Birkhäuser

Progress in Mathematics

Volume 326

More information about this series at http://www.springer.com/series/4848

Victor G. Kac • Vladimir L. Popov
Editors

Lie Groups, Geometry, and Representation Theory

A Tribute to the Life and Work of Bertram Kostant

Birkhäuser

Editors
Victor G. Kac
Department of Mathematics
Massachusetts Institute of Technology
Cambridge, MA, USA

Vladimir L. Popov
Steklov Mathematical Institute
Moscow, Russia

National Research University
Higher School of Economics
Moscow, Russia

ISSN 0743-1643 ISSN 2296-505X (electronic)
Progress in Mathematics
ISBN 978-3-030-02190-0 ISBN 978-3-030-02191-7 (eBook)
https://doi.org/10.1007/978-3-030-02191-7

Library of Congress Control Number: 2018965937

This book is published under the imprint Birkhäuser, www.birkhauser-science.com by the registered company Springer Nature Switzerland AG.
The registered company address is: Gewerbestrasse 11, 6330 Cham, Switzerland

This volume is dedicated to the memory of the great American mathematician Bertram Kostant (May 24, 1928–February 2, 2017).

Victor G. Kac
Vladimir L. Popov

Photo submitted courtesy of Ann Kostant

Contents

A Tribute to Bertram Kostant

Anthony Joseph

The scientific career of Bertram Kostant, to whom this volume is dedicated, filled an astonishing six decades. In his more active years he typically wrote one major paper each year often of one hundred pages or more. These papers were characterized by being intensely exciting, yet easy to read even by a beginning graduate student. They would invariably involve several strikingly simple constructions which would either establish an entirely new result or represent a simplification in earlier work as exemplified by his proof of the Jacobson–Morozov theorem and of the Amitsur–Levitski identity, for which the simplification, using only the commutativity of quadratic elements of the Grassmann algebra was particularly dramatic. Moreover they were often of such originality that he could leave them unpublished for many decades without the danger of rediscovery by anyone else. Naturally many of these constructions influenced and inspired his followers.

The Hochschild–Kostant–Rosenberg isomorphism gave a way to describe Hochschild homology with eventual applications to cyclic homology developed to extend DeRham theory to a non-commutative context.

Kostant refined the Chevalley theorem on Weyl group invariants by establishing a simple rule for their degrees obtained from eigenvalues arising in a principal three-dimensional subalgebra of a semisimple Lie algebra. This was inspired partly by a result of Coleman on the relation between eigenvalues of a Coxeter element and degrees of invariants. Then, in an aside, Kostant noted how one can "complete" a regular nilpotent element so that it becomes a Coxeter element. This eventually led Kazhdan and Lusztig to construct a map from nilpotent orbits to conjugacy classes in the Weyl group.

The Kostant separation theorem for the symmetric algebra of a semisimple Lie algebra and the ideas behind its proof were of prime importance. Thus he constructed a section for the coadjoint action of the group by exploiting his above mentioned estimate on degrees of Weyl group invariants. This result had important applications to the study of Whittaker modules as well as to the dynamics of the Toda lattice, which is an exactly integrable dynamical system extended by Kostant from $\mathrm{sl}(3)$ to an arbitrary semisimple Lie algebra.

Through the section he constructed, Kostant showed that the nilpotent cone of a semisimple Lie algebra is a complete intersection, a result almost unparalleled in such a complex situation. From this, he described a separation of variables theorem of the symmetric algebra, into invariants and the Kostant harmonic space. For the latter he computed multiplicities using the Kostant partition function, the q-analogue of which also computes the graded multiplicities. This led to a precise description of what became known as the Kostant–Brylinski filtration of the symmetric algebra defined by the action of a regular nilpotent element.

Much later the Kostant section had a significant influence on invariant theory, especially concerning biparabolic and stabilizer subalgebras. Again an analogue of the Kostant section can often be derived for a simple Lie algebra acting on a module when the invariants form a polynomial algebra. Thus Popov and Vinberg found a section for the action of $\mathrm{SL}(3)$ on its ten-dimensional representation thereby giving an elegant derivation of Weierstrass canonical form for elliptic curves.

From the Kostant separation of variables theorem, one may identify and compute an important determinant. Actually Kostant had already considered the corresponding determinant in the much more complicated real case. This led to a still useful criterion for unitarity for which he was awarded the Steele prize. On the other hand, the separation theorem is also of some importance in studying primitive ideals. In particular, the identification of a minimal primitive quotient with a principal series module, via Kostant's estimate of multiplicities, leads to the description of any primitive ideal as the annihilator of a simple highest weight module.

Kostant is particularly known for the orbit method. This relates coadjoint orbits to unitary representations, a procedure he developed from the work of Kirillov. This realized a process inaugurated by Dirac, termed "geometric quantization" by Souriau.[1] It led partly through his work with Auslander to the extension of Kirillov's results to real solvable groups. Kostant was a strong advocate of extending the orbit method to an arbitrary finite-dimensional Lie algebra and indeed for these it has some important consequences. Perhaps the most striking is the Duflo isomorphism from the invariants of the symmetric algebra to the centre of the enveloping algebra, which results purely combinatorially only through a horrendously complicated analysis of the Baker–Campbell–Hausdorff formula, though this is still needed to establish isomorphism for the higher cohomologies.

Kostant established a major extension of the Golden–Thompson rule with consequences for the decomposition of a real Lie group.

Kostant was one of the first to consider a Clifford algebra as what one might call a quantum deformation of a Grassmann algebra. Here one of his notable results was that the Clifford algebra of a semisimple Lie algebra is a free algebra over its invariants, just like the symmetric algebra, a result quite impossible for the Grassmann algebra. This led to a formula for the multiplicity of the adjoint representation in each graded component of the Grassmann algebra.

[1] Kostant, in his comments in Vol. II of his *Collected Works,* states that "the general topic and terminology of the material in his paper *'Quantization and Unitary Representations,'* now goes by the name of geometric quantization, given by Souriau, who discovered a number of these ideas independently."

The Kostant section pertains to a regular nilpotent orbit. However it generalizes as a slice for any nilpotent orbit and notably, for the subregular orbit, from which Slodowy was able to detail the proof of Breiskorn's description of Kleinian singularities.

The quantization of this slice in the regular case was carried out by Kostant in obtaining Whittaker modules, a construction generalized by one of his students to the not very different "even" case. The general case was first considered, using the now generally accepted construction, by Moeglin. The resulting algebras were described in detail by Premet and became known as finite W-algebras. Eventually this led to affine W-algebras with important applications to integrable Hamiltonian systems and to conformal field theory developed notably by Kac. Finite W-algebras themselves were found to have significant applications to the quantization of a nilpotent orbit obtained through the construction of a one-dimensional representation of the corresponding finite W-algebra, with significant contributions by Brundan, Goodwin and Losev. Their importance lies in determining the scale factors in the Goldie rank polynomials.

Kostant's shorter papers also initiated new themes. He was the first to construct (in $\mathrm{SO}(4,2)$) a non-trivial example of a unitarizable highest weight module, from which Enright, Howe and Wallach and independently Jacobson eventually found a complete classification. He established a surprisingly general result on tensoring by finite-dimensional modules. This eventually led to the all-important translation principle. It was an initial step in the Bernstein–Gelfand description of projective functors and their algebraization of Zhelobenko's classification of Harish-Chandra modules for complex Lie groups. Kostant found a characterization of co-commutative Hopf algebras, which has been the cornerstone of much recent work. Again the Kostant form of an enveloping algebra of a semisimple Lie algebra, already important in the modular representation theory, was later generalized to quantized enveloping algebras. This gave rise to quantum Frobenius splitting, which itself led Littelmann to a complete description of standard monomial bases.

Even unpublished work of Kostant has had a significant echo. Thus he showed that the algebra of functions on the minimal non-zero nilpotent orbit is quadratic. Later Bezrukavnikov showed this algebra to be Koszul and this in turn implied that Gerstenhaber's horrendous infinite system of conditions for (quantum) deformation reduced to just three which could be easily satisfied. Kostant proved a special case of a conjecture of Guillemin, Quillen and Sternberg on the integrability of characteristics. A few years later the general case was established through microlocal analysis (by Kashiwara and Kawai) and later purely algebraically (by Gabber). Yet even the special case established by Kostant is a starting point in the theory of generalized Harish-Chandra modules as it showed that the set of elements in the Lie algebra, acting locally finitely on a module, form a Lie subalgebra. Kostant identified a remarkable system of strongly orthogonal roots called the Kostant cascade which notably has been used to compute the index of a biparabolic subalgebra.

A notable aspect of Kostant's method of research involved attentive participation in research lectures. In this he was extraordinarily quick to realize and exploit nascent ideas. He frequently attended conferences in theoretical physics and his insights, into their often unintelligible peregrinations made him a much sought after participant. Many of these contributions relate to Schwinger terms, Lie superalgebras, dynamical systems (particularly the Toda lattice), Dirac cohomology and ideals in the nilradical of a Borel. For these and many other contributions to mathematical physics he was awarded the Wigner medal.

The work of Bertram Kostant will surely continue to influence mathematicians for many more generations.

Poisson Structures and Potentials

Anton Alekseev, Arkady Berenstein, Benjamin Hoffman, and Yanpeng Li

To the memory of Bertram Kostant

Abstract We introduce the notion of weakly log-canonical Poisson structures on positive varieties with potentials. Such a Poisson structure is log-canonical up to terms dominated by the potential. To a compatible real form of a weakly log-canonical Poisson variety, we assign an integrable system on the product of a certain real convex polyhedral cone (the tropicalization of the variety) and a compact torus.

We apply this theory to the dual Poisson-Lie group G^* of a simply-connected semisimple complex Lie group G. We define a positive structure and potential on G^* and show that the natural Poisson-Lie structure on G^* is weakly log-canonical with respect to this positive structure and potential. For $K \subset G$ the compact real form, we show that the real form $K^* \subset G^*$ is compatible and prove that the corresponding integrable system is defined on the product of the decorated string cone and the compact torus of dimension $\frac{1}{2}(\dim G - \operatorname{rank} G)$.

Keywords: Poisson structures • Poisson-Lie groups • Potentials • Tropicalization

Mathematics Subject Classification (2010): 20G42, 53D17

A.B. expresses his gratitude for hospitality and support during his visits to Geneva in 2016 and in 2017. A.A. and Y.L. were supported in part by the ERC project MODFLAT and by the grants 165666 and 159581 of the Swiss National Science Foundation. B.H. was supported by the National Science Foundation Graduate Research Fellowship under Grant Number DGE-1650441.

Anton Alekseev
Section of Mathematics, University of Geneva, 2-4 rue du Lièvre, c.p. 64, 1211 Genève 4, Switzerland, e-mail: Anton.Alekseev@unige.ch

Arkady Berenstein
Department of Mathematics, University of Oregon, Eugene, OR 97403, USA
e-mail: arkadiy@uoregon.edu

Benjamin Hoffman
Department of Mathematics, Cornell University, 310 Malott Hall, Ithaca, NY 14853, USA
e-mail: bsh68@cornell.edu

Yanpeng Li
Section of Mathematics, University of Geneva, 2-4 rue du Lièvre, c.p. 64, 1211 Genève 4, Switzerland, e-mail: yanpeng.li@unige.ch

V. G. Kac, V. L. Popov (eds.), *Lie Groups, Geometry, and Representation Theory*, Progress in Mathematics 326, https://doi.org/10.1007/978-3-030-02191-7_1

1 Introduction

Varieties with potentials play a crucial role in mirror symmetry and mathematical physics. A potential is a rational function on an algebraic variety. We are interested in algebraic varieties with positive structures, that is, an open embedding of a split algebraic torus to the variety, so that the potential pulls back to a subtraction-free rational function. Assume for simplicity that the potential can be written as a Laurent polynomial

$$\Phi(x_1,\ldots,x_m) = \sum_{\mathbf{a}\in\mathbb{Z}^n\setminus\{0\}} c_{\mathbf{a}} x^{\mathbf{a}},$$

where $\mathbf{a} = (a_1,\ldots,a_m)$ and $c_{\mathbf{a}} \geq 0$. To such a potential, one can assign a rational convex polyhedral cone $\mathscr{C}_\Phi \subset \mathbb{Z}^m$ given by inequalities

$$\sum_{k=1}^{m} a_k\xi_k < 0 \text{ for all } \mathbf{a} \text{ with } c_{\mathbf{a}} > 0,$$

where the ξ_i are coordinates on $\mathbb{Z}^m$. One can also define this cone using the tropicalized potential $\Phi^t : \mathbb{Z}^m \to \mathbb{Z}$ given by

$$\Phi^t(\xi) = \max_{\mathbf{a};\, c_{\mathbf{a}}>0} \sum_{k=1}^{m} a_k\xi_k.$$

Then, $\mathscr{C}_\Phi = \{\xi \in \mathbb{Z}^m; \Phi^t(\xi) < 0\}$. We will denote by $\mathscr{C}_\Phi(\mathbb{R})$ the open convex polyhedral cone

$$\mathscr{C}_\Phi(\mathbb{R}) = \{\xi \in \mathbb{R}^m; \Phi^t(\xi) < 0\} \subset \mathbb{R}^m$$

defined by the same inequalities. We say that a positive function f on X is dominated by the potential Φ if $f^t(\xi) < 0$ for all $\xi \in \mathscr{C}_\Phi$. Positive functions dominated by the potential Φ form a semiring under addition and multiplication. If a rational function f can be decomposed as $f = f_+ - f_-$, where f_+ and f_- are both dominated by Φ, then we say f is weakly dominated by Φ. Rational functions weakly dominated by Φ form a ring under addition and multiplication.

Important examples of positive varieties with potentials are complete and partial flag varieties for semisimple algebraic groups [5]. An important example for us is the Borel subgroup $B \subset G$. For instance, for $G = \mathrm{SL}_2$, elements of B are upper triangular matrices

$$b = \begin{pmatrix} b_{11} & b_{12} \\ 0 & b_{11}^{-1} \end{pmatrix}.$$

In this case, the positive structure is defined by the matrix entries of b, and the potential [4, 17] is given by

$$\Phi(b) = \frac{b_{11} + b_{11}^{-1}}{b_{12}}.$$

The potential cone is the following in $\mathbb{Z}^2$:

(1) $$\mathscr{C}_\Phi = \{(\xi_{11}, \xi_{12}) \in \mathbb{Z}^2; \xi_{12} > \xi_{11} > -\xi_{12}\}.$$

One can show that regular functions dominated by the potential are all non-zero Laurent polynomials of the form

$$f(b_{11}, b_{12}) = \sum_{k\geq 1} \sum_{l=-k}^{k} c_{kl} b_{11}^l b_{12}^{-k}$$

with $c_{kl} \geq 0$. Note that the cone (1) is the $n = 2$ instance of the Gelfand-Zeitlin cone [2, 3] for SL_n. In fact, for any n, there is a positive stucture and the potential on $B \subset \mathrm{SL}_n$ with potential cone is equal to the corresponding Gelfand-Zeitlin cone.

Varieties arising in applications of mirror symmetry are often symplectic or Poisson. A Poisson structure is a bivector which induces a Poisson bracket on the ring of regular functions on the variety. The Poisson bracket on a positive variety is called log-canonical if

$$\{x_i, x_j\} = c_{ij} x_i x_j,$$

where $x_1, \ldots, x_m$ are toric coordinates (defined by the positive structure) and c_{ij} is a constant matrix. Important examples of Poisson varieties with log-canonical Poisson structures are cluster varieties, and in particular double Bruhat cells, including complete and partial flag varieties [14]. For instance, in the example of $B \subset \mathrm{SL}_2$, the log-canonical Poisson bracket is of the form

$$\{b_{11}, b_{12}\} = b_{11} b_{12}.$$

It turns out that this Poisson structure is multiplicative and gives B the structure of a Poisson algebraic group.

The condition of a Poisson structure to be log-canonical is very restrictive. On a positive variety with potential we can generalize it to a notion of *weakly log-canonical* Poisson structures. This means that the Poisson bracket is given by the formulas

$$\{x_i, x_j\} = x_i x_j (c_{ij} + f_{ij}(x)),$$

where $f_{ij}(x)$ are functions weakly dominated by the potential Φ.

Given a weakly log-canonical Poisson structure π_X on a smooth complex variety X, consider the real form $(K, \pi_K) \subset (X, \pi_X)$ defined by the equations $x_i \in \mathbb{R}$ in the toric chart. Then, to such a structure we assign a constant Poisson bracket on the space

(2) $$\mathscr{C} \times \mathbb{T},$$

where $\mathscr{C}$ is a subcone of $\mathscr{C}_\Phi(\mathbb{R})$ and $\mathbb{T} \cong (S^1)^r$ is a compact torus of dimension r, where $2r$ is the maximal rank of π_K. This Poisson bracket has the form

(3) $$\{\xi_i, \xi_j\} = 0, \quad \{\phi_i, \phi_j\} = 0, \quad \{\xi_i, \phi_j\} = d_{ij},$$

where $d_{ij} \in \mathbb{R}$ is determined by the log-canonical part c_{ij} of the weakly log-canonical bracket π_X. Here ξ_i's are coordinates on the cone $\mathscr{C}(\mathbb{R})$ and ϕ_i's are coordinates on the torus $\mathbb{T}$. We refer to the space (2) as a *partial tropicalization* of the weakly log-canonical Poisson variety. Up to a change of variables, the Poisson bracket (3) defines an integrable system on the partial tropicalization.

Our prime example is the dual Poisson-Lie group G^* of a semisimple complex Lie group G endowed with the standard Poisson-Lie structure. For $G = \mathrm{SL}_2(\mathbb{C})$, the group G^* is of the form

$$G^* = \{(b^+, b^-) \in B \times B_-; b^+_{11} b^-_{11} = 1\}.$$

The positive structure is defined by the following parametrization:

$$b^+ = \begin{pmatrix} b_{11} & b_{12} \\ 0 & b_{11}^{-1} \end{pmatrix}, \qquad (b^-)^{-1} = \begin{pmatrix} b_{11} & 0 \\ b_{21} & b_{11}^{-1} \end{pmatrix},$$

and the potential is given by

$$\Phi_{G^*} = \frac{b_{11} + b_{11}^{-1}}{b_{12}} + \frac{b_{11} + b_{11}^{-1}}{b_{21}}.$$

To $K \subset G$, the compact real form of G, one associates the real form $K^* \subset G^*$. In the case of $K = \mathrm{SU}(2) \subset \mathrm{SL}_2(\mathbb{C})$, this real form is defined by equations $b_{11} \in \mathbb{R}_{>0}$ and $b_{21} = \overline{b_{12}}$. The canonical real Poisson bracket on K^* is given by [15, 18]:

$$\{b_{11}, b_{12}\} = i b_{11} b_{12}, \quad \{b_{11}, b_{21}\} = -i b_{11} b_{21}, \quad \{b_{12}, b_{21}\} = i(b_{11}^2 - b_{11}^{-2}).$$

Note that the first two expressions are log-canonical on the nose whereas the third expression has no log-canonical part and the corresponding function $f(x)$ is of the form

$$i\left(\frac{b_{11}^2}{b_{12}b_{21}} - \frac{b_{11}^{-2}}{b_{12}b_{21}}\right).$$

This expression is weakly dominated by the potential Φ_{G^*}. The corresponding partial tropicalization is the product of the Gelfand-Zeitlin cone and the circle S^1:

$$\{(\xi_{11}, \xi_{12}) \in \mathbb{R}^2; \xi_{12} > \xi_{11} > -\xi_{12}\} \times S^1$$

with Poisson bracket

$$\{\xi_{11}, \xi_{12}\} = 0, \quad \{\xi_{11}, \phi\} = 1, \quad \{\xi_{12}, \phi\} = 0,$$

which is the $n = 2$ Gelfand-Zeitlin integrable system.

We will now describe our main results. Let G be a semisimple complex Lie group with the standard Poisson-Lie structure and G^* the dual Poisson-Lie group. The standard positive structures on B and B_- and the respective potentials Φ_+ and Φ_- give rise to a positive structure θ_{G^*} and a potential Φ_{G^*} on G^*. We prove the following theorem:

Theorem 1.1. *In the coordinates given by θ_{G^*}, the standard Poisson-Lie structure on G^* is weakly log-canonical with respect to the potential Φ_{G^*}.*

Furthermore, for the compact real form $K \subset G$ we consider the partial tropicalization of the real form $K^* \subset G^*$. Our next result is as follows:

Theorem 1.2. *The partial tropicalization of K^* is of the form $\mathscr{C} \times \mathbb{T}$, where $\mathscr{C}$ is the extended string cone and $\mathbb{T}$ is a real torus of dimension $d = \dim(N)$. This space is equipped with an integrable system.*

In the case of $G = \mathrm{SL}_n(\mathbb{C})$, these results (in one of the toric charts) were obtained in [2]. We establish them in full generality: for all equivalent positive structures on $\mathrm{SL}_n^*(\mathbb{C})$ and for all semisimple complex Lie groups.

In the next several paragraphs we describe the future directions which give an additional context and motivation for our results:

Dual Poisson-Lie groups K^* of compact connected Poisson-Lie groups are Poisson manifolds with very special properties. In particular, the Ginzburg-Weinstein isomorphism theorem [13] states that (K^*, π_{K^*}) is isomorphic as a Poisson manifold to the dual of the Lie algebra $\mathfrak{k}^*$ with the linear Kirillov-Kostant-Souriau (KKS) Poisson bracket π_{KKS}. Since the bracket π_{KKS} is linear, the scaling map $D_s : x \mapsto sx$ is a Poisson map $D_s : (\mathfrak{k}^*, \pi_{\mathrm{KKS}}) \to (\mathfrak{k}^*, s\pi_{\mathrm{KKS}})$. Following [1, 3], this observation gives rise to a family of Ginzburg-Weinstein maps

$$\mathrm{gw}_s : (\mathfrak{k}^*, \pi_{\mathrm{KKS}}) \to (K^*, s\pi_{K^*}).$$

We would like to think of the partial tropicalization $\mathscr{C} \times T$ of K^* as the $s \to \infty$ limit of the Poisson space $(K^*, s\pi_{K^*})$.

In the forthcoming paper we plan to give the following evidence in support of this idea: the symplectic leaves in $\mathscr{C} \times T$ are labeled by the elements $\lambda \in \overset{\circ}{W}_+$ of the interior of the positive Weyl chamber and their symplectic volumes coincide with symplectic volumes of the corresponding coadjoint orbits $\mathscr{O}_\lambda \subset \mathfrak{k}^*$. The proof makes use of the relation between Kashiwara crystals for finite-dimensional modules of G and the Langlands dual group $G^\vee$.

A more ambitious project is to define the $s \to \infty$ limit of the Ginzburg-Weinstein map gw_s. This will be a Poisson map

$$\mathrm{gw}_\infty : (\mathfrak{k}^*, \pi_{\mathrm{KKS}}) \to (\mathscr{C} \times T, \pi_\infty),$$

where π_∞ is the constant Poisson bracket on the partical tropicalization. Among other things, such a map will define interesting completely integrable systems on the top-dimensional coadjoint orbits $\mathscr{O}_\lambda$. For the case of $G = \mathrm{SL}_n(\mathbb{C})$, the map gw_∞ can be constructed by combining results of [3] and [2]. For the general case of G semisimple, this question remains open.

The structure of the paper is as follows: in Section 2 we review the theory of positive varieties, in Section 3 we introduce potentials and tropicalizations, in Section 4 we apply this theory to double Bruhat cells and prove the main technical results on functions dominated by the potentials, in Section 5 we discuss weakly log-canonical Poisson brackets and prove Theorem 1.1. In Section 6 we consider real forms of positive varieties, define partial tropicalization, and prove Theorem 1.2.

Acknowledgements. We are grateful to I. Davydenkova who shared with us her unpublished results on the case of $G = \mathrm{Sp}(4)$, to M. Podkopaeva who participated in the initial stage of this project, to G. Koshevoy for his useful comments on the string cone, and to J. Lu for her helpful comments on an earlier draft. This work became possible thanks to the NCCR SwissMAP, which organized a series of meetings in Les Diablerets where A.B. participated and the Master Class in Geneva where B.H. took part during the academic year 2016–2017.

2 Positivity theory

In this section we review the theory of positive varieties and positive maps. One of the central notions is tropicalization, which assigns a lattice to a positive variety. We mainly follow the work of Berenstein and Kazhdan, see [4] and [5].

2.1 Algebraic tori and positive maps

Let $\mathbb{G}_a$ be the additive group and $\mathbb{G}_m$ the multiplicative group defined over $\mathbb{Q}$. By definition, the coordinate algebra $\mathbb{Q}[\mathbb{G}_a]$ is the polynomials over $\mathbb{Q}$ in one variable $\mathbb{Q}[x]$ and the coordinate algebra $\mathbb{Q}[\mathbb{G}_m]$ is the Laurent polynomials $\mathbb{Q}[x, x^{-1}]$.

Consider a split algebraic torus $S \cong \mathbb{G}_m^n$. Denote by $S_t = \mathrm{Hom}(S, \mathbb{G}_m)$ the character lattice of S and by $S^t = \mathrm{Hom}(\mathbb{G}_m, S)$ the cocharacter lattice. The lattices S_t and S^t are in natural duality, in particular $S_t \cong \mathrm{Hom}(S^t, \mathbb{Z})$. The coordinate ring $\mathbb{Q}[S]$ is the group algebra (over $\mathbb{Q}$) of the lattice S_t. That is, every element $f \in \mathbb{Q}[S]$ can be written as

$$f = \sum_{\chi \in S_t} c_\chi \chi, \tag{4}$$

where only a finite number of coefficients c_χ are non-zero.

Denote by $P[S]$ the set of non-zero elements of the form (4) in which all c_χ are non-negative rational numbers. By construction, $P[S] \subset \mathbb{Q}[S]$ is a sub-semiring without zero.

Denote by $\mathbb{Q}(S)$ the field of rational functions on S, and by $P(S) \subset \mathbb{Q}(S)$ the subset of elements of $\mathbb{Q}(S)$ represented as fractions f/g with $f, g \in P[S]$. By construction, $P(S)$ is a semifield without zero.

Example 2.1. Note that the polynomial $x^2 - x + 1 = (x^3 + 1)/(x + 1)$ belongs to $P(\mathbb{G}_m)$ but does not belong to $P[\mathbb{G}_m]$, and the polynomial $x^2 - 2x + 1 = (x - 1)^2$ belongs neither to $P[\mathbb{G}_m]$ nor to $P(\mathbb{G}_m)$.

Definition 2.2. A rational map $\phi : S \to S'$ between split algebraic tori is positive if for every character $\chi : S' \to \mathbb{G}_m$, the composition $\chi \circ \phi \in P(S)$ is a positive rational function on S. ◁

Example 2.3. Let $S = S' = \mathbb{G}_m^3$. The rational map $\phi : S \to S'$ defined by

$$\phi(x_1, x_2, x_3) = \left(\frac{x_2 x_3}{x_1 + x_3}, x_1 + x_3, \frac{x_1 x_2}{x_1 + x_3} \right)$$

is positive.

Proposition 2.4. *A rational map $\phi : S \to S'$ is positive if and only if the pullback ϕ^* restricts to a semifield homomorphism on positive rational functions:*

$$\phi^* : P(S') \to P(S).$$

Proof. Let $y_1, \ldots, y_m \in P(S')$ be the standard basis of the character lattice S'_t of S' given by the splitting of S', and let $\phi_i = y_i \circ \phi \in \mathbb{Q}(S)$ be the components of ϕ. Assume ϕ is positive, then $\phi_i \in P(S)$ for each i. We show that $\phi^* f \in P(S)$ whenever $f \in P(S')$. Indeed, $\phi^* f(y_1, \ldots, y_m) = f(\phi_1, \ldots, \phi_m) \in P(S)$ is a subtraction-free rational polynomial in $\phi_1, \ldots, \phi_m$.

Conversely, assume $\phi^* f \in P(S)$ whenever $f \in P(S')$. Then for each basic character y_i on S', we have $\phi^* y_i = y_i \circ \phi = \phi_i \in P(S)$. □

Denote by **PosTori** the category with objects split algebraic tori and arrows positive rational maps. The previous proposition shows $P(\cdot)$ defines a functor from **PosTori**^{op} to the category of semifields. However, the next example shows the situation is not so straightforward when we take positive regular functions $P[S]$ of a split algebraic torus S; see also Remark 2.20 below.

Example 2.5. Define a rational map $F : \mathbb{G}_m^2 \to \mathbb{G}_m^2$ by $F(t_1, t_2) = \left(\frac{t_1 t_2}{t_1 + t_2}, \frac{t_2^2}{t_1 + t_2} \right)$. Clearly, F is invertible and F^{-1} is given by $F^{-1}(t_1, t_2) = \left(\frac{t_1}{t_2} \cdot (t_1 + t_2), t_1 + t_2 \right)$. In particular, F is an isomorphism in the category **PosTori**. Let $f := \frac{(t_1^3 + t_2^3)(t_1 + t_2)}{t_2^2} \in P[\mathbb{G}_m^2]$. Then

$$f \circ F = \frac{\left(\left(\frac{t_1 t_2}{t_1 + t_2} \right)^3 + \left(\frac{t_2^2}{t_1 + t_2} \right)^3 \right) \left(\frac{t_1 t_2}{t_1 + t_2} + \frac{t_2^2}{t_1 + t_2} \right)}{\left(\frac{t_2^2}{t_1 + t_2} \right)^2} = \frac{t_1^3 + t_2^3}{t_1 + t_2} = t_1^2 - t_1 t_2 + t_2^2 .$$

Thus, $f \circ F \notin P[\mathbb{G}_m^2]$.

2.2 Tropicalization of positive maps

Following [5], to each positive rational map $\phi : S \to S'$, we will associate a *tropicalized map* $\phi^t : S^t \to (S')^t$ as follows.

Step 1. Let $\phi : S \to \mathbb{G}_m$ be a positive regular function, that is, $\phi \in P[S]$ and it admits the form (4) with $c_\chi \geq 0$. Define $\phi^t : S^t \to \mathbb{G}_m^t = \mathbb{Z}$ by formula

$$\phi^t(\xi) = \max_{\chi; c_\chi > 0} \langle \chi, \xi \rangle,$$

where $\langle \cdot, \cdot \rangle : S_t \times S^t \to \mathbb{Z}$ is the canonical pairing.

Step 2. Let $\phi : S \to \mathbb{G}_m$ be a positive rational function. That is, $\phi = f/g$ with $f, g \in P[S]$, then

$$\phi^t = f^t - g^t.$$

One can show that the right-hand side is independent of the presentation of the positive fraction. Note that the assignment $\phi \mapsto \phi^t$ is a homomorphism of semifields mapping $P(S)$ to the semifield of $\mathbb{Z}$-valued functions on S^t. This semifield has operations of tropical addition $(f,g) \mapsto \max\{f,g\}$ and tropical multiplication $(f,g) \mapsto f+g$.

Example 2.6. Consider $S = \mathbb{G}_m$ and $\phi = (x^3+1)/(x+1)$. Then, $\phi^t(\xi) = \max(3\xi, 0) - \max(\xi, 0) = 2\max(\xi, 0)$. Note that for any $a,b,c,d \in \mathbb{Q}_{>0}$ the function $\phi' = (ax^3+b)/(cx+d)$ has the same tropicalization $(\phi')^t = \phi^t$.

Step 3. Let $\phi : S \to S'$ be a positive rational map. Define $\phi^t : S^t \to (S')^t$ as the unique map such that for every character $\chi \in S'_t$ and for every cocharacter $\xi \in S^t$, we have

$$\langle \chi, \phi^t(\xi) \rangle = (\chi \circ \phi)^t(\xi).$$

Let $\phi_1, \ldots, \phi_m$ be the components of ϕ given by the splitting $S' \cong (\mathbb{G}_m)^m$. Then, in the induced coordinates on $(S')^t$, we have

$$\phi^t = (\phi_1^t, \ldots, \phi_m^t).$$

Example 2.7. The positive rational map from Example 2.3

$$\phi : \mathbb{G}_m^3 \to \mathbb{G}_m^3 \; : \; (x_1, x_2, x_3) \mapsto \left(\frac{x_2 x_3}{x_1 + x_3}, x_1 + x_3, \frac{x_1 x_2}{x_1 + x_3} \right)$$

has tropicalization

$$\begin{aligned} &\phi^t : (\mathbb{G}_m^3)^t \cong \mathbb{Z}^3 \to (\mathbb{G}_m^3)^t \cong \mathbb{Z}^3; \\ &\quad (\xi_1, \xi_2, \xi_3) \mapsto (\xi_2 + \xi_3 - \max\{\xi_1, \xi_3\}, \max\{\xi_1, \xi_3\}, \xi_1 + \xi_2 - \max\{\xi_1, \xi_3\}). \end{aligned}$$

Note that ϕ^t is linear on the chambers $\xi_1 < \xi_3$ and $\xi_1 > \xi_3$.

Recall that **PosTori** is the category with objects split algebraic tori and arrows positive rational maps. Denote by **PLSpaces** the category with objects finite rank lattices and arrows piecewise $\mathbb{Z}$-linear maps. The following is shown in Section 2.4 of [4]:

Proposition 2.8. *The assignment* $S \mapsto S^t$, $\phi \mapsto \phi^t$ *defines a functor from the category* **PosTori** *to the category* **PLSpaces**.

Note that all piecewise linear maps ϕ^t obtained by applying the tropicalization functor are homogeneous in the following sense:

$$\phi^t(n\xi) = n\phi^t(\xi) \tag{5}$$

for every $n \in \mathbb{Z}_{\geq 0}$. In particular, $\phi^t(0) = 0$.

2.3 Positive varieties

Definition 2.9. Let X be an irreducible variety over $\mathbb{Q}$. A *toric chart* is an open embedding $\theta : S \to X$ from a split algebraic torus S to X. ◁

Since θ is an open map, it induces an inclusion of coordinate rings. Hence, the coordinate ring of X identifies with a subalgebra of $\mathbb{Q}[S]$.

Example 2.10. For $X = S$, the identity map $\mathrm{Id} : S \to S$ is a toric chart.

Example 2.11. The inclusion $\mathbb{G}_m \to \mathbb{G}_a$ is a toric chart on $\mathbb{G}_a$. The corresponding homomorphism of coordinate algebras is the natural embedding $\mathbb{Q}[x] \to \mathbb{Q}[x, x^{-1}]$.

Example 2.12. Let X be a toric variety, and let $\theta : S \hookrightarrow X$ pick out the open torus of X. Then θ is a toric chart.

Example 2.13. Let $N \subset \mathrm{SL}_3$ be the group of unipotent upper-triangular matrices. Define $\theta : S = (\mathbb{G}_m)^3 \to N$ by

$$\theta(x_1, x_2, x_3) = \begin{pmatrix} 1 & x_1 & 0 \\ 0 & 1 & 0 \\ 0 & 0 & 1 \end{pmatrix} \begin{pmatrix} 1 & 0 & 0 \\ 0 & 1 & x_2 \\ 0 & 0 & 1 \end{pmatrix} \begin{pmatrix} 1 & x_3 & 0 \\ 0 & 1 & 0 \\ 0 & 0 & 1 \end{pmatrix} = \begin{pmatrix} 1 & x_1 + x_3 & x_1 x_2 \\ 0 & 1 & x_2 \\ 0 & 0 & 1 \end{pmatrix}.$$

Then θ is a toric chart on N.

Example 2.14. Let $N \subset \mathrm{SL}_3$ be as in the previous example. Define $\theta' : (\mathbb{G}_m)^3 \to N$ by

$$\theta'(y_1, y_2, y_3) = \begin{pmatrix} 1 & 0 & 0 \\ 0 & 1 & y_1 \\ 0 & 0 & 1 \end{pmatrix} \begin{pmatrix} 1 & y_2 & 0 \\ 0 & 1 & 0 \\ 0 & 0 & 1 \end{pmatrix} \begin{pmatrix} 1 & 0 & 0 \\ 0 & 1 & y_3 \\ 0 & 0 & 1 \end{pmatrix} = \begin{pmatrix} 1 & y_2 & y_2 y_3 \\ 0 & 1 & y_1 + y_3 \\ 0 & 0 & 1 \end{pmatrix}.$$

Then θ' is a toric chart on N.

Example 2.15. Let A be an upper cluster algebra, let $X = \operatorname{Spec} A$, and let $x_1, \dots, x_k$ be a cluster. By the Laurent phenomenon, $A \subset \mathbb{Q}[x_1^{\pm}, \dots, x_k^{\pm}]$, and the corresponding map of varieties

$$\mathbb{G}_m^k = \operatorname{Spec} \mathbb{Q}[x_1^{\pm}, \dots, x_k^{\pm}] \to X$$

is a toric chart. More details can be found in [6].

Definition 2.16. Let $\theta : S \to X$ and $\theta' : S' \to X$ be toric charts on an irreducible variety X. If $(\theta)^{-1} \circ \theta' : S' \to S$ and $(\theta')^{-1} \circ \theta : S \to S'$ are positive rational maps, then θ and θ' are *positively equivalent* toric charts. We define a *positive variety* to be a pair (X, Θ_X), where Θ_X is a positive equivalence class of toric charts. If $\theta \in \Theta_X$, we sometimes write $\Theta_X = [\theta]$. $\lhd$

Example 2.17. Let $N \subset \mathrm{SL}_3$ be upper triangular unipotent matrices. Consider the toric charts θ, θ' on N from Examples 2.13 and 2.14. Then $(\theta')^{-1} \circ \theta$ is the positive rational map $\phi : \mathbb{G}_m^3 \to \mathbb{G}_m^3$ from Example 2.3. Its inverse has a similar positive expression, so θ and θ' are positively equivalent and $[\theta] = [\theta']$.

Remark 2.18. If a birational map $\phi : S \to S'$ is positive, its inverse is not necessarily positive. For instance, the birational map $\mathbb{G}_m \to \mathbb{G}_m$ given by $x \mapsto x + 1$ does not have a positive inverse.

Definition 2.19. A *positive map* of positive varieties $\phi : (X, \Theta_X) \to (Y, \Theta_Y)$ is a rational map $\phi : X \to Y$ so that for some (equivalently any) $\theta_X \in \Theta_X$ and $\theta_Y \in \Theta_Y$, the rational map $\theta_Y^{-1} \circ \phi \circ \theta_X : S \to S'$ is positive. We denote by **PosVar** the category of positive varieties over $\mathbb{Q}$, with positive rational maps as arrows. $\lhd$

In particular, setting $Y = \mathbb{G}_m$, we see that a rational function $f \in \mathbb{Q}(X)$ is positive if $f \circ \theta_X : S \to \mathbb{G}_m$ belongs to $P(S)$. Denote the set of positive rational functions on (X, Θ_X) by $P(X, \Theta_X)$. It is clear that $P(X, \Theta_X)$ is a semifield.

Remark 2.20. Example 2.5 shows that we cannot similarly define a "positive regular semiring," $P[X, \Theta_X]$ since the definition will not be independent of the choice of toric chart. However, we indicate in Remark 4.12 that in some situations there is an alternate approach.

Definition 2.21. Let $\mathbf{PosVar}^{\bullet}$ be the category with objects *framed positive varieties* (X, θ), where θ is a toric chart on X. An arrow from (X, θ_X) to (Y, θ_Y) is a rational map $\phi : X \to Y$ such that $\theta_Y^{-1} \circ \phi \circ \theta_X : S_X \to S_Y$ is a positive map of tori. $\lhd$

Let

$$F : \mathbf{PosVar}^{\bullet} \to \mathbf{PosVar}$$

be the forgetful functor sending $(X, \theta) \mapsto (X, [\theta])$; then it is shown in Claim 3.17 in [5] that F is an equivalence of categories. We may define an adjoint equivalence

$$G : \mathbf{PosVar} \to \mathbf{PosVar}^{\bullet}$$

by simultaneously choosing a representative $\theta \in \Theta_X$ for each positive variety (X, Θ_X). In fact, any adjoint equivalence to F arises this way.

Tropicalization extends in the obvious way to framed positive varieties: if (X,θ) is a framed positive variety, with $\theta : S \to X$ a toric chart, set

$$(X,\theta)^t := \mathrm{Hom}(\mathbb{G}_m, S) = S^t.$$

A positive rational map $\phi : (X,\theta_X) \to (Y,\theta_Y)$ has a tropicalization $\phi^t := (\theta_Y^{-1} \circ \phi \circ \theta_X)^t : (X,\theta)^t \to (Y,\theta)^t$, and by Proposition 2.8, tropicalization respects the composition of positive rational maps. We then have:

Proposition 2.22. *Tropicalization defines a functor from the category* **PosVar**$^\bullet$ *to the category* **PLSpaces**.

Example 2.23. Let (X,Θ) be a positive variety. Then for every $\theta, \theta' \in \Theta$, the identity map of X induces a map $(X,\theta) \to (X,\theta')$ in **PosVar**$^\bullet$. In particular, recall Example 2.17. There, we considered $X = N$ with two toric charts. The tropicalization of the induced map was given in Example 2.7.

In general, if $\theta, \theta' : \mathbb{G}_m^k \to X$ are positively equivalent charts, the transition map $(\theta^{-1} \circ \theta')^t : \mathbb{Z}^k \to \mathbb{Z}^k$ is a piecewise $\mathbb{Z}$-linear bijection.

Precomposing the functor $(\cdot)^t$: **PosVar**$^\bullet$ $\to$ **PLSpaces** with the equivalence G : **PosVar** $\to$ **PosVar**$^\bullet$ gives a tropicalization of positive varieties. Choosing a different adjoint equivalence G', we see $(G(X,\Theta))^t$ differs from $(G'(X,\Theta))^t$ by a piecewise linear bijection.

3 Potentials

In this section we introduce the notion of potentials on positive varieties. Potentials are positive functions, and from our perspective their main role is to define interesting cones on tropicalized varieties.

Definition 3.1. Let (X,θ) be a framed positive variety. We distinguish a set of positive rational functions $\Phi := \{\varphi_1, \dots, \varphi_m\} \subset P(X,[\theta])$, called a *set of potentials on* (X,θ). The triple (X,θ,Φ) is called a *framed positive variety with potential.*

Since the semiring $P(X,[\theta])$ depends only on the positive structure $[\theta]$ and not the toric chart θ, we define similarly a *positive variety with potentials* as a triple $(X,[\theta],\Phi)$.

We write $\Phi^t = \{\phi_1^t, \dots, \phi_m^t\}$. For $\xi \in (X,\theta)^t$, we write $\Phi^t(\xi) < 0$ whenever $\phi_i^t(\xi) < 0$ for all $\phi_i^t \in \Phi^t$. ◁

Definition 3.2. Let (X,θ) be a framed positive variety. For $\Phi \subset P(X,[\theta])$, we define the cone

$$\mathscr{C}_\Phi(X,\theta) := \{\xi \in (X,\theta)^t \mid \Phi^t(\xi) < 0\} \subset (X,\theta)^t.$$

We write $\mathscr{C}_\Phi$ when the variety and positive structure (X,θ) are evident, and we use $\mathscr{C}_f$ if $\Phi = \{f\}$. If Φ is distinguished as a potential on (X,θ), we call $\mathscr{C}_\Phi$ the (*strict*) *potential cone.* ◁

Remark 3.3. (1) In [4, 5, 8], the authors consider the cone

$$\mathscr{C}^{\leq}_{\Phi}(X,\theta) := \{\xi \in (X,\theta)^t \mid \varphi_i^t(\xi) \leq 0 \text{ for all } \varphi_i \in \Phi\} \subset (X,\theta)^t$$

given by non-strict inequalities, whereas we primarily consider the strict potential cone $\mathscr{C}_\Phi$. In Section 5, we consider the open real cone

$$\mathscr{C}_\Phi(\mathbb{R}) = \{\xi \in (X,\theta)^t \otimes \mathbb{R} \mid \Phi^t(\xi) < 0\}$$

defined by strict inequalities.

(2) When Φ restricts to a set of regular functions on the toric chart $\theta : S \hookrightarrow X$, the cone $\mathscr{C}_\Phi(\mathbb{R})$ is polyhedral and hence convex.

(3) We sometimes consider the function 0 as a potential on a positive variety (X,θ). By convention, $0^t = -\infty$ and $\mathscr{C}_0(X,\theta) = (X,\theta)^t$.

Definition 3.4. Let (X,θ) be a framed positive variety and let $\Phi,\Psi \subset P(X,[\theta])$. Then Ψ is *dominated* by Φ if, for all $\xi \in (X,\theta)^t$, if $\Phi^t(\xi) < 0$, then $\Psi^t(\xi) < 0$. ◁

In other words, Ψ is dominated by Φ if and only if $\mathscr{C}_\Phi(X,\theta) \subset \mathscr{C}_\Psi(X,\theta)$.

Lemma 3.5. *Let $f : (X,\theta_X) \to (Y,\theta_Y)$ be a positive rational map. If $\Phi,\Psi \subset P(Y,[\theta_Y])$ and Ψ is dominated by Φ, then, $f^*\Psi$ is dominated by $f^*\Phi$. Here, we write*

$$f^*\Phi := \{f^*\phi_1,\ldots,f^*\phi_n\}, \text{ where } \Phi = \{\phi_1,\ldots,\phi_n\}.$$

Proof. Without loss of generality we assume $\Phi = \{\phi\}$ and $\Psi = \{\psi\}$; see Proposition 3.14 below. Since tropicalization is a functor, we know $(f^*(\phi))^t = \phi^t \circ f^t$. By assumption,

$$\phi^t(\xi) < 0 \Rightarrow \psi^t(\xi) < 0 \text{ for all } \xi \in (Y,\theta_Y)^t.$$

So,

$$(f^*(\phi))^t = \phi^t(f^t(\xi)) < 0 \Rightarrow \psi^t(f^t(\xi)) = (f^*(\psi))^t(\xi) < 0 \text{ for all } \xi \in (X,\theta_X)^t,$$

and thus $f^*\Psi$ is dominated by $f^*\Phi$. □

As a consequence of Lemma 3.5, domination is preserved by positive rational equivalences.

Definition 3.6. Let (X,Θ_X,Φ_X) and (Y,Θ_Y,Φ_Y) be positive varieties with potential. A positive rational map $f : (X,\Theta_X) \to (Y,\Theta_Y)$ is a *map of positive varieties with potential* if for some (equivalently, any) $\theta_X \in \Theta_X$ and $\theta_Y \in \Theta_Y$, we have $f^*\Phi_Y$ is dominated by Φ_X.

Positive varieties with potential and their maps form a category **PosVarPot**. ◁

Definition 3.7. For $(X,[\theta_X],\Phi)$ a positive variety with potential, let $P_\Phi(X,[\theta_X])$ be the set of positive rational functions on X that are dominated by Φ, which we call the *dominated semiring*. It is a semiring , with the usual operations $+$ and $\cdot$. Note that $\mathscr{C}_1(X,\theta_X) = \emptyset$, so $P_\Phi(X,[\theta_X])$ has no multiplicative unit in general.

Let $\tilde{P}_\Phi(X,[\theta_X]) \subset \mathbb{Q}(X)$ be the ring (with the usual operations) generated by $P_\Phi(X,[\theta_X])$, which we call the *ring of weakly dominated functions.* $\triangleleft$

Proposition 3.8. *The assignments $P_\bullet$ and $\tilde{P}_\bullet$ define functors from* **PosVarPot**^{op} *to the category of semirings* **SemiRing** *and $\mathbb{Q}$-algebras* $\mathbf{Alg}_\mathbb{Q}$*, respectively.*

Proof. Follows from Lemma 3.5. □

Example 3.9. Consider the framed positive variety (N,θ) from Example 2.13. Define the set of potentials:

$$\Phi := \left\{ \frac{1}{x_1}, \frac{1}{x_3}, \frac{x_1+x_3}{x_1x_2}, \frac{x_1+x_3}{x_2x_3} \right\}.$$

Tropically, we have

$$\Phi^t = \{-\xi_1, -\xi_3, \max\{\xi_1,\xi_3\} - (\xi_1+\xi_2), \max\{\xi_1,\xi_3\} - (\xi_2+\xi_3)\}.$$

The potential cone is

$$\begin{aligned}\mathscr{C}_\Phi = \{\xi_1,\xi_2,\xi_3 \in \mathbb{Z}^3 \mid \xi_1 > 0,\ \xi_3 > 0, \\ \max\{\xi_1,\xi_3\} < \xi_1+\xi_2,\ \max\{\xi_1,\xi_3\} < \xi_2+\xi_3\}.\end{aligned}$$

Let $f := \frac{1}{x_1x_2}$; then

$$\mathscr{C}_f = \{\xi_1,\xi_2,\xi_3 \in \mathbb{Z}^3 \mid \xi_1+\xi_2 > 0\}.$$

The function f is dominated by Φ: if a point is in $\mathscr{C}_\Phi$, it is in $\mathscr{C}_f$.

Definition 3.10. Let (X,Θ,Φ) be a positive variety with potential. Let

$$P_\Phi^{\text{est}}(X,\Theta) := \{f \mid \exists g \in P(X,\Theta) \text{ such that } f+g \in \text{Poly}^+\Phi\},$$

where $\text{Poly}^+\Phi$ is the semiring of polynomials in $\Phi = \{\phi_1,\dots,\phi_n\}$ with positive rational coefficients, and no constant terms. We call $P_\Phi^{\text{est}}(X,\Theta)$ the semiring of *estimate-dominated* functions. We define the *ring of weakly estimate-dominated functions* $\tilde{P}_\Phi^{\text{est}}(X,\Theta)$ as the ring generated by $P_\Phi^{\text{est}}(X,\Theta)$. $\triangleleft$

In summary, we then have the following diagram:

$$\begin{array}{ccc} P_\Phi(X,\theta) & \hookrightarrow & \tilde{P}_\Phi(X,\theta) \\ \uparrow & & \uparrow \\ P_\Phi^{\text{est}}(X,\theta) & \hookrightarrow & \tilde{P}_\Phi^{\text{est}}(X,\theta). \end{array}$$

Example 3.11. Let (N,θ), Φ, and $f = \frac{1}{x_1x_2}$ be as in Example 3.9. We showed that f is dominated by Φ, but it is in fact estimate-dominated: We have $\left\{\frac{1}{x_1}, \frac{x_1+x_3}{x_2x_3}\right\} \subset \Phi$,

and
$$\left(\frac{1}{x_1}\right)\left(\frac{x_1+x_3}{x_2x_3}\right)-\frac{1}{x_1x_2}=\frac{1}{x_2x_3}\in P(N,\theta).$$

Remark 3.12. In general, the inclusion $P_\Phi^{\mathrm{est}}(X,\Theta)\hookrightarrow P_\Phi(X,\Theta)$ is strict. For instance, let $(X,\Theta)=(\mathbb{G}_m,[\mathrm{Id}])$ and $\Phi=\{x^2\}$. Then $\mathscr{C}_x(\mathbb{G}_m,[\mathrm{Id}])=\mathscr{C}_\Phi(\mathbb{G}_m,[\mathrm{Id}])$, but there is no positive rational function g on $\mathbb{G}_m$ so that $x+g$ is a polynomial in x^2.

The same example shows that the assignments $P_\bullet^{\mathrm{est}}$ and $\tilde{P}_\bullet^{\mathrm{est}}$ are not functorial. The identity map
$$\mathrm{Id}:(\mathbb{G}_m,[\mathrm{Id}],x^2)\to(\mathbb{G}_m,[\mathrm{Id}],x)$$
is an isomorphism of positive varieties with potential, but we saw $P_x^{\mathrm{est}}(\mathbb{G}_m,[\mathrm{Id}])\neq P_{x^2}^{\mathrm{est}}(\mathbb{G}_m,[\mathrm{Id}])$.

Remark 3.13. In [5], the authors consider a single potential function $\Phi=\{f\}$. For our purposes it is convenient to consider sets of potentials. This is equivalent to considering a single potential, as shown in the following proposition.

Proposition 3.14. *Let $(X,[\theta])$ be a positive variety with set of potentials $\Phi=\{\varphi_1,\dots,\varphi_m\}$, and let $\Phi^+=\{\varphi_1+\cdots+\varphi_m\}$. Then $P_\Phi(X,[\theta])=P_{\Phi^+}(X,[\theta])$, and $P_\Phi^{est}(X,[\theta])=P_{\Phi^+}^{est}(X,[\theta])$.*

Proof. Since $\mathscr{C}_{\{f,f'\}}(X,\theta)=\mathscr{C}_{f+f'}(X,\theta)$ for $f,f'\in P(X,\theta)$, we have $P_\Phi(X,\theta)=P_{\Phi^+}(X,\theta)$.

Since it is obvious that $P_\Phi^{\mathrm{est}}(X,[\theta])\supseteq P_{\Phi^+}^{\mathrm{est}}(X,[\theta])$, all we need to show is $P_\Phi^{\mathrm{est}}(X,[\theta])\subset P_{\Phi^+}^{\mathrm{est}}(X,[\theta])$. For $f\in P_\Phi^{\mathrm{est}}(X,[\theta])$, there exists some $g\in P(X,[\theta])$ such that $f+g=p$, where $p\in\mathrm{Poly}^+\Phi$. It suffices to show that there is an $h\in P(X,[\theta])$ such that $p(\varphi_1,\dots,\varphi_m)+h=q$ for $q\in\mathrm{Poly}^+\Phi^+$; then $f+g+h=q$. Without loss of generality we may assume that $p=\varphi_1^{n_1}\cdots\varphi_m^{n_m}$ is a non-constant monomial in the φ_i's. Set $q:=(\varphi_1+\cdots+\varphi_m)^{n_1+\cdots+n_m}$. Then $q-p\in P(X,[\theta])$, which proves the proposition. □

4 Potentials on double Bruhat cells

In this section we recall that double Bruhat cells in semisimple Lie groups are examples of positive varieties. In particular, we focus on the double Bruhat cell G^{e,w_0} and study the semifield of functions weakly dominated by the Berenstein-Kazhdan potentials. Our main technical result is Theorem 4.13, which provides a source of weakly dominated functions to be used in the next section.

4.1 Semisimple groups

Let G be a simply connected semisimple algebraic group over $\mathbb{Q}$ with Lie algebra $\mathfrak{g}$ of rank r. Choosing a Cartan subalgebra $\mathfrak{h}\subset\mathfrak{g}$ gives a root system $R\subset\mathfrak{h}^*$; for a root

$\alpha \in R$ we denote by $\mathfrak{g}_\alpha \subset \mathfrak{g}$ the associated root space. Fixing a choice of positive roots $R^+ \subset R$ gives a Cartan decomposition $\mathfrak{g} = \mathfrak{n}_- \oplus \mathfrak{h} \oplus \mathfrak{n}$, where

$$\mathfrak{n}_- = \bigoplus_{-\alpha \in R^+} \mathfrak{g}_\alpha \qquad \mathfrak{n}_+ = \bigoplus_{\alpha \in R^+} \mathfrak{g}_\alpha.$$

Let N_-, H, and N be closed subgroups of G with Lie algebras $\mathfrak{n}_-, \mathfrak{h}$, and $\mathfrak{n}$ respectively. Then H is a Cartan subgroup of G and N_-, N are a pair of opposite maximal unipotent subgroups of G. Let F_i, H_i, E_i be the Chevalley generators of $\mathfrak{g}$, indexed by the simple roots $\alpha_1, \ldots, \alpha_r \in \mathfrak{h}^*$ of $\mathfrak{g}$. For each i, the triple F_i, H_i, E_i defines an embedding $\phi_i : \mathrm{SL}_2 \to \mathrm{G}$. We define

$$f_i(t) := \phi_i \begin{pmatrix} 1 & 0 \\ t & 1 \end{pmatrix}, \quad h_i(t) := \phi_i \begin{pmatrix} t & 0 \\ 0 & -t \end{pmatrix}, \quad e_i(t) := \phi_i \begin{pmatrix} 1 & t \\ 0 & 1 \end{pmatrix}.$$

Let $x \mapsto x^T$ be the involutive *transpose* anti-automorphism of G defined by

$$h_i(t)^T = h_i(t), \;\; e_i(t)^T = f_i(t), \;\; f_i(t)^T = e_i(t).$$

We denote by W the Weyl group $N_G(H)/H$ of G. There is a faithful linear action of W on $\mathfrak{h}^*$: identifying W with its image under this action; the group W is generated by the simple reflections $s_1, \ldots, s_r$, where $s_i(\omega) = \omega - \omega(H_i)\alpha_i$. For the simple reflection $s_i \in W$, set the lift

$$\overline{s}_i = \phi_i \begin{pmatrix} 0 & -1 \\ 1 & 0 \end{pmatrix}$$

of s_i to G. It is known that the $\overline{s}_i$'s satisfy the Coxeter relations in W, thus any reduced word of $w \in W$ gives the same lift $\overline{w} \in G$.

Consider the opposite Borel subgroups $B = HN$ and $B_- = HN_-$, with Lie algebras $\mathfrak{b} = \mathfrak{h} \oplus \mathfrak{n}_+$ and $\mathfrak{b}_- = \mathfrak{h} \oplus \mathfrak{n}_-$, respectively. Then G has a decomposition into *double Bruhat cells*:

$$G = \bigsqcup_{u,v} G^{u,v}, \qquad \text{where } G^{u,v} := BuB \cap B_- vB_-.$$

The expression on the right does not depend on the lift of $u, v \in W$ to $N_G(H) \subset G$.

Example 4.1. We introduce an example which we will reference in future sections. Let $G = \mathrm{SL}_3$; then the Lie algebra of G is $\mathfrak{sl}_3 \subset \mathrm{Mat}(3 \times 3)$, traceless 3×3-matrices. Let $E_{ij} \in \mathrm{Mat}(3 \times 3)$ denote the matrix with all entries zero except for the ij-th entry:

$$(E_{ij})_{kl} = \delta_{ik}\delta_{jl}.$$

For $i \in \{1, 2\}$, the simple coroots are $H_i = E_{ii} - E_{i+1,i+1}$. We make the standard choice of E_{12} and E_{23} as simple root vectors of $\mathfrak{sl}_3$. Then $B \subset \mathrm{SL}_3$ is the subgroup of upper triangular matrices and $B_- \subset \mathrm{SL}_3$ is the subgroup of lower triangular matrices. In this case, the transpose map $x \mapsto x^T$ defined above is the usual matrix transpose. The Weyl group of SL_3 is isomorphic to the symmetric group S_3, with simple reflections corresponding to the transpositions $s_1 = (12)$ and $s_2 = (23)$, respectively.

The *weight lattice* $P(\mathfrak{g})$ of $\mathfrak{g}$ is the set of $\omega \in \mathfrak{h}^*$ which have $\omega(H_i) \in \mathbb{Z}$ for all i. There is a basis of *fundamental weights* ω_i of P, where $\omega_i(H_j) = \delta_{ij}$. Any multiplicative character $\xi : H \to \mathbb{G}_m$ induces a weight $\omega = d\xi \in \mathfrak{h}^*$; by assumption G is simply-connected and so we may identify multiplicative characters $\mathrm{Hom}(H, \mathbb{G}_m)$ with the weight lattice $P(\mathfrak{g})$. We write h^ω for the value at $h \in H$ of the character corresponding to $\omega \in P(\mathfrak{g})$. For the fundamental weight ω_i, we have $h_j(t)^{\omega_i} = t^{\delta_{ij}}$.

Let $G_0 = N_-HN \subset G$ be the open subset of G of elements which admit a Gaussian decomposition. For an element $x \in G_0$ we write $x = [x]_-[x]_0[x]_+$. Following [8], for $u, v \in W$ and a fundamental weight ω_i, we define the *generalized minor* $\Delta_{u\omega_i, v\omega_i}$ as the regular function on G whose restriction to the open subset $\overline{u}G_0\overline{v}^{-1}$ of G is given by

$$\Delta_{u\omega_i, v\omega_i}(x) = [\overline{u}^{-1}x\overline{v}]_0^{\omega_i}.$$

It is shown in [11] that the right-hand side depends only on the weights $u\omega_i$ and $v\omega_i$, and not on the choice of lifts $\overline{u}$ and $\overline{v}$. If $a_1, a_2 \in H$ and $x \in G$, then

$$\Delta_{u\omega_i, v\omega_i}(a_1 x a_2) = a_1^{u\omega_i} a_2^{v\omega_i} \Delta_{u\omega_i, v\omega_i}. \tag{6}$$

If $G = \mathrm{SL}_n$, the generalized minors are minors.

4.2 Positive structures on double Bruhat cells

Let G be as in the previous section, and let W be its Weyl group. Let $(u, v) \in W \times W$. Recall that W is generated by the simple reflections s_i corresponding to the simple roots of $\mathfrak{g}$, and that a word u in the simple reflections of W has a length $l(u)$. We write $s_{-1}, \ldots, s_{-r}$ for the simple reflections in the first copy of W and $s_1, \ldots, s_r$ for the simple reflections in the second copy of W. A *double word* $\mathbf{i}$ for (u, v) is a shuffle of a word for u in the simple reflections $s_{-1}, \ldots, s_{-r}$, and a word for v in the simple reflections $s_1, \ldots, s_r$. If the words for u and v are both reduced, then $\mathbf{i}$ is a *double reduced word* for (u, v). The *length* of $\mathbf{i}$ is $l(\mathbf{i}) := l(u) + l(v)$. We write

$$\mathbf{i} = (i_1, \ldots, i_{l(u)+l(v)}),$$

where $i_j \in \{-r, \ldots, -1, 1, \ldots, r\}$.

A double word defines a map $\theta_{\mathbf{i}} : \mathbb{G}_m^{l(\mathbf{i})+r} \to G$. The map is given by

$$(t_1, \ldots, t_{l(\mathbf{i})}, t_{l(\mathbf{i})+1}, \ldots, t_{l(\mathbf{i})+r}) \mapsto e_{i_1}(t_1) \cdots e_{i_{l(\mathbf{i})}}(t_{l(\mathbf{i})}) \cdot h_1(t_{l(\mathbf{i})+1}) \cdots h_r(t_{l(\mathbf{i})+r}), \tag{7}$$

where we interpret $e_{-i}(t) := f_i(t)$, for $i > 0$.

Example 4.2. Let $G = \mathrm{SL}_3$ as in Example 4.1, and fix $(u, v) = (e, w_0)$, where $w_0 = (13)$ is the longest element of S_3, the Weyl group of G. A double reduced word for (e,w_0) is $\mathbf{i} = s_1 s_2 s_1$. Then $\theta_{\mathbf{i}} : \mathbb{G}_m^3 \times \mathbb{G}_m^2 \to \mathrm{SL}_3$ has

$$\theta_{\mathbf{i}}(t_1,t_2,t_3,t_4,t_5) = \theta(t_1,t_2,t_3)\cdot h_1(t_4)\cdot h_2(t_5),$$

where $\theta : \mathbb{G}_m^3 \to \mathrm{SL}_3$ is given as in Example 2.13.

Theorem 4.3 (Theorem 1.2 of [11]). *For any $(u,v) \in W \times W$ and any double reduced word $\mathbf{i}$ for (u,v), the map $\theta_{\mathbf{i}}$ restricts to a biregular isomorphism between $\mathbb{G}_m^{l(\mathbf{i})+r}$ and a Zariski open subset of $G^{u,v}$.*

As a consequence of the previous theorem and Theorem 3.1 from [8], we have the following:

Theorem 4.4. *Let $\mathbf{i}$ and $\mathbf{i}'$ be double reduced words for $(u,v) \in W \times W$. Then $\theta_{\mathbf{i}}$ and $\theta_{\mathbf{i}'}$ are positively equivalent toric charts on $G^{u,v}$.*

For the remainder of this article, we restrict our attention to the case of $(u,v) = (e,w_0)$, where w_0 is the longest element of W. A double reduced word for (e,w_0) is simply a reduced word for w_0. We have the following relationship between generalized minors and the toric charts $\theta_{\mathbf{i}}$. From Theorem 5.8(i) and formula (4.2), both of [9], we have

Theorem 4.5. *Let γ and δ be two weights in the W-orbit of the same fundamental weight ω_i of $\mathfrak{g}$, and let $\mathbf{i} = (i_1,\ldots,i_m)$ be any word in $\{1,\ldots,r\}$ and define $\theta_{\mathbf{i}} : \mathbb{G}_m^{m+r} \to G$ as in (7). Then,*

$$\Delta_{\gamma,\delta}(\theta_{\mathbf{i}}(t_1,\ldots,t_{m+r})) = \Delta_{\gamma,\delta}(\theta_{\mathbf{i}}(t_1,\ldots,t_m,1,\ldots,1))\cdot(h_1(t_{m+1})\cdots h_r(t_{m+r}))^{\delta},$$

where the first term is a polynomial in $t_1,\ldots,t_m$ with positive integer coefficients and the second term is a Laurent monomial in $t_{m+1},\ldots,t_{m+r}$. In particular, by Theorem 4.3, we get $\Delta_{\gamma,\delta} \in P(G^{e,w_0},[\theta_{\mathbf{i}}])$.

4.3 Cluster variables on double Bruhat cells

Given a reduced word $\mathbf{i}$ for $w_0 \in W$, there is a set $F(\mathbf{i}) \subset P(G^{e,w_0},\theta_{\mathbf{i}})$ of generalized minors

$$\Delta(-r;\mathbf{i}),\ldots,\Delta(-1;\mathbf{i}),\Delta(1;\mathbf{i}),\ldots,\Delta(l(w_0);\mathbf{i}),$$

which are the *cluster variables* associated to the *initial seed* of $\mathbb{Q}[G^{e,w_0}]$ determined by $\mathbf{i}$; details are given in [6].

Example 4.6. Let $G = \mathrm{SL}_3$ as in Example 4.1 and 4.2. Consider the reduced word $\mathbf{i} = s_1 s_2 s_1$ for w_0, where $w_0 = (13)$ is the longest element of of S_3. Then the set $F(\mathbf{i})$ of cluster variables for SL_3^{e,w_0} is

$$F(\mathbf{i}) = \{\Delta_{12,23},\Delta_{1,3},\Delta_{1,2},\Delta_{12,12},\Delta_{1,1}\},$$

where $\Delta_{I,J} \in \mathbb{Q}[\mathrm{SL}_3]$ is the minor with rows I and columns J.

Remark 4.7. For any choice of the reduced word $\mathbf{i}$ for w_0, we always have $\Delta(-k,\mathbf{i}) = \Delta_{\omega_k, w_0\omega_k} \in F(\mathbf{i})$. Also, we always have the *principal* generalized minors $\Delta_{\omega_k,\omega_k} \in F(\mathbf{i})$, for all k. Together, these are the *frozen variables* of the upper cluster algebra structure on $\mathbb{Q}[G^{e,w_0}]$.

Define the rational map

$$\Delta_{\mathbf{i}} : G^{e,w_0} \to \mathbb{G}_m^{r+l(w_0)},$$
$$g \mapsto (\Delta(-r;\mathbf{i})(g), \ldots, \Delta(-1;\mathbf{i})(g), \Delta(1;\mathbf{i})(g), \ldots, \Delta(l(w_0);\mathbf{i})(g)).$$

Proposition 4.8 (Lemma 2.12 of [6]). *The map $\Delta_{\mathbf{i}}$ is biregular isomorphism from Zariski open subset of G^{e,w_0} to $\mathbb{G}_m^{r+l(w_0)}$.*

We then have, for each reduced word $\mathbf{i}$ for w_0, a toric chart

$$\Delta_{\mathbf{i}}^{-1} : \mathbb{G}_m^{r+l(\mathbf{i})} \to G^{e,w_0}.$$

Proposition 4.9. *For any reduced word $\mathbf{i}$ for w_0, the maps $\Delta_{\mathbf{i}}^{-1}$ and $\theta_{\mathbf{i}}$ are positively equivalent toric charts on G^{e,w_0}.*

Proof. By Theorem 4.5, the change of coordinates $\Delta_{\mathbf{i}} \circ \theta_{\mathbf{i}}$ is positive and birational. We need to show its inverse $\theta_{\mathbf{i}}^{-1} \circ \Delta_{\mathbf{i}}^{-1}$ is positive. In [11], the authors introduce an involutive biregular *twist map* $\xi : G^{e,w_0} \to G^{e,w_0}$, and Theorem 1.9 in [11] gives that $\theta_{\mathbf{i}}$ and $\xi \circ \Delta_{\mathbf{i}}^{-1}$ are positively equivalent toric charts on G^{e,w_0}. So,

$$\theta_{\mathbf{i}}^{-1} \circ \Delta_{\mathbf{i}}^{-1} = (\theta_{\mathbf{i}}^{-1} \circ \xi \circ \Delta_{\mathbf{i}}^{-1}) \circ (\Delta_{\mathbf{i}} \circ \theta_{\mathbf{i}}) \circ (\theta_{\mathbf{i}}^{-1} \circ \xi \circ \Delta_{\mathbf{i}}^{-1})$$

is positive. □

In summary, for any two reduced words $\mathbf{i}, \mathbf{i}'$ for w_0, the following positive structures on G^{e,w_0} are all equal:

$$[\theta_{\mathbf{i}}] = [\theta_{\mathbf{i}'}] = [\Delta_{\mathbf{i}}^{-1}] = [\Delta_{\mathbf{i}'}^{-1}]. \tag{8}$$

4.4 Weakly estimate-dominated functions on double Bruhat cells

Let $\mathbf{i}$ be a reduced word for w_0. In this section we consider G^{e,w_0} with the positive structure determined by $\theta_{\mathbf{i}}$.

Recall that the Lie algebra $\mathfrak{g}$ of G has a left action on $\mathbb{Q}[G]$; in particular for the elements E_i, F_i of the Chevalley basis of $\mathfrak{g}$, we have for $\Delta \in \mathbb{Q}[G]$,

$$F_i \cdot \Delta(g) = \frac{d}{dt}\Big|_{t=0} \Delta(f_i(t)g), \quad E_i \cdot \Delta(g) = \frac{d}{dt}\Big|_{t=0} \Delta(e_i(t)g).$$

If $X \in \mathfrak{n}$, then X is everywhere tangent to the double Bruhat cell G^{e,w_0}. Define the potentials

$$\varphi_i := \frac{E_i \cdot \Delta_{\omega_i, w_0\omega_i}}{\Delta_{\omega_i, w_0\omega_i}} = \frac{\Delta_{s_i\omega_i, w_0\omega_i}}{\Delta_{\omega_i, w_0\omega_i}},$$

and let $\Phi := \{\varphi_1, \dots, \varphi_r\}$. The following proposition implies that $\Phi \subset P(G^{e,w_0}, [\theta_{\mathbf{i}}])$.

Proposition 4.10 (Proposition 4.11.(ii) in [9]). *Let* $\mathbf{i} = (i_1, \dots, i_m)$ *be a reduced word for* w_0. *Then*

$$\frac{E_{i_1} \cdot \Delta_{\omega_{i_1}, w_0\omega_{i_1}}(\theta_{\mathbf{i}}(t_1, \dots, t_m, t_{m+1}, \dots, t_{m+r}))}{\Delta_{\omega_{i_1}, w_0\omega_{i_1}}(\theta_{\mathbf{i}}(t_1, \dots, t_m, t_{m+1}, \dots, t_{m+r}))} = \frac{1}{t_1}.$$

Note that on the toric chart $\Delta_{\mathbf{i}}^{-1}$, the potentials φ_j restrict to regular functions and so, by Remark 3.3, the cone $\mathscr{C}_\Phi(G^{e,w_0}, \Delta_{\mathbf{i}}^{-1})(\mathbb{R})$ is convex. Since $\mathbb{Q}[G]$ is a $\mathfrak{g}$ module, it is a $U_\mathfrak{g}$ module. Next we will estimate the action of $U_\mathfrak{g}$. In fact, we have

Lemma 4.11. *For a word* $\mathbf{j} = (j_1, \dots, j_n)$ *in the simple roots of* $\mathfrak{g}$, *choose a reduced word* $\mathbf{i} = (i_1, \dots, i_m)$ *for* w_0 *such that* $i_1 = j_1$. *Suppose that* $E_{j_2} \cdots E_{j_n} \Delta_{\omega_i, w\omega_i} \neq 0$. *Then for any* w *in the Weyl group* W *and any fundamental weight* ω_i,

$$\frac{E_{j_1} \cdots E_{j_n} \Delta_{\omega_i, w\omega_i}}{E_{j_2} \cdots E_{j_n} \Delta_{\omega_i, w\omega_i}} \in P_\Phi^{est}(G^{e,w_0}, [\theta_{\mathbf{i}}]).$$

Proof. For a word $\mathbf{j} = (j_1, \dots, j_n)$, let $\partial_{\mathbf{j}} := \frac{d}{dq_1}\Big|_0 \cdots \frac{d}{dq_n}\Big|_0$ be the differential at zero. In the following, let $\mathbf{j}' := (j_2, \dots, j_n)$ and let $\theta_{\mathbf{i}}$ be $\theta_{\mathbf{i}}(t_1, \dots, t_m, t_{m+1}, \dots, t_{m+r})$ as in (7). By Theorem 4.5, we get

$$\begin{aligned} &E_{j_2} \cdots E_{j_n} \Delta_{\omega_i, w\omega_i}(\theta_{\mathbf{i}}) \\ &\quad = \partial_{\mathbf{j}'} \Delta_{\omega_i, w\omega_i}(e_{j_n}(q_n) \cdots e_{j_2}(q_2)\theta_{\mathbf{i}}(t_1, \dots, t_{m+r})) \\ &\quad = \partial_{\mathbf{j}'} \Delta_{\omega_i, w\omega_i}(e_{j_n}(q_n) \cdots e_{j_2}(q_2) e_{i_1}(t_1) \cdots e_{i_m}(t_m) h_1(t_{m+1}) \cdots h_r(t_{m+r})) \\ &\quad = \sum_{k=0} f_k(t_2, \dots, t_{m+r}) t_1^k, \end{aligned}$$

where f_k are products of polynomials in $t_2, \dots, t_m$ with positive integer coefficients and Laurent monomials in $t_{m+1}, \dots, t_{m+r}$. Similarly, since $j_1 = i_1$,

$$\begin{aligned} E_{j_1} \cdots E_{j_n} \Delta_{\omega_i, w\omega_i}(\theta_{\mathbf{i}}) &= \partial_{\mathbf{j}} \Delta_{\omega_i, w\omega_i}(e_{j_n}(q_n) \cdots e_{j_1}(q_1)\theta_{\mathbf{i}}(t_1, \dots, t_{m+r})) \\ &= \partial_{\mathbf{j}} \Delta_{\omega_i, w\omega_i}\big(e_{j_n}(q_n) \cdots e_{j_2}(q_2) e_{i_1}(q_1 + t_1) e_{i_2}(t_2) \cdots \\ &\qquad \cdots e_{i_m}(t_m) h_1(t_{m+1}) \cdots h_r(t_{m+r})\big) \\ &= \frac{d}{dq_1}\Big|_0 \sum_{k=0} f_k(t_2, \dots, t_{m+r})(q_1 + t_1)^k \\ &= \sum_{k=1} k f_k(t_2, \dots, t_{m+r}) t_1^{k-1}. \end{aligned}$$

By Proposition 4.10, we know $\varphi_{i_1}(\theta_{\mathbf{i}}(t_1, \dots, t_{m+r})) = 1/t_1$. Direct calculation gives

$$\frac{\alpha}{t_1}-\frac{E_{j_1}\cdots E_{j_n}\Delta_{\omega_i,w_0\omega_i}(\theta_{\mathbf{i}}(t_1,\ldots,t_{m+r}))}{E_{j_2}\cdots E_{j_n}\Delta_{\omega_i,w_0\omega_i}(\theta_{\mathbf{i}}(t_1,\ldots,t_{m+r}))}=\frac{\alpha f_0+\sum_{k=1}(\alpha-k)f_kt_1^k}{t_1\sum_{k=0}f_kt_1^k}.$$

Let α be a sufficiently large positive integer, then the right-hand side is a positive function. □

Remark 4.12. In fact, the arguments in the proof of Lemma 4.11 are applicable in a much larger context. We say that a function f on G^{e,w_0} is *totally positive* if for every word $\mathbf{j}=(j_1,\ldots,j_n)$ the expression $E_{j_1}\ldots E_{j_n}f$ is either a positive rational function or zero. The functions $\Delta_{\omega_i,w_0\omega_i}$ satisfy this property, but there are many other examples: for G simply laced, all elements of the dual canonical basis have this property, and for G semisimple, all cluster variables on G^{e,w_0} are totally positive.

If f is totally positive, $E_{j_2}\ldots E_{j_n}f$ (and all such derivatives) are products of polynomials with positive coefficients in variables $t_1,\ldots,t_m$ and of Laurent monomials in $t_{m+1},\ldots,t_{m+r}$. Hence, the proof of Lemma 4.11 applies and the logarithmic derivative

$$\frac{E_{j_1}E_{j_2}\ldots E_{j_n}f}{E_{j_2}\ldots E_{j_n}f}$$

is dominated by the potential $\phi_{j_1}=1/t_1$.

The theory of totally positive functions and its applications to functions dominated by potentials will be explored elsewhere.

Theorem 4.13. *For any nonzero vector $X\in\mathfrak{g}_\alpha$ in a positive root space of $\mathfrak{g}$ and any $w\in W$, we have $\frac{X\cdot\Delta_{\omega_i,w\omega_i}}{\Delta_{\omega_i,w\omega_i}}\in\tilde{P}^{est}_{\Phi}(G^{e,w_0},[\theta_{\mathbf{i}}])$.*

Proof. Without loss of generality, we represent X as a nested commutator of simple roots $X=[E_{k_1}[\cdots[E_{k_{n-1}},E_{k_n}]\cdots]]$. Since $\tilde{P}^{est}_{\Phi}(G^{e,w_0},[\theta_{\mathbf{i}}])$ is a ring, it suffices to show that for any word $\mathbf{j}=(j_1,\ldots,j_n)$ in the simple roots of $\mathfrak{g}$,

$$\frac{E_{j_1}\cdot E_{j_2}\cdots E_{j_n}\Delta_{\omega_i,w\omega_i}}{E_{j_2}\cdots E_{j_n}\Delta_{\omega_i,w\omega_i}}\in P^{est}_{\Phi}(G^{e,w_0},[\theta_{\mathbf{i}}]).$$

Since any choice of a reduced word $\mathbf{i}'$ for w_0 defines toric chart $\theta_{\mathbf{i}'}$ on G^{e,w_0} which is positively equivalent to $\theta_{\mathbf{i}}$, this follows immediately from Lemma 4.11. □

An analogous statement holds for the standard right action of $\mathfrak{g}$ on $\mathbb{C}[G]$. For the simple root $\alpha_i'=-w_0\alpha_i$, write $E_{i'}$ for the corresponding root vector. Define the potentials $\psi_i:=\frac{\Delta_{\omega_i,w_0\omega_i}\cdot E_{i'}}{\Delta_{\omega_i,w_0\omega_i}}=\frac{\Delta_{\omega_i,w_0s_i\omega_i}}{\Delta_{\omega_i,w_0\omega_i}}$, and let $\Psi:=\{\psi_1,\ldots,\psi_r\}$. Then $\Psi\subset P(X,[\theta_{\mathbf{i}}])$, and we have the following.

Theorem 4.14. *For any nonzero vector $X\in\mathfrak{g}_\alpha$ in a positive root space of $\mathfrak{g}$ and any $w\in W$, we have $\frac{\Delta_{\omega_i,w\omega_i}\cdot X}{\Delta_{\omega_i,w\omega_i}}\in\tilde{P}^{est}_{\Psi}(G^{e,w_0},[\theta_{\mathbf{i}}])$.*

The proof is symmetric to the proof of Theorem 4.13.

Definition 4.15. For any reduced word $\mathbf{i}$ for w_0, consider the positive variety with potential $(G^{e,w_0},[\theta_{\mathbf{i}}],\Phi_{BK})$, where

$$\Phi_{BK} := \{\varphi_1, \dots, \varphi_r, \psi_1, \dots, \psi_r\}$$

with φ_i, ψ_i are as above. We call Φ_{BK} the *Berenstein-Kazhdan Potential* on G^{e,w_0}. For a certain toric chart $\theta_z \in [\theta_{\mathbf{i}}]$ on G^{e,w_0}, the cone $\mathscr{C}^{\leq}_{\Phi_{BK}} \subset (G^{e,w_0}, \theta_z)^t$ is called the *extended string cone*. Abusing terminology slightly, we call $\mathscr{C}_{\Phi_{BK}} \subset (G^{e,w_0}, \theta)^t$ the *strict extended string cone* for any choice of $\theta \in [\theta_{\mathbf{i}}]$; any such cone is related to $\mathscr{C}_{\Phi_{BK}} \subset (G^{e,w_0}, \theta_z)^t$ by a piecewise linear bijection. See [5] for more details, including the connection between $\mathscr{C}^{\leq}_{\Phi_{BK}}$ and parameterizations of crystal bases. The potential Φ_{BK} was also discovered independently by Rietsch, see [17]. For $G = \mathrm{SL}_n$, for a certain choice of reduced word $\mathbf{i}$ for w_0, the cone $\mathscr{C}^{\leq}_{\Phi_{BK}} \subset (G^{e,w_0}, \theta_{\mathbf{i}})^t$ is the Gelfand-Zeitlin cone [7]. $\triangleleft$

5 Positive Poisson varieties

In this section we consider Poisson structures on positive varieties. We are interested in weakly log-canonical charts. These are toric charts in which the Poisson structure is log-canonical modulo terms weakly dominated by potentials. Examples are given by dual Poisson algebraic groups G^* for G a semisimple algebraic group.

5.1 Definition of positive Poisson varieties

Let (X, π) be an irreducible Poisson variety over $\mathbb{Q}$, and let Θ be a positive structure on X. Suppose $\theta : \mathbb{G}_m^n \to X$ is a toric chart with $\theta \in \Theta$ and let $(z_1, \dots, z_n)$ be the standard coordinates on $\mathbb{G}_m^n$. Fix a set of potentials $\Phi \subset P(X, [\theta])$ on X. Then θ is a *weakly log-canonical chart* for π if, in the coordinates z_i the bracket is of the form

$$\{z_i, z_j\} = z_i z_j (\pi_{ij} + f_{ij}), \tag{9}$$

where $\pi_{ij} \in k$ are constant and $f_{ij} \in \tilde{P}_{\Phi}(X, [\theta])$ are weakly dominated by Φ. We call π_{ij} the *log-canonical part* of the bracket.

If a Poisson variety (X, π) has a weakly log-canonical chart θ, we define $[\theta]_\pi$ to be the collection of all $\theta' \in [\theta]$ that is weakly log-canonical chart for π. We call $[\theta]_\pi$ a *π-compatible positive structure* on X.

We define a *positive Poisson variety* to be a quadruple $(X, \pi, \Theta_\pi, \Phi)$, where Θ_π is a π-compatible positive structure on X. Note that $[\theta]_\pi \subset [\theta]$ but in general $[\theta] \neq [\theta]_\pi$.

A *positive Poisson map* of positive Poisson varieties

$$\phi : (X_1, \pi_1, \Theta_{\pi_1}, \Phi_1) \to (X_2, \pi_2, \Theta_{\pi_2}, \Phi_2)$$

is a Poisson map $\phi : (X_1, \pi_1) \to (X_2, \pi_2)$, which is also a map of positive varieties with potential. We denote by **PosPoiss** the category of positive Poisson varieties.

Example 5.1. In [12], the authors define the *cluster manifold* $X(A)$ associated to a cluster algebra A. The cluster algebra structure on A gives $X(A)$ a Poisson structure π and a family of positively equivalent toric charts θ on X which are log-canonical for π. Cluster manifolds are then examples of positive Poisson varieties, with potential $\Phi = 0$; see Remark 3.3.

We record the following observation for future reference.

Proposition 5.2. *Let $\theta : \mathbb{G}_m^n \to X$ be a weakly log-canonical chart for $(X, \pi, \Theta_\pi, \Phi)$. If M and N are Laurent monomials in the standard coordinates z_j, then the bracket $\{M,N\}$ is weakly log-canonical; i.e.*

$$\{M,N\} = MN(\pi_{MN} + f_{MN}),$$

where $\pi_{MN} \in k$ and $f_{MN} \in \tilde{P}_\Phi(X, [\theta])$.

Proof. Assume M, N, and L are Laurent monomials in $z_1, \ldots, z_n$ and that

$$\{M,N\} = MN(\pi_{MN} + f_{MN}) \text{ and } \{M,L\} = ML(\pi_{ML} + f_{ML})$$

are weakly log-canonical. The proposition follows by induction, using the following two facts. First,

$$\{M,N^{-1}\} = -N^{-2}\{M,N\} = N^{-2}MN(-\pi_{MN} - f_{MN}) = MN^{-1}(-\pi_{MN} - f_{MN})$$

is weakly log-canonical. Second,

$$\{M,NL\} = \{M,N\}L + N\{M,L\} = MNL(\pi_{MN} + f_{MN}) + MNL(\pi_{ML} + f_{ML})$$

is weakly log-canonical. □

5.2 Poisson algebraic groups

Most of the material in this section is based on [10, 15]. Let G be a simply connected semisimple algebraic group over $\mathbb{Q}$ with Lie algebra $\mathfrak{g}$, as in Section 4.1. Let X_i be an orthonormal basis for $\mathfrak{h}$ under the Killing form. For positive roots $\alpha \in R^+$, choose root vectors $E_\alpha \in \mathfrak{g}_\alpha$ and $E_{-\alpha} \in \mathfrak{g}_{-\alpha}$, so that under the Killing form $K(\cdot,\cdot)$ we have $K(E_\alpha, E_{-\alpha}) = 1$. Consider the standard quasitriangular r-matrix

$$r_G := \frac{1}{2} \sum_{i=1}^{\operatorname{rank}\mathfrak{h}} X_i \otimes X_i + \sum_{\alpha \in R^+} E_\alpha \otimes E_{-\alpha}.$$

Let r_G^λ, r_G^ρ be the left- and right-invariant 2-tensor fields on G, respectively, which have $r_G^\lambda(e) = r_G^\rho(e) = r_G$. Then r_G satisfies the classical Yang-Baxter equation, and so $\pi_G = r_G^\lambda - r_G^\rho$ is a Poisson bivector field and (G, π_G) is a Poisson algebraic group.

The simply connected Drinfeld double of (G,π_G) is $(D,\pi_D)=(G\times G,\pi_D)$, where $\pi_D=r_D^\lambda-r_D^\rho$ and

$$r_D=\frac{1}{2}\sum_{i=1}^{\mathrm{rank}\,\mathfrak{h}}(X_i,X_i)\otimes(X_i,-X_i)+\sum_{\alpha\in R^+}(E_\alpha,E_\alpha)\otimes(0,-E_{-\alpha})+(E_{-\alpha},E_{-\alpha})\otimes(E_\alpha,0)$$

is the standard quasitriangular r-matrix. The diagonal embedding $(G,\pi_G)\to(D,\pi_D)$, sending $g\mapsto(g,g)$ is Poisson.

Recall that $[\cdot]_0:G_0\to H$ is the projection of Gaussian decomposable elements to H. Then the subgroup G^* of $G\times G$ given by

$$G^*=\{(b^+,b^-)\in B\times B_-\mid[b^+]_0[b^-]_0=1\}\subset G\times G$$

is a Poisson algebraic subgroup of (D,π_D), and $(G^*,-\pi_D|_{G^*})$ is a Poisson group dual to (G,π_G).

Proposition 5.3. *The Borel subgroups B and B_- are Poisson algebraic subgroups of G, and the projections*

$$\mathrm{pr}_1:G^*\to B,\qquad \mathrm{pr}_2:G^*\to B_-$$

are anti-Poisson.

Proof. Follows from the expression of the r-matrices r_G and r_D; see also [14]. □

Definition 5.4 (Notation). Let $f\in\mathbb{Q}[B]$ and $g\in\mathbb{Q}[B_-]$ be regular functions. Denote $f_1:=f\circ\mathrm{pr}_1\in\mathbb{Q}[G^*]$ and $g_2:=g\circ\mathrm{pr}_2\in\mathbb{Q}[G^*]$. ◁

For future reference, we note the following computation, which follows directly from the description of π_{G^*} above.

Proposition 5.5. *Let $f\in\mathbb{Q}[B]$ and $g\in\mathbb{Q}[B_-]$. Then in the notation of Definition* 5.4,

$$\{f_1,g_2\}_{G^*}=\frac{1}{2}\sum_{i=1}^{\mathrm{rank}\,\mathfrak{h}}(X_i\cdot f)_1(-X_i\cdot g)_2-(f\cdot X_i)_1(g\cdot(-X_i))_2 \\ +\sum_{\alpha\in R^+}(E_\alpha\cdot f)_1(-E_{-\alpha}\cdot g)_2-(f\cdot E_\alpha)_1(g\cdot(-E_{-\alpha}))_2. \tag{10}$$

5.3 The positive Poisson variety G^*

In this section we endow G^* with the structure of a positive Poisson variety. Recall the transpose $(\cdot)^T:G\to G$ anti-involution defined in Section 4.1. We introduce the Lie group involution $\tau:G\to G$, with $\tau(g)=(g^T)^{-1}=(g^{-1})^T$. From the expression for the r-matrix r_G, we have the following.

Proposition 5.6. *The map $\tau : (G, \pi_G) \to (G, \pi_G)$ is anti-Poisson. Also, τ restricts to the Lie group isomorphisms $\tau : B \to B_-$ and $\tau : B_- \to B$, and induces the automorphism*

$$\tau : G^* \to G^* \; : \; (b^+, b^-) \mapsto (((b^-)^T)^{-1}, ((b^+)^T)^{-1}).$$

Recall from Section 4.3, we have, for each reduced word $\mathbf{i}$ for w_0, a set of $r + l(w_0)$ cluster variables $F(\mathbf{i})$ which includes the generalized minors $\Delta_{\omega_k,\omega_k}$. For each $\mathbf{i}$ we defined a birational map $\Delta_{\mathbf{i}} : G^{e,w_0} \to \mathbb{G}_m^{r+l(w_0)}$, which had as its components the cluster variables $F(\mathbf{i})$. Let $\eta : \mathbb{G}_m^{r+l(w_0)} \to \mathbb{G}_m^{l(w_0)}$ be the projection parallel to the coordinates $\Delta_{\omega_k,\omega_k}$ of $\Delta_{\mathbf{i}}$. Let

$$\hat{\Delta}_{\mathbf{i}} := \eta \circ \Delta_{\mathbf{i}} : G^{e,w_0} \to \mathbb{G}_m^{l(w_0)}, \tag{11}$$

and let $\Delta_0 : G^{e,w_0} \to \mathbb{G}_m^r$ be given by

$$\Delta_0(g) = (\Delta_{\omega_1,\omega_1}(g), \ldots, \Delta_{\omega_r,\omega_r}(g)). \tag{12}$$

Recall that $G^{e,w_0} \subset B$ is a Zariski open subset. Consider the rational map

$$\begin{aligned} \mathbb{A}_{\mathbf{i}} : G^* &\to \mathbb{G}_m^{l(w_0)} \times \mathbb{G}_m^{l(w_0)} \times \mathbb{G}_m^r; \\ (b^+, b^-) &\mapsto (\hat{\Delta}_{\mathbf{i}}(b^+), \hat{\Delta}_{\mathbf{i}}(\tau(b^-)), \Delta_0(b^+)). \end{aligned} \tag{13}$$

The following proposition follows from the definition of G^* and Proposition 4.8.

Proposition 5.7. *The map $\mathbb{A}_{\mathbf{i}}$ is a biregular isomorphism from a Zariski open subset of G^* to $\mathbb{G}_m^{r+2l(w_0)}$.*

Example 5.8. Let $G = \mathrm{SL}_3$ as in Examples 4.1, 4.2, and 4.6 above. For $b \in G^{e,w_0}$, we have

$$\begin{aligned} \Delta_{\mathbf{i}}(b) &= (\Delta_{12,23}(b), \Delta_{1,3}(b), \Delta_{1,2}(b), \Delta_{12,12}(b), \Delta_{1,1}(b)), \\ \hat{\Delta}_{\mathbf{i}}(b) &= (\Delta_{12,23}(b), \Delta_{1,3}(b), \Delta_{1,2}(b)), \\ \Delta_0(b) &= (\Delta_{1,1}(b), \Delta_{12,12}(b)). \end{aligned}$$

The birational map $\mathbb{A}_{\mathbf{i}} : G^* \to \mathbb{G}_m^8$ is then given:

$$\begin{aligned} \mathbb{A}_{\mathbf{i}}(b^+, b^-) = &(\Delta_{12,23}(b^+), \Delta_{1,3}(b^+), \Delta_{1,2}(b^+), \\ &\Delta_{12,23}(\tau(b^-)), \Delta_{1,3}(\tau(b^-)), \Delta_{1,2}(\tau(b^-)), \Delta_{1,1}(b^+), \Delta_{12,12}(b^+)). \end{aligned}$$

Consider the toric chart $\mathbb{A}_{\mathbf{i}}^{-1} : \mathbb{G}_m^{r+2l(w_0)} \to G^*$ on G^*. By (8), another reduced word $\mathbf{i}'$ for w_0 gives a toric chart $\mathbb{A}_{\mathbf{i}'}^{-1}$ which is positively equivalent to $\mathbb{A}_{\mathbf{i}}^{-1}$.

We define the set of potentials on G^*, following the notation of Section 4.4:

$$\Phi_{G^*} := \left\{ \frac{(E_i \cdot \Delta_{\omega_i, w_0 \omega_i})_1}{(\Delta_{\omega_i, w_0 \omega_i})_1}, \frac{(\Delta_{\omega_i, w_0 \omega_i} \cdot E_{i'})_1}{(\Delta_{\omega_i, w_0 \omega_i})_1}, \right.$$
$$\left. \frac{((E_i \cdot \Delta_{\omega_i, w_0 \omega_i}) \circ \tau)_2}{(\Delta_{\omega_i, w_0 \omega_i} \circ \tau)_2}, \frac{((\Delta_{\omega_i, w_0 \omega_i} \cdot E_{i'}) \circ \tau)_2}{(\Delta_{\omega_i, w_0 \omega_i} \circ \tau)_2} \right\}_i,$$

where as usual i indexes the simple roots of G. Note that $\Phi_{G^*} \subset P(G^*, [\mathbb{A}_{\mathbf{i}}^{-1}])$ by Proposition 4.10. Recalling Definition 4.15, we see that Φ_{G^*} comes from the Berenstein-Kazhdan potential Φ_{BK} on G^{e,w_0}. Note that Φ_{G^*} restricts to regular functions on the toric chart $\mathbb{A}_{\mathbf{i}}^{-1}$, and so by Remark 3.3 the cone $\mathscr{C}_{\Phi_{G^*}}(G^*, \mathbb{A}_{\mathbf{i}}^{-1})(\mathbb{R})$ is convex.

Theorem 5.9. *For any reduced word* $\mathbf{i}$ *of* w_0*, the quadruple* $(G^*, \pi_{G^*}, [\mathbb{A}_{\mathbf{i}}^{-1}]_{\pi_{G^*}}, \Phi_{G^*})$ *is a positive Poisson variety.*

Proof. We check that π_{G^*} is weakly log-canonical in the coordinates given by $\mathbb{A}_{\mathbf{i}}$. Let $\Delta, \Delta' \in F(\mathbf{i})$. We consider three types of brackets. Recall the notation of Definition 5.4.

(a) Bracket of type $\{\Delta_1, \Delta_1'\}_{G^*}$. By Proposition 5.3,

$$\{\Delta_1, \Delta_1'\}_{G^*} = -\{\Delta, \Delta'\}_B \circ \mathrm{pr}_1 .$$

By Theorem 2.6 of [14], the bracket $\{\Delta \circ \xi, \Delta' \circ \xi\}_B$ is log-canonical on the open subset G^{e,w_0}, where $\xi : G^{e,w_0} \to G^{e,w_0}$ is the twist map of Proposition 4.9. By Theorem 3.1 of [12], the twist map ξ is anti-Poisson.

(b) Bracket of type $\{(\Delta \circ \tau)_2, (\Delta' \circ \tau)_2\}_{G^*}$. Once we note $\tau : B_- \to B$ is anti-Poisson by Proposition 5.6, the argument is the same as the previous case.

(c) Bracket of type $\{\Delta_1, (\Delta' \circ \tau)_2\}_{G^*}$. By the definition of τ and (10), we have

$$(14) \quad \{\Delta_1, (\Delta' \circ \tau)_2\}_{G^*}$$
$$= \frac{1}{2} \sum_{i=1}^{\mathrm{rank}\, \mathfrak{h}} (X_i \cdot \Delta)_1 ((X_i \cdot \Delta') \circ \tau)_2 - (\Delta \cdot X_i)_1 ((\Delta' \cdot X_i) \circ \tau)_2$$
$$+ \sum_{\alpha \in R^+} (E_\alpha \cdot \Delta)_1 ((E_\alpha \cdot \Delta') \circ \tau))_2 - (\Delta \cdot E_\alpha)_1 ((\Delta' \cdot E_\alpha) \circ \tau)_2.$$

Write $\Delta = \Delta_{\omega_j, u\omega_j}$ and $\Delta' = \Delta_{\omega_k, v\omega_k}$. By (6), the first sum

$$\frac{1}{2} \sum_{i=1}^{\mathrm{rank}\, \mathfrak{h}} (X_i \cdot \Delta)_1 ((X_i \cdot \Delta') \circ \tau)_2 - (\Delta \cdot X_i)_1 ((\Delta' \cdot X_i) \circ \tau)_2$$
$$= \frac{1}{2} \left(\sum_{i=1}^{\mathrm{rank}\, \mathfrak{h}} \omega_j(X_i) \omega_k(X_i) - u\omega_j(X_i) v\omega_k(X_i) \right) \Delta_1 (\Delta' \circ \tau)_2$$
$$= \frac{1}{2} (K(\omega_j, \omega_k) - K(u\omega_j, v\omega_k)) \Delta_1 (\Delta' \circ \tau)_2$$

is log-canonical. For future reference, we note that here the coefficient of $\Delta_1(\Delta'\circ\tau)_2$ is rational. We then write

$$\{\Delta_1,(\Delta'\circ\tau)_2\}_{G^*}=\Delta_1(\Delta'\circ\tau)_2(c+f), \tag{15}$$

where

$$f=\sum_{\alpha\in R^+}\frac{(E_\alpha\cdot\Delta)_1}{\Delta_1}\frac{((E_\alpha\cdot\Delta')\circ\tau)_2}{(\Delta'\circ\tau)_2}-\frac{(\Delta\cdot E_\alpha)_1}{\Delta_1}\frac{((\Delta'\cdot E_\alpha)\circ\tau)_2}{(\Delta'\circ\tau)_2}.$$

By Theorem 4.13 and Theorem 4.14, f is weakly estimate dominated by Φ_{G^*}, and thus weakly dominated by Φ_{G^*}. □

Remark 5.10. Theorem 5.9 describes a number of toric charts (labeled by reduced words of w_0) making the dual Poisson algebraic group G^* into a positive Poisson variety. In fact, there are many more charts with this property. In particular, every cluster chart on G^{e,w_0} gives rise to a positive Poisson variety structure on G^*. Note that the number of charts labeled by reduced words is always finite whereas the number of cluster charts is usually infinite.

It turns out that the proof of Theorem 5.9 applies in the case of arbitrary cluster charts: Poisson brackets of the type $\{f_1,f'_1\}_{G^*}$ reduce to Poisson brackets on B, and they are log-canonical in all cluster charts (by the standard properties of Poisson cluster varieties). And Poisson brackets of the type $\{f_1,(f'\circ\tau)_2\}_{G^*}$ are weakly dominated by the potentials in Remark 4.12.

This viewpoint will be explored elsewhere.

Example 5.11. Let $G=\mathrm{SL}_3$ as in Examples 4.1, 4.2, and 4.6. The set of potentials on SL_3^* is

$$\Phi_{G^*}=\left\{\frac{(\Delta_{13,23})_1}{(\Delta_{12,23})_1},\frac{(\Delta_{2,3})_1}{(\Delta_{1,3})_1},\frac{(\Delta_{12,13})_1}{(\Delta_{12,23})_1},\frac{(\Delta_{1,2})_1}{(\Delta_{1,3})_1},\right.$$
$$\left.\frac{(\Delta_{13,23}\circ\tau)_2}{(\Delta_{12,23}\circ\tau)_2},\frac{(\Delta_{2,3}\circ\tau)_2}{(\Delta_{1,3}\circ\tau)_2},\frac{(\Delta_{12,13}\circ\tau)_2}{(\Delta_{12,23}\circ\tau)_2},\frac{(\Delta_{1,2}\circ\tau)_2}{(\Delta_{1,3}\circ\tau)_2}\right\}.$$

From Theorem 5.9, we know that, with this choice of potentials Φ_{G^*}, the chart $\Delta_{\mathbf{i}}^{-1}$ is weakly log-canonical for $\pi_{\mathrm{SL}_3^*}$. As an example, we compute the bracket $\{(\Delta_{1,3})_1,(\Delta_{1,2}\circ\tau)_2\}$ in SL_3^*; see (14) above:

$$\begin{aligned}
&\{(\Delta_{1,3})_1, (\Delta_{1,2}\circ\tau)_2\} \\
&= \frac{1}{2}(K(\omega_1,\omega_1) - K(w_0\omega_1, s_1\omega_1))(\Delta_{1,3})_1(\Delta_{1,2}\circ\tau)_2 \\
&\quad + \sum_{ij\in\{12,23,13\}} (E_{ij}\cdot\Delta_{1,3})_1((E_{ij}\cdot\Delta_{1,2})\circ\tau))_2 - (\Delta_{1,3}\cdot E_{ij})_1((\Delta_{1,2}\cdot E_{ij})\circ\tau)_2. \\
&= c(\Delta_{1,3})_1(\Delta_{1,2}\circ\tau)_2 + (\Delta_{2,3})_1(\Delta_{2,2}\circ\tau)_2 + (\Delta_{3,3})_1(\Delta_{3,2}\circ\tau)_2. \\
(16)\quad &= (\Delta_{1,3})_1(\Delta_{1,2}\circ\tau)_2\left(c + \frac{(\Delta_{2,3})_1(\Delta_{2,2}\circ\tau)_2}{(\Delta_{1,3})_1(\Delta_{1,2}\circ\tau)_2} + \frac{(\Delta_{3,3})_1(\Delta_{3,2}\circ\tau)_2}{(\Delta_{1,3})_1(\Delta_{1,2}\circ\tau)_2}\right).
\end{aligned}$$

One can check that the two terms $\frac{(\Delta_{2,3})_1(\Delta_{2,2}\circ\tau)_2}{(\Delta_{1,3})_1(\Delta_{1,2}\circ\tau)_2} + \frac{(\Delta_{3,3})_1(\Delta_{3,2}\circ\tau)_2}{(\Delta_{1,3})_1(\Delta_{1,2}\circ\tau)_2}$ of (16) are indeed weakly estimate-dominated by Φ_{G^*}.

6 Tropicalization of Poisson structures

In this section, we pass to complex points of our varieties, which have thus far been defined over $\mathbb{Q}$. Specifically, a *complex positive variety* $X(\mathbb{C})$ is the complex points of the positive variety $(X,[\theta])$. A toric chart $\theta : \mathbb{G}_m^n \to X$ induces an open embedding $(\mathbb{C}^\times)^n \to X(\mathbb{C})$, which we also call a toric chart, and the definitions of the previous sections extend similarly under base change.

In the remainder of this paper, we consider only complex positive varieties and their real forms. For simplicity we write $X = X(\mathbb{C})$. In particular, taking complex points of the Poisson algebraic groups of Section 5.2 gives complex Poisson-Lie groups $G = G(\mathbb{C})$ and $G^* = G^*(\mathbb{C})$.

We define (under some extra assumptions) a tropicalization $\mathscr{C}\times\mathbb{T}$ of a complex positive Poisson variety X. Here $\mathscr{C}$ is a cone, $\mathbb{T}$ is a real torus, the tropicalization carries a constant Poisson structure. In particular, for a dual Poisson-Lie group $G^* = G^*(\mathbb{C})$ we obtain that $\mathscr{C}$ is the strict extended string cone.

6.1 *Real forms of Poisson structures*

In this section we introduce real forms of holomorphic Poisson structures. Much of this material is well known; see for instance [20].

Let (X,π) be a complex manifold with holomorphic Poisson structure $\pi\in\Lambda^2(T^{1,0}X)$. Let $\pi = \pi_R + i\pi_I$ be the decomposition of π into real and imaginary parts; it is well known that $\pi_R,\pi_I\in\Gamma(\Lambda^2(TX))$ are (real) Poisson bivectors. Let $\overline{\tau} : X\to X$ be an anti-holomorphic involution of X, which satisfies $\overline{\tau}(\pi_R) = \pi_R$. In this case, we say $\overline{\tau}$ is *Poisson*.

Remark 6.1. Equivalently, $\overline{\tau}$ is Poisson if $\overline{\tau}(\pi_I) = -\pi_I$. Extending $\overline{\tau}$ conjugate-linearly to $TX\otimes\mathbb{C}$, this is equivalent to the condition $\overline{\tau}(\pi) = \pi$.

Let $K \subset \mathrm{Fix}(\overline{\tau})$ be a (real) open submanifold of the fixed points of $\overline{\tau}$. For any $p \in K$, decompose T_pX as $T_pX = (T_pX)^{\overline{\tau}} + (T_pX)^{-\overline{\tau}}$, where

$$(T_pX)^{\overline{\tau}} = \{v \in T_pX : \overline{\tau}(v) = v\}, \quad (T_pX)^{-\overline{\tau}} = \{v \in T_pX : \overline{\tau}(v) = -v\}.$$

As shown in [20], π_R can be decomposed as $\pi_R(p) = \pi_R^{\overline{\tau}}(p) + \pi_R^{-\overline{\tau}}(p)$, where

$$\pi_R^{\overline{\tau}}(p) \in \Lambda^2(T_pX)^{\overline{\tau}}, \quad \pi_I^{-\overline{\tau}}(p) \in \Lambda^2(T_pX)^{-\overline{\tau}}.$$

Lemma 6.2 ([20]). *Using the notation above, $\pi_R^{\overline{\tau}}$ is a Poisson bivector on K.*

The pair $(K, \pi_R^{\overline{\tau}})$ is called a *real form* of X.

Example 6.3. Consider any holomorphic Poisson structure

$$\pi = \sum_{i,j} \pi_{ij}(z)\partial_{z_i} \wedge \partial_{z_j}$$

on $\mathbb{C}^n$, where $\pi_{ij}(z)$ are holomorphic functions. Let $\overline{\tau}$ be the anti-holomorphic involution of $\mathbb{C}^n$ given by $\overline{\tau}(z) = \overline{z}$. Then the set of fixed points of $\overline{\tau}$ is $K = \mathbb{R}^n \subset \mathbb{C}^n$. Thus $\overline{\tau}(\pi_R) = \pi_R$ if and only if $\pi_{ij}(\overline{z}) = \overline{\pi_{ij}(z)}$. Write

$$\partial_{z_i} = \frac{1}{2}(\partial_{x_i} - \sqrt{-1}\partial_{y_i}),$$

and direct calculation shows that

$$\pi_R^{\overline{\tau}} = \frac{1}{4}\sum_{i,j} \pi_{ij}(x)\partial_{x_i} \wedge \partial_{x_j}. \tag{17}$$

Lemma 6.4. *Let X be a complex manifold with holomorphic Poisson structure π. Let $\overline{\tau}$ be an anti-holomorphic Poisson involution of (X, π). Let $(K, \pi_R^{\overline{\tau}})$ be the corresponding real form. Then we have*

$$\{f_1|_K, f_2|_K\}_{\pi_R^{\overline{\tau}}} = \frac{1}{4}\{f_1, f_2\}_\pi|_K,$$

where the f_i are holomorphic functions on an open subset $U \subset X$ satisfying $f_i(\overline{\tau}(z)) = \overline{f_i(z)}$.

Proof. We only need to show the lemma in a neighborhood of each fixed point of $\overline{\tau}$. Choose holomorphic local coordinates $z_1, \ldots, z_n$ such that $\overline{\tau}$ is given by $\overline{\tau}(z) = \overline{z}$. Then in these coordinates, the f_i satisfy $f_i(\overline{z}) = \overline{f_i(z)}$. Set $z_j = x_j + \sqrt{-1}y_j$ and let $x = (x_1, \ldots, x_n)$ and $y = (y_1, \ldots, y_n)$. Write $f_i(z) = u_i(x,y) + \sqrt{-1}v_i(x,y)$, where $u_i, v_i \in C^\infty(X)$ are smooth real-valued functions on X. Since $f_i(\overline{z}) = \overline{f_i(z)}$, we know $u_i(x,y) = u_i(x,-y)$ and $v_i(x,y) = -v_i(x,-y)$. Thus $\partial_{y_j} u_i|_{y=0} = 0$. Then by the Cauchy-Riemann equations, we have

$$\left(\partial_{z_j} f_i(z)\right)\Big|_K = \frac{1}{2}\left(\partial_{x_j} u_i + \partial_{y_j} v_i - \sqrt{-1}\partial_{y_j} u_i + \sqrt{-1}\partial_{x_j} v_i\right)\Big|_{y=0}$$
$$= \frac{1}{2}\left(\partial_{x_j} u_i + \partial_{y_j} v_i\right)\Big|_{y=0} = \partial_{x_j} f_i(x)$$

By Equation (17), we get the conclusion. □

Proposition 6.5. *Let $X = (\mathbb{C}^\times)^n$ with holomorphic Poisson bivector π, let $\tau : X \to X$ be an algebraic involution of X, and let $\overline{\tau} : X \to X$ be the anti-holomorphic involution given by $\overline{\tau}(z) := \tau(\bar{z}) = \overline{\tau(z)}$. Assume $\overline{\tau}$ is Poisson, and consider the real form $(K, \pi_R^{\overline{\tau}})$. Let f_1, f_2 be holomorphic functions on an open subset $U \subset X$. Then*

$$\{f_1|_K, f_2|_K\}_{\pi_R^{\overline{\tau}}} = \frac{1}{4}\{f_1, f_2\}_\pi|_K.$$

Proof. Let $g_i := f_i + f_i \circ \tau$ and $h_i := f_i - f_i \circ \tau$. Then g_i and $\sqrt{-1}h_i$ satisfy the condition from Lemma 6.4:

$$g_i(\overline{\tau}(z)) = \overline{g_i(z)}, \qquad \sqrt{-1}h_i(\overline{\tau}(z)) = \overline{\sqrt{-1}h_i(z)}.$$

We then compute:

$$\begin{aligned}
\{f_1|_K, f_2|_K\}_{\pi_R^{\overline{\tau}}} &= \frac{1}{4}\{g_1|_K + h_1|_K, g_2|_K + h_2|_K\}_{\pi_R^{\overline{\tau}}} \\
&= \frac{1}{4}\Big(\{g_1|_K, g_2|_K\}_{\pi_R^{\overline{\tau}}} - \sqrt{-1}\{g_1|_K, \sqrt{-1}h_2|_K\}_{\pi_R^{\overline{\tau}}} \\
&\quad - \sqrt{-1}\{\sqrt{-1}h_1|_K, g_2|_K\}_{\pi_R^{\overline{\tau}}} - \{\sqrt{-1}h_1|_K, \sqrt{-1}h_2|_K\}_{\pi_R^{\overline{\tau}}}\Big) \\
&= \frac{1}{16}\Big(\{g_1, g_2\}_\pi|_K - \sqrt{-1}\{g_1, \sqrt{-1}h_2\}_\pi|_K \\
&\quad - \sqrt{-1}\{\sqrt{-1}h_1, g_2\}_\pi|_K - \{\sqrt{-1}h_1, \sqrt{-1}h_2\}_\pi|_K\Big) \\
&= \frac{1}{16}\{g_1 + h_1, g_2 + h_2\}_\pi|_K = \frac{1}{4}\{f_1, f_2\}_\pi|_K.
\end{aligned}$$

□

6.2 Real forms of positive Poisson varieties

In this section we define real forms of positive Poisson varieties, and give as an example the real form K^* of the positive Poisson variety G^*.

Let $(X, [\theta])$ be a complex positive variety, with $\theta : (\mathbb{C}^\times)^n \to X$. Then complex conjugation $\overline{(\cdot)} : (\mathbb{C}^\times)^n \to (\mathbb{C}^\times)^n$ defines an anti-holomorphic involution on the open subvariety $\theta((\mathbb{C}^\times)^n) \subset X$. Since the transition maps between charts in $[\theta]$ are positive, their components have rational coefficients and they commute with complex conjugation of $(\mathbb{C}^\times)^n$. So $\overline{(\cdot)}$ extends to the open subvariety $\bigcup_{\theta'\in[\theta]} \theta'((\mathbb{C}^\times)^n)$. In particular, if the charts in $[\theta]$ cover X, the involution $\overline{(\cdot)}$ is defined on all of X.

Definition 6.6. Let $(X,\pi,[\theta]_\pi,\Phi)$ be a complex positive Poisson variety, with $\theta:(\mathbb{C}^\times)^n \to X$ a π-compatible chart. Let $\tau: X\to X$ be an involution, which is also a map of positive varieties with potential. Then $\overline{\tau}: X\to X$ is an anti-holomorphic involution of X, where $\overline{\tau}(x) := \tau(\bar{x}) = \overline{\tau(x)}$. Assume $\overline{\tau}$ is Poisson.

We impose the additional conditions:

(a) In the standard coordinates $z_1,\dots,z_n$ on $(\mathbb{C}^\times)^n$, the positive transformation τ is monomial.
(b) The log-canonical part π_{ij} of the bracket of coordinate functions $\{z_i,z_j\}$ is pure imaginary.

In this case we call the tuple (X,π,θ,Φ,τ) a *framed positive Poisson variety with real form*. Maps from $(X_1,\pi_1,\Theta_{\pi_1},\Phi_1,\tau_1)$ to $(X_2,\pi_2,\Theta_{\pi_2},\Phi_2,\tau_2)$ are maps of complex positive Poisson varieties which intertwine τ_1 and τ_2. The corresponding category is denoted $\mathbf{PosPoiss}^{\bullet}_{\mathbb{R}}$. ◁

Theorem 6.7. *Let (G^*,π_{G^*}) be the dual Poisson-Lie group given by complex points of the Poisson algebraic group G^* introduced in Section 5.2, and let τ be the involution as in Proposition 5.6:*

$$\tau: G^*\to G^* \;:\; (b^+,b^-)\mapsto (((b^-)^T)^{-1},((b^+)^T)^{-1}).$$

Then for a reduced word $\mathbf{i}$ for w_0, the tuple $(G^*,\sqrt{-1}\pi_{G^*},\Delta_{\mathbf{i}}^{-1},\Phi_{G^*},\tau)$ is a framed positive Poisson variety with real form.

Proof. From the expression of the r-matrix r_D, we see that τ is a holomorphic anti-Poisson involution of (G^*,π_{G^*}). It follows that $\overline{\tau}$ is an anti-holomorphic Poisson involution of $(G^*,\sqrt{-1}\pi_{G^*})$. In the toric coordinates given by $\Delta_{\mathbf{i}}^{-1}$, the involution τ simply permutes the coordinates and thus τ satisfies condition (a) of Definition 6.6. Also, $\tau^*\Phi_{G^*} = \Phi_{G^*}$, so τ is a map of positive varieties with potential. By the proof of Theorem 5.9, the weakly log-canonical part of the bracket $\sqrt{-1}\pi_{G^*}$ is a rational multiple of $\sqrt{-1}$, and therefore satisfies condition (b) of Definition 6.6. □

We next want to specify a canonical choice for the real form of (X,π,θ,Φ,τ); note the real form K was not uniquely determined in Section 6.1. We first introduce some notation.

Definition 6.8 (Notation). Let $(X,\pi,\theta:(\mathbb{C}^\times)^n\to X,\Phi,\tau)$ be a framed positive Poisson variety with real form. We set $(X,\theta)^t(\mathbb{R}) := (X,\theta)^t\otimes_{\mathbb{Z}}\mathbb{R}$ to be the extension by scalars. For simplicity, sometimes we use $X^t(\mathbb{R})$ for $(X,\theta)^t(\mathbb{R})$. Let

$$L_{X,\theta,\tau} := \{\xi\in X^t(\mathbb{R}) \mid \tau^t(\xi)=\xi\}$$

be the fixed points of the tropical involution τ^t, and let

$$\mathbb{T}_{X,\theta,\tau} := \{g\in (S^1)^n\subset(\mathbb{C}^\times)^n \mid \tau(\bar{g}) = g\}_0$$

be the identity component of the fixed points of the anti-holomorphic involution $\overline{\tau}|_{(S_1)^n}$. Sometimes we write $L_X = L_{X,\theta,\tau}$ and $\mathbb{T}_X = \mathbb{T}_{X,\theta,\tau}$ for brevity. Finally, let $s > 0$ be a real parameter. Define the map

$$E_{X,\theta,s} : (X,\theta)^t(\mathbb{R}) \times (S^1)^n \to (\mathbb{C}^\times)^n;$$
$$(\xi_1,\dots,\xi_n,e^{\sqrt{-1}v_1},\dots,e^{\sqrt{-1}v_n}) \mapsto (e^{s\xi_1+\sqrt{-1}v_1},\dots,e^{s\xi_n+\sqrt{-1}v_n}).$$

Recall our assumption that τ is a monomial transformation in the toric coordinates given by θ. Thus $\overline{\tau}$ preserves the compact torus $(S^1)^n \subset (\mathbb{C}^\times)^n$, and $L_{X,\theta,\tau} \subset X^t(\mathbb{R})$ is a linear subspace. ◁

Pulling back $\overline{\tau}$ by $\theta \circ E_{X,\theta,s}$, we find:

Proposition 6.9. *Let (X,π,θ,Φ,τ) be a framed positive Poisson variety with real form. Then for all $s,s' > 0$, we have*

$$(\theta \circ E_{X,\theta,s})(L_X \times \mathbb{T}_X) = (\theta \circ E_{X,\theta,s'})(L_X \times \mathbb{T}_X) \subset \mathrm{Fix}(\overline{\tau}) \cap \theta((\mathbb{C}^\times)^n) \subset X.$$

Definition 6.10. Let (X,π,θ,Φ,τ) be a framed positive Poisson variety with real form. Then the *real form* of (X,π,θ,Φ,τ) is $(\mathfrak{R}(X,\pi,\theta,\Phi,\tau),\pi_{\mathfrak{R}(X,\theta,\Phi,\tau)})$, where

$$\mathfrak{R}(X,\pi,\theta,\Phi,\tau) = (\theta \circ E_{X,\theta,s'})(L_X \times \mathbb{T}_X) \subset \mathrm{Fix}(\overline{\tau}) \cap \theta((\mathbb{C}^\times)^n), \quad \pi_{\mathfrak{R}(X,\theta,\Phi,\tau)} = 4\pi_R^{\overline{\tau}}.$$

We usually write $(\mathfrak{R}(X),\pi_{\mathfrak{R}(X)})$ for simplicity. Note that the definition does not depend on the choice of $s > 0$, and that, (up to a scalar multiple of the bracket), the Poisson manifold $(\mathfrak{R}(X),\pi_{\mathfrak{R}(X)})$ is a real form in the sense of the previous section. In summary, we have the following diagram:

$$\begin{array}{ccccc}
(X,\theta)^t(\mathbb{R}) \times (S^1)^n & \xrightarrow[\sim]{E_{X,\theta,s}} & (\mathbb{C}^\times)^n & \xhookrightarrow{\theta} & X \\
\uparrow & & & & \uparrow \\
L_{X,\theta,\tau} \times \mathbb{T}_{X,\theta,\tau} & & \xrightarrow[\sim]{\theta \circ E_{X,\theta,s}} & & \mathfrak{R}(X).
\end{array}$$

◁

Remark 6.11. By Proposition 6.5, we have $\{z_i|_{\mathfrak{R}(X)}, z_j|_{\mathfrak{R}(X)}\}_{\pi_R^{\overline{\tau}}} = \frac{1}{4}\{z_i,z_j\}_\pi|_{\mathfrak{R}(X)}$. The factor of 4 in the definition of $\pi_{\mathfrak{R}(X)}$ is simply to make formulas simpler.

Proposition 6.12. *Let $f : (X_1,\pi_1,\theta_1,\Phi_1,\tau_1) \to (X_2,\pi_2,\theta_2,\Phi_2,\tau_2)$ be a map of framed positive Poisson varieties with real forms. Then $\theta_2^{-1} \circ f \circ \theta_1$ restricts to a Poisson map*

$$(\theta_1^{-1}(\mathfrak{R}(X_1)),\theta_1^*\pi_{\mathfrak{R}(X_1)}) \to (\theta_2^{-1}(\mathfrak{R}(X_2)),\theta_2^*\pi_{\mathfrak{R}(X_2)})$$

whenever it is defined.

Proof. Since f is a map of framed positive Poisson varieties with real forms, we have $f \circ \overline{\tau_1} = \overline{\tau_2} \circ f$. Because f is a positive rational map, it takes $\theta_1(\mathbb{T}_{X_1})$ to $\theta_2(\mathbb{T}_{X_2})$, and so it takes the component $\mathfrak{R}(X_1)$ of $\mathrm{Fix}(\overline{\tau}_1)$ to the component $\mathfrak{R}(X_2)$ of $\mathrm{Fix}(\overline{\tau}_2)$. That the map is Poisson follows from Proposition 6.5. □

Remark 6.13. Let $K^* \subset G^*$ be the identity component of the fixed points of $\overline{\tau}$. From Proposition 6.7 we know $(G^*, \sqrt{-1}\pi_{G^*}, \Delta_{\mathbf{i}}^{-1}, \Phi_{G^*}, \tau)$ is a positive Poisson variety with real form. The real form given by Definition 6.10 is an open dense submanifold of K^*, and its Poisson structure extends to the Poisson structure $\pi_{K^*} = 4(\sqrt{-1}\pi_{G^*})_R^{\overline{\tau}}$ on K^*. We then think of the real form of $(G^*, \sqrt{-1}\pi_{G^*}, \Delta_{\mathbf{i}}^{-1}, \Phi_{G^*}, \tau)$ as a coordinate neighborhood on K^*.

The Poisson-Lie group (G, π_G) has a compact real form (K, π_K); see [2]. It is shown, see for example [10, 20], that (K^*, π_{K^*}) is a dual Poisson-Lie group of (K, π_K). From the Iwasawa decomposition $G = KAN$ of G, we may identify $K^* = AN$.

Example 6.14. Let $G = \mathrm{SL}_3(\mathbb{C})$ as in Examples 4.1, 4.2, 4.6, and 5.11 above. It is easy to see that τ is a positive involution of $(\mathrm{SL}_3^*, [\Delta_{\mathbf{i}}^{-1}])$, and τ^t preserves $\mathscr{C}_{\Phi_{G^*}}$. In the toric coordinates $z_1, \dots, z_8$ given by $\Delta_{\mathbf{i}}^{-1}$, the anti-holomorphic involution $\overline{\tau} : (\mathbb{C}^\times)^8 \to (\mathbb{C}^\times)^8$ becomes

$$\overline{\tau}(z_1, \dots, z_8) = (\bar{z}_4, \bar{z}_5, \bar{z}_6, \bar{z}_1, \bar{z}_2, \bar{z}_3, \bar{z}_7, \bar{z}_8).$$

Fixed points are those of the form

$$(z_1, z_2, z_3, \bar{z}_1, \bar{z}_2, \bar{z}_3, x_7, x_8), \qquad \text{where } x_7, x_8 \in \mathbb{R} \setminus \{0\}.$$

Projecting parallel to the fourth, fifth, and sixth coordinates and restricting to the positive part of the last two coordinates gives coordinates on the usual presentation of K^* as the group of complex upper-triangular matrices, with positive real entries along the diagonal, as in [1, 3].

6.3 Partial tropicalization

In this section we construct, for each framed positive Poisson variety with real form $(X, \pi, \theta, \Phi, \tau)$, a real manifold $PT(X, \pi, \theta, \Phi, \tau)$ with constant Poisson bracket π_{PT} called its *partial tropicalization*. We extend partial tropicalization to a functor

$$PT : \mathbf{PosPoiss}_{\mathbb{R}}^{\bullet} \to \mathbf{PTrop}$$

to the category of partial tropicalizations, defined below. We show in Theorem 6.18 below that partial tropicalization can be thought of as the limit of a 1-parameter family of coordinates on the real form $\mathfrak{R}(X)$.

Definition 6.15 (Notation). Let $(X, \theta_X : (\mathbb{C}^\times)^n \to X, \Phi_X)$ be a framed complex positive variety with potential. Let

$$\mathscr{C}_\Phi(X, \theta_X)(\mathbb{R}) = \{\xi \in (X, \theta_X)^t(\mathbb{R}) \mid \Phi^t(\xi) < 0\};$$

see also Remark 3.3. For simplicity, sometimes we use $\mathscr{C}_\Phi(\mathbb{R})$ for $\mathscr{C}_\Phi(X,\theta_X)(\mathbb{R})$. ◁

Definition 6.16. Let (X,π,θ,Φ,τ) be a framed positive Poisson variety with real form. Define

$$PT(X,\pi,\theta,\Phi,\tau) := (\mathscr{C}_\Phi(\mathbb{R}) \cap L_{X,\theta,\tau}) \times \mathbb{T}_{X,\theta,\tau},$$

where $L_{X,\theta,\tau}$ and $\mathbb{T}_{X,\theta,\tau}$ are given in Definition 6.8.

Let $z_1,\dots,z_n$ be the standard coordinate functions on the toric chart $\theta\colon (\mathbb{C}^\times)^n \to X$. Consider induced involution τ^* of the field of rational functions $\mathbb{C}(z_1,\dots,z_n)$. Recall from Proposition 5.2 that $\{z_i,\tau^*(z_j)\}$ is weakly log-canonical. Denote by $\pi_{i\tau(j)}$ the log-canonical part of the bracket, and note that by assumption (b) of Definition 6.6 and the proof of Proposition 5.2, we have $\pi_{i,\tau(j)} \in \sqrt{-1}\mathbb{R}$ is pure imaginary. Also, recall that the coordinates $z_1,\dots,z_n$ on $(\mathbb{C}^\times)^n$ induce coordinates $\xi_1,\dots,\xi_n$ on $X^t(\mathbb{R})$. Writing $(e^{\sqrt{-1}\nu_1},\dots,e^{\sqrt{-1}\nu_n}) \in (S^1)^n$ gives us local coordinates $\nu_1,\dots,\nu_n$ on $(S^1)^n \subset (\mathbb{C}^\times)^n$. We define a constant real bracket on $X^t(\mathbb{R}) \times (S^1)^n$ as follows:

$$\begin{aligned}
\{\xi_i,\nu_j\} &:= \frac{-\sqrt{-1}}{2}(\pi_{ij} - \pi_{i\tau(j)}),\\
\{\nu_i,\xi_j\} &:= \frac{-\sqrt{-1}}{2}(\pi_{ij} + \pi_{i\tau(j)}),\\
\{\xi_i,\xi_j\} &:= 0,\\
\{\nu_i,\nu_j\} &:= 0.
\end{aligned}$$

We set π_{PT} to be the restriction of this bracket to $PT(X,\pi,\theta,\Phi,\tau) \subset X^t(\mathbb{R}) \times (S^1)^n$. The pair $(PT(X,\pi,\theta,\Phi,\tau),\pi_{PT})$ is the *partial tropicalization* of (X,π,θ,Φ,τ). ◁

We will see π_{PT} is skew-symmetric and well defined after giving a scaling interpretation for partial tropicalization in Theorem 6.18.

Lemma 6.17. *Let $f \in \tilde{P}_\Phi(X,[\theta])$ be a function which is weakly dominated by Φ, and assume f restricts to a regular function on the toric chart $\theta : (\mathbb{C}^\times)^n \to X$. Then, for every $\delta > 0$ and every point in $\mathscr{C}_\Phi(\mathbb{R}) \times (S^1)^n$, the function $f \circ \theta \circ E_{X,\theta,s}$ is bounded,*

$$|f \circ \theta \circ E_{X,\theta,s}| \le e^{-s\delta}$$

for s sufficiently large.

Proof. It suffices to show the estimate for $f \in P_\Phi(X,[\theta])$. By restricting to our toric chart, we may assume $(X,\theta) = ((\mathbb{C}^\times)^n,\mathrm{Id})$. Then by assumption $f \in \mathbb{C}[z_1^{\pm 1},\dots,z_n^{\pm 1}]$ is regular on $(\mathbb{C}^\times)^n$, and so by the triangle inequality we may assume $\nu_i = 0$ for all i. We write $(f \circ E_s)(\xi_1,\dots,\xi_n)$ for $(f \circ E_{(\mathbb{C}^\times)^n,\mathrm{Id},s})(\xi_1,\dots,\xi_n,1,\dots,1)$. Now, by a standard bounding argument as in [16], we get

$$\lim_{s\to\infty}\frac{1}{s}\log(f\circ E_s)(\xi_1,\dots,\xi_n)=f^t(\xi_1,\dots,\xi_n).$$

For $(\xi_1,\dots,\xi_n)\in\mathscr{C}_\Phi$, we have $f^t(\xi_1,\dots,\xi_n)<0$ since f is dominated. Then for sufficiently large s,

$$\frac{1}{s}\log(f\circ E_s)(\xi_1,\dots,\xi_n)\le-\delta<0.$$

Therefore, since the exponential is monotonic,

$$0<e^{s\frac{1}{s}\log(f\circ E_s)(\xi_1,\dots,\xi_n)}\le e^{-s\delta}.$$

Thus, $(f\circ E_s)(\xi_1,\dots,\xi_n)=e^{s\frac{1}{s}\log(f\circ E_s)(\xi_1,\dots,\xi_n)}\to 0$ exponentially quickly as $s\to\infty$. □

We introduce the scaled Poisson bivector $\pi_s:=s\pi_{\mathfrak{R}(X)}$ on the real form $\mathfrak{R}(X)=(\theta\circ E_{X,\theta,s})(L_{X,\theta,\tau}\times\mathbb{T}_{X,\theta,\tau})$.

Theorem 6.18 (Partial tropicalization as a limit). *Under the change of coordinates $\theta\circ E_{X,\theta,s}$, in the limit $s\to\infty$, the bivector π_s converges to the constant bivector π_{PT} on $PT(X,\pi,\theta,\Phi,\tau)\subset L_{X,\theta,\tau}\times\mathbb{T}_{X,\theta,\tau}$.*

Proof. By restricting to our toric chart, we may assume $(X,\theta)=((\mathbb{C}^\times)^n,\mathrm{Id})$. Note that since θ is an open embedding, regular functions on X restrict to Laurent polynomials in the coordinates z_i on $(\mathbb{C}^\times)^n$. We compute:

$$\begin{aligned}\{z_i,z_j\}_s&=\{e^{s\xi_i+\sqrt{-1}\nu_i},e^{s\xi_j+\sqrt{-1}\nu_j}\}_s\\ &=se^{s\xi_i+\sqrt{-1}\nu_i}e^{s\xi_j+\sqrt{-1}\nu_j}(\pi_{ij}+f_{ij}),\end{aligned}\tag{18}$$

where we write $f_{ij}=f_{ij}(e^{s\xi_1+\sqrt{-1}\nu_1},\dots,e^{s\xi_n+\sqrt{-1}\nu_n})$. On the other hand, $\{\cdot,\cdot\}_s$ is a biderivation, so

$$\begin{aligned}\{e^{s\xi_i+\sqrt{-1}\nu_i},e^{s\xi_j+\sqrt{-1}\nu_j}\}_s=e^{s\xi_i+\sqrt{-1}\nu_i}e^{s\xi_j+\sqrt{-1}\nu_j}\Big(&s^2\{\xi_i,\xi_j\}_s\\ &+\sqrt{-1}s(\{\xi_i,\nu_j\}_s+\{\nu_i,\xi_j\}_s)-\{\nu_i,\nu_j\}\Big).\end{aligned}\tag{19}$$

Combining (18) and (19) gives

$$s^2\{\xi_i,\xi_j\}_s+\sqrt{-1}s(\{\xi_i,\nu_j\}_s+\{\nu_i,\xi_j\}_s)-\{\nu_i,\nu_j\}_s=s(\pi_{ij}+f_{ij}).\tag{20}$$

The condition that π_s is a real bivector on $\mathfrak{R}(X)$ implies that for complex-valued functions $f,g\in C^\infty(\mathfrak{R}(X),\mathbb{C})$, the bracket respects complex conjugation $\{\overline{f},\overline{g}\}_s=\overline{\{f,g\}_s}$. So, similarly to (20) we find from considering the bracket $\{e^{s\xi_i-\sqrt{-1}\nu_i},e^{s\xi_j-\sqrt{-1}\nu_j}\}_s$, that

$$s^2\{\xi_i,\xi_j\}_s-\sqrt{-1}s(\{\xi_i,\nu_j\}_s+\{\nu_i,\xi_j\}_s)-\{\nu_i,\nu_j\}_s=s(\overline{\pi}_{ij}+\overline{f}_{ij}).\tag{21}$$

Recall we have assumed the log-canonical part π_{ij} of the bracket $\{z_i, z_j\}$ is pure imaginary. Putting together (20) and (21), we get

$$s^2\{\xi_i, \xi_j\}_s - \{\nu_i, \nu_j\}_s = s(\text{w.d.t.}), \tag{22}$$

$$\{\xi_i, \nu_j\}_s + \{\nu_i, \xi_j\}_s = -\sqrt{-1}\pi_{ij} + \text{w.d.t.}, \tag{23}$$

where w.d.t. stands for weakly dominated terms.

Restricting to the fixed locus of $\overline{\tau}$, we have the relation $\overline{z}_i = \tau^*(z_j)$. Therefore, on $\mathfrak{R}(X) \subset \text{Fix}(\overline{\tau})$ we have

$$\{z_i, \overline{z}_j\}_{\mathfrak{R}(X)} = \{z_i, \tau^*(z_j)\}_{\mathfrak{R}(X)} = z_i\tau^*(z_j)(\pi_{i\tau(j)} + f_{i\tau(j)}) = z_i\overline{z}_j(\pi_{i\tau(j)} + f_{i\tau(j)}). \tag{24}$$

Repeating calculations similar to those before (22) and (23) gives

$$s^2\{\xi_i, \xi_j\}_s + \{\nu_i, \nu_j\}_s = s(\text{w.d.t.}), \tag{25}$$

$$-\{\xi_i, \nu_j\}_s + \{\nu_i, \xi_j\}_s = -\sqrt{-1}\pi_{i\tau(j)} + \text{w.d.t.} \tag{26}$$

Combining (22), (23), (25), and (26), and applying Lemma 6.17 gives the result. □

Corollary 6.19. *The constant bracket π_{PT} is well defined on $PT(X, \pi, \theta, \Phi, \tau)$. It is skew-symmetric, and hence Poisson.*

Proof. Follows from Theorem 6.18 and the fact that $\Lambda^2(T_p\mathfrak{R}(X))$ is a (closed) linear subspace of $T_p\mathfrak{R}(X) \otimes T_p\mathfrak{R}(X)$, for any $p \in \mathfrak{R}(X)$. □

Definition 6.20. We define the category **PTrop** as follows. Objects are pairs (M, π), where $M = \mathscr{C} \times \mathbb{T} \subset \mathbb{R}^n \times (S^1)^n$ is the product of an open real cone and a subtorus of the compact torus, and π is a constant Poisson bivector on M which restricts to 0 on $\mathscr{C}$ and $\mathbb{T}$. We consider $\mathbb{R}^n = \mathbb{Z}^n \otimes_{\mathbb{Z}} \mathbb{R}$ to be the extension of the integer lattice, and have the exponential map $\exp(2\pi\sqrt{-1}\cdot) : \mathbb{R}^n \to (S^1)^n$. Arrows in **PTrop** from $\mathscr{C} \times \mathbb{T}$ to $\mathscr{C}' \times \mathbb{T}' \subset \mathbb{R}^m \times (S^1)^m$ are continuous piecewise linear maps $f : \mathbb{R}^n \to \mathbb{R}^m$ which are homogeneous in the sense of (5), have $f(\mathbb{Z}^n) \subset \mathbb{Z}^m$, and are subject to the following conditions.

(1) $f(\mathscr{C}) \subset \mathscr{C}'$.
(2) On each open linearity chamber $C \subset \mathbb{R}^n$ of f, note that there is a unique Lie group homomorphism $\exp(f|_C) : (S^1)^n \to (S^1)^m$ induced by $\exp(2\pi\sqrt{-1}\cdot)$. When $C \cap \mathscr{C} \neq \emptyset$, we require that the induced maps $\exp(f|_C) : (S^1)^n \to (S^1)^m$ restrict to maps of the subgroups $\mathbb{T} \to \mathbb{T}'$.
(3) On each open linearity chamber C, the restriction of the map $f \times \exp(f|_C)$: $(C \cap \mathscr{C}) \times \mathbb{T} \to \mathscr{C}' \times \mathbb{T}'$ is Poisson.

Note that an arrow in **PTrop** induces a Poisson map $(\mathscr{C} \times \mathbb{T}, \pi) \to (\mathscr{C}' \times \mathbb{T}', \pi')$, defined on an open subset of the domain. ◁

Definition 6.21. Let $f : (X,\pi_X,\theta_X,\Phi_X,\tau_X) \to (Y,\pi_Y,\theta_Y,\Phi_Y,\tau_Y)$ be a map of framed positive Poisson varieties with real form. Define the partial tropicalization $PT(f)$ of f to be the continuous piecewise linear map $f^t : X^t(\mathbb{R}) \to Y^t(\mathbb{R})$. ◁

After proving the following lemma, we will show in Theorem 6.23 that the assignment PT is functorial.

Lemma 6.22. *Let $(X,\theta : (\mathbb{C}^\times)^n \to X)$ be a framed complex positive variety. Let $f : (\mathbb{C}^\times)^n \to \mathbb{C}^\times$ be a positive rational function, and let $C \subset (X,\theta)^t(\mathbb{R})$ be an open linearity chamber of f^t. Then for $f_s = E^{-1}_{\mathbb{C}^\times,Id,s} \circ f \circ E_{X,\theta,s}$, we have*

$$\lim_{s\to\infty} f_s|_C = f^t|_C \times \exp(f^t|_C) : C \times (S^1)^n \to \mathbb{R} \times S^1. \tag{27}$$

In particular, for a point $(\xi, e^{\sqrt{-1}\nu}) \in C \times (S^1)^n$, for large enough s the value of $f_s(\xi, e^{\sqrt{-1}\nu})$ is always defined.

Proof. For $\xi := (\xi_1,\dots,\xi_n) \in C$ and $e^{\sqrt{-1}\nu} := (e^{\sqrt{-1}\nu_1},\dots,e^{\sqrt{-1}\nu_n}) \in (S^1)^n$ we have

$$f_s(\xi, e^{\sqrt{-1}\nu}) = \left(\frac{1}{s}\log|f(e^{s\xi+\sqrt{-1}\nu})|, \frac{f(e^{s\xi+\sqrt{-1}\nu})}{|f(e^{s\xi+\sqrt{-1}\nu})|}\right).$$

Write $f = A/B$, where A and B are positive polynomials in $z_1,\dots,z_n$. Then

$$A(e^{s\xi+\sqrt{-1}\nu}) = \sum_j e^{sM_j(\xi)+\sqrt{-1}M_j(\nu)}, \quad B(e^{s\xi+\sqrt{-1}\nu}) = \sum_k e^{sN_k(\xi)+\sqrt{-1}N_k(\nu)},$$

where M_j and N_k are linear polynomials with positive integer coefficients. Without loss of generality, assume $M_1(\xi) > M_j(\xi)$ for $j > 1$ and $N_1(\xi) > N_k(\xi)$ for $k > 1$. Note this uses that ξ is in an open linearity chamber of f^t. Also, note that $f^t|_C$ is given by the linear polynomial $M_1 - N_1$.

We then have

$$\lim_{s\to\infty} \frac{A(e^{s\xi+\sqrt{-1}\nu})}{e^{sM_1(\xi)+\sqrt{-1}M_1(\nu)}} = 1, \quad \lim_{s\to\infty} \frac{B(e^{s\xi+\sqrt{-1}\nu})}{e^{sN_1(\xi)+\sqrt{-1}N_1(\nu)}} = 1,$$

and so, dividing we get

$$\lim_{s\to\infty} \frac{f(e^{s\xi+\sqrt{-1}\nu})}{e^{s(M_1-N_1)(\xi)+\sqrt{-1}(M_1-N_1)(\nu)}} = 1. \tag{28}$$

Thus,

$$\lim_{s\to\infty} \frac{f(e^{s\xi+\sqrt{-1}\nu})}{|f(e^{s\xi+\sqrt{-1}\nu})|} = \lim_{s\to\infty} \frac{f(e^{s\xi+\sqrt{-1}\nu})}{e^{s(M_1-N_1)(\xi)}} = e^{\sqrt{-1}(M_1-N_1)(\nu)} = \exp(f^t|_C)(e^{\sqrt{-1}\nu}),$$

which gives the second component of (27).

Now, from (28) we find

$$\lim_{s\to\infty} \log|f(e^{s\xi+\sqrt{-1}\nu})| - s(M_1 - N_1)(\xi) = 0.$$

Therefore,

$$\begin{aligned}\lim_{s\to\infty} \frac{1}{s}\log|f(e^{s\xi+\sqrt{-1}\nu})| &= (M_1 - N_1)(\xi)\\ &= f^t(\xi),\end{aligned}$$

which gives the first component of (27). □

Theorem 6.23. *Partial tropicalization is a functor*

$$PT : \mathbf{PosPoiss}^{\bullet}_{\mathbb{R}} \to \mathbf{PTrop}.$$

Proof. Recall that after tropicalizing, maps $f : (X, \theta_X, \Phi_X) \to (Y, \theta_Y, \Phi_Y)$ of framed positive varieties with potential preserve the integer lattice and send cones into cones:

$$f^t(\mathscr{C}_{\Phi_X}) \subset \mathscr{C}_{\Phi_Y}$$

We then have condition 1 of Definition 6.20, and functoriality will then follow from functoriality of tropicalization, Proposition 2.22. It remains to check, for a map $f : (X, \pi_X, \theta_X, \Phi_X, \tau_X) \to (Y, \pi_Y, \theta_Y, \Phi_Y, \tau_Y)$, that $PT(f)$ satisfies conditions 2 and 3 of Definition 6.20.

Extending our notation from above, let $f_s := (\theta_Y \circ E_{Y,\theta_Y,s})^{-1} \circ f \circ (\theta_X \circ E_{X,\theta_X,s})$ and let $f_s^{\mathfrak{R}} := f_s|_{L_X \times \mathbb{T}_X}$ be the restriction to the real form. Then considering the component functions of f and applying Lemma 6.22, we have for open linearity chambers C of f^t,

$$\lim_{s\to\infty} f_s|_C = f^t|_C \times \exp(f^t|_C) : C \times (S^1)^n \to Y^t(\mathbb{R}) \times (S^1)^m, \tag{29}$$

where n and m are the dimensions of X and Y, respectively. On each open linearity chamber C of f^t, by Proposition 6.12 we have in our new coordinates the restricted map

$$f_s^{\mathfrak{R}}|_C : (C \cap L_X) \times \mathbb{T}_X \to L_Y \times \mathbb{T}_Y, \tag{30}$$

has its image in the real form of Y. By Lemma 6.22, the map $f_s^{\mathfrak{R}}$ is defined at each point of C for large enough s. The description of the limit (29) gives condition 2. By Proposition 6.12, the map $\theta_Y^{-1} \circ f|_{\mathfrak{R}(X)} \circ \theta_X$ is a Poisson map of real forms when it is defined. Restricting the limit $\lim_{s\to\infty} f_s^{\mathfrak{R}(X)}|_C$ of (30) to $\mathscr{C}_{\Phi_X}(\mathbb{R}) \times \mathbb{T}_X$ and applying Theorem 6.18, we find the map

$$\left(\lim_{s\to\infty} f_s^{\mathfrak{R}(X)}|_C\right)\Big|_{\mathscr{C}_{\Phi_X}(\mathbb{R})\times\mathbb{T}_X} : (C \cap (\mathscr{C}_{\Phi_X}(\mathbb{R}) \cap L_X)) \times \mathbb{T}_X \to (\mathscr{C}_{\Phi_Y}(\mathbb{R}) \cap L_Y) \times \mathbb{T}_Y.$$

preserves the partially tropicalized Poisson structures. The description of the limit (29) gives condition 3. □

6.4 Partial tropicalization of K^*

Fix a simply-connected semisimple complex Lie group G, and let $\mathbf{i}$ be a reduced word for w_0. As shown in Theorem 6.7, the tuple $(G^*, \sqrt{-1}\pi_{G^*}, \mathbb{A}_{\mathbf{i}}^{-1}, \Phi_{G^*}, \tau)$ is a framed positive Poisson variety with real form, which by Remark 6.13 is an open dense subset of the Poisson-Lie group (K^*, π_{K^*}). We summarize our results in the following theorem.

Theorem 6.24. *The partial tropicalization* $PT(G^*, \sqrt{-1}\pi_{G^*}, \mathbb{A}_{\mathbf{i}}^{-1}, \Phi_{G^*}, \tau)$ *is of the form* $\mathscr{C}_{\Phi_{BK}}(\mathbb{R}) \times \mathbb{T}$*, where* $\mathscr{C}_{\Phi_{BK}}$ *is the strict extended string cone of Definition* 4.15*, and* $\mathbb{T}$ *is a real torus of dimension* $d = \dim(N)$*. This space is equipped with a constant Poisson bracket and an integrable system.*

Proof. Let $L = L_{G^*, \mathbb{A}_{\mathbf{i}}^{-1}, \tau}$. By Theorem 6.23, it suffices to describe the cone $L \cap \mathscr{C}_{\Phi_{G^*}}(G^*, \mathbb{A}_{\mathbf{i}}^{-1})(\mathbb{R})$ for a single reduced word $\mathbf{i}$ of w_0. Write $\xi_1, \ldots, \xi_{l(w_0)}$ for the coordinates on $(G^*, \mathbb{A}_{\mathbf{i}})^t(\mathbb{R})$ coming from $(\hat{\Delta}_{\mathbf{i}})_1$; see (11)–(13) above. Similarly, write $\xi_1^\tau, \ldots, \xi_{l(w_0)}^\tau$ for the coordinates coming from $(\hat{\Delta}_{\mathbf{i}} \circ \tau)_2$, and $\xi_1^0, \ldots, \xi_r^0$ for the coordinates coming from Δ_0. In these coordinates, we have the following expression for $\tau^t : (G^*, \mathbb{A}_{\mathbf{i}}^{-1})^t \to (G^*, \mathbb{A}_{\mathbf{i}}^{-1})^t$:

$$\tau^t(\xi_1, \ldots, \xi_{l(w_0)}, \xi_1^\tau, \ldots, \xi_{l(w_0)}^\tau, \xi_1^0, \ldots, \xi_r^0)$$
$$= (\xi_1^\tau, \ldots, \xi_{l(w_0)}^\tau, \xi_1, \ldots, \xi_{l(w_0)}, \xi_1^0, \ldots, \xi_r^0).$$

Projecting parallel to the coordinates $\xi_1^\tau, \ldots, \xi_{l(w_0)}^\tau$ gives an isomorphism

$$L \cong (G^{e,w_0}, \Delta_{\mathbf{i}}^{-1})^t$$

which takes $\mathscr{C}_{\Phi_{G^*}}$ to $\mathscr{C}_{\Phi_{BK}}$.

A similar analysis in coordinates of $\mathbb{T} = \mathbb{T}_{G^*, \mathbb{A}_{\mathbf{i}}^{-1}, \tau} \subset (S^1)^{r+2l(w_0)}$ gives that the dimension of $\mathbb{T}$ is $l(w_0) = \dim(N)$. □

Remark 6.25. Integrable systems of Theorem 6.24 can be described in more detail. In particular, the constant Poisson bracket π_{PT} is of maximal rank $2d$, symplectic leaves of π_{PT} are in one-to-one correspondence with generic symplectic leaves in K^*, and symplectic volumes of the leaves of π_{PT} coincide with symplectic volumes of the corresponding leaves in K^*. This picture will be explored in detail in our forthcoming work.

In particular, we have the following.

Theorem 6.26. *For* $G = \mathrm{SL}_n(\mathbb{C})$*, the partial tropicalization*

$$(PT(G^*, \sqrt{-1}\pi_{G^*}, \mathbb{A}_{\mathbf{i}}^{-1}, \Phi_{G^*}, \tau), \pi_{PT})$$

is isomorphic to the Gelfand-Zeitlin integrable system.

Proof. In [2], this was shown for a specific choice of $\mathbf{i}$. Fix another reduced word $\mathbf{i}'$ for w_0. Let

$$\Delta_{\mathbf{i}'} \circ \Delta_{\mathbf{i}}^{-1} : (\mathbb{C}^\times)^n \to (\mathbb{C}^\times)^n$$

be the positive equivalence from the toric chart $\Delta_{\mathbf{i}}^{-1}$ to $\Delta_{\mathbf{i}'}^{-1}$, and let $\{C_j\}_j$ be the set of open linearity chambers of

$$(\Delta_{\mathbf{i}'} \circ \Delta_{\mathbf{i}}^{-1})^t : (G^*, \Delta_{\mathbf{i}}^{-1})^t(\mathbb{R}) \to (G^*, \Delta_{\mathbf{i}'}^{-1})^t(\mathbb{R}).$$

Let

$$D = \left(G^*, \Delta_{\mathbf{i}}^{-1})^t(\mathbb{R})\right) \setminus \bigcup_j C_j$$

be the complement of the open linearity chambers C_j, and let

$$\mathscr{C} = \left(\mathscr{C}_{\Phi_{G^*}}(G^*, \Delta_{\mathbf{i}}^{-1})(\mathbb{R})\right) \cap L,$$

where as before we write $L = L_{G^*, \Delta_{\mathbf{i}}^{-1}, \tau}$. By Theorem 6.23, the partial tropicalization $PT(\Delta_{\mathbf{i}} \circ \Delta_{\mathbf{i}'}^{-1})$ determines a Poisson map

$$(PT(G^*, \sqrt{-1}\pi_{G^*}, \Delta_{\mathbf{i}}^{-1}, \Phi_{G^*}, \tau), \pi_{PT}) \to (PT(G^*, \sqrt{-1}\pi_{G^*}, \Delta_{\mathbf{i}'}^{-1}, \Phi_{G^*}, \tau), \pi_{PT})$$

away from $D \cap \mathscr{C} \times \mathbb{T}_{G^*, \Delta_{\mathbf{i}}^{-1}, \tau}$, so we just need to show that $\mathscr{C} \not\subset D$. We will show $D \cap \mathscr{C}$ is of positive codimension in $\mathscr{C}$.

First, we note that D is a union of finitely many subspaces D_k of $(G^*, \Delta_{\mathbf{i}}^{-1})^t(\mathbb{R})$. So it suffices to show each $D_k \cap \mathscr{C}$ has positive codimension in $\mathscr{C}$.

Second, the cone $\mathscr{C}_{\Phi_{G^*}}(G^*, \Delta_{\mathbf{i}}^{-1})(\mathbb{R})$ is open in $(G^*, \Delta_{\mathbf{i}}^{-1})^t(\mathbb{R})$, so $\mathscr{C}$ is open in L. So if some $D_k \cap \mathscr{C}$ has codimension 0 in $\mathscr{C}$, then $L \subset D_k$.

We then show that L is not contained in any D_k. For the reduced word $\mathbf{i}$, consider the coordinates $\xi_1, \ldots, \xi_{l(w_0)}$ and $\xi_1^\tau, \ldots, \xi_{l(w_0)}^\tau$ introduced in the proof of Theorem 6.24. From the definition of $\Delta_{\mathbf{i}}$, we see that D_k is given by some linear equalities of the form

$$\sum_{p=1}^{l(w_0)} a_p \xi_p = 0, \qquad \sum_{p=1}^{l(w_0)} b_p \xi_p^\tau = 0, \qquad a_p, b_p \in \mathbb{Q}.$$

On the other hand, L is given by the equations

$$\xi_p = \xi_p^\tau, \qquad p = 1, \ldots, l(w_0).$$

From this description it is evident that D_k does not contain L. □

References

[1] A. Alekseev, *On Poisson actions of compact Lie groups on symplectic manifolds*, J. Differential Geom. 45 (1997), no. 2, 241–256.

[2] A. Alekseev, I. Davydenkova, *Inequalities from Poisson brackets*, Indag. Math. 25 (5) (2014) 846–871.
[3] A. Alekseev, E. Meinrenken, *Ginzburg-Weinstein from Gelfand-Zeitlin*, J. Differential Geom. 76 (2007), no. 1, 1–34.
[4] A. Berenstein and D. Kazhdan, *Geometric and unipotent crystals*, Geom. Funct. Anal., Special Volume, Part I (2000), 188–236.
[5] A. Berenstein, D. Kazhdan, *Geometric and unipotent crystals II: From unipotent bicrystals to crystal bases quantum groups*, Contemp. Math., vol. 433, Amer. Math. Soc., Providence, RI, 2007, pp. 13–88.
[6] A. Berenstein, S. Fomin, A. Zelevinsky, *Cluster algebras III. Upper bounds and double Bruhat cells*, Duke Math. J. 126 (1) (2005) 1–52.
[7] A. Berenstein, A. Zelevinsky, *Canonical bases for the quantum group of type A_r and piecewise linear combinatorics*, Duke Math. J. 82 (1996), 473–502.
[8] A. Berenstein, A. Zelevinsky, *Total positivity in Schubert varieties*, Comment. Math. Helv. 72 (1997), 128–166. CMP 97:14.
[9] A. Berenstein, A. Zelevinsky: *Tensor product multiplicities, canonical bases and totally positive varieties*. Invent. Math. 143, 77–128 (2001).
[10] P. Etingof, O. Schiffmann, *Lectures on Quantum Groups*, 2nd edition, International Press, 2002.
[11] S. Fomin, A. Zelevinsky, *Double Bruhat Cells and total positivity*, J. Amer. Math. Soc. 12 (1999) 335–380.
[12] M. Gekhtman, M. Shapiro, and A. Vainshtein, *Cluster algebras and Poisson geometry*, Mosc. Math. J. 3 (2003), no. 3, 899–934, 1199.
[13] V. Ginzburg, A. Weinstein, *Lie-Poisson structure on some Poisson Lie groups*, J. Amer. Math. Soc. 5 (2) (1992) 445–453.
[14] M. Kogan, A. Zelevinsky, *On symplectic leaves and integrable systems in standard complex semisimple Poisson-Lie groups*, Int. Math. Res. Not. 32 (2002) 1685–1702.
[15] J. H. Lu, A. Weinstein, *Poisson-Lie groups, dressing transformations and Bruhat decompositions*, J. Differential Geom. 31 (1990), no. 2, 501–526.
[16] G. Mikhalkin, *Tropical geometry and its applications*, International Congress of Mathematicians. Vol. II, 827–852, Eur. Math. Soc., Zürich, 2006.
[17] K. Rietsch, *A mirror symmetric construction of $qH_T^*(G/P)_{(q)}$*, Adv. Math. 217 (2008), no. 6, 2401–2442.
[18] M. A. Semenov-Tian-Shansky, *What is a classical r-matrix?*, Functional Analysis and Its Applications 17 (1983) no. 4, 259–272.
[19] M. A. Semenov-Tian-Shansky, *Dressing transformations and Poisson group actions*, Publ. Res. Inst. Math. Sci. 21 (1985), no. 6, 1237–1260.
[20] P. Xu, *Dirac submanifolds and Poisson involutions*, Ann. Scient. Éc. Norm. Sup., 36 (2003), 403–430

Quasi-lisse Vertex Algebras and Modular Linear Differential Equations

Tomoyuki Arakawa and Kazuya Kawasetsu

Dedicated to the great mathematician Bertram Kostant

Abstract We introduce the notion of quasi-lisse vertex algebras, which generalizes admissible affine vertex algebras. We show that the normalized character of an ordinary module over a quasi-lisse vertex operator algebra has a modular invariance property, in the sense that it satisfies a modular linear differential equation. As an application we obtain the explicit character formulas of simple affine vertex algebras associated with the Deligne exceptional series at level $-h^\vee/6-1$, which express the homogeneous Schur indices of 4d SCFTs studied by Beem, Lemos, Liendo, Peelaers, Rastelli and van Rees, as quasimodular forms.

Keywords: Vertex algebras • Modular linear differential equations • Quasimodular forms • Affine Kac-Moody algebras • Affine $\mathscr{W}$-algebras • Associated varieties • Deligne exceptional series • Schur limit of superconformal index

Mathematics Subject Classification (2010): 17B69, 17B67, 11F22, 81R10

1 Introduction

The vertex algebra V is called *lisse*, or C_2*-cofinite*, if the dimension of the associated variety X_V is zero. For instance, a simple affine vertex algebra V associated with an

---

The first author is partially supported by JSPS KAKENHI Grant No. 20340007 and No. 23654006. The second author was partially supported by JSPS KAKENHI Grant No. 14J09236.

Tomoyuki Arakawa
Research Institute for Mathematical Sciences, Kyoto University, Kyoto 606-8502 Japan
e-mail: arakawa@kurims.kyoto-u.ac.jp
Department of Mathematics, MIT, 77 Massachusetts Ave, Cambridge MA 02139 USA

Kazuya Kawasetsu
School of Mathematics and Statistics, The University of Melbourne, 3010 Australia
e-mail: kazuya.kawasetsu@unimelb.edu.au

V. G. Kac, V. L. Popov (eds.), *Lie Groups, Geometry, and Representation Theory*,
Progress in Mathematics 326, https://doi.org/10.1007/978-3-030-02191-7_2

affine Kac-Moody algebra $\widehat{\mathfrak{g}}$ is lisse if and only if V is an integrable representation as a $\widehat{\mathfrak{g}}$-module. Thus, the lisse property generalizes the integrability condition to an arbitrary vertex algebra.

It is known that a lisse vertex operator algebra V has nice properties, such as the modular invariance of characters [Z, Miy], and most theories of vertex operator algebras have been built under this finiteness condition (see e.g., [DLM, Hua]). However, there do exist significant vertex algebras that do not satisfy the lisse condition. For instance, admissible affine vertex algebras do not satisfy the lisse condition unless they are integrable, but nevertheless their representations are semisimple in category $\mathscr{O}$ ([AdMi, A5]) and have the modular invariance property ([KW2, AvE]). Moreover, there are a huge number of vertex algebras constructed in [BLL$^+$] from four-dimensional $N = 2$ superconformal field theories (SCFTs), whose character coincides with the Schur limit of the superconformal index of corresponding four-dimensional theories. These vertex algebras also fail to satisfy the lisse property in general.

In this paper we propose the *quasi-lisse* condition that generalizes the lisse condition. More precisely, we call a conformal vertex algebra V quasi-lisse if its associated variety X_V has finitely many symplectic leaves. For instance, a simple affine vertex algebra V associated with $\widehat{\mathfrak{g}}$ is quasi-lisse if and only if X_V is contained in the nilpotent cone $\mathscr{N}$ of $\mathfrak{g}$. Therefore, by [FM, A3], all admissible affine vertex algebras are quasi-lisse. Moreover, the W-algebras obtained from quasi-lisse affine vertex algebras by the quantized Drinfeld-Sokolov reduction ([FF, KRW]) are quasi-lisse as well. The vertex algebras constructed from 4d SCFTs are also expected to be quasi-lisse, since their associated varieties conjecturally coincide with Higgs branches of the corresponding four-dimensional theories ([BR]).

We show that the normalized character of an ordinary representation of a quasi-lisse vertex operator algebra has a modular invariance property, in the sense that it satisfies a *modular linear differential equation* (MLDE) (cf. [MMS], [KZ], [Mas], [Mil], [KNS] and [AKNS]). This seems to be new even for an admissible affine vertex algebra. Moreover, using MLDE, we obtain the explicit character formulas of simple affine vertex algebras associated with the Deligne exceptional series $A_1 \subset A_2 \subset G_2 \subset D_4 \subset F_4 \subset E_6 \subset E_7 \subset E_8$ ([D]) at level $-h^\vee/6-1$. These vertex algebras arose in [BLL$^+$] as $2d$ chiral algebras constructed from 4d SCFTs[1]. Thus [BLL$^+$], our result expresses the homogeneous Schur indices of the corresponding 4d SCFTs as *(quasi)modular forms*. This result is rather surprising especially for types D_4, E_6, E_7 and E_8 (non-admissible cases), since the characters of these vertex algebras are written [KT] in terms of *non-trivial* Kazhdan-Lusztig polynomials as their highest weights are not regular dominant.

We note that in [CS] the authors have obtained a conjectural expression of Schur indices in terms of Kontsevich-Soibelman wall-crossing invariants, which we hope to investigate in future works.

---

[1] For types G_2 and F_4, the connection with 4d SCFTs is conjectural.

Acknowledgments

The first named author thanks Victor Kac, Anne Moreau, Hiraku Nakajima, Takahiro Nishinaka, Leonardo Rastelli, Shu-Heng Shao, Yuji Tachikawa and Dan Xie for valuable discussions. He thanks Christopher Beem for pointing out an error in the first version of this article. Some part of this work was done while he was visiting Academia Sinica, Taiwan, in August 2016, for the "Conference in Finite Groups and Vertex Algebras". He thanks the organizers of the conference. The second named author would like to thank Hiroshi Yamauchi for helpful advice.

This research was supported in part by the Perimeter Institute for Theoretical Physics. Research at the Perimeter Institute is supported by the Government of Canada through Industry Canada and by the Province of Ontario through the Ministry of Economic Development & Innovation.

2 Quasi-lisse vertex algebras

Let V be a conformal vertex algebra, $R_V = V/C_2(V)$, Zhu's C_2-algebra of V ([Z]), where $C_2(V) = \langle a_{(-2)}b \mid a,b \in V\rangle_{\mathbb{C}}$. The space R_V is a Poisson algebra by

$$\bar{a}\cdot\bar{b} = \overline{a_{(-1)}b}, \quad \{\bar{a},\bar{b}\} = \overline{a_{(0)}b}.$$

Here $\bar{a}$ denotes the image of $a \in V$ in R_V, and

$$a(z) = \sum_{n\in\mathbb{Z}} a_{(n)}z^{-n-1} \in (\operatorname{End} V)[[z,z^{-1}]]$$

is the quantum field corresponding to $a \in V$. In this paper we assume V is finitely strongly generated, that is, R_V is finitely generated.

The associated variety [A2] of a vertex algebra V is the finite-dimensional algebraic variety defined by

$$X_V = \operatorname{Specm}(R_V).$$

Since R_V is a Poisson algebra, we have a finite partition

$$X_V = \bigsqcup_{k=0}^{r} X_k,$$

where X_k are smooth analytic Poisson varieties (see e.g., [BG]). Thus, for any point $x \in X_k$, there is a well-defined symplectic leaf $\mathcal{L}_x \subset X_k$ through it.

Definition 2.1. A finitely strongly generated vertex algebra V is called *quasi-lisse* if X_V has only finitely many symplectic leaves.

Let V be a quasi-lisse vertex algebra. Finiteness of the symplectic leaves implies [BG] that the symplectic leaf $\mathcal{L}_x$ at $x \in X_V$ coincides with the regular locus of the zero variety of the maximal Poisson ideal contained in the maximal ideal corresponding to x. Thus, every leaf $\mathcal{L}_x$ is a smooth connected locally-closed *algebraic* subvariety in X_V. In particular, every irreducible component of X_V is the closure of a symplectic leaf ([Gin, Corollary 3.3]).

For us, the importance of the finiteness of the symplectic leaves is in the following fact that has been established by Etingof and Schedler.

Theorem 2.2 ([ES]). *Let R be a finitely generated Poisson algebra, and suppose that* $\mathrm{Specm}(R)$ *has finitely many symplectic leaves. Then*

$$\dim R/\{R,R\} < \infty.$$

3 A necessary condition for the quasi-lisse property

A finitely strongly generated vertex algebra V is called *conical* if it is conformal, and L_0 gives a $\frac{1}{m}\mathbb{Z}_{\geq 0}$-grading

$$V = \bigoplus_{\Delta \in \frac{1}{m}\mathbb{Z}_{\geq 0}} V_\Delta$$

on V for some $m \in \mathbb{N}$, $\dim V_\Delta < \infty$ for all Δ, and $V_0 = \mathbb{C}$, where $V_\Delta = \{v \in V \mid L_0 v = \Delta v\}$. Note that if V is a vertex operator algebra, that is, if V is integer-graded, then a conical vertex operator algebra is the same as a vertex operator algebra of CFT type.

Let V be a conical vertex algebra. The L_0-grading induces the grading

$$R_V = \bigoplus_{\Delta \in \frac{1}{m}\mathbb{Z}_{\geq 0}} (R_V)_\Delta, \quad (R_V)_0 = \mathbb{C},$$

on R_V. In other words, the L_0-grading induces a contracting $\mathbb{C}^*$-action on the associated variety X_V.

Remark 3.1. The associated variety of a simple conical quasi-lisse vertex algebra is conjecturally irreducible ([AM2]). The validity of this conjecture implies that the associated variety of a quasi-lisse vertex algebra is actually *symplectic*, that is, X_V is the closure of a symplectic leaf.

Remark 3.2. Conical lisse (C_2-cofinite) conformal vertex algebras are quasi-lisse, since X_V is a point in this case.

Proposition 3.3. *Let V be a conical quasi-lisse vertex algebra. Then the image $[\omega]$ of the conformal vector ω of V is nilpotent in Zhu's C_2-algebra R_V of V.*

Proof. Since V is conical, the $\mathbb{C}^*$-action ρ on X_V induced by the conformal grading contracts to a point, say 0, that is, $\lim_{t\to 0} \rho(t)x = 0$ for all $x \in X_V$.

Set $z = [\omega]$. It is sufficient to show that the value of z at any closed point x is zero. Pick an irreducible component Y of X_V containing x. Note that Y is $\mathbb{C}^*$-invariant, and hence $0 \in Y$. On the other hand, there exists a symplectic leaf $\mathcal{L} \subset X_V$ such that $Y = \overline{\mathcal{L}}$, the Zariski closure of $\mathcal{L}$. Since it belongs to the Poisson center of R_V, z belongs to the Poisson center of $\mathcal{O}(\mathcal{L})$. Hence z is constant on $\mathcal{L}$ as $\mathcal{L}$ is symplectic, and so it is constant on Y as well. Therefore the value of z at x is the same as the one at 0, which is clearly zero. □

4 Finiteness of ordinary representations

Recall that a weak V-module (M, Y_M) is called *ordinary* if L_0 acts semi-simply on M, any L_0-eigenspace M_Δ of M of eigenvalue $\Delta \in \mathbb{C}$ is finite-dimensional, and for any $\Delta \in \mathbb{C}$, we have $M_{\Delta-n} = 0$ for all sufficiently large $n \in \mathbb{Z}$.

Theorem 4.1. *Let V be a quasi-lisse conformal vertex algebra. Then the number of simple ordinary V-modules is finite.*

Proof. Let $A(V)$ be Zhu's algebra of V. By Zhu's theorem [Z], it is sufficient to show that the number of simple finite-dimensional $A(V)$-modules is finite.

The algebra $A(V)$ is naturally filtered: There is a natural filtration of $G_\bullet A(V)$ induced by the filtration $\bigoplus_{\Delta \leq p} V_\Delta$ of V ([Z]) that makes the associated graded $\mathrm{gr}_G A(V)$ a Poisson algebra. Moreover, there is a surjection map

$$R_V \to \mathrm{gr}\, A(V)$$

of Poisson algebras [DSK, ALY]. Therefore, $\mathrm{Specm}(\mathrm{gr}_G A(V))$ is a Poisson subvariety of X_V, and hence has finitely many symplectic leaves. Hence, thanks to Theorem 1.4 of [ES] that follows from Theorem 2.2, we conclude that $A(V)$ has only finitely many simple finite-dimensional representations. □

5 Modular linear differential equations

Let ϑ_k denote the Serre derivation of weight k,

$$\vartheta_k(f) = q\frac{d}{dq}f - \frac{k}{12}E_2 f,$$

where $E_n(\tau)$ is the normalized Eisenstein series of weight $n \geq 2$. Let

$$\vartheta_k^i = \vartheta_{k+2(i-1)} \circ \cdots \circ \vartheta_{k+2} \circ \vartheta_k$$

be the i-th iterated Serre derivation of weight k with $\vartheta_k^0 = 1$. Recall that a modular linear differential equation (MLDE) of weight k is a linear differential equation

$$\vartheta_k^n f + \sum_{j=0}^{n-1} P_j \vartheta_k^j f = 0$$

with a classical modular function P_j of weight $2n-2j$ for $0 \leq j \leq n-1$.

In this section, we prove the following theorem. Let $\mathbb{H}$ denote the complex upper half-plane.

Theorem 5.1. *Let V be a quasi-lisse vertex operator algebra, $c \in \mathbb{C}$ the central charge of V. Then the normalized character*

$$\chi_V(\tau) = \mathrm{tr}_V\left(e^{2\pi i\tau(L_0 - c/24)}\right) \qquad (\tau \in \mathbb{H})$$

satisfies a modular linear differential equation of weight 0.

Let $(V, Y(\cdot,z), \omega)$ be a quasi-lisse conformal vertex operator algebra with the weight grading $V = \bigoplus_{\Delta=0}^{\infty} V_\Delta$. Y. Zhu introduced a second vertex operator algebra $(V, Y[\cdot,z], \tilde{\omega})$ associated to V ([Z]), where the vertex operator $Y[\cdot,z]$ is defined by linearly extending the assignment

$$Y[v,z] = Y(v, e^z - 1)e^{z\Delta} = \sum_{n \in \mathbb{Z}} v[n] z^{-n-1}, \quad (v \in V_\Delta,\ \Delta \geq 0),$$

and $\tilde{\omega} = \omega - c/24$. We write $Y[\tilde{\omega}, z] = \sum_{n\in\mathbb{Z}} L[n] z^{-n-2}$ and

$$V_{[\Delta]} = \{v \in V \,|\, L[0]v = \Delta v\} \qquad \text{for every } \Delta \in \mathbb{Z}_{\geq 0}.$$

Set $A = \mathbb{C}[\tilde{G}_4(q), \tilde{G}_6(q)]$, $V_A = V \otimes A$. Here, q is a formal variable and $\tilde{G}_n(q) = \sum_{j=0}^{\infty} a_{n,j} q^j$, where $\{a_{n,j}\}_{j\geq 0}$ $(n \geq 2)$ are the Fourier coefficients of the Eisenstein series $G_n(\tau)$ of weight n, that is, $G_n(\tau) = \sum_{j=0}^{\infty} a_{n,j} e^{2\pi i\tau j}$. Let $O_q(V)$ be the A-submodule of V_A generated by

$$a[-2]b + \sum_{k=2}^{\infty} (2k-1)\tilde{G}_{2k}(q) a[2k-2]b \qquad (a, b \in V).$$

Let $[V_A, V_A]$ denote the A-span of elements $a[0]b$, $a, b \in V$, in V_A.

Proposition 5.2. *The A-module*

$$V_A / ([V_A, V_A] + O_q(V))$$

is finitely generated.

Proof. We set

$$U := [V_A, V_A] + O_q(V).$$

Also, put $R = R_V = V/C_2(V)$, $C_2(V)_A = C_2(V) \otimes_{\mathbb{C}} A$, and

$$R_A = R \otimes_{\mathbb{C}} A = V_A / C_2(V)_A.$$

Define an increasing filtration $G_\bullet V_A$ of the A-module V_A by

$$G_p V_A = \bigoplus_{\Delta \le p} V_\Delta \otimes A.$$

This induces the filtration on U and V_A/U:

$$G_p U = U \cap G_p V_A, \quad \mathrm{gr}_G U = \bigoplus_p G_p U / G_{p-1} U, \quad G_p(V_A/U) = G_p V_A / G_p U,$$

and

$$\mathrm{gr}_G(V_A/U) = \bigoplus_p G_p(V_A/U)/G_{p-1}(V_A/U) = V_A / \mathrm{gr}_G U.$$

Since

$$\mathrm{gr}_G U \supset \{R_A, R_A\} + C_2(V)_A,$$

we have a surjective map of A-modules

$$R_A/\{R_A, R_A\} \twoheadrightarrow V_A / \mathrm{gr}_G U = \mathrm{gr}_G(V_A/U).$$

As $R_A/\{R_A, R_A\} = (R/\{R,R\}) \otimes_{\mathbb{C}} A$, the assertion follows from Theorem 2.2. □

Since A is a Noetherian ring, it follows from Proposition 5.2 that $V_A/([V_A, V_A] + O_q(V))$ is a Noetherian A-module. Hence, we have the following lemma.

Lemma 5.3. *For an element a of V, there exist $s \in \mathbb{Z}_{\ge 0}$ and $g_i(q) \in A$ $(0 \le i \le s-1)$ such that*

$$L[-2]^s a + \sum_{i=0}^{s-1} g_i(q) L[-2]^i a \in [V_A, V_A] + O_q(V).$$

Let M be an ordinary V-module. Define the zero-mode action $o(\cdot) : V \to \mathrm{End}(M)$ by linearly extending the assignment

$$o(a) = \mathrm{Res}_z Y^M(a,z) z^{\Delta - 1} dz : M \to M, \quad (a \in V_\Delta,\ \Delta \ge 0),$$

For any $a \in V$, define the formal 1-point function $\tilde{\chi}_M(a,q)$ by

$$\tilde{\chi}_M(a,q) = \mathrm{tr}|_M o(a) q^{L_0 - c/24}.$$

For the q-series $f(q)$ and $k \ge 0$, define the formal Serre derivation ∂_k of weight k by

$$\partial_k f = q \frac{d}{dq} f(q) + k \tilde{G}_2(q) f(q).$$

Let Δ and ℓ be non-negative integers. For $a \in V_{[\Delta]}$ and $f(q) \in \mathbb{C}[\tilde{G}_2(q), \tilde{G}_4(q), \tilde{G}_6(q)]$ of weight n, define the formal iterated Serre derivation ∂^i by

$$\partial^i(f(q)\tilde{\chi}_M(a,q)) = \partial_{\Delta + \ell + 2i - 2}(\partial^{i-1}(f(q)\tilde{\chi}_M(a,q))) \quad (i \ge 1),$$

and $\partial^0 = \mathrm{id}$. Here, $f(q)$ is said to be of weight n if $f(q)$ is a homogeneous element of weight n of the graded algebra $\mathbb{C}[\tilde{G}_2(q), \tilde{G}_4(q), \tilde{G}_6(q)]$, where the weight of $\tilde{G}_k(q)$ is k for $k = 2,4,6$.

Lemma 5.4 (cf. [DLM, (5.9)]). *Let v be an element of V and M an ordinary V-module. We have*

$$\tilde{\chi}_M(L[-2]v,q) = \partial \tilde{\chi}_M(v,q) + \sum_{\ell=2}^{\infty} \tilde{G}_{2\ell}(q)\tilde{\chi}_M(L[2\ell-2]v,q).$$

Proof. The assertion follows by [Z, Proposition 4.3.5] with $a = \omega$ and $b = v$. □

Let a be a primiary vector of V of weight Δ, that is, $L[0]a = \Delta a$ and $L[n]a = 0$ for $n \geq 1$.

Lemma 5.5 (cf. [DLM, Lemma 6.2]). *For every $i \geq 1$, there exist elements $f_j(q) \in A$ $(0 \leq j \leq i-1)$ such that for any ordinary V-module M,*

$$\tilde{\chi}_M(L[-2]^i a,q) = \partial^i \tilde{\chi}_M(a,q) + \sum_{j=0}^{i-1} f_j(\tau)\partial^j \tilde{\chi}_M(a,q). \tag{1}$$

Proof. The proof is similar to that of [DLM, Lemma 6.2]. We prove the assertion by induction on i. When $i = 1$, it follows from Lemma 5.4 that $\tilde{\chi}_M(L[-2]a,q) = \partial \tilde{\chi}_M(a,q)$, and therefore (1) follows. Suppose that $i \geq 2$. Then by Lemma 5.4, we see that

$$\tilde{\chi}_M(L[-2]^i a,q) = \partial \tilde{\chi}_M(L[-2]^{i-1}a,q) + \sum_{k=2}^{\infty} \tilde{G}_{2k}(q)\tilde{\chi}_M(L[2k-2]L[-2]^{i-1}a,q).$$

Using the relation of the Virasoro algebra, we have

$$L[2k-2]L[-2]^{i-1}a = c_k \cdot L[-2]^{i-k}a$$

with a scalar c_k for $2 \leq k \leq i$ and $L[2k-2]L[-2]^{i-1}a = 0$ if $k \geq i+1$. Therefore,

$$\tilde{\chi}_M(L[-2]^i a,q) = \partial \tilde{\chi}_M(L[-2]^{i-1}a,q) + \sum_{k=2}^{i} c_k \tilde{G}_{2k}(q)\tilde{\chi}_M(L[-2]^{i-k}a,q).$$

By the induction hypothesis, we have (1), which completes the proof. □

Let u and v be elements of V.

Lemma 5.6 ([Z, Proposition 4.3.6]). *For every ordinary V-module M,*

$$\tilde{\chi}_M(u[0]v,q) = 0,$$

$$\tilde{\chi}_M(u[-2]v,q) + \sum_{k=2}^{\infty} (2k-1)\tilde{G}_{2k}(q)\tilde{\chi}_M(u[2k-2]v,q) = 0.$$

Theorem 5.7. *Let V be a quasi-lisse vertex operator algebra, $a \in V$ primary with $L[0]a = \Delta a$. For an ordinary V-module M, the series $\tilde{\chi}_M(a,q)$ converges absolutely and uniformly on every closed subset of the domain $\{q \mid |q| < 1\}$, and the limit function $\chi_M(a,q)$ has the form $q^h f(q)$ with some analytic function $f(q)$ on $\{q \mid |q| < 1\}$. Moreover, the space spanned by $\chi_M(a,q)$ for all ordinary V-modules M is a subspace of the space of solutions of a modular linear differential equation of weight Δ.*

Proof. The proof is similar to those of [Z, Theorem 4.4.1] and [DLM, Lemma 6.3]. By Lemma 5.3, we have $L[-s]^s a + \sum_{i=0}^{s-1} g_i(q) L[-2]^i a \in O_q(V)$ where $s \in \mathbb{Z}_{\geq 0}$ and $g_i(q) \in A$ for $0 \leq i \leq s-1$. It then follows by the definition of $O_q(V)$ and Lemma 5.6 that $\tilde{\chi}_M(L[-2]^s a + \sum_{i=0}^{s-1} g_i(q) L[-2]^i a, q) = 0$. By Lemma 5.5, we obtain a differential equation

$$\partial^s \tilde{\chi}_M(a,q) + \sum_{i=0}^{s-1} h_i(q) \partial^i \tilde{\chi}_M(a,q) = 0, \tag{2}$$

for the formal series $\tilde{\chi}_M(a,q)$ with $h_i(q) \in A$. Since $h_i(q)$ converges absolutely and uniformly on every closed subset of $\{q \mid |q| < 1\}$, and (2) is regular, it follows that $\tilde{\chi}_M(a,q)$ converges uniformly on every closed subset of $\{q \mid |q| < 1\}$. By using (2) again, we see that the space, spanned by $\chi_M(a,q)$ for all ordinary V-modules M, is a subset of the space of solutions of the MLDE

$$\vartheta_\Delta^s \chi_M(a,q) + \sum_{i=0}^{s-1} p_i(q) \vartheta_\Delta^i \chi_M(a,q) = 0,$$

where $p_i(q) \in \mathbb{C}[G_4(\tau), G_6(\tau)]$ is the limit function of $h_i(q)$ with $q = e^{2\pi i \tau}$ and $\tau \in \mathbb{H}$. The remainder of the theorem is clear. □

Theorem 5.1 follows from Theorem 5.7 with $a = |0\rangle$, the vacuum vector of V.

6 Examples of quasi-lisse vertex algebras

Let $V^k(\mathfrak{g})$ be the universal affine vertex algebra associated with a simple Lie algebra $\mathfrak{g}$ at level $k \in \mathbb{C}$, and let $V_k(\mathfrak{g})$ be the unique simple graded quotient of $V^k(\mathfrak{g})$. We have $X_{V^k(\mathfrak{g})} = \mathfrak{g}^*$, where $\mathfrak{g}^*$ is equipped with the Kirillov-Kostant-Souriau Poisson structure, and $X_{V_k(\mathfrak{g})}$ is a conic, G-invariant, Poisson subvariety of $X_{V^k(\mathfrak{g})} = \mathfrak{g}^*$, where G is the adjoint group of $\mathfrak{g}$ (see [A2]).

Let $\mathcal{N} = \{x \in \mathfrak{g} \mid \operatorname{ad} x \text{ is nilpotent}\}$, the nilpotent cone of $\mathfrak{g}$, which is identified with the zero locus of the augmentation ideal $\mathbb{C}[\mathfrak{g}^*]_+^G$ of the invariant ring $\mathbb{C}[\mathfrak{g}^*]^G$ via the identification $\mathfrak{g} = \mathfrak{g}^*$. It is well-known since Kostant [Kos63] that the number of G-orbits in $\mathcal{N}$ is finite.

Lemma 6.1. *The affine vertex algebra $V_k(\mathfrak{g})$ is quasi-lisse if and only if the associated variety $X_{V_k(\mathfrak{g})} \subset \mathcal{N}$.*

Proof. The "if" part is clear since the symplectic leaves in $\mathfrak{g}^*$ are the coadjoint orbits of G. Conversely, suppose that $X_{V_k(\mathfrak{g})} \not\subset \mathcal{N}$. Since $X_{V_k(\mathfrak{g})}$ is closed, there exists a nonzero semisimple element s in $X_{V_k(\mathfrak{g})}$. As it is conic, $X_{V_k(\mathfrak{g})}$ contains infinitely many orbits of the form $G.\lambda s$, $\lambda \in \mathbb{C}^*$. □

Recall that $V_k(\mathfrak{g})$ is called *admissible* if it is an admissible representation ([KW2]) as a module over the affine Kac-Moody algebra $\widehat{\mathfrak{g}}$ associated with $\mathfrak{g}$. All admissible affine vertex algebras are quasi-lisse, since their associated varieties are contained in $\mathcal{N}$ ([FM, A3]). In fact, the associated variety of an admissible affine vertex algebra $V_k(\mathfrak{g})$ is irreducible, that is, $X_{V_k(\mathfrak{g})} = \overline{\mathbb{O}}$ for some nilpotent orbit $\mathbb{O}$ of $\mathfrak{g}$ (see [A3] for the explicit description of the orbit $\mathbb{O}$).

Highest weight representations of an admissible affine vertex algebra $V_k(\mathfrak{g})$ are exactly the admissible representations $L(\lambda)$ of $\widehat{\mathfrak{g}}$ of level k, whose integral Weyl groups are of the form $y\left(\widehat{\Delta}(k\Lambda_0)\right)$, where y is an element of the extended affine Weyl group and $\widehat{\Delta}(k\Lambda_0)$ is the integral Weyl group of $V_k(\mathfrak{g})$ ([AdMi, A5]). Let $\mathfrak{h}$ be the Cartan subalgebra of $\mathfrak{g}$. The modular invariance of the normalized *full* characters

$$e^{2\pi kt}\,\mathrm{tr}_{L(\lambda)}\left(q^{L_0-c/24}e^{2\pi ix}\right), \qquad (\tau,x,t) \in Y,$$

of those representations, where Y is some domain in $\mathbb{H}\times\mathfrak{h}\times\mathbb{C}$, has been known due to Kac and Wakimoto [KW1, KW2], and was extended in [AvE] to that of general (full) trace functions. Here it is essential to consider their full characters, since an admissible representation is not an ordinary representation in general, and thus, the normalized character $\mathrm{tr}_V\left(e^{2\pi i\tau(L_0-c/24)}\right)$ is not always well defined.

Theorem 5.7 states the modular invariance of the normalized character (instead of the normalized full character) of an admissible representation that is ordinary. As far as the authors know, this fact is new.

Here are more examples of quasi-lisse affine vertex algebras.

Theorem 6.2 ([AM1]). *Assume that $\mathfrak{g}$ belongs to Deligne's exceptional series*

$$A_1 \subset A_2 \subset G_2 \subset D_4 \subset F_4 \subset E_6 \subset E_7 \subset E_8,$$

and let $k = -h^\vee/6 - 1$. Then $X_{V_k(\mathfrak{g})} \cong \overline{\mathbb{O}_{\min}}$, where $\mathbb{O}_{\min}$ is the minimal nilpotent orbit of $\mathfrak{g}$.

In Theorem 6.2, the affine vertex algebra $V_{-h^\vee/6-1}(\mathfrak{g})$ is admissible for types A_1, A_2, G_2, F_4, and so the statement is contained in [A3]. However, $V_{-h^\vee/6-1}(\mathfrak{g})$ is not admissible for types D_4, E_6, E_7, E_8. These non-admissible quasi-lisse affine vertex algebras have appeared in [BLL$^+$] as principal examples of chiral algebras coming from 4d SCFTs. In fact, the labels D_4, E_6, E_7, E_8 also appear in Kodaira's classification of isotrivial elliptic fibrations, and the corresponding 4d SCFTs are obtained by applying F-theory to these isotrivial elliptic fibrations. By construction [BLL$^+$], the character of the above non-admissible quasi-lisse affine vertex algebras are the (homogeneous) Schur indices of these 4d SCFTs obtained from elliptic fibrations. In mathematics, such a non-admissible affine vertex algebra was first extensively studied in [Per13].

In the next section we derive the explicit form of the characters of these non-admissible quasi-lisse affine vertex algebras.

Now let us give examples of quasi-lisse vertex algebras outside of the affine vertex algebras. Let $\mathcal{W}^k(\mathfrak{g},f)$ be the W-algebra associated with $\mathfrak{g}$ and a nilpotent element $f \in \mathfrak{g}$ at level k, defined by the quantized Drinfeld-Sokolov reduction

$$\mathcal{W}^k(\mathfrak{g},f) = H^0_{DS,f}(V^k(\mathfrak{g})),$$

where $H^{\bullet}_{DS,f}(M)$ denotes the cohomology of the BRST complex with coefficient M associated with the Drinfeld-Sokolov reduction with respect to f. This definition was discovered by Feigin and Frenkel [FF] in the case that f is principal as a generalization of Kostant's Whittaker model of the center of $U(\mathfrak{g})$ ([Kos78]), and was generalized to an arbitrary f by Kac and Wakimoto ([KRW]).

By [A3], the natural surjection $V^k(\mathfrak{g}) \to V_k(\mathfrak{g})$ induces a surjective homomorphism $\mathcal{W}^k(\mathfrak{g},f) \twoheadrightarrow H^0_{DS,f}(V_k(\mathfrak{g}))$ of vertex algebras, and moreover,

$$X_{H^0_{DS,f}(V_k(\mathfrak{g}))} \cong X_{V_k(\mathfrak{g})} \cap \mathcal{S}_f,$$

where $\mathcal{S}_f$ is the Slodowy slice at f, that is, $\mathcal{S}_f = f + \mathfrak{g}^e$. Here $\{e,f,h\}$ is an $\mathfrak{sl}_2$-triple and $\mathfrak{g}^e$ is the centralizer of e in $\mathfrak{g}$. Therefore, we have the following assertion.

Lemma 6.3. *Let k be non-critical and suppose that $V_k(\mathfrak{g})$ is quasi-lisse, that is, $X_{V_k(\mathfrak{g})} \subset \mathcal{N}$. For any $f \in X_{V_k(\mathfrak{g})}$, $H^0_{DS,f}(V_k(\mathfrak{g}))$ is quasi-lisse, and hence, so is the simple quotient $\mathcal{W}_k(\mathfrak{g},f)$ of $\mathcal{W}^k(\mathfrak{g},f)$.*

We note that $H^0_{DS,f}(V_k(\mathfrak{g})) \cong \mathcal{W}_k(\mathfrak{g},f)$ if $G.f \subset X_{V_k(\mathfrak{g})}$ for type A ([A1]) and this conjecturally holds for any $\mathfrak{g}$ ([KW3]).

Lemma 6.3 implies that admissible affine vertex algebras produce many quasi-lisse W-algebras by applying the Drinfeld-Sokolov reduction. For instance, if k is a non-degenerate admissible number (see [A4]), then $X_{V_k(\mathfrak{g})} = \mathcal{N}$, and hence,

$$X_{H^0_{DS,f}(V_k(\mathfrak{g}))} \cong \mathcal{N} \cap \mathcal{S}_f,$$

which is irreducible and therefore symplectic ([Pre]). In particular, if f is a subregular nilpotent element in types ADE, $X_{H^0_{DS,f}(V_k(\mathfrak{g}))}$ has the simple singularity of the same type as $\mathfrak{g}$ ([Slo]). In type A, it has been recently shown by Genra [Gen] that the subregular W-algebra $\mathcal{W}^k(\mathfrak{sl}_n, f_{\text{subreg}})$ is isomorphic to Feigin-Semikhatov's $W_n^{(2)}$-algebra ([FS]) at level k .

See [AM1, AM3] for more examples of quasi-lisse vertex algebras, and see e.g., [XYY] for more examples of vertex algebras obtained from 4d SCFTs.

7 The characters of affine vertex algebras associated with the Deligne exceptional series

In this section, we give explicit character formulas of the quasi-lisse affine vertex algebras associated with Deligne's exceptional series that appeared in Theorem 6.2 using MLDEs.

Let $\mathfrak{g}$ be a Lie algebra in Deligne's exceptional series and V the simple affine vertex algebra $V_k(\mathfrak{g})$ with $k = -h^\vee/6 - 1$. The Deligne dimension formula (7) below implies that the central charge c of V is given by

$$c = -2h^\vee - 2.$$

Lemma 7.1. *The square of the Virasoro element ω of V is 0 in Zhu's C_2-algebra R_V.*

Proof. Let I be the ideal of $R_{V^k(\mathfrak{g})} = S(\mathfrak{g})$ generated by the image of the maximal submodule of $V^k(\mathfrak{g})$, so that $R_{V_k(\mathfrak{g})} = S(\mathfrak{g})/I$. We need to show that $\Omega^2 \in I$, where Ω is the Casimir element of $S(\mathfrak{g})$.

If $\mathfrak{g}$ is not of type A, then this result has been already stated in Lemma 2.1 of [AM1], see (the proof of) Theorem 3.1 of [AM1]. So let $\mathfrak{g}$ be of type A, in which case the maximal submodule of $V^k(\mathfrak{g})$ is generated by a singular vector, say v ([KW1]). For $\mathfrak{g} = \mathfrak{sl}_2$, the assertion follows immediately from a result in [FM], which says that the image of v in I coincides with Ωe up to nonzero constant multiplication, see the proof of Theorem 4.2.1 of [FM]. Finally let $\mathfrak{g} = \mathfrak{sl}_3$. Then the vector v has degree 2, cf. [Per]. Let V be the $\mathfrak{g}$-submodule of $S^2(\mathfrak{g})$ generated by the image $[v]$ of v in I. Proposition 3.3 of [GS] (which is valid for type A cases as well) says that $\mathfrak{g} \cdot V \subset S^3(\mathfrak{g})$ contains a submodule isomorphic to $\mathfrak{g}$. On the other hand, Kostant's Separation Theorem ([Kos63], cf. Proposition 3.2 of [GS]) implies that $\mathfrak{g} \cdot \Omega$ is the unique submodule of $S^3(\mathfrak{g})$ isomorphic to $\mathfrak{g}$. Thus, $\mathfrak{g} \cdot \Omega \subset \mathfrak{g} \cdot V$, and the assertion follows. □

As $(V, Y[\cdot, z])$ is isomorphic to $(V, Y(\cdot, z))$, it follows that

$$L[-2]^2|0\rangle = \sum_{i=1}^{\ell} b_i[-2]c_i$$

for some $\ell \geq 0$, where b_i and c_i, $1 \leq i \leq \ell$, are $L[0]$-homogeneous elements of V such that $L[0](b_i[-2]c_i) = 4b_i[-2]c_i$. On the other hand, it follows from the definition of $O_q(V)$ that

$$b_i[-2]c_i \equiv -\sum_{k=2}^{\infty} f_k(q) b_i[2k-2]c_i \pmod{O_q(V)}$$

for $1 \leq i \leq \ell$. However, $b_i[2]c_i \in \mathbb{C}|0\rangle$ and $b_i[2k-2]c_i = 0$ for $k \geq 3$ as the $L[0]$-weight of $b_i[-2]c_i$ is 4. Therefore, we get that

$$L[-2]^2|0\rangle + g(q)|0\rangle \in O_q(V)$$

with $g(q) \in A$. Using Lemma 5.5, we see that the formal characters $\tilde{\chi}_M(|0\rangle, q)$ of all ordinary V-modules M satisfy a second-order differential equation of the form $\partial^2 \tilde{\chi}_M(|0\rangle, q) + f_1(q)\partial \tilde{\chi}_M(|0\rangle, q) + f_2(q)\tilde{\chi}_M(|0\rangle, q) = 0$ with $f_1(q), f_2(q) \in A$. Hence, the characters $\chi_M(\tau)$ of ordinary V-modules M satisfy a second-order MLDE $L(f) = 0$ of weight 0.

The second-order MLDEs (of weight 0) have the form

$$f''(\tau) - \frac{1}{6}E_2(\tau)f'(\tau) - \frac{k(k+2)}{144}E_4(\tau)f(\tau) = 0 \tag{3}$$

with $k \in \mathbb{C}$. Here,

$$' = q\frac{d}{dq} = \frac{1}{2\pi i}\frac{d}{d\tau}.$$

A function $f(\tau)$ is called of *vacuum type* if f has the form $f(\tau) = q^{-\alpha/24}(1 + \sum_{n=1}^{\infty} a_n q^n)$ with $\alpha \in \mathbb{Q}$ and $a_n \in \mathbb{Z}_{\geq 0}$ for each $n \geq 1$, where $q = e^{2\pi i \tau}$. Let f be a solution of (3) of the form $f(\tau) = q^{-\alpha/24}(1 + O(q))$ with $\alpha \in \mathbb{Q}$. Then by substituting f into (3), we see that $\alpha = -k/12$ or $(k+2)/12$. If $\alpha = -k/12$, it follows that k is one of the following numbers [KNS, (3.12)]:

$$k = \frac{1}{5}, \frac{1}{2}, 1, \frac{7}{5}, 2, \frac{13}{4}, 3, \frac{7}{2}, \frac{19}{5}, 4. \tag{4}$$

On the other hand, if $\alpha = (k+2)/12$, we have [KNS, (3.16)]

$$k = \frac{1}{5}, \frac{1}{2}, 1, 2, 3, 5, 8, 11, 17, 23, 29, 53. \tag{5}$$

Since V is of CFT-type with the central charge $-h^\vee - 2$, the character $\chi_V(\tau)$ is of vacuum type and has the form $\chi_V(\tau) = q^{(h^\vee+1)/12}(1 + O(q))$. Therefore, (4) and (5) imply that the MLDE $L(f) = 0$ must be the following:

$$f''(\tau) - \frac{1}{6}E_2(\tau)f'(\tau) - \frac{(h^\vee - 1)(h^\vee + 1)}{144}E_4(\tau)f(\tau) = 0. \tag{6}$$

The vacuum type solutions of (6) are also given in [KK] and [KNS]. As a result, we conclude that

$$\chi_{V_{-4/3}(A_1)} = \frac{\eta(3\tau)^3}{\eta(\tau)^3}, \quad \chi_{V_{-3/2}(A_2)} = \frac{\eta(2\tau)^8}{\eta(\tau)^8}, \quad \chi_{V_{-5/3}(G_2)} = \frac{E_1^{(3)}(\tau)\eta(3\tau)^6}{\eta(\tau)^8},$$

$$\chi_{V_{-2}(D_4)} = \frac{E_4'(\tau)}{240\eta(\tau)^{10}}, \quad \chi_{V_{-5/2}(F_4)} = \frac{E_2^{(2)}(\tau)\eta(2\tau)^{24}}{\eta(\tau)^{28}},$$

$$\chi_{V_{-3}(E_6)} = -\frac{1}{462}\left(\frac{E_6(\tau)E_4'(\tau)}{240\eta(\tau)^{22}} - \eta(\tau)^2\right),$$

$$\chi_{V_{-4}(E_7)} = \frac{1}{204204}\left(\Delta(\tau)P_2\left(\frac{E_6(\tau)}{\sqrt{\Delta(\tau)}}\right)\frac{E_4'(\tau)}{240\eta(\tau)^{34}} - \Delta(\tau)\frac{E_6(\tau)}{\eta(\tau)^{34}}\right),$$

$$\chi_{V_{-6}(E_8)} = \frac{1}{38818159380}\left(\Delta(\tau)^2 P_4\left(\frac{E_6(\tau)}{\sqrt{\Delta(\tau)}}\right)\frac{E_4'(\tau)}{240\eta(\tau)^{58}} - \Delta(\tau)^{5/2} Q_4\left(\frac{E_6(\tau)}{\sqrt{\Delta(\tau)}}\right)\frac{1}{\eta(\tau)^{58}}\right).$$

Here, $\eta(\tau) = q^{1/24}\prod_{n=1}^{\infty}(1-q^n)$, $\Delta(\tau) = \eta(\tau)^{24}$, $E_2^{(2)}(\tau) = 2E_2(2\tau) - E_2(\tau)$,

$$E_1^{(3)}(\tau) = 1 + 6\sum_{n=1}^{\infty}\left(\sum_{d|n}\left(\frac{d}{3}\right)(n/d)^2\right)q^n$$

with the Legendre symbol $\left(\frac{\cdot}{\cdot}\right)$, $P_2(x) = x^2 + 462$, $P_4(x) = x^4 + 1341x^2 + 201894$, and $Q_4(x) = x^3 + 879x$.

In particular, it follows that the characters of $V_{-4/3}(A_1)$, $V_{-3/2}(A_2)$, $V_{-5/3}(G_2)$, and $V_{-5/2}(F_4)$ are modular forms, while those of $V_{-2}(D_4)$, $V_{-3}(E_6)$, $V_{-4}(E_7)$ and $V_{-6}(E_8)$ are quasimodular forms of positive depths ([KNS, pp.450]). Moreover, if $h^\vee$ is the dual Coxeter number of D_4, E_6, E_7 or E_8, then MLDE (6) has a solution with a logarithmic term (see [KK, section 5] and [KNS, Remark 3.8]). Note that the above formula for A_1 and A_2 follows also from the recent remark [KW4] by Kac and Wakimoto.

Remark 7.2. It should be noted that the coefficient of $E_4(\tau)f(\tau)$ in (6) is a non-constant rational function in $h^\vee$, as such phenomena are often observed for the Deligne exceptional series. In fact, the dimensions of specific modules over any Lie algebra in the Deligne exceptional series satisfy the so-called *Deligne dimension formulas*, which are rational functions in $h^\vee$. For example,

(7) $$\dim \mathfrak{g} = \frac{2(h^\vee+1)(5h^\vee-6)}{h^\vee+6},$$

and $\dim L(2\theta) = 5(h^\vee)^2(2h^\vee+3)(5h^\vee-6)/(h^\vee+12)(h^\vee+6)$ ([CdM], [D] and [LM]). Here, $L(2\theta)$ is the irreducible highest weight module of weight 2θ over $\mathfrak{g}$. In the vertex algebra setting, the first example of such phenomena was observed in [T], where coefficients of the MLDEs, which the characters of the affine vertex algebras $V_1(\mathfrak{g})$ at level 1 associated with the Deligne exceptional series satisfy, are expressed as rational functions in $h^\vee$. The second example of such phenomena was found in [K2], where the minimal affine $\mathscr{W}$-algebras associated with the Deligne exceptional series at level $-h^\vee/6$ were shown to be lisse and rational.

Remark 7.3. It follows from a result of [KNS] that there is a vacuum type solution of (6) with $h^\vee = 24$. Although the Coxeter number $h^\vee$ of any Lie algebra in the Deligne exceptional series does not coincide with 24, the number "$h^\vee = 24$" appears in many studies of the Deligne exceptional series (see e.g., [CdM] and [K1]).

References

[AdMi] Dražen Adamović and Antun Milas, Vertex operator algebras associated to modular invariant representations for $A_1^{(1)}$, *Math. Res. Lett.*, 2(5): 563–575, 1995.

[A1] Tomoyuki Arakawa, Representation theory of W-algebras, II. In *Exploring new structures and natural constructions in mathematical physics*, Volume 61, *Adv. Stud. Pure Math.*, pages 51–90, Math. Soc. Japan, Tokyo, 2011.

[A2] Tomoyuki Arakawa, A remark on the C_2 cofiniteness condition on vertex algebras, *Math. Z.*, 270(1–2): 559–575, 2012.

[A3] Tomoyuki Arakawa, Associated varieties of modules over Kac-Moody algebras and C_2-cofiniteness of W-algebras, *Int. Math. Res. Not.*, 2015: 11605–11666, 2015.

[A4] Tomoyuki Arakawa, Rationality of W-algebras: principal nilpotent cases, *Ann. Math.*, 182(2): 565–694, 2015.

[A5] Tomoyuki Arakawa, Rationality of admissible affine vertex algebras in the category $\mathcal{O}$, *Duke Math. J.*, 165(1): 67–93, 2016.

[ALY] Tomoyuki Arakawa, Ching Hung Lam, and Hiromichi Yamada. Zhu's algebra, C_2-algebra and C_2-cofiniteness of parafermion vertex operator algebras. *Adv. Math.*, 264: 261–295, 2014.

[AM1] Tomoyuki Arakawa and Anne Moreau, Joseph ideals and lisse minimal W-algebras, *J. Inst. Math. Jussieu*, to be published.

[AM2] Tomoyuki Arakawa and Anne Moreau, Sheets and associated varieties of affine vertex algebras, *Adv. Math.*, 320: 157–209, 7 November 2017.

[AM3] Tomoyuki Arakawa and Anne Moreau, On the irreducibility of associated varieties of W-algebras, to appear in the special issue of *J. Algebra* in honor of Efim Zelmanov on the occasion of his 60th birthday.

[AvE] Tomoyuki Arakawa and Jethro van Ekeren, Modularity of relatively rational vertex algebras and fusion rules of regular affine W-algebras. arXiv:1612.09100[math.RT].

[AKNS] Yusuke Arike, Masanobu Kaneko, Kiyokazu Nagatomo and Yuichi Sakai, Affine Vertex Operator Algebras and Modular Linear Differential Equations, *Lett. Math. Phys.*, 106(5): 693–718, 2016.

[BLL$^+$] Christopher Beem, Madalena Lemos, Pedro Liendo, Wolfger Peelaers, Leonardo Rastelli, and Balt C. van Rees, Infinite chiral symmetry in four dimensions, *Comm. Math. Phys.*, 336(3): 1359–1433, 2015.

[BR] Christopher Beem and Leonardo Rastelli, Vertex operator algebras, Higgs branches, and modular differential equations, arXiv:1707.07679[hep-th].

[BG] Kenneth A. Brown and Iain Gordon, Poisson orders, symplectic reflection algebras and representation theory, *J. Reine Angew. Math.*, 559: 193–216, 2003.

[CdM] Arjeh M. Cohen and Ronald de Man, Computational evidence for Deligne's conjecture regarding exceptional Lie groups, *Comptes rendus de l'Académie des Sciences, Série 1, Mathématique* 322.5 427–432, 1996.

[CS] Clay Córdova and Shu-Heng Shao, Schur indices, BPS particles, and Argyres-Douglas theories, *J. High Energy Phys.*, (1): 040, front matter+37, 2016.

[D] Pierre Deligne, La série exceptionnelle de groupes de Lie, *Comptes Rendus de l'Académie des Sciences-Série I-Mathématique* 322.4 (1996): 321–326.

[DSK] Alberto De Sole and Victor G. Kac, Finite vs affine W-algebras, *Japan. J. Math.*, 1(1): 137–261, 2006.

[DLM] Chongying Dong, Haisheng Li, and Geoffrey Mason. Modular-Invariance of Trace Functions in Orbifold Theory and Generalized Moonshine, *Comm. Math. Phys.* 214.1: 1–56, 2000.

[ES] Pavel Etingof and Travis Schedler, Poisson traces and D-modules on Poisson varieties, *Geom. Funct. Anal.*, 20(4): 958–987, 2010. With an appendix by Ivan Losev.

[FF] Boris Feigin and Edward Frenkel, Quantization of the Drinfel′d-Sokolov reduction, *Phys. Lett. B*, 246(1–2): 75–81, 1990.

[FM] Boris Feigin and Fyodor Malikov, Modular functor and representation theory of $\widehat{\mathrm{sl}}_2$ at a rational level. In *Operads: Proceedings of Renaissance Conferences (Hartford, CT/Luminy, 1995)*, Volume 202, *Contemp. Math.*, pages 357–405, Amer. Math. Soc., Providence, RI, 1997.

[FS] B. L. Feigin and A. M. Semikhatov, $\mathcal{W}_n^{(2)}$ algebras, *Nuclear Phys. B*, 698(3): 409–449, 2004.

[GS] Wee Teck Gan and Gordan Savin, Uniqueness of Joseph ideal, *Math. Res. Lett.*, 11(5–6): 589–597, 2004.

[Gen] Naoki Genra, Screening operators for W-algebras, *Selecta Math. (N.S.)*, 23(3): 2157–2202, 2017, https://doi.org/10.1007/s00029-017-0315-9 .

[Gin] Victor Ginzburg, On primitive ideals, *Selecta Math. (N.S.)*, 9(3): 379–407, 2003.

[Hua] Yi-Zhi Huang, Vertex operator algebras and the Verlinde conjecture, *Commun. Contemp. Math.*, 10(1): 103–154, 2008.

[KRW] Victor Kac, Shi-Shyr Roan, and Minoru Wakimoto, Quantum reduction for affine superalgebras, *Comm. Math. Phys.*, 241(2–3): 307–342, 2003.

[KW1] Victor G. Kac and Minoru Wakimoto, Modular invariant representations of infinite-dimensional Lie algebras and superalgebras, *Proc. Nat. Acad. Sci. U.S.A.*, 85(14): 4956–4960, 1988.

[KW2] V. G. Kac and M. Wakimoto, Classification of modular invariant representations of affine algebras. In *Infinite-dimensional Lie algebras and groups (Luminy-Marseille, 1988)*, Volume 7, *Adv. Ser. Math. Phys.*, pages 138–177, World Sci. Publ., Teaneck, NJ, 1989.

[KW3] Victor G. Kac and Minoru Wakimoto, On rationality of W-algebras, *Transform. Groups*, 13(3–4): 671–713, 2008.

[KW4] Victor G. Kac and Minoru Wakimoto. A remark on boundary level admissible representations. *Comp. Rend. Math.* 355.2: 128–132, 2017.

[KK] Masanobu Kaneko and Masao Koike, On modular forms arising from a differential equation of hypergeometric type, *The Ramanujan J.* 7.1–3: 145–164, 2003.

[KNS] Masanobu Kaneko, Kiyokazu Nagatomo, and Yuichi Sakai. Modular forms and second order ordinary differential equations: applications to vertex operator algebras. *Lett. Math. Phys.* 103.4: 439–453, 2013.

[KZ] Masanobu Kaneko, and Don Zagier. Supersingular j-invariants, hypergeometric series, and Atkin's orthogonal polynomials. *AMS/IP Stud. Adv. Math.* 7: 97–126, 1998.

[KT] Masaki Kashiwara and Toshiyuki Tanisaki, Characters of irreducible modules with non-critical highest weights over affine Lie algebras. In *Representations and quantizations (Shanghai, 1998)*, pages 275–296. China High. Educ. Press, Beijing, 2000.

[K1] Kazuya Kawasetsu, The intermediate vertex subalgebras of the lattice vertex operator algebras, *Lett. Math. Phys.* 104.2: 157–178, 2014.

[K2] Kazuya Kawasetsu, $\mathscr{W}$-algebras with non-admissible levels and the Deligne exceptional series, *Int. Math. Res. Not.* rnw240, 2016.

[Kos63] Bertram Kostant, Lie group representations on polynomial rings, *Amer. J. Math.*, 85: 327–404, 1963.

[Kos78] Bertram Kostant, On Whittaker vectors and representation theory, *Invent. Math.*, 48(2): 101–184, 1978.

[LM] Joseph M Landsberg and Laurent Manivel. Triality, exceptional Lie algebras and Deligne dimension formulas, *Adv. Math.* 171.1: 59–85, 2002.

[Mas] Geoffrey Mason, Vector-valued modular forms and linear differential operators, *Int. J. of Number Theory* 3.03: 377–390, 2007.

[Mil] Antun Milas, On certain automorphic forms associated to rational vertex operator algebras. (English summary) *Moonshine: the first quarter century and beyond* (*Edinburgh 2004*), Volume 372, *London Math. Soc. Lecture Note Ser.*, pages 330–357, Cambridge Univ. Press, Cambridge, 2010.

[Miy] Masahiko Miyamoto, Modular invariance of vertex operator algebras satisfying C_2-cofiniteness, *Duke Math. J.*, 122(1): 51–91, 2004.

[MMS] Samir D. Mathur, Sunil Mukhi, and Ashoke Sen. On the classification of rational conformal field theories, *Phys. Lett. B* 213.3: 303–308, 1988.

[Per] Ozren Perše, Vertex operator algebras associated to certain admissible modules for affine Lie algebras of type A, *Glas. Mat. Ser. III*, 43(63)(1): 41–57, 2008.

[Per13] Ozren Perše, A note on representations of some affine vertex algebras of type D, *Glas. Mat. Ser. III*, 48(68)(1): 81–90, 2013.

[Pre] Alexander Premet, Special transverse slices and their enveloping algebras, *Adv. Math.*, 170(1): 1–55, 2002, With an appendix by Serge Skryabin.

[Slo] Peter Slodowy, *Simple singularities and simple algebraic groups*, Volume 815, *Lecture Notes in Mathematics*, Springer, Berlin, 1980.

[T] Michael P. Tuite, The Virasoro algebra and some exceptional Lie and finite groups, *Symmetry, Integrability and Geometry: Methods and Applications (SIGMA)*. 3: 008, 13 pages, 2007.

[XYY] Dan Xie, Wenbin Yan, Shing-Tung Yau, Chiral algebra of Argyres-Douglas theory from M5 brane, *preprint*, arXiv:1604.02155[hep-th].

[Z] Yongchang Zhu, Modular invariance of characters of vertex operator algebras, *J. Amer. Math. Soc.* 9.1: 237–302, 1996.

On Dimension Growth of Modular Irreducible Representations of Semisimple Lie Algebras

Roman Bezrukavnikov and Ivan Losev

To the memory of Bertram Kostant

Abstract In this paper we investigate the growth with respect to p of dimensions of irreducible representations of a semisimple Lie algebra $\mathfrak{g}$ over $\overline{\mathbb{F}}_p$. More precisely, it is known that for $p \gg 0$, the irreducibles with a regular rational central character λ and p-character χ are indexed by a certain canonical basis in the K_0 of the Springer fiber of χ. This basis is independent of p. For a basis element, the dimension of the corresponding module is a polynomial in p. We show that the canonical basis is compatible with the two-sided cell filtration for a parabolic subgroup in the affine Weyl group defined by λ. We also explain how to read the degree of the dimension polynomial from a filtration component of the basis element. We use these results to establish conjectures of the second author and Ostrik on a classification of the finite-dimensional irreducible representations of W-algebras, as well as a strengthening of a result by the first author with Anno and Mirkovic on real variations of stabilities for the derived category of the Springer resolution.

Keywords: Dimension polynomials • Localization theorem • W-algebras • Real variation of stabilities

Mathematics Subject Classification (2010): 17B45, 17B50, 14F05

The work of R. B. was partially supported by the NSF under the grant DMS-1601953.

The work of I. L. was partially supported by the NSF under the grant DMS-1501558.

Roman Bezrukavnikov
Department of Mathematics, MIT, Cambridge MA 02139 USA, e-mail: bezrukav@math.mit.edu

Ivan Losev
Department of Mathematics, Northeastern University, Boston MA 02115 USA & International Laboratory of Representation theory and Mathematical Physics, NRU-HSE, Moscow, Russia, e-mail: i.loseu@neu.edu

V. G. Kac, V. L. Popov (eds.), *Lie Groups, Geometry, and Representation Theory*, Progress in Mathematics 326, https://doi.org/10.1007/978-3-030-02191-7_3

1 Introduction

In this paper we study the representation theory of semisimple Lie algebras over algebraically closed fields of large positive characteristic. More precisely, let G be a semisimple algebraic group (of adjoint type) over $\mathbb{C}$ and let $\mathfrak{g}$ be its Lie algebra. Then $\mathfrak{g}$ is defined over $\mathbb{Z}$, so for an algebraically closed field $\mathbb{F}$ of characteristic p, we can define the form $\mathfrak{g}_{\mathbb{F}}$ over $\mathbb{F}$. The universal enveloping algebra $U(\mathfrak{g}_{\mathbb{F}})$ is finite over its center, namely, we have a central algebra embedding $S(\mathfrak{g}_{\mathbb{F}}^{(1)}) \hookrightarrow U(\mathfrak{g}_{\mathbb{F}})$, $x \mapsto x^p - x^{[p]}$, where the superscript (1) indicates the Frobenius twist and the superscript $[p]$ stands for the restricted p-th power map $\mathfrak{g}_{\mathbb{F}}^{(1)} \to \mathfrak{g}_{\mathbb{F}}$. The image is known as the *p-center*. In particular, all irreducible representations of $\mathfrak{g}_{\mathbb{F}}$ are finite-dimensional. Below we will assume that $p \gg 0$ (although some statements hold under weaker assumptions).

Let $\mathfrak{h}$ denote a Cartan subalgebra of $\mathfrak{g}$. We have an identification $U(\mathfrak{g}_{\mathbb{F}})^{G_{\mathbb{F}}} \xrightarrow{\sim} \mathbb{F}[\mathfrak{h}^*]^W$ (the Harish-Chandra isomorphism), and the central subalgebra $U(\mathfrak{g}_{\mathbb{F}})^{G_{\mathbb{F}}} \subset U(\mathfrak{g}_{\mathbb{F}})$ is known as the *Harish-Chandra center*. Fix $\lambda \in \mathfrak{h}^*$ and consider the corresponding central reduction $\mathscr{U}_{\lambda,\mathbb{F}}$ of the algebra $U(\mathfrak{g}_{\mathbb{F}})$. Further, for $\chi \in \mathfrak{g}_{\mathbb{F}}^{(1)*}$, we can consider the further central reduction $\mathscr{U}_{\lambda,\mathbb{F}}^{\chi}$, which is a finite-dimensional algebra. Obviously, every irreducible representation of $U(\mathfrak{g}_{\mathbb{F}})$ factors through exactly one irreducible quotient $U_{\lambda,\mathbb{F}}^{\chi}$; some of these quotients are zero.

The study of the representation theory of the algebras $\mathscr{U}_{\lambda,\mathbb{F}}^{\chi}$ can be easily reduced to the case when the element χ is nilpotent. Here the algebra $\mathscr{U}_{\lambda,\mathbb{F}}^{\chi}$ is nonzero if and only if $\lambda \in \mathfrak{h}_{\mathbb{F}_p}^*$. We recall some results of the first author and collaborators on the representation theory of $\mathscr{U}_{\lambda,\mathbb{F}}^{\chi}$.

Consider the flag variety $\mathscr{B}$ for $\mathfrak{g}$ (over $\mathbb{C}$). Let e be a nilpotent element in $\mathfrak{g}$ in the orbit corresponding to that of χ (since $p \gg 0$, there is a natural bijection between the nilpotent orbits in $\mathfrak{g}$ and in $\mathfrak{g}_{\mathbb{F}}^{(1)}$). Consider the corresponding Springer fiber $\mathscr{B}_e$. In [BMR2], for a regular λ, the authors have constructed identifications

(1) $$K_0(\mathscr{U}_{\lambda,\mathbb{F}}^{\chi}\text{-mod}) \xrightarrow{\sim} K_0(\mathrm{Coh}(\mathscr{B}_e)) \xrightarrow{\sim} H_*(\mathscr{B}_e,\mathbb{C});$$

in the present paper all K_0-groups will be over $\mathbb{C}$ but, in fact, the first isomorphism holds over $\mathbb{Z}$.

There is a way to identify classes of simples under this isomorphism, conjectured by Lusztig and proved in [BM]. The space $K_0(\mathrm{Coh}(\mathscr{B}_e))$ admits a q-deformation, the equivariant K-theory group $K_0(\mathrm{Coh}^{\mathbb{C}^\times}(\mathscr{B}_e))$ for a contracting action of $\mathbb{C}^\times$ on $\mathscr{B}_e$, [Lu2, Section 6]. Then, according to [BM], there is a canonical basis $\mathfrak{B}$ in $K_0(\mathrm{Coh}^{\mathbb{C}^\times}(\mathscr{B}_e))$ such that the classes of simples in $K_0(\mathrm{Coh}(\mathscr{B}_e))$ are specializations of the elements of $\mathfrak{B}$ to $q = 1$. The only thing that we need to know about $\mathfrak{B}$ is that it is independent of p, and depends not on λ itself but on its *p-alcove*, so we will not need this.

A big problem with this canonical basis is that it is implicit. For example, it is unclear how to compute the dimensions of the irreducible modules. The goal of this paper is to get more explicit information about the canonical bases elements and about dimensions of the corresponding simple modules. More precisely, we want to understand the dependence of the dimensions on p.

First, let us recall that $K_0(\mathrm{Coh}^{\mathbb{C}^\times}(\mathscr{B}_e))$ is a module over the affine Hecke algebra $\mathscr{H}_q(W^a)$. Here and below we write W^a for the affine Weyl group of $\mathfrak{g}$, i.e., $W^a = W \ltimes Q$, where Q is the root lattice.

Now choose a finite localization R of $\mathbb{Z}$ and a dominant regular element $\lambda \in \mathfrak{h}^*_R$. Then for $p \gg 0$, we can reduce λ to an element in $\mathfrak{h}^*_{\mathbb{F}_p}$. Further, pick $b \in \mathfrak{B}$, and let $V_{\lambda,p}(b)$ denote the corresponding simple in $\mathscr{U}^\chi_{\lambda,\mathbb{F}}$-mod. Then (for λ and b fixed) $\dim V_{\lambda,p}(b)$ is known to be a polynomial in p, assuming p satisfies some congruence conditions depending on λ. Our first goal is to determine the degree of this polynomial.

Note that λ determines a proper standard parabolic subgroup $W_{[\lambda]} \subset W^a$. Namely, we consider the action of W^a on $\mathfrak{h}^*_{\mathbb{Q}}$. Let λ° be the intersection of $W^a\lambda$ with the fundamental alcove. For $W_{[\lambda]}$, we take the standard parabolic subgroup generated by the simple reflections corresponding to the walls containing λ°. For example, when $\lambda \in Q$, we have $W_{[\lambda]} = W$, as a standard parabolic subgroup of W^a.

Consider the partition of $W_{[\lambda]}$ into two-sided cells. This partition also determines a partition of the irreducible $W_{[\lambda]}$-modules (or $\mathscr{H}_q(W_{[\lambda]})$-modules for generic q) into families. We filter the module $K_0(\mathrm{Coh}^{\mathbb{C}^\times}(\mathscr{B}_e))$ according to two-sided cells for $W_{[\lambda]}$. Namely, given a two-sided $\mathbf{c}$ for $W_{[\lambda]}$, let $K_0(\mathrm{Coh}^{\mathbb{C}^\times}(\mathscr{B}_e))_{\leqslant \mathbf{c}}$ denote the intersection of $K_0(\mathrm{Coh}^{\mathbb{C}^\times}(\mathscr{B}_e))$ with the sum of all irreducible $\mathscr{H}_q(W_{[\lambda]})$-submodules in the localized K_0 that belong to families indexed by two-sided cells $\mathbf{c}' \leqslant \mathbf{c}$.

The following is the main result of the paper. Let us recall that from a two-sided cell $\mathbf{c}$ in $W_{[\lambda]}$, we can recover a nilpotent orbit $\mathbb{O}_\mathbf{c}$ in $\mathfrak{g}$; see Section 2.2 for more details.

Theorem 1.1. *Under the above assumptions (including $p \gg 0$) we have:*

(1) *For any regular $\lambda \in \mathfrak{h}^*_R$, the basis $\mathfrak{B}$ of $K_0(\mathrm{Coh}^{\mathbb{C}^\times}(\mathscr{B}_e))$ is compatible with the filtration $K_0(\mathrm{Coh}^{\mathbb{C}^\times}(\mathscr{B}_e))_{\leqslant \mathbf{c}}$.*
(2) *Let $b \in \mathfrak{B}$ lie in $K_0(\mathrm{Coh}^{\mathbb{C}^\times}(\mathscr{B}_e))_{\leqslant \mathbf{c}}$ but not in smaller filtration pieces. Then the degree of the polynomial $\dim V_{\lambda,p}(b)$ in p equals $\dim \mathbb{O}_\mathbf{c}/2$.*

Remark 1.2. The Kac-Weisfeiler conjecture [KW] proved by Premet [P5] (see also [BMR2] for an alternative proof under more restrictive assumptions) implies that the polynomial $\dim V_{\lambda,p}(b)$ is divisible by $p^{d_e/2}$, where d_e is the dimension of the orbit of e.

Remark 1.3. There is a classical analog of (2) for categories in characteristic 0 such as category $\mathscr{O}$. There the result is that the Gelfand-Kirillov dimension of the module corresponding to a canonical basis element equals $\dim \mathbb{O}_\mathbf{c}/2$. So part (2) means that the degree of the dimension polynomial is the modular analog of the Gelfand-Kirillov dimension. Heuristically, this can justified as follows: a module of Gelfand-Kirillov dimension d has "the same size" as the space of sections of a coherent sheaf on $\mathfrak{g}^*$ with support of dimension d, while a module in characteristic

p whose dimension D is expressed by a polynomial in p of degree d has the same size as the space of sections of such a coherent sheaf restricted to the Frobenius neighborhood of a point; see also Remark 4.6.

We can also reformulate (2) as follows. We will see below that there is a unique primitive ideal $\mathscr{J} \in \mathscr{U}$ such that the simple corresponding to b is annihilated by the reduction of $\mathscr{J}$ mod p. We will see that $\overline{\mathbb{O}}_{\mathbf{c}}$ is the associated variety of $\mathscr{J}$, so that the degree of the dimension polynomial is $\frac{1}{2}\,\mathrm{GK\text{-}dim}(\mathscr{U}/\mathscr{J})$. We expect that an analog of this result holds in a much greater generality, for example, for quantizations of symplectic singularities.

Let us discuss some applications of Theorem 1.1. First, it allows us to prove conjectures of the second author and Ostrik on the classification of finite-dimensional irreducible modules over the finite W-algebra $\mathscr{W}$ for $(\mathfrak{g}, e)$, see [LO, Section 7.6]. This is Theorem 5.2 in the paper. In particular, this theorem implies that the K_0 of the finite-dimensional representations of $\mathscr{W}$ with central character λ coincides with $\bigoplus_{\mathbf{c}} K_0(\mathrm{Coh}(\mathscr{B}_e))_{\leqslant \mathbf{c}}$, where the sum is taken over all two-sided cells in $W_{[\lambda]}$ such that $\mathbb{O}_{\mathbf{c}} = Ge$. In fact, for such $\mathbf{c}$, we have $K_0(\mathrm{Coh}(\mathscr{B}_e))_{<\mathbf{c}} = 0$. The first author and Kazhdan plan to use part (1) and the result mentioned in the previous sentence to study invariants of the pro-unipotent radical of a parahoric subgroup in a unipotent irreducible representation of p-adic group.

Another application that motivated the main result is a stronger version of the result of [ABM]. The central point of that work is the definition of *real variation of stabilities*, a concept partly inspired by the notion of the Bridgeland stability condition on a triangulated category, and a theorem asserting that the categories of $\mathscr{U}^{\chi}_{\lambda,\mathbb{F}}$-modules give rise to such a structure. Let us describe it in more detail. The above identification $K_0(\mathscr{U}^{\chi}_{\lambda,\mathbb{F}}\text{-mod}) \xrightarrow{\sim} K_0(\mathrm{Coh}(\mathscr{B}_{e,\mathbb{F}}))$ comes from an equivalence of triangulated categories $\mathscr{L}_{\tilde{\lambda}} : D^b(\mathscr{U}_{\lambda,\mathbb{F}}\text{-mod}^{\chi}) \to D^b(\mathrm{Coh}_{\mathscr{B}_e}(T^*\mathscr{B}_{\mathbb{F}}))$ (the definition of $\mathscr{L}_{\tilde{\lambda}}$ is recalled below after Lemma 2.2). Here $\tilde{\lambda}$ is an element of the root lattice, such that $\lambda = \tilde{\lambda} \mod p$; $\mathrm{Coh}_{\mathscr{B}_e}(T^*\mathscr{B}_{\mathbb{F}})$ denotes the category of coherent sheaves on $T^*\mathscr{B}_{\mathbb{F}}$ set-theoretically supported on the closed subvariety $\mathscr{B}_e$, while $\mathscr{U}_{\lambda,\mathbb{F}}\text{-mod}^{\chi}$ is the category of modules over $\mathscr{U}_{\lambda,\mathbb{F}}$ where the kernel of χ acts nilpotently. The image of the abelian category $\mathscr{U}_{\lambda,\mathbb{F}}\text{-mod}^{\chi}$ under the equivalence $\mathscr{L}_{\tilde{\lambda}}$, i.e., the corresponding t-structure on $D^b(\mathrm{Coh}_{\mathscr{B}_\chi}(T^*\mathscr{B}_{\mathbb{F}}))$ depends only on the p-alcove of $\tilde{\lambda}$, not on λ itself. Thus we get a collection of t-structures on the derived category of coherent sheaves indexed by alcoves; although the above construction applies to varieties of large finite characteristic only, the t-structures admits a canonical lift to $D^b(\mathrm{Coh}_{\mathscr{B}_e}(T^*\mathscr{B}_{\mathbb{C}}))$.

It turns out to be part of a real variation of stability conditions; the content of this statement is as follows: for two neighboring alcoves sharing a codimension-one face, the derived equivalence between the corresponding abelian categories is a *perverse equivalence* governed by a certain polynomial map $Z\colon \mathfrak{t}^*_{\mathbb{R}} \to K_0(\mathrm{Coh}(\mathscr{B}_e))^*$ called the central charge map.

A conjecture stated in [ABM, Remark 6] asserts that a similar property should hold for two alcoves symmetric relative to a higher-codimension face of the affine coroot stratification of $\mathfrak{t}^*_{\mathbb{R}}$. In Section 6 we deduce (a statement essentially equivalent to) that conjecture from Theorem 1.1. Again, we expect a similar statement to hold for all, or at least for a wide class of, symplectic singularities.

Acknowledgements. We would like to thank Pavel Etingof for his kind permission to use his results in this paper, see Section 4.3, and Victor Ostrik for numerous discussions related to the project. We are also grateful to Jim Humphreys for helpful comments on the text.

2 Preliminaries

2.1 Harish-Chandra bimodules and primitive ideals

Let us write $\mathscr{U}$ for $U(\mathfrak{g})$ and $\mathscr{U}_\lambda$ for the central reduction of $\mathscr{U}$ at $\lambda \in \mathfrak{h}^*$.

Recall that by a Harish-Chandra (shortly, HC) $\mathscr{U}$-bimodule, one means a finitely generated $\mathscr{U}$-bimodule with locally finite adjoint action of $\mathfrak{g}$. In this paper we will only consider the bimodules where the adjoint $\mathfrak{g}$-action integrates to an action of $G := \mathrm{Ad}(\mathfrak{g})$. Every HC-bimodule admits a so-called *good filtration*, i.e., a G-stable filtration such that the associated graded space is finitely generated as a module over $S(\mathfrak{g})$ (since the filtration is G-stable the left and the right actions of $\mathfrak{g}$ on the associated graded space coincide).

We write $\mathrm{HC}(\mathscr{U})$ for the category of HC $\mathscr{U}$-bimodules and $D^b_{HC}(\mathscr{U}\text{-bimod})$ for the full subcategory of $D^b(\mathscr{U}\text{-bimod})$ of all objects with HC homology. We note that $D^b_{HC}(\mathscr{U}\text{-bimod})$ is closed under taking the derived tensor products.

Inside $\mathrm{HC}(\mathscr{U})$ we will consider three kinds of subcategories defined by the central character conditions. Fix $\lambda, \lambda' \in \mathfrak{h}^*$. We consider the subcategory ${}^1_\lambda\mathrm{HC}^1_{\lambda'}$ of all HC-bimodules with genuine central characters λ on the left and λ' on the right. When $\lambda = \lambda'$, we write $\mathrm{HC}(\mathscr{U}_\lambda)$ for ${}^1_\lambda\mathrm{HC}^1_\lambda$. Note that $\mathrm{HC}(\mathscr{U}_\lambda)$ is a monoidal category. We can also consider the larger subcategory ${}^\infty_\lambda\mathrm{HC}^\infty_{\lambda'}$, where the central characters on the left and on the right are generalized. Finally, there is an intermediate category ${}^\infty_\lambda\mathrm{HC}^1_{\lambda'}$. Note that ${}^1_\lambda\mathrm{HC}^1_{\lambda'}$ contains all simple objects in ${}^\infty_\lambda\mathrm{HC}^\infty_{\lambda'}$ and all objects in ${}^\infty_\lambda\mathrm{HC}^\infty_{\lambda'}$ have finite length. Because of this, the K_0's of these categories are the same.

Now suppose that λ is regular. Let $\mu \in W\lambda$ be anti-dominant, meaning that $\langle \alpha^\vee, \mu \rangle \notin \mathbb{Z}_{\geqslant 0}$ for any positive coroot $\alpha^\vee$. Fix this μ; it is not unique unless λ is integral. Consider the block $\mathscr{O}(\mu)$ of the BGG category $\mathscr{O}$ spanned by the simples $L(u\mu)$ with highest weight $u\mu - \rho$, where u is in the integral Weyl group $W_{\mu,\mathrm{int}}$ of μ. Recall that this group is generated by all reflections s_α such that $\langle \alpha^\vee, \mu \rangle \in \mathbb{Z}$.

Then there is the Bernstein-Gelfand equivalence ${}^\infty_\lambda\mathrm{HC}^1_\lambda \xrightarrow{\sim} \mathscr{O}(\mu)$ given by $\mathscr{M} \mapsto \mathscr{M} \otimes_{\mathscr{U}_\lambda} \Delta(w_0\mu)$, where w_0 is the longest element in $W_{\mu,\mathrm{int}}$, so that $\Delta(w_0\mu)$ is projective in $\mathscr{O}(\mu)$. In particular, the simples in $\mathrm{HC}(\mathscr{U}_\lambda)$ are labelled by $u \in W_{\mu,\mathrm{int}}$. Note that there is a natural isomorphism $W_{[\mu]} \xrightarrow{\sim} W_{\mu,\mathrm{int}}$: namely, let $w_1 \in W^a$ be the minimal length element such that (in the notation of the introduction) $\mu = w_1\mu^\circ$.

Then an isomorphism $W_{[\mu]} \xrightarrow{\sim} W_{\mu,\mathrm{int}}$ is given by $w \mapsto \mathrm{pr}(w_1^{-1}ww_1)$, where we write pr for the projection $W^a \twoheadrightarrow W$. This defines a bijection $\mathrm{Irr}(\mathrm{HC}(\mathscr{U}_\lambda)) \xrightarrow{\sim} W_{[\lambda]}$. Let us write $\mathscr{M}_w$ for the simple HC $\mathscr{U}_\lambda$-bimodule corresponding to $w \in W_{[\lambda]}$.

Let $\mathscr{M}$ be a HC $\mathscr{U}$-bimodule. By the associated variety $\mathrm{V}(\mathscr{M})$, we mean the support of $\mathrm{gr}\,\mathscr{M}$ in $\mathfrak{g}$, where the associated graded space is taken with respect to any good filtration. We note that $\mathrm{V}\big(\mathrm{Tor}_i^{\mathscr{U}}(\mathscr{M}_1,\mathscr{M}_2)\big) \subset \mathrm{V}(\mathscr{M}_1) \cap \mathrm{V}(\mathscr{M}_2)$.

Let us fix a nilpotent orbit $\mathbb{O} \subset \mathfrak{g}$. We can consider the subcategories $\mathrm{HC}_{\partial\mathbb{O}}(\mathscr{U}) \subset \mathrm{HC}_{\overline{\mathbb{O}}}(\mathscr{U})$ of all $\mathscr{M} \in \mathrm{HC}(\mathscr{U})$ with $\mathrm{V}(\mathscr{M}) \subset \partial\mathbb{O}$ (resp., $\mathrm{V}(\mathscr{M}) \subset \overline{\mathbb{O}}$). These are tensor ideals in $\mathrm{HC}(\mathscr{U})$. So we can form the quotient category $\mathrm{HC}_{\mathbb{O}}(\mathscr{U})$ that also carries the tensor product. Let $\mathrm{HC}^{ss}_{\mathbb{O}}(\mathscr{U})$ denote the full subcategory of semisimple objects in $\mathrm{HC}_{\mathbb{O}}(\mathscr{U})$. One can show, using, for example, [L2, Corollary 1.3.2], that the subcategory $\mathrm{HC}^{ss}_{\mathbb{O}}(\mathscr{U})$ is closed under the tensor products. Moreover, it is a rigid monoidal category. This has the following corollary.

Corollary 2.1. *For simple HC-bimodules $\mathscr{M},\mathscr{M}' \in \mathrm{HC}_{\overline{\mathbb{O}}}(\mathscr{U})$, there are $M_1 \subset M_2 \subset \mathscr{M} \otimes_{\mathscr{U}} \mathscr{M}'$ with $M_1, \mathscr{M} \otimes_{\mathscr{U}} \mathscr{M}'/M_2 \subset \mathrm{HC}_{\partial\mathbb{O}}(\mathscr{U})$, while M_2/M_1 is the sum of simple HC-bimodules with associated variety $\overline{\mathbb{O}}$.*

We proceed to primitive ideals (=annihilators of irreducible representations). We write $\mathrm{Prim}(\mathscr{U}_\lambda)$ for the set of primitive ideals in $\mathscr{U}_\lambda$. By the Duflo theorem, every primitive ideal in $\mathscr{U}_\lambda$ is the annihilator $\mathscr{J}(\lambda')$ of some irreducible module $L(\lambda')$ with $\lambda' \in W\lambda$. Inside $\mathrm{Prim}(\mathscr{U}_\lambda)$ we can consider the subset $\mathrm{Prim}_{\mathbb{O}}(\mathscr{U}_\lambda)$ of all $\mathscr{J}$ such that $\mathrm{V}(\mathscr{U}_\lambda/\mathscr{J}) = \mathbb{O}$.

Suppose that λ is regular. We have a surjection $W_{[\lambda]} \twoheadrightarrow \mathrm{Prim}(\mathscr{U}_\lambda)$ that sends $w \in W_{[\lambda]}$ to the left annihilator of $\mathscr{M}_w$; let us denote it by $\mathscr{J}_w$. We have $\mathscr{J}_w = \mathscr{J}(w\mu)$ in our previous notation. The right annihilator of $\mathscr{M}_w$ is $\mathscr{J}_{w^{-1}}$.

Now suppose λ_0 is singular (and dominant). Choose a strictly dominant element μ in the root lattice and let $\lambda = \lambda_0 + \mu$ so that, in particular, λ is regular dominant. We have $\mathscr{J}(w'\lambda_0) = \mathscr{J}(w\lambda_0)$ provided $\mathscr{J}(w'\lambda) = \mathscr{J}(w\lambda)$; see [Ja, Section 5.4-5.8]. This gives the embedding $\mathrm{Prim}(\mathscr{U}_{\lambda_0}) \hookrightarrow \mathrm{Prim}(\mathscr{U}_\lambda)$ whose image consists of the primitive ideals $\mathscr{J}(w\lambda)$, where w is longest in wW_{λ_0}. The embedding sends $\mathrm{Prim}_{\mathbb{O}}(\mathscr{U}_{\lambda_0})$ to $\mathrm{Prim}_{\mathbb{O}}(\mathscr{U}_\lambda)$.

2.2 Hecke algebras, cells, and HC-bimodules

For a Weyl group W, we can consider its Hecke algebra $\mathscr{H}_q(W)$ which comes with the distinguished basis c_w, $w \in W$, known as the Kazhdan-Lusztig basis (we use the convention, where the elements c_w are sign-positive with respect to the standard basis T_w). This basis allows us to define the so-called *two-sided pre-order* on the basis elements. Namely, consider the two-sided based (=spanned by basis elements as a $\mathbb{C}[q^{\pm 1}]$-module) ideal I_w. Set $w \leqslant w'$ if $I_w \subset I_{w'}$. The equivalence classes for this pre-order are known as *two-sided cells*. The induced order on the set of two-sided cells will also be denoted by $\leqslant$. Similarly, we can consider left based ideals, and have the pre-order $\leqslant^L$ and the equivalence relation $\sim^L$ on W. The equivalences classes are known as the *left cells*.

The two-sided cells and left cells naturally define subquotients of $\mathscr{H}_q(W)$ that are bimodules and left modules, respectively (called two-sided and left cell modules). The two-sided cell modules allow us to partition irreducible representations of $\mathscr{H}_q(W)$, and of $\mathbb{C}W$ when W is of finite type, into subsets called families.

Now let us discuss a connection between the Hecke algebras and HC-bimodules. Let W be the Weyl group of $\mathfrak{g}$. The category $D^b_{HC}(\mathscr{U}_\lambda\text{-bimod})$ is monoidal with respect to $\bullet \otimes^L_{\mathscr{U}_\lambda} \bullet$. This monoidal structure equips $K_0(\mathrm{HC}(\mathscr{U}_\lambda))$ with an algebra structure. The resulting algebra is $\mathbb{C}W_{[\lambda]}$. The class $\mathscr{M}_w$ corresponds to the specialization of $c_{w^{-1}}$ to $q=1$. The simple reflections in $\mathbb{C}W_{[\lambda]}$ correspond to the so-called wall-crossing bimodules in $K_0(\mathrm{HC}(\mathscr{U}_\lambda))$.

We recall the definition of these bimodules. For $\lambda \in \mathfrak{h}^*$, let us write $\mathscr{D}^\lambda_{\mathscr{B}}$ for the sheaf of $(\lambda-\rho)$-twisted differential operators. Choose $w \in W_{[\lambda]}$ and view it as an element of $W_{\mu,\mathrm{int}}$ as before. Set $\psi = w_0\mu - w^{-1}w_0\mu$, which is an element of the root lattice. Let $\mathscr{W}\mathscr{C}_w$ denote the $\mathscr{D}^{w_0\mu}_{\mathscr{B}}$-$\mathscr{D}^{w_0\mu-\chi}_{\mathscr{B}}$-bimodule quantizing the line bundle $\mathscr{O}(\psi)$ on $T^*\mathscr{B}$. Then the wall-crossing bimodule WC_w is the global sections of $\mathscr{W}\mathscr{C}_w$.

Moreover, we get a homomorphism $\mathrm{Br}_{W_{[\lambda]}} \to D^b_{HC}(\mathscr{U}_\lambda\text{-bimod})$ sending the natural generators of the braid groups to the wall-crossing bimodules, see [M, Section L.3] or [BMR1, Section 2] (that treats the positive characteristic case).

Let us now discuss the representation theoretic meaning of cells. We have $\mathscr{J}_w \subset \mathscr{J}_{w'}$ if and only if $w \leqslant^L w'$ and hence $\mathscr{J}_w = \mathscr{J}_{w'}$ if and only if $w \sim^L w'$; this follows from combining [Lu4, Lemma 7.4] and [Jo1, Theorem 3.10]. So if w, w' are in the same two-sided cell, then $\mathrm{V}(\mathscr{M}_w) = \mathrm{V}(\mathscr{M}_{w'})$ (and the converse is true for integral λ). So to a two-sided cell we can assign a nilpotent orbit in $\mathfrak{g}$, denoted by $\mathbb{O}_{\mathbf{c}}$. It is easy to see that $\mathbf{c} < \mathbf{c}'$ implies $\mathbb{O}_{\mathbf{c}} \subsetneq \mathbb{O}_{\mathbf{c}'}$.

Recall that every left cell contains a so-called distinguished (a.k.a. Duflo) involution, say d. The corresponding simple HC-bimodule $\mathscr{M}_d$ is the socle of $\mathscr{U}_\lambda/\mathscr{J}_d$. Moreover, the quotient $\mathrm{V}((\mathscr{U}_\lambda/\mathscr{J}_d)/\mathscr{M}_d) \subset \partial\mathbb{O}_{\mathbf{c}}$.

Now let us discuss asymptotic Hecke algebras. To any Weyl group W, Lusztig assigned the asymptotic Hecke algebra $\mathsf{J} = \mathsf{J}(W)$ that is a unital associative algebra (say, over $\mathbb{C}$) together with a distinguished basis t_w, $w \in W$. The unit in J is the element $\sum_d t_d$, where the sum is taken over all distinguished involutions in W. There is a homomorphism $\mathbb{C}W \to \mathsf{J}$ which is known to be an isomorphism when W is of finite type.

Note that we have $t_w t_{w'} = 0$ when w, w' lie in two different two-sided cells. So we get a decomposition $\mathsf{J} = \bigoplus_{\mathbf{c}} \mathsf{J}_{\mathbf{c}}$, where $\mathsf{J}_{\mathbf{c}}$ is the ideal in J with basis t_w, $w \in \mathbf{c}$. Note that for W of finite type, the irreducible W-modules that belong to a two-sided cell $\mathbf{c}$ are precisely the modules obtained by pullback $\mathsf{J}_{\mathbf{c}}$. Moreover, if σ is a left cell in W with distinguished involution d, then the left cell module $[\sigma]$ is $\mathsf{J}(W)t_d$.

Now let us give a categorical interpretation of the algebra $\mathsf{J}(W_{[\lambda]})$. Consider the rigid monoidal category $\bigoplus_{\mathbb{O}} \mathrm{HC}^{ss}_{\overline{\mathbb{O}}}(\mathscr{U}_\lambda)$, where the sum is taken over all nilpotent orbits in $\mathfrak{g}$; some summands may be zero. Note that $\mathrm{HC}^{ss}_{\overline{\mathbb{O}}}(\mathscr{U}_\lambda)$ splits as $\bigoplus_{\mathbf{c}} \mathrm{HC}^{ss}_{\mathbf{c}}(\mathscr{U}_\lambda)$, where the summand $\mathrm{HC}^{ss}_{\mathbf{c}}(\mathscr{U}_\lambda)$ is spanned by the $\mathscr{M}_w$'s with $w \in \mathbf{c}$. So our category can be written as $\bigoplus_{\mathbf{c}} \mathrm{HC}^{ss}_{\mathbf{c}}(\mathscr{U}_\lambda)$, where the sum is taken over all two-sided cells in $W_{[\lambda]}$. Then, by the work of Joseph, e.g., [Jo2], see also [BFO], it is known that $K_0(\mathrm{HC}^{ss}_{\mathbf{c}}(\mathscr{U}_\lambda)) = \mathrm{J}_{\mathbf{c}}(W_{[\lambda]})$.

2.3 Localization in characteristic p

We now explain results of [BMR2, BM] related to the localization in characteristic $p \gg 0$.

Pick a regular dominant element $\lambda \in \mathfrak{h}^*_{\mathbb{Q}}$. Let x be the least common denominator of the coefficients of the simple roots in λ. In what follows, we assume that $p+1$ is divisible by x, so that $(p+1)\lambda$ lies in the root lattice.

Let $\mathbb{F}$ be an algebraically closed field of large enough characteristic p. Recall that $\mathscr{B}_{\mathbb{F}}$ stands for the flag variety for G over $\mathbb{F}$. Then we have the sheaf $\mathscr{D}^\lambda_{\mathscr{B}_{\mathbb{F}}}$ which is an Azumaya algebra on $T^*\mathscr{B}^{(1)}_{\mathbb{F}}$. Note that the categories $\mathrm{Coh}(\mathscr{D}^\lambda_{\mathscr{B}_{\mathbb{F}}})$, $\mathrm{Coh}(\mathscr{D}^{\lambda'}_{\mathscr{B}_{\mathbb{F}}})$ are abelian equivalent, via twist with a line bundle $\mathscr{O}(\mu)$, where μ is a weight congruent to $\lambda' - \lambda$ mod p.

We have $R\Gamma(\mathscr{D}^\lambda_{\mathscr{B}_{\mathbb{F}}}) = \mathscr{U}_{\lambda,\mathbb{F}}$. It was shown in [BMR2, Section 3.2] that the derived global section functor $R\Gamma\colon D^b(\mathrm{Coh}(\mathscr{D}^\lambda_{\mathscr{B}_{\mathbb{F}}})) \to D^b(\mathscr{U}_{\lambda,\mathbb{F}}\text{-mod})$ is an equivalence. Further, it was checked in [BMR2, Section 5.4] that $\mathscr{D}^\lambda_{\mathscr{B}_{\mathbb{F}}}$ splits in the formal neighborhood $\mathscr{B}^{(1)\wedge}_{\mathbb{F},\chi}$ of the Springer fiber in $T^*\mathscr{B}^{(1)}_{\mathbb{F}}$.

Choose a splitting bundle $\mathscr{V}_{\chi,\mathbb{F}}$. This gives rise to the abelian equivalence

$$\mathscr{V}_{\chi,\mathbb{F}} \otimes \bullet :\ \mathrm{Coh}_\chi(T^*\mathscr{B}^{(1)}_{\mathbb{F}}) \xrightarrow{\sim} \mathrm{Coh}_\chi(\mathscr{D}^\lambda_{\mathscr{B}_{\mathbb{F}}}),$$

where the subscript χ refers to the subcategory of sheaves set-theoretically supported at the Springer fiber. So we arrive at the derived equivalence

$$D^b_\chi(\mathrm{Coh}(T^*\mathscr{B}^{(1)}_{\mathbb{F}})) \xrightarrow{\sim} D^b_\chi(\mathscr{U}_{\lambda,\mathbb{F}}\text{-mod})$$

given by $M \mapsto R\Gamma(\mathscr{V}_{\chi,\mathbb{F}} \otimes M)$. The following was shown in [BMR2, Lemma 6.2.5]:

Lemma 2.2. *Fix λ' in the root lattice such that $\lambda' = \lambda \mod p$. Then there exists a canonical choice of the splitting bundle $\mathscr{V}_{\chi,\mathbb{F}}$ (recall it is defined up to a twist with a line bundle) such that the class $[\mathscr{V}_{\chi,\mathbb{F}}] \in K_0(\mathscr{B}^{(1)}_{\chi,\mathbb{F}})$ is the pull-back of $[(\mathrm{Fr}_{\mathscr{B}_{\mathbb{F}}})_*\mathscr{O}((p-1)\rho+\lambda')]$.*

The resulting equivalence is denoted by $\mathscr{L}_{\lambda'}$.

Below we always choose $\mathscr{V}_{\chi,\mathbb{F}}$ as in the lemma.

Proposition 2.3. *The following is true:*

(1) *The images of the classes of simple $\mathscr{U}^\chi_{\lambda,\mathbb{F}}$-modules in $K_0(\mathrm{Coh}(\mathscr{B}_\chi))$ are independent of p as long as $p \gg 0$.*

(2) *The dimensions of the simple* $\mathscr{U}^{\chi}_{\lambda,\mathbb{F}}$*-modules are polynomials in* p*, provided* $p+1$ *is divisible by* x*.*

Proof. (1) for general λ follows from the case $\lambda = 0$, which is (a) in [BM, Corollary 5.1.8]. (2) follows easily from (1) and Lemma 2.2; compare to [BMR2, Section 6.2]. □

We now discuss actions of algebras of interest on the above Grothendieck groups. Recall that $K_0(\mathrm{Coh}^{\mathbb{C}^\times}(\mathscr{B}_e))$ is a module over the affine Hecke algebra $\mathscr{H}_q(W^a)$. In particular, $K_0(\mathrm{Coh}(\mathscr{B}_e)) \cong H_*(\mathscr{B}_e, \mathbb{C})$ acquires an action of W^a.

As was shown in [BMR1, Section 2], the latter action is categorified by an action of the affine braid group B_{aff} on $D^b(\mathscr{U}_{\lambda,\mathbb{F}}\text{-mod}^{\chi}) \cong D^b(\mathrm{Coh}_{\mathscr{B}_\chi}(T^*\mathscr{B}_{\mathbb{F}}))$, while the former is categorified by a compatible action on the derived category of a *graded version* of $\mathscr{U}_{\lambda,\mathbb{F}}\text{-mod}^{\chi}$, which is derived equivalent to $\mathrm{Coh}^{\mathbb{G}_m}_{\mathscr{B}_\chi}(T^*\mathscr{B}_{\mathbb{F}})$ (see [BM, 5.3.1, 5.3.2]).

For future reference we mention a standard property of this action. For a simple reflection α, we let $\tilde{s}_\alpha$ denote the corresponding generator of the affine braid group.

Lemma 2.4. *For a simple reflection* α *and an irreducible module* $L \in \mathscr{U}_{\lambda,\mathbb{F}}\text{-mod}^{\chi}$*, the object* $\tilde{s}_\alpha(L)$ *either lies in the abelian category* $\mathscr{U}_{\lambda,\mathbb{F}}\text{-mod}^{\chi}$ *or is isomorphic to* $L[1]$*.*

Proof. Consider the full embedding $\mathscr{U}_{\lambda,\mathbb{F}}\text{-mod}^{\chi} \to \mathscr{U}_{\mathbb{F}}\text{-mod}^{\chi}_{\lambda}$, where the target category consists of all those $\mathfrak{g}_{\mathbb{F}}$-modules where the kernel of the central ideal corresponding to (λ,χ) acts nilpotently. By [BR, Theorem 1.3.1], we have a compatible B_{aff} action on $D^b(\mathscr{U}_{\mathbb{F}}\text{-mod}^{\chi}_{\lambda})$, and it suffices to check the same statement in $D^b(\mathscr{U}_{\mathbb{F}}\text{-mod}^{\chi}_{\lambda})$. We have the exact reflection functor Ξ_α acting on $\mathscr{U}_{\mathbb{F}}\text{-mod}^{\chi}_{\lambda}$ and an exact triangle

$$L \mapsto \Xi_\alpha(L) \to \tilde{s}_\alpha(L).$$

Recall that $\Xi_\alpha = T_{\mu\to\lambda} \circ T_{\lambda\to\mu}$ is a composition of two biadjoint translation functors for a weight μ on the α-wall. If $T_{\lambda\to\mu}(L) = 0$, then $\Xi_\alpha(L) = 0$ and $\tilde{s}_\alpha(L) \cong L[1]$. If $T_{\lambda\to\mu}(L) \neq 0$, then the adjunction arrow $L \to T_{\mu\to\lambda} \circ T_{\lambda\to\mu}(L)$ is nonzero, hence it is injective provided that L is irreducible. Thus in this case $\tilde{s}_\alpha(L) \cong \Xi_\alpha(L)/L$ is concentrated in homological degree zero. □

Remark 2.5. It is natural to expect that the aforementioned action of the affine braid group on the derived categories of coherent sheaves factors through the standard categorification of the affine Hecke algebra; the latter can be defined either using constructible sheaves on the affine flag variety, or using the theory of Soergel bimodules. For a base field of characteristic zero, this follows from the main result of [B]; see also [BY] for the relation to Soergel bimodules. For a base field of positive characteristic (which is the setting related to $\mathfrak{g}$-modules in positive characteristic as explained above) this question is still open, to the authors' knowledge.

3 Lengths

This section contains a number of results that will be used to prove Theorem 1.1.

3.1 Reduction of HC-bimodules to characteristic p

The proof of Proposition 3.3 will be based on considering reductions of HC-bimodules to characteristic p.

Let us start by discussing R-forms of Harish-Chandra bimodules. The category of HC-bimodules is defined over $\mathbb{Q}$, the Bernstein-Gelfand equivalence shows that ${}^{\infty}_{\lambda}\mathrm{HC}^{1}_{\lambda}$ is split over $\mathbb{Q}$ because the category $\mathcal{O}$ is split over the rationals. Recall that an abelian category equivalent to a category of modules over a finite-dimensional algebra over a field is called split if the endomorphism algebras of all simples coincide with the field.

Clearly, there is a finite localization R of $\mathbb{Z}$, such that the tensor category $D^b_{HC}(\mathscr{U}_\lambda\text{-bimod})$ is defined over R. All simples are defined over R as well; let us fix some R-lattices $\mathscr{M}_{w,R}$, $w \in W_{[\lambda]}$. Note that we can still talk about HC $\mathscr{U}_{\lambda,R}$-bimodules: these are bimodules $\mathscr{M}$ that admit a bounded-from-below good filtration, such that the left and the right actions of $R[\mathscr{N}]$ on $\operatorname{gr}\mathscr{M}$ coincide and the $R[\mathscr{N}]$-module $\operatorname{gr}\mathscr{M}$ is finitely generated—here $\mathscr{N}$ stands for the nilpotent cone of $\mathfrak{g}$. In particular, every HC $\mathscr{U}_{\lambda,R}$- (or $\mathscr{U}_R$-) bimodule becomes flat over R after a finite localization. Note also that any Tor of any two HC $\mathscr{U}_{\lambda,R}$-bimodules is again HC.

For a primitive ideal $\mathscr{J} \subset \mathscr{U}_\lambda$, we set $\mathscr{J}_R := \mathscr{U}_{\lambda,R} \cap \mathscr{J}$.

Let $V = V(\mu)$ denote the irreducible G-module with highest weight μ. For $m \in \mathbb{Z}_{>0}$, we write

$$V^m := \bigoplus_{\mu \mid \langle \rho^\vee, \mu\rangle \leqslant m} V(\mu)^{\dim V(\mu)}.$$

We will also impose the following conditions which we can achieve by a finite localization of R (in (c3), (c4), we fix m and then further localize R). Here (c2) follows from Corollary 2.1, which is an analogous statement over $\mathbb{C}$.

(c1) For every distinguished involution d, we have an inclusion

$$\mathscr{M}_{d,R} \hookrightarrow \mathscr{U}_{\lambda,R}/\mathscr{J}_{d,R}$$

and the quotient is filtered by bimodules $\mathscr{M}_{w,R}$ with $w <^L d$.

(c2) For every $w_1, w_2 \in \mathbf{c}$, there are $\mathscr{U}_{\lambda,R}$-sub-bimodules

$$M_1 \subset M_2 \subset \mathscr{M}_{w_1,R} \otimes_{\mathscr{U}_R} \mathscr{M}_{w_2,R}$$

such that

- both M_1 and $\mathscr{M}_{w_1,R} \otimes_{\mathscr{U}_{\lambda,R}} \mathscr{M}_{w_2,R}/M_2$ are filtered by $\mathscr{M}_{w,R}$'s, where the elements $w \in W_{[\lambda]}$ lie in cells strictly less than $\mathbf{c}$, and

– the quotient M_2/M_1 is isomorphic to the direct sum of $\mathscr{M}_{w,R}$'s for $w \in \mathbf{c}$.

(c3) Both $\mathrm{pr}_\lambda(V^m \otimes \mathscr{M}_w)$ and its complement in $V^m \otimes \mathscr{M}_w$ are defined over R for all $w \in W_{[\lambda]}$.
(c4) The projection $\mathrm{pr}_\lambda(V^m \otimes \mathscr{M}_w)_R$ is filtered by $\mathscr{M}_{w',R}$'s. Moreover, there is a quotient of $\mathrm{pr}_\lambda(V^m \otimes \mathscr{M}_w)_R$ isomorphic to the direct sum of $\mathscr{M}_{w,R}$'s that gives $\mathsf{head}(\mathrm{pr}_\lambda(V^m \otimes \mathscr{M}_w))$ after base change to $\mathbb{C}$ (recall that by the head we mean the maximal semisimple quotient).
(c5) The wall-crossing bimodules are defined over R and define a homomorphism $\mathrm{Br}_{W_{[\lambda]}} \to D^b_{HC}(\mathscr{U}_{\lambda,R}\text{-bimod})$.
(c6) All $\mathrm{Tor}^i_{\mathscr{U}_{\lambda,R}}(\mathscr{M}_{w_1,R}, \mathscr{M}_{w_2,R})$ are filtered by $\mathscr{M}_{w,R}$'s; note that after a finite localization of R only finitely many of these Tor's are nonzero because $\mathscr{U}_\lambda$ has finite homological dimension. The analogous result is true for $\mathrm{Tor}^i_{\mathscr{U}_{\lambda,R}}(\mathscr{M}_{w_1,R}, \mathscr{U}_{\lambda,R}/\mathscr{J}_{w_2,R})$.

Now let $\mathbb{F}$ be an algebraically closed field and an R-algebra. We have an action of $D^b_{HC}(\mathscr{U}_{\lambda,R}\text{-bimod})$ on $D^b(\mathscr{U}_{\lambda,\mathbb{F}}\text{-mod}^\chi)$ that gives rise to an action of $W_{[\lambda]}$ on $K_0(\mathscr{U}^\chi_{\lambda,\mathbb{F}}\text{-mod})$.

Lemma 3.1. *Under the identification $K_0(\mathrm{Coh}(\mathscr{B}_\chi)) \cong K_0(\mathscr{U}_{\lambda,\mathbb{F}}\text{-mod}^\chi)$, the two actions of $W_{[\lambda]}$ (i.e., the one defined above and the one restricted from the W^a-action in Section 2.3) coincide.*

Proof. The W^a-action on $K_0(\mathrm{Coh}(\mathscr{B}_\chi))$ corresponds to the action on $K_0(\mathscr{U}_{\lambda,\mathbb{F}}\text{-mod}^\chi)$ by the wall-crossing functors, [R, Section 5.4]. By [BMR1, Theorem 2.1.4], the wall-crossing functors through the walls defined by the simple roots for $W_{[\lambda]}$ are given by taking the derived tensor products with the wall-crossing bimodules. □

Below we are also going to use the following lemma.

Lemma 3.2. *Let L be a simple $\mathscr{U}^\chi_{\lambda,\mathbb{F}}$-module and let d be a distinguished involution such that $\mathscr{J}_d$ is a maximal primitive ideal with $\mathscr{J}_{d,R}L = 0$. Then the following is true:*

(1) *For any $w \in W$, all simple constituents of $\mathrm{Tor}_*^{\mathscr{U}_{\lambda,R}}(\mathscr{M}_{w,R}, L)$ are annihilated by $\mathscr{J}_{w,R}$.*
(2) *Let L' be a simple constituent of $\mathrm{Tor}_*^{\mathscr{U}_{\lambda,R}}(\mathscr{M}_{w,R}, L)$ and let $\mathscr{J}_{d'}$ be a maximal primitive ideal such that $\mathscr{J}_{d',R}L' = 0$. Then the two-sided cell of d' is less than or equal to that of d.*
(3) *Assume, in addition, that $d \not\sim^L w^{-1}$. Let L' be a simple constituent of $\mathrm{Tor}_*^{\mathscr{U}_{\lambda,R}}(\mathscr{M}_{w,R}, L)$ and let $d' \in W$ be as in (2). Then $d' <^L w$.*

Proof. Let us take a resolution of L by free $\mathscr{U}_{\lambda,R}$-modules:

$$\cdots \to \mathscr{U}_{\lambda,R}^{\oplus n_2} \to \mathscr{U}_{\lambda,R}^{\oplus n_1} \to L \to 0.$$

Then $\operatorname{Tor}_*^{\mathcal{U}_{\lambda,R}}(\mathcal{M}_{w,R}, L)$ is the homology of the complex

$$\cdots \to \mathcal{M}_{w,R}^{\oplus n_2} \to \mathcal{M}_{w,R}^{\oplus n_1} \to 0.$$

The individual terms of this complex are annihilated by $\mathcal{J}_{w,R}$, hence so is the homology. This proves (1).

Let us prove (2). Since $\mathcal{J}_{d,R}$ annihilates L, we have

$$\mathcal{M}_{w,R} \otimes^L_{\mathcal{U}_{\lambda,R}} L = \left(\mathcal{M}_{w,R} \otimes^L_{\mathcal{U}_{\lambda,R}} (\mathcal{U}_{\lambda,R}/\mathcal{J}_{d,R})\right) \otimes^L_{\mathcal{U}_{\lambda,R}/\mathcal{J}_{d,R}} L.$$

Note that all simples $\mathcal{M}_{w'}$ appearing in $\operatorname{Tor}^*_{\mathcal{U}_\lambda}(\mathcal{M}_w, \mathcal{U}_\lambda/\mathcal{J}_d)$ satisfy $w' \leqslant^L w$, $w'^{-1} \leqslant^L d$. This, together with (c6), implies (2). In (3), since $d \not\leqslant^L w^{-1}$, we have $w' <^L w$. (3) follows. □

3.2 Results on growth of lengths

Recall that $\lambda \in \mathfrak{h}^*_{\mathbb{Q}}$ is regular and p is large enough, in particular, λ is well defined and regular mod p. Let L be a simple in $\mathcal{U}^\chi_{\lambda,\mathbb{F}}$-mod.

Given m, we always choose p large enough so that two conditions hold:

(1) $V(\mu)$ is irreducible mod p provided that $\langle \rho^\vee, \mu\rangle \leqslant m$, and
(2) $V^m_{\mathbb{F}} \otimes V^m_{\mathbb{F}}$ is semisimple.

Let us write $\operatorname{pr}_\lambda$ for the projection $\mathcal{U}_{\mathbb{F}}\text{-mod}^\chi \twoheadrightarrow \mathcal{U}_{\mathbb{F}}\text{-mod}^\chi_\lambda$. For a module $M \in \mathcal{U}_{\mathbb{F}}\text{-mod}^\chi$, let us write $\ell(M)$ for its length. We want to understand the behavior of the length $\ell(\operatorname{pr}_\lambda(V^m_{\mathbb{F}} \otimes L))$ as a function of m.

Proposition 3.3. *Let L be an irreducible $\mathcal{U}^\chi_{\lambda,\mathbb{F}}$-module. Let $\mathcal{J} \subset \mathcal{U}_\lambda$ be a maximal primitive ideal with the following property: L is annihilated by $\mathcal{J}_R$. Let $\overline{\mathbb{O}}$ be the associated variety of $\mathcal{U}_\lambda/\mathcal{J}$. Then there are real numbers c, C with $0 < c < C$, such that for all $m \in \mathbb{Z}_{>0}$ and $p \gg m$, we have*

$$c < \frac{\ell(\operatorname{pr}_\lambda(V^m_{\mathbb{F}} \otimes L))}{m^{\dim \mathbb{O}}} < C.$$

We will deduce this from an analogous result for Harish-Chandra bimodules. Namely, let $\mathcal{M} \in {}^\infty_\lambda\mathrm{HC}^1_\lambda$. Then we can consider the HC-bimodule $\operatorname{pr}_\lambda(V^m \otimes \mathcal{M}) \in {}^\infty_\lambda\mathrm{HC}^1_\lambda$ and its length $\ell(\operatorname{pr}_\lambda(V^m \otimes \mathcal{M}))$.

Proposition 3.4. *Let $\mathcal{M} \in {}^\infty_\lambda\mathrm{HC}^1_\lambda$ with $\mathrm{V}(\mathcal{M}) = \overline{\mathbb{O}}$. Let $m \in \mathbb{Z}_{>0}$. Then there are $0 < c < C$, such that for all $m \in \mathbb{Z}_{>0}$ we have*

$$c < \frac{\ell(\operatorname{pr}_\lambda(V^m \otimes \mathcal{M}))}{m^{\dim \mathbb{O}}} < C.$$

3.3 Lengths for HC-bimodules

In this section we will prove Proposition 3.4 and work over $\mathbb{C}$. We are going to bound $\ell(\mathrm{pr}_\lambda(V^m \otimes \mathscr{M}))$ by two degree-$\dim \mathbb{O}$ polynomials in m, where $\mathrm{V}(\mathscr{M}) = \overline{\mathbb{O}}$.

Under the Bernstein-Gelfand equivalence ${}^{\infty}_{\lambda}\mathrm{HC}^1_\lambda \xrightarrow{\sim} \mathscr{O}(\mu)$ (in the notation of Section 2.1) $\mathscr{U}_\lambda$ maps to the indecomposable projective $\Delta(w_0\mu)$. Every indecomposable projective in ${}^{\infty}_{\lambda}\mathrm{HC}^1_\lambda$ appears as a summand in an object of the form $\mathrm{pr}_\lambda(V_0 \otimes \mathscr{U}_\lambda)$ for a suitable finite-dimensional G-module V_0 that we fix from now on.

Lemma 3.5. *Let $\mathscr{M} \in {}^{\infty}_{\lambda}\mathrm{HC}(\mathscr{U})^1_\lambda$ have associated variety $\overline{\mathbb{O}}$. Then*

$$\dim \mathrm{Hom}_{\mathrm{bimod}}(\mathscr{U}_\lambda, V^m \otimes \mathscr{M})$$

is a degree-$\dim \mathbb{O}$ *polynomial in m.*

Proof. Note that

$$\mathrm{Hom}_{\mathrm{bimod}}(\mathscr{U}_\lambda, V^m \otimes \mathscr{M}) = \mathrm{Hom}_{\mathrm{bimod}}(U(\mathfrak{g}), V^m \otimes \mathscr{M})$$

because $V^m \otimes \mathscr{M}$ has a genuine central character λ on the right. Also

$$\mathrm{Hom}_{\mathrm{bimod}}(U(\mathfrak{g}), V^m \otimes \mathscr{M}) = (V^m \otimes \mathscr{M})^G = \mathrm{Hom}_G(V^{m*}, \mathscr{M}).$$

So

$$\dim \mathrm{Hom}_{\mathrm{bimod}}(\mathscr{U}_\lambda, V^m \otimes \mathscr{M}) = \dim \mathrm{Hom}_G(V^{m*}, \mathscr{M}).$$

Set $\mathscr{M}' := \mathrm{gr}\, \mathscr{M}$ with respect to some good filtration; this is a finitely generated G-equivariant $\mathbb{C}[\mathfrak{g}]$-module. Clearly, $\dim \mathrm{Hom}_G(V^{m*}, \mathscr{M}) = \dim \mathrm{Hom}_G(V^{m*}, \mathscr{M}')$. Consider the filtration $\mathscr{M}'_{\leqslant i}$ on $\mathscr{M}'$ given by $\mathscr{M}'_{\leqslant i}$ being the sum of the isotypic components of $\mathscr{M}'$ with $\langle \rho^\vee, \mu\rangle \leqslant i$. This filtration is compatible with the similarly defined filtration on $\mathbb{C}[\mathfrak{g}]$.

It is well known that for any finitely generated commutative G-algebra A, the algebra $\mathrm{gr} A$, for the filtration $A = \bigcup_i A_{\leqslant i}$, is finitely generated and for any finitely generated G-equivariant A-module M, the $\mathrm{gr} A$-module $\mathrm{gr} M$ is finitely generated. This is because $\mathrm{gr} A = (\mathbb{C}[G/U \times G/U]^T \otimes A)^G$ and a similar equality holds for $\mathrm{gr} M$; here U is a maximal unipotent subgroup of G.

It follows that the GK dimensions of M, $\mathrm{gr} M$ are the same. Since

$$\dim \mathrm{Hom}_G(V_m^*, \mathscr{M}') = \dim \mathscr{M}'_{\leqslant m} = \dim \mathrm{gr}\, \mathscr{M}'_{\leqslant m},$$

the left-hand side of this equality is the Hilbert polynomial of the graded module $\mathrm{gr}\, \mathscr{M}'$. But the GK-dimension of $\mathscr{M}'$ is $\dim \mathbb{O}$ and our claim follows. □

Proof (*of Proposition* 3.4). Let us prove that $\ell(\mathrm{pr}_\lambda(V^m \otimes \mathscr{M})) \geqslant Q(m)$, where Q is a degree-$\dim \mathbb{O}$ polynomial. Note that $\dim \mathrm{Hom}_{\mathrm{bimod}}(\mathscr{U}_\lambda, V^m \otimes \mathscr{M})$ is the multiplicity of the simple bimodule covered by $\mathscr{U}_\lambda$ in $V^m \otimes \mathscr{M}$. By Lemma 3.5, this multiplicity is a degree-dim$\mathbb{O}$ polynomial in m, say $Q(m)$. On the other hand, it is clear that $\ell(\mathrm{pr}_\lambda(V^m \otimes \mathscr{M})) \geqslant \dim \mathrm{Hom}_{\mathrm{bimod}}(\mathscr{U}_\lambda, V^m \otimes \mathscr{M}) = Q(m)$.

We prove that $\ell(\mathrm{pr}_\lambda(V^m \otimes \mathscr{M})) \leqslant \widetilde{Q}(m)$, where $\widetilde{Q}$ is also a degree-dim$\mathbb{O}$ polynomial. Let V_0 be as in the second paragraph of Section 3.3. Then

$$\begin{aligned}\ell(\mathrm{pr}_\lambda(V^m \otimes \mathscr{M})) &\leqslant \dim \mathrm{Hom}_{\mathrm{bimod}}(V_0 \otimes \mathscr{U}_\lambda, V^m \otimes \mathscr{M}) \\ &= \dim \mathrm{Hom}_{\mathrm{bimod}}(\mathscr{U}_\lambda, V^m \otimes (V_0^* \otimes \mathscr{M})) =: \widetilde{Q}(m).\end{aligned}$$

Applying Lemma 3.5 to $V_0^* \otimes \mathscr{M}$, we see that $\widetilde{Q}(m)$ is a degree-dim$\mathbb{O}$ polynomial in m. □

In the proof of Proposition 3.3, we will also need the following lemma.

Lemma 3.6. *There is a constant $0 < c_0 < 1$, such that, for any object $\mathscr{M} \in {}^{\infty}_{\lambda}\mathrm{HC}(\mathscr{U})^1_\lambda$, we have $\ell(\mathsf{head}\mathscr{M}) \geqslant c_0 \ell(\mathscr{M})$.*

Proof. Note that ${}^{\infty}_{\lambda}\mathrm{HC}^1_\lambda$ is equivalent to the category of modules over a finite-dimensional algebra. We claim that in any such category

$$\ell(\mathsf{head}(M)) > \frac{1}{L}\ell(M) \tag{2}$$

for every module M, where L is the maximum of lengths of the indecomposable projectives. Suppose that we know (2) for all M' with $\ell(M') < \ell(M)$. Choose a simple constituent L in $\mathsf{head}(M)$. Let P_L be the projective cover of L. We have a homomorphism $\varphi\colon P_L \to M$ whose composition with $M \twoheadrightarrow \mathsf{head}(M)$ coincides with $P_L \twoheadrightarrow L \hookrightarrow \mathsf{head}(M)$. Clearly, $\mathsf{head}(\mathrm{coker}\varphi) = \mathsf{head}(M)/L$. Applying the induction hypothesis to $\mathrm{im}\,\varphi$ and $\mathrm{coker}\,\varphi$, we finish the proof of this lemma. □

3.4 Lengths in characteristic p

To prove Proposition 3.3, we will need the following technical lemma.

Lemma 3.7. *Let L be an irreducible $\mathscr{U}^\chi_{\lambda,\mathbb{F}}$-module, such that $\mathscr{J} = \mathscr{J}_{w^{-1}}$ is a maximal primitive ideal with $\mathscr{J}_R L = 0$. Then the following is true:*

(1) $\mathscr{M}_{w,R} \otimes_{\mathscr{U}_R} L \neq 0$,
(2) *there is $c_1 > 1$ independent of p, such that $\ell(\mathscr{M}_{w,R} \otimes_{\mathscr{U}_R} L) < c_1$.*

We will first deduce Proposition 3.3 from this lemma and then prove it.

Proof (of Proposition 3.3). Let $\mathscr{J}$ be a maximal primitive ideal in $\mathscr{U}_\lambda$, such that L is annihilated by $\mathscr{J}_R$. By (c3), $\mathrm{pr}_\lambda(V_R^m \otimes_R L) = \mathrm{pr}_\lambda(V^m \otimes \mathscr{U}/\mathscr{J})_R \otimes_{\mathscr{U}_R} L$. Thanks to Proposition 3.4, what we need to prove is that there are constants $0 < c < C$, such that $c \leqslant \ell(\mathscr{M}_\mathbb{F} \otimes_{\mathscr{U}_\mathbb{F}} L)/\ell(\mathscr{M}) \leqslant C$ for any $\mathscr{M} \in {}^{\infty}_{\lambda}\mathrm{HC}(\mathscr{U}_R)^1_\lambda$ whose right annihilator is $\mathscr{J}_R$. By (1) of Lemma 3.7 combined with Lemma 3.6 and (c4), we can set $c := c_0$ from Lemma 3.6. By Lemma 3.7, we can set $C := c_1$. □

Proof (*of Lemma* 3.7). Let us prove (1). Pick w' such that $\mathscr{M}_{w'} \otimes_{\mathscr{U}_\lambda} \mathscr{M}_w$ has $\mathscr{M}_d$ as a direct summand in $\mathrm{HC}_{\mathbf{c}}(\mathscr{U}_\lambda)$ (w' exists because $\mathrm{HC}^{ss}_{\mathbf{c}}(\mathscr{U}_\lambda)$ is a rigid monoidal category). Set $w_1 = w'$, $w_2 = w$ in (c2) and let $M_1 \subset M_2$ be as in (c2). The equality $\mathscr{M}_{w,R} \otimes_{\mathscr{U}_R} L \neq 0$ will follow once we show that

$$\left(\left(\mathscr{M}_{w',R} \otimes_{\mathscr{U}_R} \mathscr{M}_{w,R}\right)/M_1\right) \otimes_{\mathscr{U}_R} L \neq 0.$$

First, let us check that $\mathscr{M}_{d,R} \otimes_{\mathscr{U}_R} L \neq 0$. By the choice of d, we have

$$(\mathscr{U}_{\lambda,R}/\mathscr{J}_{d,R}) \otimes_{\mathscr{U}_{\lambda,R}} L \neq 0.$$

So $\mathscr{M}_{d,R} \otimes_{\mathscr{U}_{\lambda,R}} L \neq 0$ as long as $\mathrm{Tor}^i_{\mathscr{U}_{\lambda,R}}((\mathscr{U}_{\lambda,R}/\mathscr{J}_{d,R})/M_{d,R}, L)$ does not have L in its Jordan-Hoelder series for $i = 0,1$. This is a consequence of (3) Lemma 3.2. We conclude that $\mathscr{M}_{d,R} \otimes_{\mathscr{U}_{\lambda,R}} L \neq 0$.

From here we deduce that $(M_2/M_1) \otimes_{\mathscr{U}_{\lambda,R}} L \neq 0$. Similarly to the previous paragraph, this implies $\left(\left(\mathscr{M}_{w',R} \otimes_{\mathscr{U}_R} \mathscr{M}_{w,R}\right)/M_1\right) \otimes_{\mathscr{U}_{\lambda,R}} L \neq 0$. This finishes the proof of (1).

Let us prove (2). It is enough to prove this statement with $\mathscr{M}_{w,R}$ replaced by a bimodule that covers it, e.g., $\mathrm{pr}_\lambda(V_{0,R} \otimes \mathscr{U}_{\lambda,R})_R$, where V_0 is as in the beginning of Section 3.3. Note that bimodule is projective as a right module. On the level of K_0, the operator $\mathrm{pr}_\lambda(V_{0,R} \otimes \mathscr{U}_{\lambda,R}) \otimes_{\mathscr{U}_{\lambda,R}} \bullet$ is the multiplication by some element, say y, of $\mathbb{C}W_{[\lambda]}$ independent of p. For $b \in \mathfrak{B}$, we can expand $yb = \sum_{b' \in \mathfrak{B}} m_{bb'} b'$. Then $c_1 = \max_{b \in \mathfrak{B}} \sum_{b'} m_{bb'}$ satisfies the conditions of (2). □

4 Proof of Theorem 1.1

In this section we will prove Theorem 1.1. Part (1) is proved in Section 4.1, while the proof of part (2) occupies the remainder of the section. We will describe the main steps of the proof in Section 4.2.

4.1 Proof of part (1) of Theorem 1.1

For a two-sided cell $\mathbf{c}$ in $W_{[\lambda]}$, consider the full subcategory $\mathscr{U}_{\lambda,\mathbb{F}}\text{-mod}^\chi_{\leqslant \mathbf{c}}$ which is the Serre span of all simples annihilated by $\mathscr{J}_{w,R}$ with $w \in \mathbf{c}$. Note that $D^b(\mathscr{U}_{\lambda,\mathbb{F}}\text{-mod}^\chi)_{\leqslant \mathbf{c}}$ is a submodule category for the action of $D^b_{HC}(\mathscr{U}_{\lambda,R}\text{-mod})$; this follows from (2) of Lemma 3.2.

We have the following result.

Proposition 4.1. *All irreducible representations of $W_{[\lambda]}$ occurring in*

$$K_0(\mathscr{U}_{\lambda,\mathbb{F}}\text{-mod}^\chi_{\leqslant \mathbf{c}})/K_0(\mathscr{U}_{\lambda,\mathbb{F}}\text{-mod}^\chi_{< \mathbf{c}})$$

belong to the two-sided cell $\mathbf{c}$.

Proof. We need to prove two statements:

(1) $\mathbb{C}W_{[\lambda],<\mathbf{c}}K_0(\mathscr{U}_{\lambda,\mathbb{F}}\text{-mod}^{\chi}_{\leqslant\mathbf{c}})\subset K_0(\mathscr{U}_{\lambda,\mathbb{F}}\text{-mod}^{\chi}_{<\mathbf{c}})$,
(2) $\mathbb{C}W_{[\lambda],\leqslant\mathbf{c}}K_0(\mathscr{U}_{\lambda,\mathbb{F}}\text{-mod}^{\chi}_{\leqslant\mathbf{c}})=K_0(\mathscr{U}_{\lambda,\mathbb{F}}\text{-mod}^{\chi}_{\leqslant\mathbf{c}})$.

Statement (1) follows from (1) of Lemma 3.2. Let us prove statement (2). Let $L\in\mathscr{U}_{\lambda,\mathbb{F}}\text{-mod}^{\chi}_{\leqslant\mathbf{c}}$ be a simple object annihilated by $\mathscr{J}_{d,R}$, where $d\in\mathbf{c}$. Tautologically $(\mathscr{U}_{\lambda,R}/\mathscr{J}_{d,R})\otimes^L_{\mathscr{U}_{\lambda,R}/\mathscr{J}_{d,R}}L=L$. Let us write V for $\mathrm{Span}_{w\leqslant^L d}(\mathscr{M}_{w^{-1}})$. Note that $\mathbb{C}W_{[\lambda],\leqslant\mathbf{c}}V=V$. Indeed, all W-irreducibles appearing in V belong to the families indexed by two-sided cells $\leqslant\mathbf{c}$.

So we have $[\mathscr{U}_\lambda/\mathscr{J}_d]=\sum c_{w_1,w_2}[\mathscr{M}_{w_1}][\mathscr{M}_{w_2}]$ (an equality in V), where w_1 runs over $\mathbf{c}$ and w_2 over the elements such that $w_2^{-1}\leqslant^L d$. It follows that $[L]$ belongs to the linear span of the classes of the form

$$\left[(\mathscr{M}_{w_1,R}\otimes^L_{\mathscr{U}_\lambda,R}\mathscr{M}_{w_2,R})_{\mathscr{U}_{\lambda,R}/\mathscr{J}_{d,R}}L\right]=[\mathscr{M}_{w_1}]\left[\mathscr{M}_{w_2,R}\otimes^L_{\mathscr{U}_{\lambda,R}}L\right].$$

Note that $\mathscr{M}_{w_2,R}\otimes^L_{\mathscr{U}_{\lambda,R}}L\in D^b(\mathscr{U}_{\lambda,\mathbb{F}}\text{-mod}^{\chi})_{\leqslant\mathbf{c}}$ by (1) of Lemma 3.2. We deduce that $[L]\in\mathbb{C}W_{[\lambda],\leqslant\mathbf{c}}K_0(\mathscr{U}_{\lambda,\mathbb{F}}\text{-mod}^{\chi}_{\mathbb{F}})$. This implies (2). □

Proof (*of* (1) *of Theorem* 1.1). Proposition 4.1 and an easy induction on $\mathbf{c}$ show that

$$K_0(\mathscr{U}_{\lambda,\mathbb{F}}\text{-mod}^{\chi})_{\leqslant\mathbf{c}}=K_0(\mathscr{U}_{\lambda,\mathbb{F}}\text{-mod}^{\chi}_{\leqslant\mathbf{c}}).$$

This establishes the claim of part (1) at $q=1$.

To prove the full claim, one uses the graded lifts mentioned in Section 2.3. Namely, let us write $\mathscr{C}^{gr}$ for the graded lift of $\mathscr{C}:=\mathscr{U}_{\lambda,\mathbb{F}}\text{-mod}^{\chi}$. We still have the two-sided cell filtration $\mathscr{C}^{gr}_{\leqslant\mathbf{c}}$ on $\mathscr{C}^{gr}$ that is closed under the grading shifts and lifts the filtration $\mathscr{U}_{\lambda,\mathbb{F}}\text{-mod}^{\chi}_{\leqslant\mathbf{c}}$.

We claim that $D^b(\mathscr{C}^{gr}_{\leqslant\mathbf{c}})$ is invariant under the braid group $B_{[\lambda]}$, hence its Grothendieck group is invariant under the corresponding Hecke algebra. It is enough to check that for a simple module $L\in\mathscr{C}^{gr}_{\leqslant\mathbf{c}}$ and a generator $\tilde{s}_\alpha\in B_{[\lambda]}$, we have $\tilde{s}_\alpha(L)\in D^b(\mathscr{C}^{gr}_{\leqslant\mathbf{c}})$. In view of Lemma 2.4, this follows from $s_\alpha([L])\in K^0(\mathscr{C}_{\leqslant\mathbf{c}})$, which has already been proved.

So $K_0(\mathscr{C}^{gr}_{\leqslant\mathbf{c}})/K_0(\mathscr{C}^{gr}_{<\mathbf{c}})$ is an $\mathscr{H}_q(W_{[\lambda]})$-module flat over $\mathbb{C}[q^{\pm1}]$ that specializes to

$$K_0(\mathscr{U}_{\lambda,\mathbb{F}}\text{-mod}^{\chi}_{\leqslant\mathbf{c}})/K_0(\mathscr{U}_{\lambda,\mathbb{F}}\text{-mod}^{\chi}_{<\mathbf{c}})$$

at $q=1$. The latter factors through the quotient corresponding to $\mathbf{c}$, hence so is the former. □

4.2 Outline of the proof of (2) of Theorem 1.1

Below we will prove part (2) of Theorem 1.1. We will start with the $\chi = 0$ case.

Proposition 4.2. *Part* (2) *of Theorem* 1.1 *holds when* $\chi = 0$.

Here we will use an easy adaptation of an argument due to Etingof that relates the degrees of dimension polynomials with the GK dimensions of simples in the category $\mathscr{O}$ in characteristic 0. We will reduce the case of general χ to $\chi = 0$ by using the degeneration map $K_0(\mathscr{U}^{\chi}_{\lambda,\mathbb{F}}\text{-mod}) \to K_0(\mathscr{U}^{0}_{\lambda,\mathbb{F}}\text{-mod})$. This map can be shown to be independent of p. Hence it preserves the dimension polynomials. The most nontrivial step is to show that the degeneration of a simple module that lies in $K_0(\mathscr{U}^{\chi}_{\lambda,\mathbb{F}}\text{-mod})_{\leqslant \mathbf{c}}$, but not in the lower filtration terms, lies in $K_0(\mathscr{U}^{0}_{\lambda,\mathbb{F}}\text{-mod})_{\leqslant \mathbf{c}}$ (this is straightforward) but not in the lower filtration terms (this is harder, we need Proposition 3.3 to handle this part).

Before we proceed to proving part (2), let us reformulate it. For this we need the following lemma.

Lemma 4.3. *Let a simple* $L \in \mathscr{U}^{\chi}_{\lambda,\mathbb{F}}\text{-mod}_{\leqslant \mathbf{c}}$ *but not in lower filtration terms. Then there is a unique maximal primitive ideal* $\mathscr{J} = \mathscr{J}_d \in \operatorname{Prim}(\mathscr{U}_\lambda)$ *with* $\mathscr{J}_R L = 0$. *We have* $d \in \mathbf{c}$.

Proof. Let $\mathscr{J}_1, \mathscr{J}_2$ be two maximal primitive ideals with the required property. Assume, in addition, that the two-sided cell corresponding to $\mathscr{J}_1$ is maximal possible. By (c6), $\mathscr{U}_{\lambda,R}/(\mathscr{J}_{1,R} + \mathscr{J}_{2,R})$ is filtered by $\mathscr{M}_{w,R}$ for w lying in two-sided cells smaller than $\mathbf{c}$. If $\mathscr{M}_{w,R} \otimes_{\mathscr{U}_{\lambda,R}} L \neq 0$, then $\mathscr{J}_{w^{-1},R} L = 0$. The two-sided cell of $\mathscr{J}_{w^{-1}}$ is strictly smaller than those of $\mathscr{J}_1, \mathscr{J}_2$. A contradiction. The inclusion $d \in \mathbf{c}$ follows from the definition of $\mathscr{U}^{\chi}_{\lambda,\mathbb{F}}\text{-mod}_{\leqslant \mathbf{c}}$. □

Remark 4.4. So (2) of Theorem 1.1 says that the dimension polynomial of L has degree equal to $\frac{1}{2}\text{GK-dim}(\mathscr{U}_\lambda/\mathscr{J})$, where $\mathscr{J}$ is the maximal primitive ideal in $\mathscr{U}_\lambda$ such that $\mathscr{J}_R L = 0$. This statement makes sense for other classes of quantizations, e.g., for those of symplectic singularities and we expect it to hold in this setting.

4.3 Etingof's construction

We will prove Proposition 4.2 by adapting an argument due to Etingof from categories $\mathscr{O}$ for type A rational Cherednik algebras to BGG categories $\mathscr{O}$.

Let us explain this argument. Let R be a finite localization of $\mathbb{Z}$ such that $\lambda \in \mathfrak{h}^*_R$. Then we can consider the Verma module $\Delta_R(\lambda)$ with highest weight $\lambda + \rho$. Note that $\Delta_R(\lambda)$ is naturally graded with highest vector in degree 0 and the operators f_α have degree 1 for a simple root α. Let p be a prime number invertible in R. Set $\Delta_{\mathbb{F}_p}(\lambda) := \mathbb{F}_p \otimes_R \Delta_R(\lambda)$, $\Delta_{\mathbb{Q}}(\lambda) := \mathbb{Q} \otimes_R \Delta_R(\lambda)$. Let $L_{\mathbb{Q}}(\lambda)$ be the unique irreducible quotient of $\Delta_{\mathbb{Q}}(\lambda)$. The module $\Delta_{\mathbb{F}_p}(\lambda)$ has a unique graded simple quotient, denoted by $L_{\mathbb{F}_p}(\lambda)$. The modules $L_{\mathbb{F}_p}(\lambda')$ for $\lambda' \in W\lambda$ are absolutely irreducible and pairwise non-isomorphic, and so their base changes to $\mathbb{F}$ form a complete collection of the irreducibles in $\mathscr{U}^0_{\lambda,\mathbb{F}}$-mod. Note that $L_{\mathbb{Q}}(\lambda)$ and $L_{\mathbb{F}_p}(\lambda)$ are graded quotients. Let $L^i_{\mathbb{Q}}(\lambda)$, $L^i_{\mathbb{F}_p}(\lambda)$ denote the ith graded component.

The following lemma is due to Etingof (in the Cherednik case).

Lemma 4.5. *Suppose that p is sufficiently large. Then there is a positive integer N_λ, independent of p, such that $\dim L^i_{\mathbb{F}_p}(\lambda) = \dim L^i_{\mathbb{Q}}(\lambda)$ for $i < p/N_\lambda$.*

Proof. Recall that for any $\lambda' \in \mathfrak{h}^*_R$, the module $\Delta_R(\lambda')$ has a unique (up to scaling by elements of R) contravariant form $B_{\lambda'}$, and different graded components are orthogonal with respect to $B_{\lambda'}$. We assume that B is nondegenerate on the highest weight component.

We can also consider a one-parameter deformation $\Delta_{R[t]}(\lambda + t\rho)$, where t is an independent variable. The module $\Delta_{R[t]}(\lambda + t\rho)$ comes with a contravariant form $B_{\lambda+t\rho}$. Let us write B^i_λ for the restriction of B_λ to $\Delta^i_R(\lambda)$. The notation $B^i_{\lambda+t\rho}$ has a similar meaning.

We note that $L_{\mathbb{Q}}(\lambda)$ is the quotient of $\Delta_{\mathbb{Q}}(\lambda)$ by the radical of $B_{\lambda,\mathbb{Q}}$, the specialization of B_λ to $\mathbb{Q}$ and, similarly, $L_{\mathbb{F}_p}(\lambda)$ is the quotient of $\Delta_{\mathbb{F}_p}(\lambda)$ by the radical of $B_{\lambda,\mathbb{F}_p}$. So we need to check that for $i < p/N_\lambda$, the radicals of $B^i_{\lambda,\mathbb{F}_p}$ and of $B^i_{\lambda,\mathbb{Q}}$ have the same dimension.

Consider the finitely generated $\mathbb{K}[t]$-module $\Delta^i_{\mathbb{K}[t]}(\lambda + t\rho)$, where $\mathbb{K} = \mathbb{Q}$ or $\mathbb{F}_p$. The form $B^i_{\lambda+t\rho,\mathbb{K}}$ defines a descending filtration $F^j\Delta^i_{\mathbb{K}}(\lambda)$ (the Jantzen filtration): by definition, $F^0\Delta^i_{\mathbb{K}}(\lambda) = \Delta^i_{\mathbb{K}}(\lambda)$, and $F^j\Delta^i_{\mathbb{K}}(\lambda)$ is the radical of the well-defined form $t^{-j}B^i_{\lambda+t\rho,\mathbb{K}}|_{t=0}$ on $F^{j-1}\Delta^i_{\mathbb{K}}(\lambda)$. It is clear that $\dim F^j\Delta^i_{\mathbb{Q}}(\lambda) \geqslant \dim F^j\Delta^i_{\mathbb{F}_p}(\lambda)$ for all p. The equality $\dim F^j\Delta^i_{\mathbb{Q}}(\lambda) = \dim F^j\Delta^i_{\mathbb{F}_p}(\lambda)$ for all j (that will immediately imply what we need, which is the $j = 1$ case) for $i < p/N_\lambda$ will follow if we check that the order of vanishing of $f^i_p(t) := \det B^i_{\lambda+t\rho,\mathbb{F}_p}$ at $t = 0$ coincides with the order of vanishing of $f^i(t) := \det B^i_{\lambda+t\rho,\mathbb{Q}}$ at $t = 0$—for example, the former equals $\sum_j j\dim F^j\Delta^i_{\mathbb{F}_p}(\lambda)$. Clearly, $f^i_p(t)$ is obtained from $f(t)$ by reduction mod p.

The polynomial $f^i(t)$ can be decomposed as $C^i t^{n_0}\prod_{j=1}^k (t - z^i_j)^{n_j}$, where $C^i \in R$ and z^i_j are nonzero elements that lie, a priori, in the algebraic closure of $\mathbb{Q}(t)$. In fact, $f^i(z) = 0$ means that there is a singular vector in $\Delta^{i'}_{\lambda+z\rho}(\lambda)$, where $0 < i' \leqslant i$. In particular, there is $w \in W$, $w \neq 1$, such that $\langle \lambda + z\rho - w(\lambda + z\rho), \rho^\vee\rangle = i'$. This is equivalent to

$$z = \langle \rho - w\rho, \rho^\vee\rangle^{-1}(i' - \langle \lambda - w\lambda, \rho\rangle). \quad (3)$$

In particular, $f^i(z) = 0$ implies $z \in R$.

Therefore what we need to prove is that, for $i < p/N_\lambda$, we have $C^i, z^i_j \neq 0$ modulo p.

Let us show that C^i is nonzero mod p. Consider the baby Verma module $\underline{\Delta}_{\mathbb{F}_p[t]}(\lambda + t\rho) = \Delta_{\mathbb{F}_p[t]}(\lambda + t\rho)/(\mathfrak{n}^-_{\mathbb{F}_p})^{(1)}\Delta_{\mathbb{F}_p[t]}(\lambda + t\rho)$. Note that $\underline{\Delta}^i_{\mathbb{F}_p[t]}(\lambda + t\rho) = \Delta^i_{\mathbb{F}_p[t]}(\lambda + t\rho)$ as long as $i < p$. It is well known that for a generic $z \in \mathbb{F}$ the module

$\underline{\Delta}^i_{\mathbb{F}}(\lambda + z\rho)$ is irreducible. In particular, $f^i_p(t) \neq 0$ as a polynomial, so $C^i \neq 0$ mod p.

Now let us show that $z^i_j \neq 0$ mod p using (3). Recall that x stands for the least common multiple of the denominators of the coordinates of λ in the basis of simple roots. Then $xi' - x\langle\lambda - w\lambda, \rho\rangle$ is a nonzero multiple of p. Clearly as long as p is large enough, there is N_λ such that for $i < p/N_\lambda$, we have $0 < xi' - x\langle\lambda - w\lambda, \rho\rangle < p$. □

4.4 Proof of Proposition 4.2

Let us deduce Proposition 4.2 from Lemma 4.5.

Proof (*of Proposition* 4.2). The GK dimension of $L(\lambda)$ is equal to

$$\frac{1}{2}\text{GK-dim}(\mathcal{U}_\lambda / \operatorname{Ann} L(\lambda))$$

and the latter coincides with $\dim \mathbb{O}_{\mathbf{c}}/2$. Therefore $\sum_{j=0}^i \dim L^j(\lambda)$ is a polynomial in i of degree $\dim \mathbb{O}_{\mathbf{c}}/2$. So $\dim L_{\mathbb{F}}(\lambda) \geqslant \sum_{j=0}^i \dim L^j_{\mathbb{F}_p}(\lambda) = \sum_{j=0}^i \dim L^j(\lambda)$, where the equality follows from Lemma 4.5, as long as $i < p/N_\lambda$. This shows that $\dim L_{\mathbb{F}}(\lambda)$ is bounded below by a degree-$\dim \mathbb{O}_{\mathbf{c}}/2$ polynomial in p.

Now let us show that $\dim L_{\mathbb{F}}(\lambda)$ is bounded from above by a degree-$\dim \mathbb{O}_{\mathbf{c}}/2$ polynomial in p. Note that $L_{\mathbb{F}}(\lambda)$ is a quotient of the baby Verma module $\underline{\Delta}_{\mathbb{F}}(\lambda)$. We have $\underline{\Delta}^j_{\mathbb{F}}(\lambda) = 0$ for $j \geqslant 2\langle\rho, \rho^\vee\rangle p$, hence $L^j_{\mathbb{F}}(\lambda) = 0$ for such j. We claim that for any $m_1 > 1 > m_2$, there is a constant M such that

$$\sum_{j=0}^{\lfloor m_1 p \rfloor} \dim L^j_{\mathbb{F}}(\lambda) \leqslant M \sum_{j=0}^{\lfloor m_2 p \rfloor} \dim L^j_{\mathbb{F}}(\lambda); \tag{4}$$

this will establish the upper bound thanks to Lemma 4.5.

Let us write $U_{\leqslant k}(\mathfrak{n}^-_{\mathbb{F}})$ for the kth filtration term with respect to the PBW filtration on $U(\mathfrak{n}^-_{\mathbb{F}})$ and $U^{\leqslant i}(\mathfrak{n}^-_{\mathbb{F}})$ for $\bigoplus_{j=0}^i U^j(\mathfrak{n}^-_{\mathbb{F}})$. For any i, we have

$$\bigoplus_{j=0}^{i} L^j_{\mathbb{F}}(\lambda) = U^{\leqslant i}(\mathfrak{n}^-_{\mathbb{F}}) L^0_{\mathbb{F}}(\lambda). \tag{5}$$

Note that the filtrations $U^{\leqslant i}(\mathfrak{n}^-_{\mathbb{F}})$ and $U_{\leqslant k}(\mathfrak{n}^-_{\mathbb{F}})$ are compatible in the sense that there are constants $c_1 < 1 < c_2$, such that $U_{\leqslant c_1 i}(\mathfrak{n}^-_{\mathbb{F}}) \subset \mathcal{U}^{\leqslant i}(\mathfrak{n}^-_{\mathbb{F}}) \subset U_{\leqslant c_2 i}(\mathfrak{n}^-_{\mathbb{F}})$. So, thanks to (5), (4) will follow if we show that for any $m \in \mathbb{Z}_{>1}$, there is M such that for any i, we have

$$\dim U_{\leqslant mi}(\mathfrak{n}^-_{\mathbb{F}}) \leqslant M \dim U_{\leqslant i}(\mathfrak{n}^-_{\mathbb{F}}). \tag{6}$$

Let $x_1,\dots,x_n$ be a basis of $\mathfrak{n}^-_{\mathbb{F}}$. (6) will follow if we show that every element of $U_{\leqslant mi}(\mathfrak{n}^-_{\mathbb{F}})$ can be written as a sum of elements of the form PQ, where P is an ordered monomial in $x_1^{i+1},\dots,x_n^{i+1}$ of degree $\leqslant m$ and Q is an element of $U_{\leqslant i}(\mathfrak{n}^-_{\mathbb{F}})$. The latter claim follows from the analogous one on the associated graded level, which is straightforward. □

Remark 4.6. Let us mention an alternative way to the prove the upper bound for $\dim L_{\mathbb{F}}(\lambda)$ established above. Let M be an object in category $\mathscr{O}$ over $\mathbb{C}$ of Gelfand-Kirillov dimension d. We can find that a $\mathfrak{g}$-module M_R is defined over R with $M \cong M_R \otimes_R \mathbb{C}$, and consider its base change $M_{\mathbb{F}}$ to a field of almost any prime characteristic. Let $\overline{M_{\mathbb{F}}}$ be the reduction of $M_{\mathbb{F}}$ by the zero p-central character. Then one can check that:

(i) $\dim \overline{M_{\mathbb{F}}} = O(p^d)$.
(ii) The space $K_0(\mathscr{U}_{\lambda,\mathbb{F}}\text{-mod}^{\chi}_{\leqslant \mathbf{c}})$ is spanned by classes $\dim \overline{L_{\mathbb{F}}}$, where L runs over the set of irreducible modules in category $\mathscr{O}$ belonging to cells $\mathbf{c}' \leqslant \mathbf{c}$.

These two properties clearly imply the upper bound.

4.5 Degeneration map

We have a one-parameter subgroup $\gamma\colon \mathbb{F}^\times \to G_{\mathbb{F}}$ with $\gamma(t)\chi = t^2\chi$. Via γ, the group $\mathbb{F}^\times$ acts on the sheaf of algebras $\mathscr{U}_{\mathbb{F}}|_{\mathbb{F}\chi}$, where the action on the base $\mathbb{F}\chi$ is by dilations. This gives rise to the degeneration map $\delta\colon K_0(\mathscr{U}^{\chi}_{\mathbb{F}}\text{-mod}) \to K_0(\mathscr{U}^{0}_{\mathbb{F}}\text{-mod})$. Since $\mathbb{F}^\times$ acts trivially on the Harish-Chandra center, we see that the map restricts to

$$\delta\colon K_0(\mathscr{U}^{\chi}_{\lambda,\mathbb{F}}\text{-mod}) \to K_0(\mathscr{U}^{0}_{\lambda,\mathbb{F}}\text{-mod}). \tag{7}$$

The following standard lemma summarizes basic properties of the degeneration map (7).

Lemma 4.7. *The following are true.*

(1) *Under the identifications*

$$K_0(\mathscr{U}^{\chi}_{\lambda,\mathbb{F}}\text{-mod}) \cong H_*(\mathscr{B}_e,\mathbb{C}), K_0(\mathscr{U}^{0}_{\lambda,\mathbb{F}}\text{-mod}) \cong H_*(\mathscr{B},\mathbb{C}),$$

the map δ coincides with the push-forward map $H_*(\mathscr{B}_e,\mathbb{C}) \to H_*(\mathscr{B},\mathbb{C})$. In particular, it is independent of p and W^a-equivariant.
(2) *The map δ intertwines the endomorphisms $\mathrm{pr}_\lambda(V(\mu)_{\mathbb{F}}\otimes\bullet)$ for $p \gg \langle \rho^\vee,\mu\rangle$.*
(3) *The map δ preserves the dimension polynomials.*

Proof. (2) and (3) are straightforward; let us prove (1). We can consider the categories $\mathrm{Coh}(\mathscr{D}^{\lambda}_{\mathscr{B}_{\mathbb{F}}})^{t\chi}$ (here $t = 0,1$) of all coherent sheaves of $\mathscr{D}^{\lambda}_{\mathscr{B}_{\mathbb{F}}}$-modules supported at the pre-image of $t\chi$ and the corresponding derived category $D^b(\mathrm{Coh}(\mathscr{D}^\lambda_{\mathscr{B}_\mathbb{F}}))^{t\chi}$. The derived equivalence $R\Gamma$ induces identifications

$$K_0(\mathrm{Coh}(\mathscr{D}^\lambda_{\mathscr{B}_\mathbb{F}})^{t\chi}) \cong K_0(\mathscr{U}_{\lambda,\mathbb{F}}\text{-mod}^{t\chi})$$

that are compatible with the degeneration maps. Also note that the Chern character isomorphisms intertwine the degeneration maps. So it remains to show that the identifications

$$K_0(\mathrm{Coh}(\mathscr{B}^{(1)}_{t\chi})) \xrightarrow{\sim} K_0(\mathrm{Coh}(\mathscr{D}^\lambda_{\mathscr{B}_\mathbb{F}})^{t\chi})$$

intertwine the degeneration maps. This is a consequence of Lemma 2.2 and the projection formula. □

4.6 Proof of (2) of Theorem 1.1

We start with the following proposition.

Proposition 4.8. *Let L be a simple in $\mathscr{U}^\chi_{\lambda,\mathbb{F}}$-mod$_{\leqslant \mathbf{c}}$, but not in the smaller filtration terms. Then the projection of $\delta([L])$ to the $\mathbf{c}$-isotypic component of $K_0(\mathscr{U}^0_{\lambda,\mathbb{F}}\text{-mod})_{\leqslant \mathbf{c}}$ is nonzero.*

Proof. Assume the contrary. Let $\delta[L] = \sum_{i=1}^k [L_i]$, where L_i are simples in $\mathscr{U}^0_{\lambda,\mathbb{F}}$-mod. Note that k is independent of p by (1) of Lemma 4.7 and the fact that the basis of simples is independent of p. By our assumption, $L_i \in \mathscr{U}^0_{\lambda,\mathbb{F}}\text{-mod}_{<\mathbf{c}}$.

Pick a dominant weight μ and suppose that p is very large. Let M_μ denote the functor $\mathrm{pr}_\lambda(V(\mu)_\mathbb{F} \otimes \bullet)$. By (2) of Lemma 4.7, we have

(8) $$\delta([M_\mu L]) = \sum_{i=1}^k [M_\mu L_i].$$

Note that for $N \in \mathscr{U}^\chi_\mathbb{F}\text{-mod}_\lambda$, we have

(9) $$\ell([N]) \leqslant \ell(\delta[N]).$$

By Proposition 3.3 combined with Lemma 4.3,

(10) $$\sum_{\mu, \langle \rho^\vee, \mu\rangle \leqslant m} \dim V(\mu)\ell(M_\mu L)$$

grows (with respect to m) faster than

(11) $$\sum_{i=1}^k \sum_{\mu, \langle \rho^\vee, \mu\rangle \leqslant m} \dim V(\mu)\ell(M_\mu L_i).$$

But combining (8) with (9), we see that (11)$\geqslant$(10). A contradiction. □

Proof (*of* (2) *of Theorem* 1.1). By (3) of Lemma 4.7, the map

$$\delta\colon K_0(\mathscr{U}^{\chi}_{\lambda,\mathbb{F}}\text{-mod}) \to K_0(\mathscr{U}^{0}_{\lambda,\mathbb{F}}\text{-mod})$$

preserves the dimension polynomials. For an irreducible $L \in \mathscr{U}^{\chi}_{\lambda,\mathbb{F}}\text{-mod}_{\leqslant \mathbf{c}}$ (but not in smaller terms of the filtration), we have $\delta([L]) = \sum_{i=1}^{k}[L_i]$, where k is independent of p, the simple L_i belongs to a two-sided cell $\mathbf{c}_i \leqslant \mathbf{c}$, and there is i such that $\mathbf{c}_i = \mathbf{c}$. The dimension polynomial for L is the sum of dimension polynomials for the L_i's. Note that $\mathbb{O}_{\mathbf{c}_i} \subset \overline{\mathbb{O}}_{\mathbf{c}}$. By Proposition 4.2, the dimension polynomial of L has degree $\dim \mathbb{O}_{\mathbf{c}}/2$. □

5 Application to W-algebras

In this section we will use Theorem 1.1 to prove conjectures from [LO, Section 7.6] on the classification of finite-dimensional irreducible representations of W-algebras.

5.1 Background on W-algebras

Finite W-algebras (below we omit the adjective "finite") were introduced by Premet in [P1] (with alternative constructions later given by the second author). These are associative algebras constructed from pairs $(\mathfrak{g}, e)$, where $\mathfrak{g}$ is a semisimple Lie algebra over $\mathbb{C}$ and $e \in \mathfrak{g}$ is a nilpotent element. Such a W-algebra is a quantization of the transverse Slodowy slice to the adjoint orbit $\mathbb{O}$ of e. The reader is referred to the survey article [L1] for details.

We recall Premet's definition. Include e into an $\mathfrak{sl}_2$-triple (e,h,f). The element h induces the grading on $\mathfrak{g}$ by eigenvalues of $\mathrm{ad}(h)$: $\mathfrak{g} = \bigoplus_{i\in\mathbb{Z}} \mathfrak{g}(i)$. Let, as before, $\chi = (e,\cdot)$. The form $\omega(x,y) = \langle \chi, [x,y]\rangle$ is symplectic on $\mathfrak{g}(-1)$. Let us pick a lagrangian subspace $\ell \subset \mathfrak{g}(-1)$. Form a subalgebra $\mathfrak{m} \subset \mathfrak{g}$ by $\mathfrak{m} = \bigoplus_{i\leqslant -2}\mathfrak{g}(i) \oplus \ell$. Note that χ is the character of $\mathfrak{m}$ and that $\dim \mathfrak{m} = \frac{1}{2}\dim \mathbb{O}$, where we write $\mathbb{O}$ for the orbit of e. Then, by definition, the W-algebra $\mathscr{W}$ is the quantum Hamiltonian reduction $[U(\mathfrak{g})/U(\mathfrak{g})\{x - \langle \chi, x\rangle | x \in \mathfrak{m}\}]^{\mathrm{ad}\,\mathfrak{m}}$.

Let us list some important properties of the W-algebra.

(1) The algebra $\mathscr{W}$ is naturally independent of the choice of ℓ, as was demonstrated in [GG]. Moreover, it comes with a Hamiltonian action of the group $Q = Z_G(e,h,f)$ by automorphisms.

(2) Next, $\mathscr{W}$ comes with a filtration induced from the filtration on $U(\mathfrak{g})$, where $\deg \mathfrak{g}(i) = i+2$. The associated graded ring for this filtration is $\mathbb{C}[S]$, the algebra of functions on the Slodowy slice $S = e + \mathfrak{z}_{\mathfrak{g}}(f)$.

(3) Also note that the definition of $\mathscr{W}$ via the Hamiltonian reduction yields a homomorphism $U(\mathfrak{g})^G \to \mathscr{W}$. As was checked by Ginzburg, see the footnote for

[P2, Question 5.1], this homomorphism is an isomorphism onto the center of $\mathcal{W}$. So for $\lambda \in \mathfrak{h}^*$, we can talk about the central reduction $\mathcal{W}_\lambda$.

Now let us discuss a reduction mod p for W-algebras. Note that $\mathcal{W}$ is defined over some finite localization R of $\mathbb{Z}$: we can take the Hamiltonian reduction $\mathcal{W}_R$ of $\mathcal{U}_R$ and the properties (1), (2), (3) still hold. So we can reduce mod p and get the algebra $\mathcal{W}_\mathbb{F} := \mathbb{F} \otimes_R \mathcal{W}_R$.

As Premet proved, see, for example, [P4, Theorem 2.1], one has a central inclusion $\mathbb{F}[S^{(1)}] \hookrightarrow \mathcal{W}_\mathbb{F}$. In [P3, Proposition 4.1] Premet checked that one has an isomorphism $\mathcal{U}_\mathbb{F}^\chi \cong \operatorname{End}_\mathbb{F}(\mathfrak{m}_\mathbb{F}) \otimes \mathcal{W}_\mathbb{F}^\chi$.

Remark 5.1. Consider $\mathcal{U}_\mathbb{F}|_{S^{(1)}} = \mathbb{F}[S^{(1)}] \otimes_{\mathbb{F}[\mathfrak{g}^{(1)}]} \mathcal{U}_\mathbb{F}$. One can strengthen Premet's result and show that $\mathcal{U}_\mathbb{F}/\mathcal{U}_\mathbb{F}\{x - \langle\chi, x\rangle, x \in \mathfrak{m}_\mathbb{F}\}|_{S^{(1)}}$ is a Morita equivalence bimodule between $\mathcal{U}_\mathbb{F}|_{S^{(1)}}$ and $\mathcal{W}_\mathbb{F}$. This follows from [T]. From here we see that $\mathcal{U}_{\lambda,\mathbb{F}}/\mathcal{U}_{\lambda,\mathbb{F}}\{x - \langle\chi, x\rangle, x \in \mathfrak{m}_\mathbb{F}\}|_{S^{(1)}}$ is a Morita equivalence bimodule between $\mathcal{U}_{\lambda,\mathbb{F}}|_{S^{(1)}}$ and $\mathcal{W}_{\lambda,\mathbb{F}}$.

5.2 Restriction functor for HC-bimodules

In this section we will recall results from [L2] on the restriction functor between the category of HC $\mathcal{U}$-bimodules and the category of HC $\mathcal{W}$-bimodules.

Namely in [L2] the second author has constructed a functor

$$\bullet_\dagger : \operatorname{HC}(\mathcal{U}) \to \operatorname{HC}^Q(\mathcal{W})$$

to the category of Q-equivariant HC $\mathcal{W}$-bimodules (introduced in that paper) with the following properties:

(1) The functor $\bullet_\dagger$ is exact, tensor, $\mathbb{C}[\mathfrak{h}^*]^W$-bilinear, and sends $\mathcal{U}$ to $\mathcal{W}$,
(2) it maps $\operatorname{HC}_{\overline{\mathbb{O}}}(\mathcal{U})$ to the category $\operatorname{Bimod}^Q_{\text{fin}}(\mathcal{W})$ of finite-dimensional Q-equivariant $\mathcal{W}$-bimodules,
(3) and kills $\operatorname{HC}_{\partial\mathbb{O}}(\mathcal{U})$.
(4) There is a functor $\bullet^\dagger : \operatorname{Bimod}^Q_{\text{fin}}(\mathcal{W}) \to \operatorname{HC}_{\overline{\mathbb{O}}}(\mathcal{U})$ that is right adjoint to $\bullet_\dagger$.
(5) For $\mathcal{M} \in \operatorname{HC}_{\overline{\mathbb{O}}}(\mathcal{U})$, the kernel and the cokernel of the adjunction unit $\mathcal{M} \to (\mathcal{M}_\dagger)^\dagger$ are supported on $\partial\mathbb{O}$.
(6) Let $\mathcal{M} \in \operatorname{HC}(\mathcal{U})$ and $\mathcal{N}' \subset \mathcal{M}_\dagger$ be a Q-stable sub-bimodule of finite codimension. Then there is a unique maximal sub-bimodule $\mathcal{M}' \subset \mathcal{M}$ with $\mathcal{M}'_\dagger = \mathcal{N}'$ and $\operatorname{V}(\mathcal{M}/\mathcal{M}') = \overline{\mathbb{O}}$.

We will need relative versions of (2)-(5), compare to [L5, Section 3.3.2]. Namely, let us pick an affine subspace $\mathfrak{h}^1 \subset \mathfrak{h}^*$ and write $\mathcal{U}_{\mathfrak{h}^1} := \mathbb{C}[\mathfrak{h}^1] \otimes_{\mathbb{C}[\mathfrak{h}^*]^W} \mathcal{U}$, $\mathcal{W}_{\mathfrak{h}^1} := \mathbb{C}[\mathfrak{h}^1] \otimes_{\mathbb{C}[\mathfrak{h}^*]^W} \mathcal{W}$. Then we get a $\mathbb{C}[\mathfrak{h}^1]$-bilinear exact tensor functor $\bullet_\dagger : \operatorname{HC}(\mathcal{U}_{\mathfrak{h}^1}) \to \operatorname{HC}^Q(\mathcal{W}_{\mathfrak{h}^1})$.

Let us write $\operatorname{HC}_{\overline{\mathbb{O}}}(\mathcal{U}_{\mathfrak{h}^1})$ for the full subcategory of $\operatorname{HC}(\mathcal{U}_{\mathfrak{h}^1})$ consisting of HC-bimodules $\mathcal{M}$ with $\operatorname{V}(\mathcal{M}) \cap \mathcal{N} \subset \overline{\mathbb{O}}$. The notation $\operatorname{HC}_{\partial\mathbb{O}}(\mathcal{U}_{\mathfrak{h}^1})$ has a similar meaning. We also write $\mathrm{HC}^Q_\chi(\mathscr{W}_{\mathfrak{h}^1})$ for the category of all bimodules finitely generated (as left, or equivalently, right) modules over $\mathbb{C}[\mathfrak{h}^1]$. Analogs of (2)–(5) are as follows.

(2′) $\bullet_\dagger$ maps $\mathrm{HC}_{\overline{\mathbb{O}}}(\mathscr{U}_{\mathfrak{h}^1})$ to $\mathrm{HC}^Q_\chi(\mathscr{W}_{\mathfrak{h}^1})$,
(3′) $\bullet_\dagger$ annihilates $\mathrm{HC}_{\partial\mathbb{O}}(\mathscr{U}_{\mathfrak{h}^1})$.
(4′) There is a functor $\bullet^\dagger: \mathrm{HC}^Q_\chi(\mathscr{W}_{\mathfrak{h}^1}) \to \mathrm{HC}_{\overline{\mathbb{O}}}(\mathscr{U}_{\mathfrak{h}^1})$ right adjoint to $\bullet_\dagger$.
(5′) For $\mathscr{M} \in \mathrm{HC}_{\overline{\mathbb{O}}}(\mathscr{U}_{\mathfrak{h}^1})$, the kernel and the cokernel of the adjunction unit $\mathscr{M} \to (\mathscr{M}_\dagger)^\dagger$ are supported on $\mathrm{HC}_{\partial\mathbb{O}}(\mathscr{U}_{\mathfrak{h}^1})$.

5.3 Results on finite-dimensional irreducible $\mathscr{W}$-modules

Let us state our results on the classification of finite-dimensional irreducible $\mathscr{W}$-modules. For this, we will need to recall one of the main results of [L2]. Since Q acts on $\mathscr{W}$ by automorphisms, it also acts on the set $\mathrm{Irr}_{\mathrm{fin}}(\mathscr{W})$ of the isomorphism classes of finite-dimensional irreducible $\mathscr{W}$-modules. Since the action of Q on $\mathscr{W}$ is Hamiltonian, the action on $\mathrm{Irr}_{\mathrm{fin}}(\mathscr{W})$ descends to an action of the component group $A(=A_{\mathbb{O}}) := Q/Q^\circ$.

One of the main results of [L2], see Section 1.2, was a natural identification $\mathrm{Irr}_{\mathrm{fin}}(\mathscr{W})/A \xrightarrow{\sim} \mathrm{Prim}_{\mathbb{O}}(\mathscr{U})$: it sends $\mathscr{J} \in \mathrm{Prim}_{\mathbb{O}}(\mathscr{U})$ to the A-orbit of the irreducible representations of $\mathscr{W}/\mathscr{J}_\dagger$; this is well defined and gives a bijection by (6) of Section 5.2. So to finish the classification of the finite-dimensional irreducible $\mathscr{W}$-modules, we need, for every primitive ideal $\mathscr{J} \in \mathrm{Prim}_{\mathbb{O}}(\mathscr{U})$, to compute the stabilizer $H_{\mathscr{J}}$ (defined up to conjugacy) in the A-orbit over $\mathscr{J}$.

Fix a central character λ and assume for the time being that it is regular. Let us write $\mathscr{W}_{\lambda,\mathbf{c}}$ for the semisimple finite-dimensional quotient of $\mathscr{W}_\lambda$ whose simple representations are precisely the irreducibles lying over the primitive ideals corresponding to the two-sided cell $\mathbf{c}$. Recall the Springer representation $\mathrm{Spr}_{\mathbb{O}} := H_{\mathrm{top}}(\mathscr{B}_e, \mathbb{C})$ of $W \times A$. Also recall that $W_{[\lambda]}$ can be regarded as a subgroup of W via $W_{[\lambda]} \hookrightarrow W^a \twoheadrightarrow W$. Let us write $\mathrm{Spr}_{\mathbb{O},\mathbf{c}}$ for the sum of all irreducible $W_{[\lambda]}$-submodules in the Springer representation that belong to the family of irreducible W-modules indexed by the cell $\mathbf{c}$.

Theorem 5.2. *Let* $\mathbf{c}$ *be a two-sided cell in* $W_{[\lambda]}$ *and let* $\mathbb{O} = \mathbb{O}_{\mathbf{c}}$. *Then the following is true.*

(1) *Let* $\mathscr{J} \in \mathrm{Prim}_{\mathbb{O}}(\mathscr{U}_\lambda)$ *correspond to a left cell* $\sigma \subset W_{[\lambda]}$, *and let* $H_{\mathscr{J}}$ *denote a stabilizer in the A-orbit in* $\mathrm{Irr}_{\mathrm{fin}}(\mathscr{W}_\lambda)$ *lying over* $\mathscr{J}$. *Then the A-module* $\mathrm{Hom}_{W_{[\lambda]}}([\sigma], \mathrm{Spr}_{\mathbb{O}})$ *coincides with the A-module induced from the trivial* $H_{\mathscr{J}}$*-module.*
(2) *We have an isomorphism* $K_0(\mathscr{W}_{\lambda,\mathbf{c}}\text{-mod}) \cong \mathrm{Spr}_{\mathbb{O},\mathbf{c}}$ *of* $W_{[\lambda]} \times A$*-modules.*

When λ is integral, this theorem is the main result of [LO], see Theorem 1.1 and (iii) of Theorem 7.4. Note that (1) is sufficient to determine $H_{\mathscr{J}}$ (at least in all cases but 2). Indeed, the group A is abelian for all nilpotent orbits but twelve in the exceptional Lie algebras, see, e.g., [CM, Section 8.4]. If A is abelian, then $H_{\mathcal{J}}$ is just the kernel of the A-action on $\mathrm{Hom}_{W_{[\lambda]}}([\sigma], \mathrm{Spr}_{\mathbb{O}})$. Out of these twelve cases, in ten cases we have $A = S_3$, where, clearly, the induced module determines a subgroup uniquely. In the two remaining cases, we have $A = S_4$ (in F_4) and $A = S_5$ (in E_8); we haven't checked for general λ if (1) determines $H_{\mathcal{J}}$ uniquely, though for an integral λ this is indeed the case.

To finish this section, we explain what happens for singular central characters. The situation is very similar to the integral case considered in [LO]. Let λ_0 be a singular dominant element in $\mathfrak{h}^*$. Pick a dominant element μ in the root lattice, so that $\lambda := \lambda_0 + \mu$ is strictly dominant. As was explained in Section 2.1, $\mathrm{Prim}_{\mathbb{O}}(\mathcal{U}_{\lambda_0}) \hookrightarrow \mathrm{Prim}_{\mathbb{O}}(\mathcal{U}_{\lambda})$. This gives rise to the partitions $\mathrm{Prim}_{\mathbb{O}}(\mathcal{U}_{\lambda_0}) = \bigsqcup_{\mathbf{c}} \mathrm{Prim}_{\mathbf{c}}(\mathcal{U}_{\lambda_0})$, $\mathrm{Irr}_{\mathrm{fin}}(\mathcal{W}_{\lambda_0}) = \bigsqcup_{\mathbf{c}} \mathrm{Irr}(\mathcal{W}_{\lambda_0,\mathbf{c}})$.

Corollary 5.3. *Let $\mathcal{J}_0 \subset \mathrm{Prim}_{\mathbf{c}}(\mathcal{U}_{\lambda_0})$, and let $\mathcal{J}$ be the corresponding ideal in $\mathrm{Prim}_{\mathbf{c}}(\mathcal{U}_{\lambda})$. Then the A-orbit over $\mathcal{J}_0$ coincides with $A/H_{\mathcal{J}}$ and $K_0(\mathcal{W}_{\lambda_0,\mathbf{c}}\text{-mod}) = \mathrm{Spr}_{\mathbb{O},\mathbf{c}}^{W_{\lambda_0}}$.*

5.4 Reduction of representations mod p

Now fix a dominant rational $\lambda \in \mathfrak{h}^*$. Recall, [L4, Theorem 1.3], that $\mathcal{W}_\lambda$ has a minimal ideal of finite codimension, say $\mathcal{I}$. By definition, this ideal is defined over $\mathbb{Q}$. For a finite localization R of $\mathbb{Z}$, set $\mathcal{I}_R := \mathcal{W}_{\lambda,R} \cap \mathcal{I}$. We assume that $\mathrm{gr}\,\mathcal{W}_R = R[S]$ and $\mathrm{gr}\,\mathcal{W}_{\lambda,R} = R[S \cap \mathcal{N}]$, this can be achieved after a finite localization of R.

Lemma 5.4. *After a finite localization of R, we get $\mathcal{I}_R^2 = \mathcal{I}_R$.*

Proof. Note that $\mathrm{gr}\,\mathcal{W}_{\lambda,R}/\mathcal{I}_R$ is a finitely generated commutative R-algebra. So after a finite localization of R, we can achieve that $\mathcal{W}_{\lambda,R}/\mathcal{I}_R$ is a free finite rank R-module. Note that $\mathcal{W}_{\lambda,R}$ is Noetherian because of $\mathrm{gr}\,\mathcal{W}_{\lambda,R} = R[S \cap \mathcal{N}]$. In particular, $\mathcal{I}_R$ is a finitely generated left $\mathcal{W}_{\lambda,R}$-module. It follows that $\mathcal{I}_R/\mathcal{I}_R^2$ is a finitely generated module over $\mathcal{W}_{\lambda,R}/\mathcal{I}_R$, and hence is a finite rank R-module. Note that $\mathcal{I}_{\mathbb{Q}}$ is still the minimal ideal of finite codimension in $\mathcal{W}_{\mathbb{Q}}$. So $\mathcal{I}_{\mathbb{Q}} = \mathcal{I}_{\mathbb{Q}}^2$. It follows that $\mathcal{I}_R/\mathcal{I}_R^2$ is a finitely generated torsion R-module, hence it is killed by a finite localization of R. □

This lemma shows that $\big(\mathbb{F} \otimes_R (\mathcal{W}_{\lambda,R}/\mathcal{I}_R)\big)$-mod is a Serre subcategory in $\mathcal{W}_{\lambda,\mathbb{F}}$-mod.

After replacing R with a finitely generated algebraic extension, we can assume that $\mathcal{W}_{\lambda,\mathrm{Frac}(R)}/\mathcal{I}_{\mathrm{Frac}(R)}$ is split. So there is a natural bijection $\mathrm{Irr}_{\mathrm{fin}}(\mathcal{W}_\lambda) \cong \mathrm{Irr}(\mathcal{W}_{\lambda,\mathrm{Frac}(R)}/\mathcal{I}_{\mathrm{Frac}(R)})$. So, for $L \in \mathrm{Irr}(\mathcal{W}_\lambda)$, we can talk about its reduction $L_{\mathbb{F}}$ mod p. For standard reasons, $L_{\mathbb{F}}$ is irreducible. As was checked in [BL, Section 6.5], $L_{\mathbb{F}}$ has central character χ. So we get an inclusion $\mathrm{Irr}_{\mathrm{fin}}(\mathcal{W}_\lambda) \hookrightarrow \mathrm{Irr}(\mathcal{W}_{\lambda,\mathbb{F}}^{\chi})$. Recall, Section 5.1, that the target is naturally identified with $\mathrm{Irr}(\mathcal{U}_{\lambda,\mathbb{F}}^{\chi})$.

Proposition 5.5. *For $p \gg 0$, the image of* $\mathrm{Irr}_{\mathrm{fin}}(\mathscr{W}_\lambda)$ *in* $\mathrm{Irr}(\mathscr{U}^\chi_{\lambda,\mathbb{F}})$ *consists of the simples whose dimension polynomials are of degree* $\dim \mathbb{O}/2$.

Proof. The simples in $\mathrm{Irr}(\mathscr{U}^\chi_{\lambda,\mathbb{F}})$ with degree of dimension polynomial equal to $\dim \mathbb{O}/2$ correspond to the simples in $\mathrm{Irr}(\mathscr{W}^\chi_{\lambda,\mathbb{F}})$ whose dimension is independent of p. Let d be the maximal dimension of these representations. Let $\mathscr{I}'_R$ be the ideal in $\mathscr{W}_R$ generated by the elements $\sum_{\sigma\in S_{2d}} \mathrm{sgn}(\sigma) a_{\sigma(1)}\ldots a_{\sigma(2d)}$ for $a_i \in \mathscr{W}_R$, $i = 1,\ldots,2d$. By the Amitsur-Levitsky theorem, the ideal $\mathscr{I}'_R$ vanishes on all representations of $\mathscr{W}_\mathbb{F}$ of dimension $\leqslant d$. Arguing as in [L3, Lemma 5.1], we see that $\mathscr{I}'_\mathbb{C}$ is of finite codimension. So $\mathscr{I}'_\mathbb{C} \supset \mathscr{I}$. It follows that after a finite localization of R, we have $\mathscr{I}'_R \supset \mathscr{I}_R$. So any irreducible representation of $\mathscr{W}_{\lambda,\mathbb{F}}$ of dimension $\leqslant d$ factors through $\mathscr{W}_{\lambda,\mathbb{F}}/\mathscr{I}_\mathbb{F}$ for p large enough. This finishes the proof. □

5.5 Proof of Theorem 5.2

Let us prove Theorem 5.2 in the case when λ is rational.

Proof (of Theorem 5.2 for rational λ). Let us start by proving (2). From Proposition 5.5 combined with (2) of Theorem 1.1, we know that $K_0(\mathscr{W}_{\lambda,\mathbf{c}}\text{-mod}) = H_*(\mathscr{B}_e,\mathbb{C})_\mathbf{c}$, the sum of all irreducible $W_{[\lambda]}$-submodules in $H_*(\mathscr{B}_e,\mathbb{C})_\mathbf{c}$ that belong to $\mathbf{c}$. This is with respect to the standard embedding $W_{[\lambda]} \hookrightarrow W^a$. What remains to show is that

$$H_*(\mathscr{B}_e,\mathbb{C})_\mathbf{c} \xrightarrow{\sim} H_{\mathrm{top}}(\mathscr{B}_e,\mathbb{C})_\mathbf{c}, \tag{12}$$

where now the action on the right-hand side is via $W_{[\lambda]} \hookrightarrow W^a \twoheadrightarrow W$ and the map is induced by the natural projection $H_*(\mathscr{B}_e,\mathbb{C}) \twoheadrightarrow H_{\mathrm{top}}(\mathscr{B}_e,\mathbb{C})$. According to Dodd, [D, Section 7], $K_0(\mathscr{W}_{\lambda,\mathbf{c}}\text{-mod}) \subset H_*(\mathscr{B}_e,\mathbb{C})$ projects injectively to $H_{\mathrm{top}}(\mathscr{B}_e,\mathbb{C})$. The projection $H_*(\mathscr{B}_e,\mathbb{C}) \to H_{\mathrm{top}}(\mathscr{B}_e,\mathbb{C})$ is W^a-equivariant, where on the target space W^a acts via the projection $W^a \twoheadrightarrow W$, and so intertwines the two actions of $W_{[\lambda]}$. This implies (12) and finishes the proof of (2).

Let us now deduce (1) from (2). The restriction functor $\bullet_\dagger\colon \mathrm{HC}_\mathbb{O}(\mathscr{U}_\lambda) \to \mathrm{Bimod}^Q_{\mathrm{fin}}(\mathscr{W}_\lambda)$, recalled in Section 5.3, equips $K_0(\mathscr{W}_{\lambda,\mathbf{c}}\text{-mod})$ with an action of $K_0(\mathrm{HC}^{ss}_\mathbf{c}(\mathscr{U}_\lambda)) = \mathrm{J}_\mathbf{c}(W_{[\lambda]})$. By results of Dodd, [D, Section 8], this action is the same as the $\mathrm{J}_\mathbf{c}(W_{[\lambda]})$-action on $K_0(\mathscr{U}^\chi_{\lambda,\mathbb{F}}\text{-mod})_\mathbf{c}$.

By the description of the $\mathrm{J}_\mathbf{c}$-action in the previous paragraph, the span of classes of the irreducible modules lying over $\mathscr{J}$ equals $t_d K_0(\mathscr{W}_{\lambda,\mathbf{c}}\text{-mod})$.

So $t_d K_0(\mathscr{W}_{\lambda,\mathbf{c}}\text{-mod})$ is nothing but $\mathrm{Hom}_{W_{[\lambda]}}([\sigma],\mathrm{Spr}_\mathbb{O})$. On the other hand, the span is A-stable and is the A-representation induced from the trivial representation of $H_\mathscr{J}$. This finishes the proof of (1). □

Now let us reduce the proof of Theorem 1.1 to the case when λ is rational. To state our main technical result we need some notation.

Choose a regular central character $\lambda \in \mathfrak{h}^*$. Let W_0 denote a minimal parabolic subgroup of W containing $W_{[\lambda]}$; we can conjugate λ and assume that W_0 is standard, while λ is still dominant. We can write λ as $\lambda_1 + \lambda_2$, where λ_1 lies in $(\mathfrak{h}^*)^{W_0}$ and λ_2 lies in the orthogonal complement to $(\mathfrak{h}^*)^{W_0}$. Note that λ_2 is rational.

Proposition 5.6. *For all dominant regular rational $\lambda' \in \lambda_2 + (\mathfrak{h}^*)^{W_0}$ satisfying $W_{[\lambda']} = W_{[\lambda]}$, the following is true. Let σ be a left cell in $W_{[\lambda]}$ and let $\mathcal{J}, \mathcal{J}'$ be primitive ideals in $\mathcal{U}$ with central characters λ, λ' corresponding to the left cell σ. Then $H_{\mathcal{J}} = H_{\mathcal{J}'}$.*

Proof. The proof is in several steps.

Step 1. Since λ is not rational, $W_0 \neq W$. By [LO, Proposition 5.7] for any integral dominant $\mu \in (\mathfrak{h}^*)^{W_0}$ for the ideal $\mathcal{J}^\mu$ with central character $\lambda + \mu$ (corresponding to $\mathcal{J}$ under the isomorphism $\operatorname{Prim}(\mathcal{U}_\lambda) \xrightarrow{\sim} \operatorname{Prim}(\mathcal{U}_{\lambda+\mu})$), we have $H_{\mathcal{J}^\mu} = H_{\mathcal{J}}$. So in the proof we can assume that λ_1 is Zariski generic in $(\mathfrak{h}^*)^{W_0}$. To simplify the notation, we will write $\mathfrak{h}^1$ for $\lambda_2 + (\mathfrak{h}^*)^{W_0}$.

Step 2. A standard argument, see, for example, the proof of [L3, Lemma 5.1], shows that there is an ideal $\mathcal{I} \subset \mathcal{W}_{\mathfrak{h}^1}$, such that $\mathcal{W}_{\mathfrak{h}^1}/\mathcal{I}$ is finitely generated over $\mathbb{C}[\mathfrak{h}^1]$, and for a Weil generic $\hat{\lambda} \in \mathfrak{h}^1$, the specialization $\mathcal{I}_{\hat{\lambda}}$ is the minimal ideal of finite codimension in $\mathcal{W}_{\hat{\lambda}}$. Let us write $\bar{\mathcal{J}}$ for the kernel of $\mathcal{U}_{\mathfrak{h}^1} \to (\mathcal{W}_{\mathfrak{h}^1}/\mathcal{I})^\dagger$. Note that thanks to (4′) of Section 5.2, $\mathrm{V}(\mathcal{U}_{\mathfrak{h}^1}/\bar{\mathcal{J}}) \cap \mathcal{N} = \overline{\mathbb{O}}$. So $\mathcal{W}_{\mathfrak{h}^1}/\bar{\mathcal{J}}_\dagger$ is finitely generated over $\mathbb{C}[\mathfrak{h}^1]$. Since $\bar{\mathcal{J}}_\dagger \subset \mathcal{I}$, we can replace $\mathcal{I}$ with $\bar{\mathcal{J}}_\dagger$ and assume that $\mathcal{I} = \bar{\mathcal{J}}_\dagger$.

Step 3. For a dominant regular $\hat{\lambda}$ with $W_{[\hat{\lambda}]} = W_{[\lambda]}$, let us write $\mathcal{J}_{\hat{\lambda}}$ for the primitive ideal with central character $\hat{\lambda}$ corresponding to the left cell σ. Let us prove that for a Zariski generic $\hat{\lambda} \in \mathfrak{h}^1$ with $W_{[\hat{\lambda}]} = W_{[\lambda]}$, we have $\bar{\mathcal{J}}_{\hat{\lambda}} \subset \mathcal{J}_{\hat{\lambda}}$. Note that this inclusion is automatic provided $\mathcal{I}_{\hat{\lambda}}$ is the minimal ideal of finite codimension in $\mathcal{W}_{\hat{\lambda}}$.

Choose $w \in \sigma$. For $\hat{\lambda} \in \mathfrak{h}^1$, consider the Verma module $\Delta(w\hat{\lambda})$. These Verma modules form a flat family over $\mathfrak{h}^1$, and let us denote the corresponding $\mathcal{U}_{\mathfrak{h}^1}$-module by $\Delta_{\mathfrak{h}^1}$. Inside we have a $\mathcal{U}_{\mathfrak{h}^1}$-submodule $\bar{\mathcal{J}}\Delta_{\mathfrak{h}^1}$. Consider the quotient $\Delta_{\mathfrak{h}^1}/\bar{\mathcal{J}}\Delta_{\mathfrak{h}^1}$. Its specialization to $\hat{\lambda}$ is $\Delta(w\hat{\lambda})/\bar{\mathcal{J}}_{\hat{\lambda}}\Delta(w\hat{\lambda})$. It is nonzero for a Weil generic $\hat{\lambda}$. It follows that it is nonzero for a Zariski generic $\hat{\lambda}$ as well. The claim in the first paragraph of this step follows.

Step 4. The algebra $\mathcal{W}_{\mathfrak{h}^1}/\mathcal{I}$ comes with a Q-action. Since the algebra $\mathcal{W}_{\mathfrak{h}^1}/\mathcal{I}$ is a finitely generated module over $\mathbb{C}[\mathfrak{h}^1]$, we see that the number of irreducible representations is the same for two Zariski generic specializations (this is a version of the Tits deformation argument). Moreover, for two nearby generic parameters, there is a natural bijection between the irreducibles. Being natural, this bijection preserves the stabilizers in Q. And when the parameters are not nearby, the monodromy may appear, but it does not change the stabilizers in A.

Step 5. Recall, Step 1, that we can assume that λ is Zariski generic in $\mathfrak{h}^1$. Similarly, we can assume that λ' is Zariski generic. Now the claim of the proposition follows from Step 4. □

Proof (of Theorem 5.2 for general λ). (1) for λ immediately follows from Proposition 5.6 and (1) for λ' proved above. To prove (2) we can argue as follows. Take a Weil generic $\hat{\lambda}$ with $W_{[\hat{\lambda}]} = W_{[\lambda]}$. Then we have the degeneration maps (compare to [BL, Section 11.1])

$$K_0(\mathscr{W}_{\hat{\lambda}}\text{-mod}_{\text{fin}}) \to K_0(\mathscr{W}_{\lambda'}\text{-mod}_{\text{fin}}), K_0(\mathscr{W}_{\hat{\lambda}}\text{-mod}_{\text{fin}}) \to K_0(\mathscr{W}_{\lambda}\text{-mod}_{\text{fin}}).$$

By the proof of Proposition 5.6, we see that both these maps are isomorphisms. They are also $W_{[\lambda]}$-invariant. This implies (2). □

5.6 Proof of Corollary 5.3

Let λ_0, λ be as before the statement of Corollary 5.3. Let V be a finite-dimensional G-module. We can define the endo-functor $V \otimes \bullet$ of $\mathscr{W}$-mod as the tensor product with the bimodule $(V \otimes \mathscr{U})_{\dagger}$. Using this we can define translation functors $\mathscr{T}_{\lambda_0 \to \lambda}: \mathscr{W}\text{-mod}_{\lambda_0} \to \mathscr{W}\text{-mod}_{\lambda}$, $\mathscr{T}_{\lambda \to \lambda_0}: \mathscr{W}\text{-mod}_{\lambda} \to \mathscr{W}\text{-mod}_{\lambda_0}$ in a standard way. They enjoy properties similar to those of the usual translation functors (because $\bullet_{\dagger}$ is a tensor functor):

(1) On $K_0(\mathscr{W}_{\lambda_0}\text{-mod}_{\text{fin}})$ the composition $\mathscr{T}_{\lambda \to \lambda_0} \circ \mathscr{T}_{\lambda_0 \to \lambda}$ is the multiplication by $|W_{\lambda_0}|$.
(2) On $K_0(\mathscr{W}_{\lambda}\text{-mod}_{\text{fin}})$ the composition $\mathscr{T}_{\lambda_0 \to \lambda} \circ \mathscr{T}_{\lambda \to \lambda_0}$ acts as $\sum_{w \in W_{\lambda_0}} w$.

This implies the equality

$$K_0(\mathscr{W}_{\lambda_0}\text{-mod}_{\text{fin}}) = K_0(\mathscr{W}_{\lambda}\text{-mod}_{\text{fin}})^{W_{\lambda_0}}.$$

Also, for w longest in its right W_{λ_0}-coset, the maps $[\mathscr{T}_{\lambda_0 \to \lambda}], [\mathscr{T}_{\lambda \to \lambda_0}]$ map between $K_0(\mathscr{W}/\mathscr{J}(w\lambda_0)_{\dagger}\text{-mod})$ and $K_0(\mathscr{W}/\mathscr{J}(w(\lambda))_{\dagger}\text{-mod})$, which together with (1) implies that $H_{\mathscr{J}(w\lambda_0)} = H_{\mathscr{J}(w(\lambda))}$.

6 Application to real variation of stability conditions

In this section, we use Theorem 1.1 to essentially realize the idea sketched in [ABM, Remark 6].

We now describe a simplified version of a construction of [ABM]. Let $\mathscr{A}$ be an abelian category and let $\zeta: \mathbb{C} \to K_0(\mathscr{A})^*$ be a polynomial map. We assume that for some $r > 0$, we have

$$\langle [M], \zeta(x) \rangle \in \mathbb{R}_{>0} \quad \forall\, x \in (0,r), M \in \mathscr{A}, M \neq 0. \tag{13}$$

In this situation we get a filtration on $\mathscr{A}$ by Serre subcategories, where $\mathscr{A}_{\geqslant d}$ consists of objects M such that the polynomial $x \mapsto \langle [M], \zeta(x) \rangle$ has a zero of order at least

d at zero. We say that a derived equivalence $\phi: D^b(\mathscr{A}) \to D^b(\mathscr{A}')$ is a *perverse equivalence governed by* ζ if there exists a filtration on $\mathscr{A}'$ by Serre subcategories, such that $D^b(\mathscr{A}'_{\geqslant d}) = \phi(D^b(\mathscr{A}_{\geqslant d}))$, while the functor $gr_d(\phi)\colon D^b(\mathscr{A}_{\geqslant d}/\mathscr{A}_{>d}) \to D^b(\mathscr{A}'_{\geqslant d}/\mathscr{A}'_{>d})$ sends $\mathscr{A}_{\geqslant d}/\mathscr{A}_{>d}$ to $(\mathscr{A}_{\geqslant d}/\mathscr{A}_{>d})[d]$.

We now set $\mathscr{A} = \mathscr{U}_{\lambda,\mathbb{F}}\text{-mod}^{\chi}$. Let $\xi: \mathbb{R} \to \mathfrak{t}^*_{\mathbb{R}}$ be an affine linear functional sending zero to a face F in the closure of the fundamental alcove A_0; we assume that $\xi(\mathbb{R}_{>0})$ intersects A_0.

The central charge map $Z\colon \mathfrak{t} \to K_0(\mathrm{Coh}(\mathscr{B}_e))^*$ was defined in [ABM]. We use identification (1) for a choice of λ in the fundamental p-alcove to get a map $\mathfrak{t} \to K_0(\mathscr{A})^*$ which we also denote by Z. We set $\zeta = Z \circ \xi$. Then [ABM, Proposition 1(a)] implies that the positivity condition (13) holds for some $r > 0$.

The face F determines a proper subset in the set of vertices of the affine Dynkin graph; let W_F be the corresponding finite Weyl group and let w_F be the longest element in W_F; let $\tilde{w}_F$ be the canonical (minimal length) lift of w_F to the affine braid group B_{aff}. Note that a path in the complement to affine coroot hyperplanes whose end-points are contained in $t^*_{\mathbb{R}}$ defines an element in B_{aff}; the element $\tilde{w}_F$ corresponds to the path $[0,1] \to \mathfrak{t}^*_{\mathbb{C}}$, $x \mapsto \xi(R\exp(2\pi i x))$ for a small $R > 0$.

Recall the action of B_{aff} on $D^b(\mathscr{A})$.

The main result of this section is as follows.

Theorem 6.1. *The functor $\tilde{w}_F: D^b(\mathscr{A}) \to D^b(\mathscr{A})$ is a perverse equivalence governed by ζ.*

Proof. We let $\mathscr{A}_{\geqslant d}$ be the filtration introduced earlier in this section, and we let $\mathscr{A}'_{\geqslant d}$ denote the Serre subcategory generated by $\mathscr{U}_{\lambda,\mathbb{F}}\text{-mod}^{\chi}_{\leqslant \mathbf{c}}$ for all cells $\mathbf{c}$ in W_F such that $a(\mathbf{c}) \geqslant d$.

The result clearly follows from the following two statements.

(a) We have $\mathscr{A}_{\geqslant d} = \mathscr{A}'_{\geqslant d}$.
(b) The functor $\tilde{w}_F[-d]$ induces a t-exact functor on $D^b(\mathscr{A}'_{\geqslant d}/\mathscr{A}'_{>d})$.

We claim that (a) follows from Theorem 1.1. To see this, choose a regular rational weight λ with $W_{[\lambda]} = W_F$. Furthermore, we can and will assume that λ satisfies the following assumptions: it can be written as $\lambda = \mu + \nu$, where ν is an integral weight and μ lies in the closure of the fundamental alcove A_0, while $\mu + t\nu$ lies in A_0 for small (equivalently, for some) $t > 0$ (equivalently, $\mu' + t\nu$ lies in A_0 for all $\mu' \in F$ and small $t > 0$, where the bound on t depends on μ'). Choose a large prime p, such that $(p+1)\lambda$ is an integral weight. Then $\tilde{\lambda} := (p+1)\mu + \nu$ is an integral weight satisfying: $\tilde{\lambda} = \lambda \mod p$ and $\frac{\tilde{\lambda}}{p} \in A_0$.

For $M \in \mathscr{A}$, consider the polynomial D_M, such that for an integral weight η such that $\frac{\tilde{\lambda}+\eta-\rho}{p} \in A_0$, we have $\dim(T_{\tilde{\lambda}\to\eta}(M)) = D_M(\lambda)$, where T denotes the translation functor (the existence of D_M follows from [BMR2, Theorem 6.2.1]). By Theorem 1.1 for a $M \in \mathscr{A}'_{\geqslant d}$, $M \notin \mathscr{A}'_{>d}$, we have $\deg_p(D_M) = \dim(\mathscr{B}) - d$.

On the other hand the central charge Z of [ABM] satisfies (see the proof of [ABM, Proposition 1]):

$$\left\langle Z\left(\frac{\eta+\rho}{p}\right),[M]\right\rangle = p^{-\dim(\mathscr{B})}D_M(\eta).$$

It follows that the order of zero of the polynomial $\zeta_M(t) = \langle Z(\mu+t\nu),[M]\rangle$ at $t=0$ equals $\dim(\mathscr{B}) - \deg_p(D_M)$, this proves (a).

We now sketch the proof of (b). We use the fact that the braid group B_F of the Coxeter group W_F acts the category $D^b(\mathscr{U}_{\lambda,R}\text{-mod})$ compatibly with its action on $D^b(\mathscr{U}_{\lambda,\mathbb{F}}\text{-mod})$.

Note that the action of $\tilde{w}_F$ is given by the derived tensor product with the wall-crossing $\mathscr{U}_\lambda$-bimodule WC_{w_F} corresponding to the element w_F. By [L6, Theorem 3.1], the functor $\mathscr{W}\mathscr{C}_{w_F}$ is a perverse equivalence with

$$\mathscr{U}_\lambda\text{-mod}_{\geqslant d} = \{M \in \mathscr{U}_\lambda\text{-mod} \mid \dim \mathrm{V}(\mathscr{U}_\lambda/\operatorname{Ann}(M)) \leqslant \dim \mathscr{N} - 2d\}.$$

As was shown in the proof of that theorem, the statement reduces to the vanishing of Tor's involving WC_{w_F}, and the quotients of $\mathscr{U}_\lambda$ by the minimal ideals with given dimensions of associated varieties. This vanishing was checked in the proof. Now this vanishing over $\mathbb{C}$ implies the analogous vanishing over R (after a finite localization), and hence the claim that the endo-functor $\mathscr{W}\mathscr{C}_{w_F} = \mathsf{WC}_{w_F,R} \otimes^L_{\mathscr{U}_{\lambda,R}} \bullet$ of $D^b(\mathscr{U}_{\lambda,\mathbb{F}}\text{-mod}^\chi)$ is perverse with respect to the filtration $\mathscr{A}'_{\geqslant d}$. □

References

[B] R. Bezrukavnikov, *On two geometric realizations of an affine Hecke algebra,* Publ. IHES 123(1) (2016), 1–67.

[ABM] R. Anno, R. Bezrukavnikov, I. Mirkovic, *Stability conditions for Slodowy slices and real variations of stability,* Mosc. Math. J. 15 (2015), no. 2, 187–203, 403.

[BFO] R. Bezrukavnikov, M. Finkelberg, V. Ostrik, *Character D-modules via Drinfeld center of Harish-Chandra bimodules.* Invent. Math. 188 (2012), no. 3, 589–620.

[BL] R. Bezrukavnikov, I. Losev, *Etingof conjecture for quantized quiver varieties.* http://www.northeastern.edu/iloseu/bezpaper.pdf

[BM] R. Bezrukavnikov, I. Mirkovic. *Representations of semisimple Lie algebras in prime characteristic and noncommutative Springer resolution.* Ann. Math. 178 (2013), n.3, 835–919.

[BMR1] R. Bezrukavnikov, I. Mirkovic, D. Rumynin. *Singular localization and intertwining functors for reductive Lie algebras in prime characteristic.* Nagoya Math. J. 184 (2006), 1–55.

[BMR2] R. Bezrukavnikov, I. Mirkovic, D. Rumynin. *Localization of modules for a semisimple Lie algebra in prime characteristic* (with an appendix by R. Bezrukavnikov and S. Riche), Ann. of Math. (2) 167 (2008), no. 3, 945–991.

[BR] R. Bezrukavnikov, S. Riche, *Affine braid group actions on derived categories of Springer resolutions.* Ann. Sci. Éc. Norm. Supér. (4) 45 (2012), no. 4, 535–599 (2013).

[BY] R. Bezrukavnikov, Z. Yun, *On Koszul duality for Kac-Moody groups,* Represent. Theory **17** (2013), 1–98.

[CM] D. Collingwood, W. McGovern. *Nilpotent orbits in semisimple Lie algebras.* Chapman and Hall, London, 1993.

[D] C. Dodd. *Injectivity of a certain cycle map for finite dimensional W-algebras.* Int. Math. Res. Not. 2014, no. 19, 5398–5436.

[GG] W.L. Gan, V. Ginzburg. *Quantization of Slodowy slices*. IMRN, 5(2002), 243–255.

[Ja] J.C. Jantzen. *Einhüllende Algebren halbeinfacher Lie-Algebren*. Ergebnisse der Math., Vol. 3, Springer, New York, Tokyo, 1983.

[Jo1] A. Joseph. *On the associated variety of a primitive ideal*. J. Algebra, 93 (1985), no. 2, 509–523.

[Jo2] A. Joseph. *On the cyclicity of vectors associated with Duflo involutions*. In *Lecture Notes in Mathematics*, Vol. 1243, 144–188, Springer-Verlag, Berlin, 1987.

[KW] V. Kac, B. Weisfeiler, *Coadjoint action of a semisimple algebraic group and the center of the enveloping algebra in characteristic p,* Indag. Math. 38 (1976), 136–151.

[L1] I. Losev, *Finite W-algebras*. Proceedings of the International Congress of Mathematicians Hyderabad, India, 2010, p. 1281–1307.

[L2] I.V. Losev. *Finite dimensional representations of W-algebras*. Duke Math J. 159(2011), n.1, 99–143.

[L3] I. Losev. *Derived equivalences for Rational Cherednik algebras*. Duke Math J. 166(2017), N1, 27–73.

[L4] I. Losev. *Bernstein inequality and holonomic modules* (with a joint appendix by I. Losev and P. Etingof). Adv. Math. 308 (2017), 941–963.

[L5] I. Losev. *Cacti and cells*. arXiv:1506.04400. Accepted by J. Eur. Math. Soc.

[L6] I. Losev. *Wall-crossing functors for quantized symplectic resolutions: perversity and partial Ringel dualities*. arXiv:1604.06678.

[LO] I. Losev, V. Ostrik, *Classification of finite dimensional irreducible modules over W-algebras*. Compos. Math. 150(2014), N6, 1024–1076.

[Lu1] G. Lusztig. *Characters of reductive groups over a finite field*, Ann. Math. Studies 107, Princeton University Press (1984).

[Lu2] G. Lusztig. *Equivariant K-theory and representations of Hecke algebras*. Proc. Amer. Math. Soc. 94 (1985), no. 2, 337–342.

[Lu3] G. Lusztig. *Leading coefficients of character values of Hecke algebras*, Proc. Symp. Pure Math. 47(2), Amer. Math. Soc. 1987, 235–262.

[Lu4] G. Lusztig. *Cells in affine Weyl groups, IV*. J. Fac. Sci. Univ. Tokyo Sect. IA, Math. 36 (1989), 297–328.

[M] D. Miličić. *Localization and Representation Theory of Reductive Lie Groups*, available at http://www.math.utah.edu/∼milicic.

[P1] A. Premet. *Special transverse slices and their enveloping algebras*. Adv. Math. 170(2002), 1–55.

[P2] A. Premet. *Enveloping algebras of Slodowy slices and the Joseph ideal*. J. Eur. Math. Soc, 9(2007), N3, 487–543.

[P3] A. Premet. *Primitive ideals, non-restricted representations and finite W-algebras*. Mosc. Math. J., 7 (2007) N 4, 743–762.

[P4] A. Premet, *Commutative quotients of finite W-algebras*. Adv. Math. 225 (2010), N1, 269–306.

[P5] A. Premet, *Irreducible representations of Lie algebras of reductive groups and the Kac-Weisfeiler conjecture*, Invent. Math. 121(1995), 79–117.

[R] S. Riche, *Geometric braid group action on derived categories of coherent sheaves*. Represent. Theory 12 (2008) 131–169. With appendix by R. Bezrukavnikov and S. Riche.

[T] L. Topley, *A Morita theorem for modular finite W-algebras*. Math. Z. 285 (2017), N 3–4, 685–705.

Remarks on the Asymptotic Hecke Algebra

Alexander Braverman and David Kazhdan

Abstract Let G be a split reductive p-adic group, $I \subset G$ be an Iwahori subgroup, $\mathscr{H}(G)$ be the Hecke algebra and $\mathscr{C}(G) \supset \mathscr{H}(G)$ be the Harish-Chandra Schwartz algebra. The purpose of this note is to define (in spectral terms) a subalgebra $\mathscr{J}(G)$ of $\mathscr{C}(G)$, containing $\mathscr{H}(G)$, which we consider as an algebraic version of $\mathscr{C}(G)$. We show that the subalgebra $\mathscr{J}(G)^{I\times I} \subset \mathscr{J}(G)$ is isomorphic to the Lusztig's asymptotic Hecke algebra J and explain a relation between the algebra $\mathscr{J}(G)$ and the Schwartz space of the basic affine space studied in [2].

Keywords: Hecke algebras • p-adic groups

Mathematics Subject Classification (2010): 20C11, 22D10, 22D20

1 Introduction and statement of the results

1.1 Notation

Let F be a non-archimedean local field with ring of integers $\mathscr{O}$; we shall choose a generator π of the maximal ideal of $\mathscr{O}$. Typically, we shall denote algebraic varieties over F with boldface letters (e.g. $\mathbf{G}, \mathbf{X}$ etc.) and the corresponding sets of F-points – with corresponding Roman letters (e.g. G, X etc.).

Alexander Braverman
Department of Mathematics, University of Toronto, Perimeter Institute for Theoretical Physics and Skolkovo Institute for Science and Technology

David Kazhdan
Department of Mathematics, Hebrew University, Jerusalem, Israel

V. G. Kac, V. L. Popov (eds.), *Lie Groups, Geometry, and Representation Theory*, Progress in Mathematics 326, https://doi.org/10.1007/978-3-030-02191-7_4

In what follows we fix a connected split reductive group $\mathbf{G}$ over F with a Borel subgroup $\mathbf{B}$ it unipotent radical $\mathbf{U}$, maximal split torus $\mathbf{T} = \mathbf{B}/\mathbf{U}$. Let Λ be the lattice of cocharacters of $\mathbf{T}$ and $\Lambda^\vee$ be the lattice of characters of $\mathbf{T}$.

We write $K_0 = G(\mathscr{O})$ and denote by $I \subset K_0$ an Iwahori subgroup of G. We denote by $\mathscr{H}(G)$ the full Hecke algebra of G and by $\mathscr{H}(G,I)$ the Iwahori-Hecke subalgebra; we shall also denote by $\mathscr{H}_{\mathrm{aff}}$ the corresponding algebra over $\mathbb{C}[v,v^{-1}]$ (thus $\mathscr{H}(G,I)$ is obtained from $\mathscr{H}_{\mathrm{aff}}$ by specializing v to $q^{1/2}$). We denote by $\mathscr{M}(G)$ the category of smooth representations of G. This is the same as the category of locally unital modules over $\mathscr{H}(G)$. For any smooth representation (π,V) we denote by $(\pi^\vee,V^\vee)$ the subrepresentation of smooth vectors in the representation of G on the space of linear functionals on V. For any $v \in V, \lambda \in V^\vee$ we denote by $m_{v,\lambda} : G \to \mathbb{C}$ the matrix coefficient $m_{v,\lambda}(g) := \lambda(\pi(g)(v))$.

1.2 Matrix Paley-Wiener theorem

Let P be a parabolic subgroup of G with a Levi group M. The set X_M of unramified characters of M is equal to $\Lambda_M^\vee \otimes \mathbb{C}^\times$ where $\Lambda_M^\vee \subset \Lambda^\vee$ is the subgroup of characters of $\mathbf{T}$ trivial on $\mathbf{T} \cap [\mathbf{M},\mathbf{M}]$. So X_M has a structure of a complex algebraic variety; the algebra of polynomial functions on X_M is equal to $\mathbb{C}[\Lambda_M]$ where Λ_M is the lattice dual to $\Lambda_M^\vee$. We denote by $X_{M,t} \subset X_M$ the subset of unitary characters.

For any $(\sigma,V) \in \mathscr{M}(M)$ we denote by $i_{GP}(\sigma)$ the corresponding unitarily induced object of $\mathscr{M}(G)$. As a representation of K_0 this representation is equal to $\mathrm{ind}_{P\cap K_0}^{K_0}(\sigma)$. So for any unramified character $\chi : M \to \mathbb{C}^*$ the space of the representation $i_{GP}(\sigma \otimes \chi)$ is isomorphic to the space V_χ of the representation $i_{GP}(\sigma)$ and is independent on a choice of χ. Since X_M has a structure of an algebraic variety over $\mathbb{C}$ it makes sense to say that a family $\eta_\chi \in \mathrm{End}(V_\chi)$, $\chi \in \mathbf{T}^\vee$ is a regular (or a smooth) function of χ.

We denote by $Forg\colon \mathscr{M}(G) \to Vect$ the forgetful functor, by $\widetilde{\mathscr{E}(G)} = \{e(\pi)\}$ the ring of endomorphisms of $Forg$ and define $\mathscr{E}(G) \subset \widetilde{\mathscr{E}(G)}$ as the subring of endomorphisms η_π such that

1) For any Levi subgroup M of G and $\sigma \in Ob(\mathscr{M}(M))$, the endomorphisms $\eta_{i_{GP}(\sigma\otimes\chi)}$ are regular functions of χ.

2) There exists an open compact subgroup K of G such that η_π is $K \times K$-invariant for every π.

By definition, we have a homomorphism

$$PW\colon \mathscr{H}(G) \to \mathscr{E}(G), \quad f \mapsto \pi(f).$$

The following is usually called “the matrix Paley-Wiener theorem” (cf. [1], Theorem 25):

Theorem 1.3. *The map PW is an isomorphism.*

The group $G \times G$ acts on $\mathscr{E}(G)$ in the obvious way. We denote by $\mathscr{E}^I(G) \subset \mathscr{E}(G)$ the subring $I \times I$-invariant elements. It is clear that we can interpret $\mathscr{E}^I(G)$ as a subring of the ring of endomorphisms of the forgetful functor $Rep(\mathscr{H}(G,I)) \to Vect$.

1.4 Harish-Chandra algebra

Recall that for any $g \in G$, there exists a unique dominant coweight $\lambda(g)$ of $\mathbf{T}$ such that $g \in \mathbf{G}(\mathscr{O})\pi^{\lambda(g)}\mathbf{G}(\mathscr{O})$. Let us set $\Delta(g) = q^{\langle \lambda, \rho \rangle}$. Then we say that a function $f \colon G \to \mathbb{C}$ is a Schwartz function if

(a) There exists an open compact subgroup K of G such that f is both left and right K-invariant.

(b) For any polynomial function $p \colon G \to F$ and $n > 0$, there exists a constant $C = C_{p,n} \in \mathbb{R}_{>0}$ such that

$$\Delta(g)|f(g)| \leq C \ln^{-n}(1 + |p(g)|)$$

for all $g \in G$.

We denote by $\mathscr{C}(G)$ the space of all Schwartz functions. It is known that $\mathscr{C}(G)$ has an algebra structure with respect to convolution (cf. [7], Chapter 4 or [8]).

Obviously we have $\mathscr{H}(G) \subset \mathscr{C}(G)$.

For an open compact subgroup K of G we denote by $\mathscr{C}(G,K)$ the space of $K \times K$-invariants in $\mathscr{C}(G)$.

Below we recall the spectral description of $\mathscr{C}(G)$.

1.5 Tempered representations

Let (π, V) be a representation of G of finite length with central character $\eta : Z(G) \to \mathbb{C}^*$. Recall that π is called tempered if

(1) π is unitary. In particular, η is unitary (i.e., it takes values in $S^1 \subset \mathbb{C}^*$). In this case the absolute values $|m_{v,\lambda}|$ of matrix coefficients of V are functions on $G/Z(G)$.

(2) For any $\varepsilon > 0$ and any matrix coefficient $m_{v,\lambda}$ of π we have

$$|m_{v,\lambda}| \in L^{2+\varepsilon}(G/Z(G)).$$

The following facts are well-known (cf. [7], [8]):

(F1) Let π be a tempered representation of G. Then the action of $\mathscr{H}(G)$ extends naturally to an action of $\mathscr{C}(G)$.

(F2) Let P be a parabolic subgroup of G with a Levi group M. Let σ be a tempered irreducible representation of M. Then the representation $i_{GP}(\sigma)$ is tempered.

(F3) For a generic unitary character $\chi\colon M \to S^1$ the representation $i_{GP}(\sigma \otimes \chi)$ (which is tempered by F2) is irreducible.

We denote by $\mathscr{M}_t(G) \subset \mathscr{M}(G)$ the subcategory of tempered representations. As follows from (F2), for any tempered representation σ of M and a unitary character χ of M the representations $i_{GP}(\sigma \otimes \chi)$ belong to $\mathscr{M}_t(G)$.

Let $\mathscr{E}_t(G)$ be the subring of endomorphisms $\{\eta\}$ of the forgetful functor $Forg_t\colon \mathscr{M}_t(G) \to Vect$ such that

(1t) $\eta_{i_{GP}(\sigma\otimes\chi)}$ is a smooth function of $\chi \in X_{M,t}$ for any Levi subgroup M of G and $\sigma \in Ob(\mathscr{M}_t(M))$.

(2t) η is $K \times K$-invariant for some open compact subgroup of G.

The fact F1 can be upgraded to the following version of the matrix Paley-Wiener theorem (cf. [8]):

Theorem 1.6. *The map $f \mapsto \pi(f)$ defines an isomorphism between $\mathscr{C}(G)$ and $\mathscr{E}_t(G)$.*

We denote by $\mathscr{E}_t^I \subset \mathscr{E}_t(G)$ the subring of $I \times I$-invariant elements of $\mathscr{E}_t(G)$.

1.7 Asymptotic Hecke algebra

Recall that we denote by $\mathscr{H}_{\text{aff}}$ the "algebraic" version of $\mathscr{H}(G,I)$ which is an algebra over $\mathbb{C}[v,v^{-1}]$. Let us assume that $\mathbf{G}$ is of adjoint type. In [4], G. Lusztig defined the so-called asymptotic Hecke algebra J (we are going to recall the definition in Section 2). This is an algebra over $\mathbb{C}$ [1] and there is a canonical embedding $\mathscr{H}(G,I) \hookrightarrow J \otimes \mathbb{C}[v,v^{-1}]$ which becomes an isomorphism after some completion. Moreover, one can show that the specialization of this embedding to any $q \in \mathbb{C}^*$ is also injective. Hence we get an embedding $\mathscr{H}(G,I) \hookrightarrow J$.

One of the main purposes of this paper is to formulate and prove a version of the matrix Paley-Wiener theorem for J.

Let P be a parabolic subgroup with Levi group M. We say that an unramified character $\chi\colon M \to \mathbb{C}^*$ is (non-strictly) positive if for any coroot α of $\mathbf{G}$ such that the corresponding root subgroup lies in the unipotent radical $\mathbf{U_P}$ of $\mathbf{P}$ (which in particular defines a homomorphism $\alpha\colon F^* \to Z(M)$), we have $|\chi(\alpha(x))| \geq 1$ for $|x| \geq 1$.

Let $\mathscr{E}_{\mathscr{J}}(G)$ be the ring of collections

$$\{\eta_\pi \in \mathrm{End}_{\mathbb{C}}(V) \mid \text{for tempered irreducible } (\pi,V)\}$$

which extend to a rational function $E_{i_{GP}(\sigma\otimes\chi)} \in \mathrm{End}_{\mathbb{C}}(\sigma \otimes \chi)$ for every tempered irreducible representation σ of M and which are

a) regular on the set of characters χ such that χ^{-1} is (non-strictly) positive.

b) K-invariant for some open compact subgroup K of G.

As follows from the definition, we have an embedding $\mathscr{E}^I_{\mathscr{J}}(G) \to \mathscr{E}_t^I(G)$.

---

[1] In fact J can be defined over $\mathbb{Z}$ but to simplify the notation we shall always work over $\mathbb{C}$

Theorem 1.8. (1) *Let (π,V) be an irreducible tempered representation of G. Then the action the action of $\mathscr{H}(G,I)$ on V^I extends uniquely to J.*
(2) *Let P be a parabolic subgroup of G with Levi group M, let σ be an irreducible tempered representation of M, and let χ be a (non-strictly) positive character of M and $(\pi,V)=i_{GP}(\sigma\otimes\chi^{-1})$. Then the action of $\mathscr{H}(G,I)$ on V^I extends to an action of J. This extension is unique up to isomorphism.*
(3) *The map $f\mapsto\pi(f)$ defines an isomorphism between J and ${\mathscr{E}^I}_{\mathscr{J}}(G)$.*

It follows immediately from Theorem 1.3, Theorem 1.6 and Theorem 1.8 that we have inclusions $\mathscr{H}(G,I)\subset J\subset\mathscr{C}(G,I)$. Theorem 1.8 allows giving the following

Definition 1.9. We define $\mathscr{J}(G)$ to be the preimage of $\mathscr{E}_{\mathscr{J}}(G)$ in $\mathscr{C}(G)$. Note that we have natural embeddings $\mathscr{H}(G)\subset\mathscr{J}(G)\subset\mathscr{C}(G)$.

The algebra $\mathscr{J}(G)$ can be thought of as a "beyond Iwahori" version of $\mathscr{J}(G,I)$. It follows again from Theorem 1.3 and Theorem 1.6 that we have the embeddings $\mathscr{H}(G)\subset\mathscr{J}(G)\subset\mathscr{C}(G)$. Also, this definition makes sense for any reductive G.

The proof of Theorem 1.8 is given in the next section; it is essentially an exercise on manipulating the results from [4], [5] and [6]. In Section 3 we also explain a connection between the algebra $\mathscr{J}(G)$ and the Schwartz space of the basic affine space studied in [2]. In the appendix we give some examples of elements of J (viewed as functions on G) for $G=SL(2,F)$.

1.10 An algebraic version

Let us explain a version of Theorem 1.8 which is algebraic in the sense that it doesn't use the notion of "positive real number". To emphasize this we are going to work not over $\mathbb{C}$ but over an arbitrary algebraically closed field $\mathtt{K}$ of characteristic 0. First, let us make the following definitions:

(1) A character $\chi\colon F^*\to\mathtt{K}^*$ is called special if $\chi(a)=q^{\frac{1}{2}r\cdot\mathrm{val}(a)}$ with $r\in\mathbb{Z}_{>0}$

(2) A representation of F^* on a $\mathtt{K}$-vector space V is called special if there is a non-zero vector in V on which F^* acts by a special character.

(3) Let G,P,M be as above. Then a representation σ of M is called special if there exists a positive coroot $\alpha\colon F^*\to Z(M)$ whose composition with σ is special

(4) A representation π of G is called quasi-tempered if for any parabolic P with Levi decomposition $P=MU$, the Jacquet functor $r_{GP}(\pi)$ is not a special representation of M.

(5) Let $\mathscr{E}_{\mathscr{J},\mathtt{K}}$ be the algebra of collections

$$\{\eta_\pi\in\mathrm{End}_{\mathbb{C}}(V)\mid\text{for quasi-tempered irreducible }(\pi,V)\}$$

which extend to a rational function $E_{i_{GP}(\sigma)}\in\mathrm{End}_{\mathbb{C}}(\sigma)$ for every non-special irreducible representation σ of M. For an open compact subgroup K of G we denote by $\mathscr{E}_{\mathscr{J},\mathtt{K}}(G,K)$ the algebra of $K\times K$-invariant elements in $\mathscr{E}_{\mathscr{J},\mathtt{K}}$.

With these definitions in mind we have

Theorem 1.11. *Let $J_{\mathbb{K}} = J_{\mathbb{Z}} \otimes \mathbb{K}$. Then*

(1) *Let (π, V) be an irreducible quasi-tempered representation of G. Then the action of $\mathscr{H}(G,I)$ on V extends uniquely to $J_{\mathbb{K}}$.*
(2) *Let P be a parabolic subgroup of G with Levi group M. Let σ be a non-special representation of M. Let $(\pi,V) = i_{GP}(\sigma)$. Then the action of $\mathscr{H}(G,I)$ on V^I extends uniquely to $J_{\mathbb{K}}$.*
(3) *The map $f \mapsto \pi(f)$ defines an isomorphism between $J_{\mathbb{K}}$ and $\mathscr{E}_{\mathscr{J},\mathbb{K}}(G,I)$.*

It is not difficult to show that $\mathscr{E}_{\mathscr{J},\mathbb{K}}$ is naturally isomorphic to some algebra of locally constant functions on G (where the algebra structure is given by convolution) which contains $\mathscr{H}_{\mathbb{K}}(G)$; we shall denote this algebra by $\mathscr{J}_{\mathbb{K}}(G)$. We shall not pursue the details in this paper.

1.12 Acknowledgements

We would like to thank R. Bezrukavnikov, G. Lusztig and V. Ostrik for numerous discussions and help during the preparation of this paper.

2 Proof of Theorem 1.8

In this section we shall assume that **G** is semi-simple and adjoint. Then the affine Weyl group $W_{\mathrm{aff}} = \Lambda \rtimes W$ is in general not a Coxeter group, but it still has a length function ℓ defined e.g. in Section 1.1 of [6].

2.1 Definition of J

We are keeping the notations of [4] and [6]. Let $\mathscr{A} = \mathbb{C}[v, v^{-1}]$ and $\mathscr{A}^+ = \mathbb{C}[v]$. We let as before $\mathscr{H}_{\mathrm{aff}}$ denote the corresponding affine Hecke algebra. It has bases $\{T_w, \widetilde{T}_w, C_w\}_{w \in W_{\mathrm{aff}}}$ where T_w is the standard basis of $\mathscr{H}_{\mathrm{aff}}$, $\widetilde{T}_w = v^{-\ell(w)} T_w$ and C_w is the Kazhdan-Lusztig basis which satisfies

$$C_w = \sum_{y \leq w} (-1)^{\ell(w)-\ell(y)} v^{\ell(w)-\ell(y)} P_{y,w}(v^{-2}) \widetilde{T}_y,$$

where $P_{y,w}$ is a polynomial.

For $x, y \in W_{\mathrm{aff}}$, let us set

$$C_x C_y = \sum_{z \in W_{\mathrm{aff}}} h_{x,y,z} C_z.$$

For any $z \in W_{\mathrm{aff}}$, we let $a(z)$ be the smallest integer such that $u^{a(z)} h_{x,y,z} \in \mathscr{A}^+$ for all x,y. Let $\gamma_{x,y,z} \in \mathbb{Z}$ be such that $u^{a(z)} h_{x,y,z} - \gamma_{x,y,z} \in u\mathscr{A}^+$. Then let J denote the $\mathbb{C}$-vector space with basis t_w, $w \in W_{\mathrm{aff}}$; it has a ring structure defined by

$$t_x t_y = \sum_{z \in W_{\mathrm{aff}}} \gamma_{x,y,z} t_z.$$

Let $k = \mathbb{C}((v^{-1}))$ with the natural v^{-1}-adic topology on it. Let also $\widehat{\mathscr{H}_{\mathrm{aff}}}$ be the completion of $\mathscr{H}_{\mathrm{aff}}$ consisting of all (possibly infinite) sums $\sum b_x C_x$ with $b_x \in \widehat{\mathscr{A}}$ such that $b_x \to 0$ when $\ell(x) \to \infty$ (it is clear that one can replace C_x by $\widetilde{T}_x$ in the above definition). In Section 2.4 of [4], G. Lusztig defines a homomorphism $\mathscr{H}_{\mathrm{aff}} \hookrightarrow J \otimes \mathscr{A}$ which gives rise to an isomorphism between $\widehat{\mathscr{H}_{\mathrm{aff}}}$ and $J \widehat{\otimes} k$ (the latter denotes some completed tensor product). Also, the above embedding makes $J \otimes \mathscr{A}$ into a finitely generated module over $\mathscr{H}_{\mathrm{aff}}$.

2.2 Representations of $\mathscr{H}_{\mathrm{aff}}$ and $\mathscr{H}(G,I)$

Set $J_{\mathscr{A}} = J \otimes \mathscr{A}$, $J_k = J \otimes k$, $\mathscr{H}_{\mathrm{aff},k} = \mathscr{H}_{\mathrm{aff}} \underset{\mathscr{A}}{\otimes} k$. It follows from the above that $\mathscr{H}_{\mathrm{aff},k}$ can be regarded as a subalgebra of J_k. In what follows we denote by $G^\vee$ the Langlands dual group of $\mathbf{G}$ (over $\mathbb{C}$).

Let (s,u,ρ) be a triple where

(a) $s \in G^\vee(\mathbb{C})$ is a semi-simple element,

(b) $u \in G^\vee(\mathbb{C})$ is a unipotent element such that $su = us$,

(c) ρ is an irreducible representation of the group of components of the centralizer $Z_{G^\vee}(s,u)$ of the pair (s,u).

Recall that in [3] Kazhdan and Lusztig define a representation $\mathsf{K}(s,u,\rho)$ of $\mathscr{H}_{\mathrm{aff}}$. This representation may be 0; we shall say that the triple (s,u,ρ) is admissible if $\mathsf{K}(s,u,\rho) \neq 0$. We shall denote by $\mathsf{K}(s,u,\rho,q)$ the specialization of $\mathsf{K}(s,u,\rho)$ to $v = q^{1/2}$ viewed as a representation of $\mathscr{H}(G,I)$. Moreover, if ρ is some (not necessarily irreducible) representation of the group of components of $Z_{G^\vee}(s,u)$, we denote by $\mathsf{K}(s,u,\rho,q)$ the direct sum of $\mathsf{K}(s,u,\rho_i)$ where ρ is the direct sum of irreducible representations ρ_i. Then the following facts are true:

(1) (cf. Theorem 7.12 in [3]) $\mathsf{K}(s,u,\rho,q)$ has a unique simple quotient which we shall denote by $L(u,s,\rho,q)$.
(2) (cf. Theorem 8.2 in [3]) Assume that s is compact. Then the corresponding representation $\mathsf{K}(s,u,\rho,q)$ of $\mathscr{H}(G,I)$ is tempered and irreducible. Moreover, every irreducible tempered representation of $\mathscr{H}(G,I)$ is isomorphic to $\mathsf{K}(s,u,\rho,q)$ for a unique admissible triple (s,u,ρ) with compact s.
(3) (cf. Theorem 6.2 in [3]) Let P be a parabolic subgroup of G with Levi group M and let $P^\vee$ and $M^\vee$ be the corresponding parabolic and Levi subgroups in $G^\vee$. Let also $\mathfrak{p}^\vee, \mathfrak{m}^\vee$ be their Lie algebras. Assume that $s,u \in M^\vee$. Let $Z_{M^\vee}(s,u)$

be the centralizer of (s,u) in $M^\vee$ and let $Z_{G^\vee}(s,u)$ be the centralizer of (s,u) in $G^\vee$.

Now let $(\mathfrak{g}^\vee/\mathfrak{p}^\vee)_u$ be the kernel of $1-u$ on $\mathfrak{g}^\vee/\mathfrak{p}^\vee$. Assume that q^{-i} is not an eigenvalue of s on $(\mathfrak{g}^\vee/\mathfrak{p}^\vee)_u$ for every $i>0$. Let $\mathsf{K}_M(s,u,\rho_j,q)$ denote the corresponding representation of M. Then $i_{GP}(\mathsf{K}_M(s,u,\rho,q))$ is isomorphic to $\mathsf{K}(s,u,\widetilde{\rho},q)$ where $\widetilde{\rho}=\mathrm{Ind}_{Z_M(s,u)/Z_M^0(s,u)}^{Z_G(s,u)/Z_G^0(s,u)}\rho$. In particular, this is true if s is of the form $s'\cdot\chi^{-1}$ where s' is compact and χ is a (non-strictly) positive character of M.

In [6] Lusztig proves the following result:

Theorem 2.3. *Let E be an irreducible representation of J and let $E_k=\rho\underset{\mathbb{C}}{\otimes}k$. Then there exists a unique triple (s,u,ρ) such that $E_k|_{\mathscr{H}_{\mathrm{aff},k}}$ is isomorphic to $\mathsf{K}(s,u,\rho)\underset{\mathbb{C}}{\otimes}k$ (we shall denote the latter $\mathscr{H}_{\mathrm{aff},k}$-module by $\mathsf{K}(s,u,\rho)_k$). Moreover, every admissible triple (s,u,ρ) arises in this way.*

Theorem 2.3 implies that we have a bijection between irreducible representations of J and admissible triples (s,u,ρ). For any such triple (s,u,ρ) we shall denote by $E(s,u,\rho)$ the corresponding irreducible representation of J.

Note that by specializing the embedding $\mathscr{H}_{\mathrm{aff}}\hookrightarrow J\otimes\mathscr{A}$ to $v=q^{1/2}$ we get a homomorphism $\mathscr{H}(G,I)\to J$ which is injective by Proposition 1.7 of [5]. We now claim the following:

Theorem 2.4. (1) *Let π be an irreducible tempered representation of $\mathscr{H}(G,I)$. Then π extends uniquely to an irreducible representation of J.*

(2) *Any module of the form $i_{GP}(\sigma\otimes\chi^{-1})$, where σ is an irreducible tempered representation of the Levi group M and χ is a (non-strictly) positive character of M, extends to an irreducible module over J. This extension is unique up to isomorphism.*

(3) *Let M be a Levi subgroup of G and let (s,u,ρ) be an admissible triple for u in $M^\vee$ with compact s. Then there exists a $J\otimes\mathbb{C}[\Lambda_M]$-module $\mathscr{M}(s,u,\rho)$ whose fiber at any non-strictly positive χ is isomorphic to the J-module from* (2).

In order to prove Theorem 2.4 we shall use the following result of N. Xi [9]:

Theorem 2.5. *Let (s,u,ρ) be an admissible triple. Then the restriction of $E(s,u,\rho)$ to $\mathscr{H}(G,I)$ has a unique irreducible quotient, which is isomorphic to $L(s,u,\rho,q)$. Moreover, any irreducible subquotient of the kernel of the map $E(s,u,\rho)|_{\mathscr{H}(G,I)}\to L(s,u,\rho,q)$ is not isomorphic to $L(s,u,\rho,q)$.*

This result implies the following:

Corollary 2.6. *The representation $E(s,u,\rho)|_{\mathscr{H}(G,I)}$ of $\mathscr{H}(G,I)$ is isomorphic to $\mathsf{K}(s,u,\rho,q)$.*

Proof. By definition we have an isomorphism $E(s,u,\rho)|_{\mathscr{H}(G,I)}\simeq\mathsf{K}(s,u,\rho)_k$. This obviously implies that

(a) $\dim E=\dim_k\mathsf{K}(s,u,\rho)_k=\dim_{\mathbb{C}}\mathsf{K}(s,u,\rho,q)$.

(b) There exists a non-zero homomorphism $\mathsf{K}(s,u,\rho,q) \to E(s,u,\rho)|_{\mathscr{H}(G,I)}$. According to [3], Theorem 7.12, the representation $\mathsf{K}(s,u,\rho,q)$ has a unique simple quotient $L(s,u,\rho,q)$. It now follows from Theorem 2.5 that the map $\mathsf{K}(s,u,\rho,q) \to E(s,u,\rho)|_{\mathscr{H}(G,I)}$ is surjective (indeed, otherwise it would land inside the maximal proper submodule of $E(s,u,\rho)|_{\mathscr{H}(G,I)}$ which doesn't contain $L(s,u,\rho,q)$ as a subquotient, while on the other hand, any non-zero image of $\mathsf{K}(s,u,\rho,q)$ has a quotient isomorphic to $L(s,u,\rho,q)$). Since the dimensions of these two modules are equal, it follows that the map $\mathsf{K}(s,u,\rho,q) \to E(s,u,\rho)|_{\mathscr{H}(G,I)}$ is an isomorphism. □

2.7 Proof of Theorem 2.4

Let π be an irreducible tempered representation of $\mathscr{H}(G,I)$. Then π is isomorphic to $\mathsf{K}(s,u,\rho,q)$ for some admissible triple (s,u,ρ) with compact s. Now Corollary 2.6 implies that π extends to an irreducible representation of J and any two such extensions are isomorphic as abstract J-modules. In other words, any two such extensions are conjugate by means of some $\mathrm{Aut}_{\mathscr{H}(G,I)}(\pi) = \mathbb{C}^{\times}$. Hence any two such extensions are equal. This proves assertion (1).

Assertion (2) follows in a similar way from statement (3) above Theorem 2.3. Finally, the module $\mathscr{M}(s,u,\rho)$ is constructed in the following way. Let $\mathscr{B}^{\vee}$ denote the flag variety of $G^{\vee}$ and let $\mathscr{B}^{\vee}_{s,u}$ denote the variety of (s,u)-fixed points on $\mathscr{B}^{\vee}$. Then it follows from [6] that J acts on the equivariant K-theory $K_{Z(M^{\vee})}(\mathscr{B}^{\vee}_{s,u})$ (here $Z(M^{\vee})$ denotes the center of $M^{\vee}$). Moreover, this K-theory has a natural action of the centralizer $Z_{M^{\vee}}(s,u)$ and we let $\mathscr{M}(s,u,\rho)$ denote its ρ-isotypic component with respect to $Z_{M^{\vee}}(s,u)$. It easily follows from the above that $\mathscr{M}(s,u,\rho)$ satisfies the requirements of (3).

2.8 Proof of Theorem 1.8

The first two assertions of Theorem 1.8 are exactly the two assertions of Theorem 2.4. So, it remains to prove the 3rd assertion.

It follows from part (3) of Theorem 2.4 that the map $J \to \mathscr{E}^{I}{}_{\mathscr{J}}(G)$ is well defined. We now need to prove that it is an isomorphism. First, given $h \in J$ for any P, M, σ as above, we can define $\eta_h(\chi) \in \mathrm{End}_{\mathbb{C}}(i_{GP}(\sigma \otimes \chi^{-1}))$ where χ is a positive unramified character of M; the fact that $\eta_h(\chi)$ depends rationally on χ follows immediately from the fact that the embedding $\mathscr{H}(G,I) \hookrightarrow J$ is an isomorphism over the generic point of the center of $\mathscr{H}(G,I)$. Thus we get an injective map $J \hookrightarrow \mathscr{E}^{I}{}_{\mathscr{J}}(G)$. We now want to prove that this map is also surjective.

For a unipotent element u in $G^{\vee}$ let $\mathscr{Z}_u$ denote the algebra of ad-invariant polynomial functions on the centralizer $Z_{G^{\vee}}(u)$ of u in $G^{\vee}$. Let $\mathscr{Z}$ denote the direct sum of all the $\mathscr{Z}_u$ (where the sum is taken over conjugacy classes of unipotent elements in

$G^\vee$). Then $\mathscr{Z}$ maps to the center of both J and $\mathscr{E}^I{}_{\mathscr{J}}(G)$ and both algebras are finitely generated modules over $\mathscr{Z}$.

To prove that the desired surjectivity holds, it is enough to prove that it holds modulo every maximal ideal of $\mathscr{Z}$. Let $\mathfrak{m}$ be such an ideal. Set $J_\mathfrak{m} = J/\mathfrak{m}J$, $\mathscr{E}_\mathfrak{m} = \mathscr{E}^I{}_{\mathscr{J}}(G)/\mathfrak{m}\mathscr{E}^I{}_{\mathscr{J}}(G)$. It is enough to prove that the map $J_\mathfrak{m} \to \mathscr{E}_\mathfrak{m}$ is surjective for every $\mathfrak{m}$. Let $(\pi_1, V_1), \ldots, (\pi_n, V_n)$ be all the different (non-isomorphic) representations of G which have the form $i_{GP}(\sigma \otimes \chi^{-1})$ with tempered σ and positive χ such that $V_i^I \neq 0$ and $\mathscr{Z}$ acts on V_i^I through the quotient by $\mathfrak{m}$. Then by definition $\mathscr{E}_\mathfrak{m}$ embeds into $\bigoplus_i \mathrm{End}_\mathbb{C}(V_i^I)$. On the other hand, $V_1^I, \ldots, V_n^I$ are non-isomorphic irreducible representations of $J_\mathfrak{m}$ and hence the map $J_\mathfrak{m} \to \bigoplus_i \mathrm{End}_\mathbb{C}(V_i^I)$ is surjective.

3 Connection to the Schwartz space of G/U

3.1 Digression on [2]

Let $\mathbf{U}$ be a maximal unipotent subgroup of $\mathbf{G}$ defined over F, $U = \mathbf{U}(F)$. Set $X = G/U$; it is endowed with a natural action of $G \times T$. Let us denote by $\mathscr{S}_c(X)$ the space of locally constant compactly supported functions on X, and also let $C^\infty(X)$ denote just the space of functions $f\colon X \to \mathbb{C}$ such that there exists an open compact subgroup K of G such that f is K-invariant. Let $G \times T$ act on these spaces in such a way that the action of G comes from the right action of G on X and the action of T comes from the right action of T on X twisted by the character $t \mapsto q^{\langle \mathrm{val}(t), \rho\rangle}$ of T where $\mathrm{val}\colon T \to \Lambda$ denotes the natural homomorphism. In [2] we have defined the Schwartz space $\mathscr{S}(G/U)$ of functions on the basic affine space G/U which contains $\mathscr{S}_c(G/U)$ and it is contained in $C^\infty(X)$ in the case when $\mathbf{G}$ is simply connected. Let us recall this definition.

The space X has unique up to scalar G-invariant measure and we denote by $L^2(X)$ the L^2-space with respect to this measure. When $\mathbf{G}$ is simply connected one can construct a natural action of the Weyl group W on $L^2(X)$ by unitary operators Φ_w which commute with $G \times T$. In order to define these operators it is enough to consider the case when $w = s_\alpha$ – a simple reflection (here α is a simple root of G). Let us recall this definition as it will be used in the future.

For a simple root α let $\mathbf{P}_\alpha \subset \mathbf{G}$ be the minimal parabolic subgroup of type α containing B. Let $\mathbf{B}_\alpha$ be the commutator subgroup of P_α, and denote $\mathbf{X}_\alpha := \mathbf{G}/\mathbf{B}_\alpha$. We have an obvious projection of homogeneous spaces $\pi_\alpha : \mathbf{X} \to \mathbf{X}_\alpha$. It is a fibration with the fiber $\mathbf{B}_\alpha/\mathbf{U} = \mathbb{A}^2 - \{0\}$.

Let $\overline{\pi}_\alpha\colon \overline{\mathbf{X}}^\alpha \to \mathbf{X}_\alpha$ be the relative affine completion of the morphism π_α. (So $\overline{\pi}_\alpha$ is the affine morphism corresponding to the sheaf of algebras $\pi_{\alpha*}(\mathscr{O}_\mathbf{X})$ on $\mathbf{X}_\alpha$.) Then $\overline{\pi}_\alpha$ has the structure of a 2-dimensional vector bundle; $\mathbf{X}$ is identified with the complement to the zero-section in $\overline{\mathbf{X}}^\alpha$. The $\mathbf{G}$-action on $\mathbf{X}$ obviously extends to $\overline{\mathbf{X}}^\alpha$; moreover, it is easy to see that the determinant of the vector bundle $\overline{\pi}_\alpha$ admits a canonical (up to a constant) $\mathbf{G}$-invariant trivialization, i.e., $\overline{\pi}_\alpha$ admits a unique up

to a constant $\mathbf{G}$-invariant fiberwise symplectic form ω_α. We will fix such a form for every α.

Obviously $L^2(X) = L^2(\overline{X}^\alpha)$. Thus we define $\Phi_\alpha = \Phi_{s_\alpha}$ to be equal to the Fourier transform in the fibers of $\overline{\pi}_\alpha$, corresponding to the identification of $\overline{X}^\alpha$ with the dual bundle by means of ω_α.

Then

$$\mathscr{S}(X) = \sum_{w \in W} \Phi_w(\mathscr{S}_c(X)).$$

We can extend the above definition to the case when $\mathbf{G}$ is not necessarily simply connected. First the definition of [2] works without any change in the case when $[\mathbf{G},\mathbf{G}]$ is simply connected. Now, given any connected reductive $\mathbf{G}$ there always exists an algebraic reductive group $\widetilde{\mathbf{G}}$ and a central torus $\mathbf{Z}$ in $\widetilde{\mathbf{G}}$ so that $\mathbf{G} = \widetilde{\mathbf{G}}/\mathbf{Z}$. We now denote by $\widetilde{X}$ the basic affine space for $\widetilde{G} = \widetilde{\mathbf{G}}(F)$ and we set $\mathscr{S}(X) = \mathscr{S}(\widetilde{X})^T$. With this definition most results of [2] extend word-by-word to any $\mathbf{G}$.

3.2 *Action of $\mathscr{J}(G)$ on $L^2(X)$*

By definition we have $\mathscr{S}(X) \subset L^2(X)$. We claim that $\mathscr{C}(G)$ acts on $L^2(X)$. Indeed, we have

$$L^2(X) = \bigoplus_{\theta :\, \mathbf{T}(\mathscr{O}) \to S^1} L^2(X)_\theta,$$

where $L^2(X)_\theta$ denotes the subspace of $L^2(X)$ on which $\mathbf{T}(\mathscr{O})$ acts by θ. Now, each $L^2(X)_\theta$ is a direct integral of G-representations of the form $i_{GB}(\chi)$ where χ is a unitary character of T over a compact base (isomorphic to $(S^1)^{\dim \mathbf{T}}$) and therefore it acquires a natural action of $\mathscr{C}(G)$ (since it acts on each $i_{GB}(\chi)$ with unitary χ).

In particular, the algebra $\mathscr{J}(G)$ acts on $L^2(X)$. The following conjecture provides an alternative definition of $\mathscr{S}(X)$.

Conjecture 3.3. *We have $\mathscr{S}(X) = \mathscr{J}(G) \cdot \mathscr{S}_c(X)$.*

Remark. We claim that Conjecture 3.3 is equivalent to saying that $\mathscr{S}(X) = \mathscr{J}(G)_U$ where the latter means U-coinvariants with respect to the right action of U on $\mathscr{J}(G)$ (note that $\mathscr{J}(G)$ is a G-bimodule, since it contains $\mathscr{H}(G)$ as a subalgebra). Indeed, let us assume Conjecture 3.3. Then we can define a map $\zeta : \mathscr{J}(G) \to \mathscr{S}(X)$ by sending every $f \in \mathscr{J}(G)$ to $\int_U f(gu)du$. The fact that the action of $\mathscr{J}(G)$ on $\mathscr{S}_c(X)$ is well-defined guarantees that this integral is convergent; in fact we have $\zeta(f) = f \star \delta_{K/K\cap U}$ for a sufficiently small open compact subgroup K of G (here $\delta_{K/K\cap U}$ denotes the multiple of the characteristic function of $K/K\cap U \subset G/U$ normalized by the condition that its integral over G/U is equal to 1; it is easy to see that the result is independent of the choice of K if we require that f is K-invariant). Also Conjecture 3.3 guarantees that $f \star \delta_{K/K\cap U} \in \mathscr{S}(X)$. It is clear that ζ factorizes through $\mathscr{J}(G)_U$ and the resulting map $\mathscr{J}(G)_U \to \mathscr{S}(X)$ is injective. On the other hand, the restriction of ζ to $\mathscr{H}(G)$ defines a surjective map

$\mathscr{H}(G) \to \mathscr{S}_c(X)$. Hence, by definition we have that ζ is a surjective map from $\mathscr{J}(G)$ to $\mathscr{J}(G)\cdot\mathscr{S}_c(X) = \mathscr{S}(X)$. Thus we have proved that Conjecture 3.3 implies that $\mathscr{S}(X) = \mathscr{J}(G)_U$. The converse statement obvious, since as a $\mathscr{J}(G)$-module the space $\mathscr{J}(G)_U$ is clearly generated by $\mathscr{H}(G)_U = \mathscr{S}_c(X)$.

We can prove the following weaker version of Conjecture 3.3:

Theorem 3.4. *We have* $(\mathscr{S}(X))_\theta = (\mathscr{J}(G)\cdot\mathscr{S}_c(X))_\theta$ *where either the character* $\theta\colon \mathbf{T}(\mathscr{O}) \to S^1$ *is trivial or the composition of* θ *with any coroot is non-trivial.*

Proof. Let $\mathscr{S}_c = \mathscr{S}_c(X), \mathscr{S} = \mathscr{S}(X), \mathscr{S}' = \mathscr{J}(G)\cdot\mathscr{S}_c(X)$. We want to show the equality $\mathscr{S}_\theta = \mathscr{S}'_\theta$ for θ as above.

Step 1. Let us first show that $\mathscr{S}^I = (\mathscr{S}')^I$. As before, let $G^\vee$ denote the Langlands dual group of $\mathbf{G}$ over $\mathbb{C}$ and let $\mathscr{B}$ be its flag variety. According to [4] the algebra J decomposes as a direct sum of subalgebras J_u numbered by unipotent elements $u \in G^\vee$ up to conjugacy. We denote by J_0 the summand corresponding to the unit conjugacy class. We claim the action on J on $L^2(X)^I$ factorizes through the projection on J_0. This is obvious since $L^2(X)^I$ is a torsion-free module over the center of $\mathscr{H}(G,I)$ and every J_u with non-trivial u is annihilated by a non-zero ideal of the center.

According to [10] the algebra J_0 is naturally isomorphic to the $K_{G^\vee}(\mathscr{B}\times\mathscr{B})$ (here $K_{G^\vee}(?)$ stands for the complexified Grothendieck group of $G^\vee$-equivariant coherent sheaves on ?). On the other hand, let $\mathscr{K} = K_{T^\vee\times\mathbb{C}^*}(\mathscr{B})$ and let $\mathscr{K}_q$ be its specialization at $v = q^{1/2}$ where $K_{\mathbb{C}^*}(pt) = \mathbb{C}[v,v^{-1}]$ (the action of $\mathbb{C}^*$ on $\mathscr{B}$ trivial). According to Section 5 of [2] the space $\mathscr{K}$ has a natural action of $\mathscr{H}_{\text{aff}}$ and hence $\mathscr{K}_q$ has a natural action of $\mathscr{H}(G,I)$. Moreover, we have an isomorphism $\mathscr{S}^I \simeq \mathscr{K}_q$ which identifies $\mathscr{S}^I_c$ with the submodule generated by the skyscraper κ at some $T^\vee$-invariant point $e \in \mathscr{B}^\vee$. On the other hand $J_0 \otimes \mathscr{A} = K_{G^\vee\times\mathbb{C}^*}(\mathscr{B}\times\mathscr{B})$ clearly acts on $\mathscr{K}$ (and this action is compatible with the $\mathscr{H}_{\text{aff}}$-action with respect to the homomorphism $\mathscr{H}_{\text{aff}} \to J_0$ – this is proved in [10]); moreover the action of $K_{G^\vee\times\mathbb{C}^*}(\mathscr{B}\times\mathscr{B})$ on κ defines an isomorphism $K_{G^\vee\times\mathbb{C}^*}(\mathscr{B}\times\mathscr{B}) \simeq \mathscr{K}$. Hence the same is true after specialization to $v = q^{1/2}$. We see that J_0 acts on $\mathscr{K}_q = \mathscr{S}^I$ and the latter is generated as a module by an element of $\mathscr{S}^I_c$. This implies the equality $\mathscr{S}^I = (\mathscr{S}')^I$.

Step 2. Let $\mathscr{S}_0, \mathscr{S}'_0$ denote the G-module of coinvariants of $\mathscr{S}$ (resp. $\mathscr{S}'$) with respect to $\mathbf{T}(\mathscr{O})$. Then both are subrepresentations of $C^\infty(X)_0$. Recall that if a G-module V is generated by I-fixed vectors then two G-submodules W_1 and W_2 of V coincide if and only if $W_1^I = W_2^I$. Applying this to $C^\infty(X)_0$, $W_1 = \mathscr{S}_0$, $W_2 = \mathscr{S}'_0$ and using Step 1 we get the equality $\mathscr{S}_0 = \mathscr{S}'_0$.

Step 3. Let $\theta\colon \mathbf{T}(\mathscr{O}) \to S^1$ be a character. For a T-module V let V_θ denote the corresponding space of $(\mathbf{T}(\mathscr{O}),\theta)$-coinvariants. Let us prove that $\mathscr{S}_\theta = \mathscr{S}'_\theta$ assuming that the composition of θ with any coroot is a non-trivial character of $\mathscr{O}^*$. We shall refer to such θ as "regular".

In this case it is obvious from the definition of Φ_w that for any simple coroot α the operator Φ_{s_α} defines an isomorphism between $\mathscr{S}_{c,\theta}$ and $\mathscr{S}_{c,s_\alpha(\theta)}$. Indeed,

any $(\mathbf{T}(\mathscr{O}),\theta)$-equivariant function automatically vanishes on $\overline{X}^{\alpha}\backslash X$ (recall the notation of Section 3.1) and the same is true for θ replaced with $s_\alpha(\theta)$. Since the notion of regularity is W-invariant it follows that for any $w\in W$ the operator Φ_w defines an isomorphism between $\mathscr{S}_{c,\theta}$ and $\mathscr{S}_{c,w(\theta)}$. On the other hand, we claim that $\mathscr{S}'_\theta=\mathscr{S}_{c,\theta}$. For this it is enough to prove that $\mathscr{S}_{c,\theta}$ is $\mathscr{J}(G)$-invariant. This would follow if we knew that for any character $\chi\colon T\to\mathbb{C}^*$ such that $\chi|_{\mathbf{T}(\mathscr{O})}=\theta$ the action of $\mathscr{H}(G)$ on the space $\mathscr{S}_{c,\chi}$ of (T,χ)-coinvariants on $\mathscr{S}_c$ extends to $\mathscr{J}(G)$. For any χ as above we can find an element $w\in W$ such that $w(\chi)$ is non-negative. Hence by definition $\mathscr{J}(G)$ acts on $\mathscr{S}_{c,w(\chi)}=i_{GB}(w(\chi))\simeq i_{GB}(\chi)$. □

3.5 The parabolic case

Let P be a parabolic subgroup of G with a Levi subgroup M and unipotent radical U_P. Let $X_P=G/U_P$. This space has a natural $G\times M$ action. Therefore the space $\mathscr{S}_c(X)$ of locally constant compactly supported functions on X_P becomes a $G\times M$ module; for convenience we are going to twist the M action by the square root of the absolute value of the determinant of the M-action on the Lie algebra $\mathfrak{u}_P$ of U_P.

As before, we can define the space $L^2(X_P)$. For the same reason as before it has an action of $\mathscr{C}(G)$. We now define $\mathscr{S}(X_P):=\mathscr{J}(G)\cdot\mathscr{S}_c(X_P)\subset L^2(X_P)$. Equivalently $\mathscr{S}(X_P)=\mathscr{J}(G)_{U_P}$.

Conjecture 3.6. *Let P and Q be two associate parabolic subgroups, i.e., two parabolic subgroups with the same Levi subgroup M. Then there exists a $G\times M$-equivariant unitary isomorphism $\Phi_{P,Q}\colon L^2(X_P)\tilde{\to}L^2(X_Q)$ whose restriction to $\mathscr{S}(X_P)$ defines an isomorphism between $\mathscr{S}(X_P)$ and $\mathscr{S}(X_Q)$.*

4 Appendix: an $SL(2)$-example

4.1

The purpose of this appendix is to show how the algebra J gets realized inside the locally constant functions on G for the case when $G=SL(2,F)$. Let St denote the Steinberg representation of G. Let also $\mathscr{S}$ denote the space of locally constant compactly supported functions on F^2. This space has an action of G and a commuting action of the Fourier transform Φ and of the torus $T=F^\times$. Similarly, we let $\mathscr{S}_c$ denote the space of locally constant compactly supported functions on $F^2\backslash\{0\}$. It follows from the above results that

$$J=\operatorname{End}(St^I)\oplus J_0,\quad \text{where } J_0=\operatorname{End}_{\Phi,F^\times}(\mathscr{S}^I). \tag{4.1}$$

The embedding $\mathscr{H}(G,I) \hookrightarrow J$ is given by the action of $\mathscr{H}(G,I)$ on St^I and on $\mathscr{S}^I$. Since the first summand in (4.1) is a one-dimensional subspace that does not lie in $\mathscr{H}(G,I)$, it follows that the projection of $\mathscr{H}(G,I)$ to J_0 is an embedding. Moreover, it is clear that the codimension of $\mathscr{H}(G,I)$ in J_0 is at most 1. Indeed, an element of J_0 comes from an element of $\mathscr{H}(G,I)$ if and only if it sends $\mathscr{S}_c^I$ to $\mathscr{S}_c^I$. The quotient $\mathscr{S}/\mathscr{S}_c$ is naturally isomorphic to $\mathbb{C}$ (the map is given by evaluating a function at 0) with $F^\times$-action given by the character $x \mapsto q^{|x|}$ (recall the convention about the torus action from Section 3.1). Denote this character by χ and let $\mathscr{S}_\chi$ (resp. $(\mathscr{S}_c)_\chi$) denote the space of $(F^\times, \chi)$-coinvariants on $\mathscr{S}$ (resp. on $\mathscr{S}_c$). Let U denote the image of $(\mathscr{S}_c)_\chi$ in $\mathscr{S}_\chi$. Then it is clear that the quotient $J_0/\mathscr{H}(G,I)$ embeds into $\mathrm{Hom}_T(U^I, \mathbb{C})$. However, U is actually isomorphic to St, hence $\dim U^I = 1$. This shows that $\dim J_0/\mathscr{H}(G,I)$ is either 0 or 1 and in the latter case it is isomorphic to $\mathbb{C} \otimes St^I$ as an $\mathscr{H}(G,I)$-bimodule.

We would like to show that $J_0/\mathscr{H}(G,I)$ is indeed one-dimensional. Note that this will imply that $\dim J/\mathscr{H}(G,I) = 2$. For this it is enough to find one $I \times I$-invariant function f on G such that

(a) f does not have compact support but belongs to $\mathscr{C}(G,I)$,

(b) The action of f on $\mathscr{S}^I$ is well-defined and it is given by a non-zero operator.

We are going to find such a function explicitly. In fact, the function f will be $K \times I$-invariant, where $K = SL(2,\mathscr{O})$. We begin by some generalities.

4.2 General remarks and volume calculation

Let G be a totally disconnected group, $x \in G$, and let K_1, K_2 be open compact subgroups of G; set $X = K_1 x K_2$. Let V be a smooth representation of G, and $v \in V$. Fix a Haar measure on G. Then we have an element e_X in the Hecke algebra of G. By definition, we have

$$e_X(v) = \int\limits_{g \in X} g(v)dg.$$

This formula can be rewritten in the following way, which will be crucial in the future:

$$e_X(v) = \frac{\mathrm{vol}(X)}{\mathrm{vol}(K_1)\cdot \mathrm{vol}(K_2)} \int\limits_{g_1 \in K_1, g_2 \in K_2} (g_1 x g_2)(v) dg_1 dg_2. \tag{4.2}$$

In particular, let us assume that the Haar measure is chosen in such a way that $\mathrm{vol}(K_2) = 1$ and that v is K_2-invariant. Then (4.2) takes the form

$$e_X(v) = \frac{\mathrm{vol}(X)}{\mathrm{vol}(K_1)} \int\limits_{g \in K_1} g(x(v))dg. \tag{4.3}$$

We now want to apply this formula to the case when $G = SL(2,F)$, $K = K_1 = SL(2,\mathscr{O})$, $K_2 = I$ (so, in particular, we fix the Haar measure so that the volume of I is 1). Let t be a uniformizer of F and let $x_n = \begin{pmatrix} t^n & 0 \\ 0 & t^{-n} \end{pmatrix}$ where $n \in \mathbb{Z}$ and let $X_n = Kx_nI$.

Lemma 4.3. *We have*

$$\frac{\mathrm{vol}(X_n)}{\mathrm{vol}(K)} = \begin{cases} q^{2n-1} \text{ if } n \geq 0, \\ q^{-2n} \text{ if } n < 0. \end{cases}$$

Proof. Let $H_n \subset K$ be the subgroup of all $h \in K$ such that $x_n^{-1}hx_n \in I$. Then $\mathrm{vol}(X_n) = \#(K/H_n)$. Recall that $\mathrm{vol}(K) = \#(K/I) = (q+1)$.

If $h = \begin{pmatrix} a & b \\ c & d \end{pmatrix}$ then $x_n^{-1}hx_n = \begin{pmatrix} a & t^{-2n}b \\ t^{2n}c & d \end{pmatrix}$. Hence, if $n > 0$, then $x_n^{-1}hx_n \in I$ iff $b \in t^{2n}\mathscr{O}$ and the cardinality of K/H_n is $q^{2n-1}(q+1)$. If $n \leq 0$ then the condition is $c \in t^{-2n+1}\mathscr{O}$ and the cardinality of K/H_n is $q^{-2n}(q+1)$. □

We now pass to the function f.

Proposition 4.4. *Let*

$$\gamma_n = \begin{cases} q^{2n} \text{ if } n \leq 0, \\ -q^{-2n+1} \text{ if } n > 0. \end{cases} \tag{4.4}$$

Let $f = \sum_{n\in\mathbb{Z}} \gamma_n \cdot \chi_{X_n}$ where χ_{X_n} denotes the characteristic function of X_n. Then f belongs to J_0 (and it obviously does not belong to $\mathscr{H}(G,I)$).

Corollary 4.5. *As a bimodule over $\mathscr{H}(G,I)$ the quotient $J/\mathscr{H}(G,I)$ is isomorphic to $V \otimes St^I$ where V is a 2-dimensional representation of $\mathscr{H}(G,I)$ which is a non-trivial extension of $\mathbb{C}$ (trivial representation) by St^I.*

Proof. We have already explained above that Proposition 4.4 implies that $\dim J/\mathscr{H}(G,I) = 2$. Moreover, it follows from (4.1) that this quotient contains $St^I \otimes St^I$ as an $\mathscr{H}(G,I)$-submodule and we have explained at the end of Section 4.1 that the quotient of $J/\mathscr{H}(G,I)$ by $St^I \otimes St^I$ is isomorphic to $\mathbb{C} \otimes St^I$ as an $\mathscr{H}(G,I)$-bimodule. Hence we see that $J/\mathscr{H}(G,I)$ is isomorphic to $V \otimes St^I$ where V is a 2-dimensional representation of $\mathscr{H}(G,I)$ which is an extension of $\mathbb{C}$ by St^I. If that extension were trivial, then f would have be G-invariant on the left modulo functions with compact support. This would imply that γ_n is constant for $|n| >> 0$, which is obviously not the case (and it is also clear that no element of $\mathscr{C}(G)$ could satisfy this). □

Let us now turn to the proof of Proposition 4.4. It is easy to see that f belongs to $\mathscr{C}(G)$. Also since f is K-invariant on the left, its projection to the first summand of (4.1) is 0. Hence to show that it belongs to J_0 it is enough to show that its action on $\mathscr{S}^I$ is well-defined. For this, we are going to take f of the form $\sum \gamma_n\chi_{X_n}$ with arbitrary coefficients γ_n and see what conditions we need to impose of the coefficients so the action of f on $\mathscr{S}^I$ is well-defined. To check the latter, it is enough to verify that $f \star \chi_{\mathscr{O}\oplus\mathscr{O}}$ and $f \star \chi_{\mathscr{O}\oplus t\mathscr{O}}$ are well-defined.

4.5.1 The case of $\chi_{\mathcal{O}\oplus\mathcal{O}}$

Let us first compute the action of f on $\chi_{\mathcal{O}\oplus\mathcal{O}}$. It is enough to compute the value of the action of χ_{X_n} on $\chi_{\mathcal{O}\oplus\mathcal{O}}$ at $(t^{-r},0)$ where $r\in\mathbb{Z}$. It is equal to $\frac{\mathrm{vol}(X_n)}{\mathrm{vol}(K)}\cdot\mathrm{vol}(K_{n,r})$ where $K_{n,r}=\{h\in K|\ h(t^{-r},0)\in t^n\mathcal{O}\oplus t^{-n}\mathcal{O}\}$.

Assume first that $n>0$. Then we have the following cases:

(1) $r>n$. Then the result is 0.
(2) $r\leq -n$. Then $K_{n,r}=K$ and the result is $(q+1)q^{2n-1}$.
(3) $-n<r\leq n$. In this case if $h=\begin{pmatrix} a & b\\ c & d\end{pmatrix}$ then $h\in K_{n,r}$ iff $a\in t^{n+r}\mathcal{O}$ and thus $\mathrm{vol}(K_{n,r})=q^{-n-r+1}$. Hence the value is q^{n-r}.

On the other hand, assume that $n=-m, m>0$. Then we have the following cases:

(1) $r>m$ – the value is 0
(2) $r\leq -m$. Then again $K_{n,r}=K$ and we get $(q+1)q^{2m}$.
(3) $-m<r\leq m$. Then $h\in K_{n,r}$ iff $c\in t^{r+m}\mathcal{O}$ and its volume is q^{-r-m+1}. Hence the value is q^{m-r+1}.

If $n=0$ then we just get $(q+1)\chi_{\mathcal{O}\oplus\mathcal{O}}$. Altogether, the value of $f\star\chi_{\mathcal{O}\oplus\mathcal{O}}$ at $(t^{-r},0)$ for $r>0$ is

$$\sum_{n\geq r}\gamma_n q^{n-r}+\sum_{m\geq r}\gamma_{-m}q^{-m-r+1}.$$

Hence if we want this to be 0 (which we definitely want at least for r large enough) we need

$$\gamma_r+q\gamma_{-r}=0. \tag{4.5}$$

4.5.2 The case of $\chi_{\mathcal{O}\oplus t\mathcal{O}}$

In this case we define $K'_{n,r}=\{h\in K\mid h(t^{-r},0)\in t^n\mathcal{O}\oplus t^{-n+1}\mathcal{O}\}$ and we need to repeat the above calculation with $K_{n,r}$ replaced by $K'_{n,r}$.

Let us first assume that $n>0$. Then we have the following cases:

(1) $r<n-1$ – get 0
(2) $r\leq -n$ – get $(q+1)\frac{\mathrm{vol}(X_n)}{\mathrm{vol}(K)}\gamma_n=(q+1)q^{2n-1}\gamma_n$
(3) $n<r\leq n-1$
In this case $K'_{n,r}$ is given by the condition $a\in t^{n+r}\mathcal{O}$ and the value is

$$q^{-n-r+1}q^{2n-1}\gamma_n=q^{n-r}\gamma_n.$$

On the other hand, assume that $n\leq 0$ and set $n=-m$. Then we have the following cases:

(1) $r>m$ – get 0
(2) $r\leq -m-1$ – get $(q+1)q^{2m}\gamma_{-m}$.

(3) $-m \leq r \leq m$. Then $K'_{n,r}$ is given by the condition $c \in t^{r+m+1}\mathcal{O}$ and the volume of $K'_{n,r}$ is q^{-r-m}. Thus we get $q^{-r-m}q^{2m}\gamma_{-m} = q^{m-r}\gamma_{-m}$.

Thus for $r \geq 0$ we get that the value of $f \star \chi_{\mathcal{O}\oplus t\mathcal{O}}$ at $(t^{-r}, 0)$ is equal to

$$\sum_{n \geq r+1} q^{n-r}\gamma_n + \sum_{m \geq r} q^{m-r}\gamma_{-m}.$$

This will be 0 for all $r \geq 0$ if

$$q\gamma_{r+1} + \gamma_{-r} = 0. \tag{4.6}$$

4.6 *The function f*

Altogether it is now clear that if γ_n is given by (4.4) then f satisfies both (4.5) and (4.6).

Let us now compute the action of the resulting function f on both $\chi_{\mathcal{O}\oplus\mathcal{O}}$ and $\chi_{\mathcal{O}\oplus t\mathcal{O}}$. In the first case, we already know that $f \star \chi_{\mathcal{O}\oplus\mathcal{O}}(t^{-r}, 0) = 0$ if $r > 0$. On the other hand, for $r \leq 0$ we have

$$\begin{aligned} f \star \chi_{\mathcal{O}\oplus\mathcal{O}}(t^{-r}, 0) &= (q+1) \;+ \sum_{0<n\leq -r} (q+1)q^{2n-1}\left(-q^{-2n+1}\right) \\ &\quad + \sum_{n>-r} q^{n-r}\left(-q^{-2n+1}\right) \;+ \sum_{0<m\leq -r} (q+1)q^{2m}q^{-2m} \\ &\quad + \sum_{m>-r} q^{m-r+1}q^{-2m} \\ &= q+1, \end{aligned}$$

which shows that $f \star \chi_{\mathcal{O}\oplus\mathcal{O}} = (q+1)\chi_{\mathcal{O}\oplus\mathcal{O}}$. A similar calculation shows that $f \star \chi_{\mathcal{O}\oplus t\mathcal{O}} = 0$.

References

[1] J. Bernstein, *Draft of : Representations of p-adic groups*, Fall 1992. Lectures by Joseph Bernstein, Written by Karl E. Rumelhart.

[2] A. Braverman and D. Kazhdan, *On the Schwartz space of the basic affine space*, Selecta Math. (N.S.) **5** (1999), no. 1, 1–28.

[3] D. Kazhdan and G. Lusztig, *Proof of the Deligne-Langlands conjecture for Hecke algebras*, Invent. math. **87**, 153–215 (1987)

[4] G. Lusztig, *Cells in affine Weyl groups. II.* J. Algebra **109** (1987), no. 2, 536–548.

[5] G. Lusztig, *Cells in affine Weyl groups. III.* J. Fac. Sci. Univ. Tokyo Sect. IA Math. **34** (1987), no. 2, 223–243.

[6] G. Lusztig, *Cells in affine Weyl groups. IV.* J. Fac. Sci. Univ. Tokyo Sect. IA Math. **36** (1989), no. 2, 297–328.

[7] A. J. Silberger, *Introduction to Harmonic Analysis on Reductive P-Adic Groups*, Based on Lectures by Harish-Chandra at The Institute for Advanced Study, 1971–73. Princeton University Press, 1979.

[8] J.-L. Waldspurger, *La formule de Plancherel pour les groupes p-adiques (d'après Harish-Chandra)*. J. Inst. Math. Jussieu, 2(2):235–333, 2003

[9] N. Xi, *Representations of affine Hecke algebras and based rings of affine Weyl groups.* J. Amer. Math. Soc. **20** (2007), no. 1, 211–217.

[10] N. Xi, *The based ring of the lowest two-sided cell of an affine Weyl group, III*, arXiv:1506.00476.

Autoequivalences of Tensor Categories Attached to Quantum Groups at Roots of 1

Alexei Davydov, Pavel Etingof, and Dmitri Nikshych

To the memory of Bertram Kostant

Abstract We compute the group of braided tensor autoequivalences and the Brauer-Picard group of the representation category of the small quantum group $\mathfrak{u}_q(\mathfrak{g})$, where q is a root of unity.

Keywords: tensor category • quantum group • autoequivalence • braiding • quantum Frobenius

Mathematics Subject Classification (2010): 17B37, 81R50, 20G42, 18D10

1 Introduction

Let k be an algebraically closed field of characteristic zero. Let G be a simple algebraic group over k, and let $\mathfrak{g} = \mathsf{Lie}(G)$ be its Lie algebra. Let q be a root of unity of odd order coprime to 3 if G is of type G_2, and coprime to the determinant of the Cartan matrix of G. Let $\mathfrak{u}_q(\mathfrak{g})$ be Lusztig's small quantum group attached to $\mathfrak{g}$ [Lu1]. Then $\mathfrak{u}_q(\mathfrak{g})$ is a quasitriangular Hopf algebra, so the category of its finite-

---

The work of P.E. was partially supported by NSF grant DMS-1502244.

The work of D.N. was partially supported by NSA grant H98230-16-1-0008.

Alexei Davydov
Department of Mathematics, Ohio University, Athens, OH 45701, USA,
e-mail: alexei1davydov@gmail.com

Pavel Etingof
Department of Mathematics, Massachusetts Institute of Technology, Cambridge, MA 02139, USA,
e-mail: etingof@math.mit.edu

Dmitri Nikshych
Department of Mathematics and Statistics, University of New Hampshire, Durham, NH 03824, USA, e-mail: dmitri.nikshych@unh.edu

V. G. Kac, V. L. Popov (eds.), *Lie Groups, Geometry, and Representation Theory*,
Progress in Mathematics 326, https://doi.org/10.1007/978-3-030-02191-7_5

dimensional representations $\mathsf{Rep}\,\mathfrak{u}_q(\mathfrak{g})$ is a finite braided tensor category [EGNO]. One of the main goals of this paper is to compute the Picard group $\mathsf{Pic}(\mathsf{Rep}\,\mathfrak{u}_q(\mathfrak{g}))$ of this category, i.e., the group of equivalence classes of invertible $\mathsf{Rep}\,\mathfrak{u}_q(\mathfrak{g})$-module categories. Picard groups of braided tensor categories and, in particular, Brauer-Picard groups of tensor categories play a crucial role in classification of graded extensions [ENO] and also appear as symmetry groups of three-dimensional topological field theories [FS]. It is known that $\mathsf{Pic}(\mathsf{Rep}\,\mathfrak{u}_q(\mathfrak{g}))$ is isomorphic to the group $\mathsf{Aut}^{\mathrm{br}}(\mathsf{Rep}\,\mathfrak{u}_q(\mathfrak{g}))$ of braided autoequivalences of $\mathsf{Rep}\,\mathfrak{u}_q(\mathfrak{g})$ [DN, ENO]. We show under some restrictions on q that $\mathsf{Aut}^{\mathrm{br}}(\mathsf{Rep}\,\mathfrak{u}_q(\mathfrak{g}))$ is isomorphic to the group $\mathsf{Aut}(\mathfrak{g})$ of automorphisms of $\mathfrak{g}$, i.e., $\mathsf{Aut}^{\mathrm{br}}(\mathsf{Rep}\,\mathfrak{u}_q(\mathfrak{g})) = \Gamma \ltimes G^{\mathrm{ad}}$, where G^{ad} is the adjoint group of G and $\Gamma = \Gamma_{\mathfrak{g}}$ is the automorphism group of the Dynkin diagram of $\mathfrak{g}$. Namely, we prove this when the order ℓ of q is sufficiently large and also for classical groups $G = SL_N, Sp_N, SO_N$ if $\ell > N$.

Moreover, we show that $\mathsf{Rep}\,\mathfrak{u}_q(\mathfrak{g})$ has only two braidings (the standard one and its reverse) and deduce that any tensor autoequivalence of $\mathsf{Rep}\,\mathfrak{u}_q(\mathfrak{g})$ is necessarily braided. Thus, the group of tensor autoequivalences (also known as the group of biGalois objects) of $\mathsf{Rep}\,\mathfrak{u}_q(\mathfrak{g})$ is isomorphic to $\Gamma \ltimes G^{\mathrm{ad}}$. This generalizes the result of Bichon [Bi2], who proved this fact for $\mathfrak{g} = \mathfrak{sl}_2$.

We also consider the braided tensor category $\mathcal{O}_q(G)\text{-}\mathsf{comod}$ of finite-dimensional comodules over the function algebra $\mathcal{O}_q(G)$, which is the G-equivariantization of $\mathsf{Rep}\,\mathfrak{u}_q(\mathfrak{g})$. We show that every braided autoequivalence of $\mathcal{O}_q(G)\text{-}\mathsf{comod}$ comes from a Dynkin diagram automorphism if ℓ is sufficiently large, and prove a similar result in the non-braided case. This generalizes a result of Neshveyev and Tuset [NT1, NT2], who proved this when q is not a root of unity. We also show this for the classical groups SL_N, Sp_N, SO_N if $\ell > N$.

As a tool, we introduce the notion of a finitely dominated tensor category. We show that the category of comodules over a finitely presented Hopf algebra is finitely dominated and prove that tensor autoequivalences of a finitely dominated category that preserve a tensor generator form an algebraic group. While this theory plays an auxiliary role in our paper, it may be of independent interest.

We expect that the main results of this paper extend without significant difficulties to roots of unity of arbitrary order ℓ, not necessarily satisfying the above coprimeness assumptions (at least when ℓ is sufficiently large). However, this would require some important modifications in the statements. Notably, if $\ell = 2dr$, where $d = 1,2,3$ is the the ratio of squared norms of long and short roots of G, and G is simply connected, then $\mathcal{O}_q(G)\text{-}\mathsf{comod}$ is a G^L-equivariantization (rather than a G-equivariantization) of an appropriate version of $\mathsf{Rep}\,\mathfrak{u}_q(\mathfrak{g})$, where G^L is the Langlands dual group of G, see [AG]. Therefore, we expect that in this case $\mathsf{Aut}^{\mathrm{br}}(\mathsf{Rep}\,\mathfrak{u}_q(\mathfrak{g})) \cong \Gamma \ltimes G^L$ (note that by definition $G^L = (G^L)^{\mathrm{ad}}$).

The paper is organized as follows. In Section 2 we give preliminaries and auxiliary results. In Section 3 we develop the theory of finitely dominated tensor categories and study groups of tensor autoequivalences of such categories. In Section 4 we classify tensor autoequivalences of $\mathcal{O}_q(G)\text{-}\mathsf{comod}$. Finally, in Section 5 we prove that any tensor autoequivalence of $\mathsf{Rep}\,\mathfrak{u}_q(\mathfrak{g})$ is braided and classify such autoequivalences. As a consequence, we compute the Brauer-Picard groups of $\mathsf{Rep}\,\mathfrak{u}_q(\mathfrak{g})$ and $\mathsf{Rep}\,\mathfrak{u}_q(\mathfrak{b})$.

Acknowledgements. We are very grateful to Julien Bichon, Ken Brown, David Kazhdan, George Lusztig, Sergey Neshveyev, Victor Ostrik, Noah Snyder, and Milen Yakimov for useful discussions.

2 Preliminaries and auxiliary results

Let k be an algebraically closed field. In this paper we consider tensor categories over k [EGNO, Definition 4.1.1] which we will simply refer to as *tensor categories*. The most basic example of a tensor category is the trivial tensor category Vec, i.e., the category of finite-dimensional vector spaces over k. More generally, given a Hopf k-algebra H, the category $\mathsf{Rep}\,H$ of finite-dimensional left H-modules and the category $H\text{-}\mathsf{comod}$ of finite-dimensional left H-comodules are examples of tensor categories. For a tensor category $\mathscr{C}$, let $\mathsf{Aut}(\mathscr{C})$ denote the group of isomorphism classes of tensor autoequivalences of $\mathscr{C}$.

2.1 Braided tensor categories and their Picard groups

We refer the reader to [EGNO, Chapter 8] for basic definitions related to braided tensor categories. Let $\mathscr{C}$ be a braided tensor category with braiding $c_{X,Y} : X \otimes Y \to Y \otimes X$. The *reverse* braiding of c is, by definition, $c^{\text{rev}}_{X,Y} := c^{-1}_{Y,X}$, $X, Y \in \mathscr{C}$. We will denote by $\mathscr{C}^{\text{rev}}$ the tensor category $\mathscr{C}$ with the reverse braiding. Let $\mathsf{Aut}^{\text{br}}(\mathscr{C})$ denote the group of isomorphism classes of braided tensor autoequivalences of $\mathscr{C}$.

Let $\mathscr{Z}(\mathscr{C})$ denote the Drinfeld center of $\mathscr{C}$. Then the assignment $X \mapsto (X, c_{-,X})$ is a braided embedding (i.e., a fully faithful braided tensor functor) $\mathscr{C} \hookrightarrow \mathscr{Z}(\mathscr{C})$. Similarly, the assignment $X \mapsto (X, c^{-1}_{X,-})$ is a braided embedding $\mathscr{C}^{\text{rev}} \hookrightarrow \mathscr{Z}(\mathscr{C})$. These embeddings combine together into a single braided tensor functor

$$\mathscr{C} \boxtimes \mathscr{C}^{\text{rev}} \to \mathscr{Z}(\mathscr{C}). \tag{1}$$

Assume in addition that $\mathscr{C}$ is finite (i.e., has finitely many simple objects and enough projective objects). We say that $\mathscr{C}$ is *factorizable* if the functor (1) is an equivalence.

Lemma 2.1. *Let $\mathscr{C}$ be a factorizable braided tensor category with braiding c. Suppose that $\mathscr{C}$ is not pointed and that $\mathscr{C}$ has no proper non-trivial tensor subcategories. Then $\mathscr{C}$ has exactly two braidings: c and c^{rev}.*

Proof. A braiding on $\mathscr{C}$ is the same thing as a section of the forgetful functor $F : \mathscr{Z}(\mathscr{C}) \to \mathscr{C}$, i.e., a tensor subcategory $\widetilde{\mathscr{C}} \subset \mathscr{Z}(\mathscr{C})$ such that $F|_{\widetilde{\mathscr{C}}} : \widetilde{\mathscr{C}} \to \mathscr{C}$ is an equivalence. By the hypothesis, $\widetilde{\mathscr{C}} \cap (\mathscr{C} \boxtimes \mathsf{Vec})$ is either $\mathscr{C} \boxtimes \mathsf{Vec}$ or Vec. The former case corresponds to the original braiding c. Let us deal with the latter case. We will argue that $\widetilde{\mathscr{C}} \cap (\mathsf{Vec} \boxtimes \mathscr{C}^{\text{rev}})$ is non-trivial, and, hence, $\widetilde{\mathscr{C}} = \mathsf{Vec} \boxtimes \mathscr{C}^{\text{rev}}$, which corresponds to the reverse braiding c^{rev}.

Any simple object of $\mathscr{Z}(\mathscr{C}) \cong \mathscr{C} \boxtimes \mathscr{C}^{\text{rev}}$ is of the form $X \boxtimes Y$, where X, Y are simple objects of $\mathscr{C}$. If $X \boxtimes Y$ is in $\widetilde{\mathscr{C}}$ then $(X^* \otimes X) \boxtimes \mathbf{1} \in \widetilde{\mathscr{C}}$ since it is a subquotient of $(X \boxtimes Y)^* \otimes (X \boxtimes Y)$. Thus, $(X^* \otimes X) \boxtimes \mathbf{1} \in \widetilde{\mathscr{C}} \cap (\mathscr{C} \boxtimes \mathsf{Vec})$, so X must be invertible. Similarly, $\mathbf{1} \boxtimes (Y^* \otimes Y) \in \widetilde{\mathscr{C}} \cap (\mathsf{Vec} \boxtimes \mathscr{C}^{\text{rev}})$. Since $\widetilde{\mathscr{C}}$ is non-pointed, one can choose a non-invertible Y. Therefore, $\widetilde{\mathscr{C}} \cap (\mathsf{Vec} \boxtimes \mathscr{C}^{\text{rev}}) \neq \mathsf{Vec}$, as required. □

Let $\mathscr{C}$ be a finite braided tensor category. By definition, the *Picard group* $\mathsf{Pic}(\mathscr{C})$ of $\mathscr{C}$ [ENO, DN] is the group of equivalence classes of invertible $\mathscr{C}$-module categories. When $\mathscr{C}$ is factorizable, there is a canonical group isomorphism

$$\mathsf{Pic}(\mathscr{C}) \cong \mathsf{Aut}^{\text{br}}(\mathscr{C}), \tag{2}$$

see [ENO, Theorem 6.2] and [DN, Corollary 4.6].

Thus, computing the Picard group of a factorizable braided tensor category $\mathscr{C}$ amounts to computing its group of braided tensor autoequivalences.

Corollary 2.2. *Let $\mathscr{C}$ be a factorizable braided tensor category such that $\mathscr{C} \not\cong \mathscr{C}^{\text{rev}}$. Suppose that $\mathscr{C}$ is not pointed and that $\mathscr{C}$ has no proper non-trivial tensor subcategories. Then any tensor autoequivalence of $\mathscr{C}$ is automatically braided and*

$$\mathsf{Pic}(\mathscr{C}) \cong \mathsf{Aut}(\mathscr{C}). \tag{3}$$

Proof. Follows from Lemma 2.1. □

Finally, recall that the *Müger center* of a braided tensor category $\mathscr{C}$ is the tensor subcategory of all objects $X \in \mathscr{C}$ such that for each $Y \in \mathscr{C}$, the squared braiding on $X \otimes Y$ is the identity. This subcategory is symmetric.

2.2 Algebraic group actions on categories, equivariantization and de-equivariantization

Let us recall the construction of [AG]. Let G be an abstract group. An action of G on a category $\mathscr{C}$ [EGNO, 2.7] is a collection of functors $T_g \colon \mathscr{C} \to \mathscr{C}$ attached to each $g \in G$ such that $T_1 = \mathrm{Id}_{\mathscr{C}}$, equipped with functorial isomorphisms $\gamma_{g,h} \colon T_g \circ T_h \to T_{gh}$ satisfying the 2-cocycle condition $\gamma_{f,gh} \circ T_f(\gamma_{g,h}) = \gamma_{fg,h} \circ \gamma_{f,g}$. Given such an action, a G-equivariant object in $\mathscr{C}$ is an object X together with a collection of isomorphisms $u_g \colon T_g(X) \to X$ such that $u_{gh} \circ \gamma_{g,h} = u_g \circ T_g(u_h)$, and the G-equivariantization $\mathscr{C}^G$ of $\mathscr{C}$ is the category of G-equivariant objects in $\mathscr{C}$. If $\mathscr{C}$ is monoidal or braided, then we require that the functors T_g be monoidal, respectively braided, and $\gamma_{g,h}$ preserve the tensor (respectively, braided) structure, in which case the equivariantization $\mathscr{C}^G$ inherits the same structure.

For an affine group scheme G over k, let $\mathscr{O}(G)$ denote the algebra of regular functions on G (i.e., the coordinate Hopf algebra), and let $\mathscr{O}(G)\text{-}\mathsf{mod}$ denote the category of $\mathscr{O}(G)$-modules. If $\mathscr{C}$ is artinian and G is finite, then we can represent the collection of functors $\{T_g\}$ as a single functor $T\colon \mathscr{C} \to \mathscr{O}(G)\text{-}\mathsf{mod} \boxtimes \mathscr{C}$, where the Deligne tensor product $\mathscr{O}(G)\text{-}\mathsf{mod} \boxtimes \mathscr{C}$ may be interpreted as the category of $\mathscr{O}(G)$-modules in $\mathscr{C}$. Namely, $T_g(X)$ is the fiber of $T(X)$ at $g \in G$, and $T(X) = \bigoplus_{g\in G} T_g(X)$ (note that $\mathscr{O}(G)\text{-}\mathsf{mod} \cong \mathsf{Vec}_G$ in the case of finite G). Then the isomorphisms $\gamma_{g,h}$ are also combined into a single isomorphism $\gamma\colon (\mathrm{Id} \otimes T) \circ T \cong (m^* \otimes \mathrm{Id}) \circ T$, where

$$m^*\colon \mathsf{Vec}_G = \mathscr{O}(G)\text{-}\mathsf{mod} \to \mathscr{O}(G)\text{-}\mathsf{mod} \boxtimes \mathscr{O}(G)\text{-}\mathsf{mod} = \mathscr{O}(G) \otimes \mathscr{O}(G)\text{-}\mathsf{mod} = \mathsf{Vec}_{G\times G}$$

is the functor of sheaf-theoretic pullback under the multiplication map $m\colon G \times G \to G\colon m^*(M)_{g,h} = M_{gh}$. Similarly, the morphisms u_g are combined into a single morphism $u\colon T(X) \to T_{\mathrm{triv}}(X)$, where T_{triv} is the functor attached to the trivial G-action on $\mathscr{C}$. If $\mathscr{C}$ is monoidal or braided, we require that T be a tensor (respectively, braided) functor, where the tensor product in $\mathscr{O}(G)\text{-}\mathsf{mod} \boxtimes \mathscr{C}$ is over $\mathscr{O}(G)$, and that γ preserve these structures in an appropriate sense.

In this form, the definitions of an action and equivariantization make sense when G is an *affine algebraic* group (as they formalize the requirements that the functors T_g and morphisms $\gamma_{g,h}$ depend algebraically on the group elements). Namely, in this case $\mathscr{O}(G)$ stands for the algebra of regular functions on G, Vec_G is replaced by the category of quasicoherent sheaves QCoh_G, and $m^*(M) = \mathscr{O}(G \times G) \otimes_{\mathscr{O}(G)} M$, where $\mathscr{O}(G)$ embeds into $\mathscr{O}(G \times G) = \mathscr{O}(G) \otimes \mathscr{O}(G)$ via the coproduct (induced by the product in G). Also, $T(X)$ should be required to be an $\mathscr{O}(G)$-module in the ind-completion of $\mathscr{C}$, rather than in $\mathscr{C}$ itself, and it is no longer the direct sum of T_g (since $\mathscr{O}(G)$ may be infinite dimensional and non-semisimple). With these definitions, if $\mathscr{C}$ is a finite tensor category then $\mathscr{C}^G$ is a tensor category (in general, not finite), and if $\mathscr{C}$ is braided, then so is $\mathscr{C}^G$ (provided that the G-action preserves the tensor structure and the braiding). Moreover, $\mathsf{Rep}\, G$ sits as a tensor subcategory in $\mathscr{C}^G$ (namely, the category of equivariant objects which are multiples to $\mathbf{1}$ as objects of $\mathscr{C}$), and in the braided case this subcategory is contained in the Müger center of $\mathscr{C}^G$ (i.e., the squared braiding of an object of $\mathsf{Rep}\, G$ with any object of $\mathscr{C}^G$ is the identity).

Finally, given a tensor category $\mathscr{D}$ together with a braided tensor functor $\mathsf{Rep}\, G \to \mathscr{Z}(\mathscr{D})$ which gives rise to an inclusion $\mathsf{Rep}\, G \hookrightarrow \mathscr{D}$ (for an affine algebraic group G), we can define the de-equivariantization $\mathscr{D}_G$ of $\mathscr{D}$ to be the category of finitely generated $\mathscr{O}(G)$-modules in the ind-completion of $\mathscr{D}$, where $\mathscr{O}(G)$ is the algebra of regular functions on G equipped with the action of G by left (or, alternatively, right) translations [EGNO, 8.23]. Moreover, if $\mathscr{D}$ is braided and $\mathsf{Rep}\, G$ lies in the Müger center of $\mathscr{D}$, then $\mathscr{D}_G$ is braided and carries a G-action, and we have $(\mathscr{D}_G)^G \cong \mathscr{D}$. Conversely, for a braided tensor category $\mathscr{C}$ with a G-action we have $(\mathscr{C}^G)_G \cong \mathscr{C}$, i.e., equivariantization and de-equivariantization are inverses of each other (this fact is essentially proved in [AG] and can also be obtained by adjusting arguments of [DGNO, Section 4] to the infinite setting).

2.3 Quantum groups at roots of unity

Let $\mathrm{char}(k) = 0$. Let G be a simple algebraic group over k and let $\mathfrak{g}$ be the associated Lie algebra. Let $\mathscr{O}(G)$ denote the coordinate Hopf algebra of G and let $\mathscr{O}_q(G)$ denote its quantized form, see [BG, KS].

Let ℓ be an odd integer, relatively prime to the determinant of the Cartan matrix of G and to 3 if $\mathfrak{g}$ is of type G_2. Let q be a primitive ℓ-th root of unity in k. Let $\mathfrak{u}_q(\mathfrak{g})$ be the small quantum group, i.e., the Frobenius-Lusztig kernel [Lu1]. Recall [T, XI.6.3] that $\mathfrak{u}_q(\mathfrak{g})$ is a factorizable quasitriangular Hopf algebra. Also, it was shown in [DL] that there is a cocleft central exact sequence of Hopf algebras

$$k \to \mathscr{O}(G) \to \mathscr{O}_q(G) \to \mathfrak{u}_q(\mathfrak{g})^* \to k. \tag{4}$$

Moreover, the pullback of the coquasitriangular structure of $\mathfrak{u}_q(\mathfrak{g})^*$ (dual to the universal R-matrix of $\mathfrak{u}_q(\mathfrak{g})$) to $\mathscr{O}_q(G)$ defines a coquasitriangular structure on $\mathscr{O}_q(G)$, giving $\mathscr{O}_q(G)\text{-}\mathsf{comod}$ the structure of a braided tensor category such that the forgetful functor $\mathscr{O}_q(G)\text{-}\mathsf{comod} \to \mathfrak{u}_q(\mathfrak{g})^*\text{-}\mathsf{comod} = \mathsf{Rep}\,\mathfrak{u}_q(\mathfrak{g})$ is braided.

It follows that there is a natural action of G on $\mathsf{Rep}\,\mathfrak{u}_q(\mathfrak{g}) = \mathfrak{u}_q(\mathfrak{g})^*\text{-}\mathsf{comod}$ as a braided category. Furthermore, it follows from [AG, AGP] that with respect to this action $\mathscr{O}_q(G)\text{-}\mathsf{comod}$ is equivalent to the G-equivariantization of $\mathsf{Rep}\,\mathfrak{u}_q(\mathfrak{g})$, which can, in turn, be recovered as the de-equivariantization of $\mathscr{O}_q(G)\text{-}\mathsf{comod}$:

$$\mathscr{O}_q(G)\text{-}\mathsf{comod} = (\mathsf{Rep}\,\mathfrak{u}_q(\mathfrak{g}))^G, \quad \mathsf{Rep}\,\mathfrak{u}_q(\mathfrak{g}) = (\mathscr{O}_q(G)\text{-}\mathsf{comod})_G. \tag{5}$$

More precisely, [AG] considers the case when q is a root of unity of order $2dr$, where $d = 1,2,3$ is the ratio of the squared norm of long roots of G to the squared norm of short roots. In this case the role of $\mathscr{O}(G)$ is played by $\mathscr{O}(G^L)$, where G^L is the Langlands dual group to G, but the arguments of [AG] apply without significant changes to our case.

It is easy to see that a maximal torus $T \subset G$ acts on $\mathsf{Rep}\,\mathfrak{u}_q(\mathfrak{g})$ by Hopf algebra automorphisms of $\mathfrak{u}_q(\mathfrak{g})$ (i.e., by conjugation). Hence, the center of G acts trivially on $\mathsf{Rep}\,\mathfrak{u}_q(\mathfrak{g})$ (as it is contained in T, and conjugation by a central element induces the identity automorphism). On the other hand, the action of G is non-trivial, as $\mathscr{O}_q(G)\text{-}\mathsf{comod}$ is not equivalent to $\mathsf{Rep}\,u_q(\mathfrak{g}) \boxtimes \mathsf{Rep}\,G$. Thus, we get

Proposition 2.3. *There is an inclusion* $G^{\mathrm{ad}} \hookrightarrow \mathsf{Aut}^{\mathrm{br}}(\mathsf{Rep}\,\mathfrak{u}_q(\mathfrak{g}))$.

Remark 2.4. In the case of $\mathfrak{g} = \mathfrak{sl}_n$ the inclusion $G^{\mathrm{ad}} \hookrightarrow \mathsf{Aut}(\mathsf{Rep}\,\mathfrak{u}_q(\mathfrak{g}))$ was established by Bichon [Bi2], who also proved that this inclusion is an isomorphism for $n = 2$.

We also have the following (well known) lemmas.

Lemma 2.5. $\mathfrak{u}_q(\mathfrak{g})$ *has no non-trivial Hopf quotients. In particular, it has no non-trivial central grouplike elements.*

Proof. The second statement follows since ℓ is coprime to the determinant of the Cartan matrix of $\mathfrak{g}$.

To prove the first statement, recall that $\mathfrak{u}_q(\mathfrak{g})$ is a pointed Hopf algebra, and let $H := \mathrm{gr}\,\mathfrak{u}_q(\mathfrak{g})$ be its associated graded Hopf algebra under the coradical filtration (it is defined by the same generators e_i, f_i, K_i and the same relations as $\mathfrak{u}_q(\mathfrak{g})$, except that now $[e_i, f_i] = 0$ for all i). Then H^* is a pointed Hopf algebra generated in degree 1. Hence, any proper Hopf ideal $I \neq 0$ in $\mathfrak{u}_q(\mathfrak{g})$ must contain a nonzero element of degree 1 under the coradical filtration. Hence I must contain a nonzero $(1,g)$-skew-primitive element for some grouplike element g. If it is trivial, i.e., is a multiple of $g-1$, then I must contain e_i for some i (since g is not central and thus acts on some e_i with eigenvalue $\neq 1$). Thus, in any case I contains a non-trivial $(1,g)$-skew-primitive element, say e_i. Since

$$[e_i, f_i] = \frac{K_i - K_i^{-1}}{q - q^{-1}},$$

we have $K_i - K_i^{-1} \in I$. Thus $K_i - 1 \in I$ (as ℓ is odd). Thus $f_i \in I$ and $e_j, f_j \in I$ for j connected to I in the Dynkin diagram of G (as K_i acts on these elements with eigenvalues $\neq 1$). Continuing in this way, we will get that $e_i, f_i, K_i - 1 \in I$ for all i (as the Dynkin diagram of G is connected). Hence I is the augmentation ideal. This implies the required statement. □

Lemma 2.6. *The only tensor automorphism of the identity functor of* $\mathrm{Rep}\,\mathfrak{u}_q(\mathfrak{g})$ *is the identity.*

Proof. This follows from Lemma 2.5, since any such automorphism is defined by a central grouplike element. □

Finally, we will need the following lemma. Let P be the weight lattice of G, P_+ its dominant part, and L_λ be the simple $\mathscr{O}_q(G)$-comodule with highest weight $\lambda \in P_+$. Recall that $\mathscr{O}(G)\text{-comod} = \mathrm{Rep}\,G \subset \mathscr{O}_q(G)\text{-comod}$ is the semisimple subcategory whose simple objects are $L_{\ell\lambda}$ for $\lambda \in P_+$ ([Lu2, DL]).

Lemma 2.7. *If λ is not divisible by ℓ, then the matrix elements of L_λ and L_λ^* generate $\mathscr{O}_q(G/C_\lambda)$, where C_λ is a central subgroup of G.*

Proof. Let $H \subset \mathscr{O}_q(G)$ be the Hopf subalgebra generated by the matrix elements of L_λ and L_λ^*. It is clear that the $\mathscr{O}_q(G)$-comodule $L_{\ell\lambda}$ (which is a G-module) is a subquotient of $L_\lambda^{\otimes \ell}$. Let $C = C_\lambda \subset G$ be the kernel of the action of G on $L_{\ell\lambda}$. Then the matrix elements of $L_{\ell\lambda}$ generate $\mathscr{O}(G/C)$, so $\mathscr{O}(G/C) \subset H$. Let H_1 be the fiber of the $\mathscr{O}(G/C)$-module H at $1 \in G/C$, a finite-dimensional Hopf algebra. It is clear that $H \subset \mathscr{O}_q(G/C)$, and the fiber of $\mathscr{O}_q(G/C)$ at $1 \in G/C$ is $\mathfrak{u}_q(\mathfrak{g})^*$, so H_1 is a Hopf subalgebra of $\mathfrak{u}_q(\mathfrak{g})^*$, hence H_1^* is a Hopf quotient of $\mathfrak{u}_q(\mathfrak{g})$. Also $H_1 \neq k$ since λ is not divisible by ℓ and hence $H \neq \mathscr{O}(G/C)$. Since by Lemma 2.5 $\mathfrak{u}_q(\mathfrak{g})$ has no non-trivial Hopf quotients, we get $H_1 = \mathfrak{u}_q(\mathfrak{g})^*$. Thus, the de-equivariantization $(H\text{-comod})_{G/C} \subset (\mathscr{O}_q(G/C)\text{-comod})_{G/C} = \mathrm{Rep}\,\mathfrak{u}_q(\mathfrak{g})$ is actually the entire category $\mathrm{Rep}\,\mathfrak{u}_q(\mathfrak{g})$. Hence, $H\text{-comod} = (\mathrm{Rep}\,\mathfrak{u}_q(\mathfrak{g}))^{G/C} = \mathscr{O}_q(G/C)\text{-comod}$ and $H = \mathscr{O}_q(G/C)$, as desired. □

2.4 *Compatibility of tensor functors on comodule categories with vector space dimensions*

Proposition 2.8. *Let H be a finitely generated Hopf algebra over k of slower than exponential growth (e.g., of finite GK dimension). Then*

(i) *for any fiber functor $F : H\text{-comod} \to \mathrm{Vec}$ one has $\dim F(X) \geq \dim X$;*
(ii) *for any tensor autoequivalence $E : H\text{-comod} \to H\text{-comod}$ one has $\dim E(X) = \dim X$.*

Proof. (i) We have $\mathrm{length}(X^{\otimes n}) \leq \dim F(X)^n$, so there is a simple composition factor Y_n in $X^{\otimes n}$ which has dimension $d_n \geq (\dim X / \dim F(X))^n$. The matrix elements of Y_n span a space of dimension d_n^2 and are noncommutative polynomials of degree $\leq n$ of the matrix elements of X. So if $\dim X / \dim F(X) > 1$, then H has exponential growth.

(ii) Let F be the usual fiber functor on $H\text{-comod}$. Then $F \circ E$ is another fiber functor, so by (i) $\dim E(X) = \dim F(E(X)) \geq \dim X$. Also, the same is true for E^{-1}. Hence, $\dim E(X) = \dim X$. □

Remark 2.9. If H has exponential growth, then both parts of Proposition 2.8 may fail. Indeed, Bichon showed in [Bi1] that for any integer $n \geq 2$ there exists a Hopf algebra H_n such that $H_n\text{-comod} = \mathrm{Rep}\,SL_2(k)$, so that the 2-dimensional irreducible $SL_2(k)$-module corresponds to an n-dimensional H_n-comodule (namely, $H_2 = \mathscr{O}(SL_2(k))$ but H_n has exponential growth for $n \geq 3$). Now the usual fiber functor F of $\mathrm{Rep}\,SL_2(k)$ on $H_n\text{-comod}$ for $n \geq 3$ gives a counterexample to (i), and the autoequivalence of the category $H_m \otimes H_n\text{-comod}$, $m \neq n$ switching the factors gives a counterexample to (ii).

2.5 *Basic properties of tensor autoequivalences of* $\mathscr{O}_q(G)\text{-comod}$

Let F be a tensor autoequivalence of $\mathscr{O}_q(G)\text{-comod}$. In this section we prove some basic properties of F.

Recall that a tensor category $\mathscr{C}$ is said to be *tensor-generated* by its object X if every object of $\mathscr{C}$ is a subquotient of a direct sum of tensor powers of X.

Lemma 2.10. *F induces an autoequivalence of $\mathscr{O}(G)\text{-comod} = \mathrm{Rep}\,G$.*

Proof. By Lemma 2.7, if $Y \in \mathcal{O}_q(G)\text{-comod}$ and $Y \notin \mathrm{Rep}\,G$, then $Y \oplus Y^*$ tensor-generates a nonsemisimple category. Hence, if $X \in \mathrm{Rep}\,G$ is a simple object, then $F(X) \in \mathrm{Rep}\,G$ (as $F(X) \oplus F(X)^*$ tensor-generates a semisimple category). The same holds for F^{-1}. This implies the statement. □

By the results of [NT2], the restriction of F to $\mathrm{Rep}\,G$ belongs to the group $\mathrm{Out}\,G \ltimes H^2(Z_G^\vee, k^\times)$, where $\mathrm{Out}\,G$ is the group of outer automorphisms of G and Z_G is the center of G (this uses the theorem of McMullen that any automorphism of the Grothendieck semiring of $\mathrm{Rep}\,G$ comes from an automorphism of G). On the other hand, the group $\mathrm{Out}\,G \ltimes H^2(Z_G^\vee, k^\times)$ acts naturally on $\mathcal{O}_q(G)\text{-comod}$. So composing F with an element of this group if needed, we may assume that $F|_{\mathrm{Rep}\,G} \cong \mathrm{Id}$.

Remark 2.11. If F is braided, then another way to prove Lemma 2.10 is to note that $\mathrm{Rep}\,G$ is the Müger center of $\mathcal{O}_q(G)\text{-comod}$ (since $\mathfrak{u}_q(\mathfrak{g})$ is a factorizable Hopf algebra, $\mathrm{Rep}\,\mathfrak{u}_q(\mathfrak{g})$ is a factorizable braided tensor category and so its Müger center is trivial), hence must be preserved by F.

Moreover, in this case by the uniqueness of a fiber functor of a Tannakian category [DM], $F|_{\mathrm{Rep}\,G}$ is given by an outer automorphism of G. Thus, by composing F with such an automorphism if needed, we may assume that $F|_{\mathrm{Rep}\,G} \cong \mathrm{Id}$ (i.e., we do not have to use [NT2]).

Proposition 2.12. *For any finite-dimensional $\mathcal{O}_q(G)$-comodule V we have* $\dim F(V) = \dim V$.

Proof. This follows from Proposition 2.8(ii), since $\mathcal{O}_q(G)$ has GK dimension $\dim G$ (as it is module-finite over $\mathcal{O}(G)$). □

Proposition 2.13. *If F is a tensor autoequivalence of $\mathcal{O}_q(G)\text{-comod}$ such that $F|_{\mathrm{Rep}\,G} = \mathrm{Id}$, then $F(L) \cong L$ for each simple object $L \in \mathcal{O}_q(G)\text{-comod}$.*

Proof. Let $T \subset G$ be a maximal torus, and $W = N(T)/T$ the Weyl group. The character map gives an isomorphism $\mathrm{Gr}(\mathcal{O}_q(G)\text{-comod}) \otimes k \cong k[T/W]$ (where Gr stands for the Grothendieck ring). Thus, F defines an automorphism $F^* \colon T/W \to T/W$. Moreover, let $\phi_\ell \colon T/W \to T/W$ be the map defined by raising to power ℓ on T. Then, since F acts trivially on $\mathrm{Rep}\,G$, we have $F^* \circ \phi_\ell = \phi_\ell$. Also $F^*(1) = 1$ by Proposition 2.12. But the map ϕ_ℓ defines an automorphism of the formal neighborhood of 1 in T/W. Hence, F^* acts trivially on the formal neighborhood of 1 in T/W, and hence $F^* = \mathrm{Id}$. Thus, for each L the character of $F(L)$ equals the character of L. But characters of simple objects are linearly independent, which implies that $F(L) \cong L$ for all simple objects L. □

Proposition 2.14. *There are no tensor autoequivalences of $\mathcal{O}_q(G)\text{-comod}$ which reverse the braiding.*

Proof. Suppose for the sake of contradiction that F is a braiding-reversing autoequivalence. Then F preserves the Müger center $\mathrm{Rep}\,G$, and we may assume without loss of generality that $F|_{\mathrm{Rep}\,G} \cong \mathrm{Id}$. Hence, by Proposition 2.13, $F(L) \cong L$ for all simple objects L.

Let P be the weight lattice of G and $P_+ \subset P$ be the set of dominant integral weights. The eigenvalue of the Drinfeld central element z (the double twist) on the simple comodule L_λ of highest weight $\lambda \in P_+$ is $q^{2(\lambda,\lambda+2\rho)}$. Since $F(L_\lambda) = L_\lambda$ and F reverses braiding, this eigenvalue must equal its reciprocal, so we must have $(\lambda,\lambda+2\rho) = 0$ in $\mathbb{Z}/\ell$ for all $\lambda \in P_+$. Subtracting these conditions for two weights λ,μ, we get $(\lambda-\mu,\lambda+\mu+2\rho) = 0$ in $\mathbb{Z}/\ell$. Thus, $(\nu,\beta) = 0$ in $\mathbb{Z}/\ell$ for all $\nu,\beta \in P$, which is a contradiction. □

3 Tensor autoequivalences of tensor categories

3.1 Tensor autoequivalences of a finite tensor category

One of the goals of this section is to put algebraic structure on the groups $\mathsf{Aut}(\mathscr{C})$ and their subgroups. We start with the following proposition.

Proposition 3.1. *Let $\mathscr{C}$ be a finite tensor category over k. Then $\mathsf{Aut}(\mathscr{C})$ has a natural structure of an affine algebraic group over k. Moreover, if $\mathscr{C}$ is braided, then so does $\mathsf{Aut}^{\mathrm{br}}(\mathscr{C})$.*

Proof. The idea of the proof is to express categorical data (tensor functors) entirely in terms of linear-algebraic data (linear maps, i.e., eventually, matrices).

Let P be the direct sum of the indecomposable projectives of $\mathscr{C}$, and $A := (\mathrm{End}P)^{\mathrm{op}}$. Then we have a natural identification $\mathscr{C} \cong \mathsf{Rep}A$ as abelian categories, given by $Y \mapsto \mathsf{Hom}(P,Y)$. Under this identification, the tensor product functor $\otimes\colon \mathscr{C} \boxtimes \mathscr{C} \to \mathscr{C}$ is given by tensoring over $A^{\otimes 2}$ with an $(A,A^{\otimes 2})$-bimodule T, and the associativity isomorphism is represented by an isomorphism of $(A,A^{\otimes 3})$-bimodules $\Phi\colon T \otimes_{A^{\otimes 2}} (T \otimes A) \cong T \otimes_{A^{\otimes 2}} (A \otimes T)$ satisfying the pentagon relation. Any tensor autoequivalence $F\colon \mathscr{C} \to \mathscr{C}$ can then be defined by an algebra automorphism $\alpha\colon A \to A$ together with a bimodule isomorphism $J\colon T \cong T^\alpha$ which preserves Φ. It is clear that pairs (α,J) form an affine algebraic group under the obvious composition. Denote this group by G_1. Also, let $G_2 := A^\times$, also an affine algebraic group. Then we have a homomorphism of algebraic groups $\phi\colon G_2 \to G_1$ given by $\phi(a) = (\mathrm{Ad}\,a, J_a)$, where $J_a(t) = at(a^{-1} \otimes a^{-1})$, $t \in T$. It is clear that a tensor functor F determined by (α,J) is isomorphic to the identity if and only if $(\alpha,J) = \phi(a)$ for some $a \in A^\times$. Thus, $\phi(G_2)$ is normal in G_1, and $\mathsf{Aut}(\mathscr{C}) = G_1/\phi(G_2)$ is an affine algebraic group, as claimed. □

Remark 3.2. A different (but similar) proof of Proposition 3.1 may be obtained by using [EO, Proposition 2.7] which states that any finite tensor category $\mathscr{C}$ is the representation category of a finite-dimensional weak quasi-Hopf algebra H, and representing tensor autoequivalences of $\mathscr{C}$ linear-algebraically as twisted automorphisms of H (as in [Da2]).

3.2 Tensor autoequivalences of a tensor category generated by one object

Now let $\mathscr{C}$ be a tensor category which is not necessarily finite. Then in general $\mathrm{Aut}(\mathscr{C})$ is not an algebraic or even a proalgebraic group. For instance, if $\mathscr{C} = \mathrm{Vec}_{\mathbb{Z}^2}$ then $\mathrm{Aut}(\mathscr{C}) = GL_2(\mathbb{Z}) \ltimes k^\times$. Thus, to obtain a proalgebraic group, we need to put some restrictions on the tensor autoequivalences.

Given $X \in \mathscr{C}$, let $\mathrm{Aut}_X(\mathscr{C})$ be the subgroup of elements $F \in \mathrm{Aut}(\mathscr{C})$ such that $F(X) \cong X$. The following proposition is a generalization of Proposition 3.1 to the infinite case.

Proposition 3.3. *Suppose that $\mathscr{C}$ is tensor-generated by X. Then $\mathrm{Aut}_X(\mathscr{C})$ has a natural structure of an affine proalgebraic group. Moreover, if $\mathscr{C}$ is braided, then so does $\mathrm{Aut}_X^{\mathrm{br}}(\mathscr{C})$.*

Proof. The proof is analogous to the proof of Proposition 3.1, with additional technical details to deal with the fact that $\mathscr{C}$ may not be finite.

Namely, $\mathscr{C}$ has an exhausting increasing filtration $\mathscr{C} = \bigcup_{N \geq 0} \mathscr{C}_N$, where $\mathscr{C}_N$ is the full subcategory whose objects are subquotients of $(X \oplus \mathbf{1})^{\otimes N}$. Note that $\mathscr{C}_N$ are finite, and we have compatible tensor product functors $\otimes \colon \mathscr{C}_m \times \mathscr{C}_p \to \mathscr{C}_{m+p}$. Also, if $F \colon \mathscr{C} \to \mathscr{C}$ is a tensor autoequivalence such that $F(X) \cong X$, then F preserves this filtration and these tensor products, and conversely, any autoequivalence $F \colon \mathscr{C} \to \mathscr{C}$ with these properties is a tensor autoequivalence such that $F(X) \cong X$ (indeed, $F(X \oplus \mathbf{1}) \cong X \oplus \mathbf{1}$ implies $F(X) \cong X$). Here by "F preserves the tensor products" we mean that F is equipped with an appropriate tensor structure.

Let $\mathrm{Aut}_X^N(\mathscr{C})$ be the group of isomorphism classes of autoequivalences $\mathscr{C}_N \to \mathscr{C}_N$ preserving the filtration and the tensor products $\otimes \colon \mathscr{C}_m \times \mathscr{C}_p \to \mathscr{C}_{m+p}$ for $m+p \leq N$. Then the groups $\mathrm{Aut}_X^N(\mathscr{C})$ form an inverse system (i.e., we have natural homomorphisms $\psi_N \colon \mathrm{Aut}_X^{N+1}(\mathscr{C}) \to \mathrm{Aut}_X^N(\mathscr{C})$), and $\mathrm{Aut}_X(\mathscr{C}) = \varprojlim_{N \to \infty} \mathrm{Aut}_X^N(\mathscr{C})$. Thus, it suffices to show that $\mathrm{Aut}_X^N(\mathscr{C})$ are affine algebraic groups and ψ_N are homomorphisms of algebraic groups.

Let P_N be the direct sum of the indecomposable projectives of $\mathscr{C}_N$. Then we have surjections $\eta_N \colon P_N \to P_{N-1}$, such that $\mathrm{Ker}\,\eta_N$ is the intersection of kernels of all morphisms from P_N to objects of $\mathscr{C}_{N-1}$. Let $A_N := (\mathrm{End}\,P_N)^{\mathrm{op}}$. Then any $a \in A_N$ preserves $\mathrm{Ker}\,\eta_N$ and thus descends to P_{N-1}, which defines morphisms $\xi_N \colon A_N \to A_{N-1}$. For $Y \in \mathscr{C}_{N-1}$, the map $\eta_N^* \colon \mathrm{Hom}(P_{N-1}, Y) \to \mathrm{Hom}(P_N, Y)$ is an isomorphism, which implies that ξ_N are surjective. Moreover, we have natural identifications $\mathscr{C}_N \cong \mathrm{Rep}\,A_N$ as abelian categories, given by $Y \mapsto \mathrm{Hom}(P_N, Y)$, and the inclusions $\mathscr{C}_{N-1} \hookrightarrow \mathscr{C}_N$ correspond to the surjections ξ_N. Under this identification, the tensor product functors $\otimes \colon \mathscr{C}_m \boxtimes \mathscr{C}_p \to \mathscr{C}_{m+p}$ are given by tensoring over $A_m \otimes A_p$ with an $(A_{m+p}, A_m \otimes A_p)$-bimodule T_{mp}, and the associativity isomorphism is represented by an isomorphism of $(A_{m+p+s}, A_m \otimes A_p \otimes A_s)$-bimodules

$$\Phi_{mps} \colon T_{m+p,s} \otimes_{A_{m+p} \otimes A_s} (T_{mp} \otimes A_s) \cong T_{m,p+s} \otimes_{A_m \otimes A_{p+s}} (A_m \otimes T_{ps})$$

satisfying the pentagon relation. Moreover, the bimodules T_{mp} are equipped with natural identifications $\gamma^+_{m,p}: T_{mp} \otimes_{A_m} A_{m-1} \cong \xi^*_{m+p} T_{m-1,p}$, $\gamma^-_{m,p}: T_{m,p-1} \otimes_{A_p} A_{p-1} \cong \xi^*_{m+p} T_{m,p-1}$, coming from restricting the tensor product $\otimes: \mathscr{C}_m \boxtimes \mathscr{C}_p \to \mathscr{C}_{m+p}$ to $\mathscr{C}_{m-1} \otimes \mathscr{C}_p$ and $\mathscr{C}_m \otimes \mathscr{C}_{p-1}$.

Any autoequivalence $F: \mathscr{C}_N \to \mathscr{C}_N$ preserving X and the tensor product functors can then be defined by a collection of algebra automorphisms $\alpha_m: A_m \to A_m$, $m \le N$ compatible with $\{\xi_i\}$ and a collection bimodule isomorphisms $J_{mp}: T_{mp} \cong T^{\alpha}_{mp}$, $m+p \le N$ preserving $\{\Phi_{mps}\}$ and compatible with $\{\gamma^{\pm}_{mp}\}$. It is clear that pairs (α, J) form an affine algebraic group under the obvious composition. Denote this group by G_1. Also, let $G_2 := A_N^{\times}$, also an affine algebraic group. Then we have a homomorphism of algebraic groups $\phi: G_2 \to G_1$ given by $\phi(a) = (\mathrm{Ad}\, a, J_a)$, where $J_a(t) = at(a^{-1} \otimes a^{-1})$, $t \in T_{mp}$. It is clear that a tensor functor F determined by (α, J) is isomorphic to the identity if and only if $(\alpha, J) = \phi(a)$ for some $a \in A_N^{\times}$. Thus, $\phi(G_2)$ is normal in G_1, $\mathrm{Aut}^N_X(\mathscr{C}) = G_1/\phi(G_2)$ is an affine algebraic group, and ψ_N is a morphism of algebraic groups, as claimed. □

3.3 *Tensor autoequivalences of the category of comodules over a Hopf algebra*

Let us describe the affine proalgebraic group $\mathrm{Aut}_X(\mathscr{C})$ more explicitly in the case when $\mathscr{C} = H$-comod, where H is a Hopf algebra over k. The condition that $\mathscr{C}$ is tensor-generated by X means that H is generated as an algebra by the finite-dimensional subcoalgebra $C_X \subset H$ spanned by the matrix elements of X.

Definition 3.4. A *co-twisted automorphism* of H is a pair (α, J), where $\alpha: H \to H$ is a coalgebra automorphism, and $J \in (H \otimes H)^*$ is a Hopf 2-cocycle such that $\alpha(x * y) = \alpha(x) *_J \alpha(y)$, where $*_J$ is the product twisted by J.

Remark 3.5. This notion is dual to the notion of a twisted automorphism in [Da2], which explains the terminology.

Clearly, co-twisted automorphisms form a group under the obvious composition operation. Let us denote this group by $\mathrm{TAut}(H)$. We have a natural homomorphism $\zeta: \mathrm{TAut}(H) \to \mathrm{Aut}(H\text{-comod})$ given by $\zeta(\alpha, J) = (F, J)$, where F is defined by α.

Assume from now on that H has slower than exponential growth.

Proposition 3.6. *ζ is surjective.*

Proof. Let $(F, J) \in \mathrm{Aut}(H\text{-comod})$. Recall that $H = \bigoplus_V V^*_{\text{space}} \otimes I(V)$, where V runs over simple H-comodules and $I(V)$ is the injective hull of V (here the subscript "space" indicates that we are considering V^* just as a vector space, without any actions). By Proposition 2.8, (F, J) preserves vector space dimensions of H-comodules. Hence, $F(H) \cong H$, i.e., F is induced by a coalgebra automorphism α of H. Further, since (F, J) is a tensor equivalence, J gives rise to a Hopf 2-cocycle on H

(which we denote by the same letter) such that α preserves the product up to twisting by J. Then (α, J) is a co-twisted automorphism of H and $\zeta(\alpha, J) = (F, J)$. □

Let us now describe the kernel of ζ. By definition, it consists of pairs (α, J) where α is the inner automorphism corresponding to an invertible element $a \in H^*$ and $J = da$ is the coboundary of a. Moreover, the pair (α, J) corresponding to a equals $(1,1)$ if and only if a is a central grouplike element of H^*. Thus, $\mathrm{Ker}\,\zeta = (H^*)^\times/Z_H$, where $(H^*)^\times$ is the group of invertible elements of H^* and Z_H is the group of central grouplike elements of H^*. Therefore, we get

Corollary 3.7. *There is an isomorphism of groups*

$$\overline{\zeta}\colon \mathsf{TAut}(H)/((H^*)^\times/Z_H) \cong \mathsf{Aut}(H\text{-}\mathsf{comod}).$$

Let $\mathsf{TAut}_X(H)$ be the subgroup of co-twisted automorphisms (α, J) of H for which

$$\alpha^*(X) \cong X.$$

Also note that $(H^*)^\times$ is a proalgebraic group, and Z_H is a closed normal subgroup in it. Thus, $(H^*)^\times/Z_H$ is a proalgebraic group, which sits as a closed normal subgroup in $\mathsf{TAut}_X(H)$. Therefore, we have

Corollary 3.8. *The restriction* $\zeta_X\colon \mathsf{TAut}_X(H) \to \mathsf{Aut}_X(H\text{-}\mathsf{comod})$ *is a surjective homomorphism of affine proalgebraic groups. Thus, we have an isomorphism of affine proalgebraic groups*

$$\overline{\zeta}\colon \mathsf{TAut}_X(H)/((H^*)^\times/Z_H) \cong \mathsf{Aut}_X(H\text{-}\mathsf{comod}).$$

Proof. This follows from Proposition 3.3, Proposition 3.6, and Corollary 3.7. □

3.4 Finitely dominated tensor categories and their tensor autoequivalences

Let $\mathscr{C}$ be a tensor category, $\{V_i\}$ be a collection of objects of $\mathscr{C}$, and $\{f_j\}$ be a collection of morphisms between tensor products of V_i.

Definition 3.9. We say that $\mathscr{C}$ is *dominated* by $\{V_i\}$ and $\{f_j\}$ if any tensor functor F from $\mathscr{C}$ to any tensor category $\mathscr{E}$ is determined by $\{F(V_i)\}$ and $\{F(f_j)\}$ up to an isomorphism. We say that $\mathscr{C}$ is *finitely dominated* if it is dominated by a finite collection of objects and morphisms.

Here by $F(f_j)$ we mean the morphism between tensor products of $F(V_i)$ corresponding to f_j.

Remark 3.10. If $S \subset R$ are rings, then one says that $a \in R$ is dominated by S if for a ring homomorphism $f\colon R \to T$, $f(a)$ is determined by $f|_S$, [Is]. Further, one says that R is dominated by S if this is true for each $a \in R$ (this is equivalent to saying that the inclusion $S \hookrightarrow R$ is an epimorphism in the category of rings). Note that this definition makes perfect sense if S is just a subset of R, rather than a subring (in fact, the notions of domination by a subset S and the subring $\langle S\rangle$ generated by S are obviously equivalent). Thus, it is natural to talk of a ring R dominated by a subset $S \subset R$. Note that this is a weaker condition than R being generated by S (e.g., $\mathbb{Z}[x,x^{-1}]$ is dominated by x). Our notion of domination is a categorification of this notion, which justifies the terminology.

Proposition 3.11. *Suppose that $\mathscr{C}$ is finitely dominated and tensor-generated by an object X. Then $\mathsf{Aut}_X(\mathscr{C})$ is of finite type, i.e., is an affine algebraic group.*

Proof. Let $\{V_i, f_j\}$ be a finite set of objects and morphisms dominating $\mathscr{C}$. Let $\mathscr{C}_N$ be the finite abelian subcategories of $\mathscr{C}$ defined in the proof of Proposition 3.3. Let N be so large that V_i and f_j belong to $\mathscr{C}_N$. Let $\mathsf{Aut}_X^N(\mathscr{C})$ be the affine algebraic group defined in the proof of Proposition 3.3, and let K_N be the kernel of the natural homomorphism $\mathsf{Aut}_X(\mathscr{C}) \to \mathsf{Aut}_X^N(\mathscr{C})$. Then for $F \in K_N$, we have $F(V_i) = V_i$ for all i and $F(f_j) = f_j$ for all j. Thus, $F \cong \mathrm{Id}$, i.e. $K_N = 1$. Hence, $\mathsf{Aut}_X(\mathscr{C}) \subset \mathsf{Aut}_X^N(\mathscr{C})$, so that $\mathsf{Aut}_X(\mathscr{C})$ is an affine algebraic group, as claimed. □

3.5 *Finitely presented Hopf algebras*

Let H be a Hopf algebra which is finitely generated as an algebra. This means that the category $H\text{-}\mathsf{comod}$ is tensor-generated by a single object X. Hence, we have a natural surjective homomorphism of H-bicomodule algebras $\theta\colon T(X \otimes {}^*X) \to H$.

Now assume that H is finitely presented (e.g., $H = \mathscr{O}(G)$, where G is an algebraic group). This means that for some (and, hence, any) finite set of generators for H there is a finite set of defining relations. In other words, the ideal $\operatorname{Ker}\theta$ is generated by a finite-dimensional bicomodule $R \subset T(X \otimes {}^*X)$, so that H is the cokernel of the natural H-bicomodule morphism

$$\psi\colon T(X \otimes {}^*X) \otimes R \otimes T(X \otimes {}^*X) \to T(X \otimes {}^*X).$$

We have $R \subset \bigoplus_{j=0}^m (X \otimes {}^*X)^{\otimes j}$. Let $\xi_j\colon R \to (X \otimes {}^*X)^{\otimes j}$ be the corresponding projections, $j = 0, \dots, m$.

Proposition 3.12. *Let H be a finitely presented Hopf algebra and let $\mathscr{E}$ be a tensor category over k. Then a tensor functor $F\colon H\text{-}\mathsf{comod} \to \mathscr{E}$ is determined up to an isomorphism by $F(X)$, $F(R)$, and $F(\xi_j)\colon F(R) \to F(X)^{\otimes j} \otimes X_{\text{space}}^{*\otimes j}$. That is, the tensor category $H\text{-}\mathsf{comod}$ is finitely dominated (by X, R and $\{\xi_j\}$).*

Proof. Let $\mathscr{D}$ be the diagrammatic additive monoidal category whose objects are direct sums of tensor products of symbols X and R, and morphisms are freely generated by $\xi_j\colon R \to X^{\otimes j} \otimes X_{\text{space}}^{*\otimes j}$, $j = 0, \dots, m$. Then we have a natural monoidal functor

$$K\colon \mathscr{D} \to H\text{-}\mathsf{comod},$$

and our job is to show that F is determined up to an isomorphism by the composition $F \circ K\colon \mathscr{D} \to \mathscr{E}$.

In the completion $\widehat{\mathscr{D}}$ of $\mathscr{D}$ under infinite direct sums, there is an obvious morphism $\widetilde{\psi}$ such that $K(\widetilde{\psi}) = \psi$. Hence, $F(\psi) = (F \circ K)(\widetilde{\psi})$ is determined by $F \circ K$. Thus, so is $F(H) = \operatorname{Coker} F(\psi)$. Moreover, let $a\colon H \to H$ be a morphism of left H-comodules. Then a is given by the right action of an element of H^*, which we will also call a. Let $\widetilde{a}$ be the endomorphism of $X^{\otimes j} \otimes X_{\text{space}}^{*\otimes j}$ in $\mathscr{D}$ obtained by the right action of a on the second component. By taking a direct sum over j, we get a morphism

$$\widetilde{a}\colon T(X \otimes X^*_{\text{space}}) \to T(X \otimes X^*_{\text{space}})$$

in $\widehat{\mathscr{D}}$ such that $\theta \circ K(\widetilde{a}) = a \circ \theta$ in $H\text{-}\mathsf{comod}$. This implies that $F(a)$ is determined by $F \circ K$.

Now, any $Y \in H\text{-}\mathsf{comod}$ admits an injective resolution $I_\bullet(Y)$ by multiples of H:

$$Y \to I_0 \to I_1 \to \cdots,$$

where $I_j = H \otimes W_j$, and W_j is a vector space. Indeed, for any H-comodule Z (not necessarily finite dimensional) we have a canonical inclusion $Z \hookrightarrow H \otimes Z_{\text{space}}$; so we may define I_j as $H \otimes (U_j)_{\text{space}}$, where $U_j := \operatorname{Coker}(I_{j-2} \to I_{j-1})$ (where $I_{-1} := Y$ and $I_0 := H \otimes Y_{\text{space}}$). The above argument implies that the action of F on this resolution is determined by $F \circ K$. Thus, $F(Y)$ is determined by $F \circ K$. Further, if $f\colon Z \to Y$ is a morphism in $H\text{-}\mathsf{comod}$, then it lifts to a morphism of resolutions $I_\bullet(Z) \to I_\bullet(Y)$, which implies that $F(f)$ is determined by $F \circ K$, i.e., F is determined by $F \circ K$ as a functor.

Now, the tensor structure $J_{ZY}\colon F(Z) \otimes F(Y) \to F(Z \otimes Y)$ is determined by its values at $Z = Y = H$. In turn, J_{HH} is determined by $J_{\widetilde{H}\widetilde{H}}$, where $\widetilde{H} := T(X \otimes X^*_{\text{space}})$. Finally, $J_{\widetilde{H}\widetilde{H}}$ is determined by $F \circ K$. This proves the proposition. □

Corollary 3.13. *Let H be a finitely presented Hopf algebra such that $H\text{-}\mathsf{comod}$ is tensor-generated by X. Then*

(i) $\mathsf{Aut}_X(H\text{-}\mathsf{comod})$ *is an affine algebraic group;*
(ii) $\mathsf{TAut}_X(H)/((H^*)^\times/Z_H)$ *is an affine algebraic group.*

Proof. Part (i) follows from Proposition 3.11 and Proposition 3.12. Part (ii) follows from (i) and Corollary 3.8. □

3.6 The second invariant cohomology of a tensor category

Given a tensor category $\mathscr{C}$, let $\mathsf{Aut}_0(\mathscr{C})$ be the group of isomorphism classes of tensor autoequivalences of $\mathscr{C}$ that are isomorphic to the identity as abelian equivalences, [Da1, BC, GKa]. We will call this group the *second invariant cohomology group* of $\mathscr{C}$.

Proposition 3.14. $\mathrm{Aut}_0(\mathscr{C})$ *has a natural structure of an affine proalgebraic group.*

Proof. Let S be the set of isomorphism classes of objects of $\mathscr{C}$. Let G_1 be the closed subgroup of the affine proalgebraic group $\prod_{X,Y\in S}\mathrm{Aut}(X\otimes Y)$ cut out by the functoriality condition in X and Y and the tensor structure axiom

$$J_{X\otimes Y,Z}\circ(J_{X,Y}\otimes \mathrm{Id}_Z)=J_{X,Y\otimes Z}\circ(\mathrm{Id}_X\otimes J_{Y,Z}),$$

i.e., the closed subgroup in $\mathrm{Aut}(\otimes)$ cut out by the tensor structure axiom. Also let $G_2\subset\prod_{X\in S}\mathrm{Aut}(X)$ be the closed subgroup cut out by the functoriality condition in X (i.e., $G_2=\mathrm{Aut}(\mathrm{id}_{\mathscr{C}})$, a commutative affine proalgebraic group). Let $\phi: G_2\to G_1$ be the homomorphism defined by $\phi(a)_{X,Y}=a_{X\otimes Y}\circ(a_X^{-1}\otimes a_Y^{-1})$. Then ϕ is a homomorphism of affine proalgebraic groups, $\phi(G_2)$ is a central subgroup of G_1, and $\mathrm{Aut}_0(\mathscr{C})=\mathrm{Coker}\,\phi$. This implies the statement. □

Remark 3.15. In the case $\mathscr{C}=H\text{-comod}$, where H is a Hopf algebra, Proposition 3.14 follows from Theorem 6.7 of [BKa].

Corollary 3.16. (i) *If $\mathscr{C}$ is finitely dominated, then the second invariant cohomology $\mathrm{Aut}_0(\mathscr{C})$ is an affine algebraic group (i.e., is of finite type).*

(ii) *If H is a finitely presented Hopf algebra, then the second invariant cohomology $\mathrm{Aut}_0(H\text{-comod})$ is an affine algebraic group.*

Proof. (i) It is clear that $\mathrm{Aut}_0(\mathscr{C})$ is a closed subgroup of $\mathrm{Aut}_X(\mathscr{C})$, where X is a tensor-generating object. Thus, the statement follows from Proposition 3.11.

Part (ii) follows from (i) and Corollary 3.13. □

Example 3.17. Let G be an abstract group, and $H=kG$ be its group algebra. Then H is finitely generated, respectively finitely presented, iff so is G. Also, $H\text{-comod}=\mathrm{Vec}_G$, the tensor category of finite-dimensional G-graded vector spaces. Thus, $\mathrm{Aut}_0(H\text{-comod})=H^2(G,k^\times)$. Hence, Corollary 3.16 reduces in this case to the well known result that for a finitely presented group G the group $H^2(G,k^\times)$ is an algebraic group, which holds due to the Serre-Hochschild exact sequence $\cdots\to\mathrm{Hom}(N,k^\times)^G\to H^2(G,k^\times)\to 1$, where $1\to N\to\widetilde{G}\to G\to 1$ is a finite presentation of G (i.e., $\widetilde{G}$ is a finitely generated free group). Indeed, since this presentation is finite, N is the normal closure of a finitely generated group, hence $\mathrm{Hom}(N,k^\times)^G$ is an affine algebraic group (i.e., is of finite type).

Note that if G is finitely generated but not finitely presented, then this may be false. E.g., if G is the Grigorchuk group [Gri], then $H_2(G,\mathbb{Z})$ is infinitely generated 2-elementary abelian, hence $H^2(G,k^\times)=\mathrm{Hom}(H_2(G,\mathbb{Z}),k^\times)$ is an infinite product of copies of μ_2 (a proalgebraic group of infinite type). Thus, the finite presentation assumption in Proposition 3.12 and the finite domination assumption in Proposition 3.11 cannot be dropped.

4 Tensor autoequivalences of $\mathcal{O}_q(G)\text{-comod}$

4.1 Finite presentation for $\mathcal{O}_v(G)$ over $k[v,v^{-1}]$

Let G be a simple algebraic group. Let v be an indeterminate and let $\mathcal{O}_v(G)$ denote the quantized function algebra of G over $k[v,v^{-1}]$.

Proposition 4.1. *$\mathcal{O}_v(G)$ is finitely presented as an algebra.*

Proof. Let S be a finite generating subset of the cone P_+ of dominant weights of G. Let $X=\bigoplus_{\lambda\in S}V_\lambda$ be the direct sum of all standard $\mathcal{O}_v(G)$-comodules whose highest weights are in S. Then we have an element $T=T_X\in \operatorname{End}X\otimes\mathcal{O}_v(G)$ determining the coaction of $\mathcal{O}_v(G)$ on X. By [Lu3, Proposition 3.3] the matrix elements of T_X generate $\mathcal{O}_v(G)$. Also, T_X satisfies the Faddeev-Reshetikhin-Takhtajan (FRT) relation

$$R_{XX}^{12}T^{13}T^{23}=T^{23}T^{13}R_{XX}^{12},$$

where R_{XX} is the specialization of the universal R-matrix to $X\otimes X$ (note that to define R_{XX}, one may need to adjoin $v^{1/m}$ for some m, but the FRT relations contain only integer powers of v). Let A be the algebra generated over $k[v,v^{-1}]$ by the matrix elements of T with these relations taken as defining. Then we have a surjective homomorphism $\eta: A\to\mathcal{O}_v(G)$. By [BG, Lemma I.8.17], the algebra A is Noetherian (more precisely, this lemma is proved when v is specialized to $q\in k^\times$, but it generalizes verbatim to our setting). Hence the ideal $\operatorname{Ker}\eta$ is finitely generated. This implies that $\mathcal{O}_v(G)$ is finitely presented, as desired. □

Corollary 4.2. *For any $q\in k^\times$, the algebra $\mathcal{O}_q(G):=\mathcal{O}_v(G)/(v-q)$ is finitely presented. Also, the $k(v)$-algebra $\widetilde{\mathcal{O}}_v(G):=\mathcal{O}_v(G)\otimes_{k[v,v^{-1}]}k(v)$ is finitely presented.*

Remark 4.3.(1) Proposition 4.1 also holds over $\mathbb{Z}[v,v^{-1}]$ (i.e., in the setting of [Lu3]), with the same proof.
(2) A nice finite presentation of $\widetilde{\mathcal{O}}_v(G)$ (and thus of $\mathcal{O}_q(G)$ for transcendental q) is given in [I].

4.2 Tensor autoequivalences of $\mathcal{O}_q(G)\text{-comod}$ outside finitely many roots of unity

In this subsection we classify tensor and braided autoequivalences of $\mathcal{O}_q(G)\text{-comod}$. Here we don't make any coprimeness assumptions on the order of q, and just assume that q is a root of unity of sufficiently large order ℓ.

Note that any tensor autoequivalence F of $\mathcal{O}_q(G)\text{-comod}$ naturally acts on the center Z_G of G, as $Z_G^\vee$ is the universal grading group of $\mathcal{O}_q(G)\text{-comod}$. Thus, for a subgroup $C\subset Z_G$, F defines an equivalence $\mathcal{O}_q(G/C)\text{-comod}\to\mathcal{O}_q(G/F(C))\text{-comod}$.

Let $\Gamma_G = \mathsf{Out}\, G$ (e.g., if G is simply connected, then $\Gamma_G = \Gamma$) and

$$\widetilde{\Gamma_G} := \mathsf{Out}\, G \ltimes H^2(Z_G^\vee, k^\times).$$

Theorem 4.4. *For all $q \in k^\times$ except finitely many roots of unity:*

(i) $\mathsf{Aut}(\mathscr{O}_q(G)\text{-}\mathsf{comod}) \cong \widetilde{\Gamma_G}$.
(ii) $\mathsf{Aut}^{\mathrm{br}}(\mathscr{O}_q(G)\text{-}\mathsf{comod}) \cong \Gamma_G$.

Proof. If q is not a root of unity, this is shown in [NT2] (for (i)) and [NT1] (for (ii)); more precisely, the results of [NT1, NT2] are proved for simply connected groups, but the arguments extend without significant changes to the general case. So we only have to prove the statements for roots of unity.

Let us prove (i). For a positive integer N, let Σ_N be the set of all nonzero dominant integral weights λ for G such that the irreducible representation L_λ of G with highest weight λ has dimension $\leq N$. If the order of q is large enough, these L_λ have q-analogs, $\mathscr{O}_q(G)$-comodules L_λ^q of the same dimension as L_λ, which are also irreducible.

Let $F \in \mathsf{Aut}(\mathscr{O}_q(G)\text{-}\mathsf{comod})$. We claim that for sufficiently large order ℓ of q, the functor F permutes L_λ^q, $\lambda \in \Sigma_N$. Indeed, by Steinberg's tensor product theorem for quantum groups ([Lu2, Proposition 9.2]), for large enough ℓ the only irreducible comodules over $\mathscr{O}_q(G)$ which have dimension $\leq N$, don't belong to the Müger center of $\mathscr{O}_q(G)\text{-}\mathsf{comod}$, and cannot be non-trivially decomposed as a tensor product are L_λ^q, $\lambda \in \Sigma_N$. But $F(L_\lambda^q)$ cannot belong to the Müger center of $\mathscr{O}_q(G)\text{-}\mathsf{comod}$, as it generates a subcategory of the form $\mathscr{O}_q(G/C)\text{-}\mathsf{comod}$, while $F(\mathscr{O}_q(G/C)\text{-}\mathsf{comod}) = \mathscr{O}_q(G/F(C))\text{-}\mathsf{comod}$ is not contained in the Müger center of $\mathscr{O}_q(G)\text{-}\mathsf{comod}$. So, since by Proposition 2.12 F preserves vector space dimensions, it must permute L_λ^q, $\lambda \in \Sigma_N$.

Now pick N so large that $X = X_N^q := \bigoplus_{\lambda \in \Sigma_N} L_\lambda^q$ is a tensor generator of $\mathscr{O}_q(G)\text{-}\mathsf{comod}$. Then $F(X) \cong X$, so $F \in \mathsf{Aut}_X(\mathscr{O}_q(G)\text{-}\mathsf{comod})$. Note that we have a natural inclusion $\widetilde{\Gamma_G} \subset \mathsf{Aut}(\mathscr{O}_q(G)\text{-}\mathsf{comod})$ (see [NT2]). Thus, our job is to show that for sufficiently large ℓ, this inclusion is an equality.

Let $R = k[v, v^{-1}][1/f]$, where f is a nonzero polynomial vanishing at roots of unity of low order. Since by Proposition 4.1 $\mathscr{O}_v(G)$ is finitely presented, by Corollary 3.13 the commutative algebra $\mathscr{O}(\mathsf{Aut}_X(\mathscr{O}_q(G)\text{-}\mathsf{comod}))$ is the specialization at $v = q$ of a finitely generated commutative algebra H over R (for all but finitely many roots of unity q). Indeed, we can take a finite presentation of the $k(v)$-algebra $\mathscr{O}(\mathsf{Aut}_X(\widetilde{\mathscr{O}_v}(G)\text{-}\mathsf{comod}))$ and define H by the same generators and relations over R (for a suitable choice of f). Moreover, since $\widetilde{\Gamma_G}$ acts faithfully by automorphisms of $\mathscr{O}_v(G)$, we have a surjective algebra homomorphism $\beta : H \to R[\widetilde{\Gamma_G}]$, where $R[\widetilde{\Gamma_G}]$ is the algebra of R-valued functions on $\widetilde{\Gamma_G}$. Let $K := \operatorname{Ker} \beta$. By Grothendieck's generic freeness lemma [Eis, Theorem 14.4], since H is finitely generated, we may assume without loss of generality that H is a free R-module (by choosing f appropriately). Then $H \cong K \oplus R[\widetilde{\Gamma_G}]$ as an R-module, hence K is a projective R-module.

Moreover, by the result of [NT2], β becomes an isomorphism upon tensoring with $k(v)$, hence $K \otimes_R k(v) = 0$. As K is projective, this implies that $K = 0$, i.e., β is an isomorphism. Thus, β is an isomorphism after specializing v to all roots of unity q that are not roots of f. Hence, for all such roots of unity q we have an isomorphism $\mathsf{Aut}(\mathcal{O}_q(G)\text{-comod}) \cong \widetilde{\Gamma_G}$, as desired.

Part (ii) is proved in the same way, using the group Γ_G instead of $\widetilde{\Gamma_G}$, and [NT1] instead of [NT2]. □

Remark 4.5. It would be interesting to obtain a more direct proof of Theorem 4.4 (and desirably of its stronger version, giving an explicit list of excluded roots of unity) by generalizing the arguments of [NT1, NT2] to the case when q is a root of unity.

4.3 Sharper results for classical groups

For classical groups $G = SL_N, Sp_N, SO_N$, we can use the Faddeev-Reshetikhin-Takhtajan presentations of $\mathcal{O}_q(G)$ to obtain a sharper result, i.e., one for all q of order $\ell > N$. Note that in these cases Z_G is cyclic, so $\widetilde{\Gamma_G} = \Gamma_G = \mathsf{Out}\,G$.

Theorem 4.6. *If $G = SL_N, Sp_N, SO_N$ and $\ell > N$, then*

$$\mathsf{Aut}(\mathcal{O}_q(G)\text{-comod}) \cong \mathsf{Aut}^{\mathrm{br}}(\mathcal{O}_q(G)\text{-comod}) \cong \mathsf{Out}\,G.$$

Proof. Let us first prove that $\mathsf{Aut}^{\mathrm{br}}(\mathcal{O}_q(G)\text{-comod}) \cong \mathsf{Out}\,G$. Take the tensor generator $X = V$, the defining comodule. In all three cases we have the Faddeev-Reshetikhin-Takhtajan presentation of $\mathcal{O}_q(G)$, in which $\mathcal{O}_q(G)$ is generated by the entries of $T \in \mathrm{End}\,V \otimes \mathcal{O}_q(G)$ with defining relations

$$R_{VV}^{12} T^{13} T^{23} = T^{23} T^{13} R_{VV}^{12}$$

plus some additional relations depending on which case we are considering, see [FRT, RTF, Ta, Ha].

Consider first the case $G = SL_N$. Then the additional relation is the quantum determinant relation $\det_q(T) = 1$. Thus, any braided autoequivalence of $\mathcal{O}_q(G)\text{-comod}$ is determined by $F(V)$ and the action of F on the morphism $\alpha\colon k \to V^{\otimes N}$ whose image is $\bigwedge_q^N V$, the quantum top exterior power of V. The only N-dimensional simple $\mathcal{O}_q(G)$-comodules are V, V^* and their Frobenius twists $V^{(1)}, (V^*)^{(1)}$ (since the set of weights of a comodule is Weyl group invariant). Then F maps V to V or to V^* (as $V \oplus V^*$ tensor-generates the category, while $V^{(1)} \oplus (V^*)^{(1)}$ does not). So by composing F with an element of $\mathsf{Out}\,G = \mathbb{Z}/2$, we may assume without loss of generality that $F(V) = V$. By rescaling this isomorphism it is also easy to make sure that $F(\alpha) = \alpha$, so $F \cong \mathrm{Id}$, as desired.

Now consider $G = Sp_N$ (N even). Then the additional relation says that T preserves the morphism $\beta\colon k \to V \otimes V$ which deforms the symplectic form on V^*.

Thus any braided autoequivalence F of $\mathcal{O}_q(G)$-comod is determined by $F(V)$ and $F(\beta)$. As before, the only N-dimensional simple comodules are V and the Frobenius twist $V^{(1)}$, and $F(V) \not\cong V^{(1)}$ (since V tensor-generates the category but $V^{(1)}$ does not). Thus $F(V) = V$. By rescaling this isomorphism we can also make sure that $F(\alpha) = \alpha$. Thus $F \cong \mathrm{Id}$, as desired.

Finally, consider the case $G = SO_N$, $N \geq 3$. In this case, the additional relations are that T preserves the morphism $\beta: k \to V \otimes V$ which deforms the inner product on V^*, and that it preserves the morphism $\alpha: k \to V^{\otimes N}$ (i.e., has quantum determinant 1). Thus, any braided autoequivalence F of $\mathcal{O}_q(G)$-comod is determined by $F(V)$, $F(\alpha)$, and $F(\beta)$. Moreover, the only N-dimensional simple $\mathcal{O}_q(G)$-comodules are V and $V^{(1)}$, and $F(V) \neq V^{(1)}$, since V tensor-generates the category but $V^{(1)}$ does not. So, for any braided autoequivalence of $\mathcal{O}_q(G)$-comod we have $F(V) = V$. Finally, we can rescale this isomorphism so that $F(\alpha) = \alpha$. Then $F(\beta) = \pm\beta$ since $\beta \otimes \beta$ may be expressed via $\alpha^{\otimes n}$ and hence $F(\beta) \otimes F(\beta) = \beta \otimes \beta$.

Now we need to consider separately odd and even N. If N is odd, rescaling the isomorphism $F(V) \cong V$ by -1 (which preserves the relation $F(\alpha) = \alpha$), we can make sure that $F(\beta) = \beta$, so $F \cong \mathrm{Id}$. On the other hand, if N is even, then we cannot do this, so we have two cases, $F(\beta) = \beta$ and $F(\beta) = -\beta$. But in this case we have a non-trivial involutive outer automorphism of G implemented by an element of O_N with determinant -1. So by composing with such automorphism, we can make sure that $F(\beta) = \beta$, i.e., $F \cong \mathrm{Id}$, as desired. □

Remark 4.7. 1. Note that for $G = SO_8$, we have the group S_3 acting by Dynkin diagram automorphisms (hence automorphisms of Spin_8), but only a 2-element subgroup of this S_3 descends to SO_8.

2. The proof of Theorem 4.6 is similar to the arguments of [KW] for SL_N and [TW] for SO_N and Sp_N.

5 Tensor autoequivalences of $\mathsf{Rep}\, \mathfrak{u}_q(\mathfrak{g})$

Now let us classify tensor autoequivalences of $\mathsf{Rep}\, \mathfrak{u}_q(\mathfrak{g})$. We again assume that q is a root of unity of odd order coprime to 3 if G is of type G_2, and coprime to the determinant of the Cartan matrix of G.

5.1 The connected component of the identity of $\mathsf{Aut}(\mathsf{Rep}\, \mathfrak{u}_q(\mathfrak{g}))$

Recall that for tensor categories $\mathscr{C}, \mathscr{C}'$ and a tensor functor $F: \mathscr{C} \to \mathscr{C}'$ one can define the deformation cohomology $H^i_F(\mathscr{C})$, see [Da1, Y] and [EGNO, Section 7.22]. Namely, $C^i_F(\mathscr{C}) = \mathrm{End}(F \circ \otimes^i)$ with the usual differential, and $H^i_F(\mathscr{C})$ is the i-th cohomology of the complex $C^i_F(\mathscr{C})$. Then $H^2_F(\mathscr{C})$ consists of equivalence classes of first order deformations of F as a tensor functor. Note that if $\mathscr{C} = \mathsf{Rep}(H)$, where H

is a Hopf algebra over k and $F: \mathscr{C} \to \mathsf{Vec}$ is the forgetful functor, then $C^i_F(\mathscr{C}) = H^{\otimes i}$, so if H is finite dimensional, then $H^i_F(\mathscr{C}) = H^i(H^*, k)$. In particular, for $H = \mathfrak{u}_q(\mathfrak{g})$ we get

Proposition 5.1. *One has*

$$H^2_F(\mathsf{Rep}\,\mathfrak{u}_q(\mathfrak{g})) = \mathfrak{n}_+ \oplus \mathfrak{n}_-,$$

where $\mathfrak{n}_\pm$ *are the positive and negative nilpotent subalgebras of* $\mathfrak{g}$.

Proof. As explained above, we have

$$H^2_F(\mathsf{Rep}\,\mathfrak{u}_q(\mathfrak{g})) = H^2(\mathfrak{u}_q(\mathfrak{g})^*, k).$$

On the other hand, by [GK, Proposition 2.3.1] and remark thereafter, we have

$$H^2(\mathfrak{u}_q(\mathfrak{g})^*, k) = \mathfrak{n}_+ \oplus \mathfrak{n}_-.$$

This implies the statement. □

Now we can compute the identity component of $\mathsf{Aut}(\mathsf{Rep}\,\mathfrak{u}_q(\mathfrak{g}))$.

Proposition 5.2. *One has* $\mathsf{Aut}(\mathsf{Rep}\,\mathfrak{u}_q(\mathfrak{g}))_0 = \mathsf{Aut}^{\mathrm{br}}(\mathsf{Rep}\,\mathfrak{u}_q(\mathfrak{g}))_0 = G^{\mathrm{ad}}$.

Proof. We have a natural map

$$\mathsf{Lie}\,\mathsf{Aut}(\mathsf{Rep}\,\mathfrak{u}_q(\mathfrak{g})) \to H^2_F(\mathsf{Rep}\,\mathfrak{u}_q(\mathfrak{g})). \tag{6}$$

Namely, recall that elements of $\mathsf{Aut}(\mathsf{Rep}\,\mathfrak{u}_q(\mathfrak{g}))$ are twisted automorphisms (a, J) of $\mathfrak{u}_q(\mathfrak{g})$ (in the sense of [Da2]), so the map (6) attaches to a twisted derivation (d, j) the class of infinitesimal twist j. (Here by a twisted derivation we mean a first order deformation of the identity twisted automorphism, see, e.g., [Da3]). Thus, by Proposition 5.1 we have a map

$$\phi: \mathsf{Lie}\,\mathsf{Aut}(\mathsf{Rep}\,\mathfrak{u}_q(\mathfrak{g})) \to \mathfrak{n}_+ \oplus \mathfrak{n}_-.$$

The kernel of ϕ consists of ordinary Hopf algebra derivations of $\mathfrak{u}_q(\mathfrak{g})$, and it is easy to see that they can be identified with the Cartan subalgebra $\mathfrak{t} = \mathsf{Lie}(T)$ of $\mathfrak{g}$. Thus, $\dim \mathsf{Lie}\,\mathsf{Aut}(\mathsf{Rep}\,\mathfrak{u}_q(\mathfrak{g})) \le \dim \mathfrak{t} + \dim \mathfrak{n}_+ + \dim \mathfrak{n}_- = \dim \mathfrak{g}$. Since by Proposition 2.3 we have an embedding

$$G^{\mathrm{ad}} \hookrightarrow \mathsf{Aut}^{\mathrm{br}}(\mathsf{Rep}\,\mathfrak{u}_q(\mathfrak{g})) \subset \mathsf{Aut}(\mathsf{Rep}\,\mathfrak{u}_q(\mathfrak{g})),$$

this implies the required statement. □

5.2 Tensor autoequivalences of $\mathsf{Rep}\,\mathfrak{u}_q(\mathfrak{g})$ are braided

Proposition 5.3. *As braided tensor categories,* $\mathsf{Rep}\,\mathfrak{u}_q(\mathfrak{g}) \not\cong \mathsf{Rep}\,\mathfrak{u}_q(\mathfrak{g})^{\mathrm{rev}}$.

Proof. Assume that $\overline{F}\colon \mathsf{Rep}\,\mathfrak{u}_q(\mathfrak{g}) \cong \mathsf{Rep}\,\mathfrak{u}_q(\mathfrak{g})^{\mathrm{rev}}$ is a braided equivalence. Then $\overline{F}$ induces an automorphism γ of $\mathsf{Aut}(\mathsf{Rep}\,\mathfrak{u}_q(\mathfrak{g}))$ and in particular of its connected component of the identity $\mathsf{Aut}(\mathsf{Rep}\,\mathfrak{u}_q(\mathfrak{g}))_0$, which by Proposition 5.2 is G^{ad}. Since $\mathsf{Aut}(G^{\mathrm{ad}}) = \Gamma \ltimes G^{\mathrm{ad}}$ and every element of $\Gamma \ltimes G^{\mathrm{ad}}$ is implemented by a tensor autoequivalence in $\mathsf{Aut}(\mathsf{Rep}\,\mathfrak{u}_q(\mathfrak{g}))$, by composing with such an autoequivalence, we may assume without loss of generality that $\gamma = 1$. Then $\overline{F}$ commutes with G^{ad}. Moreover, by Lemma 2.6, this commutativity is an isomorphism of actions. Hence, by Subsection 2.3, $\overline{F}$ gives rise to a braided equivalence of G-equivariantizations $F\colon \mathscr{O}_q(G)\text{-}\mathsf{comod} \to (\mathscr{O}_q(G)\text{-}\mathsf{comod})^{\mathrm{rev}}$. But this contradicts Proposition 2.14. □

Now we finally obtain

Theorem 5.4. *Every tensor autoequivalence of* $\mathsf{Rep}\,\mathfrak{u}_q(\mathfrak{g})$ *is automatically braided. In other words, we have* $\mathsf{Aut}(\mathsf{Rep}\,\mathfrak{u}_q(\mathfrak{g})) = \mathsf{Aut}^{\mathrm{br}}(\mathsf{Rep}\,\mathfrak{u}_q(\mathfrak{g})) = \mathsf{Pic}(\mathsf{Rep}\,\mathfrak{u}_q(\mathfrak{g}))$.

Proof. By Lemma 2.5, $\mathsf{Rep}\,\mathfrak{u}_q(\mathfrak{g})$ has no non-trivial tensor subcategories. By Proposition 5.3, the category $\mathscr{C} = \mathsf{Rep}\,\mathfrak{u}_q(\mathfrak{g})$ satisfies the assumptions of Corollary 2.2. Thus, Corollary 2.2 implies the result. □

5.3 Classification of tensor autoequivalences of $\mathsf{Rep}\,\mathfrak{u}_q(\mathfrak{g})$

Introduce the notation $\mathbf{P} := \mathsf{Aut}^{\mathrm{br}}(\mathsf{Rep}\,\mathfrak{u}_q(\mathfrak{g}))$. We have seen that $\mathbf{P}$ contains $\Gamma \ltimes G^{\mathrm{ad}}$, and by Proposition 5.2 we have $\mathbf{P}_0 = G^{\mathrm{ad}}$. Hence, G^{ad} is normal in $\mathbf{P}$. Given $x \in \mathbf{P}$, let x' be the element of $\Gamma \ltimes G^{\mathrm{ad}} = \mathsf{Aut}(G^{\mathrm{ad}})$ induced by x. We can view x' as an element of $\mathbf{P}$. Then $x = x'x''$, where x'' belongs to the centralizer Z of G^{ad} in $\mathbf{P}$ (a finite group). Since Γ normalizes G^{ad}, it acts on Z by conjugation. Thus, we have

Lemma 5.5. $\mathbf{P} = \Gamma \ltimes (G^{\mathrm{ad}} \times Z)$.

We can now formulate one of the main results of this paper.

Theorem 5.6. *One has* $\mathbf{P} = \Gamma \ltimes G^{\mathrm{ad}}$ *in the following cases:*

(i) *If* $\mathfrak{g}$ *is of a classical type* ($\mathfrak{sl}_N$, $\mathfrak{so}_N$, *or* $\mathfrak{sp}_N$) *and the order of* q *is bigger than* N*;*
(ii) *If* $\mathfrak{g}$ *is exceptional and the order of* q *is sufficiently large.*

Proof. By Lemma 5.5, our job is to show $Z = 1$. There is a group homomorphism from $Z \times G$ to $\mathsf{Aut}^{\mathrm{br}}(\mathsf{Rep}\,\mathfrak{u}_q(\mathfrak{g}))$. By Lemma 2.6, this homomorphism admits a unique lift to an action on $\mathsf{Rep}\,\mathfrak{u}_q(\mathfrak{g})$. Since the action of Z on $\mathsf{Rep}\,\mathfrak{u}_q(\mathfrak{g})$ commutes with that of G, we conclude that Z acts on the equivariantization $(\mathsf{Rep}\,\mathfrak{u}_q(\mathfrak{g}))^G$, which is the braided category $\mathscr{O}_q(G)\text{-}\mathsf{comod}$.

Thus, it suffices to prove that the group of braided autoequivalences of $\mathscr{O}_q(G)\text{-comod}$ coincides with Γ_G. Indeed, then given $z \in Z$, this would yield that $z \in \Gamma_G$, hence $z = 1$ (as it acts trivially on G^{ad}).

Now part (i) follows from Theorem 4.6 and part (ii) follows from Theorem 4.4(ii). □

Remark 5.7. We expect that Theorem 5.6 holds without the assumptions on the order of q.

5.4 Brauer-Picard groups

Let $\mathscr{C}$ be a finite tensor category. Recall [ENO] that the *Brauer-Picard group* $\mathsf{BrPic}(\mathscr{C})$ of $\mathscr{C}$ is the group of equivalence classes of invertible $\mathscr{C}$-bimodule categories. Recall also that there is a canonical isomorphism

$$\mathsf{BrPic}(\mathscr{C}) \cong \mathsf{Aut}^{\mathrm{br}}(\mathscr{Z}(\mathscr{C})), \tag{7}$$

see [DN] (and [ENO] in the semisimple case).

When $\mathscr{C}$ is braided, its Picard group $\mathsf{Pic}(\mathscr{C})$ is naturally identified with a subgroup of $\mathsf{BrPic}(\mathscr{C})$.

Proposition 5.8. *Let $\mathscr{C}$ be a finite tensor category.*

(i) *The group $\mathsf{BrPic}(\mathscr{C})$ has a natural structure of an affine algebraic group over k.*
(ii) *If $\mathscr{C}$ is braided, then $\mathsf{Pic}(\mathscr{C})$ has a natural structure of an affine algebraic group over k.*

Proof. Part (i) follows immediately from (7) and Proposition 3.1. To prove part (ii), recall that under isomorphism (7) $\mathsf{Pic}(\mathscr{C})$ is identified with the subgroup of classes of autoequivalences trivializable on the subcategory $\mathscr{C} \subset \mathscr{Z}(\mathscr{C})$ (i.e., those $\alpha \in \mathsf{Aut}^{\mathrm{br}}(\mathscr{Z}(\mathscr{C}))$ for which $\alpha|_{\mathscr{C}} \cong \mathrm{Id}_{\mathscr{C}}$ as tensor functors), see [DN]. This means that $\mathsf{Pic}(\mathscr{C})$ is a Zariski closed subgroup of $\mathsf{BrPic}(\mathscr{C})$. □

In this subsection we will compute the Brauer-Picard groups of $\mathsf{Rep}\,\mathfrak{u}_q(\mathfrak{g})$ and $\mathsf{Rep}\,\mathfrak{u}_q(\mathfrak{b})$, where $\mathfrak{b} \subset \mathfrak{g}$ is a Borel subalgebra.

Let $E := P/\ell P = Q/\ell Q$. Note that E has a natural quadratic form $\mathbf{q}(v) := q^{(v,v)}$. Let $O(E,\mathbf{q})$ be the orthogonal group of this quadratic form. Note that Γ acts naturally on E preserving $\mathbf{q}$ and therefore $\Gamma \hookrightarrow O(E,\mathbf{q})$.

Proposition 5.9. *Under the assumptions of Theorem 5.6 one has:*

(i) $\mathsf{BrPic}(\mathsf{Rep}\,\mathfrak{u}_q(\mathfrak{b})) \cong (\Gamma \ltimes G^{\mathrm{ad}}) \times O(E,\mathbf{q})$;
(ii) $\mathsf{BrPic}(\mathsf{Rep}\,\mathfrak{u}_q(\mathfrak{g})) \cong (\Gamma \ltimes G^{\mathrm{ad}}) \times (\Gamma \ltimes G^{\mathrm{ad}})$.

Proof. (i) It is well known that the quantum double $D(\mathfrak{u}_q(\mathfrak{b}))$ is given by

$$D(\mathfrak{u}_q(\mathfrak{b})) = \mathfrak{u}_q(\mathfrak{g}) \otimes k[E],$$

where the R-matrix is the external product of the R-matrix of $\mathfrak{u}_q(\mathfrak{g})$ with the R-matrix on $k[E]$ defined by $\mathbf{q}$. Hence, $\mathscr{Z}(\mathsf{Rep}\,\mathfrak{u}_q(\mathfrak{b}))$ is equivalent as a braided category to the category $\mathsf{Rep}\,\mathfrak{u}_q(\mathfrak{g}) \boxtimes \mathsf{Rep}\,E$, where the braiding on the second factor is defined by $\mathbf{q}$. Thus, by (7), we have

$$\mathsf{BrPic}(\mathsf{Rep}\,\mathfrak{u}_q(\mathfrak{b})) \cong \mathsf{Aut}^{\mathrm{br}}(\mathsf{Rep}\,\mathfrak{u}_q(\mathfrak{g}) \boxtimes \mathsf{Rep}\,E).$$

Now, any braided autoequivalence F of $\mathsf{Rep}\,\mathfrak{u}_q(\mathfrak{g}) \boxtimes \mathsf{Rep}\,E$ must preserve the second factor, since it is the subcategory spanned by all the invertible objects. Hence F also preserves the first factor (as it is the centralizer of the second one). Thus, we get

$$\mathsf{BrPic}(\mathsf{Rep}\,\mathfrak{u}_q(\mathfrak{b})) \cong \mathsf{Aut}^{\mathrm{br}}(\mathsf{Rep}\,\mathfrak{u}_q(\mathfrak{g})) \times \mathsf{Aut}^{\mathrm{br}}(\mathsf{Rep}\,E).$$

Since $\mathsf{Aut}^{\mathrm{br}}(\mathsf{Rep}\,E) \cong O(E,\mathbf{q})$, the result follows from Theorem 5.6.

(ii) Since the category $\mathscr{C} := \mathsf{Rep}\,\mathfrak{u}_q(\mathfrak{g})$ is factorizable, one has

$$\mathscr{Z}(\mathscr{C}) \cong \mathscr{C} \boxtimes \mathscr{C}^{\mathrm{rev}}.$$

Thus, by (7), we have

$$\mathsf{BrPic}(\mathsf{Rep}\,\mathfrak{u}_q(\mathfrak{g})) \cong \mathsf{Aut}^{\mathrm{br}}(\mathscr{C} \boxtimes \mathscr{C}^{\mathrm{rev}}).$$

It follows from Lemma 2.5 that the only non-trivial tensor subcategories of $\mathscr{C} \boxtimes \mathscr{C}^{\mathrm{rev}}$ are $\mathscr{C}$ and $\mathscr{C}^{\mathrm{rev}}$, which are not braided equivalent by Proposition 5.3. Hence, any braided autoequivalence of $\mathscr{C} \boxtimes \mathscr{C}^{\mathrm{rev}}$ must preserve both factors. So we get

$$\mathsf{BrPic}(\mathsf{Rep}\,\mathfrak{u}_q(\mathfrak{g})) \cong \mathsf{Aut}^{\mathrm{br}}(\mathscr{C}) \times \mathsf{Aut}^{\mathrm{br}}(\mathscr{C}),$$

and the result follows from Theorem 5.6. □

Let $B \subset G$ be the Borel subgroup corresponding to $\mathfrak{b} \subset \mathfrak{g}$, and B^{ad} be the image of B in G^{ad}.

Corollary 5.10. *One has* $\mathsf{Aut}(\mathsf{Rep}\,\mathfrak{u}_q(\mathfrak{b})) \cong \Gamma \ltimes B^{\mathrm{ad}}$.

Proof. Let $\mathscr{C}$ be a finite tensor category. Let $\mathsf{Inv}(\mathscr{C})$ denote the group of isomorphism classes of invertible objects of $\mathscr{C}$. There is an exact sequence

$$\mathsf{Inv}(\mathscr{Z}(\mathscr{C})) \to \mathsf{Inv}(\mathscr{C}) \to \mathsf{Aut}(\mathscr{C}) \to \mathsf{BrPic}(\mathscr{C}),$$

see [GP, MN], where the first map is induced by the forgetful functor $\mathscr{Z}(\mathscr{C}) \to \mathscr{C}$, the second one sends an invertible object χ to the conjugation functor

$$X \mapsto \chi \otimes X \otimes \chi^{-1},$$

and the third one is given by $\theta(F) = \mathscr{C}$ with the usual left action of $\mathscr{C}$ and the right action of $\mathscr{C}$ twisted by F.

Now take $\mathscr{C} = \mathsf{Rep}\,\mathfrak{u}_q(\mathfrak{b})$. Then the above exact sequence takes the form

$$E \to E \to \mathsf{Aut}(\mathscr{C}) \to \mathsf{BrPic}(\mathscr{C}),$$

where the first map is the identity. Thus, the map $\theta\colon \mathsf{Aut}(\mathscr{C}) \to \mathsf{BrPic}(\mathscr{C})$ is injective. Hence, by Proposition 5.9(i), $\mathsf{Aut}(\mathscr{C}) \subset (\Gamma \ltimes G^{\mathrm{ad}}) \times O(E, \mathbf{q})$.

It is clear that $\mathsf{Aut}(\mathscr{C})$ contains the subgroup $\Gamma_{\mathrm{diag}} \ltimes B^{\mathrm{ad}} \subset (\Gamma \ltimes G^{\mathrm{ad}}) \times O(E, \mathbf{q})$, where $\Gamma_{\mathrm{diag}} \subset \Gamma \times O(E, \mathbf{q})$ is the diagonal copy of Γ. Also, one shows similarly to the proof of Proposition 5.2 (using the results of [GK]) that $\mathsf{Lie}\,\mathsf{Aut}(\mathscr{C}) = \mathfrak{b}$, hence

$$\mathsf{Aut}(\mathscr{C})_0 = B^{\mathrm{ad}} \subset G^{\mathrm{ad}}.$$

Thus, $\mathsf{Aut}(\mathscr{C})$ must normalize B^{ad}, hence $\mathsf{Aut}(\mathscr{C}) = \Gamma' \ltimes B^{\mathrm{ad}}$, where

$$\Gamma_{\mathrm{diag}} \subset \Gamma' \subset \Gamma \times O(E, \mathbf{q}).$$

It remains to show that $\Gamma' = \Gamma_{\mathrm{diag}}$. Let $F \in \mathsf{Aut}(\mathscr{C})$, and consider the action of F on the invertible objects $\mathsf{Inv}(\mathscr{C}) = E$. First of all, F must permute the objects $g_i \in E$ corresponding to the roots $\alpha_i \in Q$, since they are the only invertible objects which have a non-trivial Ext^1 with the unit object. Also, $\mathsf{Ext}^2(\mathbf{1}, g_i g_j) = 0$ if and only if i is connected to j in the Dynkin diagram of $\mathfrak{g}$ (as this is exactly the case when there is no quadratic relation between e_i and e_j). Thus, the permutation of g_i induced by F is implemented by an element of Γ. Hence, composing F with an element of Γ_{diag} if needed, we may assume that F acts trivially on E. Then $F \in G^{\mathrm{ad}} \cap \mathsf{Aut}(\mathscr{C}) = B^{\mathrm{ad}}$. This implies the required statement. □

Remark 5.11. Corollary 5.10 allows one to describe tensor autoequivalences of $\mathsf{Rep}\,\mathfrak{u}_q(\mathfrak{g})$ in terms of induction. Namely, given a tensor category $\mathscr{C}$, its indecomposable exact module categories are in bijection with Lagrangian algebras in $\mathscr{Z}(\mathscr{C})$. This bijection is given by

$$\mathscr{M} \mapsto I_{\mathscr{M}}(\mathbf{1}), \tag{8}$$

where $I_{\mathscr{M}}\colon \mathscr{C}^*_{\mathscr{M}} \to \mathscr{Z}(\mathscr{C})$ is the right adjoint to the forgetful functor $\mathscr{Z}(\mathscr{C}) \cong \mathscr{Z}(\mathscr{C}^*_{\mathscr{M}}) \to \mathscr{C}^*_{\mathscr{M}}$. As usual, $\mathscr{C}^*_{\mathscr{M}}$ denotes the dual tensor category of $\mathscr{C}$ with respect to $\mathscr{M}$. Furthermore, correspondence (8) is equivariant with respect to the isomorphism $\mathsf{BrPic}(\mathscr{C}) \cong \mathsf{Aut}^{\mathrm{br}}(\mathscr{Z}(\mathscr{C}))$. Here the group $\mathsf{BrPic}(\mathscr{C})$ (respectively, $\mathsf{Aut}^{\mathrm{br}}(\mathscr{Z}(\mathscr{C}))$) acts on the set of module categories (respectively, Lagrangian algebras) in an obvious way. The stabilizer of $\mathscr{M}$ in $\mathsf{Aut}^{\mathrm{br}}(\mathscr{Z}(\mathscr{C}))$ is the subgroup of autoequivalences induced from $\mathsf{Aut}(\mathscr{C}^*_{\mathscr{M}})$ and the orbit of $\mathscr{M}$ consists of module categories $\mathscr{N}$ such that $\mathscr{C}^*_{\mathscr{M}} \cong \mathscr{C}^*_{\mathscr{N}}$ [MN].

In our situation $\mathscr{C} = \mathsf{Rep}\,\mathfrak{u}_q(\mathfrak{b})$, the induction

$$\mathsf{Aut}(\mathsf{Rep}\,\mathfrak{u}_q(\mathfrak{b})) \to \mathsf{Aut}^{\mathrm{br}}(\mathsf{Rep}\,\mathfrak{u}_q(\mathfrak{g}))$$

is injective, and $\mathrm{Aut}^{\mathrm{br}}(\mathrm{Rep}\,\mathfrak{u}_q(\mathfrak{g}))/\mathrm{Aut}(\mathrm{Rep}\,\mathfrak{u}_q(\mathfrak{b}))$ is identified with the flag variety $G/B = G^{\mathrm{ad}}/B^{\mathrm{ad}}$. Point stabilizers are identified with images of inductions:

$$\mathrm{Aut}((\mathrm{Rep}\,\mathfrak{u}_q(\mathfrak{b}))^*_{\mathscr{M}}) \to \mathrm{Aut}^{\mathrm{br}}(\mathrm{Rep}\,\mathfrak{u}_q(\mathfrak{g}))$$

taken over module categories $\mathscr{M}$ such that $(\mathrm{Rep}\,\mathfrak{u}_q(\mathfrak{b}))^*_{\mathscr{M}} \cong \mathrm{Rep}\,\mathfrak{u}_q(\mathfrak{b})$. Since G^{ad} coincides with the union of its Borel subgroups and all Borel subgroups are conjugate, we conclude that every (braided) tensor autoequivalence of $\mathrm{Rep}\,\mathfrak{u}_q(\mathfrak{g})$ is induced from a tensor autoequivalence of a copy of $\mathrm{Rep}\,\mathfrak{u}_q(\mathfrak{b})$ (i.e., from a central tensor functor $\mathrm{Rep}\,\mathfrak{u}_q(\mathfrak{g}) \to \mathrm{Rep}\,\mathfrak{u}_q(\mathfrak{b})$).

Remark 5.12. The induction homomorphism $\mathrm{Aut}(\mathrm{Rep}\,\mathfrak{u}_q(\mathfrak{b})) \to \mathrm{Aut}^{\mathrm{br}}(\mathrm{Rep}\,\mathfrak{u}_q(\mathfrak{g}))$ and construction of Weyl reflections in $\mathrm{Aut}^{\mathrm{br}}(\mathrm{Rep}\,\mathfrak{u}_q(\mathfrak{g}))$ are discussed in [LP].

5.5 Twists for $\mathfrak{u}_q(\mathfrak{g})$

It is an interesting problem to classify twists for $\mathfrak{u}_q(\mathfrak{g})$ up to gauge transformations, i.e., categorically speaking, to classify fiber functors $F\colon \mathrm{Rep}\,\mathfrak{u}_q(\mathfrak{g}) \to \mathrm{Vec}$ up to isomorphism. By the results of [EK1, EK2], the answer to a similar question for the quantized universal enveloping algebra $U_\hbar(\mathfrak{g})$ is given in terms of Belavin-Drinfeld triples (see e.g., [KKSP]). On the other hand, twists for $\mathfrak{u}_q(\mathfrak{g})$ associated to Belavin-Drinfeld triples were worked out in [Ne] following the method of [EN] and [ESS]. Let us call them *Belavin-Drinfeld twists*, and call the corresponding fiber functors *Belavin-Drinfeld functors*.

Question 5.13. (see also [Ne], Question 9.5) Is any fiber functor on $\mathrm{Rep}\,\mathfrak{u}_q(\mathfrak{g})$ a composition of a Belavin-Drinfeld functor with a tensor autoequivalence of $\mathrm{Rep}\,\mathfrak{u}_q(\mathfrak{g})$? In other words, is any twist for $\mathfrak{u}_q(\mathfrak{g})$ gauge equivalent to a composition of a Belavin-Drinfeld twist with one coming from a twisted automorphism of $\mathfrak{u}_q(\mathfrak{g})$?

The answer is positive for $\mathfrak{g} = \mathfrak{sl}_2$ by [Mo, Proposition 8.11]. In this case there are no non-trivial Belavin-Drinfeld functors, so every fiber functor is the composition of the standard one with a tensor autoequivalence, and tensor autoequivalences form the group $PGL_2(k)$.

References

[AGP] I. Angiono, C. Galindo, M. Pereira, *De-equivariantization of Hopf algebras*, Algebras and Representation Theory **17** (2014), no. 1, 161–180.

[AG] S. Arkhipov, D. Gaitsgory, *Another realization of the category of modules over the small quantum group*, Adv. Math. **173** (2003), no. 1, 114–143.

[Bi1] J. Bichon, *The representation category of the quantum group of a non-degenerate bilinear form*, Comm. Algebra **31** (2003), no. 10, 4831–4851.

[Bi2] J. Bichon, *The group of bi-Galois objects over the coordinate algebra of the Frobenius-Lusztig kernel of* $SL(2)$, Glasg. Math. J. **58** (2016), no. 3, 727–738.

[BC] J. Bichon, G. Carnovale, *Lazy cohomology: An analogue of the Schur multiplier for arbitrary Hopf algebras*, J. Pure and Applied Algebra **204** (2006), no. 3, 627–665.

[BKa] J. Bichon and C. Kassel, *The lazy homology of a Hopf algebra*, J. Algebra **323** (2010), no. 9, 2556–2590.

[BG] K. Brown, K. Goodearl, *Lectures on algebraic quantum groups*, Birkhäuser (2012).

[DL] C. De Concini, V. Lyubashenko, *Quantum function algebra at roots of 1*, Adv. Math. **108** (1994), no. 2, 205–262.

[DN] A. Davydov, D. Nikshych, *The Picard crossed module of a braided tensor category*, Algebra and Number Theory, **7** (2013), no. 6, 1365–1403.

[Da1] A. Davydov, *Twisting of monoidal structures*, Preprint of MPI (1995), MPI/95-123, arXiv:q-alg/9703001.

[Da2] A. Davydov, *Twisted automorphisms of Hopf algebras*, in Noncommutative structures in Mathematics and Physics, Koninklijke Vlaamse Academie Van Belgie Voor Wetenschappen en Kunsten (2010) 103–130, arXiv:0708.2757.

[Da3] A. Davydov, *Twisted derivations of Hopf algebras*, J. Pure and Applied Algebra **217** (2013), 567–582.

[DGNO] V. Drinfeld, S. Gelaki, D. Nikshych, V. Ostrik, *On braided fusion categories I*, Selecta Mathematica, **16** (2010), no. 1, 1–119.

[DM] P. Deligne, J. Milne, *Tannakian categories*, Lecture Notes in Mathematics **900** (1982).

[EK1] P. Etingof, D. Kazhdan, *Quantization of Lie bialgebras I*, Selecta Math. **2** (1996), no. 1, 1–41.

[EK2] P. Etingof, D. Kazhdan, *Quantization of Lie bialgebras II*, Selecta Math. **4** (1998), no. 2, 213–231.

[Eis] D. Eisenbud, *Commutative algebra with a view towards algebraic geometry*, Vol. 150. Springer Science and Business Media (2013).

[EGNO] P. Etingof, S. Gelaki, D. Nikshych, V. Ostrik, *Tensor categories*, Mathematical Surveys and Monographs, **205**, American Mathematical Society (2015).

[ENO] P. Etingof, D. Nikshych, and V. Ostrik. *Fusion categories and homotopy theory*, Quantum Topology, **1** (2010), no. 3, 209–273.

[EN] P. Etingof, D. Nikshych, *Dynamical quantum groups at roots of 1*, Duke Math. J. **108** (2001), no. 1, 135–168.

[EO] P. Etingof, V. Ostrik, *Finite tensor categories*, Moscow Math. J. **4** (2004), 627–654.

[ESS] P. Etingof, T. Schedler, O. Schiffmann, *Explicit quantization of dynamical r-matrices for finite-dimensional semisimple Lie algebras*, J. Amer. Math. Soc. **13** (2000), 595–609.

[FRT] L.D. Faddeev, N.Yu. Reshetikhin, and L.A. Takhtajan, *Quantization of Lie groups and Lie algebras*, in Algebraic Analysis, Vol. I (M. Kashiwara and T. Kawai, eds.), Academic Press, Boston, 1988, 129–139.

[FS] J. Fuchs, C. Schweigert, *Symmetries and defects in three-dimensional topological field theory*, String-Math 2014, 21–40, Proc. Sympos. Pure Math. **93**, American Mathematical Society (2016).

[GP] C. Galindo, J. Plavnik, *Tensor functors between Morita duals of fusion categories*, Letters in Math. Phys. **107** (2017), no. 3, 553–590.

[GK] V. Ginzburg, S. Kumar, *Cohomology of quantum groups at roots of unity*, Duke Math. J. **69** (1993), no. 1, 179–198.

[GKa] P. Guillot, C. Kassel, *Cohomology of invariant Drinfeld twists on group algebras*, Int. Math. Res. Notices (2009), 1894–1939.

[Gri] R.I. Grigorchuk, *On the system of defining relations and the Schur multiplier of periodic groups generated by finite automata*, in Groups St. Andrews 1997 in Bath, I, volume 260 of London Math. Soc. Lecture Note Ser., 290–317. Cambridge Univ. Press, Cambridge, 1999.

[Ha] T. Hayashi, *Quantum deformation of classical groups*, Publ. Research Inst. Math. Sci. (Kyoto) **28** (1992), 57–81.

[I] R. Iglesias, *Bitableaux bases of the quantum coordinate algebra of a semisimple group*, J. Algebra **301** (2006), no. 1, 308–336.

[Is] J.R. Isbell, *Epimorphisms and dominions*, Proc. Conf. Categorical Algebra (La Jolla, Calif., 1965), Springer, New York, 1966, 232–246.

[KW] D. Kazhdan, H. Wenzl, *Reconstructing monoidal categories*, I.M. Gelfand Seminar, Amer. Math. Soc., Providence, RI, 1993, 111–136.

[KS] L. Korogodsky, Y. Soibelman, *Algebras of functions on quantum groups: Part I*, Mathematical Surveys and Monographs, **56**, American Mathematical Society (1998).

[KKSP] B. Kadets, E. Karolinsky, A. Stolin, and I. Pop, *Classification of quantum groups and Belavin-Drinfeld cohomologies*, Comm. Mathematical Physics **344** (2016), no. 1, 1–24.

[LP] S. Lentner, J. Priel, *Three natural subgroups of the Brauer-Picard group of a Hopf algebra with applications* arXiv:1702.05133 [math.QA].

[Lu1] G. Lusztig, *finite-dimensional Hopf algebras arising from quantized universal enveloping algebras*, J. Amer. Math. Soc. **3** (1990), no. 1, 257–296.

[Lu2] G. Lusztig, *Modular representations and quantum groups*, Classical groups and related topics: Proceedings of a conference in honor of L.K. Hua, Contemporary Mathematics **82**, Amer. Math. Soc. (1989), 59–77.

[Lu3] G. Lusztig, *Study of a* $\mathbf{Z}$*-form of the coordinate ring of a reductive group*, J. Amer. Math. Soc. **22** (2009), no. 3, 739–769.

[MN] I. Marshall, D. Nikshych, *On the Brauer-Picard groups of fusion categories*, arXiv: 1603.04318 [math.QA].

[Mo] M. Mombelli, *Module categories over pointed Hopf algebras*, Mathematische Zeitschrift **266** (2010), no. 2, 319–344.

[Ne] C. Negron, *Small quantum groups associated to Belavin-Drinfeld triples*, arXiv: 1701.00283 [math.QA].

[NT1] S. Neshveyev, L. Tuset, *Symmetric invariant cocycles on the duals of q-deformations*, Adv. Math. **227** (2011), no. 1, 146–169.

[NT2] S. Neshveyev, L. Tuset, *Autoequivalences of the tensor category of* $U_q(\mathfrak{g})$*-modules*, Int. Math. Res. Notices (2012), no. 15, 3498–3508.

[RTF] N.Yu. Reshetikhin, L.A. Takhtajan, and L.D. Faddeev, *Quantization of Lie groups and Lie algebras*, Leningrad Math. J. **1** (1990), 193–225.

[Ta] M. Takeuchi, *Quantum orthogonal and symplectic groups and their embedding into quantum GL*, Proc. Japan Acad., Ser. A Math. Sci. **65** (1989), 55–58.

[T] V. Turaev, *Quantum invariants of knots and 3-manifolds*, W. de Gruyter (1994).

[TW] I. Tuba, H. Wenzl, *On braided tensor categories of type BCD*, Journal für die reine und angewandte Mathematik **581** (2005), 31–69.

[Y] D. Yetter. *Braided deformations of monoidal categories and Vassiliev invariants*, In: "Higher category theory", *Contemp. Math. AMS* **230** (1998), 117–134.

Nil-Hecke Algebras and Whittaker $\mathscr{D}$-Modules

Victor Ginzburg

To the memory of Bertram Kostant

Abstract Given a semisimple group G, Kostant and Kumar defined a nil-Hecke algebra that may be viewed as a degenerate version of the double affine nil-Hecke algebra introduced by Cherednik. In this paper, we construct an isomorphism of the spherical subalgebra of the nil-Hecke algebra with a Whittaker type quantum Hamiltonian reduction of the algebra of differential operators on G. This result has an interpretation in terms of geometric Satake and the Langlands dual group. Specifically, the isomorphism provides a bridge between very differently looking descriptions of equivariant Borel-Moore homology of the affine flag variety (due to Kostant and Kumar) and of the affine Grassmannian (due to Bezrukavnikov and Finkelberg), respectively.

It follows from our result that the category of Whittaker $\mathscr{D}$-modules on G, considered by Drinfeld, is equivalent to the category of holonomic modules over the nil-Hecke algebra, and it is also equivalent to a certain subcategory of the category of Weyl group equivariant holonomic $\mathscr{D}$-modules on the maximal torus.

Keywords: Hamiltonian reduction • differential operators • Hecke algebra

Mathematics Subject Classification (2010): 20G20, 17B20

This work was supported in part by NSF grant DMS-1303462.

Department of Mathematics, University of Chicago, Chicago, IL 60637, USA, e-mail: ginzburg@math.uchicago.edu

V. G. Kac, V. L. Popov (eds.), *Lie Groups, Geometry, and Representation Theory*,
Progress in Mathematics 326, https://doi.org/10.1007/978-3-030-02191-7_6

1 Introduction

1.1 Reminder on nil-Hecke algebras

In this paper, we work over $\mathbb{C}$. We use the notation $\operatorname{Sym}\mathfrak{k}$, resp. $\mathscr{U}\mathfrak{k}$, for the symmetric, resp. enveloping, algebra of a vector space, resp. Lie algebra, $\mathfrak{k}$. Let $\mathscr{T}^*X$, resp. $\mathscr{D}_X$ and $\mathscr{D}(X)$, denote the cotangent bundle, resp. the sheaf and ring of algebraic differential operators, on a smooth algebraic variety X. Throughout the paper, we fix a complex connected and simply connected semisimple group G with Lie algebra $\mathfrak{g}$.

Let $\mathfrak{h}$ be a finite-dimensional vector space, $\mathscr{R} \subset \mathfrak{h}^*$ a reduced root system with the set Σ of simple roots. An associated Coxeter group $\mathscr{W}$ acts naturally on $\mathfrak{h}$ and it is generated by a set s_α, $\alpha \in \Sigma$, of simple reflections. In their work on equivariant cohomology of flag varieties, Kostant and Kumar [KK1]–[KK2] introduced a noncommutative $\mathbb{Z}$-graded algebra $\mathscr{H}(\mathfrak{h},\mathscr{W})$ called a nil-Hecke algebra, cf. Section 7.2 for an overview. The algebra $\mathscr{H}(\mathfrak{h},\mathscr{W})$ is generated by the vector space $\mathfrak{h}$ and a collection, θ_α, $\alpha \in \Sigma_{\text{aff}}$, of Demazure elements. The elements of $\mathfrak{h}$ pairwise commute and generate a copy of the algebra $\operatorname{Sym}\mathfrak{h}$ inside $\mathscr{H}(\mathfrak{h},\mathscr{W})$. Demazure elements satisfy the braid relations. The other defining relations among the generators of $\mathscr{H}(\mathfrak{h},\mathscr{W})$ are as follows , cf. (7.1.3):

$$(\theta_\alpha)^2 = 0, \qquad \theta_\alpha \cdot s_\alpha(h) - h \cdot \theta_\alpha = \langle \alpha, h \rangle, \qquad \forall\, h \in \mathfrak{h},\, \alpha \in \Sigma.$$

Let T be the (abstract) maximal torus of G and $\mathfrak{t} = \operatorname{Lie} T$. Let $\mathbb{X}^* = \mathbb{X}^*(T)$ be the weight lattice and W the abstract Weyl group. Further, let $\mathfrak{t}_{\text{aff}}$ be the affine Cartan algebra, W_{aff} the affine Weyl group, and $\mathscr{H}(\mathfrak{t}_{\text{aff}}, W_{\text{aff}})$ the corresponding nil-Hecke algebra. It is convenient to enlarge the group W_{aff} and consider $\widetilde{W} = W \ltimes \mathbb{X}^*$, an extended affine Weyl group. Similarly, there is an enlargement, $\mathscr{H}(\mathfrak{t}_{\text{aff}}, \widetilde{W})$, of $\mathscr{H}(\mathfrak{t}_{\text{aff}}, W_{\text{aff}})$. This is a $\mathbb{Z}$-graded $\mathbb{C}[\hbar]$-algebra that may be viewed as a degeneration of the *nil-DAHA* introduced by Cherednik and studied further by Fegin and Cherednik [CF].

Below, we will mostly be interested in the algebra $\mathbb{H} := \mathscr{H}(\mathfrak{t}_{\text{aff}}, \widetilde{W})|_{\hbar=1}$, a specialization of $\mathscr{H}(\mathfrak{t}_{\text{aff}}, \widetilde{W})$ at $\hbar = 1$. The grading on $\mathscr{H}(\mathfrak{t}_{\text{aff}}, \widetilde{W})$ induces an ascending $\mathbb{Z}$-filtration on $\mathbb{H}$. By construction, one has $\operatorname{gr}\mathbb{H} = \mathscr{H}(\mathfrak{t}_{\text{aff}}, \widetilde{W})|_{\hbar=0}$, resp. $\mathbb{H}_\hbar = \mathscr{H}(\mathfrak{t}_{\text{aff}}, \widetilde{W})$, where $\operatorname{gr}A$, resp. $A_\hbar$, denotes an associated graded, resp. Rees algebra, of a filtered algebra A. The algebra $\mathbb{H}$ is a kind of (micro)-localization of the cross product $W \ltimes \mathscr{D}(T)$. In particular, there are algebra embeddings $\operatorname{Sym}\mathfrak{t} \hookrightarrow \mathscr{D}(T) \hookrightarrow \mathbb{H}$, where $\operatorname{Sym}\mathfrak{t}$ is identified with the algebra of translation invariant differential operators on the torus T. The filtration on $\mathbb{H}$ agrees with the natural filtration on $\mathscr{D}(T)$ by order of the differential operator.

Let $\mathbf{e} = \frac{1}{\#W} \sum_{w \in W} w \in \mathbb{C}W$ be the symmetrizer idempotent. The algebra $\mathbb{H}^{\text{sph}} = \mathbf{e}\mathbb{H}\mathbf{e}$, called *spherical subalgebra*, is a filtered algebra with unit $\mathbf{e}$. The embeddings above restrict to algebra embeddings $(\operatorname{Sym}\mathfrak{t})^W \hookrightarrow \mathscr{D}(T)^W \hookrightarrow \mathbb{H}^{\text{sph}}$.

1.2 Main results

Let K be a linear algebraic group and X a smooth K-variety. Given a character $\chi : \mathfrak{k} = \mathrm{Lie}\,K \to \mathbb{C}$, we write $\mathfrak{k}^\chi$ for the image of $\mathfrak{k}$ in $\mathrm{Sym}\,\mathfrak{k}$, resp. $\mathscr{U}\mathfrak{k}, \mathscr{D}(X)$, etc., under the map $k \mapsto k - \chi(k)$.

Throughout the paper, we fix a maximal unipotent subgroup N of G and a nondegenerate character $\psi : \mathfrak{n} = \mathrm{Lie}\,N \to \mathbb{C}$. Let $N \times N$ act on G by $(n_l, n_r) : g \mapsto n_l^{-1} g n_r$. We write N_l, resp. N_r, for the first, resp. second, factor of $N \times N$. (We will use subscripts 'l' and 'r' in other similar contexts.) Thus, we get a pair, $\mathfrak{n}_l^\psi$ and $\mathfrak{n}_r^\psi$, of commuting Lie subalgebras of $\mathscr{D}(G)$. We define the following quantum Hamiltonian reduction

$$\mathbb{W} := \big(\mathscr{D}(G)/\mathscr{D}(G)(\mathfrak{n}_l^\psi + \mathfrak{n}_r^\psi)\big)^{N_l \times N_r}.$$

This is an associative algebra that comes equipped with a natural ascending $\mathbb{Z}$-filtration, called Kazhdan filtration, cf. Section 3.

We identify $Z(\mathscr{U}\mathfrak{g})$, the center of $\mathscr{U}\mathfrak{g}$, with the algebra of G-bi-invariant differential operators on G. The embedding of the algebra of G-bi-invariant differential operators into $\mathscr{D}(G)$ induces an injective homomorphism $Z(\mathscr{U}\mathfrak{g}) \to \mathbb{W}$ of filtered algebras.

One of our main results reads as follows:

Theorem 1.2.1. *There is an isomorphism* $\mathbb{W} \xrightarrow{\sim} \mathbb{H}^{\mathrm{sph}}$*, of filtered algebras, that maps the subalgebra* $Z(\mathscr{U}\mathfrak{g}) \subset \mathbb{W}$ *to the subalgebra* $(\mathrm{Sym}\,\mathfrak{t})^W \subset \mathbb{H}^{\mathrm{sph}}$*. The resulting map* $Z(\mathscr{U}\mathfrak{g}) \to (\mathrm{Sym}\,\mathfrak{t})^W$ *is the Harish-Chandra isomorphism.*

Theorem 1.2.1 has a classical (a.k.a. Poisson) counterpart that involves the moment map $\mu_{\mathfrak{n}\times\mathfrak{n}} : \mathscr{T}^*G \to \mathfrak{n}^* \times \mathfrak{n}^*$, associated with the Hamiltonian $N_l \times N_r$-action on $\mathscr{T}^*G$ induced by the one on G. The variety $\mathfrak{Z} = \mu^{-1}(\psi \times \psi) /\!\!/ (N_l \times N_r)$, a classical Hamiltonian reduction, comes equipped with the structure of an integrable system. Specifically, $\mathfrak{Z}$ is a smooth symplectic algebraic variety equipped with a natural smooth Lagrangian fibration $\kappa : \mathfrak{Z} \to \mathfrak{g}^* /\!\!/ \mathrm{Ad}^* G$ whose fibers are abelian algebraic groups. The resulting group scheme on $\mathfrak{g}^* /\!\!/ \mathrm{Ad}^* G$ is known as the *universal centralizer*, cf. Section 2 for a review. Our second result, to be proved in Section 4.3, reads as follows:

Theorem 1.2.2. *There is an isomorphism* $\mathrm{gr}\,\mathbb{W} \cong \mathbb{C}[\mathfrak{Z}]$ *of graded Poisson algebras, which restricts to an isomorphism* $\mathrm{gr}(Z\mathfrak{g}) \cong \kappa^*(\mathbb{C}[\mathfrak{g}^* /\!\!/ \mathrm{Ad}^* G])$*, of maximal commutative subalgebras.*

As a consequence of the two theorems above, one obtains an isomorphism $\mathrm{gr}\,\mathbb{H}^{\mathrm{sph}} \cong \mathbb{C}[\mathfrak{Z}]$, which shows that the algebra $\mathbb{H}^{\mathrm{sph}}$ may be viewed as a quantization of the symplectic variety $\mathfrak{Z}$. A slightly modified form of the isomorphism $\mathrm{gr}\,\mathbb{H}^{\mathrm{sph}} \cong \mathbb{C}[\mathfrak{Z}]$ that does not however reveal a connection with nil-Hecke algebras has been proved earlier by Bezrukavnikov, Finkelberg, and Mirkovic [BFM, Proposition 2.8].

We remark that Theorem 1.2.2 neither implies, nor is a simple consequence of Theorem 1.2.1, due to the fact that the filtrations and gradings on the algebras involved are not bounded below.

1.3 Strategy of the proof of Theorem 1.2.1

The theorems above are, in fact, formal consequences of a combination of results (to be recalled in Section 1.4) of Kostant and Kumar, [KK1], [KK2], on the one hand, and of Bezrukavnikov and Finkelberg, [BF], on the other hand. That approach is, however, rather indirect; it relies on the geometric Satake and involves the Langlands dual group. Thus, one of our primary motivations was to find a more direct geometric approach.

The strategy of our approach to the isomorphism $\mathbb{H}^{\mathrm{sph}} \cong \mathbb{W}$ is as follows. First, we construct an algebra homomorphism $\mathscr{D}(T)^W \to \mathbb{W}$. Then, we show that this homomorphism can be extended to a homomorphism $\mathbb{H}^{\mathrm{sph}} \to \mathbb{W}$. Finally, in Section 8.2 we prove using Theorem 7.1.4 that the latter homomorphism $\mathbb{H}^{\mathrm{sph}} \to \mathbb{W}$ is an isomorphism.

We should point out that we do *not* know how to construct the homomorphism $\mathscr{D}(T)^W \to \mathbb{W}$ directly. The construction of such a homomorphism is not obvious even quasi-classically, where it amounts to a construction of Beilinson and Kazhdan [BK], to be recalled in Section 2.3. Thus, we use an indirect approach that involves the algebra $\mathscr{D}(G/\bar{N})$, where $\bar{N}$ is a maximal unipotent subgroup opposite to N. Put $\bar{\mathfrak{n}} = \operatorname{Lie}\bar{N}$. Our construction starts with the following Hamiltonian reduction

$$A := \big(\mathscr{D}(G/\bar{N})/\mathscr{D}(G/\bar{N})\mathfrak{n}_l^{\psi}\big)^{N_l} = \big(\mathscr{D}(G)/\mathscr{D}(G)(\mathfrak{n}_l^{\psi} + \bar{\mathfrak{n}}_r)\big)^{N_l \times \bar{N}_r}.$$

There is a Weyl group action on $\mathscr{D}(G/\bar{N})$ by algebra automorphisms that has been introduced by Gelfand and Graev a long time ago. The Gelfand–Graev W-action descends to A, so one has an algebra $W \ltimes A$. The key ingredient of our approach is a $(W \ltimes A, \mathbb{W})$-bimodule $\mathbb{M}$, the *Miura bimodule* defined in Section 6, an object closely related to the one introduced by D. Kazhdan and the author in [GK]. We prove (Theorem 5.3.1) that the algebra A is isomorphic to $\mathscr{D}(T)$; moreover, the isomorphism intertwines the W-action on A and the natural W-action on $\mathscr{D}(T)$, see Proposition 5.5.2. Thus, we may view $\mathbb{M}$ as a $(W \times \mathscr{D}(T), \mathbb{W})$-bimodule. Further, using a general criterion of Proposition 7.2.4, we deduce that the left action of $W \times \mathscr{D}(T)$ on $\mathbb{M}$ can be extended to an action of $\mathbb{H}$, a larger algebra. Now, the Miura bimodule comes equipped with a canonical generator $1_{\mathbb{M}} \in \mathbb{M}$. We prove that for every element $a \in \mathbb{H}^{\mathrm{sph}} \subset \mathbb{H}$ there exists a uniquely determined element $a_{\mathbb{W}} \in \mathbb{W}$, such that the equation $a1_{\mathbb{M}} = 1_{\mathbb{M}}a_{\mathbb{W}}$ holds. The assignment $a \mapsto a_{\mathbb{W}}$ yields the desired homomorphism $\mathbb{H}^{\mathrm{sph}} \to \mathbb{W}$.

1.4 Relation to equivariant homology of flag varieties

Let $\check{G}$ be the *Langlands dual* group of G and $\check{T}$ the maximal torus of $\check{G}$. Let $I \subset \check{G}((z))$ be an Iwahori subgroup. Let $\mathscr{B}_{\mathrm{aff}} = \check{G}((z))/I$ be the affine flag variety, resp. $\mathsf{Gr} = \check{G}((z))/\check{G}[[z]]$ the affine Grassmannian.

Given a group K and a K-action on a space X, let $H^\bullet_K(X)$, resp. $H^K_\bullet(X)$, denote equivariant cohomology, resp. homology, of X. One has canonical isomorphisms

$$H^\bullet_{I\rtimes\mathbb{G}_m}(\mathrm{pt}) = H^\bullet_{\check{T}\times\mathbb{G}_m}(\mathrm{pt}) = (\mathrm{Sym}\,\mathfrak{t})[\hbar], \quad \text{resp.} \quad H^\bullet_{\check{G}[[z]]\rtimes\mathbb{G}_m}(\mathrm{pt}) = (\mathrm{Sym}\,\mathfrak{t})^W[\hbar].$$

In the theorems below, the algebra structure on equivariant homology of $\mathscr{B}_{\mathrm{aff}}$, resp. Gr, is given by convolution. The multiplicative group $\mathbb{G}_m$ acts on $\mathscr{B}_{\mathrm{aff}}$, resp. Gr, by loop rotation.

Kostant and Kumar, [KK1], [KK2], [Ku] proved the following theorem which is a Kac-Moody generalization of a well-known result of Bernstein-Gelfand-Gelfand, [BGG].

Theorem 1.4.1. *There are graded algebra isomorphisms*

$$H^{I\rtimes\mathbb{G}_m}_\bullet(\mathscr{B}_{\mathrm{aff}}) \cong \mathbb{H}_\hbar, \quad H^{\check{G}[[z]]\rtimes\mathbb{G}_m}_\bullet(\mathsf{Gr}) \cong \mathbb{H}^{\mathrm{sph}}_\hbar.$$

On the other hand, one has the following result.

Theorem 1.4.2. *There are graded algebra isomorphisms*

$$H^{\check{G}[[z]]}_\bullet(\mathsf{Gr}) \cong \mathrm{gr}\,\mathbb{W}, \quad H^{\check{G}[[z]]\rtimes\mathbb{G}_m}_\bullet(\mathsf{Gr}) \cong \mathbb{W}_\hbar.$$

Here, the first isomorphism is due to Bezrukavnikov, Finkelberg, and Mirković [BFM], Theorem 2.1.2(b) and Proposition 2.8(b), and the second isomorphism is due to Bezrukavnikov and Finkelberg [BF], Theorem 3. Thus, our Theorem 1.2.1 is a formal consequence of a combination of the second isomorphism in Theorem 1.4.1 and Theorem 1.4.2, respectively.

1.5 The Whittaker category

Let A be a $\mathbb{Z}$-filtered algebra such that $\mathrm{gr}A$ is a finitely generated commutative algebra. Given a finitely generated A-module M, one can choose a good filtration on M and let $\mathrm{SS}(M) = \mathrm{supp}(\mathrm{gr}M)$. We say that M is holonomic if $\dim \mathrm{SS}(M) \le \frac{1}{2}\dim \mathrm{Spec}(\mathrm{gr}A)$. (Although this definition is certainly not reasonable in the generality of arbitrary filtered algebras A as above, it is sufficient for our limited purposes.) We write A-hol for the abelian category of holonomic left A-modules. In the case of the algebra $\mathbb{H}$, one has $\mathrm{gr}\,\mathbb{H} = W \ltimes \mathscr{A}$, where $\mathscr{A}$ is a commutative algebra, cf. Proposition 7.2.7. We define the notion of holonomicity by replacing $\mathrm{gr}\,\mathbb{H}$ by $\mathscr{A}$ in the previous definition.

Let K be a connected linear algebraic group, Ω_K the canonical bundle on K, and $\chi : \mathfrak{k} = \mathrm{Lie}\,K \to \mathbb{C}$ a character. We will work with right $\mathscr{D}$-modules and put $e^\chi\Omega_K := \mathscr{D}_K/\mathfrak{k}^\chi\mathscr{D}_K$. This is a line bundle on K equipped with a flat connection that does not necessarily have regular singularities, in general. Let X be a smooth variety equipped with a K-action $a_K : K \times X \to X$. Let pr_1, resp. pr_2, denote the first, resp. second, projection $K \times X \to X$. A (K,χ)-*Whittaker* $\mathscr{D}_X$-module is, by definition, a $\mathscr{D}_X$-module $\mathscr{F}$ equipped with an isomorphism $a_K^*\mathscr{F} \cong e^\chi \Omega_K \boxtimes \mathscr{F}$ of $\mathscr{D}_{K\times X}$-modules that satisfies an appropriate cocycle condition (here and elsewhere, we abuse notation and write $\mathscr{F}' \boxtimes \mathscr{F}$ for $pr_1^*\mathscr{F}' \otimes_{\mathscr{O}_{K\times X}} pr_2^*\mathscr{F}$). There is an abelian category $(\mathscr{D}_X, K, \chi)$-mod of (K,χ)-Whittaker $\mathscr{D}_X$-modules. In the special case where $\chi = 0$, one gets the category $(\mathscr{D}_X, K)$-mod of K-equivariant $\mathscr{D}_X$-modules. If X is affine, we may (and will) identify $\mathscr{D}_X$-modules with $\mathscr{D}(X)$-modules via the functor of global sections. If, moreover, the group K is unipotent, then a (K,χ)-Whittaker structure on $\mathscr{F}$ amounts to the property (rather than an additional structure) that, for any $k \in \mathfrak{k}$, the action of $k_{\mathscr{D}} - \chi(k)$ on $\Gamma(X,\mathscr{F})$ is locally nilpotent, where $k_{\mathscr{D}}$ denotes the action that comes from the $\mathscr{D}(X)$-action.

Following Drinfeld, we consider the category $(\mathscr{D}_G, N_l \times N_r, \psi \times \psi)$-mod and its full subcategory, to be denoted $\mathscr{W}h$ and called the *Whittaker category*, whose objects are *holonomic* $\mathscr{D}$-modules. Drinfeld raised the question of finding a description of an ℓ-adic counterpart of the Whittaker category in terms of W-equivariant sheaves on the maximal torus T. The following result provides, in particular, an answer to an analogous question in the $\mathscr{D}$-module setting.

Theorem 1.5.1. *The Whittaker category $\mathscr{W}h$ is equivalent to any of the following categories:*

(1) *The category of holonomic $\mathbb{W}$-modules;*
(2) *The category of holonomic $\mathbb{H}$-modules;*
(3) *The full subcategory, $\mathscr{C}at(T,W)$, of the category of W-equivariant holonomic $\mathscr{D}(T)$-modules whose objects M have the property that the following natural map is an isomorphism:*

$$\operatorname{Sym}\mathfrak{t} \otimes_{(\operatorname{Sym}\mathfrak{t})^W} M^W \to M; \tag{1.5.2}$$

(4) *The full subcategory of the category of holonomic $\mathscr{D}(T)^W$-modules whose objects L have the property that the map*

$$\operatorname{Sym}\mathfrak{t} \otimes_{(\operatorname{Sym}\mathfrak{t})^W} L \to \mathscr{D}(T) \otimes_{\mathscr{D}(T)^W} L,$$

induced by the inclusion $\operatorname{Sym}\mathfrak{t} \hookrightarrow \mathscr{D}(T)$, *is an isomorphism.*

Theorem 1.5.1 will be proved in Section 8.3.

Remark 1.5.3. There is a canonical projection $\mathfrak{t}^*/_{\mathrm{st}}W \to \mathfrak{t}^*/W$, of a stacky quotient of $\mathfrak{t}^*$ by W to the categorical quotient. Isomorphism (1.5.2) means that M, viewed as a quasi-coherent sheaf on $\mathfrak{t}^*/_{\mathrm{st}}W$, descends to $\mathfrak{t}^*/W$. ◊

In the recent work [Lo1], G. Lonergan established an equivalence of categories which is analogous to the equivalence $\mathscr{W}h \cong \mathscr{C}at(T,W)$ in the theorem above. However, the important holonomicity condition is not present in the setting of [Lo1]. The approach in [Lo1] is of a topological nature; it relies on the geometric Satake equivalence (via [BF]) and on some general results of Goresky, Kottwitz, and MacPherson, [GKM], cf. also [LLMSSZ]. That approach is totally different from ours.

Our proof of Theorem 1.5.1 is closely related to the theorem below that may be of independent interest and which is, in a sense, 'dual' to the statement of [BBM], Theorem 1.5(1).

Given a map $f: X \to Y$, let $\int_f^i$ denote the ith derived push-forward functor on $\mathscr{D}$-modules. Let $a_{\bar{N}}: \bar{N} \times G \to G,\ (\bar{n}, g) \mapsto g\bar{n}$ be the action.

Theorem 1.5.4. *For any $M \in (\mathscr{D}_G, N_r, \psi)$-mod, we have $\int_{a_{\bar{N}}}^j (\Omega_{\bar{N}} \boxtimes M) = 0,\ \forall j \neq 0$. In particular, the functor $(\mathscr{D}_G, N_r, \psi)$-mod $\to (\mathscr{D}_G, \bar{N}_r)$-mod, $M \mapsto \int_{a_{\bar{N}}}^0 (\Omega_{\bar{N}} \boxtimes M)$ is exact.*

The proof of Theorem 1.5.4 is given in Section 6.1.

1.6 The Whittaker functor

Let X be a smooth $G \times G$-variety, $G \subset G \times G$ the diagonal, and $(\mathscr{D}_X, G)$-mod the corresponding category of G-equivariant $\mathscr{D}_X$-modules. Write $N_r = 1 \times N \subset G \times G$. It is easy to check that one has a well-defined functor

$$\Psi:\ (\mathscr{D}_X, G)\text{-mod} \longrightarrow (\mathscr{D}_X, N_l \times N_r, \psi \times \psi)\text{-mod},$$
$$M \mapsto \int_{a_{N_r}}^0 (e^{\Psi}\Omega_N \boxtimes M). \quad (1.6.1)$$

Assume now that X is affine and form a Hamiltonian reduction

$$\mathfrak{D} := (\mathscr{D}(X)/\mathfrak{n}_r^{\psi}\mathscr{D}(X))^{N_r}.$$

The map $\mathfrak{n}_l^{\psi} \to \mathscr{D}(X)$ descends to a map $\mathfrak{n}^{\psi} \to \mathfrak{D}$. We let $(\mathfrak{D}, \mathfrak{n}_l^{\psi})$-mod be a full subcategory of the category of right $\mathfrak{D}$-modules, such that the induced action of the Lie algebra $\mathfrak{n}_l^{\psi}$ on the module is locally nilpotent. One checks that for a right $\mathscr{D}(X)$-module M, the space $M/M\mathfrak{n}_r^{\psi}$ of $\mathfrak{n}_r^{\psi}$-coinvariants, has the natural structure of a right $\mathfrak{D}$-module; furthermore, the resulting $\mathfrak{n}_l^{\psi}$-action on $M/M\mathfrak{n}_r^{\psi}$ is locally nilpotent if M is G-equivariant. Also, the space $\mathscr{D}(X)/\mathfrak{n}_r^{\psi}\mathscr{D}(X)$ has the structure of a $(\mathfrak{D}, \mathscr{D}(X))$-bimodule. We have the following functors:

$$(\mathscr{D}(X), G)\text{-mod} \xrightarrow{M \mapsto M/M\mathfrak{n}_r^{\psi}} (\mathfrak{D}, \mathfrak{n}_l^{\psi})\text{-mod}$$
$$\xrightarrow{-\otimes_{\mathfrak{D}} \mathscr{D}(X)/\mathfrak{n}_r^{\psi}\mathscr{D}(X)} (\mathscr{D}(X), N_l \times N_r, \psi \times \psi)\text{-mod}. \quad (1.6.2)$$

Theorem 1.6.3. *Let X be an affine $G \times G$-variety. Then the composite functor in* (1.6.2) *is isomorphic to Ψ and the second functor in* (1.6.2) *is an equivalence. Fur-*

thermore, for any $M \in (\mathscr{D}(X), G)$-mod *and all* $j \neq 0$*, we have* $H_j(\mathfrak{n}_r^\psi, M) = 0$*, resp.* $\int^j_{a_{N_r}}(\Omega_{N_r} \boxtimes M) = 0$. [1]

Thus, the first functor in diagram (1.6.2) *is exact and it takes holonomic modules to holonomic modules.*

Below, we are interested in a special case of the above setting where $X = G$ is viewed as a $G \times G$-variety via left and right translations. The diagonal G-action on X corresponds to the conjugation action of G on itself, so $(\mathscr{D}_X, G)$-mod is, in this case, the category of $\mathrm{Ad}\, G$-equivariant $\mathscr{D}_G$-modules. Further, there is an equivalence $(\mathscr{D}(G), N_l \times N_r, \psi \times \psi)$-mod $\cong \mathbb{W}$-mod, see Section 3.1 and Lemma 5.4.1. Also, one checks that for any $\mathscr{D}(G)$-module M, the subspace $(M/M\mathfrak{n}_r^\psi)^{\mathfrak{n}_l^\psi} \subset M/M\mathfrak{n}_r^\psi$ of $\mathfrak{n}_l^\psi$-invariants has the natural structure of a $\mathbb{W}$-module, see Section 3.1. From the above theorem, we will deduce the following result.

Theorem 1.6.4. (i) *For any* $M \in (\mathscr{D}(G), \mathrm{Ad}\, G)$-mod *and all* $j \neq 0$*, we have* $H_j(\mathfrak{n}_r^\psi, M) = 0$*, resp.* $\int^j_{a_{N_r}}(\Omega_{N_r} \boxtimes M) = 0$.

(ii) *The functor* (1.6.1) *corresponds, via the equivalence*

$$(\mathscr{D}(G), N_l \times N_r, \psi \times \psi)\text{-mod} \cong \mathbb{W}\text{-mod},$$

to the functor $(\mathscr{D}(G), \mathrm{Ad}\, G)$-mod $\to \mathbb{W}$-mod, $M \mapsto (M/M\mathfrak{n}_r^\psi)^{\mathfrak{n}_l^\psi}$*. Furthermore, the latter functor induces via the equivalence* $\mathscr{W}h \cong \mathbb{W}$-hol, *of Theorem* 1.5.1*, an exact functor* $(\mathscr{D}(G), \mathrm{Ad}\, G)$-hol $\to \mathscr{W}h$*.*

Theorem 1.6.3 will be proved in Section 4.2 and Theorem 1.6.4 will be proved in Section 5.4.

Remarks 1.6.5. (i) In the special case where $X = G$ and $M \in (\mathscr{D}(G), \mathrm{Ad}\, G)$-mod is holonomic (not necessarily regular), the vanishing of $\int^j_{a_{N_r}}(\Omega_{N_r} \boxtimes M) = 0$ for all $j \neq 0$ can be proved in a different way by adapting the proof of a result of Bezrukavnikov, Braverman, and Mirkovic, [BBM, Theorem 1.5(2)], to the $\mathscr{D}$-module setting.

(ii) One can show that the functor Ψ takes finitely generated $\mathscr{D}(G)$-modules to finitely generated $\mathbb{W}$-modules. Furthermore, it is likely that this functor respects singular supports in the sense that $\mathrm{SS}(\Psi(M))$ is obtained from $\mathrm{SS}(M)$ by a suitable classical Hamiltonian reduction.

Observe that group multiplication $G \times G \to G$ induces a monoidal structure on a suitably defined (bounded) derived category counterpart $D(\mathscr{D}_G, \mathrm{Ad}\, G)$-hol, resp. $D\mathscr{W}h$, of the abelian category $(\mathscr{D}_G, \mathrm{Ad}\, G)$-hol, resp. $\mathscr{W}h$. The functor Ψ in Theorem 1.6.4 can be upgraded to a *monoidal* functor $D(\mathscr{D}_G, \mathrm{Ad}\, G)$-hol $\to D\mathscr{W}h$, cf. Remark 3.1.6. (In the setting of ∞-categories for an affine Kac-Moody group, an analogous functor has been considered in [Be].) Using that the functor (1.5.2) is exact, it is also possible to define a derived version, $D\mathscr{C}at(T, W)$, of the abelian category $\mathscr{C}at(T, W)$. The group T being abelian, multiplication $T \times T \to T$ gives $D\mathscr{C}at(T, W)$ the structure of a *symmetric* monoidal category. We expect that there

[1] I was informed by V. Drinfeld that the last equation has also been obtained by Sam Raskin (unpublished).

is a derived counterpart of Theorem 1.5.1 that provides, in particular, a monoidal equivalence $D\mathscr{W}h \cong D\mathscr{C}at(T,W)$. Following Drinfeld, we observe that the existence of such an equivalence would imply that the monoidal structure on $D\mathscr{W}h$ is symmetric. The symmetry of the monoidal structure on $D\mathscr{W}h$ seems to be unknown at the time of writing this paper.

We defer a more detailed discussion of these topics to a separate paper.

Acknowledgments

This paper was inspired by a question of V. Drinfeld. The author is grateful to Gwyn Bellamy and Vladimir Drinfeld for helpful comments and to Tsao-Hsien Chen for useful discussions. Special thanks are due to Michael Finkelberg for generously sharing his ideas, [FT], on the relation of nil-DAHA to the topology of the affine Grassmannian. I am very much indebted to Gus Lonergan for pointing out two mistakes in the original proof of Proposition 7.2.4, suggesting how to fix one of them, and also for kind clarifications of his work [Lo2] that was used to fix the second mistake. I am grateful to Roman Bezrukavnikov for informing me about the results of Lonergan [Lo1] before they were made public.

2 Three constructions of the universal centralizer

Given an algebraic group K with Lie algebra $\mathfrak{k}$, we write Ad^* for the coadjoint action of K in $\mathfrak{k}^*$.

Recall that we have fixed a connected and simply connected semisimple group G with Lie algebra $\mathfrak{g}$. We write G_x for the stabilizer of an element $x \in \mathfrak{g}^*$ under the $\mathrm{Ad}^*\,G$-action and let $\mathfrak{g}_x = \mathrm{Lie}\,G_x$. We say that x is regular if $\dim \mathfrak{g}_x = \mathrm{rk}\,\mathfrak{g}$. Let $\mathfrak{g}^{\mathrm{reg}}$ be the set of regular elements of $\mathfrak{g}^*$.

2.1 The first construction of $\mathfrak{Z}$

One can identify the cotangent bundle on G with the variety of triples

$$\mathscr{T}^*G = \{(x',x,g) \in \mathfrak{g}^* \times \mathfrak{g}^* \times G \mid x' = \mathrm{Ad}^*\,g(x)\}. \tag{2.1.1}$$

The group G acts on itself by left and right translation. The induced $G \times G$-action on $\mathscr{T}^*G$ is given by $(g_1,g_2):\ (x',x,g) \mapsto ((\mathrm{Ad}^*\,g_1(x'),\ \mathrm{Ad}^*\,g_2(x),\ g_1 g g_2^{-1})$. This action is Hamiltonian with moment map

$$\mu_l \times \mu_r : \mathscr{T}^*G \to \mathfrak{g}^* \times \mathfrak{g}^*, \quad \mu_l(x',x,g) = x', \ \ \mu_r(x',x,g) = x. \tag{2.1.2}$$

Next, let G act on itself by conjugation and act on $\mathfrak{g}^* \times G$ diagonally. We define

$$\mathscr{Z} := \{(x,g) \in \mathfrak{g}^* \times G \mid \mathrm{Ad}^* g(x) = x\}, \quad \mathscr{Z}^{\mathrm{reg}} := \{(x,g) \in \mathscr{Z} \mid x \in \mathfrak{g}^{\mathrm{reg}}\}.$$

In terms of (2.1.1), the moment map associated with the $\mathrm{Ad}\,G$-action on $\mathscr{T}^*G$ has the form $\mu_{\mathrm{Ad}}(x',x,g) = x' - x$. Writing $\mathfrak{g}^*_{\mathrm{diag}}$ for the diagonal copy of $\mathfrak{g}^*$ in $\mathfrak{g}^* \times \mathfrak{g}^*$, one obtains the following identifications:

$$\mathscr{Z} \xrightarrow[\cong]{(x,g)\mapsto(x,x,g)} (\mu_l \times \mu_r)^{-1}(\mathfrak{g}^*_{\mathrm{diag}}) = \mu_{\mathrm{Ad}}^{-1}(0).$$

Let $\mathscr{T}^{\mathrm{reg}}G := \mu_l^{-1}(\mathfrak{g}^{\mathrm{reg}}) = \mu_r^{-1}(\mathfrak{g}^{\mathrm{reg}})$. This is a $G \times G$-stable Zariski open subset of $\mathscr{T}^*G$, and we have $\mathscr{Z}^{\mathrm{reg}} = (\mu_l \times \mu_r)^{-1}(\mathfrak{g}^{\mathrm{reg}}_{\mathrm{diag}}) = \mu_{\mathrm{Ad}}^{-1}(0) \cap \mathscr{T}^{\mathrm{reg}}G$.

The first projection $\mathfrak{g}^* \times G \to \mathfrak{g}^*$ makes $\mathfrak{g}^* \times G$ a G-equivariant group scheme on $\mathfrak{g}^*$ with fiber G. The fiber of $\mathscr{Z}^{\mathrm{reg}}$ over $x \in \mathfrak{g}^{\mathrm{reg}}$ equals G_x, which is a not necessarily connected abelian group. This makes $\mathscr{Z}^{\mathrm{reg}} \to \mathfrak{g}^{\mathrm{reg}}$ a smooth G-equivariant abelian group subscheme of the group scheme $\mathfrak{g}^{\mathrm{reg}} \times G \to \mathfrak{g}^{\mathrm{reg}}$.

Let $\theta : \mathfrak{g}^* \to \mathfrak{c} := \mathfrak{g}^* /\!/ \mathrm{Ad}^* G$ be the coadjoint quotient map. The fibers of the composite $\mathfrak{g}^{\mathrm{reg}} \hookrightarrow \mathfrak{g}^* \to \mathfrak{c}$ are the regular coadjoint orbits. It follows that the map $\mathscr{Z}^{\mathrm{reg}} \to \mathfrak{c}$, $(x,x,g) \mapsto \theta(x)$, is a G-torsor; in particular, $\mathscr{Z}^{\mathrm{reg}}$ is affine. The Kostant slice, see Section 2.2, provides a section of this G-torsor.

Definition. The *universal centralizer* is defined as

$$\mathfrak{Z} := \mathscr{Z}^{\mathrm{reg}} /\!/ G := \mathrm{Spec}(\mathbb{C}[\mathscr{Z}^{\mathrm{reg}}]^G).$$

The map $\mathscr{Z}^{\mathrm{reg}} \to \mathfrak{c}$ descends to a morphism $\mathfrak{Z} \to \mathfrak{c}$, making $\mathfrak{Z}$ an abelian group scheme on $\mathfrak{c}$.

We see from the above that the universal centralizer may be identified with a Hamiltonian reduction of $\mathscr{T}^{\mathrm{reg}}G$ over $0 \in \mathfrak{g}^*$:

$$\mathfrak{Z} \cong \big(\mu_{\mathrm{Ad}}^{-1}(0) \cap \mathscr{T}^{\mathrm{reg}}G\big) /\!/ G =: \mathscr{T}^{\mathrm{reg}}G /\!\!/\!\!/ (\mathrm{Ad}\,G, 0). \tag{2.1.3}$$

From now on, we fix a principal $\mathfrak{sl}_2$-triple $(\mathsf{e},\mathsf{h},\mathsf{f})$. Let $\mathsf{e} + \mathfrak{g}_{\mathsf{f}}$ be the Kostant slice and $\mathscr{S} \subset \mathfrak{g}^*$ the image of $\mathsf{e} + \mathfrak{g}_{\mathsf{f}}$ under the Killing form $(-,-)$. Further, let $\mathscr{Z}_{\mathscr{S}} = \{(x,g) \in \mathscr{S} \times G \mid \mathrm{Ad}^* g(s) = s\}$. Using the identification $\mathscr{Z} = (\mu_l \times \mu_r)^{-1}(\mathfrak{g}^*_{\mathrm{diag}})$, one can identify $\mathscr{Z}_{\mathscr{S}}$ with the preimage of $\mathscr{S}$ under the map $\mu_l|_{\mathscr{Z}} = \mu_r|_{\mathscr{Z}} : \mathscr{Z} \to \mathfrak{g}^*$, $(x,x,g) \mapsto x$. According to Kostant [Ko1], [Ko3], all elements of $\mathscr{S}$ are regular; furthermore, the composite $\mathscr{S} \hookrightarrow \mathfrak{g}^{\mathrm{reg}} \to \mathfrak{c}$ is an isomorphism. We deduce that $\mathscr{Z}_{\mathscr{S}}$ is a smooth closed subvariety of $\mathscr{T}^*G$ contained in $\mathscr{Z}^{\mathrm{reg}}$; moreover, the composite $\mathscr{Z}_{\mathscr{S}} \hookrightarrow \mathscr{Z}^{\mathrm{reg}} \twoheadrightarrow \mathscr{Z}^{\mathrm{reg}} /\!/ G = \mathfrak{Z}$ is an isomorphism. It follows that the group scheme $\mathfrak{Z}$ is smooth, so the Hamiltonian reduction construction in (2.1.3) provides $\mathfrak{Z}$ with the structure of a smooth symplectic variety.

We leave the proof of the following lemma to the interested reader.

Lemma 2.1.4. *The restriction of the symplectic 2-form on $\mathscr{T}^*G$ to the subvariety $\mathscr{Z}_{\mathscr{S}}$ is nondegenerate, i.e., $\mathscr{Z}_{\mathscr{S}}$ is a symplectic submanifold of $\mathscr{T}^*G$. Furthermore, the isomorphism $\mathscr{Z}_{\mathscr{S}} \xrightarrow{\sim} \mathfrak{Z}$ is a symplectomorphism.*

2.2 The second construction of $\mathfrak{Z}$

This construction uses a two-sided Whittaker Hamiltonian reduction as follows.

Let $\mathfrak{n}$ be the unique maximal nilpotent subalgebra of $\mathfrak{g}$ that contains the element f and N the corresponding unipotent subgroup. Let $\psi \in \mathfrak{g}^*$ be defined by $\psi(x) = (\mathsf{e}, x)$. We will abuse notation and also write ψ for $\psi|_{\mathfrak{n}}$. Thus, we have $\psi \in \mathscr{S} \subset pr^{-1}(\psi) = \psi + \mathfrak{n}^{\perp}$, where $pr : \mathfrak{g}^* \to \mathfrak{n}^*$ is the natural projection. According to Kostant, [Ko3], the action map $N \times \mathscr{S} \to \psi + \mathfrak{n}^{\perp}$ is an isomorphism. This easily yields the following result, see [Gi, Corollary 1.3.8].

Lemma 2.2.1. *Let X be a G-variety equipped with a G-equivariant map $f : X \to \mathfrak{g}^*$. Let $\mathscr{O}_\psi$ be the localization of $\mathbb{C}[\mathfrak{n}^*]$ at $\psi \in \mathfrak{n}^*$. Then, we have:*

(i) *Let $\mathscr{F}$ be a G-equivariant quasi-coherent $\mathscr{O}_X$-module. Then, $\mathscr{F}$ is a flat $(pr \circ f)^{\bullet}\mathscr{O}_\psi$-module. In particular, the composite map $X \xrightarrow{f} \mathfrak{g}^* \xrightarrow{pr} \mathfrak{n}^*$ is flat at $\psi \in \mathfrak{n}^*$.*

(ii) *The subscheme $f^{-1}(\psi + \mathfrak{n}^{\perp})$ is N-stable. The action of N on $f^{-1}(\psi + \mathfrak{n}^{\perp})$ yields an N-equivariant isomorphism $N \times f^{-1}(\mathscr{S}) \to f^{-1}(\psi + \mathfrak{n}^{\perp})$.*

Let X be a smooth (not necessarily affine) symplectic variety equipped with a Hamiltonian action of a (not necessarily reductive) linear algebraic group K, with moment map $\mu : X \to \mathfrak{k}^*$. Let $\chi \in \mathfrak{k}^*$ be a character and assume that χ is a regular value of μ and the K-action on $\mu^{-1}(\chi)$ admits a slice, that is, a subvariety $S \subset \mu^{-1}(\chi)$ such that the action $K \times S \to \mu^{-1}(\chi)$ is an isomorphism of algebraic varieties. In such a case, the variety S is smooth and inherits a natural symplectic structure. Furthermore, one can show that any two slices are isomorphic, and we write $X /\!\!/ K := S$.

We will especially be interested in the case where $X = \mathscr{T}^*G$ is viewed as a $G_l \times G_r$-variety and $f = \mu_l \times \mu_r$ is the moment map $\mathscr{T}^*G \to \mathfrak{g}^* \oplus \mathfrak{g}^*$. In this setting, the role of the Lie algebra $\mathfrak{n}^\psi$ is played by $\mathfrak{n}_l^\psi \oplus \mathfrak{n}_r^\psi$.

Corollary 2.2.2. *One has a natural isomorphism $\mathfrak{Z} = \mathscr{T}^*G /\!\!/ (N_l \times N_r, \psi \times \psi)$.*

Proof. Applying Lemma 2.2.1(ii) in the case of the Lie algebra $\mathfrak{n}_l^\psi \oplus \mathfrak{n}_r^\psi$, we compute

$$
\begin{aligned}
\mathscr{T}^*G /\!\!/ (N_l \times N_r, \psi \times \psi) &= (\mu_{l,\mathfrak{n}^*}^{-1}(\psi) \cap \mu_{r,\mathfrak{n}^*}^{-1}(\psi))/N_l \times N_r \\
&= \big((\mu_l^{-1}(\psi + \mathfrak{n}^{\perp}) \cap \mu_r^{-1}(\psi + \mathfrak{n}^{\perp})\big)/N_l \times N_r \\
&= \mu_l^{-1}(\mathscr{S}) \cap \mu_r^{-1}(\mathscr{S}) = \{(x,g) \in \mathscr{S} \times G \mid \mathrm{Ad}^* g(x) \in \mathscr{S}\} \\
&= \{(x,g) \in \mathscr{S} \times G \mid \mathrm{Ad}^* g(x) = x\} = \mathfrak{Z}.
\end{aligned}
$$
□

2.3 The third construction of $\mathfrak{Z}$

This construction, due to [BFM], is via affine blow-ups. Let $\mathscr{B}$ be the flag variety of all Borel subalgebras of $\mathfrak{g}$. Let $\widetilde{\mathfrak{g}} := \{(x,\mathfrak{b}) \in \mathfrak{g}^* \times \mathscr{B} \mid x \in \mathfrak{b}^\perp\}$. The first projection $\pi : \widetilde{\mathfrak{g}} \to \mathfrak{g}$, $(x,\mathfrak{b}) \mapsto x$, is the Grothendieck-Springer resolution. Let $\mathfrak{t}$ be the universal Cartan, T the universal Cartan torus, W the Weyl group, and $\mathfrak{t}^* \to \mathfrak{t}^*/W \cong \mathfrak{g}^* /\!/ \mathrm{Ad}^* G = \mathfrak{c}$ the quotient. It is known that the following map is an isomorphism:

$$\widetilde{\mathfrak{g}}^{\mathrm{reg}} := \{(x,\mathfrak{b}) \in \widetilde{\mathfrak{g}} \mid x \in \mathfrak{g}^{\mathrm{reg}}\} \xrightarrow{\sim} \mathfrak{g}^{\mathrm{reg}} \times_{\mathfrak{c}} \mathfrak{t}^*, \quad (x,\mathfrak{b}) \mapsto (x, x \bmod \mathfrak{b}^\perp). \tag{2.3.1}$$

Beilinson and Kazhdan [BK] constructed a natural map

$$\varkappa : \mathfrak{Z} \times_{\mathfrak{c}} \mathfrak{t}^* \longrightarrow T \times \mathfrak{t}^*, \tag{2.3.2}$$

of W-equivariant group schemes on $\mathfrak{t}^*$, where W acts naturally on T, resp. $\mathfrak{t}^*$, diagonally on $T \times \mathfrak{t}^*$, and trivially on $\mathfrak{Z}$. The construction is as follows.

First, it was shown by Kostant that for any $(x,\mathfrak{b}) \in \widetilde{\mathfrak{g}}^{\mathrm{reg}}$, one has an inclusion $G_x \subset B$, where B stands for the Borel subgroup with Lie algebra $\mathfrak{b}$. Therefore, there is a well-defined map

$$G_x \hookrightarrow B \twoheadrightarrow B/[B,B] = T, \quad g \mapsto g \bmod [B,B].$$

Hence, the assignment $(g,x,\mathfrak{b}) \mapsto (g \bmod [B,B], x, \mathfrak{b})$ gives a morphism $\mathscr{Z}^{\mathrm{reg}} \times_{\mathfrak{g}^{\mathrm{reg}}} \widetilde{\mathfrak{g}}^{\mathrm{reg}} \to T \times \widetilde{\mathfrak{g}}^{\mathrm{reg}}$, of group schemes on $\widetilde{\mathfrak{g}}^{\mathrm{reg}}$. Using (2.3.1), we obtain a morphism

$$\mathscr{Z}^{\mathrm{reg}} \times_{\mathfrak{g}^{\mathrm{reg}}} \widetilde{\mathfrak{g}}^{\mathrm{reg}} = \mathscr{Z}^{\mathrm{reg}} \times_{\mathfrak{g}^{\mathrm{reg}}} (\mathfrak{g}^{\mathrm{reg}} \times_{\mathfrak{c}} \mathfrak{t}^*) = \mathscr{Z}^{\mathrm{reg}} \times_{\mathfrak{c}} \mathfrak{t}^* \longrightarrow T \times \widetilde{\mathfrak{g}}^{\mathrm{reg}} = T \times (\mathfrak{g}^{\mathrm{reg}} \times_{\mathfrak{c}} \mathfrak{t}^*)$$

of $W \times G$-equivariant group schemes on $\mathfrak{g}^{\mathrm{reg}}$. Taking categorical quotients by G on each side yields the following map, which is the required map $\varkappa$ in (2.3.2):

$$\begin{aligned}\mathfrak{Z} \times_{\mathfrak{c}} \mathfrak{t}^* = \mathscr{Z}^{\mathrm{reg}} /\!/ G \times_{\mathfrak{c}} \mathfrak{t}^* &= (\mathscr{Z}^{\mathrm{reg}} \times_{\mathfrak{g}^{\mathrm{reg}}} \mathfrak{t}^*) /\!/ G \\ &\longrightarrow (T \times \mathfrak{g}^{\mathrm{reg}} \times_{\mathfrak{c}} \mathfrak{t}^*) /\!/ G = T \times (\mathfrak{g}^{\mathrm{reg}} \times_{\mathfrak{c}} \mathfrak{t}^*) /\!/ G = T \times \mathfrak{t}^*.\end{aligned}$$

The graph of $\varkappa$ gives the following variety:

$$\Lambda = \{(z,x,t) \in \mathfrak{Z} \times \mathfrak{t}^* \times T \mid \varpi(z) = x \bmod W, \ \ \varkappa(z,x) = (t,x)\},$$

where $\varpi : \mathfrak{Z} \to \mathfrak{c}$ is the canonical map. We identify $\mathfrak{t}^* \times T$ with $\mathscr{T}^*T$.

The proof of the following result is left to the reader.

Proposition 2.3.3. (i) *Λ is a smooth closed Lagrangian subvariety of $\mathfrak{Z} \times \mathscr{T}^*T$.*

(ii) *For any $t \in T$, every irreducible component of the fiber, $pr_T^{-1}(t)$, of the map $pr_T : \Lambda \to T$, $(z,x,t) \mapsto t$ is a (possibly singular) Lagrangian subvariety of $\mathfrak{Z}$.*

The variety Λ is a classical counterpart of the Miura bimodule, to be introduced in Section 6.2 below.

Let $R \subset \mathbb{X}^*$ be the set of roots and the weight lattice of G, respectively.

Example 2.3.4. Let $\Sigma \subset R$ be the set of simple roots and $S \subset \Sigma$ a subset. Let L_S be an associated standard Levi subgroup of G and $L_S^{\text{der}} = [L_S, L_S]$, the derived group of L_S. Let e_S be a principal nilpotent of the Levi subalgebra $\operatorname{Lie} L_S$ and $Z(L_S^{\text{der}}, \mathsf{e}_S)$ its centralizer in L_S^{der}. Finally, put

$$\mathfrak{t}_S^{\text{reg}} := \{\lambda \in \mathfrak{t}^* \mid \langle \lambda, \check{\alpha} \rangle = 0 \;\&\; \langle \lambda, \check{\beta} \rangle \neq 0, \;\forall \alpha \in S, \beta \in \Sigma \smallsetminus S\}.$$

Then, one has a decomposition

$$pr_T^{-1}(1) \cong \bigcup_{S \subset \Sigma} \overline{\mathfrak{t}_S^{\text{reg}}} \times Z(L_S^{\text{der}}, \mathsf{e}_S),$$

where each piece is isomorphic to a union of irreducible components of $pr_T^{-1}(1)$ permuted transitively by the center of G.

Next, identify $\operatorname{Sym} \mathfrak{t}$ with $\mathbb{C}[\mathfrak{t}^*]$. Following [BFM], let $\mathbb{C}\big[T \times \mathfrak{t}^*, \frac{t^\alpha - 1}{\check{\alpha}}, \alpha \in R\big]$ be an algebra obtained from $\mathbb{C}[T \times \mathfrak{t}^*]$ by adjoining all rational functions $\frac{t^\alpha - 1}{\check{\alpha}}$, $\alpha \in R$, where t^λ stands for an element $\lambda \in \mathbb{X}^*$ viewed as a regular function on T.

Theorem 2.3.5 ([BFM], Proposition 2.8)**.** *The algebra map* $\varkappa^* : \mathbb{C}[T \times \mathfrak{t}^*] \to \mathbb{C}[\mathfrak{Z} \times_{\mathfrak{c}} \mathfrak{t}^*]$, *induced by the morphism* $\varkappa$ *in* (2.3.2)*, extends to a* W*-equivariant algebra isomorphism*

$$\mathbb{C}\left[T \times \mathfrak{t}^*, \frac{t^\alpha - 1}{\check{\alpha}}, \alpha \in R\right] \cong \mathbb{C}[\mathfrak{Z} \times_{\mathfrak{c}} \mathfrak{t}^*].$$

Thus, we have

$$\mathfrak{Z} \cong \left(\operatorname{Spec} \mathbb{C}\left[T \times \mathfrak{t}^*, \frac{t^\alpha - 1}{\check{\alpha}}, \alpha \in R\right]\right) \Big/ W.$$

An important step in the proof of the theorem is played by the following result, see [BFM, Section 4].

Proposition 2.3.6. *The algebra* $\mathbb{C}\big[T \times \mathfrak{t}^*, \frac{t^\alpha - 1}{\check{\alpha}}, \alpha \in R\big]^W$ *is flat over* $\mathbb{C}[\mathfrak{t}^*]^W$.

A simple proof of this proposition based on nil-Hecke algebras will be given in Section 7.1 below.

3 Review of quantum Hamiltonian reduction and Kazhdan filtrations

3.1 Quantum Hamiltonian reduction

In this subsection we collect some general, mostly standard results on quantum Hamiltonian reduction.

Given a Lie algebra $\mathfrak{k}$, we write $E/E\mathfrak{k}$, resp. $\mathfrak{k}E\backslash E$ or $E/\mathfrak{k}E$, for the space of coinvariants of a right, resp. left, $\mathfrak{k}$-module E. We write $E^{\mathfrak{k}}$ for the space of invariants.

Let $\mathfrak{b}$, resp. $\bar{\mathfrak{b}}$, be the unique Borel subalgebra of $\mathfrak{g}$ such that $\mathsf{f} \in \mathfrak{n} = [\mathfrak{b},\mathfrak{b}]$, resp. $\mathsf{e} \in \bar{\mathfrak{n}} = [\bar{\mathfrak{b}},\bar{\mathfrak{b}}]$. We get a triangular decomposition $\mathfrak{g} = \mathfrak{n} \oplus \mathfrak{t} \oplus \bar{\mathfrak{n}}$, where we identify $\mathfrak{t}$ with $\mathfrak{b} \cap \bar{\mathfrak{b}}$. We have algebra embeddings $\mathscr{U}\bar{\mathfrak{n}} \hookrightarrow \mathscr{U}\bar{\mathfrak{b}} \hookrightarrow \mathscr{U}\mathfrak{g}$. It is clear that $\bar{\mathfrak{n}}(\mathscr{U}\bar{\mathfrak{b}}) = (\mathscr{U}\bar{\mathfrak{b}})\bar{\mathfrak{n}}$ is a two-sided ideal of $\mathscr{U}\bar{\mathfrak{b}}$ and we have $\mathscr{U}\bar{\mathfrak{b}}/(\mathscr{U}\bar{\mathfrak{b}})\bar{\mathfrak{n}} \cong \mathscr{U}\mathfrak{t}$. Let $Z\mathfrak{g}$ be the center of $\mathscr{U}\mathfrak{g}$ and $Q := \mathscr{U}\mathfrak{g}/(\mathscr{U}\mathfrak{g})\mathfrak{n}^{\psi}$.

The following result is well-known, cf. [Ko3].

Proposition 3.1.1. (i) *The embedding $Z\mathfrak{g} \hookrightarrow \mathscr{U}\mathfrak{g}$ induces an algebra isomorphism $Z\mathfrak{g} \xrightarrow{\sim} (\mathscr{U}\mathfrak{g}/\mathfrak{n}^{\psi})^{\mathrm{ad}\,\mathfrak{n}}$.*

(ii) *Q is free as a left $(\mathscr{U}\bar{\mathfrak{n}} \otimes Z\mathfrak{g})$-module.*

(iii) *The embedding $\mathscr{U}\mathfrak{t} \hookrightarrow \mathscr{U}\mathfrak{g}$ induces an isomorphism $\mathscr{U}\mathfrak{t} \xrightarrow{\sim} \bar{\mathfrak{n}}Q\backslash Q$ of left $\mathscr{U}\mathfrak{t}$-modules.*

(iv) *For all $i > 0$, we have $H^i(\mathfrak{n}^{\psi}, Q) = 0$; furthermore, $H^0(\mathfrak{n}^{\psi}, Q) = Q^{\mathfrak{n}^{\psi}} \cong Z\mathfrak{g}$, by* (i).

We will also use

Proposition 3.1.2. *Let M be an $(\mathscr{U}\mathfrak{g}, \mathscr{U}\mathfrak{g})$-bimodule such that the adjoint $\mathfrak{g}$-action on M is locally finite. Then, for any $j > 0$, we have $H_j(\mathfrak{n}^{\psi}, M) = 0$, where we view M as an $\mathfrak{n}^{\psi}$-module via the right action.*

Proof. Let M be as in the statement and assume, in addition, that M is finitely generated as a left, equivalently right, $\mathscr{U}\mathfrak{g}$-module. In this case, the required statement follows from [Gi, Lemma 4.4.1]. More precisely, the lemma was stated under an additional assumption that the center of $\mathscr{U}\mathfrak{g}$ acts on M by scalars both on the left and on the right. However, this additional assumption was not used in the proof of the lemma.

In the general case, we can find a family of sub $(\mathscr{U}\mathfrak{g}, \mathscr{U}\mathfrak{g})$-bimodules $M_\alpha \subset M$, such that each M_α is finitely generated as a left $\mathscr{U}\mathfrak{g}$-module and we have $M = \varinjlim M_\alpha$. The functor $H_\bullet(\mathfrak{n}^{\psi}, -) = \mathrm{Tor}_\bullet^{\mathscr{U}\mathfrak{n}}(\psi, -)$ commutes with direct limits. Hence, we have $H_\bullet(\mathfrak{n}^{\psi}, M) = \varinjlim H_\bullet(\mathfrak{n}_r^{\psi}, M_\alpha)$. This yields the proposition by the first paragraph of the proof. □

From now on, we let A be a finitely generated left noetherian associative algebra and $\mathscr{U}\mathfrak{g} \to A$ an algebra homomorphism such that

$$\text{the adjoint action } \mathrm{ad}\,x : a \mapsto xa - ax, \text{ of } \mathfrak{g} \text{ on } A, \text{ is locally finite.} \tag{3.1.3}$$

Let $\mathfrak{k}$ be a nilpotent Lie subalgebra of $\mathscr{U}\mathfrak{g}$, such as $\mathfrak{n}, \mathfrak{n}^{\psi}$, or $\bar{\mathfrak{n}}$. We write $(A,\mathfrak{k})$-mod for a full subcategory of the category of A-modules such that the induced $\mathfrak{k}$-action on the module is **locally nilpotent**. Condition (3.1.3) implies that the adjoint action of $\mathfrak{k}$ on A is locally nilpotent. In particular, one has $A/A\mathfrak{k} \in (A,\mathfrak{k})$-mod.

The space $A/\!\!/\mathfrak{k} := (A/A\mathfrak{k})^{\mathrm{ad}\,\mathfrak{k}}$, resp. $\mathfrak{k}\backslash\!\!\backslash A := (A/\mathfrak{k}A)^{\mathrm{ad}\,\mathfrak{k}}$, acquires an algebra structure. One has an algebra isomorphism $(A/\!\!/\mathfrak{k})^{op} \cong \mathfrak{k}^{op}\backslash\!\!\backslash A^{op}$. A left, resp. right, A-module structure on E induces a natural left, resp. right, $A/\!\!/\mathfrak{k}$-module structure on $E^{\mathfrak{k}}$, resp. $E/E\mathfrak{k}$. More generally, for each $j \geq 0$, the cohomology group $H^j(\mathfrak{k}, E)$, resp.

homology group $H_j(\mathfrak{k},E)$, has the structure of a left, resp. right, $A/\!\!/\mathfrak{k}$-module. Similarly, there is a left, resp. right, $\mathfrak{k}\backslash\!\!\backslash A$-module structure on $H_\bullet(\mathfrak{k},E)$, resp. $H^\bullet(\mathfrak{k},E^\mathfrak{k})$. In particular, $A/A\mathfrak{k}$ is an $(A,A/\!\!/\mathfrak{k})$-bimodule, resp. $\mathfrak{k}A\backslash A$ is an $(\mathfrak{k}\backslash\!\!\backslash A,A)$-bimodule.

In this section, we will only consider the case $\mathfrak{k}=\mathfrak{n}^\Psi$ and put $\mathfrak{A}=A/\!\!/\mathfrak{n}^\Psi$. The composite map $Z\mathfrak{g}\hookrightarrow \mathscr{U}\mathfrak{g}\to A$ descends to an algebra homomorphism $Z\mathfrak{g}\to\mathfrak{A}$. The map $\mathscr{U}\mathfrak{g}\to A$ induces a morphism $Q\to A/A\mathfrak{n}^\Psi$, of $(\mathscr{U}\mathfrak{g},Z\mathfrak{g})$-bimodules, and also a morphism $\mathscr{U}\mathfrak{t}=\bar{\mathfrak{n}}(\mathscr{U}\mathfrak{g})\backslash\mathscr{U}\mathfrak{g}/(\mathscr{U}\mathfrak{g})\mathfrak{n}^\Psi\to\bar{\mathfrak{n}}A\backslash A/A\mathfrak{n}^\Psi$, of $(\mathscr{U}\mathfrak{t},Z\mathfrak{g})$-bimodules.

Below, we work with left modules; similar results hold for right modules.

Proposition 3.1.4. (i) *For any* $M\in(A,\mathfrak{n}^\Psi)$-mod *and* $i\neq 0$, *we have* $H^i(\mathfrak{n}^\Psi,M)=0$. *Furthermore, the functor* $(A,\mathfrak{n}^\Psi)$-mod $\to\mathfrak{A}$-mod, $M\mapsto M^{\mathfrak{n}^\Psi}$, *is an equivalence and* $L\mapsto A/A\mathfrak{n}^\Psi\otimes_\mathfrak{A}L$ *is its quasi-inverse.*

(ii) *For* $L\in\mathfrak{A}$-mod *and all* $i\neq 0$, *we have* $H_i(\bar{\mathfrak{n}},A/A\mathfrak{n}^\Psi\otimes_\mathfrak{A}L)=0$*; furthermore, there are natural isomorphisms* $H_0(\bar{\mathfrak{n}},A/A\mathfrak{n}^\Psi\otimes_\mathfrak{A}L)\cong(\bar{\mathfrak{n}}A\backslash A/A\mathfrak{n}^\Psi)\otimes_\mathfrak{A}L\cong\mathscr{U}\mathfrak{t}\otimes_{Z\mathfrak{g}}L$.

(iii) *The algebra* $\mathfrak{A}$ *is left noetherian.*

(iv) *For all* $i\neq 0$, *we have* $H_i(\mathfrak{n}^\Psi,A)=0$, *where* $\mathfrak{n}^\Psi$ *acts on* A *by right multiplication.*

Proof. Put $Z=Z\mathfrak{g}$. Let L be a Z-module. View $Q\otimes_Z L$ as an $\mathfrak{n}^\Psi$-module and let $C^\bullet(\mathfrak{n}^\Psi,Q\otimes_Z L)$ be the corresponding cohomological Chevalley-Eilenberg complex. Thus, we have

$$C^\bullet(\mathfrak{n}^\Psi,Q\otimes_Z L)=(\wedge^\bullet\mathfrak{n}^\Psi)^*\otimes(Q\otimes_Z L)=((\wedge^\bullet\mathfrak{n}^\Psi)^*\otimes Q)\otimes_Z L.$$

Assume first that the Z-module L is flat. Then, we find

$$\begin{aligned}H^i(\mathfrak{n}^\Psi,Q\otimes_Z L)&=H^i(C^\bullet(\mathfrak{n}^\Psi,Q\otimes_Z L))\\&=H^i(((\wedge^\bullet\mathfrak{n}^\Psi)^*\otimes Q)\otimes_Z L)=H^i(\mathfrak{n}^\Psi,Q)\otimes_Z L.\end{aligned}$$

Proposition 3.1.1(iv) says that the cohomology group on the right equals L if $i=0$, resp. vanishes if $i>0$. Next, let L be an arbitrary, not necessarily flat, Z-module and choose a resolution $E^\bullet$, of L, by flat Z-modules. Since Q is free over Z, see Proposition 3.1.1(ii), we deduce that the complex $Q\otimes_Z E^\bullet$ is a resolution of $Q\otimes_Z L$. Since all E^i are flat, each term of the latter resolution is acyclic with respect to the functor $(-)^{\mathfrak{n}^\Psi}$, by the above. It follows that one has

$$H^i(\mathfrak{n}^\Psi,Q\otimes_Z L)=H^i((Q\otimes_Z E^\bullet)^{\mathfrak{n}^\Psi})=H^i(E^\bullet),$$
$$\text{where } H^0(E^\bullet)=L,\ \ H^i(E^\bullet)=0\ \forall i>0.$$

This implies part (i) in the special case $A=\mathscr{U}\mathfrak{g}$, cf. also [GG, Section 6.2].

To complete the proof of (i) in the general case, we view an object $M\in(A,\mathfrak{n}^\Psi)$-mod as an object of $(\mathscr{U}\mathfrak{g},\mathfrak{n}^\Psi)$-mod. This yields the cohomology vanishing of part (i) and shows that the functor $M\mapsto M^{\mathfrak{n}^\Psi}$ restricts to an exact functor $(A,\mathfrak{n}^\Psi)$-mod $\to\mathfrak{A}$-mod. Further, applying the equivalence in the special case above to $M = A/A\mathfrak{n}^\Psi$, viewed as an object of $(\mathscr{U}\mathfrak{g},\mathfrak{n}^\Psi)$-mod, and using that $(A/A\mathfrak{n}^\Psi)^{\mathfrak{n}^\Psi} = \mathfrak{A}$, we deduce that the natural map $Q\otimes_{Z\mathfrak{g}}\mathfrak{A} \to A/A\mathfrak{n}^\Psi$ is an isomorphism. This yields isomorphisms of functors

$$A/A\mathfrak{n}^\Psi \otimes_{\mathfrak{A}} (-) = (Q\otimes_Z \mathfrak{A})\otimes_{\mathfrak{A}} (-) = Q\otimes_Z(-). \qquad (3.1.5)$$

Using the special case $A = \mathscr{U}\mathfrak{g}$, we deduce that the map $L \to (Q\otimes_Z L)^{\mathfrak{n}^\Psi} = (A/A\mathfrak{n}^\Psi \otimes_{\mathfrak{A}} L)^{\mathfrak{n}^\Psi}$ is an isomorphism, for any $L \in \mathfrak{A}$-mod. It follows that the functors in (i) are quasi-inverse to each other. Further, since Q is free over $\mathscr{U}\bar{\mathfrak{n}}\otimes Z$, the isomorphism $Q\otimes_Z\mathfrak{A} \cong A/A\mathfrak{n}^\Psi$ implies that $A/A\mathfrak{n}^\Psi$ is free as a left $(\mathscr{U}\bar{\mathfrak{n}}\otimes\mathfrak{A}^{op})$-module. Therefore, one has $H_i(\bar{\mathfrak{n}}, A/A\mathfrak{n}^\Psi \otimes_{\mathfrak{A}} L) = H_i(\bar{\mathfrak{n}}, Q)\otimes_Z L$, for any free $\mathfrak{A}$-module L. Moreover, the group on the right vanishes for all $i \neq 0$, and we have $H_0(\bar{\mathfrak{n}}, Q)\otimes_Z L = Q/\bar{\mathfrak{n}}Q\otimes_Z L \cong \mathscr{U}\mathfrak{t}\otimes L$. The case of an arbitrary $\mathfrak{A}$-module L then follows by considering a resolution of L by free $\mathfrak{A}$-modules, similar to the proof of (i). This proves (ii).

Now, let L be a finitely generated left $\mathfrak{A}$-module. Then $A/A\mathfrak{n}^\Psi \otimes_{\mathfrak{A}} L$ is a finitely generated A-module, hence it is noetherian. The equivalence of part (i) implies that L is also noetherian, proving that $\mathfrak{A}$ is left noetherian. Finally, view A as a $(\mathscr{U}\mathfrak{g},\mathscr{U}\mathfrak{g})$-bimodule via left and right multiplication. This bimodule satisfies the assumptions of Proposition 3.1.2, by (3.1.3). Thus, part (iv) follows from Proposition 3.1.2. □

Remark 3.1.6. (i) Associated with the homomorphism $\mathfrak{n}^\Psi \to A$, there is a BRST complex $D\mathfrak{A} := \wedge^\bullet\big(\mathfrak{n}^\Psi \oplus (\mathfrak{n}^\Psi)^*\big)\otimes A$ that has the natural structure of a dg algebra. Using that the groups $H_i(\mathfrak{n}^\Psi, A)$ and $H^i(\mathfrak{n}^\Psi, A/A\mathfrak{n}^\Psi)$ vanish for all $i\neq 0$, see parts (i), (iv) of Proposition 3.1.4, it is immediate to deduce that $H^0(D\mathfrak{A}) \cong \mathfrak{A}$ and $H^i(D\mathfrak{A}) = 0$, for all $i \neq 0$.

(ii) Assume that the algebra A has finite global dimension, i.e., there is an integer $n = n(A) > 0$ such that one has $\mathrm{Ext}_A^{>n}(M,M') = 0$, for all left A-modules M,M'. Then, for all left $\mathfrak{A}$-modules L,L', by Proposition 3.1.4, we get $\mathrm{Ext}_{\mathfrak{A}}^{>n}(L,L') = \mathrm{Ext}_A^{>n}(A/A\mathfrak{n}^\Psi\otimes_{\mathfrak{A}} L, A/A\mathfrak{n}^\Psi\otimes_{\mathfrak{A}} L') = 0$. Hence, the algebra $\mathfrak{A}$ has finite global dimension. This can also be deduced from (i).

Let $D^b(A,\mathfrak{n}^\Psi)$-mod be a full triangulated subcategory of $D^b(A\text{-mod})$ whose objects $M^\bullet$ have the property that $H^i(M^\bullet) \in (A,\mathfrak{n}^\Psi)$-mod, $\forall i$. The proof of Proposition 3.1.4 can be easily upgraded to show that the functor $\wedge^\bullet\big(\mathfrak{n}^\Psi\oplus(\mathfrak{n}^\Psi)^*\big)\otimes(-)$ yields a triangulated equivalence of $D^b(A,\mathfrak{n}^\Psi)$-mod and the bounded derived category of dg $D\mathfrak{A}$-modules. The latter category is triangulated equivalent to $D^b(\mathfrak{A}\text{-mod})$, by (i).

3.2 Kazhdan filtrations

Several proofs in the paper will involve a 'filtered-to-graded' reduction. In this subsection we introduce the necessary definitions and notation.

Let E be a vector space equipped with a $\mathbb{Z}$-grading $E = \bigoplus_\ell E(\ell)$ and an ascending $\mathbb{Z}$-filtration $E_\leq$ by graded subspaces $E_{\leq j} = \bigoplus_\ell E_{\leq j}(\ell)$, $j\in\mathbb{Z}$. The *Kazhdan filtration* associated with the given grading and filtration is an ascending $\mathbb{Z}$-filtration

$F_\bullet E$ defined as follows:

$$F_n E := \bigoplus_{\ell\in\mathbb{Z}} F_n E(\ell), \qquad F_n E(\ell) := E_{\leq (n-\ell)/2}(\ell).$$

If E is an algebra and the $\mathbb{Z}$-grading and filtration $E_{\leq}$ are compatible with the algebra structure, then so is the Kazhdan filtration.

The adjoint action on $\mathfrak{g}$ of the semisimple element h, of the fixed $\mathfrak{sl}_2$-triple, gives a $\mathbb{Z}$-grading

$$\mathfrak{g} = \bigoplus_{\ell\in\mathbb{Z}} \mathfrak{g}(\ell) \quad \text{where} \quad \mathfrak{g}(\ell) = \{x \in \mathfrak{g} \mid [\mathsf{h}, x] = \ell\cdot x\}, \tag{3.2.1}$$

by *even* integers. The $\mathbb{Z}$-grading on $\mathfrak{g}$ induces one on $\mathscr{U}\mathfrak{g}$. We equip $\mathscr{U}\mathfrak{g}$ with the Kazhdan filtration, $F_\bullet\mathscr{U}\mathfrak{g}$, associated with that $\mathbb{Z}$-grading and the standard PBW filtration on $\mathscr{U}\mathfrak{g}$. The Kazhdan filtration will be our default filtration on $\mathscr{U}\mathfrak{g}$. The quotient filtration $F_\bullet Q$, on $Q = \mathscr{U}\mathfrak{g}/(\mathscr{U}\mathfrak{g})\mathfrak{n}^\psi$ has no negative terms, i.e., $F_{-1}Q = 0$. We write $\mathscr{U}_\hbar\mathfrak{g}$, $Z_\hbar\mathfrak{g}$, $Q_\hbar$, etc., for the corresponding Rees objects. Note that the Kazhdan and PBW filtrations on $\mathscr{U}\mathfrak{g}$ restrict to the same filtration on $Z\mathfrak{g}$.

The algebra $\mathscr{U}_\hbar\mathfrak{g}$ may be identified, in a natural way with a $\mathbb{C}[\hbar]$-algebra generated by a copy of the vector space $\mathfrak{g}$ such that the space $\mathfrak{g}(\ell) \subset \mathfrak{g}$, $\ell\in\mathbb{Z}$, is placed in degree $\ell+2$, with defining relations $x\otimes y - y\otimes x = \hbar[x,y]$, $\forall x,y\in\mathfrak{g}$. With this identification, the space $\mathfrak{n}^\psi = \{n - \psi(n),\ n\in\mathfrak{n}\}$ becomes a *graded* Lie subalgebra of $\mathscr{U}_\hbar\mathfrak{g}$; moreover, one has a natural graded algebra isomorphism $\mathscr{U}_\hbar\mathfrak{n} = \mathscr{U}(\mathfrak{n}^\psi)[\hbar]$. Furthermore, the $\mathscr{U}_\hbar\mathfrak{g}$-module $Q_\hbar$ may be identified with $\mathscr{U}_\hbar\mathfrak{g}/(\mathscr{U}_\hbar\mathfrak{g})\mathfrak{n}^\psi$.

Proposition 3.2.2. (1) *Analogues of the statements* (ii)–(iv) *of Proposition* 3.1.1 *hold with* $\mathscr{U}\mathfrak{g}$*, resp.* $Z\mathfrak{g}$*, and* Q*, being replaced by* $\operatorname{gr}\mathscr{U}\mathfrak{g}$*, resp.* $\operatorname{gr}Z\mathfrak{g}$ *and* $\operatorname{gr}Q$*.*

(2) *The functors* $M \mapsto M^{\mathfrak{n}^\psi}$ *and* $L \mapsto \mathscr{U}_\hbar\mathfrak{g}/(\mathscr{U}_\hbar\mathfrak{g})\mathfrak{n}^\psi \otimes_{Z_\hbar\mathfrak{g}} L$ *give quasi-inverse equivalences* $(\mathscr{U}_\hbar, \mathfrak{n}^\psi)\text{-grmod} \rightleftarrows (Z_\hbar\mathfrak{g})\text{-grmod}$ *of the corresponding categories of* $\mathbb{Z}$*-graded* $\mathbb{C}[\hbar]$*-modules.*

(3) *Let* $M \in (\mathscr{U}\mathfrak{g}, \mathfrak{n}^\psi)\text{-mod}$ *and let* $F_\bullet M$ *be a* $\mathbb{Z}$*-filtration which is compatible with the Kazhdan filtration on* $\mathscr{U}\mathfrak{g}$ *and such that each of the spaces* F_iM, $i\in\mathbb{Z}$ *is* $\mathfrak{n}$*-stable. Equip* $M^{\mathfrak{n}^\psi}$ *with the filtration induced from the one on* M *by restriction. Then the natural map* $\operatorname{gr}Q \otimes_{\operatorname{gr}Z\mathfrak{g}} \operatorname{gr}(M^{\mathfrak{n}^\psi}) \to \operatorname{gr}M$*, resp.* $\operatorname{gr}(M^{\mathfrak{n}^\psi}) \to (\operatorname{gr}M)^{\mathfrak{n}^\psi}$*, is an isomorphism.*

(4) *Let* M *be an* $(\mathscr{U}\mathfrak{g}, \mathscr{U}\mathfrak{g})$*-bimodule equipped with a* $\mathbb{Z}$*-filtration* $F_\bullet M$ *which is compatible with the Kazhdan filtration of* $\mathscr{U}\mathfrak{g}\otimes\mathscr{U}\mathfrak{g}$*. Assume that the* $\operatorname{ad}\mathfrak{g}$*-action on* M *is locally finite and each of the spaces* F_iM *is* $\operatorname{ad}\mathfrak{g}$*-stable. Let* $\mathbb{C}_\psi$ *denote a* 1*-dimensional* $\mathscr{U}(\mathfrak{n}^\psi)[\hbar]$*-module such that the vector space* $\mathfrak{n}^\psi$ *and the element* $\hbar$ *kill* $\mathbb{C}_\psi$*. Then* $\operatorname{Tor}_j^{\mathscr{U}(\mathfrak{n}^\psi)[\hbar]}(\mathbb{C}_\psi, M_\hbar) = 0$ *for all* $j > 0$*, where the algebra* $\mathscr{U}(\mathfrak{n}^\psi)[\hbar] = \mathscr{U}_\hbar\mathfrak{n} \subset \mathscr{U}_\hbar\mathfrak{g}$ *acts on* $M_\hbar$ *on the right.*

Proof. Part (1) is a direct consequence of results of Kostant, [Ko3]. Given (1), the proof of (2) is completely analogous to the proof of Proposition 3.1.4(i) in the case $A = \mathscr{U}\mathfrak{g}$. Now, let M be as in (3) and let $M_\hbar$ be the corresponding Rees $\mathscr{U}_\hbar\mathfrak{g}$-module. It follows from (2) that the natural map $Q_\hbar \otimes_{Z_\hbar\mathfrak{g}} (M_\hbar)^{\mathfrak{n}^\psi} \to M_\hbar$ is an isomorphism. The functor $\mathbb{C}[\hbar]/(\hbar)\otimes_{\mathbb{C}[\hbar]}(-)$ is right exact. Hence, applying this functor yields a surjection $\operatorname{gr} Q \otimes_{Z\mathfrak{g}} \operatorname{gr}(M^{\mathfrak{n}^\psi}) \twoheadrightarrow \operatorname{gr} M$ of N-equivariant $\mathbb{C}[\psi + \mathfrak{n}^\perp]$-modules. It follows from Lemma 2.2.1(ii) that taking $\mathfrak{n}$-invariants is an exact functor on the category of such modules, and that

$$(\operatorname{gr} Q \otimes_{\operatorname{gr} Z\mathfrak{g}} \operatorname{gr}(M^{\mathfrak{n}^\psi}))^{\mathfrak{n}^\psi} = \operatorname{gr}(M^{\mathfrak{n}^\psi}), \quad \text{resp. } \operatorname{gr} M = \operatorname{gr} Q \otimes_{\operatorname{gr} Z\mathfrak{g}} (\operatorname{gr} M)^{\mathfrak{n}^\psi}. \tag{3.2.3}$$

Also, since $F_i(M^{\mathfrak{n}^\psi}) := M^{\mathfrak{n}^\psi} \cap F_i M$, we have $\operatorname{gr}((M_\hbar)^{\mathfrak{n}^\psi}) = \operatorname{gr}(M^{\mathfrak{n}^\psi})$. Thus, from the surjection above, we deduce that the map

$$\operatorname{gr}(M^{\mathfrak{n}^\psi}) = (\operatorname{gr} Q \otimes_{Z\mathfrak{g}} \operatorname{gr}(M^{\mathfrak{n}^\psi}))^{\mathfrak{n}^\psi} \longrightarrow (\operatorname{gr} M)^{\mathfrak{n}^\psi}$$

is surjective. The injectivity of this map is immediate from its definition. Thus, from the isomorphism $\operatorname{gr}(M^{\mathfrak{n}^\psi}) \xrightarrow{\sim} (\operatorname{gr} M)^{\mathfrak{n}^\psi}$, using (3.2.3), we deduce that the composite map $\operatorname{gr} Q \otimes_{\operatorname{gr} Z\mathfrak{g}} \operatorname{gr}(M^{\mathfrak{n}^\psi}) \to \operatorname{gr} Q \otimes_{\operatorname{gr} Z\mathfrak{g}} (\operatorname{gr} M)^{\mathfrak{n}^\psi} = \operatorname{gr} M$ is an isomorphism.

The proof of (4) is similar to the proof of Proposition 3.1.2. □

3.3 Hamiltonian reduction for filtered algebras

Fix an algebra A as in Section 3.1. Let $A_\leq$ be a multiplicative $\operatorname{ad}\mathfrak{g}$-stable filtration such that $\bigcup_j A_{\leq j} = A$ and $A_{\leq -1} = \{0\}$. We will assume that the homomorphism $\mathscr{U}\mathfrak{g} \to A$ sends the PBW filtration on $\mathscr{U}\mathfrak{g}$ to the filtration $A_\leq$. The adjoint action of the element h gives a $\mathbb{Z}$-grading on A. Associated with this $\mathbb{Z}$-grading and filtration, one has the Kazhdan filtration $F_\bullet A$ on A. The latter induces a quotient filtration $F_\bullet(A/A\mathfrak{n}^\psi)$. By restriction, we get a $\mathbb{Z}$-filtration on the algebra $\mathfrak{A} = A/\!\!/\mathfrak{n}^\psi$. We write $A_\hbar = \sum_{n\in\mathbb{Z}} \hbar^n F_n A$, resp. $\mathfrak{A}_\hbar$, for the Rees algebra associated with the Kazhdan filtration. Note that since the Rees algebra associated with the Kazhdan filtration is canonically isomorphic as a $\mathbb{C}[\hbar]$-algebra (but not as a graded algebra), to the Rees algebra associated with the filtration $A_\leq$, one has $\bigcap_{i\in\mathbb{Z}} \hbar^i A_\hbar = 0$. It follows that the Kazhdan filtration on A is separating.

We identify $\mathfrak{n}^\psi$ with a graded Lie subalgebra of $\mathscr{U}(\mathfrak{n}^\psi)[\hbar] = \mathscr{U}_\hbar\mathfrak{n}$ as in Proposition 3.1.1(4). It follows from definitions that there is a natural graded space, resp. graded algebra, isomorphism

$$(A_\hbar/A_\hbar\mathfrak{n}^\psi)|_{\hbar=0} \cong \operatorname{gr} A/(\operatorname{gr} A)\mathfrak{n}^\psi, \quad \text{resp.} \quad (A_\hbar/\!\!/\mathfrak{n}^\psi)|_{\hbar=0} \cong (\operatorname{gr} A)/\!\!/\mathfrak{n}^\psi. \tag{3.3.1}$$

Lemma 3.3.2. *Assume that* $\operatorname{gr}_\leq A$ *is a finitely generated commutative algebra. Then,* $A_\hbar$ *is a finitely generated left noetherian algebra and* $A_\hbar/A_\hbar\mathfrak{n}^\psi$ *is flat over* $\mathbb{C}[\hbar]$.

Assume, in addition, that the Kazhdan filtration on $A/A\mathfrak{n}^\psi$ *is separating. Then, we have:*

(1) *The Kazhdan filtration on* $\mathfrak{A}$ *is separating and there are natural graded algebra isomorphisms*

$$A_\hbar/\!\!/\mathfrak{n}^\psi \cong \mathfrak{A}_\hbar, \quad \textit{resp.,} \quad (\operatorname{gr} A)/\!\!/\mathfrak{n}^\psi \cong \operatorname{gr}\mathfrak{A}.$$

(2) *For any finitely generated module* $M \in (A, \mathfrak{n}^\psi)$-mod, *one has*

$$\dim \mathrm{SS}(M) - \dim \mathrm{SS}(M^{\mathfrak{n}^\psi}) = \tfrac{1}{2}(\dim \mathrm{Spec}(\mathrm{gr} A) - \dim \mathrm{Spec}(\mathrm{gr} \mathfrak{A})).$$

Proof. The algebra $\mathrm{gr}_{\leq} A$ being finitely generated and commutative, it follows easily that the Rees algebra of A associated with the filtration $A_{\leq}$, hence also $A_\hbar$, is finitely generated and left noetherian. To prove that $A/A\mathfrak{n}^\psi$ is flat over $\mathbb{C}[\hbar]$, we apply Proposition 3.1.1(4) in the case $M = A$. We deduce that $\mathrm{Tor}_j^{\mathbb{C}[\hbar]}(\mathbb{C}[\hbar]/(\hbar),\ A_\hbar/A_\hbar\mathfrak{n}^\psi) = 0$ for all $j > 0$. It follows that $A_\hbar/A_\hbar\mathfrak{n}^\psi$ is $\hbar$-torsion free. Since $\mathbb{C}[\hbar,\hbar^{-1}] \otimes_{\mathbb{C}[\hbar]} A_\hbar/A_\hbar\mathfrak{n}^\psi \cong \mathbb{C}[\hbar,\hbar^{-1}] \otimes_{\mathbb{C}[\hbar]} A/A\mathfrak{n}^\psi$, we deduce that $A_\hbar/A_\hbar\mathfrak{n}^\psi$ is flat over $\mathbb{C}[\hbar]$.

To prove statement (1), we let $\mathscr{U}\mathfrak{n}^\psi$ act on A by right multiplication. The filtrations on A and $\mathfrak{n}^\psi$ make the corresponding Chevalley-Eilenberg complex $C_\bullet(\mathfrak{n}^\psi, A)$ a filtered complex. An associated graded complex, $\mathrm{gr}\, C(\mathfrak{n}^\psi, A)$, may be identified with $C_\bullet(\mathfrak{n}^\psi, \mathrm{gr} A)$, the Koszul complex of $\mathrm{gr} A$. Here, we view $\mathrm{gr} A$ as a $\mathbb{C}[(\mathfrak{n}^\psi)^*]$-module via the map $\mathrm{gr}\,\mathscr{U}\mathfrak{n}^\psi \to \mathrm{gr} A$ induced by the homomorphisms $\mathscr{U}\mathfrak{n}^\psi \to \mathscr{U}\mathfrak{g} \to A$. There is a standard spectral sequence for homology:

$$E_1 = H(\mathrm{gr}\, C_\bullet(\mathfrak{n}^\psi, A)) \quad \Rightarrow \quad E_\infty = \mathrm{gr}\, H(C_\bullet(\mathfrak{n}^\psi, A)). \tag{3.3.3}$$

The locally finite $\mathrm{ad}\,\mathfrak{g}$-action on A can be exponentiated to a G-action. Therefore, the map $\mathrm{gr}\,\mathscr{U}\mathfrak{g} \to \mathrm{gr} A$, induced by $\mathscr{U}\mathfrak{g} \to A$, gives a G-equivariant morphism $\mathrm{Spec}\,\mathrm{gr} A \to \mathrm{Spec}(\mathrm{gr}\,\mathscr{U}\mathfrak{g}) = \mathfrak{g}^*$. From Lemma 2.2.1, we deduce that $\mathrm{Spec}\,\mathrm{gr} A$ is flat over $\mathscr{O}_\psi$. It follows that the homology groups $H_j(C_\bullet(\mathfrak{n}^\psi, \mathrm{gr} A))$ vanish for all $i \neq 0$. Hence, our spectral sequence degenerates at the E_1-term.

Next, we claim that the above spectral sequence is convergent. This is not immediately clear since the Kazhdan filtration on A is not necessarily bounded below. To prove convergence, observe first that the action of the algebra A on itself by *left* multiplication makes $C_\bullet(\mathfrak{n}^\psi, A)$ a complex of finitely generated left A-modules. The Chevalley-Eilenberg differential respects the A-module structure. We deduce, as in the proof of [Gi], Theorem 4.14(i), that in the spectral sequence (3.3.3), one has $Z^\infty = \bigcap_{r \geq 0} Z_p^r = 0$, $\forall\ p$, i.e., the first equation in formula (4.3.5) from [Gi] holds. Further, we know that $H_j(\mathfrak{n}^\psi, A) = 0$, $\forall\, j > 0$, see Proposition 3.1.4(iv). This, combined with the assumption that the filtration on $A/A\mathfrak{n}^\psi$ is separating implies that the second equation in [Gi], formula (4.3.5), holds. Thus, the convergence criterion stated in [Gi, Lemma 4.3.7] applies in our present setting. It follows that the spectral sequence in (3.3.3) does converge. By [Gi, Lemma 4.3.3], we deduce that the canonical map $\mathrm{gr} A/(\mathrm{gr} A)\mathfrak{n}^\psi \to \mathrm{gr}(A/A\mathfrak{n}^\psi)$ is an isomorphism. Explicitly, this means that the following surjective map is an isomorphism:

$$F_n A / \textstyle\sum_i (F_{n-1-i} A)\mathfrak{n}_i^\psi \to (F_n A + A\mathfrak{n}^\psi)/(F_{n-1} A + A\mathfrak{n}^\psi), \qquad \forall\, n \in \mathbb{Z}. \tag{3.3.4}$$

Here, $\mathfrak{n}_i^\psi$ stands for the i-th homogeneous component of $\mathfrak{n}^\psi\,\mathscr{U}(\mathfrak{n}^\psi)[\hbar] = \mathscr{U}_\hbar\mathfrak{n}$, i.e., $\mathfrak{n}_i^\psi$ is a copy of the vector space $\mathfrak{n} \cap \mathfrak{g}(i+2)$ if $i < -2$ and $\mathfrak{n}_0^\psi = \{n - \psi(n),\ n \in \mathfrak{n} \cap \mathfrak{g}(-2)\}$.

The isomorphisms in (1) are immediate consequences of (3.3.4) and Proposition 3.2.2(3) applied to $M = A/A\mathfrak{n}^\psi$. Also, the assumption of the proposition clearly implies that the filtration $F_\bullet\mathfrak{A}$ is separating.

Now, let M be as in (2). Choose a finite-dimensional $\mathfrak{n}^\psi$-stable subspace $M_0 \subset M$, such that $M = A\,M_0$. The filtration on M defined by $F_iM := (F_iA)M_0$, $i \in \mathbb{Z}$, satisfies the assumptions of Proposition 3.2.2(3). Hence, we have $\operatorname{gr} M \cong \operatorname{gr} Q \otimes_{\operatorname{gr} Z\mathfrak{g}} \operatorname{gr}(M^{\mathfrak{n}^\psi})$. Using that $\operatorname{gr} Q \cong \mathbb{C}[\psi + \mathfrak{n}^\perp] \cong \mathbb{C}[N \times \mathscr{S}]$ and $\mathscr{S} \cong \mathfrak{c} \cong \operatorname{Spec}(\operatorname{gr} Z\mathfrak{g})$, we compute

$$\begin{aligned}\dim \operatorname{Supp}(\operatorname{gr} Q \otimes_{\operatorname{gr} Z\mathfrak{g}} \operatorname{gr}(M^{\mathfrak{n}^\psi}) &= \dim \operatorname{Supp} \operatorname{gr} Q - \dim \mathfrak{c} + \dim \operatorname{Supp} \operatorname{gr}(M^{\mathfrak{n}^\psi}) \\ &= \dim N + \dim \operatorname{Supp} \operatorname{gr}(M^{\mathfrak{n}^\psi}).\end{aligned}$$

In particular, for $M = A/A\mathfrak{n}^\psi$, we get

$$\dim \operatorname{Supp}((\operatorname{gr} A/(\operatorname{gr} A)^{\mathfrak{n}^\psi})) = \dim N + \dim \operatorname{Spec} \operatorname{gr} \mathfrak{A}.$$

On the other hand, since $\operatorname{Spec} \operatorname{gr} A$ is flat over $\mathscr{O}_\psi$, using part (ii) of Lemma 2.2.1, we get

$$\dim \operatorname{Supp}((\operatorname{gr} A/(\operatorname{gr} A)^{\mathfrak{n}^\psi})) = \dim \operatorname{Spec} \operatorname{gr} A - \dim \mathfrak{n}^*.$$

The required equation follows. □

4 Proofs of Theorem 1.2.2 and Theorem 1.6.3

4.1 Whittaker coinvariants of $\mathscr{D}$-modules

Let X be a smooth G-variety. The algebra of differential operators comes equipped with an algebra homomorphism $\mathscr{U}\mathfrak{g} \to \mathscr{D}(X)$. The homomorphism intertwines the PBW filtration on $\mathscr{U}\mathfrak{g}$ with the natural filtration $\mathscr{D}_\leq(X)$ by order of the differential operator. Following Section 3.3, we get a $\mathbb{Z}$-grading on $\mathscr{D}(X)$ and an associated Kazhdan filtration $F_\bullet\mathscr{D}(X)$.

Remark 4.1.1. Let $\gamma\colon \mathbb{G}_m \to T$ be a 1-parameter subgroup generated by the element h of our fixed $\mathfrak{sl}_2$-triple. We let the group $\mathbb{G}_m$ act on X by $z : x \mapsto \gamma(z)x$. The induced $\mathbb{G}_m$-action on $\mathscr{D}(X)$ by algebra automorphisms gives a $\mathbb{Z}$-grading on $\mathscr{D}(X)$ that agrees with the $\mathbb{Z}$-grading on $\mathscr{D}(X)$ considered above. ◊

Let K be a connected algebraic subgroup of G and $\chi : \mathfrak{k} = \operatorname{Lie} K \to \mathbb{C}$ a character. We have $(\mathfrak{k}^\chi)^{op} \cong \mathfrak{k}^{-\chi}$. The Lie algebra map $\mathfrak{k}^\chi \to \mathscr{D}(X)$ gives a Lie algebra map $\mathfrak{k}^{-\chi} \to \mathscr{D}(X)^{op}$. Assume, in addition, that X is affine and the quotient $X \to X//K$ is a K-torsor. Then, there is a canonical isomorphism $\mathscr{D}(X//K) \cong \mathscr{D}(X)/\!\!/\mathfrak{k}$. In the special case where K is unimodular, e.g., unipotent, and the canonical bundle on X has a K-invariant trivialization, we get $\mathscr{D}(X) \cong \mathscr{D}(X)^{op}$, resp. $\mathscr{D}(X/K) \cong \mathscr{D}(X/K)^{op}$. We deduce an isomorphism $\mathscr{D}(X)/\!\!/\mathfrak{k} \cong (\mathscr{D}(X)/\!\!/\mathfrak{k})^{op}$. More generally, with the same assumptions one can construct a natural isomorphism $\mathscr{D}(X)/\!\!/\mathfrak{k}^{-\chi} \cong (\mathscr{D}(X)/\!\!/\mathfrak{k}^{\chi})^{op}$. On the other hand, by general properties of Hamiltonian reduction there is an isomorphism $(\mathscr{D}(X)/\!\!/\mathfrak{k}^{\chi})^{op} \cong \mathfrak{k}^{-\chi}\backslash\!\!\backslash\mathscr{D}(X)^{op}$, see Section 3.1. Thus, under the same assumptions, one also has an isomorphism $(\mathscr{D}(X)/\!\!/\mathfrak{k}^{\chi})^{op} \cong \mathfrak{k}^{-\chi}\backslash\!\!\backslash\mathscr{D}(X)$.

Now, let X be a smooth $G\times G$-variety. We will abuse notation and write $G\times G = G_l \times G_r$. Thus, we have a natural algebra map $\mathscr{U}\mathfrak{g}_l \otimes \mathscr{U}\mathfrak{g}_r \to \mathscr{D}(X)$, so $\mathscr{U}\mathfrak{g}_l \otimes \mathscr{U}\mathfrak{g}_r$ acts on any (say, right) $\mathscr{D}(X)$-module M. Changing a sign of the $\mathfrak{g}$-action, we may view M as an $(\mathscr{U}\mathfrak{g}, \mathscr{U}\mathfrak{g})$-bimodule. Let $(\mathscr{D}(X), G)$-mod be the category of right $\mathscr{D}(X)$-modules equivariant with respect to the diagonal G-action. For such a module M, the $\operatorname{ad}\mathfrak{g}$-action on M is locally finite. We conclude that Proposition 3.1.2 yields the following result.

Corollary 4.1.2. *For any $M \in (\mathscr{D}(X), G)$-mod and $j \neq 0$, we have $H_j(\mathfrak{n}_r^{\psi}, M) = 0$. In particular, the functor $(\mathscr{D}(X), G)$-mod $\mapsto (\mathscr{D}(X)/\!\!/\mathfrak{n}_r^{\psi})$-mod, $M \mapsto M/M\mathfrak{n}_r^{\psi}$, is exact.* □

4.2 Proof of Theorem 1.6.3

Let X be a smooth affine $G\times G$-variety and M a right $\mathscr{D}(X)$-module. We may view M as a left $\mathscr{D}(X)^{op}$-module, in particular, an $\mathfrak{n}_r^{-\psi}$-module. The Lie algebra $\mathfrak{n}$ being unimodular, one has a canonical isomorphism of functors $H_\bullet(\mathfrak{n}, -) \cong H^{d-\bullet}(\mathfrak{n}, -)$, where $d = \dim N$. It follows that there is a canonical isomorphism $H^\bullet(\mathfrak{n}_r^{-\psi}, M) \cong H_{d-j}(\mathfrak{n}_r^{\psi}, M)$. By Section 3.1, this space has the natural structure of a left $\mathfrak{D}$-module, where $\mathfrak{D} = \mathscr{D}(X)^{op}/\!\!/\mathfrak{n}_r^{-\psi}$. We have the identification $\mathscr{D}(X)^{op}/\mathscr{D}(X)^{op}\mathfrak{n}_r^{-\psi} = \mathscr{D}(X)/\mathfrak{n}_r^{\psi}\mathscr{D}(X)$, and it follows from Proposition 3.1.4(i) that the functor

$$\mathscr{D}(X)^{op}/\mathscr{D}(X)^{op}\mathfrak{n}_r^{-\psi} \otimes_{\mathfrak{D}} (-):\ \mathfrak{D}\text{-mod} \to (\mathscr{D}(X), \mathfrak{n}_r^{\psi})\text{-mod}$$

is an equivalence. Also, as has been mentioned in Section 1.5, there is a canonical equivalence of $(\mathscr{D}(X), \mathfrak{n}_r^{\psi})$-mod $\cong (\mathscr{D}_X, N_r, \psi)$-mod. We deduce that the second functor in (1.6.2) is an equivalence. The first functor in (1.6.2) is exact, by Corollary 4.1.2.

It remains to identify the composite functor in (1.6.2) with the functor (1.6.1). It is well-known, cf. [MV], that an averaging functor $\int_{a_N}^{-d}(\Omega_K \boxtimes -)$ is a right adjoint to the natural full embedding $(\mathscr{D}(X), N_r)$-mod $\to \mathscr{D}(X)$-mod (a homological shift by d is due to a shift in adjunction formulas, e.g., [HTT], Theorem 3.2.14). Similarly, one shows that the functor $\int_{a_N}^{-d}(e^{-\psi}\Omega_N \boxtimes -)$ is a right adjoint to the natural full embedding $(\mathscr{D}(X), \mathfrak{n}_r^{\psi})$-mod $\to \mathscr{D}(X)$-mod, to be denoted $\mathbf{I}$.

On the other hand, the functor $\mathbf{I}$ can be factored as a composition

$$(\mathscr{D}(X), \mathfrak{n}_r^{\psi})\text{-mod} \to \mathfrak{D}\text{-mod} \to \mathscr{D}(X)\text{-mod},$$

where the first functor is the equivalence $M \mapsto M^{\mathfrak{n}_r^{-\psi}}$, to be denoted $\mathbf{J}$, and the second functor is $\mathscr{D}(X)^{op}/\mathscr{D}(X)^{op}\mathfrak{n}_r^{-\psi} \otimes_{\mathfrak{D}} (-)$. From the canonical isomorphisms

$$\begin{aligned}\mathrm{Hom}_{\mathscr{D}(X)^{op}}(\mathscr{D}(X)^{op}/\mathscr{D}(X)^{op}\mathfrak{n}_r^{\psi} \otimes_{\mathfrak{D}} M,\ \mathscr{F}) &\cong \mathrm{Hom}_{\mathfrak{D}}(M, \mathscr{F}^{\mathfrak{n}_r^{-\psi}})\\ &\cong \mathrm{Hom}_{\mathfrak{D}}(M, H_d(\mathfrak{n}_r^{\psi}, \mathscr{F})),\end{aligned}$$

we see that the functor $\mathbf{J}^{-1} \circ H_d(\mathfrak{n}_r^{\psi}, -)$ is a right adjoint of $\mathbf{I}$. Hence, the functor $\mathbf{J}^{-1} \circ H_d(\mathfrak{n}_r^{\psi}, -)$ must be isomorphic to the averaging functor $\int_{a_N}^{-d}(e^{-\psi}\Omega_N \boxtimes -)$. These two functors are left exact. The corresponding j-th right derived functors are $\mathbf{J}^{-1} \circ H_{d-j}(\mathfrak{n}_r^{\psi}, -)$ and $\int_{a_N}^{j-d}(e^{-\psi}\Omega_K \boxtimes -)$, respectively. It follows that these derived functors are isomorphic. The composite functor in (1.6.2) is nothing but the functor $\mathbf{J}^{-1} \circ H_0(\mathfrak{n}_r^{\psi}, -)$, proving the required isomorphism of functors. □

4.3 Proof of Theorem 1.2.2

Let $\mathbb{X}^{++} \subset \mathbb{X}^*$ be the set of dominant weights. We write $\lambda \leq \mu$ if $\mu - \lambda$ is a linear combination of simple roots with nonnegative integer coefficients.

Let V_λ be a simple G-representation with highest weight $\lambda \in \mathbb{X}^{++}$, i.e., V_λ has a unique, up to a constant factor, nonzero vector $v_\lambda \in V_\lambda$ such that $tv_\lambda = \lambda(t)v_\lambda$, $\forall t \in T$. Let V_λ^* be the contragredient representation. We have the Peter-Weyl decomposition $\mathbb{C}[G] = \bigoplus_{\mu \in \mathbb{X}^{++}} V_\mu \otimes V_\mu^*$. We put $R_\lambda := \bigoplus_{\mu \leq \lambda} V_\mu \otimes V_\mu^*$. The spaces R_λ, $\lambda \in \mathbb{X}^{++}$, form an exhaustive multiplicative filtration of the algebra $\mathbb{C}[G]$.

Now, view $X = G$ as a $G \times G$-variety where $G \times G = G_l \times G_r$ acts on G by left and right translations. Thus, we have an algebra homomorphism $\mathscr{U}(\mathfrak{g}_l \oplus \mathfrak{g}_r) = \mathscr{U}\mathfrak{g}_l \otimes \mathscr{U}\mathfrak{g}_r \to \mathscr{D}(G)$. The diagonal G-action on X corresponds to the action of G on itself by conjugation.

We have a natural embedding $\mathbb{C}[G] \hookrightarrow \mathscr{D}(G)$. For $\lambda \in \mathbb{X}^{++}$, let $P_\lambda\mathscr{D}(G) := R_\lambda \cdot \mathscr{U}(\mathfrak{g}_l \oplus \mathfrak{g}_r)$ be a right $\mathscr{U}(\mathfrak{g}_l \oplus \mathfrak{g}_r)$-submodule of $\mathscr{D}(G)$ generated by the subspace $R_\lambda \subset \mathbb{C}[G]$. We have

$$\begin{aligned}\mathscr{D}(G) = \mathbb{C}[G] \cdot \mathscr{U}\mathfrak{g}_l &= \textstyle\bigcup_{\lambda \in \mathbb{X}^{++}} R_\lambda \cdot \mathscr{U}\mathfrak{g}_l\\ &\subset \textstyle\bigcup_{\lambda \in \mathbb{X}^{++}} R_\lambda \cdot \mathscr{U}(\mathfrak{g}_l \oplus \mathfrak{g}_r) = \bigcup_{\lambda \in \mathbb{X}^{++}} P_\lambda\mathscr{D}(G).\end{aligned}$$

Thus, the spaces $P_\lambda\mathscr{D}(G)$, $\lambda \in \mathbb{X}^{++}$, form an exhaustive filtration of $\mathscr{D}(G)$.

Remark 4.3.1. The filtration $P_\lambda\mathscr{D}(G)$, $\lambda \in \mathbb{X}^{++}$ has a natural geometric interpretation in terms of the *wonderful compactification* of G^{ad}, the adjoint group. It is immediate from this interpretation that $\mathscr{U}(\mathfrak{g}_l \oplus \mathfrak{g}_r) \cdot R_\lambda = R_\lambda \cdot \mathscr{U}(\mathfrak{g}_l \oplus \mathfrak{g}_r)$ and each of the spaces $P_\lambda\mathscr{D}(G)$ is stable under the conjugation action of G on itself. Furthermore, for any $\lambda, \mu \in \mathbb{X}^{++}$, one has $P_\lambda\mathscr{D}(G) \cdot P_\mu\mathscr{D}(G) \subset P_{\lambda+\mu}\mathscr{D}(G)$.

The adjoint action of the element $\mathsf{h}_l + \mathsf{h}_r \in \mathfrak{g}_l \oplus \mathfrak{g}_r$ gives a $\mathbb{Z}$-grading on $\mathscr{U}(\mathfrak{g}_l \oplus \mathfrak{g}_r)$. The corresponding Kazhdan filtration $F_\bullet\mathscr{U}(\mathfrak{g}_l \oplus \mathfrak{g}_r)$ is a tensor product of the

Kazhdan filtrations on $\mathscr{U}\mathfrak{g}_l$ and $\mathscr{U}\mathfrak{g}_r$, respectively. Recall that the Kazhdan filtration on $Q = \mathscr{U}\mathfrak{g}/(\mathscr{U}\mathfrak{g})\mathfrak{n}^\psi$ has no components of negative degrees. It follows that the quotient Kazhdan filtration on $\mathscr{U}(\mathfrak{g}_l \oplus \mathfrak{g}_r)/\mathscr{U}(\mathfrak{g}_l \oplus \mathfrak{g}_r)\mathfrak{n}_{lr}^\psi = Q_l \otimes Q_r$ has no components of negative degree. Here, we have used simplified notation $\mathfrak{n}_{lr}^\psi = \mathfrak{n}_l^\psi \oplus \mathfrak{n}_r^\psi$.

The conjugation action of the 1-parameter subgroup γ on G induces a $\mathbb{Z}$-grading on $\mathscr{D}(G)$. Let $F_\bullet\mathscr{D}(G)$ be the corresponding Kazhdan filtration on $\mathscr{D}(G)$. By definition, one has $(P_\lambda\mathscr{D}(G))\mathfrak{n}_{lr}^\psi = R_\lambda \cdot \mathscr{U}(\mathfrak{g}_l \oplus \mathfrak{g}_r)\mathfrak{n}_{lr}^\psi$. We deduce from the above that for every λ, there exists $m = m(\lambda) \gg 0$ such that the quotient Kazhdan filtration on $(P_\lambda\mathscr{D}(G) + \mathscr{D}(G)\mathfrak{n}_{lr}^\psi)/\mathscr{D}(G)\mathfrak{n}_{lr}^\psi$ has no components of degrees $\leq -m(\lambda)$. This implies that the Kazhdan filtration on $\mathscr{D}(G)/\mathscr{D}(G)\mathfrak{n}_{lr}^\psi$ is separating.

To complete the proof of Theorem 1.2.2, we consider the setting of Section 3 in the case where $A = \mathscr{D}(G)$ and the Lie algebra $\mathfrak{g}$, resp. $\mathfrak{n}^\psi$, is replaced by $\mathfrak{g}_l \oplus \mathfrak{g}_r$, resp. $\mathfrak{n}_{lr}^\psi$. Thus, Lemma 3.3.2 is applicable. The filtration on $\mathscr{D}(G)/\mathscr{D}(G)\mathfrak{n}_{lr}^\psi$ being separating, from the second isomorphism in statement (1) of the lemma, we deduce that the natural map

$$(\mathrm{gr}^F \mathscr{D}(G))/\!\!/(N_l \times N_r, \psi \times \psi) \to \mathrm{gr}^F(\mathscr{D}(G)/\!\!/\mathfrak{n}_{lr}^\psi)$$

is an isomorphism. Further, we have

$$\mathbb{C}[\mathscr{T}^*G/\!\!/(N_l \times N_r, \psi \times \psi)] = (\mathrm{gr}^F \mathscr{D}(G))/\!\!/(N_l \times N_r, \psi \times \psi),$$

by Corollary 2.2.2. Theorem 1.2.2 follows. □

Corollary 4.3.2. *The algebra $\mathbb{W} = \mathscr{D}(G)/\!\!/(\mathfrak{n}_l^\psi + \mathfrak{n}_r^\psi)$ is simple and the Kazhdan filtration on this algebra is separating.*

Proof. The filtration on $\mathscr{D}(G)/\!\!/(\mathfrak{n}_l^\psi + \mathfrak{n}_r^\psi)$ is separating by Lemma 3.3.2. Therefore, for a nonzero two-sided ideal $I \subsetneq \mathbb{W}$, we have that $\mathrm{gr}I \subsetneq \mathrm{gr}\mathbb{W}$ is a nonzero Poisson ideal of $\mathrm{gr}\mathbb{W} = \mathbb{C}[\mathfrak{Z}]$, where $\mathbb{C}[\mathfrak{Z}]$ is viewed as a Poisson algebra. But this Poisson algebra contains no nontrivial Poisson ideals, since $\mathfrak{Z}$ is a smooth symplectic variety, see Section 2.1. □

Remark 4.3.3. Associated with the filtration R_λ on $\mathbb{C}[G]$, one has the corresponding Rees algebra $\bigoplus_{\lambda\in\mathbb{X}^{++}} R_\lambda$. This $\mathbb{X}^{++}$-graded algebra is the coordinate ring of the Vinberg semigroup associated with G. Similarly, the filtration $P_\lambda\mathscr{D}(G)$ on $\mathscr{D}(G)$ induces a quotient filtration on $\mathscr{D}(G)/\mathscr{D}(G)\mathfrak{n}_{lr}^\psi$. By restriction, one gets a filtration $P_\lambda\mathbb{W}$ on $\mathscr{D}(G)/\!\!/\mathfrak{n}_{lr}^\psi$. The corresponding Rees algebra $\bigoplus_{\lambda\in\mathbb{X}^{++}} P_\lambda\mathbb{W}$ may be viewed as a quantum counterpart of the coordinate ring of a Vinberg type deformation of the universal centralizer $\mathfrak{Z}$. This deformation, as well as an analogue of the wonderful compactification of $\mathfrak{Z}$ has been studied in [Ba1]–[Ba2].

5 Hamiltonian reduction of $\mathscr{T}^*(G/\bar{N})$ and $\mathscr{D}(G/\bar{N})$

5.1 The affine closure of $\mathscr{T}^*(G/\bar{N})$

Let B, resp. $\bar{B}$, be the Borel subgroup of G, corresponding to the Borel subalgebra $\mathfrak{b}$, resp. $\bar{\mathfrak{b}}$. Thus, N, resp. $\bar{N}$, is the unipotent radical of B, resp. $\bar{B}$, and $B \cap \bar{B}$ is a maximal torus. We identify the flag variety $\mathscr{B}$ with $G/\bar{B}$ and let $\widetilde{\mathscr{B}} := G/\bar{N}$.

The group G acts on $\widetilde{\mathscr{B}}$ on the left and the torus $T = \bar{B}/\bar{N}$ acts on $\widetilde{\mathscr{B}}$ on the right. Thus we have a $G_l \times T_r$-action on $\widetilde{\mathscr{B}}$ and an induced Hamiltonian $G_l \times T_r$-action on $\mathscr{T}^*\widetilde{\mathscr{B}}$, with moment map $\mu_{G,\mathscr{T}^*\widetilde{\mathscr{B}}} \times \mu_{T,\mathscr{T}^*\widetilde{\mathscr{B}}}: \mathscr{T}^*\widetilde{\mathscr{B}} \to \mathfrak{g}^* \times \mathfrak{t}^*$. We have a natural isomorphism $(\mathscr{T}^*\widetilde{\mathscr{B}})/T_r = \widetilde{\mathfrak{g}}$, so the quotient map $q: \mathscr{T}^*\widetilde{\mathscr{B}} \to (\mathscr{T}^*\widetilde{\mathscr{B}})/T_r = \widetilde{\mathfrak{g}}$ makes $\mathscr{T}^*\widetilde{\mathscr{B}}$ a G_l-equivariant T-torsor over $\widetilde{\mathfrak{g}}$. The moment map $\mu_{G,\mathscr{T}^*\widetilde{\mathscr{B}}} \times \mu_{T,\mathscr{T}^*\widetilde{\mathscr{B}}}$ factors as a composition $\mathscr{T}^*\widetilde{\mathscr{B}} \xrightarrow{q} \widetilde{\mathfrak{g}} \to \mathfrak{g}^* \times \mathfrak{t}^*$.

Let $z: \xi \mapsto z\cdot\xi$ denote the standard $\mathbb{G}_m$-action on a cotangent bundle by dilations along the fibers. We define a $\bullet$-action of $\mathbb{G}_m$ on $\mathscr{T}^*G$, resp. $\mathfrak{g}^*$, by $z:\ \xi \mapsto z\bullet\xi := z^{-2}\cdot \mathrm{Ad}^*\gamma(z)(\xi)$, where the $\mathrm{Ad}^*\gamma(z)$-action on $\mathscr{T}^*G$ is induced by $\gamma(z):\ g \mapsto \gamma(z)g\gamma(z)^{-1}$. The action on $\mathscr{T}^*G$ of the torus $T_l \times T_r \subset G_l \times G_r$ descends to a well-defined action on $\mathscr{T}^*\widetilde{\mathscr{B}}$. Hence, the $\bullet$-action descends to a well-defined action on $\mathscr{T}^*\widetilde{\mathscr{B}}$.

The variety $\mathscr{T}^*\widetilde{\mathscr{B}}$ is quasi-affine; furthermore, it was shown in [GR, Section 5.5] that the algebra $\mathbb{C}[\mathscr{T}^*\widetilde{\mathscr{B}}]$ is finitely generated. Thus, $(\mathscr{T}^*\widetilde{\mathscr{B}})_{\mathrm{aff}} := \operatorname{Spec}\mathbb{C}[\mathscr{T}^*\widetilde{\mathscr{B}}]$ is an affine algebraic variety that contains $\mathscr{T}^*\widetilde{\mathscr{B}}$ as a Zariski open and dense subvariety. It is clear that any action of a linear algebraic group on $\mathscr{T}^*\widetilde{\mathscr{B}}$ extends to an action on $(\mathscr{T}^*\widetilde{\mathscr{B}})_{\mathrm{aff}}$. The moment map $\mu_{G,\mathscr{T}^*\widetilde{\mathscr{B}}} \times \mu_{T,\mathscr{T}^*\widetilde{\mathscr{B}}}$ extends to a $G_l \times T_r$-equivariant morphism $\bar{\mu}_{G,\mathscr{T}^*\widetilde{\mathscr{B}}} \times \bar{\mu}_{T,\mathscr{T}^*\widetilde{\mathscr{B}}}: (\mathscr{T}^*\widetilde{\mathscr{B}})_{\mathrm{aff}} \to \mathfrak{g}^* \times \mathfrak{t}^*$.

The following result, [GR, Section 5.5], [GK], is a commutative counterpart of the Gelfand–Graev construction mentioned in the introduction.

Proposition 5.1.1. *There is a W-action on $(\mathscr{T}^*\widetilde{\mathscr{B}})_{\mathrm{aff}}$ such that:*

(1) *The T_r-action and the W-action combine together to give a $W \ltimes T$-action on $(\mathscr{T}^*\widetilde{\mathscr{B}})_{\mathrm{aff}}$.*
(2) *The $W \ltimes T$-action commutes both with the G_l-action and with the $\bullet$-action on $(\mathscr{T}^*\widetilde{\mathscr{B}})_{\mathrm{aff}}$.*
(3) *The map $\bar{\mu}^*_{G,\mathscr{T}^*\widetilde{\mathscr{B}}} \times \bar{\mu}_{T,\mathscr{T}^*\widetilde{\mathscr{B}}}$ is $W \ltimes T$-equivariant, where W acts naturally on $\mathfrak{t}^*$ and $W \ltimes T$ acts trivially on $\mathfrak{g}^*$.*

Remark 5.1.2. Let $\mathscr{T}^{\mathrm{reg}}\widetilde{\mathscr{B}} := \mu^{-1}_{G,\mathscr{T}^*\widetilde{\mathscr{B}}}(\mathfrak{g}^{\mathrm{reg}})$. We have open embeddings $\mathscr{T}^{\mathrm{reg}}\widetilde{\mathscr{B}} \subset \mathscr{T}^*\widetilde{\mathscr{B}} \subset (\mathscr{T}^*\widetilde{\mathscr{B}})_{\mathrm{aff}}$. The set $\mathscr{T}^{\mathrm{reg}}\widetilde{\mathscr{B}}$ is stable under the W-action, but the set $\mathscr{T}^*\widetilde{\mathscr{B}}$ is *not*. ◊

Lemma 5.1.3. *The morphism $\mu_{G,\mathscr{T}^*\widetilde{\mathscr{B}}}|_{\mathscr{T}^{\mathrm{reg}}\widetilde{\mathscr{B}}}:\ \mathscr{T}^{\mathrm{reg}}\widetilde{\mathscr{B}} \to \mathfrak{g}^{\mathrm{reg}}$ is flat and we have $\bar{\mu}^{-1}_{G,\mathscr{T}^*\widetilde{\mathscr{B}}}(\mathfrak{g}^{\mathrm{reg}}) = \mathscr{T}^{\mathrm{reg}}\widetilde{\mathscr{B}}$. Further, let $S \subset \mathfrak{g}^*$ be a closed subvariety such that $S \subset$*

$\mathfrak{g}^{\mathrm{reg}}$, *and let* $I_Y \subset \mathbb{C}[\mathfrak{g}^*]$ *be the corresponding ideal. Then, we have* $\mathbb{C}[\mu^{-1}_{G,\mathscr{T}^*\widetilde{\mathscr{B}}}(S)] = \mathbb{C}[\mathscr{T}^*\widetilde{\mathscr{B}}]/\mathbb{C}[\mathscr{T}^*\widetilde{\mathscr{B}}]I_S$.

Proof. To simplify the notation, we write $\mu = \mu_{G,\mathscr{T}^*\widetilde{\mathscr{B}}}$, resp. $\bar{\mu} = \bar{\mu}_{G,\mathscr{T}^*\widetilde{\mathscr{B}}}$. Let $\mathscr{A} := \mu_*\mathscr{O}_{\mathscr{T}^*\widetilde{\mathscr{B}}}$. This is a quasi-coherent sheaf of $\mathscr{O}_{\mathfrak{g}^*}$-algebras and we have

$$\mathbb{C}[(\mathscr{T}^*\widetilde{\mathscr{B}})_{\mathrm{aff}}] = \mathbb{C}[\mathscr{T}^*\widetilde{\mathscr{B}}] = \Gamma(\mathfrak{g}^*, \mathscr{A}).$$

Let $i_x : \{x\} \hookrightarrow \mathfrak{g}^*$ be a one point embedding and $I_x \subset \mathbb{C}[\mathfrak{g}^*]$ the ideal of x. Then $\bar{\mu}^{-1}(x)$ is a closed affine subvariety of $(\mathscr{T}^*\widetilde{\mathscr{B}})_{\mathrm{aff}}$ and we have

$$\mathbb{C}[\bar{\mu}^{-1}(x)] = \mathbb{C}[(\mathscr{T}^*\widetilde{\mathscr{B}})_{\mathrm{aff}}]/\mathbb{C}[(\mathscr{T}^*\widetilde{\mathscr{B}})_{\mathrm{aff}}]I_x = \Gamma(\mathfrak{g}^*, \mathscr{A})/\Gamma(\mathfrak{g}^*, \mathscr{A})I_x = i_x^*\mathscr{A}.$$

The map $\mu : \mathscr{T}^{\mathrm{reg}}\widetilde{\mathscr{B}} \to \mathfrak{g}^{\mathrm{reg}}$ factors as $\pi|_{\widetilde{\mathfrak{g}}^{\mathrm{reg}}} \circ q$, where q is a T-torsor, π is the Grothendieck-Springer map, and $\pi|_{\widetilde{\mathfrak{g}}^{\mathrm{reg}}}$ is a finite morphism. Therefore, $\mathscr{T}^{\mathrm{reg}}\widetilde{\mathscr{B}} \to \mathfrak{g}^{\mathrm{reg}}$ is a flat affine morphism. Hence, for $x \in \mathfrak{g}^{\mathrm{reg}}$, the fiber $\mu^{-1}(x)$ is affine. Writing $i : \mu^{-1}(x) \hookrightarrow \mathscr{T}^{\mathrm{reg}}\widetilde{\mathscr{B}}$ for the corresponding embedding, by base change, we get $i_x^*\mathscr{A} = \mu_* i^*\mathscr{O}_{\mathscr{T}^*\widetilde{\mathscr{B}}} = \mu_*\mathscr{O}_{\mu^{-1}(x)}$. Thus, from the previous paragraph, we deduce that

$$\mathbb{C}[\bar{\mu}^{-1}(x)] = i_x^*\mathscr{A} = \Gamma(\mu^{-1}(x), i^*\mathscr{O}_{\mathscr{T}^*\widetilde{\mathscr{B}}}) = \mathbb{C}[\mu^{-1}(x)].$$

We conclude that $\mu^{-1}(x) = \bar{\mu}^{-1}(x)$ for any $x \in \mathfrak{g}^{\mathrm{reg}}$. The proof of the last statement of the lemma is very similar. □

The remainder of Section 5.1 will not be used in the paper and is only given for completeness.

Let $\widetilde{\mathscr{B}}_{\mathrm{aff}}$ be the affine closure of $\widetilde{\mathscr{B}}$ and $\bar{\mathsf{p}} : (\mathscr{T}^*\widetilde{\mathscr{B}})_{\mathrm{aff}} \to \widetilde{\mathscr{B}}_{\mathrm{aff}}$ the canonical extension of the projection $\mathsf{p} : \mathscr{T}^*\widetilde{\mathscr{B}} \to \widetilde{\mathscr{B}}$. For every $w \in W$, let $\bar{\mathsf{p}}_w : (\mathscr{T}^*\widetilde{\mathscr{B}})_{\mathrm{aff}} \to \widetilde{\mathscr{B}}_{\mathrm{aff}}$ be a map defined by $\bar{\mathsf{p}}_w(y) = \bar{\mathsf{p}}(w(y))$. Thus, we have $\bar{\mathsf{p}} = \bar{\mathsf{p}}_1$.

Proposition 5.1.4. *For* $y \in (\mathscr{T}^*\widetilde{\mathscr{B}})_{\mathrm{aff}}$ *the following properties are equivalent:*

(i) *We have* $\bar{\mu}_{G,\mathscr{T}^*\widetilde{\mathscr{B}}}(y) \in \mathfrak{g}^{\mathrm{reg}}$;

(ii) *The* T_r*-orbit,* $T_r \cdot y$*, of* y *is a closed orbit in* $(\mathscr{T}^*\widetilde{\mathscr{B}})_{\mathrm{aff}}$ *of maximal dimension, i.e.,* $\dim(T_r \cdot y) = \dim T$;

(iii) *For every* $w \in W$ *we have* $\bar{\mathsf{p}}_w(y) \in \widetilde{\mathscr{B}}$.

Observe first that the fibers of the map $\bar{\mu}_{G,\mathscr{T}^*\widetilde{\mathscr{B}}} \times \bar{\mu}_{T,\mathscr{T}^*\widetilde{\mathscr{B}}}$ are T_r-stable. We need the following standard result.

Lemma 5.1.5. *Every fiber of the map* $(\bar{\mu}_{G,\mathscr{T}^*\widetilde{\mathscr{B}}} \times \bar{\mu}_{T,\mathscr{T}^*\widetilde{\mathscr{B}}}$ *contains exactly one closed* T_r*-orbit; furthermore this orbit is contained in the closure of any other* T_r*-orbit in that fiber.*

Proof. We have $\mathbb{C}[(\mathscr{T}^*\widetilde{\mathscr{B}})_{\mathrm{aff}}]^{T_r} = \mathbb{C}[\mathscr{T}^*\widetilde{\mathscr{B}}]^{T_r} = \mathbb{C}[\widetilde{\mathfrak{g}}]$. The algebra $\mathbb{C}[\widetilde{\mathfrak{g}}]$ may be identified with the pullback of $\mathbb{C}[\mathfrak{g}^* \times_{\mathfrak{t}^*/W} \mathfrak{t}^*]$ via $\bar{\mu}_{G,\mathscr{T}^*\widetilde{\mathscr{B}}} \times \bar{\mu}_{T,\mathscr{T}^*\widetilde{\mathscr{B}}}$. Hence, for any $s \in \mathfrak{g}^* \times_{\mathfrak{t}^*/W} \mathfrak{t}^* = \operatorname{Spec} \mathbb{C}[(\mathscr{T}^*\widetilde{\mathscr{B}})_{\text{aff}}]^{T_r}$, the fiber of the map $(\mathscr{T}^*\widetilde{\mathscr{B}})_{\text{aff}} \to \operatorname{Spec} \mathbb{C}[(\mathscr{T}^*\widetilde{\mathscr{B}})_{\text{aff}}]^{T_r}$ may be identified with $(\bar{\mu}_{G,\mathscr{T}^*\widetilde{\mathscr{B}}} \times \bar{\mu}_{T,\mathscr{T}^*\widetilde{\mathscr{B}}})^{-1}(s)$. The lemma follows since T-invariant functions on an affine variety separate closed T-orbits. $\square$

Proof (*Proposition* 5.1.4). Let $y \in (\mathscr{T}^*\widetilde{\mathscr{B}})_{\text{aff}}$ be such that $\bar{\mu}_{G,\mathscr{T}^*\widetilde{\mathscr{B}}}(y) \in \mathfrak{g}^{\text{reg}}$. Then, it follows from Lemma 5.1.3 that $y \in \mathscr{T}^{\text{reg}}\widetilde{\mathscr{B}}$. Hence, the T_r-orbit of y is a closed orbit of maximal dimension. Also, for every $w \in W$, we have $\bar{\mathsf{p}}_w(y) \in \widetilde{\mathscr{B}}$, since the set $\mathscr{T}^{\text{reg}}\widetilde{\mathscr{B}}$ is W-stable.

Next, fix $y \in (\mathscr{T}^*\widetilde{\mathscr{B}})_{\text{aff}}$ such that the element $\bar{\mu}_{G,\mathscr{T}^*\widetilde{\mathscr{B}}}(y)$ is not regular. We claim that the T_r-orbit of y is either not closed or it does not have maximal dimension. Let F be the fiber of the map $\bar{\mu}_{G,\mathscr{T}^*\widetilde{\mathscr{B}}} \times \bar{\mu}_{T,\mathscr{T}^*\widetilde{\mathscr{B}}} : (\mathscr{T}^*\widetilde{\mathscr{B}})_{\text{aff}} \to \mathfrak{g}^* \times_{\mathfrak{t}^*/W} \mathfrak{t}^*$ over the point $(\bar{\mu}_{G,\mathscr{T}^*\widetilde{\mathscr{B}}} \times \bar{\mu}_{T,\mathscr{T}^*\widetilde{\mathscr{B}}})(y)$. It suffices to show that there are no closed T_r-orbits in F of maximal dimension. To prove this, let $F' := F \cap T^*\widetilde{\mathscr{B}}$. All T_r-orbits in $T^*\widetilde{\mathscr{B}}$, in particular those in F', have maximal dimension. Further, since $\bar{\mu}_{G,\mathscr{T}^*\widetilde{\mathscr{B}}}(y)$ is not regular in $\mathfrak{g}^*$, the Springer fiber over $\bar{\mu}_{G,\mathscr{T}^*\widetilde{\mathscr{B}}}(y)$ has dimension > 0. It follows that F', hence also F, contains infinitely many T_r-orbits of maximal dimension. On the other hand, Lemma 5.1.5 says that F contains exactly one closed T_r-orbit O and, moreover, O is contained in the closure of any other T_r-orbit in F. Hence, none of the infinitely many T_r-orbits above is closed and O is contained in the closure of each of those orbits. It follows that the dimension of O is not maximal.

To complete the proof, let $y \in (\mathscr{T}^*\widetilde{\mathscr{B}})_{\text{aff}} \smallsetminus \mathscr{T}^{\text{reg}}\widetilde{\mathscr{B}}$ be as above and suppose by contradiction that $\bar{\mathsf{p}}_w(y) \in \widetilde{\mathscr{B}}$ for every $w \in W$. Since all orbits of the T_r-action on $\widetilde{\mathscr{B}}$ have maximal dimension, the T_r-orbit of y has maximal dimension. Therefore, this orbit is not closed in $(\mathscr{T}^*\widetilde{\mathscr{B}})_{\text{aff}}$, by the paragraph above. Hence, by Hilbert-Mumford, there is a 1-parameter subgroup $\tau : \mathbb{C}^\times \to T$ such that the limit $y' := \lim_{z\to 0} \tau(z) \cdot y$ exists and $y' \notin T_r \cdot y$. Hence, for every $w \in W$, we have $\lim_{z\to 0} \tau(z) \cdot \bar{\mathsf{p}}_w(y) = \bar{\mathsf{p}}_w(y')$.

Let $h \in \operatorname{Lie} T = \mathfrak{t}$ be a generator of the 1-parameter subgroup τ. Observe that the existence of a well-defined limit $\lim_{z\to 0} \tau(z) \cdot x \in \widetilde{\mathscr{B}}_{\text{aff}}$ for some $x \in \widetilde{\mathscr{B}}$ implies that $\mu(h) \geq 0$ for every dominant integral weight $\mu \in \mathfrak{t}^*$. Since the limit $\lim_{z\to 0} \tau(z) \cdot \bar{\mathsf{p}}_w(y)$ exists for every $w \in W$, it follows that we must have $\mu(h) \geq 0$ for all integral weights $\mu \in \mathfrak{t}^*$. This forces $h = 0$, a contradiction. $\square$

5.2 *Classical Hamiltonian reduction of* $\mathscr{T}^*(G/\bar{N})$

The following result goes back to Kostant.

Lemma 5.2.1. *Let $x \in \mathsf{e} + \mathfrak{b}$ and $\mathfrak{b}_+ \in \mathscr{B}$ a Borel subalgebra such that $x \in \mathfrak{b}_+$. Then the N-orbit through the point $\mathfrak{b}_+ \in \mathscr{B}$ is the (unique) open N-orbit in $\mathscr{B}$.*

Proof. Recall the $\mathbb{Z}$-grading (3.2.1) and introduce a $\mathbb{Z}$-filtration $\mathfrak{g}^{\leq k} := \bigoplus_{\ell \leq k} \mathfrak{g}(\ell)$, $k \in \mathbb{Z}$, on $\mathfrak{g}$. We have $\mathfrak{b} = \mathfrak{g}^{\leq 0}$, resp. $\mathfrak{n} = \mathfrak{g}^{\leq -2}$, and $\bar{\mathfrak{b}} = \bigoplus_{\ell \geq 0} \mathfrak{g}(\ell)$. Also, we have $\mathsf{f} \in \mathfrak{g}(-2)$, resp. $\mathsf{e} \in \mathfrak{g}(2)$.

Now, let x and $\mathfrak{b}_+$ be as in the statement of the lemma. Conjugating the pair $(x, \mathfrak{b}_+)$ by an element of N if necessary, we may (cf. Section 2.2) and will assume that $x \in \mathsf{e} + \mathfrak{g}_{\mathsf{f}}$. Thus, since $\mathfrak{g}_{\mathsf{f}} \subset \mathfrak{g}^{\leq -2}$, in $\operatorname{gr}\mathfrak{g}$, we have $\operatorname{gr} x = x (\operatorname{mod} \mathfrak{g}^{\leq 0}) = \mathsf{e}$. Observe next that the filtration on $\mathfrak{g}$ induces a filtration $\mathfrak{b}_+ \cap \mathfrak{g}^{\leq k}$, $k \in \mathbb{Z}$, on the Borel subalgebra $\mathfrak{b}_+$. It is clear that $\operatorname{gr}\mathfrak{b}_+$ is a solvable Lie subalgebra of $\operatorname{gr}\mathfrak{g}$ that contains the element $\mathsf{e} = \operatorname{gr} x$. Since $\dim \operatorname{gr}\mathfrak{b}_+ = \dim \mathfrak{b}_+$ and the Lie algebra $\operatorname{gr}\mathfrak{g}$ is isomorphic to $\mathfrak{g}$, we conclude that $\operatorname{gr}\mathfrak{b}_+$ is a Borel subalgebra of $\operatorname{gr}\mathfrak{g}$. But the only Borel subalgebra that contains the element e is the algebra $\bar{\mathfrak{b}}$. Therefore, we must have that $\operatorname{gr}\mathfrak{b}_+ = \bar{\mathfrak{b}}$. It follows that $\operatorname{gr}_i \mathfrak{b}_+ = 0$ for all $i < 0$. We deduce that $\mathfrak{b}_+ \cap \mathfrak{n} = \mathfrak{b}_+ \cap \mathfrak{g}^{\leq -2} = 0$, and the lemma follows. □

Lemma 5.2.2. *Let $Y := \mu^{-1}_{G, \mathscr{T}^* \widetilde{\mathscr{B}}}(\psi + \mathfrak{n}^{\perp})$. Then the image of Y under the projection $\mathsf{p} : \mathscr{T}^* \widetilde{\mathscr{B}} \to \widetilde{\mathscr{B}}$ is contained in the open set $B\bar{N}/\bar{N} \subset G/\bar{N} = \widetilde{\mathscr{B}}$. Furthermore, the map*

$$\mu_{T, \mathscr{T}^* \widetilde{\mathscr{B}}} \times \mathsf{p} : \; Y \longrightarrow \mathfrak{t}^* \times B\bar{N}/\bar{N} = N \times T \times \mathfrak{t}^* = N \times \mathscr{T}^* T$$

is an N-equivariant isomorphism.

Proof. We identify $\mathfrak{g}^*$ with $\mathfrak{g}$ and $\mathscr{B}$ with $G/\bar{B}$, respectively. Then, Lemma 5.2.1 translates into the first statement of Lemma 5.2.2. Further, by Lemma 2.2.1 one has an N-equivariant isomorphism $N \times \mu^{-1}_{G, \mathscr{T}^* \widetilde{\mathscr{B}}}(\mathscr{S}) \xrightarrow{\sim} Y$. Since $q : \mathscr{T}^* \widetilde{\mathscr{B}} \to \widetilde{\mathfrak{g}}$ is a T-torsor, we deduce that $\mu^{-1}_{G, \mathscr{T}^* \widetilde{\mathscr{B}}}(\mathscr{S})$ is a T-torsor over $\pi^{-1}(\mathscr{S})$, where π is the Grothendieck-Springer map. Recall that $\mathscr{S} \subset \mathfrak{g}^{\mathrm{reg}}$ and $\widetilde{\mathfrak{g}}^{\mathrm{reg}} \cong \mathfrak{g}^{\mathrm{reg}} \times_{\mathfrak{c}} \mathfrak{t}^*$. The map π corresponds to the first projection $\mathfrak{g}^{\mathrm{reg}} \times_{\mathfrak{c}} \mathfrak{t}^* \to \mathfrak{g}^{\mathrm{reg}}$. Hence, we obtain, $\pi^{-1}(\mathscr{S}) = \pi^{-1}(\mathscr{S}) \cap (\mathfrak{g}^{\mathrm{reg}} \times_{\mathfrak{c}} \mathfrak{t}^*) = \mathscr{S} \times_{\mathfrak{c}} \mathfrak{t}^* \cong \mathfrak{t}^*$. We conclude that $\mu^{-1}_{G, \mathscr{T}^* \widetilde{\mathscr{B}}}(\mathscr{S})$ is a T-torsor over $\mathfrak{t}^*$. The result follows. □

Let $\mathscr{Y} := \mathscr{T}^* \widetilde{\mathscr{B}} /\!\!/ (N, \psi)$. This means, according to the definition given in Section 2.2 and Lemma 5.2.2, that the map $\mu_{T, \mathscr{T}^* \widetilde{\mathscr{B}}} \times \mathsf{p}$ yields an isomorphism $p_{\mathscr{Y}} : \mathscr{Y} \xrightarrow{\sim} \mathscr{T}^* T$. Furthermore, from Lemma 5.1.3 in the case $S = \psi + \mathfrak{n}^{\perp}$, we find that

$$\mathbb{C}[\mathscr{Y}] = \mathbb{C}[Y]^N = \mathbb{C}[\mu^{-1}_{G, \mathscr{T}^* \widetilde{\mathscr{B}}}(\psi + \mathfrak{n}^{\perp})]^N$$
$$= \big(\mathbb{C}[\mathscr{T}^* \widetilde{\mathscr{B}}] / \mathbb{C}[\mathscr{T}^* \widetilde{\mathscr{B}}]\mathfrak{n}^{\psi}\big)^N = \mathbb{C}[\mathscr{T}^* \widetilde{\mathscr{B}}] /\!\!/ \mathfrak{n}^{\psi}.$$

Thus, we have $\mathscr{Y} = \operatorname{Spec}(\mathbb{C}[\mathscr{T}^* \widetilde{\mathscr{B}}] /\!\!/ \mathfrak{n}^{\psi})$.

The subset $\psi + \mathfrak{n}^{\perp} \subset \mathfrak{g}^*$ is stable under the $\bullet$-action. It follows from Proposition 5.1.1(2) that the $W \ltimes T_r$-action, resp. $\bullet$-action, on $\mathbb{C}[\mathscr{T}^* \widetilde{\mathscr{B}}]$ by algebra automorphisms descends to an action on the algebra $\mathbb{C}[\mathscr{T}^* \widetilde{\mathscr{B}}] /\!\!/ \mathfrak{n}^{\psi}$. Therefore, this action induces a $W \ltimes T_r$-action, resp. $\bullet$-action, on $\mathscr{Y}$.

Let W act on $\mathscr{T}^* T = \mathfrak{t}^* \times T$ diagonally, resp. T act by translations along the second factor.

Lemma 5.2.3. *The map $p_{\mathscr{Y}}$ is a $T \ltimes W$-equivariant symplectomorphism $\mathscr{Y} \xrightarrow{\sim} \mathscr{T}^*T$. Furthermore, this map intertwines the $\bullet$-action on Y with the $\mathbb{G}_m$-action on $\mathscr{T}^*T$ by dilations by z^{-2} along the factor $\mathfrak{t}^*$.*

Proof. All statements, except for the W-equivariance, are straightforward and are left for the reader. Compatibility of the W-actions is a consequence of properties of the W-action on $\mathbb{C}[\mathscr{T}^*\widetilde{\mathscr{B}}]$, see [GR, Lemma 5.2.5]. □

5.3 Quantum Hamiltonian reduction of $\mathscr{D}(\widetilde{\mathscr{B}})$

One has a chain of algebra isomorphisms

$$\mathscr{D}(NT\bar{N}/\bar{N})/\!\!/\mathfrak{n}^{\Psi} \cong \mathscr{D}(N\times T)/\!\!/\mathfrak{n}^{\Psi} \cong (\mathscr{D}(N)/\!\!/\mathfrak{n}^{\Psi})\otimes\mathscr{D}(T) \cong \mathbb{C}\otimes\mathscr{D}(T) \cong \mathscr{D}(T).$$

Let $j: NT\bar{N}/\bar{N} \hookrightarrow G/\bar{N} = \widetilde{\mathscr{B}}$ be the open embedding.

Theorem 5.3.1. (i) *The following functors are mutually inverse equivalences:*

$$(\mathscr{D}_{\widetilde{\mathscr{B}}}, N, \psi)\text{-mod} \underset{\mathscr{D}_{\widetilde{\mathscr{B}}}\otimes_{\mathscr{D}(\widetilde{\mathscr{B}})}(-)}{\overset{\Gamma(\widetilde{\mathscr{B}},-)}{\rightleftarrows}} (\mathscr{D}(\widetilde{\mathscr{B}}), N, \psi)\text{-mod}$$

$$\underset{\mathscr{D}(\widetilde{\mathscr{B}})/\mathscr{D}(\widetilde{\mathscr{B}})\mathfrak{n}^{\Psi}\otimes_{\mathscr{D}(\widetilde{\mathscr{B}})/\!\!/\mathfrak{n}^{\Psi}}(-)}{\overset{(-)^{\mathfrak{n}^{\Psi}}}{\rightleftarrows}} (\mathscr{D}(\widetilde{\mathscr{B}})/\!\!/\mathfrak{n}^{\Psi})\text{-mod}. \qquad (5.3.2)$$

(ii) *The map $j^*: \mathscr{D}(\widetilde{\mathscr{B}})/\!\!/\mathfrak{n}^{\Psi} \to \mathscr{D}(NT\bar{N}/\bar{N})/\!\!/\mathfrak{n}^{\Psi} \cong \mathscr{D}(T)$, induced by restriction of differential operators via j, is an algebra isomorphism.*

We begin with the following result.

Lemma 5.3.3. (i) *For any $\mathscr{F} \in (\mathscr{D}_{\widetilde{\mathscr{B}}}, N, \psi)$-mod the canonical map $\mathscr{F} \to j_{\bullet}j^{\bullet}\mathscr{F}$ is an isomorphism.*

(ii) *The functor $\mathscr{D}_T$-mod $\to (\mathscr{D}_{\widetilde{\mathscr{B}}}, N, \psi)$-mod, $M \mapsto j_{\bullet}(e^{\Psi}\Omega_N \boxtimes M)$, is an equivalence. An inverse is given by the functor $\mathscr{F} \mapsto \big(pr_{\bullet}j^{\bullet}\mathscr{F}\big)^{\mathfrak{n}^{\Psi}}$, where $pr: B\bar{N}/\bar{N} = N\times T \to T$ is the second projection.*

Proof. The argument is an adaptation of the argument from [MS] to the setting of not necessarily holonomic $\mathscr{D}$-modules. Given a point $x \in \widetilde{\mathscr{B}}$, let $i_x: \{x\} \hookrightarrow \widetilde{\mathscr{B}}$ denote the embedding, N_x the stabilizer of x in N, and ψ_x the restriction of ψ to the Lie algebra $\mathfrak{n} \cap \operatorname{Lie} N_x$. Note that if $\psi_x \neq 0$, then the $\mathscr{D}_{N_x}$-module $e^{\psi_x}\Omega_{N_x}$ cannot be obtained as a pullback via the map $N_x \to \mathrm{pt}$ of a $\mathscr{D}$-module on a point. Also, it is clear that for any $\mathscr{E} \in (\mathscr{D}_{\widetilde{\mathscr{B}}}, N, \psi)$-mod and $x \in \widetilde{\mathscr{B}}$, the $\mathscr{D}$-module $R^k i_x^! \mathscr{E}$, on $\{x\}$, is an (N_x, ψ_x)-Whittaker $\mathscr{D}$-module. It follows that if $\psi_x \neq 0$, then $R^k i_x^! \mathscr{E} = 0$ for all $k \in \mathbb{Z}$.

Now let $\mathscr{E} \in (\mathscr{D}_{\widetilde{\mathscr{B}}}, N, \psi)$-mod be a nonzero $\mathscr{D}_{\widetilde{\mathscr{B}}}$-module such that $\operatorname{Supp}\mathscr{E} \subset \widetilde{\mathscr{B}} \smallsetminus B\bar{N}/\bar{N}$. Then, for a sufficiently general point x in the support of $\mathscr{E}$, there exists k such that $R^k i_x^! \mathscr{E} \neq 0$. Write $x = g\bar{N}/\bar{N}$, so we have $\operatorname{Lie} N_x = \operatorname{Ad} g(\bar{\mathfrak{n}})$. Lemma 5.2.2(ii) implies that $\mathfrak{n} \cap \operatorname{Lie} N_x = \operatorname{Ad}^* g(\bar{\mathfrak{n}})^\perp \cap (\psi + \mathfrak{n}^\perp) = \emptyset$. This means that $\psi_x \neq 0$. It follows from the above that $\mathscr{E} = 0$, a contradiction. Thus, there are no nonzero objects $\mathscr{E} \in (\mathscr{D}_{\widetilde{\mathscr{B}}}, N, \psi)$-mod such that $\operatorname{Supp}\mathscr{E} \subset \widetilde{\mathscr{B}} \smallsetminus B\bar{N}/\bar{N}$.

To complete the proof of (i) observe that, for any $\mathscr{F} \in (\mathscr{D}_{\widetilde{\mathscr{B}}}, N, \psi)$-mod, the kernel, resp. cokernel, of the map $\mathscr{F} \to j_\bullet j^\bullet \mathscr{F}$ is an object of $\mathscr{E} \in (\mathscr{D}_{\widetilde{\mathscr{B}}}, N, \psi)$-mod supported on $\widetilde{\mathscr{B}} \smallsetminus B\bar{N}/\bar{N}$. Hence $\mathscr{E} = 0$, proving (i). Further, it is immediate from definitions that the functors $e^\Psi \Omega_N \boxtimes (-)$ and $(pr_\bullet(-))^{\mathfrak{n}^\psi}$ give mutually inverse equivalences $\mathscr{D}(T)\text{-mod} \leftrightarrows (\mathscr{D}_{N\times T}, N, \psi)\text{-mod}$. Therefore, part (ii) follows from (i). □

Proof (*Theorem* 5.3.1). First, we apply Proposition 3.1.4 in the case where $A = \mathscr{D}(\widetilde{\mathscr{B}})$. It follows from part (i) of the proposition that the functors $(-)^{\mathfrak{n}^\psi}$ and $\mathscr{D}(\widetilde{\mathscr{B}})/\mathscr{D}(\widetilde{\mathscr{B}})\mathfrak{n}^\psi \otimes_{\mathscr{D}(\widetilde{\mathscr{B}})/\!\!/\mathfrak{n}^\psi} (-)$, on the right of diagram (5.3.2), are mutually inverse equivalences.

Next, let $\mathscr{F} \in (\mathscr{D}_{\widetilde{\mathscr{B}}}, N, \psi)$-mod. The variety $B\bar{N}/\bar{N} \cong T$ being affine, from Lemma 5.3.3(i), we deduce $H^i(\widetilde{\mathscr{B}}, \mathscr{F}) = H^i(\widetilde{\mathscr{B}}, j_\bullet j^\bullet \mathscr{F}) = H^i(B\bar{N}/\bar{N}, j^\bullet \mathscr{F}) = 0$, for all $i \neq 0$. It follows that $\Gamma(\widetilde{\mathscr{B}}, -)$ is an exact functor on $(\mathscr{D}_{\widetilde{\mathscr{B}}}, N, \psi)$-mod. Furthermore, we have

$$\begin{aligned}\Gamma(\widetilde{\mathscr{B}}, \mathscr{F})^{\mathfrak{n}^\psi} &= \Gamma(\widetilde{\mathscr{B}}, j_\bullet j^\bullet \mathscr{F})^{\mathfrak{n}^\psi} = \Gamma(B\bar{N}/\bar{N}, j^\bullet \mathscr{F})^{\mathfrak{n}^\psi} \\ &= \Gamma(T, pr_\bullet j^\bullet \mathscr{F})^{\mathfrak{n}^\psi} = \Gamma(T, (pr_\bullet j^\bullet \mathscr{F})^{\mathfrak{n}^\psi}). \quad (5.3.4)\end{aligned}$$

We see that the action of the algebra $\mathscr{D}(\widetilde{\mathscr{B}})/\!\!/\mathfrak{n}^\psi$ on $\Gamma(\widetilde{\mathscr{B}}, \mathscr{F})^{\mathfrak{n}^\psi}$ factors through the restriction map $j^* : \mathscr{D}(\widetilde{\mathscr{B}})/\!\!/\mathfrak{n}^\psi \to \mathscr{D}(T)$. Moreover, the resulting functor $\Gamma(\widetilde{\mathscr{B}}, \mathscr{F})^{\mathfrak{n}^\psi} : (\mathscr{D}_{\widetilde{\mathscr{B}}}, N, \psi)\text{-mod} \to \mathscr{D}(T)\text{-mod}$ is an equivalence, by Lemma 5.3.3(ii), and an inverse equivalence is the functor $j_\bullet(e^\Psi \Omega_N \boxtimes -)$.

We apply $\Gamma(\widetilde{\mathscr{B}}, -)^{\mathfrak{n}^\psi}$, an exact functor on $(\mathscr{D}_{\widetilde{\mathscr{B}}}, N, \psi)$-mod, to the morphism $\mathscr{D}_{\widetilde{\mathscr{B}}} \otimes \mathfrak{n}^\psi \to \mathscr{D}_{\widetilde{\mathscr{B}}}$ given by right multiplication by $\mathfrak{n}^\psi$. The cokernel of this morphism equals $\mathscr{D}_{\widetilde{\mathscr{B}}}/\mathscr{D}_{\widetilde{\mathscr{B}}}\mathfrak{n}^\psi$. We deduce that $\Gamma(\widetilde{\mathscr{B}}, \mathscr{D}_{\widetilde{\mathscr{B}}}/\mathscr{D}_{\widetilde{\mathscr{B}}}\mathfrak{n}^\psi)^{\mathfrak{n}^\psi} = (\mathscr{D}(\widetilde{\mathscr{B}})/\mathscr{D}(\widetilde{\mathscr{B}})\mathfrak{n}^\psi)^{\mathfrak{n}^\psi} = \mathscr{D}(\widetilde{\mathscr{B}})/\!\!/\mathfrak{n}^\psi$. On the other hand, it is clear that $j^\bullet(\mathscr{D}_{\widetilde{\mathscr{B}}}/\mathscr{D}_{\widetilde{\mathscr{B}}}\mathfrak{n}^\psi) = e^\Psi \Omega_N \boxtimes \mathscr{D}_T$. Hence, $(pr_\bullet j^\bullet(\mathscr{D}_{\widetilde{\mathscr{B}}}/\mathscr{D}_{\widetilde{\mathscr{B}}}\mathfrak{n}^\psi))^{\mathfrak{n}^\psi} = (pr_\bullet j^\bullet(e^\Psi \Omega_N \boxtimes \mathscr{D}_T)^{\mathfrak{n}^\psi} = \mathscr{D}_T$. Thus, using (5.3.4), we obtain

$$\begin{aligned}\mathscr{D}(\widetilde{\mathscr{B}})/\!\!/\mathfrak{n}^\psi &= \Gamma(\widetilde{\mathscr{B}}, \mathscr{D}_{\widetilde{\mathscr{B}}}/\mathscr{D}_{\widetilde{\mathscr{B}}}\mathfrak{n}^\psi)^{\mathfrak{n}^\psi} \\ &= \Gamma(T, pr_\bullet j^\bullet(\mathscr{D}_{\widetilde{\mathscr{B}}}/\mathscr{D}_{\widetilde{\mathscr{B}}}\mathfrak{n}^\psi))^{\mathfrak{n}^\psi} = \Gamma(T, \mathscr{D}_T) = \mathscr{D}(T).\end{aligned}$$

Thus, we have shown that the restriction map $j^\bullet : \mathscr{D}(\widetilde{\mathscr{B}})/\!\!/\mathfrak{n}^\psi \to \mathscr{D}(T)$, in Theorem 5.3.1(ii), is an algebra isomorphism and the functor

$$\Gamma(\widetilde{\mathscr{B}},-)^{\mathfrak{n}^\psi} : (\mathscr{D}_{\widetilde{\mathscr{B}}}, N, \psi)\text{-mod} \to \mathscr{D}(T)\text{-mod}$$

is an equivalence. We conclude that both the functor $(-)^{\mathfrak{n}^\psi}$ and the composite functor $\Gamma(\widetilde{\mathscr{B}},-)^{\mathfrak{n}^\psi}$, in diagram (5.3.2), are equivalences. It follows that the functor $\Gamma(\widetilde{\mathscr{B}},-)$, in diagram (5.3.2), is also an equivalence. This proves part (i) of Theorem 5.3.1. □

5.4 Proof of Theorem 1.6.4

We begin with the following result.

Lemma 5.4.1. *The following natural maps are algebra isomorphisms*

$$(\mathscr{D}(G)/\!\!/\mathfrak{n}_l^\psi)/\!\!/\bar{\mathfrak{n}}_r \xrightarrow{u} \mathscr{D}(G)/\!\!/(\mathfrak{n}_l^\psi+\bar{\mathfrak{n}}_r) \xleftarrow{v} (\mathscr{D}(G)/\!\!/\bar{\mathfrak{n}}_r)/\!\!/\mathfrak{n}_l^\psi;$$
$$(\mathscr{D}(G)/\!\!/\mathfrak{n}_l^\psi)/\!\!/\mathfrak{n}_r^\psi \to \mathscr{D}(G)/\!\!/(\mathfrak{n}_l^\psi+\mathfrak{n}_r^\psi) \leftarrow (\mathscr{D}(G)/\!\!/\mathfrak{n}_r^\psi)/\!\!/\mathfrak{n}_l^\psi;$$
$$\bar{\mathfrak{n}}_r\backslash\!\!\backslash\big(\mathscr{D}(G)/\!\!/\mathfrak{n}_l^\psi\big) \to \big(\mathscr{D}(G)/(\bar{\mathfrak{n}}_r\mathscr{D}(G)+\mathscr{D}(G)\mathfrak{n}_l^\psi)\big)^{N_l\times\bar{N}_r} \leftarrow \big(\bar{\mathfrak{n}}_r\backslash\!\!\backslash\mathscr{D}(G)\big)/\!\!/\mathfrak{n}_l^\psi.$$

Proof. Let $N=N_l$, resp. $\bar{N}=\bar{N}_r$. First, we prove that the map v is an isomorphism. To prove this, consider an open embedding $j_G : NT\bar{N}/\bar{N} \hookrightarrow G/\bar{N}$. We have a commutative diagram of algebra homomorphisms

$$\begin{array}{ccccc}
\mathscr{D}(G/\bar{N})/\!\!/\mathfrak{n}^\psi & & \xrightarrow{\quad v \quad} & & \mathscr{D}(G)/\!\!/(\mathfrak{n}_l^\psi+\bar{\mathfrak{n}}_r) \\
\Big\downarrow{\scriptstyle j^*} & & & & \Big\downarrow{\scriptstyle j_G^*} \\
\mathscr{D}(NT\bar{N}/\bar{N})/\!\!/\mathfrak{n}^\psi & = & \mathscr{D}(T) & = & \mathscr{D}(NT\bar{N}/\bar{N})/\!\!/(\mathfrak{n}_l^\psi+\bar{\mathfrak{n}}_r).
\end{array}$$

By commutativity, it suffices to show that the map j_G^* on the right is an algebra isomorphism. The latter is proved by mimicking the arguments of Section 5.3.

Next, we prove that that the map u is an isomorphism. We simplify the notation and write $\mathscr{D}=\mathscr{D}(G)$, resp. $M=\mathscr{D}/\mathscr{D}\mathfrak{n}_l^\psi$, and $D_l=\mathscr{D}(G)/\!\!/\mathfrak{n}_l^\psi=M^{\mathfrak{n}_l^\psi}$. The action of $\mathfrak{n}_r^\psi$ on $\mathscr{D}$ by right multiplication descends to M. Observe that M is an object of the category $(\mathscr{D},\mathfrak{n}_l^\psi)$-mod. In that category, one has an exact sequence

$$M\otimes\mathfrak{n}_r^\psi \xrightarrow{a} M \to M/M\mathfrak{n}_r^\psi \to 0, \tag{5.4.2}$$

where the map a is given by the $\mathfrak{n}_r^\psi$-action on M. We apply the functor $(-)^{\mathfrak{n}_l^\psi}$, which is an exact functor on $(\mathscr{D},\mathfrak{n}_l^\psi)$-mod, to (5.4.2). We have

$$(M/M\mathfrak{n}_r^\psi)^{\mathfrak{n}_l^\psi} = \big((\mathscr{D}/\mathscr{D}\mathfrak{n}_l^\psi)/(\mathscr{D}/\mathscr{D}\mathfrak{n}_l^\psi)\mathfrak{n}_r^\psi\big)^{\mathfrak{n}_l^\psi} = \big(\mathscr{D}/\mathscr{D}(\mathfrak{n}_l^\psi+\mathfrak{n}_r^\psi)\big)^{\mathfrak{n}_l^\psi},$$
$$(M\mathfrak{n}_r^\psi)^{\mathfrak{n}_l^\psi} = (\operatorname{Im} a)^{\mathfrak{n}_l^\psi} = \operatorname{Im}[(M\otimes\mathfrak{n}_r^\psi)^{\mathfrak{n}_l^\psi} \to M^{\mathfrak{n}_l^\psi}] = (M^{\mathfrak{n}_l^\psi})\mathfrak{n}_r^\psi = D_l\mathfrak{n}_r^\psi.$$

Thus, from the exact sequence (5.4.2), we obtain the following exact sequence in $(\mathscr{D}, \mathfrak{n}_l^\psi + \mathfrak{n}_r^\psi)$-mod:

$$0 \to D_l \mathfrak{n}_r^\psi \to D_l \to \left(\mathscr{D}/\mathscr{D}(\mathfrak{n}_l^\psi + \mathfrak{n}_r^\psi)\right)^{\mathfrak{n}_l^\psi} \to 0. \tag{5.4.3}$$

We now apply the functor $(-)^{\mathfrak{n}_r^\psi}$, which is an exact functor on the category $(\mathscr{D}, \mathfrak{n}_l^\psi + \mathfrak{n}_r^\psi)$-mod, to (5.4.3). It is immediate to see that the exactness of the resulting sequence implies that the map u is an isomorphism. All other isomorphisms of the lemma are proved in a similar way. □

Proof (*Theorem* 1.6.4). Let $D_r := \mathscr{D}(G)/\!\!/\mathfrak{n}_r^\psi$. The embedding $\mathfrak{n}_l^\psi \hookrightarrow \mathscr{D}(G)$ induces a map $\mathfrak{n}^\psi \to D_r$. The functor $(D_r, \mathfrak{n}^\psi)$-mod $\to (D_r/\!\!/\mathfrak{n}^\psi)$-mod, $M \mapsto M^{\mathfrak{n}^\psi}$, is an equivalence, by Proposition 3.1.4. Further, by Lemma 5.4.1, we have an isomorphism $D_r/\!\!/\mathfrak{n}^\psi \cong \mathbb{W}$. The result now follows from Theorem 1.6.3. □

5.5 *The Gelfand–Graev action*

Let ρ be the half-sum of positive roots. Let W act on $\operatorname{Sym}\mathfrak{t}$ via the dot-action: $w \cdot a := w(a) + \langle w(a) - a, \rho\rangle$, $w \in W$, $a \in \mathfrak{t}$. We will identify $Z\mathfrak{g}$ with $\operatorname{Sym}\mathfrak{t}^{W\cdot}$ via the Harish-Chandra isomorphism. The differential of the $G \times T$-action on $\widetilde{\mathscr{B}}$ gives an algebra map $\mathscr{U}\mathfrak{g} \to \mathscr{D}(\widetilde{\mathscr{B}})$, $a \mapsto a_l$, resp. $\mathscr{U}\mathfrak{t} = \operatorname{Sym}\mathfrak{t} \to \mathscr{D}(\widetilde{\mathscr{B}})$, $a \mapsto a_r$. It is known that the map $a \otimes a' \mapsto a_l \otimes a'_r$ factors through a homomorphism $\mathscr{U}\mathfrak{g} \otimes_{Z\mathfrak{g}} \operatorname{Sym}\mathfrak{t} \to \mathscr{D}(\widetilde{\mathscr{B}})$, cf. [BB].

Restricting the $G \times T$-action on $\widetilde{\mathscr{B}}$ to the subgroup $T \times T \subset G \times T$ gives a weight decomposition

$$\mathbb{C}[\widetilde{\mathscr{B}}] = \bigoplus_{\mu,\lambda \in \mathbb{X}^*} {}^{(\mu)}\mathbb{C}[\widetilde{\mathscr{B}}]^{(\lambda)}, \quad \text{resp. } \mathbb{C}[\mathscr{T}^*\widetilde{\mathscr{B}}] = \bigoplus_{\mu,\lambda \in \mathbb{X}^*} {}^{(\lambda)}\mathbb{C}[\mathscr{T}^*\widetilde{\mathscr{B}}]^{(\mu)}.$$

Further, the $\mathbb{G}_m$-action on $\widetilde{\mathscr{B}}$ given by $\gamma(z) : g\bar{N} \mapsto \gamma(z) g \gamma(z)^{-1}\bar{N}$ induces a $\mathbb{Z}$-grading on $\mathscr{D}(\widetilde{\mathscr{B}})$. This $\mathbb{Z}$-grading may be expressed in terms of the weight decomposition as follows:

$$\mathscr{D}(\widetilde{\mathscr{B}}) = \bigoplus_{\ell \in \mathbb{Z}} \mathscr{D}(\widetilde{\mathscr{B}})(\ell), \quad \mathscr{D}(\widetilde{\mathscr{B}})(\ell) := \bigoplus_{\{\mu,\lambda \in \mathbb{X}^* \mid \langle \lambda, \mathsf{h}\rangle + \langle \mu, \mathsf{h}\rangle = \ell\}} {}^{(\mu)}\mathscr{D}(\widetilde{\mathscr{B}})^{(\lambda)}.$$

Let $F_\bullet\mathscr{D}(\widetilde{\mathscr{B}})$ be the Kazhdan filtration associated with the $\mathbb{Z}$-grading and filtration $\mathscr{D}_\leq(\widetilde{\mathscr{B}})$, as in Section 4.1.

We have maps $\mathfrak{n}^\psi = \mathfrak{n}_l^\psi \to \mathscr{U}\mathfrak{g}_l \to \mathscr{D}(\widetilde{\mathscr{B}})$. The right T-action on $\mathscr{D}(\widetilde{\mathscr{B}})$ survives the Hamiltonian reduction by $\mathfrak{n}^\psi$. We get an $\mathbb{X}^*$-grading

$$\mathscr{D}(\widetilde{\mathscr{B}})/\!\!/\mathfrak{n}^\psi = \bigoplus_{\mu \in \mathbb{X}^*} (\mathscr{D}(\widetilde{\mathscr{B}})/\!\!/\mathfrak{n}^\psi)^{(\mu)}.$$

The Kazhdan filtration induces a $\mathbb{Z}$-filtration $F_\bullet(\mathscr{D}(\widetilde{\mathscr{B}})/\!\!/\mathfrak{n}^\Psi)$ which is compatible with the $\mathbb{X}^*$-grading. Thus, we have

$$F_n(\mathscr{D}(\widetilde{\mathscr{B}})/\!\!/\mathfrak{n}^\Psi) = \bigoplus_{\mu\in\mathbb{X}^*} F_n(\mathscr{D}(\widetilde{\mathscr{B}})/\!\!/\mathfrak{n}^\Psi)^{(\mu)}.$$

The (Kazhdan shifted) principal symbol map $\mathrm{gr}^F_\bullet\, \mathscr{D}(\widetilde{\mathscr{B}}) \to \mathbb{C}[\mathscr{T}^*\widetilde{\mathscr{B}}]/\!\!/\mathfrak{n}^\Psi$ descends to a map $\mathrm{gr}^F(\mathscr{D}(\widetilde{\mathscr{B}})/\!\!/\mathfrak{n}^\Psi) \to \mathbb{C}[\mathscr{T}^*\widetilde{\mathscr{B}}]/\!\!/\mathfrak{n}^\Psi$, of $\mathbb{Z}\times\mathbb{X}^*$-graded algebras.

Proposition 5.5.1. *The map* $\mathrm{gr}^F\, \mathscr{D}(\widetilde{\mathscr{B}})/\!\!/\mathfrak{n}^\Psi \to \mathbb{C}[\mathscr{T}^*\widetilde{\mathscr{B}}]/\!\!/\mathfrak{n}^\Psi$ *is an isomorphism.*

Proof. According to [GR, Corollary 3.6.1], the principal symbol map $\mathrm{gr}\,\mathscr{D}(\widetilde{\mathscr{B}}) \to \mathbb{C}[\mathscr{T}^*\widetilde{\mathscr{B}}]$ is a graded algebra isomorphism, where $\mathrm{gr}(-)$ is taken with respect to the order filtration on differential operators. This isomorphism respects the $\mathbb{X}$-gradings. Therefore, it gives a graded algebra isomorphism $\mathrm{gr}^F\, \mathscr{D}(\widetilde{\mathscr{B}}) \cong \mathbb{C}[\mathscr{T}^*\widetilde{\mathscr{B}}]$, where the grading on $\mathbb{C}[\mathscr{T}^*\widetilde{\mathscr{B}}]$ comes from the $\bullet$-action. Further, an argument similar to the one used in Section 4.3 shows that the Kazhdan filtration on $\mathscr{D}(\widetilde{\mathscr{B}})$ is separating. The statement of the proposition now follows from the isomorphism of Lemma 3.3.2(1) in the case where $A = \mathscr{D}(\widetilde{\mathscr{B}})$. □

In the 1960's, Gelfand and Graev constructed, using partial Fourier transforms, a Weyl group action on $\mathscr{D}(\widetilde{\mathscr{B}})$ by algebra automorphisms. The Gelfand–Graev action on $\mathscr{D}(\widetilde{\mathscr{B}})$ commutes with the G-action, in particular with the N-action, by left translations. Therefore, the Gelfand–Graev action descends to a W-action on $\mathscr{D}(\widetilde{\mathscr{B}})/\!\!/\mathfrak{n}^\Psi$ by algebra automorphisms.

We view the function t^ρ as a zero order differential operator on T and define a W-action on the algebra $\mathscr{D}(T)$ by the formula $w\cdot u = w(t^\rho \circ u \circ t^{-\rho})$. Here $w(-)$ denotes the *natural* action of $w\in W$ on $\mathscr{D}(T)$ induced by the action of w on T. Note that $w\cdot u = w(u)$ for any $u\in\mathbb{C}[T]\subset\mathscr{D}(T)$. Also, the embedding $\mathrm{Sym}\,\mathfrak{t} = \mathscr{U}\mathfrak{t} \hookrightarrow \mathscr{D}(T)$ intertwines the dot-actions on $\mathrm{Sym}\,\mathfrak{t}$ and $\mathscr{D}(T)$.

It is clear that the restriction map $j^* : \mathscr{D}(\widetilde{\mathscr{B}})/\!\!/\mathfrak{n}^\Psi \to \mathscr{D}(B\bar{N}/\bar{N})/\!\!/\mathfrak{n}^\Psi = \mathscr{D}(T)$ takes $F_n(\mathscr{D}(\widetilde{\mathscr{B}})/\!\!/\mathfrak{n}^\Psi)$ to $\mathscr{D}_{\leq n}(T)$ for all n.

Proposition 5.5.2. *The isomorphism* $j^* : \mathscr{D}(\widetilde{\mathscr{B}})/\!\!/\mathfrak{n}^\Psi \to \mathscr{D}(T)$ *of Theorem* 5.3.1 *induces an isomorphism* $\mathrm{gr}^F j^* : \mathrm{gr}^F(\mathscr{D}(\widetilde{\mathscr{B}})/\!\!/\mathfrak{n}^\Psi) \xrightarrow{\sim} \mathrm{gr}\,\mathscr{D}(T) = \mathbb{C}[\mathscr{T}^*T]$ *of graded algebras. Furthermore, the map* j^* *intertwines the* W*-action on* $\mathscr{D}(\widetilde{\mathscr{B}})/\!\!/\mathfrak{n}^\Psi$ *and the dot-action on* $\mathscr{D}(T)$.

To prove the proposition, recall that $v_\lambda\in V_\lambda$ denotes a nonzero $\bar{N}$-fixed vector of the irreducible representation V_λ such that $tv_\lambda = \lambda(t)v_\lambda$, $\forall t\in T$. Let $v^*_{-\lambda}\in V^*_\lambda$ be a nonzero N-fixed vector and $f^\lambda(g) := \langle v^*_{-\lambda}, gv_\lambda\rangle$. This function is right $\bar{N}$-invariant, hence it descends to a function f^λ on $\widetilde{\mathscr{B}}$, such that $f^\lambda(n\cdot t\cdot\bar{N}/\bar{N}) = \lambda(t)$, for all $t\in T, n\in N, \bar{n}\in\bar{N}$.

To simplify the notation, put $\mathsf{A} = \mathscr{D}(\widetilde{\mathscr{B}})/\!\!/\mathfrak{n}^\Psi$. The function f^λ, viewed as an element of $\mathscr{D}(\widetilde{\mathscr{B}})$, survives the Hamiltonian reduction giving an element $f^\lambda_{\mathsf{A}}\in\mathsf{A}$.

Let F_w, $w \in W$ denote the Gelfand–Graev automorphisms of the algebra $\mathscr{D}(\widetilde{\mathscr{B}})$, resp. A.

Proof (*Proposition* 5.5.2). The map $\operatorname{Sym}\mathfrak{t} \to \mathscr{D}(\widetilde{\mathscr{B}})$, $u \mapsto u_r$ survives the Hamiltonian reduction by $\mathfrak{n}^\psi \subset \mathscr{D}(\widetilde{\mathscr{B}})$, giving an algebra map $\operatorname{Sym}\mathfrak{t} \to \mathsf{A}$, $u \mapsto u_{\mathsf{A}}$. It is clear that for any $u \in \operatorname{Sym}\mathfrak{t}$, resp. f^λ_{A}, $\mathbb{X}^{++}$, we have $j^*(u_{\mathsf{A}}) = u$, resp. $j^*(f^\lambda_{\mathsf{A}}) = t^\lambda$. The elements $\{u, t^\mu \mid u \in \operatorname{Sym}\mathfrak{t},\ \mu \in \mathbb{X}^{++}\}$ generate the algebra $\mathscr{D}(T)$. Therefore, the elements $\{u_{\mathsf{A}}, f^\mu_{\mathsf{A}} \mid u \in \operatorname{Sym}\mathfrak{t},\ \mu \in \mathbb{X}^{++}\}$ generate the algebra A, by Theorem 5.3.1.

Let $\operatorname{ord} u$ denote the order of a differential operator $u \in \mathscr{D}(\widetilde{\mathscr{B}})$. Clearly, one has $f^\mu \in {}^{-\mu}\mathscr{D}_{\leq 0}(\widetilde{\mathscr{B}})^\mu$. Thus, we have

$$f^\mu_{\mathsf{A}} \in F_0\mathsf{A}^{(\mu)}, \quad \forall \mu \in \mathbb{X}^{++}, \quad \text{resp.} \quad u_{\mathsf{A}} \in F_{2n}\mathsf{A}^{(0)}, \quad \forall u \in \operatorname{Sym}^n \mathfrak{t}.$$

Furthermore, $\operatorname{gr}_0(f^\mu_{\mathsf{A}}) \neq 0$, resp. $\operatorname{gr}_{2n}(u_{\mathsf{A}}) \neq 0$. Let symb be the isomorphism of Proposition 5.5.1, and recall the isomorphism $p_{\mathscr{Y}}$ from Section 5.2. One has the following diagram of graded algebra maps:

$$\mathbb{C}[\mathscr{T}^*T] \xleftarrow[\cong]{p_{\mathscr{Y}}} \mathbb{C}[\mathscr{T}^*\widetilde{\mathscr{B}}]/\!\!/\mathfrak{n}^\psi \xleftarrow[\cong]{\text{symb}} \operatorname{gr}\mathsf{A}$$

$$\xrightarrow{\operatorname{gr} j^*} \operatorname{gr}\mathscr{D}(T) = \mathbb{C}[\mathscr{T}^*T]. \tag{5.5.3}$$

Let $p_{\mathsf{A}} = p_{\mathscr{Y}} \circ \text{symb}$. It is immediate from the construction of the maps that, for any element a from our generating set $\{u_{\mathsf{A}}, f^\mu_{\mathsf{A}} \mid u \in \operatorname{Sym}\mathfrak{t},\ \mu \in \mathbb{X}^{++}\}$ of A, one has $p_{\mathsf{A}}(\operatorname{gr} a) = (\operatorname{gr} j^*)(\operatorname{gr} a)$. It follows that the map $(\operatorname{gr} j^*) \circ p_{\mathsf{A}}^{-1} : \mathbb{C}[\mathscr{T}^*T] \to \mathbb{C}[\mathscr{T}^*T]$ is the identity. We deduce that the map $\operatorname{gr} j^*$ is a graded algebra isomorphism, proving the first statement of the proposition.

For any, not necessarily dominant, $\lambda \in \mathbb{X}^*$, we let $f^{(\lambda)}_{\mathsf{A}}$ be the preimage of t^λ under the isomorphism j^* of Theorem 5.3.1. This agrees with the notation f^λ_{A} for dominant λ.

It is known that the map $\operatorname{Sym}\mathfrak{t} \to \mathscr{D}(\widetilde{\mathscr{B}})$, $u \mapsto u_r$, intertwines the dot-action on $\operatorname{Sym}\mathfrak{t}$ and the Gelfand–Graev action on $\mathscr{D}(\widetilde{\mathscr{B}})$, cf. [BBP]. It follows that the map $u \mapsto u_{\mathsf{A}}$ intertwines the dot-action on $\operatorname{Sym}\mathfrak{t}$ and the W-action on A. Thus, to prove that the map j^* is W-equivariant, it remains to show that for any $\mu \in \mathbb{X}^{++}$ and $w \in W$, one has $\mathsf{F}_w(f^\mu_{\mathsf{A}})) = f^{w(\mu)}_{\mathsf{A}}$. First, it is immediate from the construction of the action that for any $\lambda \in \mathbb{X}^*$ and $w \in W$, the map F_w yields an isomorphism $\mathscr{D}(\widetilde{\mathscr{B}})^{(\lambda)} \xrightarrow{\sim} \mathscr{D}(\widetilde{\mathscr{B}})^{(w(\lambda))}$, cf. [BBP], [GR]. Also, it was shown in the course of the proof of [BBP, Lemma 3.18], that one has

$$\operatorname{ord}\mathsf{F}_w(u) = \operatorname{ord} u + \tfrac{1}{2}\langle\lambda, \mathsf{h}\rangle - \tfrac{1}{2}\langle w(\lambda), \mathsf{h}\rangle, \qquad \forall \lambda \in \mathbb{X}^*, u \in \mathscr{D}(\widetilde{\mathscr{B}})^{(\lambda)},\ w \in W.$$

It follows that the automorphisms F_w, of $\mathscr{D}(\widetilde{\mathscr{B}})$, respect the Kazhdan filtration. Hence, the induced automorphisms of the algebra A have similar properties. In particular, the map $\mathsf{F}_w : \mathsf{A}^{(\lambda)} \to \mathsf{A}^{(w(\lambda))}$ is an isomorphism and the W-action on A induces a W-action on $\mathrm{gr}^F H$ by $\mathbb{Z}$-graded algebra automorphisms.

Next, we observe that for any $\lambda \in \mathbb{X}^*$, the space $\mathscr{D}(T)^{(\lambda)}$ is a rank-one free (say, left) $\mathrm{Sym}\,\mathfrak{t}$-module with generator t^λ. Therefore, the space $\mathsf{A}^{(\lambda)}$ is a rank-one free $\mathrm{Sym}\,\mathfrak{t}$-module with generator f_{A}^{λ}. On the other hand, we know that the map $\mathsf{F}_w : \mathsf{A}^{(\mu)} \to \mathsf{A}^{(w(\mu))}$ is an isomorphism and, for all $u \in \mathrm{Sym}\,\mathfrak{t}$, $\mu \in \mathbb{X}^{++}$, we have $\mathsf{F}_w(f_{\mathsf{A}}^{\mu} u_{\mathsf{A}}) = \mathsf{F}_w(f_{\mathsf{A}}^{\mu})(w \cdot u)_{\mathsf{A}}$. It follows that the element $\mathsf{F}_w(f_{\mathsf{A}}^{\mu})$ is a generator of $\mathsf{A}^{(w(\mu))}$ as a $\mathrm{Sym}\,\mathfrak{t}$-module. The generator of a rank-one free $\mathrm{Sym}\,\mathfrak{t}$-module is determined uniquely up to a nonzero constant factor. We deduce that $\mathsf{F}_w(f_{\mathsf{A}}^{\mu}) = c_{w,\mu} \cdot f_{\mathsf{A}}^{w(\mu)}$, where $c_{w,\mu} \in \mathbb{C}$ is a nonzero constant. To complete the proof of W-equivariance of the map j^*, we must show that $c_{w,\mu} = 1$ for all $w \in W$ and $\mu \in \mathbb{X}^{++}$. We know that the elements $\mathsf{F}_w(f_{\mathsf{A}}^{\mu})$ and $f_{\mathsf{A}}^{w(\mu)}$ both belong to $F_0\mathsf{A}^{(w(\mu))}$. Hence, in $\mathrm{gr}_0\,\mathsf{A}$, we have an equation $\mathrm{gr}(\mathsf{F}_w(f_{\mathsf{A}}^{\mu})) = c_{w,\mu} \cdot \mathrm{gr}\, f_{\mathsf{A}}^{w(\mu)}$. Further, it follows from [GR, Proposition 5.4.1] that the isomorphism $p_{\mathsf{A}} = p_{\mathscr{Y}} \circ \mathrm{symb}$, in (5.5.3), respects the W-actions. Therefore, we compute

$$
\begin{aligned}
p_{\mathsf{A}}(\mathrm{gr}(\mathsf{F}_w(f_{\mathsf{A}}^{\mu}))) &= p_{\mathsf{A}}(\mathsf{F}_w(\mathrm{gr}\, f_{\mathsf{A}}^{\mu})) = w \cdot p_{\mathsf{A}}(\mathrm{gr}\, f_{\mathsf{A}}^{\mu}) \\
&= w \cdot t^{\mu} = t^{w(\mu)} = (\mathrm{gr}\, j^*)(\mathrm{gr}\, f_{\mathsf{A}}^{w(\mu)}) = p_{\mathsf{A}}(\mathrm{gr}\, f_{\mathsf{A}}^{w(\mu)}).
\end{aligned}
$$

It follows that $\mathrm{gr}(\mathsf{F}_w(f_{\mathsf{A}}^{\mu})) = \mathrm{gr}\, f_{\mathsf{A}}^{w(\mu)}$, hence $c_{w,\mu} = 1$, and we are done. □

6 The Miura bimodule

6.1 Proof of Theorem 1.5.4

In this subsection, we consider the actions of N and $\bar{N}$ on G by right translations. Let $D_r = \mathscr{D}(G)/\!\!/\mathfrak{n}_r^{\psi}$. The natural projection $p : G \to G/\bar{N}_r$ is an affine morphism. Therefore, for any open affine $U \subset G/\bar{N}_r$ the set $p^{-1}(U)$ is affine. Let $A_U := \Gamma(p^{-1}(U), \mathscr{D}_G)$. Given a left D_r-module F, put $F_U = A_U/\!\!/\mathfrak{n}_r^{\psi} \otimes_{D_r} F$. We have $A_U \otimes_{\mathscr{D}(G)} (\mathscr{D}(G)/\mathscr{D}(G)\mathfrak{n}_r^{\psi} \otimes_{D_r} F) = A_U/A_U\mathfrak{n}_r^{\psi} \otimes_{A_U/\!\!/\mathfrak{n}_r^{\psi}} F_U$, since A_U is flat over $\mathscr{D}(G)$. Applying Proposition 3.1.4(ii) in the case $A = A_U$ we deduce that $H_i(\bar{\mathfrak{n}}_r,\, A_U/A_U\mathfrak{n}_r^{\psi} \otimes_{A_U/\!\!/\mathfrak{n}_r^{\psi}} F_U) = 0$, $\forall i \neq 0$, and $H_0(\bar{\mathfrak{n}}_r,\, A_U/A_U\mathfrak{n}_r^{\psi} \otimes_{A_U/\!\!/\mathfrak{n}_r^{\psi}} F_U) = \mathscr{U}\mathfrak{t} \otimes_{Z\mathfrak{g}} F_U$.

We have a chain of equivalences of categories of left modules:

$$
D_r\text{-mod} = \mathscr{D}(G)/\!\!/\mathfrak{n}_r^{\psi}\text{-mod} \cong (\mathscr{D}(G), N_r, \psi) \cong (\mathscr{D}_G, N_r, \psi)\text{-mod}. \qquad (6.1.1)
$$

Let $\mathscr{F} \in (\mathscr{D}_G, N_r, \psi)$-mod be the object that corresponds to an object $F \in D_r$-mod via the above equivalences. Thus, $\Gamma(p^{-1}(U), \mathscr{F}) = A_U/A_U\mathfrak{n}_r^{\psi} \otimes_{A_U/\!\!/\mathfrak{n}_r^{\psi}} F_U$. For any $j \in \mathbb{Z}$, by the definition of the functor

$$\int_p^{\text{derived}} : \mathscr{D}_G\text{-mod} = \mathscr{D}_G^{op}\text{-mod} \longrightarrow \mathscr{D}_{\widetilde{\mathscr{B}}}^{op}\text{-mod} = \mathscr{D}_{\widetilde{\mathscr{B}}}\text{-mod},$$

one has $\Gamma(U, \int_p^j \mathscr{F}) = H_{-j}(\bar{\mathfrak{n}}_r, \Gamma(p^{-1}(U), \mathscr{F}))$. We deduce that $\Gamma(U, \int_p^j \mathscr{F}) = 0$, $\forall j \neq 0$, and there is an isomorphism $\Gamma(U, \int_p^0 \mathscr{F}) \cong \mathscr{U}\mathfrak{t} \otimes_{Z\mathfrak{g}} F_U$. In particular, the functor $\int_p^0 : D_r\text{-mod} \to \mathscr{D}_{G/\bar{N}_r}\text{-mod}$ is exact.

To complete the proof, recall that the averaging functor $\mathscr{F} \mapsto \int_{a_{\bar{N}_r}}^{\text{derived}} (\Omega_{\bar{N}_r} \times \mathscr{F})$ on the derived category of $\mathscr{D}_G$-modules is isomorphic to the functor $p^* \circ \int_p^{\text{derived}}$, cf. [MV, Section 2.5]. The functor $p^* : \mathscr{D}_{G/\bar{N}_r}\text{-mod} \to (\mathscr{D}_G, \bar{N}_r)\text{-mod}$ is well-known to be an equivalence, and the theorem follows. □

Let $\mathscr{M} = \bar{\mathfrak{n}}_r \mathscr{D}_G \backslash \mathscr{D}_G / \mathscr{D}_G \mathfrak{n}_r^\psi$. This sheaf of coinvariants is a sheaf of vector spaces on G, to be called the *Miura sheaf*. Using an isomorphism $\mathscr{D}_G \cong \mathscr{O}_G \otimes \mathscr{U}\mathfrak{g}$ and the PBW theorem for $\mathscr{U}\mathfrak{g}$, it is easy to show that for any affine open $V \subset G$, one has $\Gamma(V, \mathscr{M}) = \bar{\mathfrak{n}}_r \mathscr{D}(V) \backslash \mathscr{D}(V) / \mathscr{D}(V) \mathfrak{n}_r^\psi$. According to Section 3.1, the sheaf $p_\bullet \mathscr{M}$ has the natural structure of a $(\bar{\mathfrak{n}}_r \backslash\!\backslash (p_\bullet \mathscr{D}_G), \mathscr{D}(G) /\!\!/ \mathfrak{n}_r^\psi)$-bimodule. Recall that $\mathscr{D}(G) /\!\!/ \mathfrak{n}_r^\psi = D_r$. Also, by Section 3.1, one has algebra isomorphisms $\bar{\mathfrak{n}}_r \backslash\!\backslash (p_\bullet \mathscr{D}_G) \cong (p_\bullet \mathscr{D}_G^{op}) /\!\!/ \bar{\mathfrak{n}}_r^{op} \cong (p_\bullet \mathscr{D}_G) /\!\!/ \bar{\mathfrak{n}}_r \cong \mathscr{D}_{\widetilde{\mathscr{B}}}$. Thus, we can view $p_\bullet \mathscr{M}$ as a $(\mathscr{D}_{\widetilde{\mathscr{B}}}, D_r)$-bimodule. By Proposition 3.1.4(ii), there is a canonical isomorphism

$$p_\bullet \mathscr{M} \cong \mathscr{U}\mathfrak{t} \otimes_{Z\mathfrak{g}} (p_\bullet \mathscr{D}_G) /\!\!/ \mathfrak{n}_r^\psi, \tag{6.1.2}$$

of sheaves of $(\mathscr{U}\mathfrak{t}, D_r)$-bimodules. Furthermore, the proof of Theorem 1.5.4 shows that on the category D_r-mod, there is an isomorphism of functors

$$\int_p (\mathscr{D}_G / \mathscr{D}_G \mathfrak{n}_r^\psi \otimes_{D_r} -) \cong p_\bullet \mathscr{M} \otimes_{D_r} (-). \tag{6.1.3}$$

6.2 From $\mathbb{W}$-modules to $\mathscr{D}(T)$-modules

We are going to mimic formulas (6.1.2)-(6.1.3) in the setting where the ring $\mathscr{D}(G)$ is replaced by the ring $D_l := \mathscr{D}(G) /\!\!/ \mathfrak{n}_l^\psi$.

View D_l as an $(\mathscr{U}\bar{\mathfrak{n}}_r, \mathscr{U}\mathfrak{n}_r)$-bimodule, where the left, resp. right, action is provided by left, resp. right, multiplication inside the algebra D_l by the elements of $\bar{\mathfrak{n}}_r$, resp. $\mathfrak{n}_r$. The analogue of the Miura sheaf $\mathscr{M}$ is played by $\mathbb{M} := \bar{\mathfrak{n}}_r D_l \backslash D_l / D_l \mathfrak{n}_r^\psi$. By Section 3.1, this space of coinvariants has the structure of a $(\bar{\mathfrak{n}}_r \backslash\!\backslash D_l, D_l /\!\!/ \mathfrak{n}_r^\psi)$-bimodule. By Lemma 5.4.1, we have $D_l /\!\!/ \mathfrak{n}_r^\psi \cong \mathbb{W}$, and there is a chain of algebra isomorphisms

$$\begin{aligned}\bar{\mathfrak{n}}_r \backslash\!\backslash D_l = \bar{\mathfrak{n}}_r \backslash\!\backslash (\mathscr{D}(G) /\!\!/ \mathfrak{n}_l^\psi) &= (\bar{\mathfrak{n}}_r \backslash\!\backslash \mathscr{D}(G)) /\!\!/ \mathfrak{n}_l^\psi \\ &= (\mathscr{D}(G) /\!\!/ \bar{\mathfrak{n}}_r) /\!\!/ \mathfrak{n}_l^\psi = \mathscr{D}(\widetilde{\mathscr{B}}) /\!\!/ \mathfrak{n}_l^\psi = \mathscr{D}(T),\end{aligned}$$

where the last isomorphism comes from Theorem 5.3.1(i). We conclude that $\mathbb{M}$ has the structure of a $(\mathscr{D}(T), \mathbb{W})$-bimodule. Further, according to Proposition 3.1.4(ii), one has natural isomorphisms

$$\mathbb{M} = \bar{\mathfrak{n}}_r D_l \backslash D_l / D_l \mathfrak{n}_r^{\psi} \cong \operatorname{Sym}\mathfrak{t} \otimes_{Z\mathfrak{g}} (D_l /\!\!/ \mathfrak{n}_r^{\psi}) \cong \operatorname{Sym}\mathfrak{t} \otimes_{Z\mathfrak{g}} \mathbb{W}.$$

We let the group W act on $\operatorname{Sym}\mathfrak{t} \otimes_{Z\mathfrak{g}} \mathbb{W}$ by $w(a \otimes h) = (w \cdot a) \otimes h$, where $a \mapsto w \cdot a$ is the dot-action. It follows from the construction that the left action $\mathscr{D}(T) \otimes \mathbb{M} \to \mathbb{M}$ is a W-equivariant map. Hence, this action can be extended to a left action of $W \ltimes \mathscr{D}(T)$ on $\mathbb{M}$. Thus, $\mathbb{M}$ acquires the structure of a $(W \ltimes \mathscr{D}(T), \mathbb{W})$-bimodule, to be called the *Miura bimodule*.

For any $\mathbb{W}$-module F, the W-action on $\mathbb{M}$ gives the $\mathscr{D}(T)$-module $\mathbb{M} \otimes_{\mathbb{W}} F$ a W-equivariant structure. Thus we obtain a functor $\mathbb{W}\text{-mod} \to W \ltimes \mathscr{D}(T)\text{-mod}$, $F \mapsto \mathbb{M} \otimes_{\mathbb{W}} F$. This functor is exact since $\mathbb{M} \otimes_{\mathbb{W}} (-) = (\operatorname{Sym}\mathfrak{t} \otimes_{Z\mathfrak{g}} \mathbb{W}) \otimes_{\mathbb{W}} (-) = \operatorname{Sym}\mathfrak{t} \otimes_{Z\mathfrak{g}} (-)$, and $\operatorname{Sym}\mathfrak{t}$ is free over $Z\mathfrak{g}$.

Proposition 6.2.1. *The functor* $\mathbb{W}\text{-mod} \to W \ltimes \mathscr{D}(T)\text{-mod}$, $F \mapsto \mathbb{M} \otimes_{\mathbb{W}} F$, *takes holonomic* $\mathbb{W}$*-modules to holonomic* $\mathscr{D}(T)$*-modules.*

Proof. Using an analogue of the composition of the chain of equivalences in (6.1.1), we obtain an equivalence $(\mathscr{D}_G, N_l \times N_r, \psi \times \psi)\text{-mod} \xrightarrow{\sim} \mathbb{W}\text{-mod}$, to be denoted $\mathbf{I}$. Further, Theorem 5.3.1(ii) yields an equivalence $(\mathscr{D}_{\widetilde{\mathscr{B}}}, N_l, \psi)\text{-mod} \xrightarrow{\sim} \mathscr{D}(T)\text{-mod}$, which we denote by $\mathbf{J}$. Recall the notation $D_l = \mathscr{D}(G) /\!\!/ \mathfrak{n}_l^{\psi}$, resp. $D_r = \mathscr{D}(G) /\!\!/ \mathfrak{n}_r^{\psi}$, and the projection $p : G \to \widetilde{\mathscr{B}} = G/\bar{N}_r$.

We consider the following diagram of functors where $\mathscr{D}$ stands for $\mathscr{D}(G)$ and horizontal inclusions in the middle of the diagram are the natural full embeddings:

$$\begin{array}{ccccccc}
\mathbb{W}\text{-mod} & \overset{\mathbf{I}}{=\!=\!=} & (\mathscr{D}, N_l \times N_r, \psi \times \psi)\text{-mod} & \hookrightarrow & (\mathscr{D}, N_r, \psi)\text{-mod} & =\!=\!= & D_r\text{-mod} \\
\downarrow{\scriptstyle \mathbb{M} \otimes_{\mathbb{W}} (-)} & & \downarrow{\scriptstyle \int_p^0} & & {\scriptstyle \int_p^0}\downarrow & & {\scriptstyle p_\bullet \mathscr{M} \otimes_{D_r} (-)}\downarrow \\
\mathscr{D}(T)\text{-mod} & \overset{\mathbf{J}}{=\!=\!=} & (\mathscr{D}, N_l \times \bar{N}_r, \psi \times 0)\text{-mod} & \hookrightarrow & (\mathscr{D}, \bar{N}_r)\text{-mod} & =\!=\!= & \mathscr{D}_{G/\bar{N}_r}\text{-mod}
\end{array}$$

Using that $\mathbf{J}$ is an equivalence and formula (6.1.3), it is easy to verify that two composite functors $\mathbb{W}\text{-mod} \to \mathscr{D}_{G/\bar{N}_r}\text{-mod}$, along the perimeter of the diagram, are isomorphic. It follows that the functor $\int_p^0 \circ \mathbf{I}$ is isomorphic to the functor $\mathbf{J} \circ \mathbb{M} \otimes_{\mathbb{W}} (-)$. This implies the statement of the proposition since the functor $\int_p^0$ sends holonomic $\mathscr{D}$-modules to holonomic $\mathscr{D}$-modules. □

Remark 6.2.2. For $\mathscr{F} \in (\mathscr{D}_G, N_l \times N_r, \psi \times \psi)\text{-mod}$, the W-equivariant structure on the $\mathscr{D}(T)$-module $\mathbf{J}^{-1}\big(\int_p^0 \mathscr{F}\big)$ is not immediately visible from the definition of the functor $\mathbf{J}^{-1} \circ \int_p^0$. ◊

The proof of the following result is left for the reader.

Lemma 6.2.3. *View $\mathbb{M}$ as a $\mathscr{D}(T)\otimes \mathbb{W}^{op}$-module. Then, we have* $\mathrm{SS}(\mathbb{M})=\Lambda$, *the Lagrangian subvariety from Proposition* 2.3.3. *In particular, $\mathbb{M}$ is a holonomic $\mathscr{D}(T)\otimes \mathbb{W}^{op}$-module.*

7 Nil-Hecke algebras

7.1 Degenerate nil-Hecke algebras

Let $\Sigma\subset\mathscr{R}\subset\mathfrak{h}^*$ be a reduced root system, where Σ denotes the set of simple roots. Let $\mathscr{W}$ be the corresponding Coxeter group, $\{s_\alpha,\ \alpha\in\Sigma\}$ the set of generators of $\mathscr{W}$, and $\ell:\mathscr{W}\to\mathbb{Z}_{\geq 0}$ the length function. For $\alpha,\beta\in\Sigma$, let $m_{\alpha,\beta}$ denote the order of the element $s_\alpha s_\beta\in\mathscr{W}$. The *nil-Hecke algebra* $\mathscr{H}(\mathscr{W})$ is defined as a $\mathbb{C}$-algebra with generators $\mathbf{t}_\alpha,\ \alpha\in\Sigma$, subject to the relations

$$(\mathbf{t}_\alpha)^2=0,\quad (\mathbf{t}_\alpha\mathbf{t}_\beta)^{m_{\alpha,\beta}}=(\mathbf{t}_\beta\mathbf{t}_\alpha)^{m_{\alpha,\beta}},\quad \forall\alpha,\beta\in\Sigma. \tag{7.1.1}$$

For each $w\in\mathscr{W}$, there is an element $\mathbf{t}_w\in\mathscr{W}$ defined as a product $\mathbf{t}_{\alpha_{i_1}}\cdots\mathbf{t}_{\alpha_{i_k}}$ for a reduced factorization $w=s_{\alpha_{i_1}}\cdots s_{\alpha_{i_k}}$ into simple reflections. It is known that $\mathbf{t}_w$ is independent of such a factorization and one has $\mathbf{t}_w\mathbf{t}_y=\mathbf{t}_{wy}$ if $\ell(w)+\ell(y)=\ell(wy)$ and $\mathbf{t}_w\mathbf{t}_y=0$, otherwise, see [Ku]. Moreover, the set $\{\mathbf{t}_w,\ w\in\mathscr{W}\}$ is a $\mathbb{C}$-basis of $\mathscr{H}(\mathscr{W})$.

Let $\mathbb{C}(\mathfrak{h}^*)$ be the field of fractions of the algebra $\operatorname{Sym}\mathfrak{h}=\mathbb{C}[\mathfrak{h}^*]$. For every $\alpha\in R$, define an element $\theta_\alpha=\frac{1}{\check\alpha}(s_\alpha-1)\in W\ltimes\mathbb{C}(\mathfrak{h}^*)$. There is a natural algebra map $\mathscr{H}(\mathscr{W})\hookrightarrow\mathbb{C}(\mathfrak{h}^*)\rtimes\mathscr{W}$ given on the generators by $\mathbf{t}_\alpha\mapsto\theta_\alpha,\ \alpha\in\Sigma$. Let $\mathscr{H}(\mathfrak{h},\mathscr{W})$ be a free left $\operatorname{Sym}\mathfrak{h}$-submodule of $\mathbb{C}(\mathfrak{h}^*)\rtimes\mathscr{W}$ with basis $\theta_w,\ w\in\mathscr{W}$. It is immediate to check that $\mathscr{H}(\mathfrak{h},\mathscr{W})$ is a subalgebra of $\mathbb{C}(\mathfrak{h}^*)\rtimes\mathscr{W}$ and that $\mathscr{H}(\mathfrak{h},\mathscr{W})$ is free as a $\operatorname{Sym}\mathfrak{h}$-module via right multiplication, [Ku].

Proposition 7.1.2 ([Ku], Theorem 11.1.2)**.** *The algebra $\mathscr{H}(\mathfrak{h},\mathscr{W})$ is generated by the algebras $\mathscr{H}(\mathscr{W})$ and* $\operatorname{Sym}\mathfrak{h}$ *subject to the following commutation relations:*

$$\theta_\alpha\cdot s_\alpha(h)-h\cdot\theta_\alpha=\langle\alpha,h\rangle,\qquad \forall\, h\in\mathfrak{h},\ \alpha\in\Sigma. \tag{7.1.3}$$

The subspace $\operatorname{Sym}\mathfrak{h}\subset\mathbb{C}(\mathfrak{h}^*)$ is stable under the action of the subalgebra $\mathscr{H}(\mathfrak{h},\mathscr{W})\subset\mathbb{C}(\mathfrak{h}^*)\rtimes\mathscr{W}$ on $\mathbb{C}(\mathfrak{h}^*)$. Conversely, one has

Theorem 7.1.4 ([Ku], Section 11.2)**.** *The $\mathscr{H}(\mathfrak{h},\mathscr{W})$-action on* $\operatorname{Sym}\mathfrak{h}$ *is faithful, and we have*

$$\mathscr{H}(\mathfrak{h},\mathscr{W})=\{u\in\mathbb{C}(\mathfrak{h}^*)\rtimes\mathscr{W}\mid u(\operatorname{Sym}\mathfrak{h})\subseteq\operatorname{Sym}\mathfrak{h}\}.$$

We equip $\operatorname{Sym}\mathfrak{h}$ with the natural grading and extend it to a grading on $\mathscr{H}(\mathfrak{h},\mathscr{W})$ by placing θ_w in degree $-\ell(w)$. This makes $\mathscr{H}(\mathfrak{h},\mathscr{W})$ a $\mathbb{Z}$-graded algebra, resp. $\operatorname{Sym}\mathfrak{h}$ a $\mathbb{Z}_{\geq 0}$-graded $\mathscr{H}(\mathfrak{h},\mathscr{W})$-module.

The assignment $s_\alpha\mapsto\check\alpha\cdot\theta_\alpha+1,\ \alpha\in\Sigma$, extends to an algebra embedding $\mathbb{C}\mathscr{W}\hookrightarrow\mathscr{H}(\mathfrak{h},\mathscr{W})$. We will identify $\mathbb{C}\mathscr{W}$ with its image, which is contained in the degree zero homogeneous component of $\mathscr{H}(\mathfrak{h},\mathscr{W})$. It is clear that for any $w \in \mathscr{W}$ and $\alpha \in \mathscr{R}$, inside $\mathscr{W} \ltimes \mathbb{C}(\mathfrak{h}^*)$, one has $\theta_{w(\alpha)} = w \cdot \theta_\alpha \cdot w^{-1}$. Taking α to be a simple root, we see that the element θ_β belongs to $\mathscr{H}(\mathfrak{h},\mathscr{W})$ for any root β.

Given a $\operatorname{Sym}\mathfrak{h} \rtimes \mathscr{W}$-module M and $\alpha \in \Sigma$, let $M^\pm = \{m \in M \mid s_\alpha(m) = \pm m\}$. It is clear that the action of $\check{\alpha} \in \mathfrak{h} \subset \operatorname{Sym}\mathfrak{h}$ on M sends $M^\pm$ to $M^\mp$. Observe also that the action of $\mathscr{H}(\mathfrak{h},\mathscr{W})$ on any $\mathscr{H}(\mathfrak{h},\mathscr{W})$-module is $(\operatorname{Sym}\mathfrak{h})^{\mathscr{W}}$-linear since the center of $\mathscr{H}(\mathfrak{h},\mathscr{W})$ equals $(\operatorname{Sym}\mathfrak{h})^{\mathscr{W}}$.

Lemma 7.1.5. *Assume that the group $\mathscr{W}$ is finite. Then the $\mathscr{H}(\mathfrak{h},\mathscr{W})$-action on* $\operatorname{Sym}\mathfrak{t}$ *yields an algebra isomorphism*

$$\mathscr{H}(\mathfrak{h},\mathscr{W}) \cong \operatorname{End}_{(\operatorname{Sym}\mathfrak{h})^{\mathscr{W}}} \operatorname{Sym}\mathfrak{h}. \tag{7.1.6}$$

Furthermore, for a $\operatorname{Sym}\mathfrak{h} \rtimes \mathscr{W}$*-module M, the following are equivalent:*

(i) *The natural map* $\operatorname{Sym}\mathfrak{h} \otimes_{(\operatorname{Sym}\mathfrak{h})^{\mathscr{W}}} M^{\mathscr{W}} \to M$ *is an isomorphism.*

(ii) *The action of* $\mathscr{W} \ltimes \operatorname{Sym}\mathfrak{h}$ *on M admits a (necessarily unique) extension to an $\mathscr{H}(\mathfrak{h},\mathscr{W})$-action on M.*

(iii) *For every $\alpha \in \Sigma$, the action map $\check{\alpha} : M^+ \to M^-$ is a bijection.* [2]

Proof. Recall that $\operatorname{Sym}\mathfrak{h}$ is a free $(\operatorname{Sym}\mathfrak{h})^{\mathscr{W}}$-module of rank $\#\mathscr{W}$. Isomorphism (7.1.6) is a simple consequence of Theorem 7.1.4, cf. also [GKV, Proposition 2.3]. The equivalence of (i) and (ii) now follows from Morita equivalence of the algebras $(\operatorname{Sym}\mathfrak{h})^{\mathscr{W}}$ and $\operatorname{End}_{(\operatorname{Sym}\mathfrak{h})^{\mathscr{W}}} \operatorname{Sym}\mathfrak{h}$. The equivalence of (ii) and (iii) easily follows from the formula $s_\alpha = \check{\alpha} \cdot \theta_\alpha + 1$. □

7.2 Degenerate nil-DAHA

Let $\Sigma \subset R \subset \mathfrak{t}^*$ be a finite reduced root datum, and $\Sigma_{\text{aff}} = \Sigma \sqcup \{\alpha_0\} \subset R_{\text{aff}} \subset \mathfrak{t}^*_{\text{aff}}$ an associated affine root datum. Let $\mathbb{X}^*$ and Q be the weight and root lattice of R, respectively. Let $\widetilde{W} = W \ltimes \mathbb{X}^* \supset W_{\text{aff}} = W \ltimes Q$ be the extended affine Weyl group. Thus, $\widetilde{W} = (\mathbb{X}^*/Q) \ltimes W_{\text{aff}}$.

Similar to the case of affine Hecke algebras, we define $\mathscr{H}(\mathfrak{t}_{\text{aff}},\widetilde{W}) := \mathbb{X}^* \ltimes_Q \mathscr{H}(\mathfrak{t}_{\text{aff}},W_{\text{aff}})$. To avoid confusion, we write elements of $\mathbb{X}^*$, viewed as a subgroup of $\widetilde{W}$, in the form e^μ and also use the same symbol for the image of that element under the algebra embedding $\mathbb{C}\widetilde{W} \hookrightarrow \mathscr{H}(\mathfrak{t}_{\text{aff}},\widetilde{W})$. Thus, for any root $\alpha \in R$ in $\mathscr{H}(\mathfrak{t}_{\text{aff}},\widetilde{W})$, there is an element θ_{e^α} ($= \theta_w$ for $w = e^\alpha$), and also a different element, θ_α, the Demazure element associated with α viewed as a (not necessarily simple) root in R_{aff}.

Let $\hbar \in \mathfrak{t}_{\text{aff}}$ be the minimal imaginary coroot. We identify $\mathfrak{t}_{\text{aff}}$ with $\mathfrak{t} \oplus \mathbb{C}\hbar$, resp. $\operatorname{Sym}(\mathfrak{t}_{\text{aff}})$ with $\mathbb{C}[\mathfrak{t}^*][\hbar]$. The natural linear action of $\widetilde{W}$ on $\mathfrak{t}_{\text{aff}}$ induces a $\widetilde{W}$-action on $\mathbb{C}[\mathfrak{t}^*][\hbar]$, by algebra automorphisms. Explicitly, for $\mu \in \mathbb{X}^*$, the action of e^μ on

[2] The equivalence of (ii) and (iii) was pointed out to me by Gwyn Bellamy.

$\mathbb{C}[\mathfrak{t}^*][\hbar]$ is given by the formula $(e^\mu f)(x,\hbar) = f(x-\hbar\mu,\hbar)$. The action of the algebra $\mathscr{H}(\mathfrak{t}_{\text{aff}}, W_{\text{aff}})$ on $\mathbb{C}[\mathfrak{t}^*][\hbar]$ extends to a faithful $\mathscr{H}(\mathfrak{t}_{\text{aff}}, \widetilde{W})$-action which agrees with the $\widetilde{W}$-action, where $\widetilde{W}$ is viewed as a subset of $\mathscr{H}(\mathfrak{t}_{\text{aff}}, \widetilde{W})$ via the canonical embedding. From the above formula for the action of e^μ, one finds the following commutation relations in $\mathscr{H}(\mathfrak{t}_{\text{aff}}, \widetilde{W})$:

$$\xi \cdot e^\mu = e^\mu \cdot (\xi + \langle \mu, \xi \rangle \cdot \hbar), \qquad \mu \in \mathbb{X}^*,\ \xi \in \mathfrak{t}. \tag{7.2.1}$$

Let $\mathscr{D}_\hbar(T)$ be the Rees algebra of the algebra $\mathscr{D}(T)$ of differential operators, equipped with the filtration by order of the differential operator. Recall the notation $t^\mu \in \mathbb{C}[T]$ for the function on T associated with $\mu \in \mathbb{X}^*$. For $\xi \in \mathfrak{t}$, let ∂_ξ denote the corresponding translation invariant vector field on T. We may view t^μ and ∂_ξ as elements of $\mathscr{D}_\hbar(T)$ placed in degrees 0 and 1, respectively. The commutation relations (7.2.1) are identical to the commutation relations in the algebra $\mathscr{D}_\hbar(T)$ between the elements t^μ and ∂_ξ. Further, it follows from (7.2.1) that the set $\mathrm{S} := (\operatorname{Sym} \mathfrak{t}_{\text{aff}}) \smallsetminus \{0\}$ is an Ore subset of the algebra $\mathscr{D}_\hbar(T)$. The corresponding noncommutative localization $\mathrm{S}^{-1} \cdot \mathscr{D}_\hbar(T)$ may be viewed as a kind of microlocalization of $\mathscr{D}_\hbar(T)$. We obtain algebra embeddings

$$W \ltimes \mathscr{D}_\hbar(T) \xhookrightarrow{w\mapsto w,\ t^\mu \mapsto e^\mu,\ \xi \mapsto \partial_\xi} \mathbb{H} \hookrightarrow \widetilde{W} \ltimes \mathrm{Q}(\mathfrak{t}_{\text{aff}})$$
$$= W \ltimes \mathrm{S}^{-1} \cdot \mathscr{D}_\hbar(T). \tag{7.2.2}$$

Given a $\mathbb{C}[\hbar]$-algebra A and $c \in \mathbb{C}$, we let $A|_{\hbar=c} := A/(\hbar - c)A$ denote its specialization at $\hbar = c$. We have $\mathscr{D}_\hbar(T)|_{\hbar=1} = \mathscr{D}(T)$, resp. $\mathscr{D}_\hbar(T)|_{\hbar=0} = \mathbb{C}[\mathscr{T}^*T]$. We define $\mathbb{H} = \mathscr{H}(\mathfrak{t}_{\text{aff}}, \widetilde{W})|_{\hbar=1}$. The graded algebra embedding $W \ltimes \mathscr{D}_\hbar(T) \hookrightarrow \mathbb{H}$ induces an embedding $W \ltimes \mathscr{D}(T) \hookrightarrow \mathbb{H}$, resp. $W \ltimes \mathbb{C}[\mathscr{T}^*T] \hookrightarrow \operatorname{gr}\mathbb{H}$, of specializations at $\hbar = 1$, resp. $\hbar = 0$.

Recall from the introduction, the notation $\mathbf{e} = \frac{1}{\#W}\sum_{w\in W} w \in \mathbb{C}W$ and the spherical algebra $\mathbf{e}\mathscr{H}(\mathfrak{t}_{\text{aff}}, \widetilde{W})\mathbf{e}$, resp. $\mathscr{H}(\mathfrak{t}_{\text{aff}}, \widetilde{W})^{\text{sph}} := \mathbf{e}\mathbb{H}\mathbf{e}$. The space $\mathscr{H}(\mathfrak{t}_{\text{aff}}, \widetilde{W})\mathbf{e}$, resp. $\mathbb{H}\mathbf{e}$, has the natural structure of an $(\mathscr{H}(\mathfrak{t}_{\text{aff}}, \widetilde{W}), \mathbf{e}\mathscr{H}(\mathfrak{t}_{\text{aff}}, \widetilde{W})\mathbf{e})$-bimodule, resp. $(\mathbb{H}, \mathbb{H}^{\text{sph}})$-bimodule. The embedding $W \ltimes \mathscr{D}_\hbar(T) \hookrightarrow \mathbb{H}$ induces an embedding $\mathscr{D}(T)^W \hookrightarrow \mathbb{H}^{\text{sph}}$, resp. $\mathbb{C}[\mathscr{T}^*T]^W \hookrightarrow \operatorname{gr}\mathbb{H}^{\text{sph}}$.

Lemma 7.2.3. *Let $c \in \mathbb{C}$ and put $\mathscr{H}_c = \mathscr{H}(\mathfrak{t}_{\text{aff}}, \widetilde{W})|_{\hbar=c}$. The algebras $\mathscr{H}(\mathfrak{t}_{\text{aff}}, \widetilde{W})$ and $\mathbf{e}\mathscr{H}(\mathfrak{t}_{\text{aff}}, \widetilde{W})\mathbf{e}$, resp. $\mathscr{H}_c$ and $\mathbf{e}\mathscr{H}_c\mathbf{e}$, are Morita equivalent and the action map induces an algebra isomorphism*

$$\mathscr{H}(\mathfrak{t}_{\text{aff}}, \widetilde{W}) \xrightarrow{\sim} \operatorname{End}_{\mathbf{e}\mathscr{H}(\mathfrak{t}_{\text{aff}}, \widetilde{W})\mathbf{e}} \mathscr{H}(\mathfrak{t}_{\text{aff}}, \widetilde{W})\mathbf{e}, \quad \text{resp.} \quad \mathscr{H}_c \xrightarrow{\sim} \operatorname{End}_{\mathbf{e}\mathscr{H}_c\mathbf{e}} \mathscr{H}_c\mathbf{e}.$$

Proof. It is immediate from (7.1.6) that $\mathbf{e}\mathscr{H}(W, \mathfrak{t})\mathbf{e} = (\operatorname{Sym}\mathfrak{t})^W$ and, moreover, there exist elements $h_i', h_i'' \in \mathscr{H}(\mathfrak{t}, W)$ such that one has $1 = \sum_i h_i' \cdot \mathbf{e} \cdot h_i''$. Since $\mathscr{H}(W, \mathfrak{t})$ is a subalgebra of $\mathscr{H}(\mathfrak{t}_{\text{aff}}, \widetilde{W})$, this equation may be viewed as an equation in $\mathscr{H}(\mathfrak{t}_{\text{aff}}, \widetilde{W})$. Thus, we have $\mathscr{H}(\mathfrak{t}_{\text{aff}}, \widetilde{W})\mathbf{e}\mathscr{H}(\mathfrak{t}_{\text{aff}}, \widetilde{W}) = \mathscr{H}(\mathfrak{t}_{\text{aff}}, \widetilde{W})$, which is known to imply all statements of the lemma involving $\mathscr{H}(\mathfrak{t}_{\mathrm{aff}},\widetilde{W})$. The case of specializations at $\hbar = c$ is similar. □

Proposition 7.2.4. (i) *Let M be a $W \ltimes \mathscr{D}_\hbar(T)$-module. The action of $W \ltimes \mathscr{D}_\hbar(T)$ on M can be extended (necessarily uniquely) to an $\mathscr{H}(\mathfrak{t}_{\mathrm{aff}},\widetilde{W})$-action on M if and only if the natural map $\mathbb{C}[\mathfrak{t}^*]\otimes_{\mathbb{C}[\mathfrak{t}^*]^W} M^W \to M$ is an isomorphism. A similar statement holds in the case of specializations at $\hbar = c$, for any $c \in \mathbb{C}$.*

(ii) *Let L be a $\mathscr{D}(T)^W$-module. The action of $\mathscr{D}(T)^W$ on L can be extended (necessarily uniquely) to an $\mathbb{H}^{\mathrm{sph}}$-action on L if and only if the map $\operatorname{Sym}\mathfrak{t}\otimes_{(\operatorname{Sym}\mathfrak{t})^W} L \to \mathscr{D}(T)\otimes_{\mathscr{D}(T)^W} L$, induced by the inclusion $\operatorname{Sym}\mathfrak{t} \hookrightarrow \mathscr{D}(T)$, is an isomorphism.*

Proof. The simple reflection $s_{\alpha_0} : \mathfrak{t}_{\mathrm{aff}} \to$ is a reflection with respect to the hyperplane $\delta(x) = \hbar$, where $\delta \in R$ is the highest root. The corresponding Demazure operator acts on $\mathbb{C}[\mathfrak{t}^*][\hbar]$ as follows:

$$(\theta_{\alpha_0} f)(x,\hbar) = \frac{f(s_{\alpha_0}(x,\hbar)) - f(x,\hbar)}{\langle x,\check{\delta}\rangle - \hbar} = \frac{f(x-(\langle x,\check{\delta}\rangle-\hbar)\delta)-f(x,\hbar)}{\langle x,\check{\delta}\rangle-\hbar}. \tag{7.2.5}$$

For any weight $\mu \in \mathbb{X}^*$, we compute

$$\begin{aligned}(e^{\mu}\circ\theta_{\delta}\circ e^{-\mu} f)(x,\hbar) &= (e^{\mu}\circ\theta_{\delta} f)(x+\hbar\mu,\hbar)\\ &= e^{\mu}\Big(\frac{f(x-\langle x,\check{\delta}\rangle\delta+\hbar\mu,\hbar)-f(x+\hbar\mu,\hbar)}{\langle x,\check{\delta}\rangle}\Big)\\ &= \frac{f(x-\hbar\mu-\langle x-\hbar\mu,\check{\delta}\rangle\delta+\hbar\mu,\hbar)-f(x-\hbar\mu+\hbar\mu,\hbar)}{\langle x-\hbar\mu,\check{\delta}\rangle}\\ &= \frac{f(x-(\langle x,\check{\delta}\rangle-\hbar\langle\mu,\check{\delta}\rangle)\delta)-f(x,\hbar)}{\langle x,\check{\delta}\rangle-\hbar\langle\mu,\check{\delta}\rangle}.\end{aligned} \tag{7.2.6}$$

Now let M be as in (i). By Lemma 7.1.5, the action of $W \ltimes \operatorname{Sym}\mathfrak{t}$ on M can be extended to an action of the algebra $\mathscr{H}(\mathfrak{t},W)$. Next, we use the action of the subalgebra $\mathbb{C}[T] \subset \mathscr{D}_\hbar(T)$ on M. Specifically, since any root is W-conjugate to a simple root, there exists a weight $\mu \in \mathbb{X}^*$ such that we have $\langle\mu,\check{\delta}\rangle = 1$. For such a μ, formula (7.2.6) shows that in $\mathscr{H}(\mathfrak{t}_{\mathrm{aff}},\widetilde{W})$, one has $\theta_{\alpha_0} = e^{\mu}\circ\theta_{\delta}\circ e^{-\mu}$. Accordingly, we let θ_{α_0} act on M by the operator $e^{\mu}\circ\theta_{\delta}\circ e^{-\mu}$.[3] We claim that the action of θ_{α_0} thus defined and the actions of $\mathscr{H}(\mathfrak{t},W)$ and $\mathbb{C}[T]$ on M combine together to give M the structure of an $\mathscr{H}(\mathfrak{t}_{\mathrm{aff}},\widetilde{W})$-module. Thus, we must check that the relations (7.1.1) and (7.1.3) hold. The relations which do not involve θ_{α_0} hold by construction. The equation $(\theta_{\alpha_0})^2 = 0$ is clear. Also, using (7.2.1) and (7.2.6) it is straightforward to check that (7.1.3) holds for $\alpha = \alpha_0$.

To prove the braid relations, we use [Lo2]. Specifically, let $\mathscr{I}(\mathfrak{t}_{\mathrm{aff}})$ be the algebra associated, as in [Lo2, Section 2.7], with the Coxeter group $\Gamma := W_{\mathrm{aff}}$ and its reflection representation $\mathfrak{h} := \mathfrak{t}_{\mathrm{aff}}$. The relations we have checked so far say that M has the

[3] This correction of an original, incorrect, construction of θ_{α_0} was suggested to me by Gus Lonergan.

structure of an $\mathscr{I}(\mathfrak{t}_{\mathrm{aff}})$-module. The algebra $\mathscr{H}(\mathfrak{t}_{\mathrm{aff}}, W_{\mathrm{aff}})$ is a quotient of $\mathscr{I}(\mathfrak{t}_{\mathrm{aff}})$ by a two-sided ideal. Furthermore, the main result of [Lo2] is essentially equivalent, see [Lo2, Section 2.7], to the statement that the canonical map $\mathscr{I}(\mathfrak{h}) \twoheadrightarrow \mathscr{H}(\mathfrak{h},\Gamma)$ is, in fact, an isomorphism, which is what we want. We remark that in [Lo2] the Coxeter group is assumed to be finite. The braid relation $(\theta_\alpha\theta_\beta)^{m_{\alpha,\beta}} = (\theta_\beta\theta_\alpha)^{m_{\alpha,\beta}}$ was proved in [Lo2, Section 3] under that assumption. However, in the case of an infinite Coxeter group with two generators, $\mathbf{t}_\alpha, \mathbf{t}_\beta$, the element $\mathbf{t}_\alpha\mathbf{t}_\beta$ has an infinite order, i.e., one has $m_{\alpha,\beta} = \infty$. Therefore, the corresponding braid relation is, in that case, vacuous. This proves that the map $\mathscr{I}(\mathfrak{h}) \twoheadrightarrow \mathscr{H}(\mathfrak{h},\Gamma)$ is an isomorphism for any, not necessarily finite, Coxeter group Γ.

Thus, we have extended the action of $W \ltimes \mathscr{D}_\hbar(T)$ on M to an $\mathscr{H}(\mathfrak{t}_{\mathrm{aff}}\widetilde{W})$-action. Note that if such an extension exists, then it is unique since each of the operators θ_α, $\alpha \in \Sigma_{\mathrm{aff}}$, is defined uniquely due to the Morita equivalence (7.1.6) in the special case of the rank-one nil-Hecke algebra associated with the root system $\{\alpha,-\alpha\}$. Part (i) of Proposition 7.2.4 follows.

Combining part (i) for the specialization at $\hbar = 1$ with Morita equivalence of $W \ltimes \mathscr{D}(T)$ and $\mathscr{D}(T)^W$ yields part (ii). □

By definition, one has $\mathscr{H}(\mathfrak{t}_{\mathrm{aff}}, \widetilde{W})|_{\hbar=0} = \mathrm{gr}\,\mathbb{H}$, resp. $\mathbf{e}\mathscr{H}(\mathfrak{t}_{\mathrm{aff}}, \widetilde{W})\mathbf{e}|_{\hbar=0} = \mathrm{gr}\,\mathbb{H}^{\mathrm{sph}}$. Note that the last statement of the proposition below implies Proposition 2.3.6.

Proposition 7.2.7. *There is an algebra isomorphism*

$$\mathrm{gr}\,\mathbb{H} \cong W \ltimes \mathbb{C}\left[T \times \mathfrak{t}^*, \frac{t^\alpha - 1}{\check{\alpha}},\ \alpha \in R\right], \ \textit{resp.}\ \mathrm{gr}\,\mathbb{H}^{\mathrm{sph}} \cong \mathbb{C}\left[T \times \mathfrak{t}^*, \frac{t^\alpha - 1}{\check{\alpha}},\ \alpha \in R\right]^W.$$

Furthermore, we have

$$(\mathrm{gr}\,\mathbb{H})\mathbf{e} \cong \mathbb{C}\left[T \times \mathfrak{t}^*, \frac{t^\alpha - 1}{\check{\alpha}},\ \alpha \in R\right],$$

and each of the above objects is flat over $\mathbb{C}[\mathfrak{t}^*]^W$.

Proof. It is clear from (7.2.1) that the elements $e^\mu \in \mathscr{H}(\mathfrak{t}_{\mathrm{aff}}, \widetilde{W})|_{\hbar=0}$, $\mu \in \mathbb{X}^*$, and $\xi \in \mathfrak{t}$, generate a copy of the algebra $\mathbb{C}[\mathscr{T}^*T]$ inside $\mathscr{H}(\mathfrak{t}_{\mathrm{aff}}, \widetilde{W})|_{\hbar=0}$. For $w \in \widetilde{W}$, let $\bar{\theta}_w$ denote the image of the element $\theta_w \in \mathscr{H}(\mathfrak{t}_{\mathrm{aff}}, \widetilde{W})$ in $\mathscr{H}(\mathfrak{t}_{\mathrm{aff}}, \widetilde{W})|_{\hbar=0}$. Then, the commutation relations (7.1.3) imply easily that the assignment

$$\bar{\theta}_\alpha \mapsto \tfrac{1}{\check{\alpha}}(s_\alpha - 1), \quad \bar{\theta}_{\alpha_0} \mapsto \tfrac{1}{\delta}(s_\delta - 1), \quad \alpha \in \Sigma,$$

extends by multiplicativity to a well-defined algebra homomorphism

$$\Phi:\ \mathrm{gr}\,\mathbb{H} = \mathscr{H}(\mathfrak{t}_{\mathrm{aff}}, \widetilde{W})|_{\hbar=0} \longrightarrow W \ltimes \mathbb{C}\left[T \times \mathfrak{t}^*, \frac{t^\alpha - 1}{\check{\alpha}},\ \alpha \in R\right]. \tag{7.2.8}$$

For any $\mu \in \mathbb{X}^*$, using that $t^\mu s_\alpha t^{-\mu} s_\alpha = t^{\langle \mu, \check{\alpha}\rangle \alpha}$, we find that

$$t^{\mu}\tfrac{1}{\check\alpha}(s_\alpha-1)t^{-\mu}s_\alpha+\tfrac{1}{\check\alpha}(s_\alpha-1)=\tfrac{1}{\check\alpha}(t^{\langle\mu,\check\alpha\rangle\alpha}-s_\alpha)+\tfrac{1}{\check\alpha}(s_\alpha-1)$$
$$=\frac{t^{\langle\mu,\check\alpha\rangle\alpha}-1}{\check\alpha}. \quad (7.2.9)$$

Letting μ run through the set of fundamental weights, we see that the image of the map Φ contains all elements $\frac{t^\alpha-1}{\check\alpha}$, $\alpha\in\Sigma$. The algebra on the right of (7.2.8) is generated by these elements and its subalgebra $W\ltimes\mathbb{C}[T\times\mathfrak{t}^*]$. It follows that the map Φ is surjective. Also, it is clear that applying $\mathbb{C}(\mathfrak{t}^*)\otimes_{\mathbb{C}[\mathfrak{t}^*]}(-)$, a localization functor, to the map Φ yields an isomorphism between the corresponding localized $\mathbb{C}(\mathfrak{t}^*)$-modules. Hence, $\operatorname{Ker}\Phi$ is a torsion $\operatorname{Sym}\mathfrak{t}$-module.

Further, the algebra $\mathscr{H}(\mathfrak{t}_{\text{aff}},\widetilde{W})$ is a free $\mathbb{C}[\mathfrak{t}^*_{\text{aff}}]$-module with basis θ_w, $w\in\widetilde{W}$, by definition. It follows that $\mathscr{H}(\mathfrak{t}_{\text{aff}},\widetilde{W})|_{\hbar=0}$ is a free $\mathbb{C}[\mathfrak{t}^*]$-module with basis $\bar\theta_w$, $w\in\widetilde{W}$, in particular, this module is torsion free. Therefore, we have $\operatorname{Ker}\Phi=0$, so the map Φ is an isomorphism. Since $\mathbb{C}[\mathfrak{t}^*]$ is free over $\mathbb{C}[\mathfrak{t}^*]^W$ and $\mathscr{H}(\mathfrak{t}_{\text{aff}},\widetilde{W})|_{\hbar=0}$ is free over $\mathbb{C}[\mathfrak{t}^*]$, we deduce that $W\ltimes\mathbb{C}[T\times\mathfrak{t}^*,\frac{t^\alpha-1}{\check\alpha},\ \alpha\in R]$ is a free $\mathbb{C}[\mathfrak{t}^*]^W$-module.

Finally, the map Φ being an isomorphism, it follows that its restriction to the spherical subalgebra yields an isomorphism $\operatorname{gr}\mathbb{H}^{\text{sph}}|_{\hbar=0}\xrightarrow{\sim}\mathbb{C}[T\times\mathfrak{t}^*,\frac{t^\alpha-1}{\check\alpha},\ \alpha\in R]^W$. Viewed as a $\mathbb{C}[\mathfrak{t}^*]^W$-module, the algebra $\operatorname{gr}\mathbb{H}^{\text{sph}}$ is a direct summand of $\operatorname{gr}\mathbb{H}$, hence it is flat over $\mathbb{C}[\mathfrak{t}^*]^W$. This implies that $\mathbb{C}[T\times\mathfrak{t}^*,\frac{t^\alpha-1}{\check\alpha},\ \alpha\in R]^W$ is a flat $\mathbb{C}[\mathfrak{t}^*]^W$-module, completing the proof. □

8 Spherical degenerate nil-DAHA via Hamiltonian reduction

8.1 The action of $\mathbb{H}$ on the Miura bimodule $\mathbb{M}$

The following result provides the link between nil-Hecke algebras and Hamiltonian reduction.

Proposition 8.1.1. *The left action of $W\ltimes\mathscr{D}(T)$ on the Miura bimodule $\mathbb{M}$ can be extended to an $\mathbb{H}$-action, making $\mathbb{M}$ an $(\mathbb{H},\mathbb{W})$-bimodule.*

Proof. This follows from Proposition 7.2.4 since $\mathbb{M}^W=\mathbb{W}$, so the map

$$\operatorname{Sym}\mathfrak{t}\otimes_{(\operatorname{Sym}\mathfrak{t})^W}\mathbb{M}^W\to\mathbb{M}=\operatorname{Sym}\mathfrak{t}\otimes_{(\operatorname{Sym}\mathfrak{t})^W}\mathbb{W}$$

is an isomorphism. □

Recall that the algebra $\mathbb{W}$ comes equipped with the Kazhdan filtration. We equip $\operatorname{Sym}\mathfrak{t}$ with the natural filtration induced by the grading and equip $\mathbb{M}=\operatorname{Sym}\mathfrak{t}\otimes_{\mathfrak{S}^W_\hbar}\mathbb{W}$ with a tensor product filtration. Let $\mathbb{M}_\hbar$ be an associated Rees module. It is straightforward to see from the construction that the $(\mathbb{H},\mathbb{W})$-bimodule structure on $\mathbb{M}$ is compatible with the filtrations. Therefore, the left $\mathbb{H}$-action on the Miura bimodule $\mathbb{M}$ can be lifted to an $\mathbb{H}_\hbar$-action on $\mathbb{M}_\hbar$. Thus, $\mathbb{M}_\hbar$ acquires the structure of a

$\mathbb{Z}$-graded $(\mathbb{H}_\hbar, \mathbb{W}_\hbar)$-bimodule. We will identify $\mathbb{M}_\hbar$ with $(\mathrm{Sym}\,\mathfrak{t})[\hbar] \otimes_{(\mathrm{Sym}\,\mathfrak{t})^W[\hbar]} \mathbb{W}_\hbar$, resp. $\mathbb{M}_\hbar^W$ with $\mathbb{W}_\hbar$. Let $1_{\mathbb{M}} = 1 \otimes 1 \in (\mathrm{Sym}\,\mathfrak{t})[\hbar] \otimes_{(\mathrm{Sym}\,\mathfrak{t})^W[\hbar]} \mathbb{W}_\hbar$ be the generator. Recall that $\mathbb{H}_\hbar = \mathscr{H}(\mathfrak{t}_{\mathrm{aff}}, \widetilde{W})$.

The following theorem and its corollary are an extended version of Theorem 1.2.1 from the introduction.

Theorem 8.1.2. (i) *The map* $\mathscr{H}(\mathfrak{t}_{\mathrm{aff}}, \widetilde{W}) \to \mathbb{M}_\hbar$, $u \mapsto u(1 \otimes 1)$, *induces an isomorphism* $\mathscr{H}(\mathfrak{t}_{\mathrm{aff}}, \widetilde{W})\mathbf{e} \xrightarrow{\sim} \mathbb{M}_\hbar$, *of graded left* $\mathscr{H}(\mathfrak{t}_{\mathrm{aff}}, \widetilde{W})$*-modules.*

(ii) *The map* $\mathbf{e}\mathscr{H}(\mathfrak{t}_{\mathrm{aff}}, \widetilde{W})\mathbf{e} \to \mathbb{M}_\hbar^W = \mathbb{W}_\hbar$, $\mathbf{e}u\mathbf{e} \mapsto \mathbf{e}u(1 \otimes 1)$, *yields a graded algebra isomorphism*

$$\mathbf{e}\mathscr{H}(\mathfrak{t}_{\mathrm{aff}}, \widetilde{W})\mathbf{e} \xrightarrow{\sim} \mathrm{End}_{\mathbb{W}_\hbar^{op}} \mathbb{W}_\hbar = \mathbb{W}_\hbar,$$

such that the following diagram commutes:

$$\begin{array}{ccc} (\mathrm{Sym}\,\mathfrak{t})^W[\hbar] & = & Z_\hbar\mathfrak{g} \\ \downarrow & & \downarrow \\ \mathbf{e}\mathscr{H}(\mathfrak{t}_{\mathrm{aff}}, \widetilde{W})\mathbf{e} & \longrightarrow & \mathbb{W}_\hbar. \end{array}$$

From this theorem, combined with Lemma 7.2.3, we deduce

Corollary 8.1.3. *The action of* $\mathbb{H}$ *on* $\mathbb{M}$ *yields the following isomorphisms of filtered algebras:*

$$\mathbb{H} \xrightarrow{\sim} \mathrm{End}_{\mathbb{W}^{op}} \mathbb{M}, \quad \text{resp.} \quad \mathbb{H}^{\mathrm{sph}} \xrightarrow{\sim} \mathbb{W}.$$

8.2 *Proof of Theorem 8.1.2*

To simplify the notation, we put

$$\mathscr{H} = \mathscr{H}(\mathfrak{t}_{\mathrm{aff}}, \widetilde{W}) = \mathbb{H}_\hbar, \quad \text{resp.} \quad \mathscr{H}^{\mathrm{sph}} = \mathbf{e}\mathscr{H}(\mathfrak{t}_{\mathrm{aff}}, \widetilde{W})\mathbf{e} = \mathbb{H}_\hbar^{\mathrm{sph}}.$$

Also, write $\mathrm{S}_\hbar$ for $(\mathrm{Sym}\,\mathfrak{t})[\hbar] = \mathbb{C}[\mathfrak{t}^*_{\mathrm{aff}}]$, resp. $\mathbb{C}(\mathfrak{t}^*_{\mathrm{aff}})$ for the field of fractions of $\mathrm{S}_\hbar$.

The natural $\mathscr{H}$-action on $\mathrm{S}_\hbar$ makes $\mathrm{S}_\hbar$ a cyclic left $\mathscr{H}$-module with generator $1_{\mathrm{S}_\hbar}$. The annihilator of $1_{\mathrm{S}_\hbar}$ is a left ideal $J \subset \mathscr{H}$ generated by the elements θ_w, $w \in \widetilde{W}$. Since $\mathbf{e}1_{\mathrm{S}_\hbar} = 1_{\mathrm{S}_\hbar}$, the map $\mathscr{H} \to \mathrm{S}_\hbar$, $h \mapsto h1_{\mathrm{S}_\hbar}$, descends to a surjection $\mathscr{H}\mathbf{e} \twoheadrightarrow \mathrm{S}_\hbar$ with kernel $J\mathbf{e}$. This gives an $\mathscr{H}^{\mathrm{sph}}$-module surjection $\mathscr{H}^{\mathrm{sph}} = \mathbf{e}\mathscr{H}\mathbf{e} \twoheadrightarrow \mathrm{S}_\hbar^W = \mathbf{e}\mathrm{S}_\hbar$, with kernel $J^{\mathrm{sph}} := \mathbf{e}J\mathbf{e}$, a left ideal of $\mathscr{H}^{\mathrm{sph}}$.

The action of $\mathscr{H}$ on $\mathbb{M}_\hbar$ gives a map $\tau : \mathbf{e}\mathscr{H}\mathbf{e} \to \mathbb{W}_\hbar = \mathbb{M}_\hbar^W$, $h \mapsto h1_{\mathbb{W}_\hbar}$. The image of $J^{\mathrm{sph}} \subset \mathscr{H}^{\mathrm{sph}}$ under the map τ is a left $\mathscr{H}^{\mathrm{sph}}$-submodule $\tau(J^{\mathrm{sph}}) \subset \mathbb{W}_\hbar$. This submodule is not, a priori, stable under left multiplication by $\mathbb{W}_\hbar$. We let $K := \mathbb{W}_\hbar \cdot \tau(J^{\mathrm{sph}})$ be a left ideal of the algebra $\mathbb{W}_\hbar$ generated by $\tau(J^{\mathrm{sph}})$. Also, we let $\bar{K} = \bigcap_{i \in \mathbb{Z}} (K + \hbar^i \mathbb{W}_\hbar)$ be the 'closure' of $K \subset \mathbb{W}_\hbar$ in $\hbar$-adic topology.

Lemma 8.2.1. *One has a decomposition* $\mathbb{W}_\hbar = Z_\hbar\mathfrak{g} \oplus \bar{K}$ *as a direct sum of* $Z_\hbar\mathfrak{g}$*-stable graded subspaces.*

Proof. We first prove an analogue of the statement of the lemma at $\hbar = 0$. To this end, let $\mathscr{H}_0 := \mathscr{H}|_{\hbar=0}$, resp. $\mathscr{H}_0^{\text{sph}} = \mathbf{e}\mathscr{H}_0\mathbf{e}$. We have a chain of isomorphisms

$$(\operatorname{gr}\mathbb{H})\mathbf{e} = \mathscr{H}_0\mathbf{e} \cong \mathbb{C}\left[T \times \mathfrak{t}^*, \frac{t^{\check{\alpha}}-1}{\check{\alpha}}, \ \alpha \in R\right]$$
$$\cong \mathbb{C}[\mathfrak{Z} \times_{\mathfrak{c}} \mathfrak{t}^*] \cong \operatorname{Sym}\mathfrak{t} \otimes_{(\operatorname{Sym}\mathfrak{t})^W} \operatorname{gr}\mathbb{W}, \quad (8.2.2)$$

where the first, resp. second, and third, isomorphism is Proposition 7.2.7, resp. Theorem 2.3.5 and Theorem 1.2.2.

Let $J_0 \subset \mathscr{H}_0\mathbf{e}$ be the image of J in $\mathscr{H}_0 = \mathscr{H}/\hbar\mathscr{H}$ and let $I \subset \mathbb{C}\big[T \times \mathfrak{t}^*, \frac{t^{\check{\alpha}}-1}{\check{\alpha}}, \alpha \in R\big]$, resp. $L \subset \operatorname{Sym}\mathfrak{t} \otimes_{(\operatorname{Sym}\mathfrak{t})^W} \operatorname{gr}\mathbb{W}$, be an ideal that corresponds to J_0 via the first, resp. the composite, isomorphism in (8.2.2). The proof of Proposition 7.2.7, combined with (7.2.8)-(7.2.9), shows that the ideal I is generated by the elements $\frac{t^{\check{\mu}}-1}{\check{\mu}}$, $\mu \in \mathbb{X}^*$. It is clear from this description, cf. also (7.2.9), that one has a direct sum decomposition $\mathbb{C}\big[T \times \mathfrak{t}^*, \frac{t^{\alpha}-1}{\check{\alpha}}, \alpha \in R\big] = \mathbb{C}[\mathfrak{t}^*] \oplus I$, of graded $\operatorname{Sym}\mathfrak{t}$-stable subspaces. Hence, one has a decomposition $\operatorname{Sym}\mathfrak{t} \otimes_{(\operatorname{Sym}\mathfrak{t})^W} \operatorname{gr}\mathbb{W} = \operatorname{Sym}\mathfrak{t} \oplus L$. Taking W-invariants and writing $J_0^{\text{sph}} = J_0^W$, we deduce the following isomorphism that respects the direct sum decompositions:

$$\mathscr{H}_0^{\text{sph}} = (\operatorname{Sym}\mathfrak{t})^W \oplus J_0^{\text{sph}} \cong \operatorname{gr}\mathbb{W} = (\operatorname{Sym}\mathfrak{t})^W \oplus L^W. \quad (8.2.3)$$

It is straightforward to check that the isomorphism $\mathscr{H}_0^{\text{sph}} \xrightarrow{\sim} \operatorname{gr}\mathbb{W}$ in (8.2.3) agrees with $\tau_0 : \mathscr{H}_0^{\text{sph}} \xrightarrow{\sim} \operatorname{gr}\mathbb{W}$, the specialization of the map $\tau : \mathscr{H}^{\text{sph}} \to \mathbb{W}_\hbar$ at $\hbar = 0$. The image of the ideal J_0^{sph} equals $\tau_0(J_0^{\text{sph}})$, by definition. We conclude that $L^W = \tau_0(J_0^{\text{sph}})$. Thus, the decomposition on the right of (8.2.3) reads

$$\mathbb{W}_\hbar/\hbar\mathbb{W}_\hbar = (Z_\hbar\mathfrak{g}/\hbar{\cdot}Z_\hbar\mathfrak{g}) \oplus \tau_0(J_0^{\text{sph}}). \quad (8.2.4)$$

This is equivalent to a pair of equations

$$\mathbb{W}_\hbar = Z_\hbar\mathfrak{g} + \tau(J^{\text{sph}}) + \hbar\mathbb{W}_\hbar, \qquad Z_\hbar\mathfrak{g} \cap (\tau(J^{\text{sph}}) + \hbar\mathbb{W}_\hbar) = \hbar\mathbb{W}_\hbar. \quad (8.2.5)$$

Remark 8.2.6. The decomposition in (8.2.4) does not necessarily imply that $\mathbb{W}_\hbar = Z_\hbar\mathfrak{g} \oplus \tau(J^{\text{sph}})$, since the $\mathbb{Z}$-grading on $\mathbb{W}_\hbar$ is not bounded below. ◊

It is immediate from definitions that $\bar{K}$ is a graded left ideal of $\mathbb{W}_\hbar$. Furthermore, the ideals $\bar{K}$ and K have the same image in $\operatorname{gr}\mathbb{W} = \mathbb{W}_\hbar/\hbar\mathbb{W}_\hbar$. Also, the image of K in $\operatorname{gr}\mathbb{W}$ equals $\operatorname{gr}\mathbb{W} \cdot \tau_0(J_0^{\text{sph}})$, by definition. We have shown that $L^W = \tau_0(J_0^{\text{sph}})$, where L^W is an ideal of $\operatorname{gr}\mathbb{W}$. Therefore, $\tau_0(J_0^{\text{sph}})$ is an ideal. Hence, inside $\operatorname{gr}\mathbb{W}$, we have $J_0^{\text{sph}} 1_{\operatorname{gr}\mathbb{W}} = \operatorname{gr}\mathbb{W} \cdot \tau_0(J_0^{\text{sph}})$. Combining these observations together, we deduce that

$$\tau(J^{\text{sph}}) + \hbar\mathbb{W}_\hbar = K + \hbar\mathbb{W}_\hbar = \bar{K} + \hbar\mathbb{W}_\hbar. \tag{8.2.7}$$

Using (8.2.7), we see that equations (8.2.5) take the following form:

$$\mathbb{W}_\hbar = Z_\hbar\mathfrak{g} + \bar{K} + \hbar\mathbb{W}_\hbar, \qquad Z_\hbar\mathfrak{g} \cap (\bar{K} + \hbar\mathbb{W}_\hbar) = \hbar\mathbb{W}_\hbar. \tag{8.2.8}$$

We claim that the equations in (8.2.8) imply the required direct sum decomposition $\mathbb{W}_\hbar = Z_\hbar\mathfrak{g} \oplus \bar{K}$. Indeed, let $a \in \mathbb{W}_\hbar$. Separating homogeneous components, one may assume without loss of generality that a is homogeneous of some degree $n \in \mathbb{Z}$. Note that $Z_\hbar\mathfrak{g}$ has no homogeneous components of negative degree. Hence, in the case $n < 0$, the first equation in (8.2.8) implies that there exist $k_1 \in \bar{K}$ and $a_1 \in \mathbb{W}_\hbar$, of degrees n and $n-1$, respectively, such that $a = k_1 + \hbar a_1$. Applying the same argument to a_1, one gets $a_1 = k_2 + \hbar a_2$, and so on. Thus, for any $m \geq 1$, one can find $k_1, \dots, k_m \in K$ such that one has $a = \hbar^{m+1}a_{m+1} + \sum_{i=1}^m \hbar^{i-1}k_i$. Since K is an ideal, we have $\sum_{i=1}^m \hbar^{i-1}k_i \in K$ so $a \in \hbar^{m+1}\mathbb{W}_\hbar + K$. We conclude that $a \in \bigcap_{m\in\mathbb{Z}} (\hbar^{m+1}\mathbb{W}_\hbar + K) = \bar{K}$. In the case $n \geq 0$, one shows by induction on n using a similar argument that $a \in Z_\hbar\mathfrak{g} + \bar{K}$. Finally, from the second equation in (8.2.8), one deduces that $Z_\hbar\mathfrak{g} \cap \bar{K} = 0$. □

Proof (*Theorem* 8.1.2). We know that $\mathscr{H}^{\text{sph}}$ and $\mathscr{H}\mathbf{e}$ are torsion free $\mathrm{S}_\hbar^W$-modules. Therefore, the map $\mathscr{H}^{\text{sph}} \to \mathbb{C}(\mathfrak{t}^*_{\text{aff}})^W \otimes_{\mathrm{S}_\hbar^W} \mathscr{H}^{\text{sph}}$ is injective. Further, the algebra $\mathfrak{Z}$ is a flat scheme over $\mathfrak{c}$, see Section 2.1. Hence, the algebra $\operatorname{gr}\mathbb{W} \cong \mathbb{C}[\mathfrak{Z}]$ is a torsion free $\mathrm{S}_\hbar^W$-module. It follows, since the filtration on $\mathbb{W}$ is separating by Lemma 4.3.2, that the algebra $\mathbb{W}_\hbar$ is torsion free over $\mathrm{S}_\hbar^W = Z_\hbar\mathfrak{g}$. We conclude that the map $\mathbb{W}_\hbar \to \mathbb{C}(\mathfrak{t}^*_{\text{aff}})^W \otimes_{\mathrm{S}_\hbar^W} \mathbb{W}_\hbar$ is injective. It follows that the map $\tau : \mathscr{H}^{\text{sph}} \to \mathbb{W}_\hbar$ is injective.

To prove that τ is surjective, recall from Section 7.2 that $\mathscr{H}$ is a subalgebra of $\widetilde{W}_{\text{aff}} \ltimes \mathbb{C}(\mathfrak{t}^*_{\text{aff}})$. The latter algebra acts faithfully on $\mathbb{C}(\mathfrak{t}^*_{\text{aff}})$. We have a decomposition $\mathbb{C}(\mathfrak{t}^*_{\text{aff}}) = \mathbf{e}\mathbb{C}(\mathfrak{t}^*_{\text{aff}}) \oplus (1-\mathbf{e})\mathbb{C}(\mathfrak{t}^*_{\text{aff}})$. The action of the subalgebra $\mathbf{e}(W \ltimes \mathbb{C}(\mathfrak{t}^*_{\text{aff}}))\mathbf{e}$ respects this decomposition, moreover, the action of this subalgebra kills the second direct summand. Further, we have $\mathscr{H}^{\text{sph}} \subset \mathbf{e}(W \ltimes \mathbb{C}(\mathfrak{t}^*_{\text{aff}}))\mathbf{e}$, so the action of $\mathscr{H}^{\text{sph}}$ kills $(1-\mathbf{e})\mathbb{C}(\mathfrak{t}^*_{\text{aff}})$ and $\mathrm{S}_\hbar^W = \mathbf{e}\mathrm{S}_\hbar \subset \mathbf{e}\mathbb{C}(\mathfrak{t}^*_{\text{aff}})$ is an $\mathscr{H}^{\text{sph}}$-stable subspace.

Next, we use the map τ to pullback the $\mathscr{H}^{\text{sph}}$-action to a $\mathbb{W}_\hbar$-action on $\mathbf{e}\mathbb{C}(\mathfrak{t}^*_{\text{aff}}) = \mathbb{C}(\mathfrak{t}^*_{\text{aff}})^W$. Thanks to the injectivity of τ, the resulting $\mathbb{W}_\hbar$-module is faithful. We claim that the subspace $\mathrm{S}_\hbar^W \subset \mathbb{C}(\mathfrak{t}^*_{\text{aff}})^W$ is stable under the $\mathbb{W}_\hbar$-action. To see this, we use Lemma 8.2.1. Specifically, it follows from the lemma that the map τ induces a bijection $\mathscr{H}^{\text{sph}}/J^{\text{sph}} \cong \mathrm{S}_\hbar^W \xrightarrow{\sim} Z_\hbar\mathfrak{g} \cong \mathbb{W}_\hbar/\bar{K}$. Via this bijection, the action of the algebra $\mathscr{H}^{\text{sph}}$ on $\mathrm{S}_\hbar^W$ can be extended to a $\mathbb{W}_\hbar$-action on $\mathrm{S}_\hbar^W \cong \mathbb{W}_\hbar/\bar{K}$. Is is immediate from the construction that the action so defined agrees with the above defined $\mathbb{W}_\hbar$-action on $\mathbb{C}(\mathfrak{t}^*_{\text{aff}})^W$. The claim follows.

We now apply Theorem 7.1.4 to conclude that inside $\operatorname{End}_{\mathbb{C}}\mathbb{C}(\mathfrak{t}^*_{\text{aff}})$, one has an inclusion $\mathbb{W}_\hbar \subseteq \tau(\mathscr{H}^{\text{sph}})$. Thus τ is surjective, hence, it is an isomorphism. All other statements of the theorem easily follow from this. □

8.3 Proof of Theorem 1.5.1

It follows from Proposition 3.1.4 and the isomorphism $(\mathscr{D}(G)/\!\!/\mathfrak{n}_l^\psi)/\!\!/\mathfrak{n}_r^\psi$ of Lemma 5.4.1, that the functor $M \mapsto M^{\mathfrak{n}_l^\psi \times \times \mathfrak{n}_r^\psi}$ gives an equivalence of categories $(\mathscr{D}_G, N_l \times N_r, \psi \times \psi)$-mod $\cong \mathbb{W}$-mod. The equation in Lemma 3.3.2(3) shows that this functor induces, by restriction to holonomic modules, an equivalence $\mathscr{W}h \cong \mathbb{W}$-hol. Also, by Theorem 1.2.1, we have an equivalence $\mathbb{W}$-hol $\cong \mathbb{H}^{\mathrm{sph}}$-hol. The corresponding equivalence $\mathbb{W}$-hol $\cong \mathbb{H}$-hol follows from this by Morita equivalence, see Lemma 7.2.3, since the latter equivalence takes holonomic modules to holonomic modules, by definition.

Let $\mathscr{C}$ be the category of not necessarily holonomic $W \ltimes \mathscr{D}(T)$-modules M such that (1.5.2) is an isomorphism. Further, let $\mathscr{C}^{\mathrm{sph}}$ be the category of not necessarily holonomic $\mathscr{D}(T)^W$-modules L such that the map $\mathrm{Sym}\,\mathfrak{t} \otimes_{(\mathrm{Sym}\,\mathfrak{t})^W} L \to \mathscr{D}(T) \otimes_{\mathscr{D}(T)^W} L$ is an isomorphism. It follows from Morita equivalence of the algebras $\mathscr{D}(T)^W$ and $W \ltimes \mathscr{D}(T)$ that the functor $\mathscr{D}(T) \otimes_{\mathscr{D}(T)^W} (-)$ provides an equivalence $\mathscr{C}^{\mathrm{sph}} \xrightarrow{\sim} \mathscr{C}$.

Further, we have a functor $\mathbb{H}$-mod $\to \mathscr{C}$ induced by the algebra embedding $W \ltimes \mathscr{D}(T) \hookrightarrow \mathbb{H}$. The proof of Proposition 7.2.4 shows that this functor is fully faithful. Thus, this functor is an equivalence. Using Morita equivalences above, we deduce that the functor $\mathbb{H}^{\mathrm{sph}}$-mod $\to \mathscr{C}^{\mathrm{sph}}$, induced by the algebra embedding $\mathscr{D}(T)^W \hookrightarrow \mathbb{H}^{\mathrm{sph}}$, is also an equivalence.

To complete the proof of the theorem we must show that an object $L \in \mathscr{C}^{\mathrm{sph}}$ is holonomic as a $\mathscr{D}(T)^W$-module if and only if it is holonomic as an $\mathbb{H}^{\mathrm{sph}}$-module. Assume first that L is holonomic as an $\mathbb{H}^{\mathrm{sph}}$-module, equivalently, as a $\mathbb{W}$-module. Then, Proposition 6.2.1 implies that $\mathbb{M} \otimes_{\mathbb{W}} L = \mathrm{Sym}\,\mathfrak{t} \otimes_{(\mathrm{Sym}\,\mathfrak{t})^W} L$ is a holonomic $\mathscr{D}(T)$-module. Hence, L is holonomic as a $\mathscr{D}(T)^W$-module.

Conversely, assume that L is holonomic as a $\mathscr{D}(T)^W$-module L. Choose a good filtration on L, and equip the module $M := \mathrm{Sym}\,\mathfrak{t} \otimes_{(\mathrm{Sym}\,\mathfrak{t})^W} L$, resp. $\mathbb{H}\mathbf{e} \otimes_{\mathscr{D}(T)^W} L$ and $\mathbb{W} \otimes_{\mathscr{D}(T)^W} L$, with the tensor product filtrations. The map $\mathscr{T}^*T \to (\mathscr{T}^*T)/W$ being finite, we deduce that $\mathrm{Supp}\,\mathrm{gr} M$ is a Lagrangian subvariety of $\mathscr{T}^*T$. We have the following diagram, cf. Proposition 2.3.3:

$$\mathfrak{Z} = \mathrm{Spec}(\mathrm{gr}\,\mathbb{W}) = \mathrm{Spec}(\mathrm{gr}\,\mathbb{H}^{\mathrm{sph}}) \xleftarrow{p_{\mathfrak{Z}}} \Lambda \xrightarrow{p_{\mathscr{T}^*T}} \mathscr{T}^*T = \mathrm{Spec}(\mathrm{gr}\,\mathscr{D}(T)).$$

where $p_{\mathfrak{Z}}$, resp. $p_{\mathscr{T}^*T}$, is the first, resp. second, projection and we have used Theorem 1.2.1 to identify $\mathrm{gr}\,\mathbb{W}$ with $\mathrm{gr}\,\mathbb{H}^{\mathrm{sph}}$, resp. Theorem 1.2.2 to identify $\mathrm{Spec}\,\mathrm{gr}\,\mathbb{W}$ with $\mathfrak{Z}$. Thus, we have

$$\mathrm{Supp}\,\mathrm{gr}(\mathbb{W} \otimes_{\mathscr{D}(T)^W} L) = \mathrm{Supp}(\mathrm{gr}\,\mathbb{W} \otimes_{\mathrm{gr}\,\mathscr{D}(T)^W} \mathrm{gr} L) \subseteq p_{\mathfrak{Z}}(p_{\mathscr{T}^*T}^{-1}(\mathrm{Supp}\,\mathrm{gr} M)).$$

By Proposition 2.3.3, Λ is a Lagrangian subvariety of $\mathfrak{Z} \times \mathscr{T}^*T$. It follows that $p_{\mathfrak{Z}}(p_{\mathscr{T}^*T}^{-1}(\mathrm{Supp}\,\mathrm{gr} M))$ is an isotropic subvariety of $\mathfrak{Z}$. We deduce that $\mathbb{W} \otimes_{\mathscr{D}(T)^W} L$ is a holonomic $\mathbb{W}$-module. On the other hand, the action of the algebra $\mathbb{H}^{\mathrm{sph}} \cong \mathbb{W}$ on L gives a canonical surjection $\mathbb{W} \otimes_{\mathscr{D}(T)^W} L \twoheadrightarrow L$, of $\mathbb{H}^{\mathrm{sph}}$-modules. This implies that L is holonomic as an $\mathbb{W}$-module. $\square$

References

[Ba1] A. Balibanu, *The Peterson variety and the wonderful compactification.* Represent. Theory 21 (2017), 132–150.

[Ba2] ______, *The partial compactification of the universal centralizer.* arXiv:1710.06327.

[BK] A. Beilinson, D. Kazhdan, *Flat projective connections.* (1991), Unpublished manuscript. http://www.math.stonybrook.edu/ kirillov/manuscripts.html

[Be] D. Beraldo, *Loop group actions on categories and Whittaker invariants.* arXiv:1310.5127.

[BGG] J. Bernstein, I. Gelfand, S. Gelfand, *Schubert cells, and the cohomology of the spaces G/P.* (Russian) Uspehi Mat. Nauk 28 (1973), no. 3 (171), 3–26.

[BBP] R. Bezrukavnikov, A. Braverman, L. Positselskii, *Gluing of abelian categories and differential operators on the basic affine space.* J. Inst. Math. Jussieu 1 (2002), 543–557.

[BBM] ______, ______, I. Mirkovic, *Some results about geometric Whittaker model.* Adv. Math. 186 (2004), 143–152.

[BF] ______, M. Finkelberg, *Equivariant Satake category and Kostant-Whittaker reduction.* Mosc. Math. J. 8 (2008), 39–72, 183.

[BFM] ______, ______, I. Mirkovic, *Equivariant homology and K-theory of affine Grassmannians and Toda lattices.* Compos. Math. 141 (2005), 746–768.

[BB] W. Borho, J.-L. Brylinski, *Differential operators on homogeneous spaces. I. Irreducibility of the associated variety for annihilators of induced modules.* Invent. Math. 69 (1982), 437–476.

[CF] I. Cherednik, B. Feigin, *Rogers-Ramanujan type identities and Nil-DAHA.* Adv. Math. 248 (2013), 1050–1088.

[FT] M. Finkelberg, A. Tsymbalyuk, *Multiplicative slices, relativistic Toda and shifted affine quantum groups.* arXiv:1708.01795.

[GG] W.L. Gan, V. Ginzburg, *Almost-commuting variety, D-modules, and Cherednik Algebras.* IMRP Int. Math. Res. Pap. 2006, 26439, 1–54.

[Gi] V. Ginzburg, *Harish-Chandra bimodules for quantized Slodowy slices.* Represent. Theory 13 (2009), 236–271.

[GKV] ______, M. Kapranov, E. Vasserot, *Residue construction of Hecke algebras.* Adv. Math. 128 (1997), 1–19.

[GK] ______, D. Kazhdan, *Differential operators on G/U and the Miura bimodule.* In preparation.

[GR] ______, S. Riche, *Differential operators on G/U and the affine Grassmannian.* J. Inst. Math. Jussieu 14 (2015), 493–575.

[GKM] M. Goresky, R. Kottwitz, R. MacPherson, *Equivariant cohomology, Koszul duality, and the localization theorem.* Invent. Math. 131 (1998), 25–83.

[HTT] R. Hotta, K. Takeuchi, T. Tanisaki, *D-modules, perverse sheaves, and representation theory.* Progress in Mathematics, 236. Birkhäuser Boston, Inc., Boston, MA, 2008.

[Ko1] B. Kostant, *The principal three-dimensional subgroup and the Betti numbers of a complex simple Lie group.* Amer. J. Math. 81 (1959) 973–1032.

[Ko2] ______, *Lie group representations on polynomial rings.* Amer. J. Math. 85 (1963), 327–404.

[Ko3] ______, *On Whittaker vectors and representation theory.* Invent. Math. 48 (1978), 101–184.

[KK1] ______, S. Kumar, *T-equivariant K-theory of generalized flag varieties.* J. Differential Geom. 32 (1990), 549–603.

[KK2] ______, ______, *The nil Hecke ring and cohomology of G/P for a Kac-Moody group G.* Adv. in Math. 62 (1986), 187–237.

[Ku] S. Kumar, *Kac-Moody groups, their flag varieties and representation theory.* Progress in Mathematics, 204. Birkhäuser Boston, Inc., Boston, MA, 2002.

[LLMSSZ] T. Lam, L. Lapointe, J. Morse, A. Schilling, M. Shimozono, M. Zabrocki, *k-Schur functions and affine Schubert calculus. Fields Institute Monographs*, 33. Springer, New York; Fields Institute for Research in Mathematical Sciences, Toronto, ON, 2014.

[Lo1] G. Lonergan, *The Fourier transform for the quantum Toda lattice.* arXiv:1706.05344.

[Lo2] ______, *A remark on descent for Coxeter groups.* arXiv:1707.01156.

[MS] D. Miličić, W. Soergel, *Twisted Harish-Chandra sheaves and Whittaker modules: the nondegenerate case.* Developments and retrospectives in Lie theory, 183–196, Dev. Math., 37, Springer, Cham, 2014.

[MV] I. Mirković, K. Vilonen, *Characteristic varieties of character sheaves.* Invent. Math. 93 (1988), 405–418.

Spectral Properties of Semi-classical Toeplitz Operators

V. Guillemin, A. Uribe, and Z. Wang

To Bertram Kostant, a beloved friend and esteemed role model of the senior author (Victor Guillemin) of this paper

Abstract The main results of this paper are an asymptotic expansion in powers of $\hbar$ for the spectral measure $\mu_\hbar$ of a semi-classical Toeplitz operator, $Q_\hbar$, and an equivariant version of this result when $Q_\hbar$ admits an n-torus as a symmetry group. In addition we discuss some inverse spectral consequences of these results.

Keywords: Toeplitz operators • spectral asymptotics

Mathematics Subject Classification (2010): 58J50 (Primary), 47B35, 81Q20

1 Introduction

Before we attempt to explain the meaning of "semi-classical" in the title above, we will first review a few basic facts about the classical theory of Toeplitz operators (the topic of the Princeton series monograph [BG]). To define these operators, let X be a compact manifold of dimension n and Σ a closed symplectic cone in $T^*X \setminus 0$. To this cone one can attach a Hilbert subspace $H^2(X)$ of $L^2(X)$ with the property that the orthogonal projection

---

Z.W. is partially supported by the National Science Foundation of China, Grant No. 11571131 and Grant No. 11526212.

V. Guillemin
Mathematics Department, Massachusetts Institute of Technology, Cambridge, MA 02138,
e-mail: vwg@math.mit.edu

A. Uribe
Mathematics Department, University of Michigan, Ann Arbor, Michigan 48109,
e-mail: uribe@umich.edu

Z. Wang
School of Mathematical Sciences, University of Science and Technology of China, Hefei, Anhui 230026, P. R. China, e-mail: wangzuoq@ustc.edu.cn

V. G. Kac, V. L. Popov (eds.), *Lie Groups, Geometry, and Representation Theory*,
Progress in Mathematics 326, https://doi.org/10.1007/978-3-030-02191-7_7

$$\Pi : L^2(X) \to H^2(X) \tag{1.1}$$

is a Fourier integral operator of Hermite type (in the sense of [BG], §3) with micro-support on the diagonal Δ_Σ in $\Sigma \times \Sigma$. The prototypical example of such an operator is the classical *Szegö projector*. Namely, in this example X is the boundary of a strictly pseudoconvex domain, $H^2(X)$ is the L^2-closure of the kernel of the $\bar{\partial}_b$ operator, and Σ is the characteristic variety of this operator. (For a slightly simplified version of this example see Section 2 below.)

The Toeplitz operators we will be concerned with in this paper are operators of the form

$$Q = \Pi P \Pi, \tag{1.2}$$

where P is a classical pseudodifferential operator of order d on X whose leading symbol will be denoted by p. One of the main results of [BG], a result which we will make heavy use of below, is that in defining these operators one can assume without loss of generality that P commutes with Π. (Thus in particular these operators form an algebra.)

To return to the "semi-classical" in the title of this article, suppose that there is a free action of the circle group S^1 on X, and that the induced action on $L^2(X)$ preserves $H^2(X)$ and commutes with P. Then if for each integer N we denote by $L^2(X)_N$ the subspace of functions $f \in L^2(X)$ satisfying

$$f\left(e^{\sqrt{-1}\theta}x\right) = e^{\sqrt{-1}N\theta} f(x), \tag{1.3}$$

P preserves this space and also preserves the intersection,

$$H^2(X)_N := L^2(X)_N \cap H^2(X).$$

Moreover, the operator

$$P_\hbar = \hbar^d P|_{L^2(X)_N}, \qquad \hbar = \frac{1}{N} \tag{1.4}$$

can be viewed as a zeroth order semi-classical pseudodifferential operator[1], and the operator

$$Q_\hbar = P_\hbar|_{H^2(X)_N} \tag{1.5}$$

as a zeroth order semi-classical Toeplitz operator.

In the result we are about to describe, the symbols of these operators play a major role, and to describe these symbols, let

$$\phi : T^*X \to \mathbb{R}$$

[1] One can also study operators of the form $\hbar^{\tilde{d}} P|_{L^2(X)_N}$, which is a semi-classical pseudodifferential operator of order $d - \tilde{d}$.

be the moment map associated with the lifted action of S^1 on T^*X. We will assume that 1 is a regular value of ϕ and that S^1 acts freely on $\phi^{-1}(1)$. Then $\phi^{-1}(1)$ is a smooth submanifold of T^*X, and

$$(T^*X)_{\text{red}} = \phi^{-1}(1)/S^1 \tag{1.6}$$

is the symplectic reduction of T^*X at $\phi = 1$. Since $p|_{\phi^{-1}(1)}$ is S^1-invariant it is the pull-back to $\phi^{-1}(1)$ of a function

$$p_{\text{red}} \colon (T^*X)_{\text{red}} \to \mathbb{R}, \tag{1.7}$$

and this reduced symbol is the semi-classical symbol of $P_\hbar$. Moreover, the quotient

$$\Sigma_{\text{red}} = \phi^{-1}(1) \cap \Sigma / S^1 \tag{1.8}$$

sits inside $(T^*X)_{\text{red}}$ and the restriction of p_{red} to Σ_{red} is the semi-classical symbol of $Q_\hbar$.

A few final assumptions: We will henceforth assume P is a self-adjoint elliptic pseudodifferential operator of order d with positive leading symbol p. Therefore, since X is compact, these assumptions imply that the spectrum of P is discrete, bounded from below and has no finite points of accumulation, and hence that the same is true of $P_\hbar$ and $Q_\hbar$.

We will now describe the main results of this paper. Let $\mu_\hbar$ be the spectral measure

$$\mu_\hbar(f) := \operatorname{trace} f(Q_\hbar), \qquad \forall f \in C_0^\infty(\mathbb{R}). \tag{1.9}$$

We will prove:

Theorem 1.1. *As $\hbar = \frac{1}{N} \to 0$, the spectral measure $\mu_\hbar$ admits an asymptotic expansion*

$$\mu_\hbar(f) \sim (2\pi\hbar)^{-r} \sum_{i=0}^{\infty} c_i(f)\hbar^i \tag{1.10}$$

with $r = \frac{1}{2}\dim \Sigma_{\text{red}}$ and

$$c_0(f) = \int_{\Sigma_{\text{red}}} (p_{\text{red}})^* f d\sigma, \tag{1.11}$$

σ being the symplectic volume form on Σ_{red}.

We will use this result to prove a slightly more general result. Suppose one has an action on X of an m-torus G, and that this action commutes with Π. Then if the operator P in (1.2) is G-invariant and α is an element of the weight lattice of G, the action of G on $H^2(X)$ preserves the space of functions in $H^2(X)$ which transform under this action by the weight $N\alpha$. Let us denote the space of these functions by $H^2(X)_{N\alpha}$ and by $P_{N\alpha}$ the restriction of P to this space. We will prove an analogue of Theorem 1.1 for the spectral measure of $P_{N\alpha}$: Since the action of G preserves Π, it preserves the microsupport Δ_Σ of Π and hence induces a Hamiltonian action on Σ with moment map

$$\phi : \Sigma \to \mathfrak{g}^*.$$

Assume now that α is a regular value of this moment map and that G acts freely on the level set $\phi^{-1}(\alpha)$ (these are the assumptions required for the symplectic reduction of Σ at α,

$$\Sigma_\alpha = \phi^{-1}(\alpha)/G, \tag{1.12}$$

to be well-defined). We will prove:

Theorem 1.2. *The spectral measure of $\Pi P_{N\alpha}\Pi$, $N = \frac{1}{\hbar}$, admits an asymptotic expansion*

$$(2\pi\hbar)^{-r}\sum_{i=0}^{\infty}\mu_{\alpha,i}\hbar^i \tag{1.13}$$

as $\hbar \to 0$, where $r = \frac{1}{2}\dim\Sigma_\alpha$ and

$$\mu_{\alpha,0}(f) = \int_{\Sigma_\alpha} p_\alpha^* f\, d\sigma, \tag{1.14}$$

σ being the symplectic volume form on Σ_α and p_α the "reduced" semi-classical symbol of P, i.e., the map $p_\alpha : \Sigma_\alpha \to \mathbb{R}$ is defined by

$$\gamma^* p_\alpha = p|_{\phi^{-1}(\alpha)}, \tag{1.15}$$

where γ is the projection of $\phi^{-1}(\alpha)$ onto Σ_α.

Remark 1.3. There are two interesting antecedents for the results we have just described: one is a theorem of Borthwick-Paul-Uribe having to do with Toeplitz operators on line bundles over projective varieties (see [BPU]), and the second an analogue of the theorem above for semi-classical pseudodifferential operators (see [GS] Chapter 12, Theorem 12.13.1.)

We conclude this introduction by giving a quick overview of the contents of Sections 2 through 8: In Section 2 we will recall how classical Szegö projectors and Toeplitz operators are defined and in Section 3 we will describe the symbol calculus of these operators. Then in Section 4 we will transition to the semi-classical setting and give a brief account of the "alternative approach" to defining Toeplitz operators that was described in Subsection 3.4 of [GUW], and use this approach to show that the $P_\hbar$'s above are operators of this type. Then in Section 5 we will prove Theorem 1.1 and in Section 6 deduce Theorem 1.2 from Theorem 1.1.

Finally in Sections 7 and 8 we will discuss some applications of Theorem 1.2 to inverse problems involving the "equivariant spectrum" of the Toeplitz operator (1.2), i.e., the spectral invariants of this operator associated with the asymptotic expansion

(1.13). More explicitly, in Section 7 we will show that for Toeplitz operators on the boundary of the unit ball in $\mathbb{C}^m$ the symbols of these operators are determined by the equivariant spectrum, or *e-spectrally determined* in abbreviation, and in Section 8 we will show by a reduction-in-stages argument that this result implies an analogous result for Toeplitz operators on arbitrary toric varieties.

A concluding remark: Though the results of this paper are officially theorems in semi-classical analysis, the theorems in Sections 7 and 8 can also be interpreted as illustrations of the "quantization commutes with reduction" principle for G-actions on symplectic manifolds, a principle which had its origins in the work of Bert Kostant and his collaborators, and for this reason we deem it an honor to be able to contribute this paper to the Kostant memorial volume.

2 The classical theory of Toeplitz operators

The Toeplitz operators that one studies in complex analysis are the operators that we alluded to in the introduction (i.e., Toeplitz operators on the boundary of a strictly pseudoconvex domain), and the "standard model" that one usually chooses for this class of operators is the algebra of Toeplitz operators on the unit ball B^m in $\mathbb{C}^m$. However, since we are concerned in this article not only with the Toeplitz operators that come up in complex analysis, but also with Toeplitz operators that come up in other contexts as well, we will choose a somewhat simpler "standard model". Namely we will choose for the manifold X in Section 1 the product space $\mathbb{R}^k \times \mathbb{T}^l$, where $\mathbb{T}^l$ is the standard l-dimensional torus

$$S^1 \times \cdots \times S^1 \qquad (l \text{ factors})$$

and $k+l=n$, and define our Hardy space $H^2(X)$ to be the space of L^2 solutions of the system of equations

$$\frac{1}{\sqrt{-1}}\left(\frac{\partial}{\partial y_j}+y_j|D_\theta|\right)f=0, \qquad j=1,\ldots,k, \tag{2.1}$$

where $|D_\theta|$ is the square root of the Laplace operator on $\mathbb{T}^l$ and $(y_1,\ldots,y_k)$ are cartesian coordinates in $\mathbb{R}^k$. For every weight $0 \neq \alpha$ in the weight lattice $\mathbb{Z}^l \subset \mathfrak{t}^*$ of $\mathbb{T}^l$, the function

$$f_\alpha(y,\theta)=e^{-\frac{|y|^2|\alpha|}{2}}\left(\frac{|\alpha|}{\pi}\right)^k e^{\sqrt{-1}\alpha(\theta)} \tag{2.2}$$

is a solution of (2.1) and these functions form an orthogonal basis of $H^2(X)$. In particular, the map

$$R\colon L^2(\mathbb{T}^l)\to L^2(X) \tag{2.3}$$

mapping $e^{\sqrt{-1}\alpha(\theta)}$ onto f_α is an orthonormal embedding of $L^2(\mathbb{T}^l)$ onto $L^2(X)$, and if we denote by Π the orthogonal projection

$$\Pi\colon L^2(X) \to H^2(X), \tag{2.4}$$

the map R satisfies the identities

$$R^t R = \mathrm{Id} \tag{2.5}$$

and

$$RR^t = \Pi. \tag{2.6}$$

Finally we note that if η and τ are dual variables to y and θ in the cotangent bundle of $\mathbb{R}^k \times \mathbb{T}^l$, then the symbol of the operators in (2.1) are

$$\frac{1}{\sqrt{-1}}\eta_j + y_j|\tau|, \quad j = 1,\ldots,k, \tag{2.7}$$

and hence the characteristic variety of the system of equations defining $H^2(X)$ is

$$y = \eta = 0. \tag{2.8}$$

In other words, the characteristic variety is the symplectic cone

$$\Sigma = \{(0,0)\} \times T^*\mathbb{T}^l \tag{2.9}$$

in the cotangent bundle $T^*(\mathbb{R}^k \times \mathbb{T}^l)$, and hence the microsupport of the projection Π can be viewed as the diagonal

$$\Delta_\Sigma = T^*\mathbb{T}^l \tag{2.10}$$

in the product $T^*\mathbb{T}^l \times T^*\mathbb{T}^l$.

Turning to the topic of this section: "Toeplitz operators". These are by definition operators on $H^2(X)$ of the form $\Pi P \Pi$ where P is a classical pseudodifferential operator on X. Alternatively they can be described as operators of the form

$$R\widetilde{P}R^t,$$

where $\widetilde{P}\colon C^\infty(\mathbb{T}^l) \to C^\infty(\mathbb{T}^l)$ is a classical pseudodifferential operator on $\mathbb{T}^l$ or as operators of the form $P\Pi$, where P is a classical pseudodifferential operator on X that commutes with Π. (For the equivalence of these three definitions, see [BG]. We note that by the third definition, these operators form an algebra, and that by the second definition, this algebra is isomorphic to the algebra of classical pseudodifferential operators on $\mathbb{T}^l$.)

With these operators as our "standard model", we will define Toeplitz operators in general to be operators which are microlocally isomorphic to the operators in this standard model. More explicitly, let X be an n-dimensional manifold and $\Sigma \subset$

$T^*X \setminus 0$ a symplectic cone. We will define a "Toeplitzification" of Σ to be a self-adjoint linear operator $\Pi\colon L^2(X) \to L^2(X)$ with the properties

(1) $\Pi^2 = \Pi$;
(2) The microsupport of Π is the diagonal Δ_Σ in $\Sigma \times \Sigma$;
(3) At every point $(x,\xi,x,\xi) \in \Delta_\Sigma$, Π is microlocally isomorphic to the "Π" in the canonical model above.

Moreover, if one has an S^1 action on X and the induced action on $L^2(X)$ commutes with Π, we will call this an *equivariant Toeplitz structure* if the isomorphism above conjugates this S^1 action to the action of a circle subgroup of $\mathbb{T}^l$ on the canonical model.

3 The symbol calculus for Toeplitz operators

The Toeplitz operators $\Pi P \Pi$ that we defined in Section 2 are "Fourier integral operators of Hermite type". The symbol calculus for general Hermite FIOs is fairly complicated. (In particular, it involves "symplectic spinors", objects which we will not attempt to define except to say that they are sections of a vector bundle over Σ whose fiber at $(x,\xi) \in \Sigma$ is an infinite-dimensional vector space.) Fortunately however, for Toeplitz operators there is an alternative symbol calculus that is much simpler and is based on the following result:

Theorem 3.1. *Let $P\colon C^\infty(X) \to C^\infty(X)$ be a classical pseudodifferential operator of order d and let $Q = \Pi P \Pi$. Then if $p\colon T^*X \setminus 0 \to \mathbb{C}$ is the leading symbol of P, its restriction*

$$p|_\Sigma = q \tag{3.1}$$

is an intrinsically defined symbol of Q, i.e., does not depend on the extension P of Q to $L^2(X)$. Moreover, if this symbol vanishes there exists a pseudodifferential operator $P'\colon C^\infty(X) \to C^\infty(X)$ of order $d-1$ with $Q = \Pi P' \Pi$.

Proof. It suffices to prove this for the "standard model" described in Section 2. However, for the standard model, the symbol of Q we have just described is the symbol of the classical pseudodifferential operator RQR^t and hence only depends on the restriction of p to Σ. □

4 The semi-classical theory of Toeplitz operators

The approach to this subject that we will describe below is motivated by the alternative approach to the theory of semi-classical pseudodifferential operators developed by T. Paul and A. Uribe in [PU]. To describe this approach, let Y be a manifold, let X be the product manifold $Y \times S^1$, and let

$$P\colon C^\infty(X) \to C^\infty(X) \tag{4.1}$$

be a classical pseudodifferential operator which is invariant with respect to the action of S^1 on $Y \times S^1$. Its symbol then has the form $p(y,\eta,\tau)$, where τ is the dual cotangent variable to the angle variable θ and η is the dual to y. Now let

$$L^2(X) = \bigoplus_N L^2(X)_N \tag{4.2}$$

be the orthogonal decomposition of $L^2(X)$ into the sum of the subspaces, $L^2(X)_N$, of functions in $L^2(X)$ which transform under the action of S^1 by the recipe

$$f\left(e^{\sqrt{-1}\theta} \cdot x\right) = e^{\sqrt{-1}N\theta} f(x). \tag{4.3}$$

Then the operator

$$P_\hbar = \hbar^d P|_{L^2(X)_N} \tag{4.4}$$

can be thought of as a semi-classical pseudodifferential operator of order 0. Moreover, via the identification

$$L^2(Y) \to L^2(X)_N \tag{4.5}$$

mapping $f(y)$ to $f(y)e^{\sqrt{-1}N\theta}$, it can be regarded as an operator on Y. (For instance if X is equipped with a Riemannian metric $\sum g_{ij}(y)dy_i \otimes dy_j + V(y)d\theta \otimes d\theta$, then for $d = 2$ the Laplace-Beltrami operator on X gets converted by this process into the Schrödinger operator

$$\hbar^2 \Delta_Y + V(y) \tag{4.6}$$

on Y.) Another key definition in this approach to the theory of semi-classical pseudodifferential operators is the notion of symbol for the operators above: The semi-classical symbol of (4.4) is the function

$$p(y,\eta,1), \tag{4.7}$$

where $p(y,\eta,\tau)$ is the symbol of the classical S^1-invariant pseudodifferential operator P.

Turning finally to the topic of this paper, this alternative approach to the semi-classical theory of pseudodifferential operators can also be applied to Toeplitz operators by mimicking the definition above of semi-classical pseudodifferential operators. Namely, if Q is the classical Toeplitz operator $\Pi P \Pi$, its semi-classical counterpart is the analogue

$$Q_\hbar = \hbar^d Q|_{H^2(X)_N}, \tag{4.8}$$

where $H^2(X)_N$ is the intersection of $H^2(X)$ with $L^2(X)_N$.

Finally we note that the theory above can be generalized slightly by allowing the manifold $X = Y \times S^1$ to be replaced by a manifold X on which one has a free action of the circle group S^1. In this case the semi-classical operators $P_\hbar$ and $Q_\hbar$ can no longer be thought of as operators on Y, but one can get around this problem either by thinking of them as operators on sections of $\mathbb{L}^N$, where $\mathbb{L}^N$ is the line bundle

$$X \times \mathbb{C}/S^1, \tag{4.9}$$

S^1 acting on $\mathbb{C}$ by multiplication by $e^{\sqrt{-1}N\theta}$, or by thinking of them in terms of local trivialization of $\mathbb{L}$ over open subsets of X. (In the next two sections of this paper we will for the most part adopt this second approach.)

5 The proof of Theorem 1.1

Let $\{\mathscr{U}_j, j = 1, \ldots, J\}$ be a covering of Σ by conic open subsets of $T^*X \setminus 0$ having the property that on $\mathscr{U}_j$ the Toeplitz projector Π is conjugated by an invertible S^1-equivariant Fourier integral operator to the Toeplitz projector in the canonical model described in Section 2. In addition let

$$\{P_j, j = 1, \ldots, J\}$$

be an S^1-equivariant microlocal partition of unity subordinate to this cover (i.e., each P_j is a classical zeroth order pseudodifferential operator which is S^1-invariant and has microsupport on $\mathscr{U}_j$; and the sum of these P_j's is microlocally equal to the identity operator on a conic open neighborhood of Σ.) Thus to compute the trace of $\Pi f(P_\hbar)\Pi$, it suffices to compute the traces of each of the summands in the sum

$$\sum_j \Pi P_j f(P_\hbar)\Pi; \tag{5.1}$$

and to compute the trace of $\Pi P_j f(P_\hbar)\Pi$ it suffices, by conjugation by an invertible S^1-equivariant Fourier integral operator, to compute the trace of the operators corresponding to $\Pi P_j f(P_\hbar)\Pi$ in the canonical model. However, in this canonical model, we can assume that the P_j's and $f(P_\hbar)$ commute with Π, and hence that this trace is identified with the trace of the semi-classical pseudodifferential operator

$$\widetilde{P}_j f(\widetilde{P}_\hbar), \tag{5.2}$$

where $\widetilde{P}_j = R^t P_j R$, and $\widetilde{P}_\hbar = R^t P_\hbar R$, R being the "Fourier integral operator of Hermite type" defined by (2.3). However, the $\widetilde{P}_j$'s are classical pseudodifferential operators and $\widetilde{P}_\hbar$ is a semi-classical pseudodifferential operator. Therefore since $\widetilde{P}_j$ is S^1-equivariant, $\widetilde{P}_j f(\widetilde{P}_\hbar)$ is a semi-classical pseudodifferential operator as well; so to compute the trace of the jth summand of (5.1) we are reduced to computing the trace of this pseudodifferential operator. Fortunately, however, the pseudodifferential version of the theorem we are trying to prove is well-known; i.e., one has an asymptotic expansion of this trace in the power of $\hbar$ which is identical with the expansions (1.10) and (1.11), except for the fact that Σ is now the cotangent bundle of $\mathbb{T}^l$ rather than the factor $T^*\mathbb{T}^l$ in the product $T^*\mathbb{R}^k \times T^*\mathbb{T}^l$ (and except for the fact that one has to insert a factor ρ_j in the integral (1.11), ρ_j being the symbol of P_j.) Thus, conjugating the operators $\Pi P_j f(P_\hbar)\Pi$ in the jth summand of (5.1) into the canonical model and summing over j, we get the asymptotic expansions (1.10) and (1.11).

Remark 5.1. For a proof of the pseudodifferential version of Theorem 1.1, see [GS], §12.12, page 330–331. Also, for a somewhat simpler proof of an enlightening special case of this theorem, namely when $P_\hbar$ is a Schrödinger operator $\hbar^2\Delta + V$, see [DGS] (and for some inverse spectral applications of this result see [GW]).

6 The proof of Theorem 1.2

We will briefly describe how the expansion (1.13) can be deduced from the expansion (1.10). Let $\mathfrak{g}$ be the Lie algebra of G. For each weight α of G, let

$$\mathfrak{g}_\alpha = \{v \in \mathfrak{g} : \alpha(v) = 0\}.$$

$\mathfrak{g}_\alpha$ is the Lie algebra of a codimension one subtorus G_α of G. Assuming that this subtorus acts freely on X, the quotient manifold

$$Y = X/G_\alpha$$

is well defined and the projection

$$\pi : X \to Y \tag{6.1}$$

is a smooth fibration. Let π^*T^*Y be the "horizontal subbundle" of T^*X with respect to this projection. Then one gets from (6.1) a fibration

$$\pi^*T^*Y \to T^*Y \tag{6.2}$$

mapping the intersection $\Sigma \cap \pi^*T^*Y$ onto a symplectic cone $\Sigma_{\alpha,0}$ in T^*Y. This cone can easily be seen to be the symplectic reduction of Σ with respect to the action of G_α on the zero level set of its G_α moment map.

Moreover, this reduced space has a residual action of the circle group G/G_α, and it is easily seen that the symplectic reduction of $\Sigma_{\alpha,0}$ with respect to the action of this circle group at the "one" level set of its moment map is Σ_α.

We will next describe the quantum analogue of this picture. Let $H^2(Y)$ be the space of G_α-invariant functions in $H^2(X)$. Then the orthogonal projection

$$\pi_Y : L^2(Y) \to H^2(Y) \tag{6.3}$$

is a Toeplitz projector with microsupport on the diagonal in $\Sigma_{\alpha,0} \times \Sigma_{\alpha,0}$. Moreover, there is a residual action of the circle group G/G_α on this space and the corresponding decomposition into weight spaces

$$H^2(Y) = \bigoplus_N H^2(Y)_N \tag{6.4}$$

coincides with the sum

$$\bigoplus_N H^2(X)_{N\alpha}. \tag{6.5}$$

As a consequence of these observations, the asymptotic formula (1.13) can easily be deduced from the asymptotic formula (1.10) with X replaced by Y.

7 The basic example

As we mentioned in the introduction to Section 2 the basic example of a Szegö projector for complex analysts is the example associated with the boundary S^{2m-1} of the unit ball B^m in $\mathbb{C}^m$; i.e., in this example, $H^2(S^{2m-1})$ is the subspace of $L^2(S^{2m-1})$ spanned by the functions

$$f|_{S^{2m-1}}, \quad f \in \mathscr{O}(B^m) \cap C^\infty(\overline{B}^m). \tag{7.1}$$

The symplectic cone Σ in T^*S^{2m-1} can be identified with $\mathbb{C}^m \setminus 0$, the symplectic form on this cone being the standard Kähler form

$$\omega = \sqrt{-1} \sum_i dz_i \wedge d\bar{z}_i. \tag{7.2}$$

Moreover, ω is preserved by the action on $\mathbb{C}^m \setminus 0$ of the m-torus

$$\mathbb{T}^m = S^1 \times \cdots \times S^1 \tag{7.3}$$

and this action is a Hamiltonian action with moment map

$$\phi(z_1, \ldots, z_m) = (|z_1|^2, \ldots, |z_m|^2). \tag{7.4}$$

Hence one gets other interesting examples by reduction by subgroups G of $\mathbb{T}^m$. We will focus in this section on the simplest of these examples, reduction by $\mathbb{T}^m$ itself, and in this case we will show that the asymptotic formula (1.13) in Theorem 1.2 is essentially a formula for the eigenvalues of the operator $Q_\hbar = \Pi P_\hbar \Pi$.

The moment map associated with the $\mathbb{T}^m$ action on $\mathbb{C}^m$ is the mapping (7.4). Therefore, for $\alpha = (\alpha_1, \ldots, \alpha_m) \in \mathbb{Z}_+^m$, $\phi^{-1}(z) = \alpha$ if and only if $|z_i|^2 = \alpha_i$. Hence if the α_i's are all non-zero, $\mathbb{T}^m$ acts freely on the space $\phi^{-1}(\alpha)$ and the reduced space

$$\phi^{-1}(\alpha)/\mathbb{T}^m \tag{7.5}$$

consists of a single point A_α. Moreover, the quantum analogue $H^2(X)_\alpha$ of this reduced space is the set of holomorphic functions $f \in \mathscr{O}(B^m)$ which transform under the action of $\mathbb{T}^m$ by the weight α, i.e., satisfy

$$f(e^{\sqrt{-1}\theta_1}z_1,\dots,e^{\sqrt{-1}\theta_m}z_m) = e^{\sqrt{-1}\alpha(\theta)}f(z). \tag{7.6}$$

This is just the one-dimensional subspace of $\mathscr{O}(B^m)$ spanned by $z_1^{\alpha_1}\cdots z_m^{\alpha_m}$. However, these are exactly the eigenspaces of $Q_\hbar$; so the asymptotic formula (1.13) in Theorem 1.2 is just an asymptotic formula for the equivariant eigenvalues of $Q_\hbar$.

As one application of this formula, we will prove the following inverse spectral result.

Theorem 7.1. *The symbol of the Toeplitz operator Q is e-spectrally determined.*

Proof. The leading term in the asymptotic expansion (1.13) determines the symbol of $Q_\hbar$ on the set (7.5); however, this set consists of the single point A_α; so it in effect determines the symbol of $Q_\hbar$ at A_α and the symbol p of P on the set $\Phi^{-1}(\alpha)$. Moreover, since P is a classical pseudodifferential operator on S^{2m-1} of order zero, p is a homogeneous function of order zero on the punctured cotangent bundle $T^*S^{2m-1}\setminus 0$, and its restriction to the symplectic cone $\Sigma = \mathbb{C}^m\setminus 0$ is a homogeneous function of degree zero on Σ. Thus if z is an element of $\phi^{-1}(\alpha)$, i.e.,

$$(|z_1|^2,\dots,|z_m|^2) = \alpha,$$

then p takes the same value at z and at tz, so p is spectrally determined on the set of z's satisfying

$$(|z_1|^2,\dots,|z_m|^2) = t\alpha. \tag{7.7}$$

Since this contains the set of points $(z_1,\dots,z_m)\in\mathbb{C}^m\setminus 0$ for which the $|z_i|$'s are rational numbers it is a dense subset of $\mathbb{C}^m\setminus 0$, so it follows that the symbol of $Q = \Pi P \Pi$ is e-spectrally determined. □

Remark 7.2. In some interesting cases this becomes an e-spectral result for Q itself. For instance this is the case for Toeplitz operators of the form $\Pi M_F \Pi$, where $M_F : L^2(S^{2m-1}) \to L^2(S^{2m-1})$ is the operator "multiplication by a function $F \in C^\infty(S^{2m-1})$".

8 Toric varieties

By a reduction-in-stages argument we will show in this section how to extract from the example above inverse spectral results for a much larger class of examples: toric varieties. To describe these examples, let G be a subtorus of $\mathbb{T}^m$ and let

(8.1) $$\phi\colon \mathbb{C}^m \setminus 0 \to \mathfrak{g}^*$$

be the moment map associated with the action of G on $\mathbb{C}^m \setminus 0$. Then if α is a regular value of this map and G acts freely on $\phi^{-1}(\alpha)$, the symplectic reduction

(8.2) $$M = \phi^{-1}(\alpha)/G$$

is well defined and inherits from $\mathbb{C}^m \setminus 0$ a Kähler structure. Moreover, from the action of $\mathbb{T}^m$ on $\mathbb{C}^m \setminus 0$, M inherits a Hamiltonian action of the quotient torus $K = \mathbb{T}^m/G$.

Lemma 8.1. *The K-action on M is a toric action.*

Proof. Since the Lie algebra $\mathfrak{k}$ of K is a quotient algebra of the Lie algebra $\mathfrak{t}_m$ of $\mathbb{T}^m$ one has an inclusion of $\mathfrak{k}^*$ into $\mathfrak{t}_m^*$, and if $\beta \in \mathfrak{k}^*$ is a regular value of the K moment map of M into $\mathfrak{k}^*$, its image δ in $\mathfrak{t}_m^*$ is a regular value of the $\mathbb{T}^m$ moment map of $\mathbb{C}^m \setminus 0$ into $\mathfrak{t}_m^*$. Moreover, the symplectic reduction of M with respect to β coincides with the symplectic reduction of $\mathbb{C}^m \setminus 0$ with respect to δ. However, since the action of $\mathbb{T}^m$ on $\mathbb{C}^m \setminus 0$ is toric, the second of these reduced spaces is just a point and hence so is the first. □

Thus, by a theorem of Delzant, the moment map

(8.3) $$\psi\colon M \to \mathfrak{k}^*$$

defined by this action has the property that if β is a regular value of ψ, the reduced space $\phi^{-1}(\beta)/K$ is just a point A_0. In fact the image of the map (8.3) is the moment polytope Δ_M of M and $\beta \in \Delta_M$ is a regular value of ψ if and only if β is in the interior $\mathring{\Delta}_M$ of Δ_M. Thus if M_0 is the set of points in M at which K acts freely, one gets from (8.3) a bijective map

(8.4) $$M_0/K \to \mathring{\Delta}_M.$$

Hence one can identify the points A_0 above with elements of $\mathring{\Delta}_M$.

Suppose now that α in definition (8.2) is an element of the weight lattice of G, and let $\mathfrak{g}_\alpha$ be the annihilator of α in $\mathfrak{g}$. Then $\mathfrak{g}_\alpha$ is the Lie algebra of a subtorus G_α of G. Now let

(8.5) $$\tilde{\varphi}_\alpha\colon T^*S^{2m-1} \to \mathfrak{g}_\alpha^*$$

be the moment map associated to the lifted Hamiltonian action of G_α on T^*S^{2m-1}. Then its restriction

(8.6) $$\varphi_\alpha\colon \mathbb{C}^m \setminus 0 \to \mathfrak{g}_\alpha^*$$

is, by definition, the moment map associated with the action of G_α on this cone. The symplectic reduction $\tilde{\varphi}_\alpha^{-1}(0)/G_\alpha$ is the cotangent space of the quotient manifold

$$Y = S^{2m-1}/G_\alpha$$

and the reduced symplectic cone

$$C = \varphi_\alpha^{-1}(0)/G_\alpha \tag{8.7}$$

sits inside T^*Y as a symplectic sub-cone. One has a residual action of the group $S^1 = G/G_\alpha$ on this cone, and if

$$\Phi\colon C \to \mathbb{R} \tag{8.8}$$

is the moment map associated with this action, the symplectic reduction

$$\Phi^{-1}(1)/S^1 \tag{8.9}$$

is another description of M. More generally, the symplectic reduction

$$\Phi^{-1}(N)/S^1 \tag{8.10}$$

is the manifold M but now with its symplectic form ω replaced by $N\omega$.

Now let $H^2(S^{2m-1})$ be the Hardy space (7.1) and let $H^2(Y)$ be the space of G_α-invariant elements in $H^2(S^{2m-1})$. Let

$$H^2(Y) = \bigoplus_N H^2(Y)_N \tag{8.11}$$

be the decomposition of this space into weight spaces with respect to the residual action of S^1 on $H^2(Y)$. For every N, $H^2(Y)_N$ can be regarded as a quantization of the symplectic manifold $(M, N\omega)$, and the quantum analogue of the assertion that the action of K on M is *toric* is as follows:

Lemma 8.2. *The action of K on each summand of* (8.11) *is* multiplicity free, *that is, the weight spaces of the representation of the torus K on each $H^2(Y)_N$ are one-dimensional.*

Proof. By definition K is the quotient group $\mathbb{T}^m/G$, so one has a natural inclusion map $\iota\colon \mathbb{Z}_K \hookrightarrow \mathbb{Z}^m$ of the weight lattice of K into the weight lattice of $\mathbb{T}^m$. Moreover, for $\beta \in \mathbb{Z}_K$ and $\delta = \iota(\beta)$ the weight space $H^2(S^{2m-1})_\delta$ coincides with the weight space $H^2(Y)_\beta$, and we showed in §7 that the first of these spaces is one-dimensional. Hence so is the second. □

As a consequence we can deduce:

Theorem 8.3. *Let Π be the orthogonal projection of $L^2(Y)$ onto $H^2(Y)$, let $P\colon C^\infty(Y) \to C^\infty(Y)$ be a K-invariant zeroth order pseudodifferential operator, and let $Q\colon H^2(Y) \to H^2(Y)$ be the associated Toeplitz operator $Q = \Pi P \Pi$. Then the symbol of Q is e-spectrally determined.*

Proof. Since the action of K on the symplectic manifold $(M, N\omega)$ is a toric action, the symplectic manifolds (1.12) in Theorem 1.2 are just points. Moreover, these points can be identified with the lattice points of the moment polytope $N\Delta_M$. Since the action of K on $H^2(Y)_N$ is multiplicity free, the asymptotic expansions (1.13) are just asymptotic expansions for the eigenvalues of the semi-classical Toeplitz operator $Q_\hbar$, $\hbar = 1/N$. Since the symplectic quotients (1.12) are just points the leading terms (1.14) in the expansions (1.13) determine the restrictions of p to the pre-image with respect to the moment map of the lattice points in $N\mathring{\Delta}_M$. However, the union of these $N\mathring{\Delta}_M$'s is the entire weight lattice $\mathbb{Z}_K$ of K. The symbol p is a homogeneous function of order zero, so it is e-spectrally determined on the preimage of $\mathbb{R}^+\mathbb{Z}_K$ and, since $\mathbb{R}^+\mathbb{Z}_K$ is dense in $\mathfrak{k}^*$, p is therefore spectrally determined. □

References

[BPU] D. Borthwick, T. Paul and A. Uribe, Legendre distributions with applications to relative Poincaré series, *Invent. Math.* 122 (1993), 359–402.

[BG] L. Boutet de Monvel and V. Guillemin, *The Spectral Theory of Toeplitz Operators*, Annals of Mathematics Studies, vol. 99. Princeton University Press, Princeton (1981).

[DGS] E. Dryden, V. Guillemin and R. Sena Dias, Semi-classical weights and equivariant spectral theory, *Advances in Mathematics* 299 (2014), 202–246.

[GS] V. Guillemin and S. Sternberg, *Semiclassical Analysis*, International Press, Boston, MA, 2013.

[GUW] V. Guillemin, A. Uribe and Z. Wang, Semiclassical states associated to isotropic submanifolds of phase space. *Lett Math Phys* 106 (2016), 1695–1728.

[GW] V. Guillemin and Z. Wang, The generalized Legendre transform and its applications to inverse spectral problems. *Inverse Problems* 32 (2016), 015001 (22pp).

[PU] T. Paul and A. Uribe, A construction of quasimodes using coherent states. *Annales de l'I.H.P., Section A* 59 (1993), 357–381.

Dual Kashiwara Functions for the $B(\infty)$ Crystal

Anthony Joseph

In memory of Bertram Kostant
We heard the words that will generations inspire

Abstract Let $\mathfrak{g}$ be a Kac-Moody algebra. For each sequence J of reduced Weyl group elements, Kashiwara constructed a crystal B_J which as a set identifies with the free $\mathbb{N}$ module of rank $|J|$ and showed that it contains a "highest weight" subcrystal $B_J(\infty)$ having some remarkable combinatorial properties. The goal of the present work is to exhibit $B_J(\infty)$ as an explicit polyhedral subset of B_J by constructing for each simple root α, a set of dual Kashiwara functions which are linear functions on B_J and whose maximum restricted to $B_J(\infty)$ determines the dual Kashiwara parameter $\varepsilon^{\star}_{\alpha}$. Up to a natural conjecture concerning identities in the Demazure modules, it is shown that these functions are given through rather explicitly determined "trails" with respect to J in the fundamental module of lowest weight $-\varpi^{\vee}_{\alpha}$ for the Langlands dual of $\mathfrak{g}$. The proof uses Kashiwara duality extended to the non-symmetrizable case and the theory of S-graphs developed by the author.

Keywords: Crystals • Kac-Moody algebras

Mathematics Subject Classification (2010): 17B35

1 Preamble

Formally this paper is a sequel to [7], [10], [8], [9], though in fact it had been written previous to these works and served as their motivation. Partly as a consequence, the present work is practically self-contained and needs little knowledge of the above papers which serve mainly to attest to its interest.

Let $\mathfrak{g}$ be a symmetrizable Kac-Moody algebra with I, an index set of its simple roots. Recall that the Kashiwara $B(\infty)$ crystal can be viewed as giving bases for all

Donald Frey Professional Chair, Department of Mathematics, The Weizmann Institute of Science, Rehovot, 76100, Israel, e-mail: anthony.joseph@weizmann.ac.il

V. G. Kac, V. L. Popov (eds.), *Lie Groups, Geometry, and Representation Theory*, Progress in Mathematics 326, https://doi.org/10.1007/978-3-030-02191-7_8

the simple integrable highest weight $\mathfrak{g}$ modules and as such has an intricate combinatorial structure which would be valuable to understand.

A construction of $B(\infty)$ was given by Kashiwara [13]. It involves choosing a countable sequence J of elements of I in which every element occurs sufficiently many times. Then a crystal B_J is defined. It is the tensor product of elementary crystals following the choice of J. It identifies with $\mathbb{N}^{|J|}$ as a set. Then $B_J(\infty)$ is defined as the subcrystal of B_J generated by the element b_∞ in which every entry is zero. After Kashiwara, as a *crystal* $B_J(\infty)$ is independent of J. This was generalized to all Kac-Moody algebras in [6, Sect. 2], through the Littelmann path model. As part of this construction and via results of Littelmann-Mathieu-Kumar, one obtains for all dominant integral weights λ a combinatorially determined subcrystal $B(\lambda)$ of $B(\infty)$ which has the same formal character as the largest integrable quotient of the corresponding Verma module, for details see [6, Sect. 3]. (In the symmetrizable case the latter is known to be its unique simple quotient.)

Our goal was to describe $B_J(\infty)$ as a subset B_J. In particular to show that it is a polyhedral subset and to determine the defining linear inequalities. For this it suffices to show that the "dual Kashiwara parameters" $\varepsilon_t^\star : t \in I$ are given as the maxima of certain linear functions on B_J which we called dual Kashiwara functions. Underlying this study is the belief that the set Z_t of dual Kashiwara functions defined by $t \in I$ must possess some remarkable combinatorial properties.

In this paper it is explained that Z_t must be almost invariant with respect to the action of the Kashiwara operators $\{e_s, f_s : s \in I\}$.

Invariance of Z_t under a fixed pair $e_s, f_s : s \in I$, motivated a Preparation Theorem [7, Thm. 8.6] in which the notion of an S-set (of type s) was introduced. This itself was new and gave rise to a new presentation of the Catalan numbers [10], a convexity result [8] and a surprising connection with identities in Demazure modules [9]. Invariance under the entire set of Kashiwara operators leads to the notion of a t-semi-invariant set to be constructed inductively on $j \in J$ from a "driving function" z_t^1, associated to a given element $t \in I$ using S-sets of every type. Here we describe how this driving function should be chosen and why the resulting t-semi-invariant set Z_t gives the dual Kashiwara functions with their required properties (Prop. 5.1).

In order for our construction of a t-semi-invariant set to succeed, we need to have some knowledge of the functions our inductive construction produces. This problem was studied in [9]. For this we followed the construction given by Berenstein and Zelevinsky [2], when $\mathfrak{g}$ is finite dimensional, of $\mathbf{i}$-trails in a fundamental module for the Langlands dual of $\mathfrak{g}$. We call these simply trails. In the work of [2], these were not defined combinatorially and particularly difficult to compute. Under the hypothesis of there being no "false trails", we gave a precise combinatorial recipe for the set of all trails [9]. Here we note that the functions they define on B_J gives the t-semi-invariant set, again under the hypothesis that there are no false trails.

The construction of all trails is seemingly a question of immense difficulty and certainly more difficult than establishing the "Demazure property," blithely assumed by Demazure [3] in his work on Schubert varieties and not settled till some twenty years later first by Andersen [1] and then more concretely (and over $\mathbb{Z}$) through the work of Kashiwara [13].

Acknowledgements. I would like to thank Shmuel Zelikson for the many ideas he contributed to this work without which this paper could not have come into existence.

2 Notation and preliminaries

2.1 Let $\mathfrak{g}$ be a Kac-Moody Lie algebra of rank ℓ. It is defined by a Cartan subalgebra $\mathfrak{h}$, a set $\pi = \{\alpha_i\}_{i=1}^{\ell} \subset \mathfrak{h}^*$ (resp. $\pi^\vee = \{\alpha_i^\vee\}_{i=1}^{\ell} \subset \mathfrak{h}$) of simple roots (resp. coroots) and their scalar products $\alpha_i^\vee(\alpha_j)$ which form the Cartan matrix $\mathbf{C}$ of $\mathfrak{g}$. Here we recall that the diagonal entries of $\mathbf{C}$ are all equal to 2, whilst its off-diagonal entries are non-positive integers satisfying $\alpha_i^\vee(\alpha_j) \neq 0 \Leftrightarrow \alpha_j^\vee(\alpha_i) \neq 0$. We set $I = \{1,2,\ldots,\ell\}$. Then the Weyl group W is generated by the simple reflections $\{s_i\}_{i\in I}$, defined as linear transformations on $\mathfrak{h}^*$ given by $s_i\lambda = \lambda - \alpha_i^\vee(\lambda)\alpha_i$. Given $w \in W$ let $\ell(w)$ denote its reduced length. Set $P := \{\lambda \in \mathfrak{h}^* | \alpha_i^\vee(\lambda) \in \mathbb{Z}\}$. Obviously $\mathbb{Z}\pi \subset P$.

2.2 The notion of a crystal was introduced by Kashiwara [12], inspired by the work of Lusztig [16] by taking a $q \to 0$ limit for highest weight modules over the quantized enveloping algebra $U_q(\mathfrak{g})$. For the latter to be defined, it was necessary to assume that $\mathbf{C}$ is symmetrizable. However the Littelmann path model [14] only requires the Bruhat order on the Weyl group to define the corresponding "highest weight" crystals, and moreover this construction is purely combinatorial. The Kashiwara $B(\infty)$ crystal is then realized as a limit of highest weight crystals as the highest weight becomes infinite in the precise sense defined in [6, 2.5.6].

2.3 The above construction of $B(\infty)$ for a given choice of the Cartan matrix $\mathbf{C}$ was explained in [6], as well as the rather more delicate task of constructing an involution $\star$ on $B(\infty)$. This again was purely combinatorial and extended the Kashiwara involution (depending on the $q \to 0$ limit). For our own convenience, we shall use [6] as a basic reference, though the results in [12] would suffice for the symmetrizable case. Just for the moment it will not be necessary to recall all the properties of crystals, it being only important for the moment to understand that one can form their tensor product. This enabled Kashiwara to present $B(\infty)$ as a tensor product of very simple "elementary" crystals $B_i : i \in I$ based on $\mathfrak{sl}(2)$ and whose definition can be found in [6, 2.4.1]. As a set each B_i identifies with $\mathbb{N}$, the elements of which we refer to as entries. This presentation is the combinatorial analogue of the Bott-Samelson desingularisation of the flag and Schubert varieties and is extremely deep. It realises the consequence of Bott's dream of having a large torus acting on a simple highest weight module, so that weight spaces become multiplicity-free (see [5, 6.5.6] for further details).

2.4 A crystal B is not merely a set but admits an action of Kashiwara operators $e_i, f_i : i \in I$, which are essentially inverse to one another. [To be precise we need to introduce the notion of a zero element 0 (distinct from the zero entry described above). Then $e_i, f_i : B \to B \sqcup \{0\}$ satisfy the following rule. If $b, e_i b$ (resp. $b', f_i b'$) are both different from 0, then $f_i e_i b = b$ (resp. $e_i f_i b' = b'$).] This will be made more precise in 3.2 for the crystals defined below.

Let $J = (\dots, i_j, i_{j-1}, \dots, i_1)$ be a sequence of elements of I in which each element of I appears infinitely many times. One may define [6, 2.4.2] a crystal B_J which as a set is the Cartesian product $\dots \times B_{i_j} \times B_{i_{j-1}} \times \dots \times B_{i_1}$, in which all but finitely many entries are zero. Let b_∞ be the element of B_J in which every entry is zero. Then $B_J(\infty)$ is defined to be the subcrystal generated by b_∞. In this one may assume without loss of generality that for all $j \in \mathbb{N}^+$ the product of simple reflections $s_{i_j} s_{i_{j-1}} \dots s_{j_1}$ is a reduced decomposition of some Weyl group element. Indeed the rules given in 3.2 imply that neither e_i nor f_i would enter to the immediate right of a place labeled by the same element of I as its predecessor, Added to this, the Kashiwara operators satisfy the braid relations on $B_J(\infty)$. The first statement is easy to check, the second follows from the Littelmann path model. For the symmetrizable case, it had also been a result of Kashiwara. For details see [6, 3.3]. Actually one may also consider the present problem without this restriction on J, but then it is less clear that the proposed solution is a valid one.

2.5 As a crystal $B_J(\infty)$ is independent of J and the resulting crystal is denoted by $B(\infty)$. In the Kashiwara theory, $B(\infty)$ had already been defined as the crystal [12] associated to a Verma module for $U_q(\mathfrak{g})$. Then Kashiwara in [13, 2.2] showed that it could be described in the above form through an embedding theorem $B(\infty) \hookrightarrow B(\infty) \otimes B_i : i \in I$.

In [6, 2.5] we constructed $B(\infty)$ as the limit of "highest weight" crystals obtained from the Littelmann path model. Then the embedding theorem became a consequence of the Littelmann tensor product decomposition. As such, it does not require $\mathbf{C}$ to be symmetrizable.

Our goal is to compute $B_J(\infty)$ as a subset of $\mathbb{N}^{|J|}$. The result is very much dependent on the choice of J.

2.6 In the case of B_J the action of the e_i, f_i is given by a set of Kashiwara functions $r_i^k : i \in I, k \in \mathbb{N}^+$ to be defined in 4.1 below. They are notably *linear* in the entries of B_J. Thus, it is hence natural to conjecture that $B_J(\infty)$ is a polyhedral subset of B_J that is, given by a set of *linear* inequalities. Indeed this was shown in type A by Gleizer and Postnikov [4] and by Berenstein and Zelevinsky [2] whenever W is finite. The former used some special properties of type A, whilst the latter used the unique longest element w_π of W and notably the fact that for each $i \in I$, the longest element admits a reduced decomposition starting on the right from s_i.

2.7 In addition to the Kashiwara operators mentioned above, a crystal B admits "evaluation maps" $\varepsilon_i, \varphi_i : B \to \mathbb{Z} \cup \{-\infty\}$ and a weight map $\mathrm{wt} : B \to P$, the latter satisfying

$$\mathrm{wt}(e_i b) = \mathrm{wt}(b) + \alpha_i, \text{ if both } b, e_i b \neq 0,$$

with a similar rule for f_i which can be deduced from 2.6.

The evaluation and weight maps are related by the rule

$$\varphi_i(b) = \varepsilon_i(b) + \alpha_i^\vee(\mathrm{wt}\, b). \tag{1}$$

The evaluation maps satisfy

$$\varphi_i(e_i b) = \varphi_i(b) + 1, \text{if both } b, e_i b \neq 0, \varphi_i(f_i b) = \varphi_i(b) - 1, \text{if both } b, f_i b \neq 0, \tag{2}$$

with a similar rule for ε_i which can be deduced from (1).

These additional maps are necessary to describe the tensor product of a pair of crystals.

A crystal is said to be upper normal if for all $b \in B$, $i \in I$, one has $\varepsilon_i(b) = \max\{j \mid e_i^j b \neq 0\}$. A rather non-trivial fact is that $B(\infty)$ is upper normal [6, Prop. 2.5.6], a result of Kashiwara in the symmetrizable case.

In view of this, we shall refer to ε_i as the Kashiwara parameter of type $i \in I$.

In the special case of B_J one may remark that $\varepsilon_i(b) = \max_k r_i^k(b)$, for all $b \in B_J, i \in I$.

2.8 Our approach to showing that $B_J(\infty)$ is a polyhedral subset of B_J is to use the Kashiwara duality $\star$ operation on $B(\infty)$. It was obtained by Kashiwara [12] by taking a $q \to 0$ limit starting from a rather trivial antiautomorphism of the "positive root" subalgebra of $U_q(\mathfrak{g})$. Yet it is highly non-trivial and only obtained for arbitrary $\mathbf{C}$ after some efforts [6, 2.5.11-2.5.25].

It leads to dual Kashiwara operators, and notably for all $i \in I$, a dual Kashiwara parameter $\varepsilon_i^\star$ of type i.

The goal of this paper is to show that there exist "dual Kashiwara functions" $z_i^k : i, k \in \mathbb{N}^+$ defined on B_J, which are *linear* on B_J and satisfy

$$\varepsilon_i^\star(b) = \max_k z_i^k(b), \forall b \in B_J(\infty), i \in I. \tag{3}$$

Then it is rather easy to deduce that $B_J(\infty)$ is polyhedral through the upper normality of $B(\infty)$; see Section 7.

To allay possible confusion on the part of the reader, we emphasize that in presenting $B(\infty)$ as $B_J(\infty)$, we are building the latter using the Kashiwara operators $e_i, f_i : i \in I$ and then computing $\varepsilon_i^\star(b) : b \in B_J(\infty)$ using $\{e_i^\star\}_{i\in I}$ together with upper normality. Had we already used the starred operators to parametrize $B(\infty)$, then linearity would be immediate. Passing between different parametrizations is already complicated when W is finite and much worse in general. In particular it gives no hint as to whether linearity is preserved. (Indeed it does not even make sense to talk of $\star$ being a linear operator on $B(\infty)$, as the latter is not even a free $\mathbb{N}$ module. It would have made sense to talk of $\star$ as being a linear operator on B_J had it been defined on this set.) Again whereas the dual Kashiwara parameter $\varepsilon_i^\star(b) : b \in B_J(\infty)$ is well defined, the dual Kashiwara functions given by the right-hand side of (3) are not uniquely determined. Notably outside of type A, they are over-determined by trails (see the example coming from Table 2).

2.9 The construction of (the) dual Kashiwara functions is by no means obvious. Yet a basic fact, again due to Kashiwara [13, Cor. 2.2.2] in the symmetrizable case and extended in [6, 2.5.15] to the general case, is that the Kashiwara and dual Kashiwara operators almost commute. Consequently $\varepsilon_i^\star(f_j b) = \varepsilon_i^\star(b)$ if $i \neq j$ and this is almost true if $i = j$. It cannot be completely true, otherwise $\varepsilon_i^\star$ would vanish on $B(\infty)$. This slight lack of invariance is achieved for each $i \in I$ through an initial "driving function". It is denoted as z_i^1 and will be given by a simple explicit expression 4.7. Then for each step in a reduced decomposition, one must produce additional functions expressing the required invariance. These additional functions then provide the driving functions for the next induction step.

As noted in [7, Acknowledgements], Zelikson pointed out to me (already four years ago) that the wiring diagrams of Gleizer-Postnikov [4] can be used to obtain, rather easily and explicitly, the dual Kashiwara functions in type A. Here this result is stated following the captions to Figures 1,2, with the proof indicated. Actually the present theory can be easily used to compute Z_t (see 5.1) whenever the corresponding fundamental coweight $\varpi_t^\vee$ is minuscule. (This result will be presented elsewhere; but apart from being much simpler it does not add anything new to the result of Berenstein and Zelevinsky [2, Thm. 3.9], noted above.) In [15], Littelmann also discusses the structure of $B(\infty)$. In this he cites a result of Berenstein and Zelevinsky (see [15, Thm. 1.7]) but this was just before [2], so he was unaware of their precise results. In any case there is practically no intersection of his results with those of the present paper, except as they pertain to the consequences of [2]. In type A, for the particular words adapted to quivers in type, S. Zelikson [18] has noted that the set of dual Kashiwara functions has a rather special structure. On its generalization to finite type, we are preparing a joint paper.

3 The Kashiwara functions

3.1 For all $i \in I$, let B_i denote the ith elementary crystal (introduced by Kashiwara) and defined for example in [6, 2.4.1]. The elements of B_i are enumerated in the form $b_i(-m) : m \in \mathbb{N}$. Then $\varepsilon_i(b_i(-m)) = m$, $e_i b_i(0) = 0$, $e_i b_i(-m) = b_i(-m+1) : m > 0$, $f_i b_i(-m) = b_i(-m-1)$. In particular B_i is upper normal with respect to the ith index.

A typical element $b \in B_J$ can be written in the form

$$\ldots \otimes b_{i_n}(-m_n) \otimes b_{i_{n-1}}(-m_{n-1}) \otimes \ldots \otimes b_{i_1}(-m_1),$$

with $b_{i_j}(-m_j) \in B_{i_j} : m_j \in \mathbb{N}$ in which all but finitely many m_j are equal to zero. We may also represent b in the form $(\dots, m_n, m_{n-1}, \dots, m_1)$, calling m_j its jth entry.

We may also consider the $\{m_j\}_{j \in J}$ as coordinate functions on B_J, and then define a locally finite function on B_J to be a finite sum of the $\{m_j\}_j \in J$.

It is convenient to represent the subscript $j \in J$ by the pair $(s,k) \in I \times \mathbb{N}^+$, where for a fixed $s \in I$, we let $k \in \mathbb{N}^+$ denote the number of times $i_j = s$ counting from the right. For this we set $m_s^k = m_j$.

Formally we have a map $\gamma : J \to I$ defined by $\gamma(j) = i_j$. We set $n(j) := |\{i \leq j \mid \gamma(i) = \gamma(j)\}|$. This gives a map $\Gamma : \mathbb{N}^+ \to I \times \mathbb{N}^+$ defined by $\Gamma(j) = (\gamma(j), n(j))$. Then $m_{\gamma(j)}^{n(j)} = m_j$.

Observe that the natural linear order on $J \subset \mathbb{N}^+$ induces a linear order on its image under Γ. Then identifying J with its image, we may write $j > \Gamma(j')$ to mean $j > j'$.

In this notation the Kashiwara function described in [6, 2.3.2, 2.4.1] can be written as

$$r_s^k(b) = m_s^k + \sum_{j \in J | j > (s,k)} \alpha_s^\vee(\alpha_{i_j}) m_j, \forall s \in I, k \in \mathbb{N}^+. \tag{4}$$

It is a linear function on B_J.

We call r_s^k the kth Kashiwara function of type $s \in I$. The Kashiwara functions are not locally finite, but the sum in (4) is finite since all but finitely many m_j are equal to zero.

When we consider m_j as the jth co-ordinate function on B_J, the expression m_j in (4) should be written as $m_j(b)$. Alternatively we may replace $r_s^k(b)$ by r_s^k on the left-hand side of (4).

Notice that the difference of successive Kashiwara functions pertaining to $s \in I$ takes the form

$$r_s^k - r_s^{k+1} = m_s^k + m_s^{k+1} + \sum_{j \in J | (s,k+1) > j > (s,k)} \alpha_s^\vee(\alpha_{i_j}) m_j, \tag{5}$$

since $\alpha_s^\vee(\alpha_s) = 2$. Moreover the sum M_k on the right-hand side evaluated at $b \in B_J$ above is an integer ≤ 0. This was the property noted in [7, 4.2]. This is not specifically used; but it being $\leq$ is one reason why the problem presently being studied is so difficult.

One may remark that the above successive difference of Kashiwara functions is not only linear, but also locally finite.

3.2 The Kashiwara functions determine the crystal structure of B_J by the following rules noted in [5, 5.2.5]. First, recall (2.7) that

$$\varepsilon_i(b) = \max_k \{r_i^k(b)\}. \tag{6}$$

Second, let $L_i(b)$ (resp. $R_i(b)$) be the leftmost (resp. rightmost) value of k such that $r_i^k(b) = \varepsilon_i(b)$. Then $e_i b$ (resp. $f_i b$) is obtained from b by just decreasing (resp.

increasing) $m_i^k(b)$ by 1 where $k = L_i(b)$ (resp. $k = R_i(b)$). For this we say that e_i (resp. f_i) enters b at the $L_i(b)^{th}$ (resp. $R_i(b)^{th}$) place.

From the above one easily checks the following fact.

(i) If e_i (resp. f_i) enters b at the kth place, then e_i (resp. f_i) enters e_ib (resp. f_ib) at the k'^{th} place with $k' \geq k$ (resp. $k' \leq k$).

We conclude from (i) that $e_i^kb = 0$, for $k \in \mathbb{N}$ sufficiently large. By contrast

(ii) For $k \in \mathbb{N}$ sufficiently large, f_i enters f_i^kb at the first place j of type i, that is when $j \in J$ satisfies $\Gamma(j) = (i,1)$. In particular f_i acts injectively on $B(\infty)$ and then through (i) one has $e_if_i = \mathrm{Id}_{B(\infty)}$, for all $i \in I$.

To summarize briefly, the set $\{r_i^k\}_{k\in\mathbb{N}^+}$ of Kashiwara functions of type $i \in I$ determine the insertion rules for the action of e_i, f_i on an element $b \in B_J$.

Let $\mathscr{E}$ (resp. $\mathscr{F}$) denote the monoid generated by the $e_i : i \in I$ (resp. $f_i : i \in I$). It is a remarkable (and non-trivial) fact [6, Lemma 2.4.2] that $\mathscr{F}b_\infty$ is $\mathscr{E}$ stable and hence is just $B_J(\infty)$ and that this subcrystal of B_J is upper normal. These results are due to Kashiwara when **C** is symmetrizable.

3.3 A crystal is called lower normal if $\varphi_i(b) = \max\{j \mid f_i^jb \neq 0\}$. However this is *not* the case of $B(\infty)$. Yet it is true of the subcrystals $B(\lambda)$, where λ is a dominant integral weight. This is made possible by a parameter shift dependent on λ; see [6, 2.5.5].

4 An algorithm for computing $\varepsilon_i^\star$

4.1 Let U^- denote the subalgebra of $U_q(\mathfrak{g})$ which is generated by the root vectors with roots in $-\pi$ (which are of course amongst the generators of $U_q(\mathfrak{g})$ in its standard presentation). It is the free subalgebra modulo the quantum Serre relations. It admits an anti-involution which extends the identity on the said generators. From this almost trivial map, Kashiwara [13, Thm. 2.1.1] constructs an involution $\star$ on $B(\infty)$ (as a set) which is *highly non-trivial* and then constructs $e_i^\star, f_i^\star : i \in I$ by transport of structure. In the general not necessarily symmetrizable case, our approach was a little different. Indeed first we construct [6, 2.5.14] the $e_i^\star, f_i^\star : i \in I$ directly on $B(\infty)$. Then by applying the $e_i^\star, f_i^\star : i \in I$ to b_∞, we obtain a map $b \mapsto b^\star$ on $B(\infty)$ which we show [6, 2.5.20] coincides with the Kashiwara involution if **C** is symmetrizable. Then we further show that $\star$ is an involution [6, Thm. 2.5.25] in general. Again as in Kashiwara [13, Cor. 2.2.2] these two sets of operators almost commute between themselves [6, Lemma 2.5.15].

Consider a presentation of $B(\infty)$ as some $B_J(\infty)$. With this we may compute $\varepsilon_i(b), \varphi_i(b) : b \in B_J(\infty)$ and the action of the Kashiwara operators e_i, f_i, for all $i \in I$, through the rules given in Section 3. However, computation of the star operators in such a presentation is far more arduous. It goes as follows. Fix $i \in I$ and choose J so that $i_1 = i$. Then by definition [6, 2.5.14], the action of $e_i^\star$ (resp. $f_i^\star$) is to decrease (resp. increase) the right-hand entry by 1. This is easy enough and one finds that $\varepsilon_{i_1}^\star = m_{i_1}$. However to obtain a formula in the general case by this method one has to change J for every choice of $i \in I$ to obtain $i = i_1$ and this is only possible if W is finite. Again the transition rules for passing from the different choices of J are difficult to compute. Nevertheless this method was successfully applied in [2, Thm. 3.9] to determine the $z_t^i : t \in I,\ i \in \mathbb{N}^+$ as linear functions on B_J given by what we are calling trails in the corresponding fundamental module of the Langlands dual of $\mathfrak{g}$ determined by t.

It is not surprising that one cannot simply read off $\varepsilon_i^\star(b)$ from the presentation of $b \in B_J(\infty)$ as in the case of $\varepsilon_i(b)$. However since $B(\infty)$ is upper normal, we immediately obtain that

(i) For all $b \in B(\infty)$, $\varepsilon_i^\star(b)$ is the largest non-negative integer such that $(e_i^\star)^k b \neq 0$.

Since $\star$ preserves weights, this also defines $\varphi_i^\star$ on $B(\infty)$, though this will not be used.

4.2 The rule described in (i) of 4.1 gives little idea of how to compute $\varepsilon_i^\star(b) : b \in B(\infty)$ from the presentation of $B(\infty)$ as some $B_J(\infty)$. Again one can ask if it is possible to define $\varepsilon_i^\star(b)$, for all $b \in B_J$, just as in the case of $\varepsilon_i(b)$ and then ask if the resulting functions are linear.

This programme could have been in principle much easier to carry out had the $\star$ involution not been so utterly complicated.

4.3 Our approach is to use the almost commutation between the two sets of Kashiwara operators noted in 4.1 to obtain an algorithm to compute the $\varepsilon_i^\star(b) : i \in I$ on $B(\infty)$. At first sight this seems disarmingly simple. However it gives little idea as to whether the suggested presentation in terms of dual Kashiwara functions is at all possible. The general principle is that these functions should be differences of successive Kashiwara functions since it is these differences which govern the generation of $B_J(\infty)$ through the action $\mathscr{F}$. Moreover this general idea was already present in [17]; but there the authors were rather indiscriminate about which differences to choose and their proposed solution to describing $B_J(\infty)$ as a polyhedral subset of B_J fell apart even in type A_3 (because of the failure of their positivity conjecture).

Nevertheless building on this principle and the known solution in type A, we were able to complete a key induction step which we designated as the Preparation Theorem [7, Thm. 8.6]. Here we shall indicate how to complete the construction of the dual Kashiwara functions, but first we describe the algorithm mentioned above.

Here we shall need the tensor product rules given below.

Let B_2, B_1 be crystals. Given $b = b_2 \otimes b_1 \in B_2 \otimes B_1$, define

$$\mathrm{wt}\, b = \mathrm{wt}\, b_2 + \mathrm{wt}\, b_1, \tag{7}$$

$$\varepsilon_i(b) = \max\{\varepsilon_i(b_2),\ \varepsilon_i(b_1) - \alpha_i^\vee(\mathrm{wt}\, b_2)\} = \max\{\varphi_i(b_2), \varepsilon_i(b_1)\} - \alpha_i^\vee(\mathrm{wt}\, b_2). \tag{8}$$

Then by (1) we obtain

$$\varphi_i(b) = \max\{\varphi_i(b_2), \varepsilon_i(b_1)\} + \alpha_i^\vee(\mathrm{wt}\ b_1). \tag{9}$$

Finally the action of the $e_i, f_i : i \in I$, is defined by

$$e_i(b_2 \otimes b_1) = \begin{cases} e_i b_2 \otimes b_1, \text{ if } \varphi_i(b_2) \geq \varepsilon_i(b_1), \\ b_2 \otimes e_i b_1, \text{ if } \varphi_i(b_2) < \varepsilon_i(b_1). \end{cases} \tag{10}$$

$$f_i(b_2 \otimes b_1) = \begin{cases} f_i b_2 \otimes b_1, \text{ if } \varphi_i(b_2) > \varepsilon_i(b_1), \\ b_2 \otimes f_i b_1, \text{ if } \varphi_i(b_2) \leq \varepsilon_i(b_1). \end{cases} \tag{11}$$

One may verify that these formulae give rise to the Kashiwara functions and the resulting insertion rules given in 3.1 and 3.2. Of course this is well known and due to Kashiwara.

4.4 Take $s,t \in I$ to be distinct. Since $e_t^\star$ and f_s commute [6, Lemma 2.5.15], it follows from 4.1(i) that

$$\varepsilon_t^\star(e_s b) = \varepsilon_t^\star(b), \quad \varepsilon_t^\star(f_s b) = \varepsilon_t^\star(b), \forall s \neq t \in I, b \in B(\infty). \tag{12}$$

Fix $t \in I$. For all $b \in B(\infty)$, set

$$N_t(b) := \varepsilon_t(b) + \varepsilon_t^\star(b) + \alpha_t^\vee(\mathrm{wt}\, b) = \varphi_t(b) + \varepsilon_t^\star(b).$$

By the embedding theorem [6, Thm. 2.5.7] and [6, 2.5.11] there exists a subset $B^t \subset B(\infty)$, so that we may write b uniquely in the form $b^t \otimes b_t(-n)$, for some $b^t \in B^t$, $n \in \mathbb{N}$. Then by definition [6, 2.5.14], of the action of $e_t^\star$, one has

$$e_t^\star(b^t \otimes b_t(-n)) = b^t \otimes b_t(-(n-1))$$

and in particular $B^t = B^{e_t^\star} := \{b \in B(\infty) \mid e_t^\star b = 0\}$. Finally by 4.1(i) we obtain

$$\varepsilon_t^\star(b) = \varepsilon_t^\star(b^t \otimes b_t(-n)) = n. \tag{13}$$

N.B. The embedding theorem is used to give the description of $B_J(\infty)$ as a subset of B_J. As a consequence, $b_t(-n)$ of the above presentation is just the rightmost factor of $b \in B_J(\infty)$ when $\gamma(1) = t$. In the symmetrizable case, this presentation was already in the work of Kashiwara [13, 2.2] and the fact that $\varepsilon_t^\star(b) = n$ was used by Berenstein-Zelevinsky in the proof of [2, Thm. 3.9].

Lemma. *For all $b \in B(\infty)$,*

(i) f_t *enters the left-hand factor of* $b = b^t \otimes b_t(-n)$ *if and only if* $N_t(b) > 0$. *In this case* $\varepsilon_t^\star(f_t b) = \varepsilon_t^\star(b)$ *and* $N_t(f_t b) = N_t(b) - 1$.

(ii) f_t *enters the right-hand factor of* $b = b^t \otimes b_t(-n)$ *if and only if* $N_t(b) = 0$. *In this case* $\varepsilon_t^\star(f_t b) = \varepsilon_t^\star(b) + 1$ *and* $N_t(f_t b) = N_t(b)$.

(iii) $N_t(b) \geq 0$.

Proof. Suppose $f_t b = f_t b^t \otimes b_t(-n)$. By (11), (1), (8), this holds if and only if

$$\varepsilon_t(b^t) > n - \alpha_t^\vee(\text{wt}\, b^t). \tag{14}$$

Moreover, by (8), the left-hand side of (14) equals $\varepsilon_t(b)$, whilst the right-hand side equals $-(n + \alpha_t^\vee(\text{wt}\, b)) = -(\varepsilon_t^\star(b) + \alpha_t^\vee(\text{wt}\, b))$. Thus (14) is equivalent to $N_t(b) > 0$. This proves the first part of (i). Finally when (14) holds $\varepsilon_t^\star(f_t b) = n = \varepsilon_t^\star(b)$, whilst $\varphi_t(f_t b) = \varphi_t(b) - 1$, by (2). This proves the second part of (i).

Suppose $f_t b = b^t \otimes f_t b_t(-n)$. By (11), (1), (8), this holds if and only if

$$\varepsilon_t(b^t) \leq n - \alpha_t^\vee(\text{wt}\, b^t). \tag{15}$$

When (15) holds, its right-hand side equals $\varepsilon_t(b)$ by (8). On the other hand, we have already seen that its right-hand side equals $-(\varepsilon_t^\star(b) + \alpha_t^\vee(\text{wt}\, b))$. In view of (i), it follows that (15) is equivalent to $N_t(b) = 0$. Yet in this case $\varepsilon_t^\star(f_t b) = n + 1$, whilst $\varphi_t(f_t b) = \varphi_t(b) - 1$, as before. Hence (ii).

Since f_t must enter b at one of its two places, (iii) follows. □

4.5 By the upper normality of $B(\infty)$, the value of $\varepsilon_t^\star(b) : b \in B(\infty)$ is independent of the presentation of $B(\infty)$ as some $B_J(\infty)$, depending only on its crystal structure. Lemma 4.4 shows that $\varepsilon_t^\star(f_t^k b) : k \in \mathbb{N}$ stays constant till $k = N_t(b)$ and then increases by 1 for each additional factor of f_t. Since $\varepsilon_t^\star(b_\infty) = 0$, by upper normality, it follows from Lemma 4.4 and (12) that $\varepsilon_t^\star(b)$ may be computed inductively for all $b \in \mathscr{F} b_\infty = B_J(\infty)$. Although this is disarmingly simple, it is not so helpful and does not actually describe $\varepsilon_t^\star(b)$ in the desired form through dual Kashiwara functions, that is by (3).

Our procedure is to fix $t \in I$ and to compute the dual Kashiwara functions $z_t^k : k \in \mathbb{N}^+$ as follows. We discern an "initial driving function" associated to $t \in I$, denoted as z_t^1, which in particular has the property that $\varepsilon_t^\star(b) \geq z_t^1(b)$ and then adjoin further functions by imposing the invariance implied by (12) and the considerations to be described in 4.8. This is carried out in detail in Section 5.

4.6 A cancellation lemma

Recall the notation of 3.2. Let the length $\ell(e)$ of an element $e \in \mathscr{E}$ in a given presentation be the number of its factors as products of the $e_i : i \in I$ in that presentation. Given $b \in B_J$, define the length $\ell(b)$ of b as the sum of the coefficients $m_j : j \in \mathbb{N}^+$ in its presentation $b = (\ldots m_j, m_{j-1}, \ldots, m_1)$. Then $b \in B_J(\infty)$ if and only there is an element $e \in \mathscr{E}$, such that $eb = b_\infty$ and in this case $\ell(e) = \ell(b)$. (This shows in particular that $\ell(e)$ is independent of the presentation.) Call a submonomial of e a product of e_j occurring in e in which some factors have been omitted (but without changing their order).

Fix positive integers $j' - 1 \geq j'' + 1$ and set

$$J' = (\dots, i_{j'}, i_{j'-1}, \dots, i_{j''+1}), \quad J'' = (i_{j''}, i_{j''-1}, \dots, i_1).$$

Given $b' \in B_{J'}, b \in B_{J''}$, consider $b := b' \otimes b''$ to belong to B_J by the concatenation of the above expressions.

Lemma. *Suppose $b \in B_J(\infty)$; then $b', b'' \in B_J(\infty)$. Moreover for all $t \in I$, one has $\varepsilon_t^\star(b) \geq \varepsilon_t^\star(b'')$.*

Proof. Recall (2.4) that $B_J(\infty)$ is the unique subcrystal of B_J containing b_∞. Thus for the first part we just have to show that b_∞ is contained in $\mathscr{E}b'$ (resp. $\mathscr{E}b''$). The proof is by induction on $\ell(b'), \ell(b'')$. It is obvious if either one of these are zero. By the hypothesis there exists $e \in \mathscr{E}$ such that $eb = b_\infty$. Let e_i be the rightmost factor of e. Then we can write $e = e'e_i$ where lengths are added. If e_i enters in b' (resp. in b''), then the length of the left-hand (resp. right-hand) factor in $e_i b$ is decreased by one. In either case the required assertion results from the induction hypothesis. One may also remark that this argument also proves that there is a submonomial e' (resp. e'') of e such that $e'b' = b_\infty$ (resp. $e''b'' = b_\infty$).

For the second part, fix $t \in I$ and recall as in 4.4 that the embedding theorem [6, 2.5.6] gives a canonical decomposition $B(\infty) = B^t \otimes B_t$, where we can identify B^t with $B(\infty)^{e_t^\star}$ and writing $b = b^t \otimes b_t(-m)$, one has $m = \varepsilon_t^\star(b)$.

Now suppose $b = b''' \otimes b_t(-n) \in B(\infty)$. By the first part of the lemma, $b''' \in B(\infty)$ and so we may write $b''' = (b''')^t \otimes b_t(-n')$, for some (uniquely determined) $(b''')^t \in B^t$. Then $b = (b''')^t \otimes b_t(-(n+n'))$. Substitution in the previous expression gives $\varepsilon_t^\star(b) = n + n' \geq n$.

Again by the first part of the lemma $b'' \in B(\infty)$, and so we may write $b'' = (b'')^t \otimes b_t(-n)$, with $(b'')^t \in B^t$ and $n = \varepsilon_t^\star(b'')$. Taking $b''' = b' \otimes (b'')^t$ in the previous paragraph, gives the required result. □

Remark. Unfortunately it is false that $\varepsilon_t^\star(b) \geq \varepsilon_t^\star(b')$. Indeed suppose $\mathfrak{g}$ is of type A_2. Then with respect to the reduced decomposition $s_1 s_2 s_1$ of the unique longest element, one has $b = b_1(-r) \otimes b_2(-s) \otimes b_1(-t) \in B(\infty)$ if and only if $r \leq s$, a result already due to Kashiwara [13, Ex. 2.2.5]. As a baby example of the present theory, one shows that $\varepsilon_2^\star(b) = \max\{s - t, r\}$. Now take $b' = b_1(-r) \otimes b_2(-s)$, $b'' = b_1(-t)$ and $r = 0$, $s = t = 1$. Then $\varepsilon_2^\star(b') = 1$, whilst $\varepsilon_2^\star(b) = 0 = \varepsilon_2^\star(b'')$.

4.7 The initial driving function

Fix $t \in I$. Extending (4) we may define the zeroth Kashiwara function of type $t \in I$ through $r_t^0(b) := -\alpha_t^\vee(\mathrm{wt}\, b)$ and the initial driving function associated to $t \in I$ through $z_t^1(b) := r_t^0(b) - r_t^1(b)$, for all $b \in B_J$.

This is justified by the following observations below.
Given $b \in B_J$, let us write

$$b = b' \otimes b_t(-n) \otimes b'', \tag{16}$$

such that b'' has no factors in B_t. We call $b_t(-n)$ the first t-place of b.

Lemma. *For all $b \in B_J$, one has $z_t^1(b) = n - \alpha_t^\vee(\mathrm{wt}\, b'') \geq n$.*

Proof. From (4) it follows that $r_t^1(b) = n - \alpha_t^\vee(\mathrm{wt}\, b')$. On the other hand, $\alpha_t^\vee(\mathrm{wt}\, b) = \alpha_t^\vee(\mathrm{wt}\, b'') + \alpha_t^\vee(\mathrm{wt}\, b') - 2n$. Summing and recalling that the off-diagonal entries of the Cartan matrix $\mathbf{C}$ are non-negative integers gives the assertion. □

4.8 Retain the hypotheses and notation of 4.7.

Set $N_t'(b) = \varphi_t(b) + z_t^1(b)$.

Lemma. *Take $b \in B_J(\infty)$, $t \in I$, and write b as in (16). Consider the passage from b to $f_t b$.*

(i) *Suppose f_t enters the left-hand factor b' of b. Then $N_t'(b) > 0$, $z_t^1(f_t b) = z_t^1(b)$ and $N_t'(f_t b) = N_t'(b) - 1$.*

(ii) *Suppose f_t enters in the factor $b_t(-n)$ of b. Then $N_t'(b) = 0$, $z_t^1(f_t b) = z_t^1(b) + 1$ and $N_t'(f_t b) = N_t'(b) = 0$.*

In particular $N_t'(b)$ is a non-negative integer.

(iii) *$N_t'(b) = 0$ if and only if f_t enters in the first t-place of b.*

(iv) *There exists a non-negative integer $d_t(b)$ such that $N_t(b) - N_t'(b) = d_t(b)$.*

(v) *$\varepsilon_t^\star(b) = z_t^1(b) + d_t(b)$.*

Proof. By (11), (1), (8), assertion (i) holds if and only if

$$\varepsilon_t(b') > n - \alpha_t^\vee(\mathrm{wt}\, b'). \tag{17}$$

Moreover by (8), the left-hand side of (17) equals $\varepsilon_t(b)$, whilst the right-hand side equals $-(n + \alpha_t^\vee(\mathrm{wt}\, b - \mathrm{wt}\, b'')) = -z_t^1(b) - \alpha_t^\vee(\mathrm{wt}\, b)$. Through (1), this proves the first part of (i). The second part of (i) follows from Lemma 4.7 and the third part follows from the second part and $\varphi_t(f_t b) = \varphi_t(b) - 1$ which is obtained from (2).

By (11), (1), (8) assertion (ii) holds if and only if

$$\varepsilon_t(b') \leq n - \alpha_t^\vee(\mathrm{wt}\, b'). \tag{18}$$

Moreover, the right-hand side (18) equals $\varepsilon_t(b)$ as well as $-z_t^1(b)-\alpha_t^\vee(\mathrm{wt}\,b)$, so we obtain that $N_t'(b)=0$. Moreover f_t enters in the first t-place of b, so the last two parts of (ii) are obtained.

(iii) is obtained from (i) and (ii).

By 3.2(ii), eventually f_t enters the first t-place of $b\in B_J$. More precisely, there exists a largest non-negative integer $\ell_t(b)$ such that f_t^k enters b' for all $k\leq \ell_t(b)$. Then by (ii), one has $N_t'(f_t^{\ell_t(b)}b)=0$ and so $N_t'(b)=\ell_t(b)$, by (i). On the other hand by Lemma 4.4(iii) there exists a non-negative integer $d_t(b)$ such that $N_t(f_t^{\ell_t(b)}b)=d_t(b)$. This proves the assertion if $\ell_t(b)=0$, so we may assume $\ell_t(b)>0$.

Suppose $d_t(b)>0$. Let k be a non-negative integer $<\ell_t(b)$. By Lemma 4.4, one has $N_t(f_t^kb)\geq N_t(f_t^{\ell_t(b)}b)>0$. Then by Lemma 4.4(i), we obtain $N_t(f_t^{k+1}b)=N_t(f_t^kb)-1$, and so $N_t(b)=d_t(b)+\ell_t(b)=N_t'(b)+d_t(b)$.

This argument breaks down if $d_t(b)=0$. However by Lemma 4.6, we obtain $b_t(-n)\otimes b''\in B_J(\infty)$. Through the embedding theorem [6, 2.5.7], we may rewrite $b_t(-n)\otimes b''$ as $b'''\otimes b_t(-m)$ and substituting into (16) gives $b=b^t\otimes b_t(-m)$ with $b^t=b'\otimes b'''$. Then the condition that $f_t^k: k\leq \ell_t(b)$ enters into the left-hand factor b' of $b=b'\otimes b_t(-n)\otimes b''$ implies the same for $b=b'\otimes b'''\otimes b_t(-m)$, and hence in particular it enters into the left-hand factor of b^t of $b=b^t\otimes b_t(-m)$. Then (i) of Lemma 4.4 applies to give $N_t(b)=N_t(f_t^{\ell_t(b)}b)+\ell_t(b)=d_t(b)+\ell_t(b)=N_t'(b)+d_t(b)$ as before. Hence (iv) and (v).

□

4.9 Retain the above notation.

Lemma. *For all $t\in I$ and $b\in B_J(\infty)$ one has $N_t(b)=0$ if and only if $\varepsilon_t^\star(b)=z_t^1(b)$ and $\varepsilon_t(b)=r_t^1(b)$.*

Proof. Indeed by definition of $N_t(b)$ and Lemma 4.8(iv),(v), one has $N_t(b)=\varphi_t(b)+\varepsilon_t^\star(b)\geq\varphi_t(b)+z_t^1(b)=\varepsilon_t(b)-r_t^1(b)\geq 0$. Equality implies that $\varepsilon_t^\star(b)=z_t^1(b)$ and $\varepsilon_t(b)=r_t^1(b)$. □

4.10 Take $b\in B_J(\infty)$ and fix $t\in I$.

Proposition.

(i) *$\varepsilon_t^\star(f_tb)\neq\varepsilon_t^\star(b)$ if and only if $\varepsilon_t^\star(b)=z_t^1(b)$ and f_t enters b at the first t-place. If either of these equivalent conditions hold, then $\varepsilon_t^\star(f_tb)=\varepsilon_t^\star(b)+1=z_t^1(b)+1=z_t^1(f_tb)$.*

(ii) *Suppose $e_tb\neq 0$. Then $\varepsilon_t^\star(e_tb)=\varepsilon_t^\star(b)$ unless $\varepsilon_t^\star(b)=z_t^1(b)$ and e_t enters b at the first t-place. In the latter case $\varepsilon_t^\star(e_tb)=\varepsilon_t^\star(b)-1=z_t^1(b)-1=z_t^1(e_tb)$.*

Proof. Consider (i). For $\Leftarrow$ observe that the condition $r_t^1(b)=\varepsilon_t(b)$ exactly means that 1 is the rightmost value of k such that $r_t^k(b)=\varepsilon_t(b)$. By the insertion rules given in 3.2, this exactly means f_t enters b at the first t-place. Then by Lemma 4.9, we obtain $N_t(b) = 0$, which by Lemma 4.4(ii) gives $\varepsilon_t^\star(f_t b) \neq \varepsilon_t^\star(b)$ as required. For $\Rightarrow$ observe that Lemma 4.4(i) forces $N_t(b) = 0$ and apply Lemma 4.9. The last assertion follows from Lemma 4.4(ii) and Lemma 4.7.

Consider (ii). Suppose that $N_t(b) > 0$ and that $b' := e_t b \neq 0$. In this case we claim that

$$\varepsilon_t^\star(e_t b) = \varepsilon_t^\star(b). \tag{19}$$

For this, first note that $f_t b' = b$. If $N_t(b') = 0$, we obtain $N_t(b) = 0$ by Lemma 4.4(ii), which contradicts our assumption. Thus $N_t(b') > 0$ and so by Lemma 4.4(i), we obtain $\varepsilon_t^\star(e_t b) = \varepsilon_t^\star(b') = \varepsilon_t^\star(f_t b') = \varepsilon_t^\star(b)$, as required.

Thus by Lemma 4.4(iii) we are reduced to the case $N_t(b) = 0$. By Lemma 4.9 this implies that $\varepsilon_t^\star(b) = z_t^1(b)$.

Suppose that e_t does not enter b at the first t-place. Then f_t does not enter $e_t b$ at the first t-place. By (i) this forces $\varepsilon_t^\star(b) = \varepsilon_t^\star(f_t e_t b) = \varepsilon_t^\star(e_t b)$, so again (19) is obtained. This proves the first part of (ii). For the last part suppose $\varepsilon_t^\star(b) = z_t^1(b)$ and e_t enters at the first t-place. Then $\varepsilon_t^\star(b) - 1 = z_t^1(b) - 1 = z_t^1(e_t b)$. Furthermore by the first part, (19) does not hold and so $\varepsilon_t^\star(e_t b) = \varepsilon_t^\star(b) - 1$ by Lemma 4.4. □

Remark. Suppose that e_t enters b at the first t-place. (This is slightly stronger than saying that $r_t^1(b) = \varepsilon_t(b)$.) Suppose in addition that $\varepsilon_t^\star(b) = z_t^1(b)$. By Lemma 4.9, this forces $N_t(b) = 0$. Then $N_t(e_t b)$ equals either 1 or 0, but one cannot say which holds. In the first case, $\varepsilon_t^\star(b) = \varepsilon_t^\star(f_t e_t b) = \varepsilon_t^\star(e_t b)$. In the second case, we obtain $\varepsilon_t^\star(b) = \varepsilon_t^\star(f_t e_t b) = \varepsilon_t^\star(e_t b) + 1$ by Lemma 4.9 and (i).

4.11 Lemma 4.10 gives a further understanding of the inequality $\varepsilon_t^\star(b) \geq z_t^1(b)$, for all $b \in B_J(\infty)$. Both are zero for $b = b_\infty$. Consider the change in both sides when $b \mapsto f_s b$. For $s \neq t$, the left-hand side remains unchanged by (12), whereas the right-hand side remains unchanged unless f_s enters b at a place strictly to the right of the first t-place and then it decreases by $-\alpha_t^\vee(\alpha_s)$. If $s = t$, then both sides remain fixed unless f_t enters b at the first t-place. Then by Proposition 4.10(i), z_t^1 increases by 1 with $\varepsilon_t^\star$ remaining fixed, until both are equal and then they both increase by 1.

5 Invariance

5.1 Fix $t \in T$. A countable set $Z_t := \{z_t^j\}_{j \in \mathbb{N}^+}$ of linear functions on B_J will be said to be t-semi-invariant if for each $s \in I \setminus \{t\}$ (resp. $s = t$), we can write Z_t (resp. $Z_t \setminus \{z_t^1\}$) as a union of finite subsets $\{Z_{t,s}^j\}_{j \in \mathbb{N}^+}$ such that $E_{t,s}^j := \max_{z \in Z_{t,s}^j} z(b)$ satisfies

$$E_{t,s}^j(f_s b) = E_{t,s}^j(b), \forall b \in B_J. \tag{20}$$

In the above we shall always take z_t^1 to be the initial driving function associated to t as defined in 4.7.

N.B. One has $Z_t \supset \{z_t^1\}$ and so is *not* the empty set. Again by (20), equality holds if and only if all the $m_s^k : s \in I \setminus \{t\}, k \in \mathbb{N}^+$ in z_t^1 have zero coefficient. The latter condition means that $z_t^1 = m_t^1$. For any J, there is always one choice of $t \in I$ such that the latter holds, for example if $i_1 = t$.

Proposition. *Let $Z_t := \{z_t^j\}_{j\in\mathbb{N}^+}$ be a t-semi-invariant countable set of linear functions on B_J. Then $\varepsilon_t^\star(b) = \max_{z\in Z_t} z(b)$, for all $b \in B_J(\infty)$.*

Proof. Set $E_t(b) = \max_{z\in Z_t}(b) : b \in B_J(\infty)$. For $s \neq t$, it is immediate from (20) that $E_t(f_s b) = E_t(b)$, for all $b \in B_J$. Again it is immediate from (20) that $E_t(f_t b) = E_t(b)$, for all $b \in B_J$, except when

$$z_t^1(b) = E_t(b), \tag{21}$$

and

$$z_t^1(f_t b) > z_t^1(b). \tag{22}$$

Suppose both (21), (22) hold. By definition of z_t^1, equation (22) holds if and only if f_t enters b at the first t-place and then $z_t^1(f_t b) = z_t^1(b) + 1$. Then by (21) we obtain $E_t(f_t b) = E_t(b) + 1$.

We conclude from the above and Proposition 4.10(i) that $E_t(b)$ and $\varepsilon_t^\star(b)$ change in exactly the same way on replacing b by $fb : f \in \mathscr{F}$.

Finally b_∞ is the zero vector in B_J when the latter is identified with $\mathbb{N}^{|J|}$. Thus linearity implies that $E_t(b_\infty) = 0$. Yet we also have $\varepsilon_t^\star(b_\infty) = 0$ by upper normality. Since $B_J(\infty)$ is generated by $\mathscr{F}$ over b_∞, if follows that $E_t = \varepsilon_t^\star$ on $B_J(\infty)$, as required. □

Remark 1. Obviously we only need (20) to hold for $b \in B_J(\infty)$. However since $B_J(\infty)$ is (for the moment) an unknown subset of B_J, it is more practical to attempt to establish (20) as presented. We have not used the finiteness of the subsets $Z_{t,s}^j$, or the countability of their number; but this is how they will be presented.

Remark 2. An important element of their construction is that the $Z_{t,s}^j : j \in \mathbb{N}^+$ consist of locally finite functions which have integer coefficients, that is, are finite integer linear combinations of the coordinate functions $m_j : j \in J$ on B_J. However this is not needed for the statement of the Proposition.

Remark 3. It should be clear that the nature of z_t^1 is crucial to this result. Had it been omitted (equivalently had it been taken to be zero), then we would have obtained simply that E_t is the zero function. This is why we call z_t^1 the driving function associated to t.

5.2 Equation (20) brings one almost inevitably to the notion of an S-set of type $s \in I$. Their construction is quite non-trivial. It was the subject of the Preparation Theorem [7]. It is one of the important by-products of this work, confirming our belief that the study of $B_J(\infty)$ inspires significant (and new) combinatorial constructions.

By definition, an S-set is the set of linear functions (eventually on B_J) associated to the vertices of an S-graph. Here an S-graph is defined by a coefficient set $\mathbf{c} = (\ldots, c_i, c_{i-1}, \ldots, c_1)$ of non-negative integers which are almost all zero. In terms of this coefficient set, an S-graph is required to possess a number of intricately related properties [7, Sect. 6] which will not be recalled here. Only the S-graphs $\mathscr{G}(\mathbf{c})$ constructed in [7, Sect. 7] will be presently considered. They have a number of additional important and surprising properties [8], [9] and [10].

The main idea behind the construction of Z_t, having the properties required for Proposition 5.1, is to adjoin S-sets of type $s \in I$ given by a process of induction on J.

The induction procedure is indicated in two baby examples given in Tables 1,2.

We start at $j = (t,1)$ and set $Z_t^{(t,1)} = \{z_t^1\}$. Here we emphasize that $Z_t^{(t,1)}$ is never an S-set of type t.

Take $j \in J$ with $j \geq (t,1)$ and set $[1,j] := \{1,2,\dots,j\}$.

We say that a linear function on B_J has support in $[1,j]$ if when expressed as a sum of the $m_{j'}$, the coefficients of $m_{j'} : j' > j$ vanish.

Assume that a finite set Z_t^j of locally finite linear functions on B_J with support in $[1,j]$ is given (eventually constructed inductively from the initial driving function z_t^1 as indicated below).

Set $s = i_{j+1}$.

For any locally finite linear function z on B_J with integer coefficients, we obtain a coefficient set $\mathbf{c}^z$ of type s whenever the coefficients of the m_s^i in z are all non-positive for all $i \in \mathbb{N}^+$ by letting $-c_i^z$ denote the coefficient of m_s^i.

For comparison with [9, 5.3.3,4], we let n^z, or simply n, be the smallest positive integer such that $c_i^z = 0$ for all $i \geq n^z$ and set $N^z = \{1,2,\dots,n^z-1\}$. Then we set $\mathbf{c}^z = (c_{n-1}^z, c_{n-2}^z, \dots, c_1^z)$ viewed as a linearly order set by lifting the partial order on the $c_i^z : i \in N$, induced by the natural total order on $\mathbb{N}$. We shall often omit the z superscript.

An element $z \in Z_t^j$, which gives rise to a coefficient set $\mathbf{c}^z$ of type s, is called a potential driving function of type s belonging to Z_t^j.

The S-set of (locally finite linear) functions (of type s) associated to a coefficient set $\mathbf{c}$ is just the set $Z(\mathbf{c})$ of functions attached to the vertices of the S-graph $\mathscr{G}(\mathbf{c})$. As shown in [10, 5.8], this set $Z(\mathbf{c})$ only depends on the coefficient set $\mathbf{c}$ with its natural order and not on the lifting to a linear order. One may remark that $Z(\mathbf{c})$ consists of the extremal points of the convex set $K(\mathbf{c}) \in \mathbb{Q}^{n-1}$ given in [8, Thm. 1.4]. (This fact is also a convenient way to understand and describe $Z(\mathbf{c})$.) Let $K_{\mathbb{Z}}(\mathbf{c})$ denote the set of integer points in $K(\mathbf{c})$. Remarkably the latter set arises in the description of trails (see [9, Prop. 7.3], a fact which explains why $\mathscr{G}(\mathbf{c})$ is the correct choice of S-graph. It also explains why the functions coming from trails are highly redundant in computing $\varepsilon_t^\star(b)$ as given by (3).

Every element of $Z(\mathbf{c}^z)$, defined by the potential driving function z, is a sum of z and a sum with *non-negative* coefficients of successive Kashiwara functions $r_s^k - r_s^{k+1} : k \in N$ of type s. Thus writing $z' > z$ whenever $z' \in z\sum_{k\in\mathbb{N}^+}\mathbb{N}(r_s^k - r_s^{k+1})$, defines a partial order on the set of potential driving functions of type s in Z_t^j.

We remark that if $z' \in K_{\mathbb{Z}}(\mathbf{c}^z)\setminus\{z\}$ is a potential driving functions of type s in Z_t^j, then $c_i^{z'} \leq c_i^z$ for all $i \in N$. This follows from [9, (37)].

Let $D(Z_t^j)$ be the minimal elements of the set of potential driving functions of type s in Z_t^j. It is called the set $D(Z_t^j)$ of driving functions of type s in Z_t^j.

The necessity to eliminate some potential driving functions in the construction below can be understood [9, 8.4] in terms of trails. An example for which this elimination was really necessary occurs in type D_5, for a particular choice of J and of $t \in I$. There are 93 trails in this case, so we will not give all the details here! It was obtained with the aid of the computer calculations of Zelikson, though the calculation can be done by hand with a cool head.

The referee thought it would be interesting to know the actual choice of J which was made in the above. Well here it is: 45321343152342321532, where the jth entry counting from the right is just $i_j \in I$. This element is *not* a power of a Coxeter element for which $\mathscr{K}_t^{BZ}$ may have better properties, as we hope to explain in a forthcoming paper.

For the above choice of J, let us use d_j in the jth entry to designate the coefficient of m_j in a function on B_J. Then 0000001 – 1101 – 201 – 100000 lies in Z_4 and is a driving function of type 3. The corresponding S-set has six elements, as computed using [7, Sect. 7]. One of these is elements is 00000010000 – 101 – 100000. It is a potential driving function of type 3 which is eliminated by the above procedure. Moreover it needs to be eliminated because the S-set it defines, if included, leads through (20) to a set not satisfying the conditions of Proposition 5.1.

The elements of $Z(\mathbf{c}^z)$ have support in $[1,(s,n^z)]$. Recall that $i_{j+1} = s$ and by our choices $(s,n^z) \leq j+1$. In accordance with [9, 7.2.3], the subset of $Z(\mathbf{c}^z)$ of elements whose coefficients in $m_s^{n^z}$ vanish, is denoted by $Z^-(\mathbf{c}^z)$. (For example z and the potential driving functions in $Z(\mathbf{c}^z)$ belong to $Z^-(\mathbf{c}^z)$. One may remark that $Z^-(\mathbf{c})$ is reduced to $\{z\}$ when the coefficients are increasing, but otherwise it can be quite large, especially when the coefficients are decreasing.) These functions have support in $[1,j]$, whereas the elements of $Z(\mathbf{c}^z) : z \in D(Z_t^j)$ have support in $[1,j+1]$.

Set

$$Z_t^{j+1} = \bigcup_{z \in D(Z_t^j)} Z(\mathbf{c}^z). \tag{23}$$

It is immediate from definitions that this recipe is the same as that using ℓ-minimal trails [9, 8.4]. Note that the construction of Z_t^{j+1} from Z_t^j is *purely combinatorial.*

5.3 It would seem a miracle if the above recipe would give the required t-semi-invariant set Z_t in the limit. However it only needs one innocent-looking criterion to be satisfied for it to work, namely we require that

$$Z_t^{j+1} \supset Z_t^j. \tag{24}$$

Then we set $Z_t = \lim_{\rightarrow} Z_t^j$.

At first sight it may seem that (24) holds by construction, in fact it is a quite non-trivial condition. For example it is not obvious that $D(Z_t^j)$ is non-empty, as required for Z_t^{j+1} to be non-empty.

One may remark that since the functions in Z_t^j have support in $[1,j]$, condition (24) is equivalent to the inclusion $Z_t^j \subset \bigcup_{z\in D(Z_t^j)} Z^-(\mathbf{c}^z)$. This is just the inclusion described in [9, (57)] and a slightly stronger version of the hypothesis of [9, Thm. 9.2]. It is a close analogue, even in some natural sense a generalization, of the Demazure property mentioned in the last sentence of the preamble. It is equivalent in the language of [9, 8.2] to the absence of false trails.

In the above notation, set

$$K_t^{j+1} = \bigcup_{z\in D(Z_t^j)} K_{\mathbb{Z}}(\mathbf{c}^z), \forall j \geq (t,1). \tag{25}$$

Further set

$$D_s = \bigcup_{j\geq(t,1)|i_{j+1}=s} D(Z_t^j), \quad \overline{\mathscr{K}}_t(s) = \bigcup_{z\in D_s} K_{\mathbb{Z}}(\mathbf{c}^z) \quad \mathscr{K}_t(s) = \overline{\mathscr{K}}_t(s) \cup \{z_t^1\}. \tag{26}$$

One may remark that $z_t^1 \in \overline{\mathscr{K}}_t(s)$, for all $s \in I \setminus \{t\}$.

Theorem. *Suppose that $K_t^{j+1} \supset Z_t^j$, for all $j \geq (t,1)$. Then $\mathscr{K}_t(s)$ is independent of $s \in I$.*

Proof. The hypothesis of the theorem is the same as that of [9, Thm. 9.2]. Thus the conclusion is obtained from [9, Thm. 9.2] combined with [9, Thm. 8.3]. □

5.4 For all $t \in I$, let $\varpi_t^\vee$ denote the corresponding fundamental coweight.

The above result can also be expressed in terms of the set $\mathscr{K}_t^{BZ}$ of all Berenstein-Zelevinsky trails (as defined in [9, 2.2]) taken with respect to J in the fundamental module of lowest weight $V(-\varpi_t^\vee)$ in the Langlands dual of $\mathfrak{g}$. (In the present paper we are not made aware of the need to pass to the Langlands dual because all the computations involving the trails themselves are tucked away in [9]. One might add that the passage to the Langlands dual means, in more mundane terms, interchanging roots and coroots in the dual Kashiwara functions. Thus it is little more than a question of book-keeping.)

To any $K \in \mathscr{K}_t^{BZ}$ there is a locally finite linear function z^K on B_J given by [9, 2.3]. In this connection, the support of K is defined to be the support of z^K. It lies in $[1,j]$ if the trail trivializes at $j \in J$ in the sense of [9, 2.2].

Fix $j_1 \in J$ and suppose that $Z_t^j \subset K_t^{j+1}$, for all $j < j_1$. Then by [9, Thm. 9.2] combined with [9, Thm. 7.6], it follows that $K_t^{j+1} : j < j_1$ identifies with the subset of trails which trivialize at $j+1$ and hence having support in $[j+1,1]$. In particular for all $j \leq j_1$, one has

$$K_t^j = \{z \in K_t^{j_1} \mid \operatorname{Supp} z \subset [j,1]\}. \tag{27}$$

Taking j_1 increasingly large, it follows that the K_t^j are increasing in j and their direct limit $\mathscr{K}_t$ identifies with the set of all trails, that is, with the set $\mathscr{K}_t^{BZ}$ defined in [9, 2.2].

Under the hypothesis of Theorem 5.3 it is shown in [9, Cor. 8.3] that $\mathscr{K}_t(s) = \mathscr{K}_t^{BZ}$, for all $s \in I$. This is the origin of the (crucial) independence of $\mathscr{K}_t(s)$ on $s \in I$.

We may summarize the above in the following:

Corollary. *Suppose $Z_t^j \subset K_t^{j+1}$, for all $j \in J$. Then*

$$\mathscr{K}_t := \varinjlim K_t^j = \mathscr{K}_t^{BZ} = \mathscr{K}_t(s), \forall s \in I.$$

Remark. Unfortunately false trails can exist. An example occurs in type C_3 with $J = 213213213$, in the notation of 6.2. One needs to establish the above independence when potential driving functions are eliminated by only using the weaker order relation $z' > z$ given $z' \in K_{\mathbb{Z}}(\mathbf{c}^z) \setminus \{z\}$ in 6.2. In simple cases (for example that above) this independence can be checked by hand. If successful this constructs all trails combinatorially.

5.5 The S condition Fix a coefficient set $\mathbf{c}$ defined by a driving function z of type $s \in I$. In the notation of 5.2, we may write $\mathbf{c} = \{c_i\}_{i \in N}$, with $N = \{1,2,\dots,n-1\}$. Set $\hat{N} := \{1,2,\dots,n\}$. Let $V(\mathscr{G}(\mathbf{c}))$ denote the set of vertices of $\mathscr{G}(\mathbf{c})$. They are labelled by the elements of $\hat{N}$. Let $V^i(\mathscr{G}(\mathbf{c}))$ be those vertices with label $i \in \hat{N}$. The edges of $\mathscr{G}(\mathbf{c})$ are labelled by the elements of $\mathbf{c}$.

To each $v \in V(\mathscr{G}(\mathbf{c}))$ there is defined a locally finite linear function z_v on B_J which is a finite sum of successive differences of Kashiwara functions with non-negative integer coefficients given by the rules [7, 6.2-6.5]. More explicitly, z_v is a sum of the initial driving function z defining $\mathbf{c}$ plus a sum of successive differences of Kashiwara functions of type s given by [7, (P_3)] through *any* path from v to the marked vertex on the corresponding S-graph. By definition $Z(\mathbf{c}) = \{z_v | v \in V(\mathscr{G}(\mathbf{c}))\}$.

Given $v \in V^i(\mathscr{G}(\mathbf{c}))$, then the coefficient of m_s^i in z_v equals zero by [7, (P_8) of Sect. 6.7].

To each edge of $\mathscr{G}(\mathbf{c})$ there is assigned an element of $\mathbf{c}$. This is done in such a way that $\mathscr{G}(\mathbf{c})$ acquires the S-property. The latter asserts [7, (P_7) of Sect. 6.7] that for every $v \in V(\mathscr{G}(\mathbf{c}))$, there exists $k \in \hat{N}$ and $v' \in V^k(\mathscr{G}(\mathbf{c}))$ such that there is an ordered path from v to v'. Here an ordered path means that the integers on edges increase along the path. One may also remark that both v' and the path are uniquely determined [10, 5.6.6,5.6.8] by the pair (v,k), though this is not needed below.

In view of property [7, (P_2), of Sect. 6.7] of an S-graph, we conclude that given the pair (v,k) there exist a positive integer u, integers $0 \leq d_1 \leq d_2 \leq \cdots \leq d_{u-1}$ and $g_1, g_2, \dots, g_u \in \hat{N}$, with $g_u = k$, such that

$$z_v - z_{v'} = \sum_{i=1}^{u-1} d_i (r_s^{g_i} - r_s^{g_{i+1}}).$$

The significance of this result is that setting $d_0 = 0$, the above expression may be rearranged in the form

$$z_{v'} - z_v = \sum_{i=1}^{u-1} (d_i - d_{i-1})(r_s^k - r_s^{g_i}). \tag{28}$$

Now suppose that f_s enters $b \in B_J$ at the kth-place. According to 3.2(i) this occurs exactly when

$$r_s^k(b) - r_s^j(b) \in \begin{cases} \mathbb{N}^+, & \text{if } k > j, \\ \mathbb{N}, & \text{if } k \leq j. \end{cases} \tag{29}$$

Substitution into (29) into (28) gives $z_{v'}(b) \geq z_v(b)$, when f_s enters b at the kth-place.

On the other hand, $m_s^k(f_sb) = m_s^k(b) + 1$, whilst $m_s^j(f_sb) = m_s^j(b); j \neq k$. Then by (4) we obtain $r_s^k(f_sb) = r_s^k(b) + 1$, whilst $r_s^j(f_sb) = r_s^j(b)$ if $j > k$ and $r_s^j(f_sb) = r_s^j(b) + 2$ if $j < k$ since $\alpha_s^\vee(\alpha_s) = 2$. Consequently

$$(r_s^k(f_sb) - r_s^j(f_sb)) - (r_s^k(b) - r_s^j(b)) = \begin{cases} -1, & \text{if } k > j, \\ 1, & \text{if } k < j. \end{cases} \tag{30}$$

We conclude that $r_s^k(f_sb) \geq r_s^j(f_sb)$, for all $j \in \mathbb{N}^+$. Substitution into (28) gives $z_{v'}(f_sb) \geq z_v(f_sb)$. Yet $z_{v'}(f_sb) = z_{v'}(b)$ since the coefficient of m_s^k of $z_{v'}$ is zero.

Since k can be any element of $\hat{N}$, we conclude that for all $b \in B_J$, one has

$$\max_{z \in Z(\mathbf{c})} z(f_sb) = \max_{z \in Z(\mathbf{c})} z(b). \tag{31}$$

Again if e_sb is non-zero, then $f_se_sb = b$, so replacing b by e_sb in (31), we conclude that for all $b \in B_J$ one has

$$\max_{z \in Z(\mathbf{c})} z(e_sb) = \max_{z \in Z(\mathbf{c})} z(b), \quad \text{given } e_sb \neq 0. \tag{32}$$

This result can also be proved directly through a similar computation to the above.

This verifies in detail what was already noted in [6, 6.7].

This result may be expressed in the following form.

Let $z \in Z_t$ be a driving function of type $s \in I$. Set $Z_{t,s}^z = \{z \in Z(\mathbf{c}^z)\}$ and $E_{t,s}^z(b) = \max_{z \in Z_{t,s}^z} z(b)$.

Proposition. *For all $b \in B_J$, one has $E_{t,s}^z(f_sb) = E_{t,s}^z(b)$.*

5.6 Assume that the hypothesis of Theorem 5.3 holds and let $\mathscr{K}_t$ be defined as in 5.4.

Theorem. *Suppose that $Z_t^j \subset K_t^{j+1}$, for all $j \geq (t,1)$. Then for all $b \in B_J(\infty)$ one has*

$$\varepsilon_t^\star(b) = \max_{z \in \mathscr{K}_t} z(b). \tag{33}$$

Proof. Choose the pair $j \in J$, $s \in I$ such that $i_{j+1} = s$. By Proposition 5.5 equation (20) holds for the $Z^j_{t,s}$ being $\bigcup_{z \in D(Z^j_t)} Z(\mathbf{c}^z)$ and hence for the $Z^j_{t,s}$ being $\bigcup_{z \in D(Z^j_t)} K_{\mathbb{Z}}(\mathbf{c}^z)$. As j runs over J, the latter union is just $\overline{\mathscr{K}}_t(s)$. Then by Corollary 5.4, the conclusion is obtained from Proposition 5.5. □

5.7 Given a subset Z of some $\mathbb{Q}^{n-1}$, let $E(Z)$ denote its set of extremal elements. Recall that [8, Thm. 1.4] asserts that $Z(\mathbf{c}) = E(K(\mathbf{c}))$.

$(*)$. Assume that the hypothesis of Theorem 5.3 holds.

By the above $E(\mathscr{K}_t(s)) \subset Z_t(s) := \bigcup_{z \in D_s} Z(\mathbf{c}^z) \cup \{z^1_t\}$. Suppose equality holds. Then by Theorem 5.3, $Z_t(s)$ is independent of $s \in I$ and this common set Z_t is defined to be a giant S-set ([9, 8.1]).

For some time we believed that $E(\mathscr{K}_t(s)) = Z_t(s)$, so that we could write $E(\mathscr{K}_t^{BZ}) = Z_t$, with Z_t a giant S-set. Indeed this is often possible. Eventually a counter-example was found to this first equality in type F_4 through computer calculations of Zelikson. In this case $|\mathscr{K}_t^{BZ}| = 272$, whilst $|E(\mathscr{K}_t^{BZ})| = 195$, so we can hardly present the details here.

Yet assuming $(*)$ holds, $\max_{z \in Z_t(s)} z(b) : b \in B_J$ is *independent* of $s \in I$.

6 The lower bound

6.1 In order that the hypothesis of Theorem 5.6 be automatically satisfied, we should show that $z \in Z^j_t \setminus K^{j+1}_t$ must violate (33), that is, there exists $b \in B_J(\infty)$ such that $z(b) > \varepsilon^\star_t(b)$ for some $b \in B_J(\infty)$. Indeed this would lead to a contradiction with the conclusion of Theorem 5.6. However the latter assumes that $\mathscr{K}_t$ can be constructed and this requires the hypothesis that we would like to establish.

The above consideration motivates proving that for all $b \in B_J(\infty)$, one has

$$\varepsilon^\star_t(b) \geq \max_{z \in Z^j_t} z(b), \forall j \geq (t,1) \tag{34}$$

by induction. Besides it would be rather strange if it were not possible to prove this directly and in particular without appealing to the existence of $\mathscr{K}_t$.

6.2 Truncation Fix $t \in I$. For all $k \in \mathbb{N}^+$, the k-truncation $\tau^k_s(b)$ of $b \in B_J$ relative to $s \in I$, or simply τ^k is defined by setting to zero the $m_j : j > (s,k)$.

By Lemma 4.6 one has

$$b \in B_J(\infty) \Rightarrow \tau^k(b) \in B_J(\infty), \tag{35}$$

and

$$\varepsilon_t^\star(\tau^k(b)) \leq \varepsilon_t^\star(b). \tag{36}$$

Since the action of e_s can only decrease entries, one has

$$\tau^k(e_s^u \tau^k(b)) = e_s^u \tau^k(b), \forall u \in \mathbb{N}. \tag{37}$$

Fix $b \in B_J(\infty)$.

Take $u \in \mathbb{N}$ maximal with the property that $e_s^u \tau^k(b) \neq 0$. Since $\tau^k(b) \in B_J(\infty)$ and $B_J(\infty)$ is upper normal, we obtain $\varepsilon_s(e_s^u \tau^k(b)) = 0$. By definition of the latter $r_s^i(e_s^u \tau^k(b) \leq 0$, for all $i \in \mathbb{N}^+$. By definition of τ^k, (4) and (37), we obtain $r_s^k(e_s^u \tau^k(b)) = m_s^k(e_s^u \tau^k(b)) \geq 0$, and so

$$0 = r_s^k(e_s^u \tau^k(b)) \geq r_s^i(e_s^u \tau^k(b)), \forall i \in \mathbb{N}^+. \tag{38}$$

As in 5.2, take $j \geq (t,1)$ and set $s = i_{j+1}$. The latter means that we can write $j = (s,k) - 1$ for some $k \in \mathbb{N}^+$.

In the notation of 5.2, take $z \in D(Z_t^j)$. It is a driving function of type s and has support in $[1, j]$. Define $n, N, \mathbf{c}$ as in 5.2 with respect to z. In particular $n+1 \leq k$.

Take $z' \in Z(\mathbf{c})$, $b \in B_J$.

One has $z' \in z + \sum_{i \in N} \mathbb{N}(r_s^i - r_s^{i+1})$. Yet for $i \in N$, one has $(s, i+1) \leq (s,k)$, so the function z' has support in $[1, j+1]$, and thus we obtain

$$z'(\tau^k(b)) = z'(b), \forall b \in B_J. \tag{39}$$

Set $b' = e_s^u \tau^k(b)$. It is an element of $B_J(\infty)$. We want to use the S-property of $\mathscr{G}(\mathbf{c})$ to obtain an upper bound on $z'(b')$.

By [6, (P_8)] a function $z_v : v \in V^{n+1}(\mathscr{G}(\mathbf{c}))$ has a zero coefficient of m_s^{n+1} and consequently has support in $[1, j]$. In particular the elements of $Z^-(\mathbf{c})$ contain those of $Z(\mathbf{c})$ corresponding to the vertices of $\mathscr{G}(\mathbf{c})$ with label $n+1$.

Now assume that $n+1 = k$.

Observe that (38) asserts that $r_s^i(b')$ takes its maximal value at $i = k$.

Through the S-property, for all $v \in V(\mathscr{G}(\mathbf{c}))$, there exists $v' \in V^{n+1}(\mathscr{G}(\mathbf{c}))$ such that the terms on the right-hand side of (28) are all ≥ 0 when evaluated on b'. Thus we obtain

$$\max_{z' \in Z(\mathbf{c})} z'(e_s^u \tau^k(b)) \leq \max_{z' \in Z^-(\mathbf{c})} z'(e_s^u \tau^k(b)). \tag{40}$$

Lemma. *Take $z \in D(Z_t^j)$. Set $j+1 = (s,k)$ and define $n, \mathbf{c}$ with respect to z as above. Assume that $n+1 = k$. Then*

$$z'(b) \leq \varepsilon_t^\star(b), \forall b \in B_J(\infty), z' \in Z^-(\mathbf{c}). \tag{41}$$

implies that

$$z'(b) \leq \varepsilon_t^\star(b), \forall b \in B_J(\infty), z' \in Z(\mathbf{c}). \tag{42}$$

Proof. Given $b \in B_J(\infty)$, define $u \in \mathbb{N}$, as in the lines following (37). As above this gives (40). By invariance, namely (32), the left-hand side of (40) equals

$$\max_{z' \in Z(\mathbf{c})} z'(\tau^k(b)). \tag{43}$$

On the other hand, (41) implies that the right-hand side of (40) is bounded above by $\varepsilon_t^\star(e_s^u \tau^k(b))$. By (12) and Lemma 4.4, the latter expression is bounded above by $\varepsilon_t^\star(\tau^k(b))$, which by (36) is bounded above by $\varepsilon_t^\star(b)$. Thus we have shown that $z'(\tau^k(b)) \leq \varepsilon_t^\star(b)$, for all $b \in B_J(\infty), z' \in Z(\mathbf{c})$. Finally by (39) the left-hand side of this inequality equals $z'(b)$. Hence (42). □

Remark. The idea of the above proof is to make a choice of $b' \in B_J(\infty)$ such that $r_s^i(b')$ takes its maximal value at k. This cannot be achieved without truncation. However we cannot choose truncation arbitrarily since we also require that $k = n+1$, that is $k-1$ is the maximal value of $i \in \mathbb{N}^+$ such that $c_i^z \neq 0$.

6.3 Return to our goal of showing that (34) holds by induction on $j \in J$. It holds for $j = (t,1)$ by Lemma 4.8(iv),(v) since $Z_t^{(t,1)} = \{z_t^1\}$.

Now as in 5.4 fix $j \in J$ with $j \geq (t,1)$ and assume that $Z_t^{j'} \subset K_t^{j'+1}$, for all $j' < j$. Then $Z_t^{j'}$ may be constructed for all $j' \leq j$. Then define Z_t^{j+1} by (23).

Proposition. *Suppose that* (34) *holds for* Z_t^j, *then it holds for* Z_t^{j+1}. *In particular if* $Z_t^{j'} \subset K_t^{j'+1}$, *for all* $j' < j$, *then*

$$z(b) \leq \varepsilon_t^\star(b), \forall z \in Z_t^{j+1}.$$

Proof. As in Lemma 6.2 we can take $j+1 = (s,k)$, for some $s \in I$, $k \in \mathbb{N}^+$. Take $z \in D(Z_t^j)$. Then $n^z + 1 \leq k$. If a strict inequality holds, then z must be a driving function of type s by definition and lie in $Z_t^{(s,k')}$ with $k' = n^z + 1$. Then $j \geq (s,k') + 1$, and so $Z(\mathbf{c}^z) \subset Z_t^j$, by the observation following (27). In this case (42) holds trivially. If equality holds, then the assertion results from Lemma 6.2 and the hypothesis of the proposition. □

7 The description of $B_J(\infty)$ through linear inequalities

7.1 Consider an element $b \in B_J(\infty)$. By definition of the latter, there exists $r \in \mathbb{N}^+$ such that $m_i = 0$, for all $i > r$, that is, we can write $b = (\ldots, 0, 0, m_r, \ldots, m_1) : m_i \in \mathbb{N}$, which we write briefly as $(m_r, \ldots, m_1)$. The question is what are the allowed values of the $m_i : i \leq r$. Our goal is to show that there are given by a set of linear inequalities determined by the dual Kashiwara functions.

In this it is necessary to calculate the set of dual Kashiwara functions not just with respect to J; but with respect to all the $J^j := \ldots, i_k, i_{k-1}, \ldots, i_j : j \in \mathbb{N}^+$, which again define sequences of reduced decompositions.

In view of the above, let $Z_t[j] : t \in I, j \in J$ be a t-semi-invariant set of linear functions on B_{J^j}.

By induction on [6, Lemma 2.5.12(iii)], it follows that b can be written as $(m_r,\dots,m_1)$ if and only if

$$e^{\star}_{i_j} f^{\star m_{j+1}}_{i_{j+1}} \cdots f^{\star m_r}_{i_r} b_\infty = 0, \forall j \in [1, r-1]. \tag{44}$$

In this the expression $f^{\star m_{j+1}}_{i_{j+1}} \cdots f^{\star m_r}_{i_r} b_\infty$ is equal to $(m_r,\dots,m_{j+1})$ *viewed* as an element of $B_{J^{j+1}}$.

We conclude that $b = (m_r,\dots,m_1) \in B_J(\infty)$ *if and only if*

$$e^{\star}_{i_j}(m_r,\dots,m_{j+1}) = 0, \forall j \in [1, r-1]. \tag{45}$$

Since $B_{J^j}(\infty)$ is upper normal [6, 2.5.6] (also with respect to the star structure) this is equivalent to

$$\varepsilon^{\star}_{i_j}(m_r,\dots,m_{j+1}) = 0, \forall j \in [1, r-1]. \tag{46}$$

By Proposition 5.1 we may rewrite (46) in the form

$$z(m_r, m_{r-1},\dots,m_{j+1}) \leq 0, \forall z \in Z_{i_j}[j+1] : j \in [1, r-1]. \tag{47}$$

In this, we recall that the inequalities

$$m_j \geq 0 : j \in [1, r], \tag{48}$$

are assumed from the start (in the definition of B_J).

Thus (47), (48) are precisely the inequalities which must be satisfied by the $m_j : j = 1,2,\dots,r$ in order to obtain $(\dots,0,0,m_r,m_{r-1},\dots,m_1) \in B_J(\infty)$.

7.2 Faces and wiring diagrams Fix $i \in I$, $k \in \mathbb{N}$. Define the linear function on B_J through

$$z^{F_i^{k+1}} := r_i^k - r_i^{k+1}. \tag{49}$$

It is called the face function for the "face" F_i^{k+1}. Faces arise naturally from wiring diagrams in the sense of [4]. From the examples in Figures 1, 2, one can easily deduce the meaning of a face and of a wiring diagram in general. In this one may note that wiring diagrams are defined for all types but are only planar if the Dynkin diagram is linear.

The driving face associated to $t \in I$ is just F_t^1. Its associated function is the driving function z_t^1.

Often we write the face function $z^{F_i^{k+1}}$ simply as F_i^{k+1}.

A significant consequence of the identity $\mathscr{K}_t = \mathscr{K}_t^{BZ}$ (obtained by assuming that there are no "false trails") is that every element of $z^K : K \in \mathscr{K}_t^{BZ}$ may be viewed as a sum of the face functions $F_s^{k+1} : s \in I, k \in \mathbb{N}^+$, added to z_t^1. From the definition [9, Sect. 2] of a trail, it is quite unclear why this should be. It should translate geometrically as a way of describing combinatorially all trails by adjoining appropriately selected faces to the driving face F_t^1.

Now suppose that $\mathfrak{g}$ is of type A.

In this case the above geometric description works out particularly well. Indeed let p be a rigorous path of type t associated to J, in the sense of [4, 5.1]. Let $\mathscr{P}_t$ denote the set of rigorous paths of type t. Let $\mathscr{F}_p$ denote the union of faces enclosed by p and set $z^p := \sum_{F \in \mathscr{F}_p} z^F$.

Lemma. *For all $p \in \mathscr{P}_t$ one has $z^p \in \mathscr{K}_t^{BZ}$ and the map $p \mapsto z^p$ is a bijection Π of $\mathscr{P}_t$ onto $\mathscr{K}_t^{BZ}$ satisfying $z^p = z^{\Pi(p)}$.*

Remark. The proof is quite straightforward using Figure 1 combined with Table 1 as a first example. We leave the details to the reader. In this case, $\mathscr{K}_t^{BZ}$ has itself the structure of a giant S-set in the sense of 5.7.

7.3 A wiring diagram can be a useful way to present a trail K as the corresponding sum of faces. This is what we shall do in the examples below.

7.3.1 A computation of Z_t in type A_3 Consider the reduced decomposition 123212 of the longest element the Weyl group for $\mathfrak{sl}(4)$. This was the first case for which the early theory of Nakashima and Zelevinsky [17] broke down.

The functions in Z_3 obtained from the present theory can be easily computed. The result (and the computation) is described through Table 1.

7.3.2 A comparison with [4] in type A_3 Retain the example of 7.3.1.

We compute $B_J(\infty)$ through 7.1. Here we can assume $r = 6$.

Let us examine the inequalities given by (47). These conditions are empty when $j = 6, 5, 4$.

Suppose $j = 3$. Then we must compute $Z_{i_3}(4) = Z_2(4)$ for the reduced decomposition defined by $J^4 = (i_6, i_5, i_4) = (1,2,3)$. The computation is similar to that given in 7.3.1 but much shorter. Indeed $Z_2(4)$ is just given by the driving function z_2^1 relevant to the given data, and so $Z_{i_3}(4) = \{z_2^1 = m_5 - m_4\}$.

Suppose $j = 2$. Then we must compute $Z_{i_2}(3) = Z_1(3)$ for the reduced decomposition defined by $J^3 = (i_6, i_5, i_4, i_3) = (1,2,3,2)$. This set is given by the driving function z_1^1 for the given data, and so $Z_{i_2}(3) = \{z_1^1 = m_6 - m_5 - m_3\}$.

Suppose $j = 1$. Then we must compute $Z_{i_1}(2) = Z_2(2)$ for the reduced decomposition defined by $J^2 = (i_6, i_5, i_4, i_3, i_2) = (1,2,3,2,1)$. This set is the driving function z_2^1 for the given data combined with the S-set of type 1 it generates, and so $Z_{i_1}(2) = \{z_2^1 = m_3 - m_2, z_2^1 + z^{F_1^2} = m_6 - m_5\}$.

This completes our description of $B_J(\infty)$ using 7.1.

To compare with [4] one notes that the inequalities given there come in three sets corresponding to the three rigorous paths in the sense of [4]. These inequalities are respectively $m_5 - m_4 \leq 0, -m_3 \leq 0$ and $m_6 - m_5 \leq 0, m_3 - m_2 \leq 0, -m_1 \leq 0$ and

$-m_6 \leq 0$. These are exactly the inequalities given by (47), (48), except that in the latter, the inequality $m_6 \leq m_5 + m_3$ is made redundant by $m_6 \leq m_5, m_3 \geq 0$.

7.3.3 Exclusion of functions Retain the example of 7.3.1.

Suppose we just want to know when $(0,0,m_4,m_3,m_2,m_1) \in B_J(\infty)$. Then it suffices to take $r = 4$ in 7.1. The conditions are empty when $j = 4,3,2$.

Suppose $j = 1$. Then we must compute $Z_{i_1}(2) = Z_2(2)$ for the reduced decomposition defined by $J^2 = (i_4,i_3,i_2) = (3,2,1)$. This set is given by the driving function z_2^1 for the given data, and so $Z_{i_1}(2) = \{z_2^1 = m_3 - m_2\}$. Thus $(0,0,m_4,m_3,m_2,m_1) \in B_J(\infty)$ if and only if $m_3 \leq m_2$ (and $m_i \geq 0 : i = 1,2,3,4$). In particular $b := (0,0,m,0,0,m)$ of $B_J(\infty)$, for all $m \in \mathbb{N}^+$.

The function $z = m_4 - m_2$ corresponds to adding $r_2^1 - r_2^2 = m_3 - m_2 + m_1$ to the driving function. It takes the value m on the element $b := (0,0,m,0,0,m)$ of B_J, which by the above belongs to $B_J(\infty)$. Yet one easily checks directly from 4.4 that the correct value of $\varepsilon_3^\star(b) = 0$.

This shows how computing $B_J(\infty)$ inductively can be used to exclude certain functions belonging to Z_t, namely z above.

In terms of faces, the elements of Z_3 correspond to following set of face sums $\{F_3^1, F_3^1 + F_2^3, F_3^1 + F_2^3 + F_2^2, F_2^3, F_3^1 + F_2^3 + F_2^2 + F_1^2\}$. On the other hand, z above corresponds to the face sum $F_3^1 + F_2^2$, which must be excluded. Indeed this face sum is not included in the S-set of type 2 defined by the driving function z_3^1. This provided our first understanding that S-sets should not be symmetric with respect to adjoining faces.

7.3.4 Rigorous paths in type A Retain the example of 7.3.1.

Recall the notion of a pseudo-line in a wiring diagram [4, 2.2] and of a rigorous path [4, 5.1].

The wiring diagram (as defined in [4, 2.2]) is given in Figure 1 below. To compute Z_3, the pseudo-lines are oriented as described in the paragraph following the caption to Figures 1,2. The above sums of faces are those unions enclosed by all rigorous paths starting from L_4 and ending in L_3.

One observes that the face union $F_3^1 + F_2^2$ corresponding to z of 7.3.3 is *not* enclosed in a rigorous path. Indeed the fragment on the right-hand side of the second figure of [4, 5.1] is not avoided. The French will appreciate this rule as a "priorité à droite", whilst the fragment on the left-hand side of the said figure is more appropriately a "priorité à gauche". They exclude certain paths (picturesquely along which one may not drive without stopping or risking "une contravention"). These rules arise through the transition maps [4, Sect. 5] of the semi-tropical calculus. In the present theory they match (a priori very surprisingly) with the asymmetry inherent in an S-set. Of course in type A, every S-set degenerates to its pointed chain (in the sense of [7, 6.3]). The general case is more complicated.

This example in which z is excluded by the inequalities defining $B(\infty)$ is a little too simplistic, in that we can easily directly check that (24) holds.

The form $(.|.)$ extends from $\mathfrak{g}$ to a non-degenerate invariant symmetric bilinear form on $\widehat{\mathfrak{g}}$ by

$$(at^m|bt^n) = \delta_{m,-n}(a|b), \quad (\mathfrak{g}[t,t^{-1}]|\mathbb{C}K + \mathbb{C}d) = 0,$$
$$(K|K) = (d|d) = 0, \quad (K|d) = 1. \tag{1.3}$$

Choosing a Cartan subalgebra $\mathfrak{h}$ of $\mathfrak{g}$, one defines the associated Cartan subalgebra of $\widehat{\mathfrak{g}}$:

$$\widehat{\mathfrak{h}} = \mathfrak{h} + \mathbb{C}d + \mathbb{C}K. \tag{1.4}$$

The restriction of the bilinear form $(.|.)$ from $\widehat{\mathfrak{g}}$ to $\widehat{\mathfrak{h}}$ is non-degenerate, and we identify $\widehat{\mathfrak{h}}$ with $\widehat{\mathfrak{h}}^*$ and $\mathfrak{h}$ with $\mathfrak{h}^*$ using this form. Then d and K are identified with elements, traditionally denoted by Λ_0 and δ respectively. We denote $q = e^{-\delta}$.

Choosing a Borel subalgebra $\mathfrak{b} = \mathfrak{h} + \mathfrak{n}_+$ of $\mathfrak{g}$, one defines the corresponding Borel subalgebra of $\widehat{\mathfrak{g}}$:

$$\widehat{\mathfrak{b}} = \widehat{\mathfrak{h}} + \mathfrak{n}_+ + \bigoplus_{n>0} \mathfrak{g}t^n.$$

Given $\Lambda \in \widehat{\mathfrak{h}}^*$, one extends it to a linear function on $\widehat{\mathfrak{b}}$ by zero on all other summands. Then there exists a unique irreducible $\widehat{\mathfrak{g}}$-module $L(\Lambda)$, which admits an eigenvector v_Λ of $\widehat{\mathfrak{b}}$ with weight Λ. Since K is a central element of $\widehat{\mathfrak{g}}$, it is represented on $L(\Lambda)$ by a scalar $\Lambda(K)$, called the level of $L(\Lambda)$ (and of Λ).

Let $\alpha_1, \ldots, \alpha_\ell$ be simple roots of $\mathfrak{g}$, θ be the highest root, and let $\bar{\Lambda}_1, \ldots, \bar{\Lambda}_\ell$ be its fundamental weights, i.e., $(\bar{\Lambda}_i|\alpha_j^\vee) = \delta_{ij}$, where $\alpha^\vee = 2\alpha/(\alpha|\alpha)$. Then $\alpha_0 = \delta - \theta, \alpha_1, \ldots, \alpha_\ell$ are simple roots of $\widehat{\mathfrak{g}}$, and the fundamental weights Λ_i of $\widehat{\mathfrak{g}}$ are defined by $(\Lambda_i|\alpha_j^\vee) = \delta_{ij}$, $\Lambda_i(d) = 0$, $i, j = 0, 1, \ldots, \ell$. Any $\Lambda \in \widehat{\mathfrak{h}}^*$ can be uniquely written in the form

$$\Lambda = \sum_{i=0}^{\ell} m_i \Lambda_i + a\delta, \text{ where } m_i, a \in \mathbb{C}. \tag{1.5}$$

A $\widehat{\mathfrak{g}}$-module can be "integrated" to the corresponding group, hence it is called integrable, if and only if all m_i are non-negative integers. In this case the level of $L(\Lambda)$ is a non-negative integer.

The character of $L(\Lambda)$ is defined as the following series, corresponding to the weight space decomposition of $L(\Lambda)$ with respect to $\widehat{\mathfrak{h}}$:

$$(\mathrm{ch} L(\Lambda))(h) = \mathrm{tr}_{L(\Lambda)} e^h, \; h \in \widehat{\mathfrak{h}}.$$

This series is convergent in the domain $\{h \in \widehat{\mathfrak{h}} \,|\, \alpha_i(h) > 0, \; i = 0, 1, \ldots, \ell\}$. Note that, adding $b\delta$ to Λ, where $b \in \mathbb{C}$, multiplies the character by q^{-b}. Thus, $\mathrm{ch} L(\Lambda)$ depends essentially only on the labels $m_0, \ldots, m_\ell$ of Λ in (1.5).

If the $\widehat{\mathfrak{g}}$-module $L(\Lambda)$ is integrable, its character is given by the Weyl-Kac character formula:

$$\widehat{R}\mathrm{ch}\,L(\Lambda) = \sum_{w\in\widehat{W}} \varepsilon(w)w(e^{\Lambda+\widehat{\rho}}) = \sum_{w\in W} \varepsilon(w)w \sum_{\gamma\in Q^\vee} t_\gamma(e^{\Lambda+\widehat{\rho}}). \tag{1.6}$$

Here $\widehat{R} = e^{\widehat{\rho}}\prod_{\alpha\in\widehat{\Delta}_+}(1-e^{-\alpha})^{\mathrm{mult}(\alpha)}$ is the affine Weyl denominator, $\widehat{\rho}$ is the affine Weyl vector:

$$\widehat{\rho} = \rho + h^\vee \Lambda_0, \tag{1.7}$$

where ρ is the Weyl vector for $\mathfrak{g}$ and $h^\vee$ is the dual Coxeter number ($=\frac{1}{2}$ the eigenvalue on $\mathfrak{g}$ of the Casimir element). Furthermore, $\widehat{W} = W \ltimes \{t_\alpha \,|\, \alpha\in Q^\vee\}$ is the affine Weyl group, where W is the Weyl group of $\mathfrak{g}$, $\varepsilon(w) = \det_{\widehat{\mathfrak{h}}^*} w$, $Q^\vee = \sum_{i=1}^{\ell}\mathbb{Z}\alpha_i^\vee$ is the coroot lattice of $\mathfrak{g}$, and the translation $t_\gamma \in \mathrm{End}\,\widehat{\mathfrak{h}}^*$ for $\gamma\in\mathfrak{h}^*$ is defined by

$$t_\gamma(\lambda) = \lambda + \lambda(K)\gamma - ((\lambda\,|\,\gamma) + \tfrac{1}{2}\lambda(K)(\gamma\,|\,\gamma))\delta,\ \lambda\in\widehat{\mathfrak{h}}^*. \tag{1.8}$$

The details of the above discussion may be found in the book [K90].

In the paper [KW88] a similar character formula was proved for admissible $L(\Lambda)$, defined by the condition that for $\alpha\in\widehat{\Delta}_+^{\mathrm{re}}$, the set of real roots of $\widehat{\mathfrak{g}}$, the number $(\Lambda+\widehat{\rho}\,|\,\alpha^\vee)$ must be a positive integer each time when it is an integer. Admissible $\widehat{\mathfrak{g}}$-modules include the integrable ones, but exclude, for example the $\widehat{\mathfrak{g}}$-modules $L(k\Lambda_0)$, where k is a negative integer.

It is known for arbitrary (non-critical, i.e., of level $\neq -h^\vee$) Λ that

$$\widehat{R}\mathrm{ch}\,L(\Lambda) = \sum_{w\in\widehat{W}} c(w)w(e^{\Lambda+\widehat{\rho}}), \tag{1.9}$$

where $c(w)$ are integers [KK79], and that $c(w)$ can be computed via the Kazhdan-Lusztig polynomials for $\widehat{W}$ [KT00]. However explicit formulas for the integers $c(w)$ are unknown in general.

In Sections 2 and 3 of this paper we find explicit character formulas for level -1 modules $L(\Lambda)$ over $\widehat{s\ell}_n$ and over $\widehat{sp}_n$ with $n\geq 3$, with highest weights $\Lambda = -(1+s)\Lambda_0 + s\Lambda_1$ and $\Lambda = -(1+s)\Lambda_0 + s\Lambda_{n-1}$, $s\in\mathbb{Z}$, and $\Lambda = -(1+s)\Lambda_0 + s\Lambda_1$, $s\in\mathbb{Z}_{\geq 0}$ and $\Lambda = -2\Lambda_0+\Lambda_2$, respectively (see Theorems 2.1, and 3.1, 3.2 respectively). In particular, we compute in both cases the character of $L(-\Lambda_0)$, which are simple affine vertex algebras of level -1. As shown in [AP12] and [AP14], the above modules are all $\mathfrak{g}$-locally finite irreducible highest weight modules over these vertex algebras. The main ingredients of the proof are the free field realizations of these modules, given in [KW01], the irreducibility theorems from [AP14], [AP12], and the affine denominator identity for affine Lie superalgebras $\widehat{s\ell}_{n|1}$ and $\widehat{spo}_{n|2}$, given in [KW94], [G11].

In Section 4 we indicate a proof, under a certain hypothesis, of an explicit character formula for certain modules $L(\Lambda)$ of negative integer level over affine Lie algebras, and conjecture that the hypothesis holds for the affine Lie algebras of Deligne series $\widehat{D}_4$, $\widehat{E}_6$, $\widehat{E}_7$ and $\widehat{E}_8$.

Throughout the paper the base field is $\mathbb{C}$.

Both of us wish to thank ESI, Vienna, where we began discussion of this paper, for its hospitality. The first named author wishes to thank IHES, where the paper was completed, for perfect working conditions. We would like to thank Maria Gorelik for important suggestions.

2 Proof of the character formulas for $\widehat{\mathfrak{g}}$, where $\mathfrak{g} = s\ell_n$

In this section we prove a character formula for certain highest weight modules $L(\Lambda)$ of level -1 over the affine Lie algebra $\widehat{s\ell}_n$. The normalized invariant bilinear form $(.|.)$ on $s\ell_n$ is the trace form. We choose as its Cartan subalgebra, as usual, the subalgebra of all diagonal traceless matrices. Then the simple roots of $s\ell_n$ are $\alpha_1 = \varepsilon_1 - \varepsilon_2, \dots, \alpha_{n-1} = \varepsilon_{n-1} - \varepsilon_n$, where $\varepsilon_1, \dots, \varepsilon_n$ is the standard basis of the dual of all diagonal matrices, and its root (= coroot) lattice is $Q = \sum_{i=1}^{n-1} \mathbb{Z}\alpha_i$.

We will also use the embedding of $s\ell_n$ in the Lie superalgebra $s\ell_{n|1}$. The trace form on $s\ell_n$ extends to the supertrace form $(.|.)$ on $s\ell_{n|1}$, and its Cartan subalgebra embeds in the Cartan subalgebra of $s\ell_{n|1}$ of supertraceless diagonal matrices. Then the simple roots of $s\ell_{n|1}$ are $\alpha_1, \dots, \alpha_{n-1}, \alpha_n = \varepsilon_n - \varepsilon_{n+1}$, where α_n is an odd root. Also, $s\ell_n$ and $s\ell_{n|1}$ have the same Weyl group, and their dual Coxeter numbers are n and $n-1$ respectively.

Theorem 2.1. *Let $n \geq 3$, and let $L(\Lambda)$ be an irreducible level -1 $\widehat{s\ell}_n$-module with highest weight*

$$\Lambda = -(1+s)\Lambda_0 + s\Lambda_{n-1} \ (resp. \ = -(1+s)\Lambda_0 + s\Lambda_1),\ s \in \mathbb{Z}_{\geq 0}.$$

Then the character of $L(\Lambda)$ is given by the following formula:

$$\widehat{R}\operatorname{ch} L(\Lambda) = \sum_{w \in W} \varepsilon(w) w \sum_{\substack{\gamma \in Q \\ (\gamma|\bar{\Lambda}_{n-1}(resp.\ \bar{\Lambda}_1)) \geq 0}} t_\gamma(e^{\Lambda + \widehat{\rho}}). \tag{2.1}$$

The proof of formula (2.1) uses the free field construction, given in [KW01], of the affine Lie superalgebra $\widehat{g\ell}_{m|n}$ of level 1 in a Fock space F in the case of $m = 0$. Note that in that paper we used the supertrace form, which is equal to the negative of the trace form on $g\ell_n = g\ell_{0|n}$. Hence we get a $\widehat{g\ell}_n$-module structure on F of level -1. Recall some properties of this module, described in [KW01].

First, we have the charge decomposition into a direct sum of $\widehat{g\ell}_n$-submodules:

$$F = \bigoplus_{s \in \mathbb{Z}} F_s. \tag{2.2}$$

Secondly, there is a Virasoro algebra acting on F, and leaving all subspaces F_s invariant, for which all fields $a(z)$, $a \in g\ell_n$, are primary of conformal weight 1, and each F_s in (2.2) has a unique, up to a constant factor, non-zero vector $|s\rangle$ with minimal L_0-eigenvalue (see Section 3 for more details). Moreover this vector is invariant with respect to the Cartan subalgebra of $\widehat{g\ell}_n$ and has the following weight:

$$\text{weight}\,|s\rangle = \begin{cases} -\Lambda_0 - \frac{s}{2}\delta + s\varepsilon_1 & \text{if } s \in \mathbb{Z}_{\geq 0}, \\ -\Lambda_0 + \frac{s}{2}\delta + s\varepsilon_n & \text{if } s \in \mathbb{Z}_{\leq 0}. \end{cases} \tag{2.3}$$

Thirdly, by formula (3.15) from [KW01], the character of the $\widehat{g\ell}_n$-module F is given by

$$\operatorname{ch} F := \sum_{s\in\mathbb{Z}} x^s \operatorname{ch} F_s = e^{-\Lambda_0} \prod_{j=1}^{n} \prod_{k=1}^{\infty} (1 - x e^{\varepsilon_j} q^{k-\frac{1}{2}})^{-1} (1 - x^{-1} e^{-\varepsilon_j} q^{k-\frac{1}{2}})^{-1}.$$

Letting in this formula $x = e^{-\varepsilon_{n+1}} q^{\frac{1}{2}}$, we obtain

$$e^{-\Lambda_0} \prod_{j=1}^{n} \prod_{k=1}^{\infty} (1 - e^{\varepsilon_j - \varepsilon_{n+1}} q^k)^{-1} (1 - e^{-(\varepsilon_j - \varepsilon_{n+1})} q^{k-1})^{-1} = \sum_{s\in\mathbb{Z}} e^{-s\varepsilon_{n+1}} q^{\frac{s}{2}} \operatorname{ch} F_s. \tag{2.4}$$

It was proved in [AP14] that all $\widehat{g\ell}_n$-modules F_s are irreducible, provided that $n \geq 3$. Therefore, using (2.3), we see that $F_s = V(\lambda^{(s)}) \otimes L(\Lambda^{(s)})$, where $V(\lambda^{(s)})$ is an irreducible $\widehat{g\ell}_1$-module with highest weight $\lambda^{(s)} \in \mathbb{C}\sum_{i=1}^{n} \varepsilon_i + \mathbb{C}\delta$ and $L(\Lambda^{(s)})$ is an irreducible $\widehat{s\ell}_n$-module with highest weight

$$\Lambda^{(s)} = -\Lambda_0 + s\bar{\Lambda}_1 \text{ (resp. } -\Lambda_0 - s\bar{\Lambda}_{n-1}) \in \mathfrak{h}^* \text{ if } s \geq 0 \text{ (resp. } s \leq 0),$$

where $\lambda^{(s)} \oplus \Lambda^{(s)} = \text{weight}\,|s\rangle$.

Hence, using that $\Lambda_i = \Lambda_0 + \bar{\Lambda}_i$, we obtain that the character of F_s is given by

$$\varphi(q) \operatorname{ch} F_s = \begin{cases} q^{\frac{s}{2}} e^{s(\varepsilon_1 - \bar{\Lambda}_1)} \operatorname{ch} L(-(1+s)\Lambda_0 + s\Lambda_1) & \text{if } s \in \mathbb{Z}_{\geq 0}, \\ q^{-\frac{s}{2}} e^{s(\varepsilon_n + \bar{\Lambda}_{n-1})} \operatorname{ch} L(-(1-s)\Lambda_0 - s\Lambda_{n-1}) & \text{if } s \in \mathbb{Z}_{\leq 0}. \end{cases} \tag{2.5}$$

Here and thereafter $\varphi(q) = \prod_{n=1}^{\infty} (1 - q^n)$. Substituting (2.5) in the RHS of (2.4), we obtain

$$\begin{aligned} \text{LHS of (2.4)} = \frac{1}{\varphi(q)} \Bigg(& \sum_{s>0} e^{s(\varepsilon_1 - \varepsilon_{n+1})} q^s e^{-s\bar{\Lambda}_1} \operatorname{ch} L(-(1+s)\Lambda_0 + s\Lambda_1) \\ & + \sum_{s\leq 0} e^{s(\varepsilon_n - \varepsilon_{n+1})} e^{s\bar{\Lambda}_{n-1}} \operatorname{ch} L(-(1-s)\Lambda_0 - s\Lambda_{n-1}) \Bigg). \end{aligned} \tag{2.6}$$

Next, we embed the Lie algebra $s\ell_n$ in the Lie superalgebra $s\ell_{n|1}$ as described above. We extend this embedding to the affine Lie (super)algebras $\widehat{s\ell}_n \hookrightarrow \widehat{s\ell}_{n|1}$. Then identitiy (2.6) can be rewritten as follows:

$$e^{-\Lambda_0}\varphi(q)\prod_{j=1}^{n}\prod_{k=1}^{\infty}(1-e^{\alpha_j+\cdots+\alpha_n}q^k)^{-1}(1-e^{-(\alpha_j+\cdots+\alpha_n)}q^{k-1})^{-1}$$
$$=\sum_{s>0}e^{s(\alpha_1+\cdots+\alpha_n)}q^s e^{-s\bar{\Lambda}_1}\mathrm{ch}\,L(-(1+s)\Lambda_0+s\Lambda_1)$$
$$+\sum_{s\le 0}e^{s\alpha_n}e^{s\bar{\Lambda}_{n-1}}\mathrm{ch}\,L(-(1-s)\Lambda_0-s\Lambda_{n-1}). \quad (2.7)$$

We denote the 0-th fundamental weight and the Weyl vector for $\widehat{s\ell}_{n|1}$ by Λ_0' and $\widehat{\rho}'$ respectively. Then, by (1.7), we have, when restricted to $\widehat{s\ell}_n$:

$$\Lambda_0'=\Lambda_0 \text{ and } \widehat{\rho}'=\widehat{\rho}-\Lambda_0. \quad (2.8)$$

Recall the formulas for the Weyl denominator $\widehat{R}$ for $\widehat{s\ell}_n$ and the Weyl superdenominator $\widehat{R}'$ for $\widehat{s\ell}_{n|1}$:

$$\widehat{R}=e^{\widehat{\rho}}\varphi(q)^{n-1}\prod_{\alpha\in\widehat{\Delta}_+^{\mathrm{re}}}(1-e^{-\alpha}), \quad (2.9)$$

$$\widehat{R}'=e^{\widehat{\rho}'}\varphi(q)^{n}\prod_{\alpha\in\widehat{\Delta}_+^{\mathrm{re}}}(1-e^{-\alpha})$$
$$\times\prod_{j=1}^{n}\prod_{k=1}^{\infty}(1-e^{\alpha_j+\cdots+\alpha_n}q^k)^{-1}(1-e^{-(\alpha_j+\cdots+\alpha_n)}q^{k-1})^{-1}. \quad (2.10)$$

Multiplying both sides of (2.7) by $\widehat{R}$, we obtain, using (2.9) and (2.10):

$$\widehat{R}'=\sum_{s>0}e^{s(\alpha_1+\cdots+\alpha_n)}q^s e^{-s\bar{\Lambda}_1}\widehat{R}\mathrm{ch}\,L(-(1+s)\Lambda_0+s\Lambda_1)$$
$$+\sum_{s\le 0}e^{s\alpha_n}e^{s\bar{\Lambda}_{n-1}}\widehat{R}\mathrm{ch}\,L(-(1-s)\Lambda_0-s\Lambda_{n-1}). \quad (2.11)$$

On the other hand, $\widehat{R}'$ can be computed by the superdenominator identity [KW94], [G11]:

$$\widehat{R}'=\sum_{w\in W}\varepsilon(w)w\sum_{\gamma\in Q}t_\gamma\frac{e^{\widehat{\rho}'}}{1-e^{-\alpha_n}}. \quad (2.12)$$

Expanding $t_\gamma\frac{e^{\widehat{\rho}'}}{1-e^{-\alpha_n}}$ in the geometric series in the domain $|e^{-\alpha_n}|<1,\ |q|<1$, we obtain for $\gamma\in Q$ (using (1.8)):

$$t_\gamma\frac{e^{\widehat{\rho}'}}{1-e^{-\alpha_n}}=e^{(n-1)\Lambda_0}\left(\sum_{\substack{p\ge 0\\(\gamma|\alpha_n)\le 0}}-\sum_{\substack{p<0\\(\gamma|\alpha_n)>0}}\right)e^{\rho+(n-1)\gamma-p\alpha_n}q^{\frac{n-1}{2}(\gamma|\gamma)+(\rho|\gamma)-p(\alpha_n|\gamma)}.$$

Using that $n\alpha_n=\sum_{i=1}^{n}\varepsilon_i-n\bar{\Lambda}_{n-1}$, and that $(\alpha_n|\gamma)=-(\bar{\Lambda}_{n-1}|\gamma)$ for $\gamma\in Q$, we can rewrite this formulas as

$$t_\gamma \frac{e^{\widehat{\rho}'}}{1-e^{-\alpha_n}} = e^{(n-1)\Lambda_0}\left(\sum_{\substack{p\geq 0 \\ (\gamma|\bar{\Lambda}_{n-1})\geq 0}} - \sum_{\substack{p<0 \\ (\gamma|\bar{\Lambda}_{n-1})<0}} \right)$$

$$\times e^{-\frac{p}{n}\sum_{i=1}^n \varepsilon_i} e^{\rho+(n-1)\gamma+n\bar{\Lambda}_{n-1}} q^{\frac{n-1}{2}(\gamma|\gamma)+(\rho|\gamma)+p(\bar{\Lambda}_{n-1}|\gamma)}.$$

Plugging this into (2.12) and using that $\sum_{i=1}^n \varepsilon_i$ is W-invariant, we obtain

$$\widehat{R}' = e^{(n-1)\Lambda_0} \sum_{\gamma\in Q} \left(\sum_{\substack{p\geq 0 \\ (\gamma|\bar{\Lambda}_{n-1})\geq 0}} - \sum_{\substack{p<0 \\ (\gamma|\bar{\Lambda}_{n-1})<0}} \right)$$

$$\times e^{-p(\alpha_n+\bar{\Lambda}_{n-1})} q^{\frac{n-1}{2}(\gamma|\gamma)+(\rho|\gamma)+p(\bar{\Lambda}_{n-1}|\gamma)} \sum_{w\in W} \varepsilon(w) w e^{\rho+(n-1)\gamma+p\bar{\Lambda}_{n-1}}. \quad (2.13)$$

Thus we obtain the identity

$$\text{RHS of (2.11)} = \text{RHS of (2.13)}.$$

Comparing the coefficient of $e^{-p\alpha_n}$ for $p\geq 0$ in this identity, we obtain

$$\widehat{R}\mathrm{ch} L(-(1+p)\Lambda_0 + p\Lambda_{n-1}) = e^{(n-1)\Lambda_0} \sum_{\substack{\gamma\in Q \\ (\gamma|\bar{\Lambda}_{n-1})\geq 0}} q^{\frac{n-1}{2}(\gamma|\gamma)+(\rho|\gamma)+p(\bar{\Lambda}_{n-1}|\gamma)}$$

$$\times \sum_{w\in W} \varepsilon(w) w e^{\rho+(n-1)\gamma+p\bar{\Lambda}_{n-1}} = \sum_{w\in W} \varepsilon(w) w \sum_{\substack{\gamma\in Q \\ (\gamma|\bar{\Lambda}_{n-1})\geq 0}} t_\gamma e^{-(1+p)\Lambda_0+p\Lambda_{n-1}+\widehat{\rho}}.$$

This establishes formula (2.1) for $\Lambda = -(1+s)\Lambda_0 + s\Lambda_{n-1}$. Formula for $\Lambda = -(1+s)\Lambda_0 + s\Lambda_1$ follows by the involution of the Dynkin diagram of $\widehat{s\ell}_n$ which keeps the 0th node fixed.

Remark 2.2. In the case $\Lambda = -\Lambda_0$, formula (2.1) can be written in a more symmetric form:

$$\widehat{R}\mathrm{ch} L(-\Lambda_0) = \sum_{w\in W} \varepsilon(w) w \sum_{\substack{\gamma\in Q \\ (\gamma|\bar{\Lambda}_1)\geq 0, (\gamma|\bar{\Lambda}_{n-1})\geq 0}} t_\gamma(e^{-\Lambda_0+\widehat{\rho}}).$$

Indeed, let $\Omega = \{\gamma\in Q \,|\, (\gamma|\bar{\Lambda}_1)\geq 0,\, (\gamma|\bar{\Lambda}_{n-1})<0\}$. Then $\gamma\in\Omega$ iff $r_\theta(\gamma)-\theta\in\Omega$. It follows that

$$\sum_{w\in W} \varepsilon(w) w \sum_{\gamma\in\Omega} t_\gamma(e^{-\Lambda_0+\widehat{\rho}}) = 0,$$

which proves the claim. We are grateful to the referee for this remark.

Remark 2.3. Let $\Lambda = -(1+s)\Lambda_0 + s\Lambda_1$ and let $L(\Lambda) = \bigoplus_{j\in\mathbb{Z}_{\geq 0}} L(\Lambda)_j$ be the eigenspace decomposition of $L(\Lambda)$ with respect to $-d$. Let $\dim_q L(\Lambda) = \sum_{j\geq 0}(\dim L(\Lambda)_j)q^j$ be the "homogeneous" q-dimension of $L(\Lambda)$. Dividing both sides of the last equality in the proof of (2.1) by $e^{(n-1)\Lambda_0}R$, where R is the Weyl denominator of $\mathfrak{g}$, and letting all elements of $\mathfrak{h}^*$ equal 0, we obtain by the usual argument:

$$\varphi(q)^{\dim\mathfrak{g}} \dim_q L(\Lambda) = \sum_{\substack{\gamma\in Q\\ (\bar{\Lambda}_1|\gamma)\geq 0}} \dim(s\bar{\Lambda}_1 + (n-1)\gamma) q^{\frac{n-1}{2}(\gamma|\gamma)+(s\bar{\Lambda}_1+\rho|\gamma)},$$

where $\dim(\lambda) = \prod_{\alpha\in\Delta_+}(\lambda+\rho|\alpha)/(\rho|\alpha)$ is the expression of the Weyl dimension formula for $\mathfrak{g}$.

Remark 2.4. For $\widehat{s\ell}_2$ the characters of the above modules are very easy (see, e.g., [KW88]): $\widehat{R}\,\mathrm{ch}\, L(-(1+s)\Lambda_0 + s\Lambda_1) = e^{-(1+s)\Lambda_0+s\Lambda_1+\widehat{\rho}}\left(1-e^{-(s+1)\alpha_1}\right)$, $s\in\mathbb{Z}_{\geq 0}$.

3 Proof of the character formulas for $\widehat{\mathfrak{g}}$, where $\mathfrak{g} = sp_n$

In order to prove the character formulas for $\widehat{sp}_n$, where $n \geq 4$, even, we use the embedding of sp_n in $s\ell_n$, obtained as the fixed point subalgebra of the involution σ, corresponding to the flip of the Dynkin diagram of $s\ell_n$. The normalized invariant bilinear form $(.|.)$ on sp_n is again the trace form. Let $n' = \frac{n}{2}$.

We will also use the embedding of the Lie algebra sp_n in the Lie superalgebra $spo_{n|2}$. The simple roots of the latter are $\alpha_*, \alpha_1, \ldots, \alpha_{n'-1}, \alpha_{n'}$, where α_* is an odd root, and $\alpha_1, \ldots, \alpha_{n'}$ are the simple roots of the former. We let $Q^\vee = \sum_{i=1}^{n'}\mathbb{Z}\alpha_i^\vee$ be the coroot ($\neq$ root) lattice of sp_n.

In this section we prove the following two theorems.

Theorem 3.1. *Let $n\geq 4$ be even, and let $L(\Lambda)$ be an irreducible $\widehat{sp}_n$-module with highest weight Λ.*

(a) *If $\Lambda = -(1+s)\Lambda_0 + s\Lambda_1$, where $s\in\mathbb{Z}_{\geq 1}$, then*

$$\mathrm{ch}\, L(\Lambda) = \widehat{R}^{-1}\sum_{w\in W}\varepsilon(w)w\sum_{\substack{\gamma\in Q^\vee\\ (\gamma|\bar{\Lambda}_1)\geq 0}} t_\gamma(e^{\Lambda+\widehat{\rho}})$$

(b) *If $\Lambda = -\Lambda_0$, then*

$$\mathrm{ch}\, L(\Lambda) = \frac{1}{2}\left(\widehat{R}^{-1}\sum_{w\in W}\varepsilon(w)w\sum_{\substack{\gamma\in Q^\vee\\ (\gamma|\bar{\Lambda}_1)\geq 0}} t_\gamma(e^{\Lambda+\widehat{\rho}}) + e^{-\Lambda_0}\frac{\varphi(q^2)}{\varphi(q)}\prod_{\alpha\in\Delta_\ell}\prod_{k\in\mathbb{Z}_{odd>0}}(1-e^\alpha q^k)^{-1}\right),$$

where Δ_ℓ is the set of long roots of sp_n.

(c) *If* $\Lambda = -2\Lambda_0 + \Lambda_2$, *then* $\mathrm{ch}\,L(\Lambda)$ *is obtained by an expression, obtained from* (b) *by dividing by* q *and replacing plus by minus between the summands.*

Theorem 3.2. *The characters of the* $\widehat{sp}_n$*-modules* $L(-\Lambda_0)$ *and* $L(-2\Lambda_0+\Lambda_2)$ *can be written in the form* (1.9) *as follows:*

(a)

$$\widehat{R}\mathrm{ch}\,L(-\Lambda_0) = \sum_{w\in W} \varepsilon(w)w \sum_{\substack{\gamma\in Q^\vee \\ (\gamma|\bar{\Lambda}_1)\geq 0 \\ (\gamma|\bar{\Lambda}_{n'})\in 2\mathbb{Z}}} t_\gamma(e^{\Lambda+\widehat{\rho}}).$$

(b)

$$\widehat{R}\mathrm{ch}\,L(-2\Lambda_0+\Lambda_2) = \sum_{w\in W} \varepsilon(w)w \sum_{\substack{\gamma\in Q^\vee \\ (\gamma|\bar{\Lambda}_2-\bar{\Lambda}_1)\geq 0 \\ (\gamma|\bar{\Lambda}_{n'})\in 2\mathbb{Z}}} t_\gamma(e^{\Lambda+\widehat{\rho}}).$$

First, letting in the character $\mathrm{ch}F$ of the $\widehat{g\ell}_n$-module F, considered in Section 2, $x = e^{\alpha_*-\varepsilon_{n'}}q^{\frac{1}{2}}$, and restricting the module F to $\widehat{sp}_n$, we obtain (cf. (1.4))

$$\begin{aligned}
& e^{-\Lambda_0}\prod_{k=1}^{\infty}(1-e^{\alpha_*}q^k)(1-e^{-\alpha_*}q^{k-1}) \\
&\quad\times\prod_{k=1}^{\infty}\prod_{i=1}^{n'}(1-e^{\alpha_*+\alpha_1+\cdots+\alpha_i}q^k)(1-e^{-\alpha_*-\alpha_1-\cdots-\alpha_i}q^{k-1}) \\
&\quad\times\prod_{k=1}^{\infty}\prod_{i=1}^{n'-1}(1-e^{\alpha_*+\alpha_1+\cdots+2\alpha_i+\cdots+2\alpha_{n'-1}+\alpha_{n'}}q^k) \\
&\qquad\qquad\times(1-e^{-\alpha_*-\alpha_1-\cdots-2\alpha_i-\cdots-2\alpha_{n'-1}-\alpha_{n'}}q^{k-1}) \\
&\qquad\qquad\qquad = \sum_{s\in\mathbb{Z}} e^{s(\alpha_*-\varepsilon_{n'})}q^{\frac{s}{2}}\mathrm{ch}\,F_s\big|_{\widehat{sp}_n}. \qquad (3.1)
\end{aligned}$$

Next, we embed the Lie algebra sp_n in the Lie superalgebra $spo_{n|2}$ as described above. We denote the 0-th fundamental weight and the Weyl vector for $\widehat{spo}_{n|2}$ by Λ_0' and $\widehat{\rho}'$ respectively. Then, by (1.7), we have again (2.8), when restricted to $\widehat{sp}_n$.

Denote by $\widehat{R}$ and $\widehat{R}'$ the Weyl denominator and Weyl superdenominator for $\widehat{sp}_n$ and $\widehat{spo}_{n|2}$ respectively. Then we have by (2.8) for sp_n :

$$\widehat{R}' = \varphi(q)\widehat{R}\times(\text{LHS of (3.1)}). \qquad (3.2)$$

On the other hand, the superdenominator identity for $\widehat{spo}_{n|2}$ reads [KW94], [G11]:

$$\widehat{R}' = \sum_{w\in\widehat{W}} \varepsilon(w)w\frac{e^{\widehat{\rho}'}}{1-e^{-\alpha_*}}, \qquad (3.3)$$

where $\widehat{W} = W \ltimes \{t_\gamma \mid \gamma \in Q^\vee\}$ is the Weyl group of $\widehat{sp}_n$.

Using (3.1)–(3.3) and applying the same argument as in Section 2, we obtain for every $s \in \mathbb{Z}_{\geq 0}$ that

$$\varphi(q)\widehat{R}(\mathrm{ch}\, F_s)\big|_{\widehat{sp}_n} = \sum_{w\in W} \varepsilon(w) w \sum_{\substack{\gamma\in Q^\vee \\ (\gamma|\bar{\Lambda}_1)\geq 0}} t_\gamma(e^{\Lambda+\widehat{\rho}}). \tag{3.4}$$

This proves claim (a) of Theorem 3.1, since, due to [AP12], the restriction of the $\widehat{s\ell}_n$-module $L(-\Lambda_0 + s\bar{\Lambda}_1)$ to $\widehat{sp}_n$ is irreducible for $s > 0$.

In order to prove claims (b) and (c), we need to study the module F more carefully. Recall that F is the unique irreducible module over the Clifford algebra Cl with generators $\varphi_k^{(i)}$ and $\varphi_k^{(i)*}$, $i = 1,\dots,n$, $k \in \frac{1}{2} + \mathbb{Z}$, with relations $\varphi_{-k}^{(i)*}\varphi_k^{(i)} - \varphi_k^{(i)}\varphi_{-k}^{(i)*} = 1$ and $= 0$ in the rest of the cases, which admits a non-zero vector $|0\rangle$, such that $\varphi_k^{(i)}|0\rangle = 0 = \varphi_k^{(i)*}|0\rangle$ for all $k > 0$, $i = 1,\dots,n$. The charge decomposition (2.2) is defined by letting charge $|0\rangle = 0$, charge $\varphi_k^{(i)} = 1 = -\mathrm{charge}\, \varphi_k^{(i)*}$.

Note that the algebra Cl carries an involution σ, defined by

$$\sigma(\varphi_k^{(i)}) = (-1)^i \varphi_k^{(n+1-i)*}.$$

This involution induces an involution of the space F_0, denoted again by σ, letting $\sigma|0\rangle = |0\rangle$, so that we have its eigenspace decomposition

$$F_0 = F_0^1 \oplus F_0^{-1}. \tag{3.5}$$

The space F_0 is spanned by monomials

$$v = \varphi_{-k_1}^{(i_1)} \cdots \varphi_{-k_m}^{(i_m)} \varphi_{-l_1}^{(j_1)*} \cdots \varphi_{-l_m}^{(j_m)*}|0\rangle. \tag{3.6}$$

Since

$$\sigma(v) = (-1)^{\sum_{p=1}^m (i_p+j_p)+m} \varphi_{-l_1}^{(n+1-j_1)} \cdots \varphi_{-l_m}^{(n+1-j_m)} \varphi_{-k_1}^{(n+1-i_1)*} \cdots \varphi_{-k_m}^{(n+1-i_m)*}|0\rangle,$$

we see that, if $v \in F_0^{-1}$, we have $(1 \leq p \leq m)$

$$n+1-j_p = i_p, \quad n+1-i_p = j_p, \quad l_p = k_p; \tag{3.7}$$

$$\sum_{p=1}^m (i_p + j_p) + m \equiv 1 \mod 2. \tag{3.8}$$

By (3.7) we have $\sum_{p=1}^m i_p = \sum_{p=1}^m (n+1) - \sum_{p=1}^m j_p$. Therefore $\sum_{p=1}^m (i_p + j_p) = m \mod 2$, which contradicts (3.8). Hence F_0^{-1} contains no monomials (3.6).

Thus, for a monomial (3.6) we have that either $\sigma(v) = v$, or v and $\sigma(v)$ are linearly independent. Denote by $F_0^\sharp$ the subspace of F_0 spanned by monomials fixed by σ. From the above discussion we obtain

$$\operatorname{ch} F_0^{\pm 1} = \frac{1}{2}(\operatorname{ch} F_0 \pm \operatorname{ch} F_0^\sharp). \tag{3.9}$$

Recall the construction of the representation of $\widehat{g\ell}_n$ of level -1 in F [KW01]. Let $\varphi^{(i)}(z) = \sum_{k\in\frac{1}{2}+\mathbb{Z}} \varphi_k^{(i)} z^{-k-\frac{1}{2}}$, $\varphi^{(i)*}(z) = \sum_{k\in\frac{1}{2}+\mathbb{Z}} \varphi_k^{(i)*} z^{-k-\frac{1}{2}}$. Then

$$e_{ij}(z) \mapsto :\varphi^{(i)}(z)\varphi^{(j)*}(z):, \; K \mapsto -1, \; d \mapsto -L_0 \tag{3.10}$$

defines a representation of $\widehat{g\ell}_n$ in F of level -1 (preserving (2.2)). Here

$$\sum_{k\in\mathbb{Z}} L_k z^{-k-2} = \frac{1}{2} \sum_{j=1}^{n} (:\partial\varphi^{(j)}(z)\varphi^{(j)*}(z): - :\partial\varphi^{(j)*}(z)\varphi^{(j)}(z):)$$

is the representation in F of the Virasoro algebra. In particular, the Heisenberg subalgebra H of $\widehat{g\ell}_n$ is represented in F as

$$\sum_{k\in\mathbb{Z}} (I_n t^k) z^{-k-1} \mapsto \sum_{k\in\mathbb{Z}} H_k z^{-k-1} = \sum_{i=1}^{n} :\varphi^{(i)}(z)\varphi^{(i)*}(z):.$$

Since the $\widehat{gl}_n$-module F_0 is irreducible [AP14], we have its following decomposition as an $H \oplus \widehat{sl}_n$-module

$$F_0 = V \otimes L(-\Lambda_0), \tag{3.11}$$

where V is an irreducible H-module with highest weight vector $|0\rangle$, i.e. $(I_n t^k)|0\rangle = 0$ for $k \geq 0$. We obviously have

$$V = \mathbb{C}[H_{-k} \mid k \in \mathbb{Z}_{>0}]\,|0\rangle \text{ and } \sigma(H_k) = -H_k.$$

Hence, in particular V is σ-invariant, so that we have the eigenspace decomposition with respect to $\sigma : V = V^1 \oplus V^{-1}$. Note that the action of $\widehat{sp}_n \subset \widehat{g\ell}_n$ on F_0 commutes with the action of σ on F_0, hence both F_0^1 and F_0^{-1} are $\widehat{sp}_n$-modules. Moreover, due to [AP12], the $\widehat{s\ell}_n$-module $L(-\Lambda_0)$, restricted to $\widehat{sp}_n$, is a direct sum of two irreducible modules, with highest weights $-\Lambda_0$ and $-\Lambda_0 + \bar{\Lambda}_2 \mod \mathbb{C}\delta$. But it is easy to see that

$$(\varphi_{-\frac{1}{2}}^{(1)} \varphi_{-\frac{1}{2}}^{(n-1)*} + \varphi_{-\frac{1}{2}}^{(2)} \varphi_{-\frac{1}{2}}^{(n)*})|0\rangle$$

is a singular vector for $\widehat{sp}_n$, and its weight is $-\Lambda_0 + \bar{\Lambda}_2 - \delta$. Thus we obtain

Lemma 3.3. *As an $H \oplus \widehat{sp}_n$-module, one has*

$$F_0^1 \simeq V^1 \otimes L(-\Lambda_0) + V^{-1} \otimes L(-\Lambda_0 + \bar{\Lambda}_2 - \delta),$$
$$F_0^{-1} = V^{-1} \otimes L(-\Lambda_0) + V^1 \otimes L(-\Lambda_0 + \bar{\Lambda}_2 - \delta).$$

It is easy to see that

$$\operatorname{ch} V^{\pm 1} = \frac{1}{2}\left(\frac{1}{\varphi(q)} \pm \frac{\varphi(q)}{\varphi(q^2)}\right), \tag{3.12}$$

hence we have

$$(\operatorname{ch} V^1)^2 - (\operatorname{ch} V^{-1})^2 = \frac{1}{\varphi(q^2)}. \tag{3.13}$$

Next, we obviously have

$$F_0^\sharp = \mathbb{C}\left[\varphi_{-k}^{(i)}\varphi_{-k}^{(n+1-i)*} \,\middle|\, 1 \le i \le n,\ k \in \frac{1}{2} + \mathbb{Z}_{\ge 0}\right]|0\rangle,$$

hence

$$\operatorname{ch} F_0^\sharp = e^{-\Lambda_0} \prod_{k \in \mathbb{Z}_{\mathrm{odd}>0}} \prod_{i=1}^{n} (1 - e^{\varepsilon_i - \varepsilon_{n+1-i}} q^k)^{-1}.$$

It follows that

$$\operatorname{ch} F_0^\sharp|_{\widehat{sp}_n} = e^{-\Lambda_0} \prod_{k \in \mathbb{Z}_{\mathrm{odd}>0}} \prod_{\alpha \in \Delta_\ell} (1 - e^\alpha q^k)^{-1}. \tag{3.14}$$

Now we are able to complete the proofs of claims (b) and (c) of Theorem 3.1. By Lemma 3.3, we have

$$\operatorname{ch} V^{\pm 1} \operatorname{ch} L(\Lambda_0) + \operatorname{ch} V^{\mp 1} \operatorname{ch}(-\Lambda_0 + \bar{\Lambda}_2 - \delta) = \operatorname{ch} F_0^{\pm 1}. \tag{3.15}$$

From (3.9), (3.13) and (3.15) we obtain

$$\frac{1}{\varphi(q^2)} \operatorname{ch} L(\Lambda_0) = \frac{1}{2}\left(\operatorname{ch} V^1 - \operatorname{ch} V^{-1}\right) \operatorname{ch} F_0|_{\widehat{sp}_n} + \frac{1}{2}(\operatorname{ch} V^1 + \operatorname{ch} V^{-1}) \operatorname{ch} F_0^\sharp|_{\widehat{sp}_n}.$$

Now claim (b) follows from (3.14). Claim (c) follows from Lemma 3.3 and claims (a), (b).

Next, we turn to the proof of Theorem 3.2. First, from the denominator identity of $A_{2n'-1}^{(2)}$ we deduce the following lemma.

Lemma 3.4. *Let $M = \{\gamma \in Q^\vee \mid (\gamma|\bar{\Lambda}_n) \in 2\mathbb{Z}\}$. Then*

$$e^{-\Lambda_0} \frac{\varphi(q^2)}{\varphi(q)} \widehat{R} \prod_{\substack{\alpha \in \Delta_\ell \\ k \in \mathbb{Z}_{odd>0}}} (1 - e^\alpha q^k)^{-1} = \sum_{w \in W} \varepsilon(w) w \sum_{\gamma \in M} t_\gamma(e^{n'\Lambda_0 + \rho}).$$

Using this lemma, we can rewrite the character formulas, given by Theorem 3.1(b) and (c) as follows:

$$\widehat{R} \operatorname{ch} L(-\Lambda_0) = \frac{1}{2} \sum_{w \in W} \varepsilon(w) w \left(\sum_{\substack{\gamma \in Q^\vee \\ (\gamma|\bar{\Lambda}_1) \ge 0}} + \sum_{\substack{\gamma \in Q^\vee \\ (\gamma|\bar{\Lambda}_{n'}) \in 2\mathbb{Z}}} \right) t_\gamma(e^{n'\Lambda_0 + \rho}). \tag{3.16}$$

$$\widehat{R}\mathrm{ch}\,(-\Lambda_0+\bar{\Lambda}_2)=\frac{1}{2q}\sum_{w\in W}\Bigg(\sum_{\substack{\gamma\in Q^\vee\\(\gamma|\bar{\Lambda}_1)\geq 0}}-\sum_{\substack{\gamma\in Q^\vee\\(\gamma|\bar{\Lambda}_{n'})\in 2\mathbb{Z}}}\Bigg)t_\gamma(e^{n'\Lambda_0+\rho}). \tag{3.17}$$

In order to rewrite these formulas into a nicer form, we introduce a different $\mathbb{Z}$-basis of $Q^\vee$:

$$\gamma_i=\alpha_i^\vee+\cdots+\alpha_{n'}^\vee,\ i=1,\ldots,n'.$$

Then, letting $\gamma=\sum_k j_k\gamma_k$, we have

$$(\gamma|\bar{\Lambda}_1)=j_1,\quad (\gamma|\bar{\Lambda}_2-\bar{\Lambda}_1)=j_2,\quad (\gamma|\bar{\Lambda}_{n'})=\sum_k j_k. \tag{3.18}$$

Using that $(-\Lambda_0+\widehat{\rho}\,|\,\delta-\theta)=0$, we obtain

Lemma 3.5. *For $\Omega\subset\mathbb{Z}^{n'}$ let*

$$\Omega'=\{(-j_1-1,j_2,\ldots,j_{n'})\,|\,(j_1,\ldots,j_{n'})\in\Omega\}.$$

Then

$$\sum_{w\in W}\varepsilon(w)w\sum_{(j_1,\ldots,j_{n'})\in\Omega}t_{\sum_k j_k\gamma_k}(e^{n\Lambda_0+\rho})=-\sum_{w\in W}\varepsilon(w)w\sum_{(j_1,\ldots,j_{n'})\in\Omega'}t_{\sum_k j_k\gamma_k}(e^{n\Lambda_0+\rho}).$$

Introduce the following shorthand notation:

$$[\text{condition } (*) \text{ on } \gamma]:=\sum_{w\in W}\varepsilon(w)w\sum_{\substack{\gamma\in Q^\vee\\ \gamma\text{ satisfies }(*)}}t_\gamma(e^{n'\Lambda_0+\rho}).$$

Applying Lemma 3.5 to the set $\Omega=\{(j_1,\ldots,j_n)\in\mathbb{Z}^{n'}\,|\,j_1\geq 0,\sum_k j_k\in\mathbb{Z}_{\mathrm{odd}}\}$, we obtain in this notation:

$$\left[(\gamma|\bar{\Lambda}_1)\geq 0,(\gamma|\bar{\Lambda}_{n'})\in 1+2\mathbb{Z}\right]=-\left[(\gamma|\bar{\Lambda}_1)<0,(\gamma|\bar{\Lambda}_{n'})\in 2\mathbb{Z}\right]. \tag{3.19}$$

In the above notation, formula (3.16) becomes

$$\widehat{R}\mathrm{ch}L(-\Lambda_0)=\frac{1}{2}([(\gamma|\bar{\Lambda}_1)\geq 0]+[(\gamma|\bar{\Lambda}_{n'})\in 2\mathbb{Z}]).$$

Using (3.19), this completes the proof of claim (c) of Theorem 3.2.

Likewise in the above notation, formula (3.17) becomes:

$$\widehat{R}\mathrm{ch}L(-\Lambda_0+\bar{\Lambda}_2)=\frac{1}{2q}([(\gamma|\bar{\Lambda}_1)\geq 0]-[(\gamma|\bar{\Lambda}_{n'})\in 2\mathbb{Z}]).$$

Using (3.19) this can be rewritten as

$$\widehat{R}\,\mathrm{ch}\,L(-\Lambda_0+\bar{\Lambda}_2) = -\frac{1}{q}\sum_{w\in W}\varepsilon(w)w \sum_{\substack{\gamma\in Q^\vee \\ (\gamma|\bar{\Lambda}_1)<0 \\ (\gamma|\bar{\Lambda}_{n'})\in 2\mathbb{Z}}} t_\gamma(e^{n'\Lambda_0+\rho}). \tag{3.20}$$

In order to rewrite this formula further, we need the following properties of roots and weights of sp_n, which are straightforward.

Lemma 3.6. *The weight $\lambda := \bar{\Lambda}_2$ satisfies the following properties:*

(a) *λ is a positive short root, given by*

$$\lambda = \alpha_1 + 2(\alpha_1+\cdots+\alpha_{n'-1}) + \alpha_{n'} = \frac{1}{2}(\gamma_1+\gamma_2).$$

(b) $(\lambda|\gamma_i) = 1$ *if* $i=1,2$, *and* $=0$ *otherwise.*

(c)

$$r_\lambda\left(\sum_{k=1}^{n'} j_k\gamma_k\right) = -j_2\gamma_1 - j_1\gamma_2 + \sum_{k=3}^{n'} j_k\gamma_k.$$

(d)

$$r_{\delta-\lambda} = r_\lambda t_{-\gamma_1-\gamma_2}.$$

(e)

$$r_{\delta-\lambda}(n'\Lambda_0+\rho) = n'\Lambda_0+\rho+\lambda-\delta = (-\Lambda_0+\lambda-\delta)+\widehat{\rho}.$$

Using Lemma 3.6, we can rewrite (3.20) as follows:

$$\widehat{R}\,\mathrm{ch}\,L(-\Lambda_0+\bar{\Lambda}_2) = -\frac{1}{q}\sum_{w\in W}\varepsilon(w)w \sum_{\substack{\gamma\in Q^\vee \\ (\gamma|\bar{\Lambda}_1)<0 \\ (\gamma|\bar{\Lambda}_{n'})\in 2\mathbb{Z}}} t_\gamma r_{\bar{\Lambda}_2} t_{-\gamma_1-\gamma_2}(e^{-\Lambda_0+\bar{\Lambda}_2+\widehat{\rho}-\delta}).$$

$$= \sum_{w\in W}\varepsilon(w)w \sum_{\substack{j_1,\dots,j_{n'}\in\mathbb{Z} \\ j_1<0 \\ \sum_k j_k\in 2\mathbb{Z}}} t_{(-j_2-1)\gamma_1+(-j_1-1)\gamma_2+\sum_{k=3}^{n'} j_k\gamma_k}(e^{-\Lambda_0+\bar{\Lambda}_2+\widehat{\rho}}).$$

Replacing in the last expression $-j_2-1$ by j_1 and $-j_1-1$ by j_2, we obtain

$$\widehat{R}\,\mathrm{ch}\,L(-\Lambda_0+\bar{\Lambda}_2) = \sum_{w\in W}\varepsilon(w)w \sum_{\substack{j_1,\dots,j_{n'}\in\mathbb{Z} \\ j_2\geq 0 \\ \sum_k j_k\in 2\mathbb{Z}}} t_{\sum_k j_k\gamma_k}(e^{-\Lambda_0+\bar{\Lambda}_2+\widehat{\rho}}).$$

Now, by (3.18), claim (b) of Theorem 3.2 follows.

4 A character formula for the Deligne series modules

In this section we prove the following simple theorem.

Theorem 4.1. *Let $\mathfrak{g}$ be a simply laced Lie algebra of rank ℓ (so that $\alpha_i^\vee = \alpha_i$), and let Λ be a weight of $\widehat{\mathfrak{g}}$ of level $k \in \mathbb{Z}_{<0}$, such that the following conditions hold:*

(i) $(\Lambda|\alpha_i) \in \mathbb{Z}_{\geq 0}$ *for* $i = 1, \ldots, \ell$,
(ii) *there exists a positive root α of $\mathfrak{g}$, such that* $(\Lambda + \widehat{\rho}|\delta - \alpha) = 0$,
(iii) *if $\beta \in \widehat{\Delta}_+$ is orthogonal to $\Lambda + \widehat{\rho}$, then $\beta = \delta - \alpha$,*
(iv) *(extra hypothesis) in the character formula* (4.2) *below, one has*

$$c(\gamma) := c(t_\gamma) = (\textit{linear function in } \gamma \in Q) + \textit{ const.}$$

Then

$$\widehat{R}\operatorname{ch}L(\Lambda) = \frac{1}{2}\sum_{w\in W}\varepsilon(w)w\sum_{\gamma\in Q}((\alpha|\gamma)+1)t_\gamma(e^{\Lambda+\widehat{\rho}}). \tag{4.1}$$

Note that Theorem 2.1 shows that the extra hypothesis fails for $\mathfrak{g} = s\ell_n$, $k = -1$. However, the comparison with [Kaw15], [AK16] indicates that the following conjecture may hold.

Conjecture 4.2. *If $\mathfrak{g} = D_4$, E_6, E_7, or E_8, then the extra hypothesis* (iv) *holds (hence the character formula* (4.1) *holds).*

Remark 4.3. If $\Lambda = k\Lambda_0$ for $k \in \mathbb{Z}_{<0}$, conditions (i)–(iii) of Theorem 4.1 hold for $k = (-\frac{h^\vee}{6} - 1) + s$, where $s = 0, 1, \ldots, b-1$ and $b = 2, 3, 4, 6$ for $\mathfrak{g} = D_4, E_6, E_7, E_8$ respectively. This explains the name "Deligne series modules", cf. [BLPR15], [Kaw15], [AK16], [AM16].

The proof of Theorem 4.1 is easy (but it is probably quite hard to verify the extra hypothesis (iv)). Indeed, by (i) $\operatorname{ch}L(\Lambda)$ is W-invariant, and by (ii), $k + h^\vee > 0$, hence (1.9) holds and it can be rewritten as

$$\widehat{R}\operatorname{ch}L(\Lambda) = \sum_{w\in W}\varepsilon(w)w\sum_{\gamma\in Q}c(\gamma)t_\gamma(e^{\Lambda+\widehat{\rho}}). \tag{4.2}$$

Since by (ii) we have $r_{\delta-\alpha}(\Lambda + \widehat{\rho}) = \Lambda + \widehat{\rho}$, and also $r_{\delta-\alpha} = r_\alpha t_{-\alpha}$, $r_\alpha t_\gamma r_\alpha = t_{r_\alpha(\gamma)}$, $\varepsilon(wr_\alpha) = -\varepsilon(w)$, we can rewrite (4.2) as

$$\widehat{R}\operatorname{ch}L(\Lambda) = -\sum_{w\in W}\varepsilon(w)w\sum_{\gamma\in Q}c(\gamma)t_{r_\alpha(\gamma)-\alpha}(e^{\Lambda+\widehat{\rho}}).$$

Replacing in this formula γ by $r_\alpha(\gamma) - \alpha$, we obtain

$$\widehat{R}\operatorname{ch}L(\Lambda) = -\sum_{w\in W}\varepsilon(w)w\sum_{\gamma\in Q}c(r_\alpha(\gamma)-\alpha)t_\gamma(e^{\Lambda+\widehat{\rho}}). \tag{4.3}$$

Taking half the sum of (4.2) and (4.3), we obtain

$$\widehat{R}\operatorname{ch}L(\Lambda) = \frac{1}{2}\sum_{w\in W}\varepsilon(w)w\sum_{\gamma\in Q}\widetilde{c}(\gamma)t_\gamma(e^{\Lambda+\widehat{\rho}}), \tag{4.4}$$

where $\widetilde{c}(\gamma) = c(\gamma) - c(r_\alpha(\gamma) - \alpha)$. The function $\widetilde{c}(\gamma)$ has the following two properties:

$$\widetilde{c}(\gamma) = -\widetilde{c}(r_\alpha(\gamma) - \alpha), \gamma \in Q, \tag{4.5}$$

$$\widetilde{c}(\gamma) = (\text{linear function in } \gamma) + \text{ const}, \gamma \in Q. \tag{4.6}$$

By (4.6), which holds due to the condition (iv), we can write for some $\beta \in \mathfrak{h}^*$, $a \in \mathbb{C}$:

$$\widetilde{c}(\gamma) = (\gamma|\beta) + a, \ \gamma \in Q. \tag{4.7}$$

Then we obtain for all $\gamma \in Q$:

$$\widetilde{c}(r_\alpha(\gamma) - \alpha) = (\gamma|\beta) - (\alpha|\beta)(\gamma|\alpha) - (\alpha|\beta) + a. \tag{4.8}$$

Since, by (4.5), one has (4.7) = -(4.8), we obtain

$$2(\gamma|\beta) - (\alpha|\beta)(\gamma|\alpha) - (\alpha|\beta) + 2a = 0, \text{ for all } \gamma \in Q.$$

Hence $\beta = \frac{1}{2}(\alpha|\beta)\alpha$ and $a = \frac{1}{2}(\alpha|\beta)$. Therefore, by (4.7), we obtain

$$\widetilde{c}(\gamma) = \text{const.} \times ((\gamma|\alpha) + 1) \text{ for all } \gamma \in Q. \tag{4.9}$$

By (4.4) and (4.9), we have

$$\widehat{R}\operatorname{ch}L(\Lambda) = \text{const.} \times \sum_{w\in W}\varepsilon(w)w\sum_{\gamma\in Q}((\gamma|\alpha)+1)t_\gamma(e^{\Lambda+\widehat{\rho}}). \tag{4.10}$$

Since the stabilizer in $\widehat{W}$ of any $\lambda \in \mathfrak{h}^*$ of positive level is generated by reflections r_α, $\alpha \in \widehat{\Delta}_+^{\text{re}}$ fixing λ [K90], by the conditions (ii) and (iii) we see that $\widehat{W}_{\Lambda+\widehat{\rho}} = \{1, r_{\delta-\alpha}\}$. It follows that const. $= \frac{1}{2}$ in (4.10), proving (4.1).

Conjecture 4.4. *If* $\mathfrak{g} = D_4$, E_6, E_7 *or* E_8 *and* $k = -1, -2, \ldots, -b$, *where* $b = 2$, 3, 4, *or* 6 *respectively, then all irreducible highest weight modules over the vertex algebra* $L(k\Lambda_0)$, *which are* $\mathfrak{g}$*-locally finite, are those from Theorem* 4.1. (*It follows from* [AM16] *that all these vertex algebras are quasilisse, hence, by* [AK16], *have only finitely many irreducible highest weight modules.*)

Example 4.5. Let $\mathfrak{g} = D_4$, $k = -1$. Then the following Λ's satisfy the conditions (i), (ii), (iii) of Theorem 4.1 (we label the branching node of the Dynkin diagram of D_4 by 2):

$$-\Lambda_0;\ -2\Lambda_0 + \Lambda_i\ (i = 1,3,4);\ -3\Lambda_0 + \Lambda_2;$$
$$-3\Lambda_0 + \Lambda_i + \Lambda_j\ ((i,j) = (1,3),(1,4),(3,4)).$$

Remark 4.6. Of course, one has a formula for homogeneous q-dimension, similar to that in Remark 2.3 in all cases, considered in Sections 3 and 4. We checked on the computer that in the case of D_4, it is compatible with the formula for q-dimension of $L(-2\Lambda_0)$ from [AK16].

Remark 4.7. Note that $((\gamma|\alpha)+1)t_\gamma(e^{\Lambda+\widehat{\rho}}) = \frac{1}{k+h^\vee} D_\alpha t_\gamma(e^{\Lambda+\widehat{\rho}})$, where D_α is the derivative in the direction α. Hence the RHS of (4.10) is a linear combination of derivatives of theta functions.

Remark 4.8. By Theorem 3.1 from [KRW03], one has

$$\widehat{R}(\mathfrak{g},f)ch_{H(\Lambda)}(\tau,h) = (\widehat{R}_n ch_\Lambda)(\tau,-\tau x+h,\tau/4), \text{ where } h \in \mathfrak{h}^f,$$

for any W-algebra $W^k(\mathfrak{g},f)$, obtained by the quantum Hamiltonian reduction of the $\widehat{\mathfrak{g}}$-module $L(\Lambda)$ of level k. Here $\widehat{R}(\mathfrak{g},f)$ is the denominator of $W^k(\mathfrak{g},f)$, $\widehat{R}_n = q^{\dim\mathfrak{g}/24}\widehat{R}$ is the normalized affine Weyl denominator, $ch_\Lambda = q^{m_\Lambda} chL(\Lambda)$ is the normalized character [K90]. In particular, if $\mathfrak{g} = D_4, E_6, E_7, E_8$, $k = -b$, f is the minimal nilpotent element of $\mathfrak{g}$, and $L(\Lambda)$ is a module of level k over the corresponding simple vertex algebra, then the simple W-algebra $W_k(\mathfrak{g},f)$ is 1-dimensional [AKMPP16], hence $ch_{H(\Lambda)} = 1$, and we get a formula relating $\widehat{R}(\mathfrak{g},f)$ to $chL(\Lambda)$ for $z = -\tau x + h$.

References

[AKMPP16] D. Adamovic, V.G. Kac, P. Moseneder-Frajria, P. Papi, O. Perse, Conformal embeddings of affine vertex algebras in minimal W-algebras I: structural results, arXiv:1602.04687

[AP12] D. Adamovic, O. Perse, The vertex algebra $M(1)^+$ and certain affine vertex algebras of level -1, *Sigma* **8** (2012), 1–16.

[AP14] D. Adamovic, O. Perse, Fusion rules and complete reducibility of certain modules for affine Lie algebras, *J. Algebra Appl.* **13** (2014), no. 1, 1350062.

[AK16] T. Arakawa, K. Kawasetsu, Quasi-lisse vertex algebras and modular linear differential equations, arXiv:1610.05865.

[AM16] T. Arakawa, A. Moreau, On the irreducibility of associated varieties of W-algebras, arXiv:1608.03142.

[BLPR15] C. Beem, M. Lemos, P. Liendo, W. Peelaers, L. Rastelli, B.C. van Rees, Infinite chiral symmetry in four dimensions, *Comm. Math. Phys.* **336** (2015), 1359–1433.

[G11] M. Gorelik, Weyl denominator identity for affine Lie superalgebras with non-zero dual Coxeter number, *J. Algebra* **337** (2011), 50–62.

[K90] V. G. Kac, Infinite-dimensional Lie algebras, Third edition, Cambridge University Press, 1990.

[KK79] V. G. Kac, D.A. Kazhdan, Structure of representations with highest weight of infinite-dimensional Lie algebras, *Adv. in Math* **34** (1979), no. 1, 97–108.

[KRW03] V. G. Kac, S.-S. Roan, M. Wakimoto, Quantum reduction of affine superalgebras, *Comm. Math. Phys.* **241** (2003), 307–342.

[KW88] V. G. Kac, M. Wakimoto, Modular invariant representations of infinite-dimensional Lie algebras and superalgebras, *Proc. Nat. Acad. Sci. USA* **85** (1988), pp 4956–4960.

[KW94] V. G. Kac, M. Wakimoto, Integrable highest weight modules over affine superalgebras and number theory, *Progress in Math.* **123**, Birkhäuser, 1994, pp 415–456.

[KW01] V. G. Kac, M. Wakimoto, Integrable highest weight modules over affine superalgebras and Appell's function, *Commun. Math. Phys.* **215** (2001), 631–682.

[KT00] M. Kashiwara, T. Tanisaki, Characters of the irreducible modules with non-critical highest weights over affine Lie algebras, in Representations and Quantizations (Shanghai, 1998), China High Education Press, Beijing, 2000.

[Kaw15] K. Kawasetsu, W-algebras with non-admissible levels and the Deligne exceptional series, arXiv:1505.06985.

Alcoved Polytopes II

Thomas Lam and Alexander Postnikov

To the memory of Bertram Kostant

Abstract This is the second of two papers where we study polytopes arising from affine Coxeter arrangements. Our results include a formula for their volumes, and also compatible definitions of hypersimplices, descent numbers and major index for all Weyl groups. We give a q-analogue of Weyl's formula for the order of the Weyl group. For A_n, C_n and D_4, we give a Gröbner basis which induces the triangulation of alcoved polytopes.

Keywords: Convex polytopes • affine Weyl group • hypersimplex

Mathematics Subject Classification (2010): Primary 52B, Secondary 05A05, 20F55

1 Introduction

This is the second of two papers in which we investigate *alcoved polytopes* arising from affine Coxeter arrangements. Let $\Phi \subset V$ be an irreducible crystallographic root system and let W be the corresponding Weyl group. Associated to Φ is an infinite hyperplane arrangement known as the affine Coxeter arrangement. This hyperplane arrangement subdivides V into simplices of the same volume which are

---

T.L. was supported by NSF DMS 0901111, and by a Sloan Fellowship.

A.P. was supported by NSF DMS-1100147 and CAREER DMS-0546209.

Thomas Lam
Department of Mathematics, University of Michigan, Ann Arbor, MI 48109, USA,
e-mail: tfylam@umich.edu

Alex Postnikov
Department of Mathematics, M.I.T., Cambridge, MA 02139, USA, e-mail: apost@math.mit.edu

V. G. Kac, V. L. Popov (eds.), *Lie Groups, Geometry, and Representation Theory*,
Progress in Mathematics 326, https://doi.org/10.1007/978-3-030-02191-7_10

called *alcoves*. We define a *proper alcoved polytope* to be a convex polytope P which is the closure of a union of alcoves.

In [API], we studied these polytopes in the special situation of the root system $\Phi = A_n$. Two motivating examples for us were the *hypersimplices* and the *alcoved matroid polytopes*. Alcoved polytopes arising from other root systems have also been studied. Payne [Pay] showed that alcoved polytopes with vertices lying in the coweight lattice are normal and Koszul in classical types. Werner and Yu [WY] studied generating sets of alcoved polytopes. Fomin and Zelevinsky's *generalized associahedra* [FZ] are examples of polytopes which can be realized as alcoved polytopes.

We prove that the volume of an alcoved polytope P is given by

$$\mathrm{Vol}(P) = \sum_{\bar{w}\in W/C} I(P_{\bar{w}}),$$

where W/C are certain cosets of the Weyl group, $P_{\bar{w}}$ are certain alcoved polytopes and $I(P)$ denotes the number of integral coweights lying in P. The group $C \subset W$ was studied previously by Verma [Ver]. The order of the group C is equal to the index of connection of Φ. For the case $\Phi = A_{n-1}$, the group C is the cyclic group generated by the long cycle $(123\cdots n)$, written in cycle notation.

Recall that the usual hypersimplex $\Delta_{k,n}$ has volume equal to the Eulerian number $A_{k,n-1}$. We define *generalized hypersimplices* Δ_k^{Φ} to be certain alcoved polytopes which generalize this property of $\Delta_{k,n}$ for each root system Φ. To this end we introduce the *circular descent number* $\mathrm{cdes} : W \to \mathbb{Z}$, so that the volume of Δ_k^{Φ} counts the number of elements of W with a fixed circular descent number. We also introduce a *circular major map* $\mathrm{cmaj} : W \to C$ that interacts in an interesting way with cdes. In particular $\{w \in W \mid \mathrm{cmaj}(w) = \mathrm{id}\}$ gives a set of coset representatives for W/C. In type A_{n-1}, the circular major map generalizes the major index of S_n, taken modulo n.

Weyl's formula for the order of the Weyl group W states that

$$|W| = f \cdot r! \cdot a_1 a_2 \cdots a_r,$$

where f is the index of connection, r is the rank of Φ and a_i are the coefficients of the simple roots in the highest root of Φ. We prove, using the geometry of alcoved polytopes, that

$$\sum_{w\in W} q^{\mathrm{cdes}(w)} e^{\mathrm{cmaj}(w)} = \left(\sum_{x\in C} e^x\right) \cdot A_r(q) \cdot [a_1]_q \cdots [a_r]_q,$$

where $A_r(q)$ is the usual Eulerian polynomial, and $[n]_q$ denotes the usual q-analogue. Here, $e^x \in \mathbb{Z}[C]$ lies in the group algebra of C.

Finally, in analogy with Sturmfels' triangulation of the hypersimplex in type A [Stu], we study the toric ideals I_P associated with an alcoved polytope P. When Φ is one of the root systems A_n, C_n, D_4, we give a Gröbner basis $\mathcal{G}_P$ for I_P which induces the triangulation of P into alcoves.

2 Root system notation

We recall standard terminology related to root systems, see [Hum] for more details. Let V be a real Euclidean space of rank r with nondegenerate symmetric inner product (λ,μ). Let $\Phi \subset V$ be an irreducible *crystallographic root system* with choice of basis of *simple roots* $\alpha_1,\dots,\alpha_r$. Let $\Phi^+ \subset \Phi$ be the corresponding set of *positive roots* and let $\Phi^- = -\Phi^+$ be the set of *negative roots*. Then Φ is the disjoint union of Φ^+ and Φ^-. We will write $\alpha > 0$, for $\alpha \in \Phi^+$; and $\alpha < 0$, for $\alpha \in \Phi^-$. The collection of *coroots* $\alpha^\vee = 2\alpha/(\alpha,\alpha) \in V$, for $\alpha \in \Phi$, forms the *dual root system* $\Phi^\vee$. The *Weyl group* $W \subset \mathrm{Aut}(V)$ is generated by the reflections $s_\alpha : \lambda \mapsto \lambda - (\lambda,\alpha^\vee)\,\alpha$ with respect to roots $\alpha \in \Phi$. The Weyl group W is actually generated by simple reflections $s_i = s_{\alpha_i}$ subject to the Coxeter relations. The *length* $\ell(w)$ of an element $w \in W$ is the length of a shortest decomposition for w in terms of simple reflections. There is a unique element $w_\circ \in W$ of maximal possible length.

The *root lattice* $L = L(\Phi) \subset V$ is the integer lattice spanned by the roots $\alpha \in \Phi$. It is generated by the simple roots α_i. The *weight lattice* is defined by $\Lambda = \Lambda(\Phi) = \{\lambda \in V \mid (\lambda,\alpha^\vee) \in \mathbb{Z},\ \text{for all } \alpha \in \Phi\}$. The weight lattice Λ contains the root lattice L as a subgroup of finite index f. The quotient group Λ/L is isomorphic to the center of the universal simply-connected Lie group $G^\vee$ associated with the root system $\Phi^\vee$. The index $f = |\Lambda/L|$ is called the *index of connection.*

The *coroot lattice* is the integer lattice $L^\vee = L^\vee(\Phi)$ spanned by the coroots $\alpha_i^\vee$. The *coweight lattice* $\Lambda^\vee$ is the integer lattice defined by $\Lambda^\vee = \Lambda^\vee(\Phi) = \{\lambda \in V \mid (\lambda,\alpha) \in \mathbb{Z},\ \text{for all } \alpha \in \Phi\}$. Let $\omega_1,\dots,\omega_r \subset V$ be the basis dual to the basis of simple roots $\alpha_1,\dots,\alpha_r$, i.e., $(\omega_i,\alpha_j) = \delta_{ij}$. The ω_i are called the *fundamental coweights*. They generate the coweight lattice $\Lambda^\vee$.

Let $\rho = \omega_1 + \cdots + \omega_r$. The *height* of a root α is the number (ρ,α) of simple roots that add up to α. Since we assumed that Φ is irreducible, there exists a unique *highest root* $\theta \in \Phi^+$ of maximal possible height. For convenience we set $\alpha_0 = -\theta$. Let $a_0 = 1$ and $a_1,\dots,a_r$ be the positive integers given by $a_i = (\omega_i,\theta)$. In other words, $\theta = a_1\alpha_1 + \cdots + a_r\alpha_r$ or, equivalently, $a_0\alpha_0 + a_1\alpha_1 + \cdots + a_r\alpha_r = 0$. The *Coxeter number* is defined as $h = (\rho,\theta) + 1 = a_0 + a_1 + \cdots + a_r$.

Lemma 2.1. *Let $\lambda \in \Lambda^\vee$ be an integral coweight and let $w \in W$ be a Weyl group element. Then*

$$w(\lambda) - \lambda \in L^\vee.$$

Proof. Since $L^\vee \subset \Lambda^\vee$ it suffices to check this for a simple reflection s_α. We compute $s_\alpha(\lambda) - \lambda = (\lambda,\alpha)\,\alpha^\vee \in L^\vee$. □

3 The affine Weyl group and alcoved polytopes

The *affine Weyl group* W_{aff} associated with the root system Φ is generated by the reflections $s_{\alpha,k} : V \to V$, $\alpha \in \Phi$, $k \in \mathbb{Z}$, with respect to the affine hyperplanes

$$H_{\alpha,k} = \{\lambda \in V \mid (\lambda, \alpha) = k\}.$$

The coweight lattice $\Lambda^\vee$ and coroot lattice $L^\vee$ act on the space V by translations. We will identify $\Lambda^\vee$ and $L^\vee$ with these groups of translations. The Weyl group W normalizes these groups.

Lemma 3.1. [Hum] *The affine Weyl group W_{aff} is the semidirect product $W \ltimes L^\vee$ of the usual Weyl group W and the coroot lattice $L^\vee$.*

The connected components of the complement to these hyperplanes $V \setminus \bigcup H_{\alpha,k}$ are called *alcoves*. Let $\mathscr{A}$ be the set of all alcoves. A *closed alcove* is the closure of an alcove. Each alcove A has the following form:

$$A = \{\lambda \in V \mid m_\alpha < (\lambda, \alpha) < m_\alpha + 1, \text{ for } \alpha \in \Phi^+\},$$

where $m_\alpha = m_\alpha(A)$ is a collection of integers associated with the alcove A.

Lemma 3.2. [Hum] *The affine Weyl group W_{aff} acts simply transitively on the collection $\mathscr{A}$ of all alcoves.*

The fundamental alcove is the simplex given by

$$\begin{aligned} A_\circ &= \{\lambda \in V \mid 0 < (\lambda, \alpha) < 1, \text{ for } \alpha \in \Phi^+\} \\ &= \{\lambda \in V \mid (\lambda, \alpha_i) > 0, \text{ for } i = 1, \dots, r; \text{ and } (\lambda, \theta) < 0\} \\ &= \{x_1\omega_1 + \cdots + x_r\omega_r \mid x_1, \dots, x_r > 0 \text{ and } a_1x_1 + \cdots + a_rx_r < 1\} \\ &= \text{Convex Hull of the points } 0,\ \omega_1/a_1,\ \dots,\ \omega_r/a_r. \end{aligned} \tag{1}$$

Lemma 3.2 implies that all alcoves are obtained from $A_\circ$ by the action of W_{aff}. In particular, all closed alcoves are simplices with the same volume. The closure of the fundamental alcove $A_\circ$ is a fundamental domain of W_{aff}. Let $\mathscr{F} \supset \mathscr{A}$ be the set of all faces of alcoves of all dimensions. We will think of elements of $\mathscr{F}$ as relatively open sets, so that the space V is the disjoint union of elements of $\mathscr{F}$.

Our main object is defined as follows.

Definition 3.3. An alcoved polytope P is a convex polytope in the space V such that P is a union of finitely many elements of $\mathscr{F}$. A proper alcoved polytope is an alcoved polytope of top dimension.

By the definition, each proper alcoved polytope comes equipped with a triangulation into closed alcoves. The following lemma is immediate from the definitions.

Lemma 3.4. *A bounded subset $P \subset V$ is an alcoved polytope if and only if P is the intersection of several half-spaces of the form $\{\lambda \in V \mid (\lambda, \alpha) \geq k\}$, for $\alpha \in \Phi$ and $k \in \mathbb{Z}$.*

Let (W, S) be a Coxeter group and $u, v \in W$. A path from u to v is a sequence $u = w_0 \to w_1 \to w_2 \to \cdots \to w_s = v$ such that $w_{i+1} = w_i s$ for some simple reflection $s \in S$. A subset $K \subset W$ is called *convex* if for every $u, v \in K$, we have that any shortest path from u to v lies in K.

Proposition 3.5. *Let $P \subset V$ be a bounded subset which is a union of closed alcoves. Then P is a convex polytope if and only if one of the following conditions hold:*

(1) *For any two alcoves $A, B \subset P$, any shortest path $A = A_0 \to A_1 \to A_2 \to \cdots \to A_s = B$ lies in P. Here $A_i \in \mathscr{A}$ are alcoves and $A' \to A''$ means that the closures of the two alcoves A' and A'' share a facet.*
(2) *The subset $W_P = \{w \in W_{\mathrm{aff}} \mid w(A_\circ) \subset P\}$ of the affine Weyl group is a convex subset.*

Proof. Suppose P is a polytope and $A = A_0 \to A_1 \to A_2 \to \cdots \to A_s = B$ is a shortest path with $A, B \in P$. Suppose that A_i lies in P but A_{i+1} lies outside of P. Then the hyperplane $H_{\alpha,k}$ which separates A_i and A_{i+1} must be a facet of P, so eventually the shortest path must cross $H_{\alpha,k}$ again. Suppose that A_j lies on the same side of $H_{\alpha,k}$ as A_{i+1} and A_{j+1} lies on the other side. Then reflecting the path $A_{i+1} \to A_{i+2} \to \cdots \to A_j$ in the hyperplane $H_{\alpha,k}$, we get another path from A to B which is shorter. Conversely, suppose condition (1) holds but P is not convex. This implies that there are alcoves $A, B \in P$ and points $a \in A$, $b \in B$ such that the straight line $\overline{ab}$ does not lie in P. Since P is compact we may assume a and b lie in the interior of A and B respectively. Thus without loss of generality, we may assume that the line $\overline{ab}$ does not intersect any face of $\mathscr{F}$ of codimension more than one. The sequence of alcoves obtained by travelling along $\overline{ab}$ must be a shortest path. This is clear as every hyperplane $H_{\alpha,k}$ that $\overline{ab}$ intersects separates A from B and so must be a separating hyperplane between some two alcoves A_i and A_{i+1} in any path from A to B.

Condition 2 follows immediately from translating the action of the simple generators of W_{aff} on alcoves. □

Thus each alcoved polytope is of the form

$$P = \{\lambda \in V \mid k_\alpha \le (\lambda, \alpha) \le K_\alpha, \text{ for } \alpha \in \Phi^+\},$$

where $k_\alpha = k_\alpha(P)$ and $K_\alpha = K_\alpha(P)$ are two collections of integers indexed by positive roots $\alpha \in \Phi^+$.

Let $\Lambda^\vee/h = \{\lambda/h \mid \lambda \in \Lambda^\vee\}$ be the coweight lattice $\Lambda^\vee$ shrunk h times. Recall that h denotes the Coxeter number.

We are grateful to Bertram Kostant for many helpful coversations over the years. We appreciate his attention to combinatorial aspects of representation theory. One little thing that we learned from him is that the center of mass of an alcove is the only point of the lattice $\Lambda^\vee/h$ in the interior of this alcove.

Lemma 3.6. [Kos], see also [LP, Section 14] *For every alcove $A \in \mathscr{A}$, there is exactly one point of the lattice $\Lambda^\vee/h$ in the interior of A. For the fundamental alcove $A_\circ$, we have $A_\circ \cap (\Lambda^\vee/h) = \{\rho/h\}$.*

Proof. Since the affine Weyl group acts simply transitively on $\mathscr{A}$ and preserves the lattice $\Lambda^\vee/h$, it is enough to prove the claim for the fundamental alcove $A_\circ$. In the basis of fundamental coweights $\omega_1, \ldots, \omega_r$, the fundamental alcove $A_\circ$ is given by the inequalities $x_1, \ldots, x_r > 0$ and $a_1x_1 + \cdots + a_rx_r < 1$; and the lattice $\Lambda^\vee/h$ is given by $x_1,\dots,x_r \in \mathbb{Z}/h$. Recall that $h = a_0 + a_1 + \cdots + a_r$. Then the intersection $A_\circ \cap (\Lambda^\vee/h)$ consists of a single point with coordinates $(1/h,\dots,1/h)$. In other words, this intersection point is $(\omega_1 + \cdots + \omega_r)/h = \rho/h$. □

For $A \in \mathscr{A}$, we call the single element of $(\Lambda^\vee/h) \cap A$ the *central point* of the alcove A. Let $Z = (\Lambda^\vee/h) \setminus \bigcup H_{\alpha,k}$ be the set of central points of all alcoves; equivalently,

$$Z = \{\lambda \in V \mid h\cdot(\alpha,\lambda) \equiv 1,\dots,h-1 \pmod h, \text{ for } \alpha \in \Phi^+\}.$$

The set Z of central points is in one-to-one correspondence with the set $\mathscr{A}$ of alcoves.

4 Weyl's formula for the order of the Weyl group

Let Vol be the volume form on the space V normalized by $\mathrm{Vol}(A_\circ) = 1$. Then the volume of any alcove is 1 and the volume $\mathrm{Vol}(P)$ of an alcoved polytope P is the number of alcoves in P. Equivalently, $\mathrm{Vol}(P) = |P \cap Z|$.

Let Π be the alcoved polytope given by

$$\begin{aligned}\Pi &= \{\lambda \in V \mid 0 \le (\lambda,\alpha_i) \le 1, \text{ for } i = 1,\dots,r\}\\ &= \{x_1\omega_1 + \cdots + x_r\omega_r \mid 0 \le x_i \le 1, \text{ for } i = 1,\dots,r\},\end{aligned}$$

i.e., Π is the parallelepiped generated by the fundamental coweights $\omega_1,\dots,\omega_r$. This polytope is a fundamental domain of the coweight lattice $\Lambda^\vee$. Since $A_\circ$ is the simplex with the vertices $0, \omega_1/a_1,\dots,\omega_r/a_r$, we have

$$\mathrm{Vol}(\Pi) = \mathrm{Vol}(\Pi)/\mathrm{Vol}(A_\circ) = r!\,a_1\cdots a_r.$$

Thus the parallelepiped Π consists of $r!\,a_1\cdots a_r$ alcoves.

Let H be the alcoved polytope given by

$$H = \{\lambda \in V \mid -1 \le (\lambda,\alpha) \le 1, \text{ for } \alpha \in \Phi^+\}.$$

The polytope H consists of all alcoves adjacent to the origin 0, i.e., it consists of the $|W|$ alcoves of the form $w(A_\circ)$, for $w \in W$. In particular, its volume is the order of the Weyl group: $\mathrm{Vol}(H) = |W|$. Lemma 3.1 implies that the polytope H is a fundamental domain of the coroot lattice $L^\vee$. Thus $\mathrm{Vol}(H)/\mathrm{Vol}(\Pi) = |\Lambda^\vee/L^\vee| = f$ is the index of connection. This implies the well-known formula for the order of the Weyl group, see [Hum, 4.9]:

$$|W| = f\cdot r!\cdot a_1\cdots a_r. \tag{2}$$

5 The group C

For an integral coweight $\lambda \in \Lambda^\vee$, the affine translation $A + \lambda$ of an alcove A is an alcove; and the affine translation $P + \lambda$ of an alcoved polytope P is an alcoved polytope.

Let us define the equivalence relation "$\sim$" on the affine Weyl group W_{aff} by

$$u \sim w \text{ if and only if } u(A_\circ) = w(A_\circ) + \lambda, \text{ for some } \lambda \in \Lambda^\vee,$$

where $u, w \in W_{\text{aff}}$. The relation "$\sim$" is invariant with respect to the left action of the affine Weyl group. According to Lemma 3.6, this equivalence relation can be defined in terms of central points of alcoves as

$$u \sim w \text{ if and only if } u(\rho/h) - w(\rho/h) \in \Lambda^\vee.$$

Let C be the subset of the usual Weyl group W given by

$$C = \{w \in W \mid w \sim 1\} = \{w \in W \mid w(\rho) - \rho \in h\Lambda^\vee\}.$$

Also let $C_{\text{aff}} = \{w \in W_{\text{aff}} \mid w \sim 1\}$. Actually, C is a subgroup in W and C_{aff} is a subgroup in W_{aff}. Indeed, $u \sim 1$ and $w \sim 1$ imply that $uw \sim u \sim 1$.

The coroot lattice $L^\vee$ is a normal subgroup in C_{aff}. The group C_{aff} is the semidirect product $C \ltimes L^\vee$ and, thus, $C \simeq C_{\text{aff}}/L^\vee$.

Equivalence classes of elements of the Weyl group (respectively, the affine Weyl group) with respect to the relation "$\sim$" are exactly cosets in W/C (respectively, $W_{\text{aff}}/C_{\text{aff}}$). Since Π is a fundamental domain of the coweight lattice $\Lambda^\vee$ and, for an alcove $A \in \mathscr{A}$, there is a translation $A + \lambda$ such that $A + \lambda = w(A_\circ)$, for some $w \in W$, we deduce that there are natural one-to-one correspondences between the followings sets:

$$W/C \simeq W_{\text{aff}}/C_{\text{aff}} \simeq \mathscr{A}/\Lambda^\vee \simeq \{\text{alcoves in } \Pi\}.$$

In particular, the number of cosets $|W/C|$ equals $\text{Vol}(\Pi) = |W|/f$ and, thus, the order of the group C is $|C| = f$.

There is a natural bijection $b : \Lambda^\vee \to C_{\text{aff}}$ given by $b(\lambda) = w$ whenever $w(A_\circ) = A_\circ + \lambda$. Notice that b may not be a homomorphism of groups. However the map $\bar{b} : \Lambda^\vee \to C$ given by the composition of b with the natural projection $C_{\text{aff}} \to C_{\text{aff}}/L^\vee \simeq C$ is a homomorphism. Indeed, let $\bar{b}(\lambda) = u$ and $\bar{b}(\mu) = w$. Then $u, w \in W$ are given by $u(A_\circ) \equiv A_\circ + \lambda \mod L^\vee$ and $w(A_\circ) \equiv A_\circ + \mu \mod L^\vee$. Then $uw(A_\circ) \equiv A_\circ + \lambda + u(\mu) \equiv A_\circ + \lambda + \mu \mod L^\vee$. The last equation follows from Lemma 2.1. The kernel of the map $\bar{b}$ is $L^\vee$. Thus $\bar{b}$ induces the natural isomorphism of groups:

$$\Lambda^\vee/L^\vee \simeq C.$$

The group C is the cyclic group $\mathbb{Z}/(n+1)\mathbb{Z}$ in type A_n, a group of order 2 for types B_n, C_n and E_7, a group of order 4 for type D_n, a group of order 3 for type E_6, and trivial for E_8, F_4, and G_2.

6 The statistic cdes

Let us say that a root $\alpha \in \Phi$ is an *inversion* of Weyl group element $w \in W$ if $w(\alpha) < 0$. Equivalently, a positive root α in an inversion of w if and only if $\ell(ws_\alpha) < \ell(w)$. Let us define

$$\mathrm{inv}_\alpha(w) = \begin{cases} 0 \text{ if } w(\alpha) > 0, \\ 1 \text{ if } w(\alpha) < 0. \end{cases}$$

Lemma 6.1. *We have* $\mathrm{inv}_\alpha(w) = -m_\alpha(w^{-1}(A_\circ))$.

Proof. Indeed, α is an inversion of w if and only if $(w(\alpha), \rho) = (\alpha, w^{-1}(\rho)) < 0$, that is, $-1 < (\lambda, \alpha) < 0$, for any $\lambda \in w^{-1}(A_\circ)$. □

Let $d_i(w) = \mathrm{inv}_{\alpha_i}(w)$, for $i = 0, \ldots, r$. If $d_i(w) = 1$, we say that w has a descent at i.

Definition 6.2. Let $w \in W$. The *circular descent number* $\mathrm{cdes}(w)$ is defined by

$$\mathrm{cdes}(w) = \sum_{i=0}^{r} a_i d_i(w).$$

Note that $\mathrm{cdes}(w)$ is always positive. Indeed if $d_i(w) = 0$ for $i \in [1, r]$, then we have $w = 1$ and $d_0(w) = 1$.

Define $\delta_w \in L^\vee$ for $w \in W$ by

$$\delta_w = \sum_{i=0}^{r} d_i(w) \cdot \omega_i$$

where for convenience we let $\omega_0 = 0$.

Lemma 6.3. *The coweight δ_w is the unique integral coweight such that $w^{-1}(A_\circ) + \delta_w \in \Pi$.*

Proof. Let $i \in [1, r]$. Then by Lemma 6.1, $-d_i(w) < (w^{-1}(\rho/h), \alpha_i) < 1 - d_i(w)$, so that $\lambda = w^{-1}(A_\circ) + \delta_w$ satisfies $0 \leq (\lambda, \alpha_i) \leq 1$. The coweight δ_w must be unique since adding or subtracting any fundamental coweight ω_i will cause λ to violate the inequality $0 < (\lambda, \alpha_i) < 1$. □

We set $S = \{\alpha_0, \ldots, \alpha_r\}$ and $S_i = \{\alpha_j \in S \mid a_j = i\}$. For convenience we set $J = \{j \in [0, r] \mid a_j = 1\}$.

Proposition 6.4. *We have the following equivalent descriptions of the group $C \subset W$.*

$$C = \{w \in W \mid \text{cdes}(w) = 1\} \tag{3}$$
$$= \{w \in W \mid w(S) = S\} \tag{4}$$
$$= \{w \in W \mid w(S_k) = S_k \text{ for all } k\}. \tag{5}$$

For any $j \in J$ there exist a unique Weyl group element $w^{(j)} \in C$ such that $w(\alpha_i) > 0$, for $i \neq j$ and $w(\alpha_j) < 0$.

Proof. Let $c \in C$. By Lemma 6.3 and the definition of C, we see that $c^{-1}(A_\circ) + \delta_c = A_\circ$. By Lemma 3.6, this implies that $c^{-1}(\rho/h) + \delta_c = \rho/h$. It is clear that $(\delta_c, \theta) = \text{cdes}(c) - d_0(c)$. We compute, using Lemma 6.1, that

$$d_0(c) - 1 \leq (c^{-1}(\rho/h), \theta) \leq d_0(c).$$

So summing, we have

$$\text{cdes}(c) - 1 \leq (\rho/h, \theta) \leq \text{cdes}(c)$$

which immediately implies that $\text{cdes}(c) = 1$ since $(\rho/h, \theta) = 1 - 1/h$. The converse follows in the same manner. This establishes the equality in (3).

Now suppose $c \in C$. We establish (4). By definition, $c(\alpha_j) < 0$ for some $j \in J$ and $c(\alpha_i) > 0$ for $i \neq j$. Let $S_{\neq j} = \{\alpha_i \in S | i \neq j\}$. We have $c^{-1}(A_\circ) + \omega_j = A_\circ$, and in particular, the set $\{0, \omega_1/a_1, \ldots, \omega_r/a_r\}$ is sent to itself under the map $\lambda \mapsto c^{-1}(\lambda) + \omega_j$. Substituting this fact into $(c(\alpha_i), \omega_k/a_k) = (\alpha_i, c^{-1}(\omega_k/a_k))$ and noting that $\{\omega_1, \ldots, \omega_n\}$ are a dual basis to $\{\alpha_1, \ldots, \alpha_n\}$, we deduce that $c(S_{\neq j}) = S_{\neq 0}$, and $c(\alpha_j) = \alpha_0$. Thus (4) holds.

We get (5) from (4) by noting that up to scalar multiplication, the relation $\sum_i a_i \alpha_i = 0$ is the only linear dependence among the roots in S.

Conversely, (5) clearly implies (3) by the definition of cdes. The last statement of the proposition also follows from this discussion. □

By property (3), we have $f = |C| = \#\{i \in [0, r] \mid a_i = 1\}$. These i's correspond to minuscule coweights ω_i. Recall that a minuscule weight is one whose weight polytope has no internal weights.

We remark that the group C was previously studied by Verma [Ver] but not in the current context of the statistic cdes. The group C is related to the statistic cdes on the whole of W in an intimate way.

Theorem 6.5. *The statistic* cdes *is constant on the double cosets $C \backslash W / C$.*

Proof. Let $w \in W$ and $c \in C$ so that $c(\alpha_j) = \alpha_0$ for $j \in J$. We need to prove that $\text{cdes}(cw) = \text{cdes}(w) = \text{cdes}(wc)$. The latter equality is immediate from condition (5) of Proposition 6.4.

Let $\alpha \in S$ and let $\beta = w(\alpha) = b_1\alpha_1 + \cdots + b_r\alpha_r$. The b_i are either all positive or all negative. The element w has a descent at α if and only if $\beta < 0$. Now

$$cw(\alpha) = c(\beta) = b_1 c(\alpha_1) + \cdots + b_r c(\alpha_r). \tag{6}$$

If $b_j = 0$, then clearly $d_\alpha(w) = d_\alpha(cw)$. If $b_j \neq 0$, then we have a term of the form

$$b_j c(\alpha_j) = b_j \alpha_0 = -b_j(a_1\alpha_1 + \cdots + a_r\alpha_r) \tag{7}$$

in equation (6). Since $-\alpha_0$ is the longest root, we have $|a_k| \geq |b_k|$. Thus substituting (7) into (6), we see that $b_j \neq 0$ implies that $d_\alpha(w) = 1 - d_\alpha(cw)$. Indeed $b_j \in \{0, 1, -1\}$, and we have $d_\alpha(w) - d_\alpha(cw) = b_j$.

Now we have

$$0 = a_0 w(\alpha_0) + \cdots + a_r w(\alpha_r),$$

and so expressing both sides in terms of the simple roots $\alpha_1, \ldots, \alpha_r$, we see that the coefficient of α_j is 0. Write $b_j(\alpha)$ for the coefficient b_j in the proof earlier, obtaining the equality

$$0 = \sum_{i=0}^{r} a_i b_j(\alpha_i) = \sum_{i=0}^{r} (d_i(cw) - d_i(w)) = \mathrm{cdes}(cw) - \mathrm{cdes}(w). \qquad \square$$

7 The map cmaj

Define the *circular major map* $\mathrm{cmaj} : W \to C$ by

$$\mathrm{cmaj}(w) = \bar{b}(\delta_w)$$

where $\bar{b}$ is the isomorphism from Section 5.

Lemma 7.1. *The map* cmaj *satisfies* $\mathrm{cmaj}(c) = c$ *for* $c \in C$.

Proof. By Lemma 6.3, we have $c(A_\circ) = A_\circ + c(\delta_c)$. But by Lemma 2.1, $c(\delta_c) = \delta_c \bmod L^\vee$, so that by definition $\bar{b}(\delta_c) = c$. $\square$

Theorem 7.2. *The map* cmaj *satisfies*

$$\mathrm{cmaj}(c_1 w c_2) = c_1 w c_2^{\mathrm{cdes}(w)}$$

for $c_i \in C$ *and* $w \in W$.

Proof. Let $w \in W$ and $c = w^{(j)} \in C$, so that $c^{-1}(\alpha_0) = \alpha_j$. Thus $c(\alpha_j) < 0$, so that $\mathrm{cmaj}(c) = \omega_j$. We first consider wc. We have

$$w^{-1}(\rho/h) + \delta_w = \mu/h$$

for some $\mu \in \Lambda$ satisfying $0 < (\mu, \alpha_i) < h$ for $i \in [1, r]$. Applying c^{-1} on the left to both sides, we obtain

$$c^{-1} w^{-1}(\rho/h) + c^{-1}(\delta_w) = c^{-1}(\mu/h).$$

Now $0 < (c^{-1}(\mu/h), c^{-1}(\alpha_i)) < 1$ and as i varies, we obtain every simple root in the form $c^{-1}(\alpha_i)$ apart from $c(\alpha_0) = \alpha_j$ for some $j \in J$. But $-\mathrm{cdes}(w) < (\mu/h, \alpha_0) < 1 - \mathrm{cdes}(w)$, so we have $0 < (c^{-1}(\mu/h) + \mathrm{cdes}(w) \, . \, \omega_j, \alpha_j) < 1$.

Thus

$$\delta_{wc} = c(\delta_w) + \mathrm{cdes}(w) \, . \, \omega_j$$
$$\mathrm{cmaj}(wc) \equiv \mathrm{cmaj}(w) + \mathrm{cdes}(w) \, . \, \omega_j \mod L^\vee .$$

Hence $\mathrm{cmaj}(wc) = \mathrm{cmaj}(w) c^{\mathrm{cdes}(w)}$. We have used Lemma 2.1.

Similarly,

$$c^{-1}(\rho/h) + \delta_c = \rho/h$$

gives

$$w^{-1}c^{-1}(\rho/h) + w^{-1}(\omega_j) = w^{-1}(\rho/h)$$
$$w^{-1}c^{-1}(\rho/h) + w^{-1}(\omega_j) + \delta_w = w^{-1}(\rho/h) + \delta_w = \rho/h.$$

This implies that

$$\delta_{cw} = w^{-1}(\omega_j) + \delta_w$$
$$\mathrm{cmaj}(cw) = \mathrm{cmaj}(w) + \omega_j \mod L^\vee .$$

Hence $\mathrm{cmaj}(cw) = c \cdot \mathrm{cmaj}(w)$. □

Theorem 7.2 shows that the map cmaj allows us to pick representatives for the right cosets W/C. For example $\{w \mid \mathrm{cmaj}(w) = \mathrm{id}\}$ is a set of right coset representatives. In type A, cmaj; has an explicit representation theoretic meaning, see Theorem 11.1.

8 Relation between volumes and numbers of lattice points

Let P be an alcoved polytope, and let $A \in \mathscr{A}$ be an alcove. Let $k_\alpha = k_\alpha(P)$, $K_\alpha = K_\alpha(P)$, and $m_\alpha = m_\alpha(A)$, for $\alpha \in \Phi^+$ as in Section 3. Let us define the alcoved polytope $P_{(A)}$ as

$$P_{(A)} = \{\lambda \in V \mid k_\alpha - m_\alpha \le (\lambda, \alpha) \le K_\alpha - m_\alpha - 1, \text{ for all } \alpha \in \Phi^+\}.$$

The following claim follows directly from the definitions.

Lemma 8.1. *For P and A as above, the set $P_{(A)} \cap \Lambda^\vee$ of lattice points in $P_{(A)}$ is exactly the set of integral coweights $\lambda \in \Lambda^\vee$ such that $A + \lambda$ is an alcove in P.*

The lemma says that lattice points in $P_{(A)}$ are in one-to-one correspondence with alcoves in P that are obtained by affine translations of A.

For $w \in W$, the definition of the polytope $P_{(w)} = P_{(w(A_\circ))}$ can be rewritten as

$$P_{(w)} = \{\lambda \in V \mid k_\alpha + d_\alpha(w^{-1}) \leq (\lambda, \alpha) \leq K_\alpha + d_\alpha(w^{-1}) - 1, \text{ for all } \alpha \in \Phi^+\}.$$

We have used Lemma 6.1.

Notice that $P_{(A+\lambda)} = P_{(A)} - \lambda$. Thus the polytopes $P_{(A)} \equiv P_{(B)}$ are equivalent modulo affine translations by elements of $\Lambda^\vee$, whenever $A \equiv B \bmod \Lambda^\vee$. This implies that the polytope $P_{(\bar{w})} = P_{(w)}$ is correctly defined modulo affine translations by coweights $\lambda \in \Lambda$, where $w \in W$ is any representative of a coset $\bar{w} \in W/C$.

Let $I(P) = |P \cap \Lambda|$ be the number of lattice points in P. The following statement establishes a relation between the volume of an alcoved polytope and the numbers of lattice points in smaller alcoved polytopes.

Theorem 8.2. *Let P be an alcoved polytope. Then*

$$\mathrm{Vol}(P) = \sum_{\bar{w} \in W/C} I(P_{(\bar{w})}).$$

Proof. According to Lemma 8.1, the total number of alcoves in P equals the sum of $I(P_{(A)})$ over representatives A of cosets $\mathscr{A}/L^\vee$. This is exactly the claim of the theorem. □

9 Generalized hypersimplices

For $k = 1, \ldots, h-1$, let us define the k-th *generalized hypersimplex* Δ_k^Φ as the alcove polytope given by

$$\Delta_k^\Phi = \{\lambda \in V \mid 0 \leq (\lambda, \alpha_i) \leq 1, \text{ for } i = 1, \ldots, r; \text{ and } k-1 \leq (\lambda, \theta) \leq k\}.$$

In other words, the generalized hypersimplices are the slices of the parallelepiped Π by the parallel hyperplanes of the form $H_{\theta,k}$, for $k \in \mathbb{Z}$. Clearly, the first generalized hypersimplex is the fundamental alcove: $\Delta_1^\Phi = A_\circ$. Also the last generalized hypersimplex is the alcove given by $\Delta_{h-1}^\Phi = w_\circ(A_\circ) + \rho$, where $w_\circ \in W$ is the longest element in W.

Lemma 9.1. *Let $w \in W$. The polytope $(\Delta_k^\Phi)_{(w)}$ consists of a single point $\lambda \in \Lambda^\vee$, if* $\mathrm{cdes}(w^{-1}) = k$*, and $(\Delta_k^\Phi)_{(w)}$ is empty, if* $\mathrm{cdes}(w^{-1}) \neq k$.

Proof. By definition, the polytope $(\Delta_k^\Phi)_{(w)}$ is given by

$$(\Delta_k^\Phi)_{(w)} = \{\lambda \in \Lambda \mid d_i(w^{-1}) = (\lambda, \alpha_i), \text{ for } i = 1, \ldots, r;\ k - d_0(w^{-1}) = (\lambda, \theta)\}.$$

The first r equations $d_i(w^{-1}) = (\lambda, \alpha_i)$ have a single solution $\lambda = \sum_{i \in D} \omega_i = \delta_{w^{-1}}$, where $D = \{i \mid d_{\alpha_i}(w^{-1}) = 1\}$. The last equation $k - d_0(w^{-1}) = (\lambda, \theta^\vee)$ for the point $\lambda = \sum_{i \in D} \omega_i$ says that $\mathrm{cdes}(w) = k$. □

Corollary 9.2. *All representatives w of a coset $\bar{w} \in W/C$ have the same generalized descent numbers* $\mathrm{cdes}(w^{-1})$.

Proof. The polytopes $(\Delta_k^{\Phi})_{(u)} \equiv (\Delta_k^{\Phi})_{(w)}$ are equivalent modulo affine translations, whenever $\bar{u} = \bar{w}$ in W/C. □

Define $\mathrm{cdes}(\bar{w}^{-1}) = \mathrm{cdes}(w^{-1})$, where $w \in W$ is any representative of a coset $\bar{w}$. Theorem 8.2 implies the following statement.

Theorem 9.3. *The volume* $\mathrm{Vol}(\Delta_k^{\Phi})$ *of k-th generalized hypersimplex Δ_k^{Φ} equals the number of cosets $\bar{w} \in W/C$ such that* $\mathrm{cdes}(\bar{w}^{-1}) = k$. *Equivalently,* $f \cdot \mathrm{Vol}(\Delta_k^{\Phi})$ *equals the number of elements $w \in W$ such that* $\mathrm{cdes}(w^{-1}) = k$.

Let $\mathscr{H}(b_1,\dots,b_r;k,K)$ be the *thick hypersimplex* given by

$$\{\lambda \in V \mid 0 \le (\lambda,\alpha_i) \le b_i, \text{ for } i = 1,\dots,r; \text{ and } k \le (\lambda,\theta) \le K\}.$$

Proposition 9.4. *We have*

$$\mathrm{Vol}(\mathscr{H}(b_1,\dots,b_r;k,K) = \sum_l \mathrm{Vol}(\Delta_l^{\Phi}) \cdot I(\mathscr{H}(b_1-1,\dots,b_r-1;l-K+1,l-k)).$$

Proof. Let $\lambda \in \Lambda^\vee$ be in the interior of Δ_l^{Φ} and $\mu \in \Lambda^\vee$. Then $0 < (\lambda,\alpha_i) < 1$ for $i \in [1,r]$ and $l-1 < (\lambda,\theta) \le l$. Thus $\lambda + \mu \in \mathscr{H}(b_1,\dots,b_r;k,K)$ if and only if we have $0 \le (\mu,\alpha_i) \le b_i - 1$ and $l-K+1 \le (\mu,\theta) \le l-k$. We conclude that for any alcove $A \in \Delta_l^{\Phi}$, we have $\mathscr{H}(b_1,\dots,b_r;k,K)_{(A)} = \mathscr{H}(b_1-1,\dots,b_r-1;l-K+1,l-k) \bmod \Lambda^\vee$. As l varies, we obtain a translate of $\mathscr{H}(b_1,\dots,b_r;k,K)_{(\bar{w})}$ for each coset $\bar{w}$ exactly once in this form. □

10 A q-analogue of Weyl's formula

Recall that for a permutation $w = w_1 \dots w_n$ in the symmetric group S_n, a *descent* is an index $i \in \{1,\dots,n-1\}$ such that $w_i > w_{i+1}$. Let $\mathrm{des}(w)$ be the number of descents of $w \in S_n$. The n-th *Eulerian polynomial* $A_n(q)$ is defined as

$$A_n(q) = \sum_{w \in S_n} q^{\mathrm{des}(w)+1},$$

for $n \ge 1$, and $A_0(q) = 1$. These polynomials can also be expressed as $A_n(q) = (1-q)^{n+1} \cdot \sum_{k \ge 0} k^n q^k$. Let $[n]_q = (1-q^n)/(1-q)$ denote the q-analogue of an integer $n \in \mathbb{Z}$.

The group algebra $\mathbb{Z}[\Lambda^\vee/L^\vee]$ has a $\mathbb{Z}$-basis of formal exponents e^x, for $x \in \Lambda^\vee/L^\vee$, with multiplication $e^x \cdot e^y = e^{x+y}$. Let $\mathbb{Z}[q][\Lambda^\vee/L^\vee] = \mathbb{Z}[q] \otimes \mathbb{Z}[\Lambda^\vee/L^\vee]$. The following theorem generalizes Weyl's formula (2) for the order of the Weyl group.

Theorem 10.1. *The following identity holds in the group algebra* $\mathbb{Z}[q][\Lambda^\vee/L^\vee]$*:*

$$\sum_{w\in W} q^{\mathrm{cdes}(w)}\, e^{\mathrm{cmaj}(w)} = \left(\sum_{x\in\Lambda^\vee/L^\vee} e^x\right)\cdot A_r(q)\cdot[a_1]_q\cdots[a_r]_q.$$

In particular, we have the following identity for polynomials in $\mathbb{Z}[q]$*:*

$$\sum_{w\in W} q^{\mathrm{cdes}(w)} = f\cdot A_r(q)\cdot[a_1]_q\cdots[a_r]_q.$$

We first establish the following generating function for the volumes of generalized hypersimplices.

Proposition 10.2. *The generating function for the volumes of generalized hypersimplices is given by*

$$\sum_{k=1}^{h-1}\mathrm{Vol}(\Delta_k^\Phi)q^k = A_r(q)\cdot[a_1]_q\cdots[a_r]_q.$$

Proof. The union of the generalized hypersimplices Δ_k^Φ for $k=1,2,\ldots,h-1$ is the fundamental parallelpiped Π. For a bounded polytope $P\subset V$, define the generating function

$$g_P(q) = \sum_k \mathrm{Vol}(P\cap\{\lambda\in V\mid k-1\le(\lambda,\theta)\le k\})\,q^k\in\mathbb{R}[q^{\pm1}].$$

Then $g_\Pi(q)=\sum_{k=1}^{h-1}\mathrm{Vol}(\Delta_k^\Phi)q^k$. We note that if $(\lambda,\theta)=a\in\mathbb{Z}$, then $g_{P+\lambda}(q)=q^a g_P(q)$. Now set Ξ to be the parallelpiped spanned by the vectors $\omega_1/a_1,\ldots,\omega_r/a_r$. Then Π is a union of translates of Ξ by integral linear combinations of the vectors ω_i/a_i, and we deduce that

$$g_\Pi(q) = g_\Xi(q)\cdot[a_1]_q\cdots[a_r]_q.$$

Since we are normalizing the fundamental alcove $A_\circ$ with vertices $\omega_1/a_1,\ldots,\omega_r/a_r$ to have $\mathrm{Vol}(A_\circ)=1$, it follows that $g_\Xi(q)$ is equal to the generating function of the normalized volumes of the usual hypersimplices:

$$g_\Xi(q) = \sum_{k=1}^{r}\mathrm{Vol}([0,1]^r\cap\{(x_1,\ldots,x_r)\in\mathbb{R}^r\mid k-1\le x_1+\cdots+x_r\le k\})\,q^k$$

which is well known to be equal to the Eulerian polynomial $A_r(q)$. This also follows from Theorem 9.3 (see Section 11) and is studied in detail in [API]. □

Proof of Theorem 10.1. Using Theorem 7.2, we let $W'=\{w\in W\mid \mathrm{cmaj}(w)=\mathrm{id}\}$ be a set of left coset representatives for $C\backslash W$. Then $(W')^{-1}$ is a set of right coset representatives for W/C. We calculate

$$\begin{aligned}
&\sum_{w\in W} q^{\mathrm{cdes}(w)}\, e^{\mathrm{cmaj}(w)} \\
&= \sum_{w\in W'} q^{\mathrm{cdes}(w)} \sum_{c\in C} e^{\mathrm{cmaj}(c)} && \text{by Theorems 6.5 and 7.2,} \\
&= \sum_{w\in (W')^{-1}} q^{\mathrm{cdes}(w^{-1})} \cdot \left(\sum_{x\in \Lambda^\vee/L^\vee} e^x \right) \\
&= \left(\sum_{k=1}^{h-1} \mathrm{Vol}(\Delta_k^{\Phi})\, q^k \right) \cdot \left(\sum_{x\in \Lambda^\vee/L^\vee} e^x \right) && \text{by Theorem 9.3,} \\
&= \left(\sum_{x\in \Lambda^\vee/L^\vee} e^x \right) \cdot A_r(q)\cdot [a_1]_q \cdots [a_r]_q && \text{by Proposition 10.2.}
\end{aligned}$$

□

Remark 10.3. We have

$$\sum_{w\in W} q^{\mathrm{cdes}(w)}\, e^{\mathrm{cmaj}(w)} = \sum_{w\in W} q^{\mathrm{cdes}(w)}\, e^{\mathrm{cmaj}(w^{-1})}.$$

This follows from the fact (Theorem 6.5) that cdes is constant on $C\backslash W/C$ double cosets. Each double coset is a disjoint union of left (resp. right) cosets $C\backslash W$ (resp. W/C) for which $e^{\mathrm{cmaj}}(w)$ (resp. $e^{\mathrm{cmaj}}(w^{-1})$) takes the values $\left(\sum_{x\in\Lambda^\vee/L^\vee} e^x\right)$.

Remark 10.4. It would be interesting to compare Theorem 10.1 with Stembridge and Waugh's Weyl group identity [SW].

The following question seems interesting.

Question 10.5. What is $\sum_{w\in W} x^{\mathrm{cmaj}(w)} y^{\mathrm{cmaj}(w^{-1})}$ in $\mathbb{Z}[\Lambda^\vee/L^\vee]\otimes\mathbb{Z}[\Lambda^\vee/L^\vee]$?

11 Example: type A

Let $\Phi = A_{n-1} \subset \mathbb{R}^n/\mathbb{R}(1,1,\ldots,1)$ throughout this section. The simple roots are $\alpha_i = e_i - e_{i+1}$ where e_i are the coordinate basis vectors of $\mathbb{R}^n$. The longest root is $\theta = e_1 - e_n$ and we have $a_i = 1$ for $i \in [0,n]$. The Weyl group $W = S_n$ is the symmetric group on n letters and $\mathrm{cdes}(w)$ is equal to the usual number of descents of w plus a descent at n if $w_n > w_1$. This is the reason for calling cdes the circular descent number. The group $C = \langle c = (123\cdots(n-1)n)\rangle$ is generated by the long cycle. The fundamental coweights are given by $\omega_i = e_1 + e_2 + \cdots + e_i$ and one can check that $\delta_{c^i} = \omega_i$. Thus $\mathrm{cmaj}(w) = c^{-\mathrm{maj}(w) \bmod n}$ where $\mathrm{maj}(w)$ denotes the usual major index of w. We can verify Proposition 7.2 directly: left multiplication by the long cycle c maps $w_1w_2\cdots w_n$ to $(w_1+1)(w_2+1)\cdots(w_n+1)$ where '$n+1$' is identified with '1'. Right multiplication by c maps $w_1w_2\cdots w_n$ to $w_2w_3\cdots w_nw_1$.

The following theorem ([KW] and [EC2, Ex. 7.88]) suggests that the map cmaj may have an explicit representation theoretic interpretation. Let χ^λ denote the irreducible character of the symmetric group S_n labeled by a partition λ.

Theorem 11.1. *Let $C_n \subset S_n$ be a cyclic subgroup of order n. Let $\rho = \mathrm{ind}_{C_n}^{S_n} e^{2\pi\sqrt{-1}j/n}$ be an induced character of S_n. Then we have*

$$\left\langle \rho, \chi^\lambda \right\rangle = \#\{\mathrm{SYT}(T) \mid \mathrm{sh}(T) = \lambda \text{ and } \mathrm{maj}(T) \equiv j \bmod n\}.$$

Here a descent of a standard young tableau (SYT) T is an index i such that the box containing $i+1$ is to the southwest of the box containing i in T. The index $\mathrm{maj}(T)$ *is defined to be the sum of all the descents of T.*

It is not hard to see that the polytopes $\Delta_k^{A_{n-1}}$ are affinely equivalent to the usual hypersimplices defined as the convex hull of the points ε_I where $\varepsilon_I = \sum_{i\in I} e_i$ and I varies over all k-subsets of $[n]$. The alcoved triangulation here is identical to the one studied in [API].

12 Example: Type C

Let $\Phi = C_n$ with $2n$ long roots $\pm 2e_i$ for $1 \le i \le n$ and $2n(n-1)$ short roots $\pm e_i \pm e_j$ for $1 \le i \ne j \le n$. A system of simple roots is given by $\alpha_1 = e_1 - e_2, \alpha_2 = e_2 - e_3, \ldots, \alpha_{n-1} = e_{n-1} - e_n, \alpha_n = 2e_n$. Then $\theta = 2e_1 = 2\alpha_1 + 2\alpha_2 + \cdots + 2\alpha_{n-1} + \alpha_n$, so that $a_0 = a_n = 1$ and $a_i = 2$ for $1 \le i \le n-1$. The fundamental coweights are given by $\omega_1 = e_1$, $\omega_2 = e_1 + e_2$, ..., $\omega_{n-1} = e_1 + \cdots + e_{n-1}$, $\omega_n = 1/2(e_1 + \cdots + e_n)$.

We identify the Weyl group W of type C_n with the group of signed permutations $w_1 w_2 \cdots w_n$ in the usual way: $w_i \in \pm\{1,2,\ldots,n\}$ and $|w_1||w_2|\cdots|w_n|$ is a usual permutation in S_n. For $i \in [1, n-1]$, a signed permutation $w = w_1 w_2 \cdots w_n$ has a descent at i if $w_i > w_{i+1}$, as usual. We have a descent at 0 if $w_1 > 0$ and a descent at n if $w_n < 0$. The group C has order two, with unique non-identity element $c = (-n - (n-1) \cdots -2 -1)$. The map $\mathrm{cmaj} : W \to C$ is given by

$$\mathrm{cmaj}(w) = \begin{cases} \mathrm{id} & \text{if } w_n > 0 \\ c & \text{if } w_n < 0. \end{cases}$$

Theorem 10.1 states in this case that

$$\sum_{w\in W} q^{\mathrm{cdes}(w)}\, e^{\mathrm{cmaj}(w)} = (e^{\mathrm{id}} + e^c) \cdot A_n(q) \cdot (1+q)^{n-1}.$$

13 Gröbner bases

In this section we will regularly refer to the results of the first paper in this series [API], in particular Appendix 19.

Let $\Phi \subset V$ be a fixed irreducible root system and P a proper alcoved polytope. We first note that the triangulations of alcoved polytopes are coherent.

Lemma 13.1. *Any polytopal subdivision arising from a hyperplane arrangement is coherent.*

Proof. As only finitely many hyperplanes will be involved in a triangulation or subdivision of a polytope, we may assume the set $\mathscr{S}$ of hyperplanes is finite.

Pick a linear functional ϕ_H for each hyperplane $H \in \mathscr{S}$ such that H is given by the vanishing of ϕ_H. Then define the piecewise linear convex function $h : V \to \mathbb{R}$ by

$$h(v) = \sum_{H \in \mathscr{S}} |\phi_H(v)|.$$

It is clear that $h(v)$ is convex, since it is a sum of convex functions. The domains of linearity are exactly the regions determined by the hyperplane arrangement. Thus the subdivision of a polytope induced by a hyperplane arrangement is coherent. □

We denote by N the set of vertices of the affine Coxeter arrangement. By [API, Theorem 19.1] the triangulation of P can be described by some appropriate term order on the polynomial ring

$$k[P] = k[x_a | a \in N \cap P].$$

Let us fix coordinates on V, so that all points in N have integer coordinates. Identify a vertex $a = (a_1, \dots, a_n) \in N$ with the coordinates $(a, 1) \in V \oplus \mathbb{R}$. Thus the triangulation is also equivalent to the reduced Gröbner basis $\mathscr{G}_P$ of the toric ideal $I_P := I_{P \cap N}$ in the notation of [API, Appendix 19]. By our choice of coordinates this toric ideal is homogeneous.

In general, the Gröbner basis $\mathscr{G}_P$ appears to be quite complicated, but many simplifications occur when N is a lattice. One can check directly that this is the case for the root systems A_n, C_n and D_4.

We assume that Φ is one of these root systems from now on. Set $c_i = \frac{\omega_i}{a_i}$. Then N is spanned by the c_i. In this case, an alcove has normalized volume 1 with respect to N. Thus by [API, Proposition 19.2], $\mathscr{G}_P$ has an initial ideal generated by square-free monomials.

Example 13.2. With the notation as in Section 12, the vertices N of the affine Coxeter arrangement of type C_n are exactly the points with all coordinates, either integers or half-integers. One can check that the lattice N is spanned by the vectors c_i.

Lemma 13.3. *Let $a,b \in N$. The midpoint $(a+b)/2$ is either a vertex or it lies on a unique edge, such that it is the midpoint of the two closest vertices lying on that edge.*

Proof. Suppose $c = (a+b)/2$ is not in N. The closed fundamental alcove $\overline{A_\circ}$ is the convex hull of the points c_i and 0. Thus there is a (affine) Weyl group element σ which takes the midpoint $c = (a+b)/2$ into $\overline{A_\circ}$. Since $\sigma(a)$ and $\sigma(b)$ are both integral linear combinations of the c_i, it is clear that $\sigma(c)$ must have the form $\frac{c_i+c_j}{2}$ or $\frac{c_i}{2}$. In the first case, $\sigma(c)$ lies on the edge given by the intersection of the hyperplanes $H_{\alpha_k,0}$ for $k \neq i,j$ satisfying $k \in [1,r]$ and $H_{\theta,1}$. In the second case the edge is given by the hyperplanes $H_{\alpha_k,0}$ for $k \neq i$. Thus c is the midpoint of $\sigma^{-1}(c_i)$ and $\sigma^{-1}(c_j)$, or the midpoint of $\sigma^{-1}(c_i)$ and $\sigma^{-1}(0)$. □

In the first case of Lemma 13.3, we set $u(a,b) = v(a,b) = (a+b)/2$. In the second case we set $u(a,b)$ and $v(a,b)$ to be the two closest vertices on the edge containing $(a+b)/2$.

Example 13.4. For type A_{n-1}, we can describe the vertices $u(a,b)$ and $v(a,b)$ in the following explicit manner ([API]). Let I,J be two k-element multi-subsets of $[n]$. Let $a_1 \leq a_2 \leq \cdots \leq a_{2k}$ be the increasing rearrangement of $I \cup J$. We define two k-element multi-subsets $U(I,J)$ and $V(I,J)$ by $U(I,J) = \{a_1,a_3,\ldots,a_{2k-1}\}$ and $V(I,J) = \{a_2,a_4,\ldots,a_{2k}\}$. For a k-element multi-subset I, we let $a_I \in \mathbb{R}^n$ be the (integer) vector with j-th coordinate $(a_I)_j$ equal to the number of occurrences of $\{1,2,\ldots,j\}$ in I. Then one can check that $u(a_I,a_J)$ and $v(a_I,a_J)$ are exactly $a_{U(I,J)}$ and $a_{V(I,J)}$.

Lemma 13.5. *Suppose $a,b \in P$ are vertices of the affine Coxeter arrangement, where P is a proper alcoved polytope. Then the vertices $u(a,b)$ and $v(a,b)$ are also in P.*

Proof. As P is convex, $c = (a+b)/2 \in P$. Assume now that c is not a vertex and suppose $u(a,b) \notin P$. Then there exists some root α and some integer k, so that $H_{\alpha,k}$ separates $u(a,b)$ and c. Here we pick $H_{\alpha,k}$ so that it may go through c but not through $u(a,b)$. The intersection of $H_{\alpha,k}$ and the edge segment joining $u(a,b)$ to $v(a,b)$ is a vertex of the affine Coxeter arrangement. But this is impossible, since there are no vertices lying between $v(a,b)$ and $u(a,b)$. □

Define a marked set $\mathscr{G}_P$ of elements which lie in I_P as follows:

$$\mathscr{G}_P = \{\underline{x_a x_b} - x_{u(a,b)}x_{v(a,b)}\}, \tag{8}$$

where a,b range over pairs of unequal vertices in P. The main result of this section is the following theorem.

Theorem 13.6. *Let Φ be one of the root systems A_n, C_n or D_4 and P a proper alcoved polytope. Then there exists a term order $\prec_P$ such that the quadratic binomials $\mathscr{G}_P$ form a (reduced) Gröbner basis of the toric ideal I_P with respect to $\prec_P$, such that the underlined monomial is the leading term.*

Proof. By Lemma 13.5, the binomials in $\mathscr{G}_P$ do indeed make sense and since $a+b = u(a,b)+v(a,b)$, they lie within I_P.

By Lemma 13.1, the triangulation is coherent and is given by the domains of linearity of the piecewise-linear function h. The same function h gives a weight vector ω as described in [API, Appendix]. By [API, Theorem A.1] the weight vector ω induces a term order $\prec_P$ such that $\Delta_{\prec_P}(I_P) = \Delta_\omega$ (we have also used the fact that the triangulation is unimodular, and [API, Proposition A.2]).

Now let $a,b \in P$ be vertices of the affine arrangement. If $x_a x_b \neq x_{u(a,b)} x_{v(a,b)}$, then clearly a and b do not belong to the same simplex of the triangulation of P. Thus $x_a x_b$ belongs to the Stanley-Reisner ideal of the alcoved triangulation of P while $x_{u(a,b)} x_{v(a,b)}$ does not. This implies that under $\prec_P$, the underlined terms in the basis above are exactly the leading terms. In other words, the set $\mathscr{G}_P$ is *marked coherently*.

Finally we check that $\mathscr{G}_P$ is indeed a Gröbner basis of I_P under $\prec$. Since $\mathscr{G}_P$ is marked coherently, it follows that the reduction of any polynomial modulo $\mathscr{G}_P$ is Noetherian (that is, it terminates). It is clear that a monomial $x_{p_1} \dots x_{p_k}$ cannot be reduced further under $\mathscr{G}_p$ if any only if all the p_i belong to the simplex of the triangulation. Thus every monomial can be reduced via $\mathscr{G}_P$ to a standard monomial and hence $\mathrm{in}_\prec(\mathscr{G}_P)$ generates $\mathrm{in}_\prec(I_P)$.

The fact that this Gröbner basis is reduced is clear. □

Corollary 13.7. *Let Φ be one of the root systems A_n, C_n or D_4 and $P \subset V$ be a convex polytope with vertices among the vertices of the affine Coxeter arrangement. Then P is alcoved if and only if the conclusion of Lemma* 13.5 *holds.*

Proof. "Only if" is the content of Lemma 13.5. For the other direction, we note that the quadratic binomials $\mathscr{G}_P$ can be defined by (8). There is some large alcoved polytope Q which contains P and since $\mathscr{G}_P \subset \mathscr{G}_Q$, this allows us to conclude that $\mathscr{G}_P$ is marked coherently. And so there is a term order $\prec_P$ which selects the marked monomial as the leading monomial in $\mathscr{G}_P$. It is easy to check that $\mathscr{G}_P$ is the Groöbner basis of I_P with respect to $\prec_P$, and the standard monomials are exactly given by monomials corresponding to faces of alcoves. Thus we obtain an alcoved triangulation of P. □

Naturally associated to the ideal I_P is the projective algebraic variety Y_P defined as

$$Y_P = \mathrm{Proj}\left(k[x_a | a \in \mathscr{P}]/I_P\right).$$

This is the projective toric variety associated to the polytope P. The following corollary is immediate from Theorem 13.6 and [API, Proposition A.2].

Corollary 13.8. *Let Φ be one of the root systems A_n, C_n or D_4 and P a proper alcoved polytope. Let Y_P be the projective toric variety defined by I_P. Then Y_P is projectively normal and its Hilbert polynomial is equal to the Erhart polynomial of P (with respect to N).*

This should be compared with the work of Payne [Pay], who showed, in types A, B, C, and D that alcoved polytopes whose vertices lie in the coweight lattice are normal with respect to the coweight lattice.

References

[FZ] S. FOMIN AND A. ZELEVINSKY, Y-systems and generalized associahedra, *Annals of Mathematics* 158 (2003), 977–1018.

[Hum] J. E. HUMPHREYS, Reflection groups and Coxeter groups, Cambridge University Press, 1990.

[Kos] B. KOSTANT, private communication.

[KW] W. KRAŚKIEWICZ AND J. WEYMAN, Algebra of coinvariants and the action of the Coxeter element, *Bayreuth. Math. Schr.* 63, (2001), 265–284.

[API] T. LAM AND A. POSTNIKOV, Alcoved Polytopes I, *Discrete Comput. Geom.* 38 (2007), no. 3, 453–478.

[LP] C. LENART AND A. POSTNIKOV, Affine Weyl groups in K-theory and representation theory, *Int. Math. Res. Not.* 2007, no. 12, Art. ID rnm038, 65 pp.

[Pay] S. PAYNE, Lattice polytopes cut out by root systems and the Koszul property, *Adv. Math.* 220 (2009), no. 3, 926–935.

[EC2] R. STANLEY, *Enumerative Combinatorics, Vol 2*, Cambridge, 1999.

[SW] J. STEMBRIDGE AND D. WAUGH, A Weyl group generating function that ought to be better known, *Indagationes Mathematicae* 9 (1998), 451–457.

[Stu] B. STURMFELS, Gröbner bases and convex polytopes, *University Lecture Series* 8, American Mathematical Society, Providence, RI, 1996.

[Ver] D. VERMA, The role of affine Weyl groups in the representation theory of algebraic Chevalley groups and their Lie algebras, Lie groups and their representations *Proc. Summer School, Bolyai János Math. Soc.*, Budapest, 1971, 653–705.

[WY] A. WERNER AND J. YU, Symmetric alcoved polytopes, *Electron. J. Combin.* 21 (2014), no. 1, Paper 1.20, 14 pp.

Representation Theory of Quantized Gieseker Varieties, I

Ivan Losev

To the memory of Bertram Kostant

Abstract We study the representation theory of quantizations of Gieseker moduli spaces. We describe the categories of finite-dimensional representations for all parameters and categories $\mathcal{O}$ for special values of parameters. We find the values of parameters, where the quantizations have finite homological dimension, and we establish the abelian localization theorem. We classify two-sided ideals. Finally, we determine the annihilators of the irreducible objects in categories $\mathcal{O}$ for some special choices of one-parameter subgroups.

Keywords: Gieseker varieties • quantizations • categories $\mathcal{O}$ • localization theorems • two-sided ideals

Mathematics Subject Classification (2010): Primary 16G99, Secondary 16G20, 53D20, 53D55

1 Introduction

Our goal is to study the representation theory of quantizations of Gieseker moduli spaces.

My work was funded by the Russian Academic Excellence Project '5-100' and was partially supported by the NSF under the grant DMS-1501558.

Department of Mathematics, Northeastern University, Boston MA 02115 USA & International Laboratory of Representation theory and Mathematical Physics, NRU-HSE, Moscow, Russia, e-mail: i.loseu@neu.edu

V. G. Kac, V. L. Popov (eds.), *Lie Groups, Geometry, and Representation Theory*, Progress in Mathematics 326, https://doi.org/10.1007/978-3-030-02191-7_11

1.1 Gieseker moduli spaces and their quantizations

We explain constructions of Gieseker moduli spaces and of their quantizations via Hamiltonian reduction.

Pick two vector spaces V, W of dimensions n, r, respectively. Consider the space $R := \mathfrak{gl}(V) \oplus \operatorname{Hom}(V,W)$ and a natural action of $G := \mathrm{GL}(V)$ on it. Then we can form the cotangent bundle T^*R; this is a symplectic vector space. Identifying $\mathfrak{gl}(V)^*$ with $\mathfrak{gl}(V)$ and $\operatorname{Hom}(V,W)^*$ with $\operatorname{Hom}(W,V)$ using the trace form, we identify T^*R with $\mathfrak{gl}(V)^{\oplus 2} \oplus \operatorname{Hom}(V,W) \oplus \operatorname{Hom}(W,V)$. The action of G on T^*R is symplectic, so we get the moment map $\mu\colon T^*R \to \mathfrak{g}$. It can be described in two equivalent ways. First, we have $\mu(A,B,i,j) = [A,B] - ji$. Second, the dual map $\mu^*\colon \mathfrak{g} \to \mathbb{C}[T^*R]$ sends $\xi \in \mathfrak{g}$ to the vector field ξ_R (the infinitesimal action of ξ) which can be viewed as a polynomial function on T^*R.

Now pick a non-trivial character θ of G and consider the open subset of θ-stable points $(T^*R)^{\theta-ss} \subset T^*R$. For example, for $\theta = \det^k$ with $k > 0$, the subset of semistable points consists of all quadruples (A,B,i,j) such that $\ker i$ does not contain nonzero A- and B-stable subspaces. Then we can form the GIT Hamiltonian reduction $\mathcal{M}^\theta(n,r) := \mu^{-1}(0)^{\theta-ss}/G$; this is the Gieseker moduli space (for all choices of θ). It is a smooth symplectic quasi-projective variety of dimension $2rn$ that is a resolution of singularities of the categorical Hamiltonian reduction $\mathcal{M}(n,r) := \mu^{-1}(0)/\!/G$. We note that the dilation action of $\mathbb{C}^\times$ on T^*R descends to both $\mathcal{M}^\theta(n,r)$, $\mathcal{M}(n,r)$; the corresponding actions will be called *contracting* below. The resulting grading on $\mathbb{C}[\mathcal{M}(n,r)]$ is positive, which means that $\mathbb{C}[\mathcal{M}(n,r)] = \bigoplus_{i\geqslant 0} \mathbb{C}[\mathcal{M}(n,r)]_i$, where $\mathbb{C}[\mathcal{M}(n,r)]_i$ denotes the ith graded component, and $\mathbb{C}[\mathcal{M}(n,r)]_0 = \mathbb{C}$.

Now we explain how to construct quantizations of $\mathcal{M}(n,r)$, meaning filtered associative unital algebras $\mathcal{A}$ with $\operatorname{gr}\mathcal{A} \xrightarrow{\sim} \mathbb{C}[\mathcal{M}(n,r)]$ (an isomorphism of graded Poisson algebras). Take $\lambda \in \mathbb{C}$ and set

$$\mathcal{A}_\lambda(n,r) := (D(R)/[D(R)\{x - \lambda \operatorname{tr} x, x \in \mathfrak{g}\}])^G.$$

This is a filtered algebra, whose filtration is induced from the Bernstein filtration on $D(R)$, where $\deg R = \deg R^* = 1$, and there is a natural epimorphism $\mathbb{C}[\mathcal{M}(n,r)] \twoheadrightarrow \operatorname{gr}\mathcal{A}_\lambda(n,r)$ which is an isomorphism because μ is flat.

We can also consider the quantization $\mathcal{A}_\lambda^\theta(n,r)$ of $\mathcal{M}^\theta(n,r)$. This is a sheaf (in conical topology) of filtered algebras of $\mathcal{M}^\theta(n,r)$, also obtained by quantum Hamiltonian reduction; we will recall how below. Its global sections coincide with $\mathcal{A}_\lambda(n,r)$ and its higher cohomology vanishes. So we have the global section functor $\Gamma_\lambda^\theta\colon \operatorname{Coh}(\mathcal{A}_\lambda^\theta(n,r)) \to \mathcal{A}_\lambda(n,r)\text{-mod}$, where we write $\operatorname{Coh}(\mathcal{A}_\lambda^\theta(n,r))$ for the category of coherent $\mathcal{A}_\lambda^\theta(n,r)$-modules and $\mathcal{A}_\lambda(n,r)$-mod for the category of finitely generated $\mathcal{A}_\lambda(n,r)$-modules.

We note that the variety $\mathcal{M}^\theta(n,r)$ comes with another torus action. Namely, let T_0 denote a maximal torus in $\mathrm{GL}(W)$. The torus $T := T_0 \times \mathbb{C}^\times$ acts on R by $(t,z).(A,i) := (zA, ti), t \in T_0, z \in \mathbb{C}^\times, A \in \mathfrak{gl}(V), i \in \operatorname{Hom}(V,W)$. The action naturally lifts to a Hamiltonian action on T^*R commuting with $\mathrm{GL}(V)$ and hence descends to $\mathcal{M}^\theta(n,r)$. Note that the T-action on $\mathcal{M}^\theta(n,r)$ commutes with the contracting action of $\mathbb{C}^\times$.

Now let $\bar{\mathcal{A}}_\lambda(n,r)$ denote the algebra $(D(\bar{R})/[D(\bar{R})\{x-\lambda\,\mathrm{tr}\,x, x\in\mathfrak{g}\}])^G$, where $\bar{R}=\mathfrak{sl}(V)\oplus\mathrm{Hom}(V,W)$. Note that $\mathcal{A}_\lambda(n,r)=D(\mathbb{C})\otimes\bar{\mathcal{A}}_\lambda(n,r)$ so most questions about the representation theory of $\mathcal{A}_\lambda(n,r)$ reduce to similar questions about $\bar{\mathcal{A}}_\lambda(n,r)$.

1.2 Classical cases

There is one case that was studied very extensively in the past fifteen years or so: $r=1$. Here the variety $\mathcal{M}^\theta(n,1)$ is the Hilbert scheme $\mathrm{Hilb}_n(\mathbb{C}^2)$ of n points on $\mathbb{C}^2$ and $\mathcal{M}(n,1)=\mathbb{C}^{2n}/\mathfrak{S}_n$. The quantization $\bar{\mathcal{A}}_\lambda(n,r)$ is the spherical subalgebra in the Rational Cherednik algebra $H_\lambda(n)$ for the pair $(\mathfrak{h},\mathfrak{S}_n)$, where $\mathfrak{h}$ is the reflection representation of $\mathfrak{S}_n$, see [GG] for details. The representation theory of $H_\lambda(n)$ (and hence of $\bar{\mathcal{A}}_\lambda(n,1)$) was studied, for example, in [BEG, GS1, GS2, R, KR, BE, L1, Wi, L5]. In particular, the following is known:

(1) when (=for which λ) the algebra $\bar{\mathcal{A}}_\lambda(n,1)$ has finite homological dimension, [BE], and hence is Morita equivalent to $H_\lambda(n)$,
(2) how to classify the finite-dimensional irreducible representations of $H_\lambda(n)$, [BEG],
(3) how to compute the characters of irreducible modules in the so-called category $\mathcal{O}$, [R, L5],
(4) how to determine the supports of these modules, [Wi],
(5) how to describe the two-sided ideals of $\bar{\mathcal{A}}_\lambda(n,1)$, [L1],
(6) when an analog of the Beilinson-Bernstein localization theorem holds, [GS1, KR].

We point out that there is an even more classical special case of the algebras $\bar{\mathcal{A}}_\lambda(n,r)$: when $n=1$. In this case $\mathcal{M}^\theta(1,r)=\mathbb{C}^2\times T^*\mathbb{P}^{r-1}$ and $\bar{\mathcal{A}}_\lambda(1,r)=D^\lambda(\mathbb{P}^{r-1})$ (the algebra of λ-twisted differential operators). (1)–(6) in this case are known and easy.

We will address analogs of (1),(2),(5),(6) for $\bar{\mathcal{A}}_\lambda(n,r)$ with general n,r. We prove some results towards (3) and (4) as well.

1.3 Main results

First, we give answers to (1) and (6).

Theorem 1.1. *The following are true.*

(1) *The algebra $\bar{\mathcal{A}}_\lambda(n,r)$ has finite homological dimension if and only if λ is not of the form $\frac{s}{m}$, where $1\leqslant m\leqslant n$ and $-rm<s<0$.*

(2) *For* $\theta = \det$, *abelian localization holds for* λ *(i.e.,* Γ_λ^θ *is an equivalence) if and only if* λ *is not of the form* $\frac{s}{m}$, *where* $1 \leqslant m \leqslant n$ *and* $s < 0$.

In fact, (2) above is a straightforward consequence of (1) and results of McGerty and Nevins, [MN2].

Let us proceed to a classification of finite-dimensional representations.

Theorem 1.2. *The algebra* $\bar{\mathscr{A}}_\lambda(n,r)$ *has a finite-dimensional representation if and only if* $\lambda = \frac{s}{n}$ *with* s *and* n *coprime and* $\lambda \notin (-r,0)$. *If this is the case, then the category* $\bar{\mathscr{A}}_\lambda(n,r)\text{-mod}_{fin}$ *of finite-dimensional representations is equivalent to* Vect, *the category of vector spaces.*

Now let us proceed to classifying two-sided ideals.

Theorem 1.3. *The following are true.*

(1) *If the algebra* $\bar{\mathscr{A}}_\lambda(n,r)$ *has infinite homological dimension, then it is simple.*
(2) *Assume that* $\bar{\mathscr{A}}_\lambda(n,r)$ *has finite homological dimension and let* m *stand for the denominator of* λ *(equal to* $+\infty$ *if* λ *is not rational). Then there are* $\lfloor n/m \rfloor$ *proper two-sided ideals in* $\bar{\mathscr{A}}_\lambda(n,r)$, *all of them are prime, and they form a chain* $\bar{\mathscr{A}}_\lambda(n,r) \supsetneq \mathscr{J}_{\lfloor n/m \rfloor} \supsetneq \cdots \supsetneq \mathscr{J}_2 \supsetneq \mathscr{J}_1 \supsetneq \{0\}$.

Finally, we explain some partial results on category $\mathscr{O}$ for $\bar{\mathscr{A}}_\lambda^\theta(n,r)$; we will recall necessary definitions below in Section 2.2. We use the notation $\mathscr{O}_\nu(\mathscr{A}_\lambda^\theta(n,r))$ for this category. Here ν is a co-character of T that is supposed to be generic, meaning that $\mathscr{M}^\theta(n,r)^{\nu(\mathbb{C}^\times)}$ is finite (in which case, $\mathscr{M}^\theta(n,r)^{\nu(\mathbb{C}^\times)}$ coincides with $\mathscr{M}^\theta(n,r)^T$, which is in a natural bijection with the set of r-multipartitions of n). The co-character ν fails to be generic precisely when it lies in a finite union of suitable hyperplanes in $\operatorname{Hom}(\mathbb{C}^\times, T)$, we will describe them explicitly below.

For now, we need to know the following about category $\mathscr{O}_\nu(\mathscr{A}_\lambda^\theta(n,r))$:

- $\mathscr{O}_\nu(\mathscr{A}_\lambda^\theta(n,r))$ is a highest weight category, so it makes sense to speak about standard objects $\Delta_\nu^\theta(p)$.
- The labeling set for standard objects is naturally identified with $\mathscr{M}^\theta(n,r)^T$, i.e., the set of r-multipartitions of n.

Theorem 1.4. *If the denominator of* λ *is bigger than* n, *then the category* $\mathscr{O}_\nu(\mathscr{A}_\lambda^\theta(n,r))$ *is semisimple. If the denominator of* λ *equals* n, *the category* $\mathscr{O}_\nu(\mathscr{A}_\lambda^\theta(n,r))$ *has only one nontrivial block. That block is equivalent to the nontrivial block of* $\mathscr{O}_\nu(\mathscr{A}_{1/nr}^\theta(nr,1))$ *(a.k.a. the category of* B*-equivariant perverse sheaves on* $\mathbb{P}^{nr-1}$*).*

In some cases, we can say which simple objects belong to the nontrivial block; we will do this below.

We can also determine the annihilators of simple modules in category $\mathscr{O}$. In this paper we treat a special co-character $\nu\colon \mathbb{C}^\times \to T$ to be called *dominant*. Namely, we pick integers $d_1,\ldots,d_r$ with $d_1 \gg d_2 \gg \cdots \gg d_r$; in fact, we can just take $d_i - d_{i+1} > n$. Consider $\nu\colon \mathbb{C}^\times \to T$ given by $t \mapsto \big((t^{d_1},\ldots,t^{d_r}),t\big)$. It is easy to see that it is generic. We also do not consider the case when λ is integral, in this case the category was described in [We].

Now take a multipartition $\tau = (\tau^{(1)}, \ldots, \tau^{(r)})$. Let m denote the denominator of λ, and we assume that $m > 1$. We also assume that $\lambda > 0$, so that the functor Γ_λ^θ is an equivalence. Divide $\tau^{(1)}$ by m with remainder: $\tau^{(1)} = m\tau' + \tau''$, where τ', τ'' are partitions with maximal possible $|\tau'|$, and the sum and the multiplication by m are defined component-wise. For example, if $m = 3$ and $\tau = (8,6,1)$, then $\tau' = (1^2)$ and $\tau'' = (5,3,1)$.

Theorem 1.5. *Assume that $m > 1$. Under the assumptions above, the annihilator of $\Gamma(L_\nu^\theta(\tau))$, where $L_\nu^\theta(\tau)$ denotes the irreducible module labeled by τ, coincides with the ideal $\mathscr{J}_{|\tau'|}$ from Theorem* 1.3.

We also give some results towards a computation of the annihilators of $\Gamma(L_\nu^\theta(\tau))$ for non-dominant ν. Namely, for each two adjacent chambers C, C' in $\operatorname{Hom}(\mathbb{C}^\times, T)$, there is a *cross-walling* bijection $\mathfrak{cw}_{C \to C'}^\lambda : \mathscr{M}^\theta(n,r)^T \to \mathscr{M}^\theta(n,r)^T$ which preserves the annihilator of the simple modules. We will define this bijection below and we plan to compute it in a subsequent paper.

Acknowledgments. I would like to thank Roman Bezrukavnikov, Dmitry Korb, Davesh Maulik, Andrei Okounkov and Nick Proudfoot for stimulating discussions. I also would like to thank Boris Tsvelikhovsky for comments on an earlier version of this text.

2 Preliminaries

2.1 Symplectic resolutions and their quantizations

Let X_0 be a normal Poisson affine variety equipped with an action of $\mathbb{C}^\times$ such that the grading on $\mathbb{C}[X_0]$ is positive (meaning that the graded component $\mathbb{C}[X_0]_i$ is zero when $i < 0$, and $\mathbb{C}[X_0]_0 = \mathbb{C}$) and there is a positive integer d such that $\{\mathbb{C}[X_0]_i, \mathbb{C}[X_0]_j\} \subset \mathbb{C}[X_0]_{i+j-d}$. By definition, a symplectic resolution of singularities of X_0 is a pair (X, ρ) of

- a smooth symplectic algebraic variety X (with form ω), and
- a morphism $\rho : X \to X_0$ of Poisson varieties that is a projective resolution of singularities.

Below we assume that (X, ρ) is a symplectic resolution of singularities. Besides, we will assume that (X, ρ) is *conical*, which means that the $\mathbb{C}^\times$-action lifts to X in such a way that ρ is equivariant. This $\mathbb{C}^\times$-action will be called *contracting* later on.

Note that $\rho^* : \mathbb{C}[X_0] \to \mathbb{C}[X]$ is an isomorphism because X_0 is normal. By the Grauert-Riemenschneider theorem, we have $H^i(X, \mathscr{O}_X) = 0$ for $i > 0$.

If X, X' are two conical symplectic resolutions of X_0, then the Picard groups of X, X' are naturally identified, see, e.g., [BPW, Proposition 2.19]. Moreover, the Chern class map defines an isomorphism $\mathbb{C}\otimes_{\mathbb{Z}}\operatorname{Pic}(X)\xrightarrow{\sim}H^2(X,\mathbb{C})$, see, e.g., [BPW, Section 2.3]. Let us write $\tilde{\mathfrak{p}}$ for $H^2(X,\mathbb{C})$ and let $\tilde{\mathfrak{p}}_{\mathbb{Z}}$ denote the image of $\operatorname{Pic}(X)$ in $H^2(X,\mathbb{C})$.

Set $\tilde{\mathfrak{p}}_{\mathbb{R}}:=\mathbb{R}\otimes_{\mathbb{Z}}\tilde{\mathfrak{p}}_{\mathbb{Z}}$. According to Namikawa [Nam], there is a linear hyperplane arrangement in $\tilde{\mathfrak{p}}_{\mathbb{R}}$ together with the action of a crystallographic reflection group W subject to the following conditions:

- The walls for W are in the arrangement.
- The conical symplectic resolutions of X are classified by W-conjugacy classes of chambers.

For θ inside a chamber, we will write X^θ for the corresponding resolution.

We will study quantizations of X_0 and X. By a quantization of X_0, we mean a filtered algebra $\mathscr{A}$ together with an isomorphism $\operatorname{gr}\mathscr{A}\cong\mathbb{C}[X_0]$ of graded Poisson algebras.

By a quantization of $X=X^\theta$, we mean a sheaf $\mathscr{A}^\theta$ of filtered algebras in the conical topology on X (in this topology, "open" means Zariski open and $\mathbb{C}^\times$-stable) that is complete and separated with respect to the filtration together with an isomorphism $\operatorname{gr}\mathscr{A}^\theta\cong\mathscr{O}_{X^\theta}$ (of sheaves of graded Poisson algebras). A result of Bezrukavnikov and Kaledin [BeKa], with variations given in [L2, Section 2.3], shows that the quantizations $\mathscr{A}^\theta$ are parameterized, up to an isomorphism, by the points in $\tilde{\mathfrak{p}}$. We write $\mathscr{A}^\theta_\lambda$ for the quantization corresponding to $\lambda\in\tilde{\mathfrak{p}}$. Note that $\mathscr{A}^\theta_{-\lambda}$ is isomorphic to $(\mathscr{A}^\theta_\lambda)^{opp}$; this follows from [L2, Section 2.3].

We set $\mathscr{A}_\lambda=\Gamma(\mathscr{A}^\theta_\lambda)$. It follows from [BPW, Section 3.3] that the algebras $\mathscr{A}_\lambda$ are independent of the choice of θ. From $H^i(X^\theta,\mathscr{O}_{X^\theta})=0$, we deduce that the higher cohomology of $\mathscr{A}^\theta_\lambda$ vanishes and that $\mathscr{A}_\lambda$ is a quantization of $\mathbb{C}[X]=\mathbb{C}[X_0]$. By the previous paragraph, $\mathscr{A}_{-\lambda}\cong\mathscr{A}^{opp}_\lambda$.

Let us compare the categories $\mathscr{A}_\lambda$-mod of finitely generated $\mathscr{A}_\lambda$-modules and $\operatorname{Coh}(\mathscr{A}^\theta_\lambda)$ of coherent sheaves of $\mathscr{A}^\theta_\lambda$-modules. We have functors $\Gamma\colon\operatorname{Coh}(\mathscr{A}^\theta_\lambda)\to\mathscr{A}_\lambda$-mod of taking global sections and its left adjoint, the localization functor Loc. When we need to indicate the dependence on (λ,θ), we write $\Gamma^\theta_\lambda,\operatorname{Loc}^\theta_\lambda$. We say that abelian (resp., derived) localization holds for (λ,θ) if the functors $\Gamma^\theta_\lambda,\operatorname{Loc}^\theta_\lambda$ are mutually inverse (resp., if the corresponding derived functors $R\Gamma^\theta_\lambda,L\operatorname{Loc}^\theta_\lambda$ are mutually inverse).

The following result was obtained in [BPW, Section 5.3].

Lemma 2.1. *Let $\chi\in\tilde{\mathfrak{p}}_{\mathbb{Z}}$ be ample for X^θ and let $\lambda\in\tilde{\mathfrak{p}}$. Then there is $n_0\in\mathbb{Z}$ such that abelian localization holds for $(\lambda+n\chi,\theta)$ for all $n\geqslant n_0$.*

2.2 Category $\mathscr{O}$

Suppose that we have a conical symplectic resolution $X=X^\theta$ which comes equipped with a Hamiltonian action of a torus T that commutes with the contracting $\mathbb{C}^\times$-action. Let $\lambda\in\tilde{\mathfrak{p}}$. The action of T on $\mathscr{O}_X$ lifts to a Hamiltonian action of T on $\mathscr{A}^\theta_\lambda$.

So we get a Hamiltonian action on $\mathscr{A}_\lambda$. By Φ we denote the quantum comoment map $\mathfrak{t} \to \mathscr{A}_\lambda$.

Let $\nu: \mathbb{C}^\times \to T$ be a one-parameter subgroup. The subgroup ν induces a grading $\mathscr{A}_\lambda = \bigoplus_{i\in\mathbb{Z}} \mathscr{A}_\lambda^{i,\nu}$. We set $\mathscr{A}_\lambda^{\geqslant 0,\nu} = \bigoplus_{i\geqslant 0} \mathscr{A}_\lambda^{i,\nu}$ and define $\mathscr{A}_\lambda^{>0,\nu}$ similarly. Further, set $\mathsf{C}_\nu(\mathscr{A}_\lambda) := \mathscr{A}_\lambda^{0,\nu} / \bigoplus_{i>0} \mathscr{A}_\lambda^{-i,\nu} \mathscr{A}_\lambda^{i,\nu}$. Note that $\mathscr{A}_\lambda / \mathscr{A}_\lambda \mathscr{A}_\lambda^{>0,\nu}$ is an $\mathscr{A}_\lambda$-$\mathsf{C}_\nu(\mathscr{A}_\lambda)$-bimodule, while $\mathscr{A}_\lambda / \mathscr{A}_\lambda^{<0,\nu} \mathscr{A}_\lambda$ is a $\mathsf{C}_\nu(\mathscr{A}_\lambda)$-$\mathscr{A}_\lambda$-bimodule.

Define the category $\mathscr{O}_\nu(\mathscr{A}_\lambda)$ as the full subcategory of $\mathscr{A}_\lambda$-mod consisting of all modules, where the action of $\mathscr{A}_\lambda^{>0,\nu}$ is locally nilpotent. We get two functors $\Delta_\nu, \nabla_\nu : \mathsf{C}_\nu(\mathscr{A}_\lambda)\text{-mod} \to \mathscr{O}_\nu(\mathscr{A}_\lambda)$ given by

$$\Delta_\nu(N) := (\mathscr{A}_\lambda / \mathscr{A}_\lambda \mathscr{A}_\lambda^{>0,\nu}) \otimes_{\mathsf{C}_\nu(\mathscr{A}_\lambda)} N, \quad \nabla_\nu(N) := \operatorname{Hom}_{\mathsf{C}_\nu(\mathscr{A}_\lambda)}(\mathscr{A}_\lambda / \mathscr{A}_\lambda^{<0,\nu} \mathscr{A}_\lambda, N).$$

Here we consider the restricted Hom with respect to the natural grading on $\mathscr{A}_\lambda / \mathscr{A}_\lambda^{<0,\nu} \mathscr{A}_\lambda$.

Now suppose that $|X^T| < \infty$. We say that a one-parameter subgroup $\nu: \mathbb{C}^\times \to T$ is *generic* if $X^{\nu(\mathbb{C}^\times)} = X^T$. Equivalently, ν is generic if and only if it does not lie in $\ker \kappa$ for any character κ of the T-action on $\bigoplus_{p\in X^T} T_p X$. The hyperplanes $\ker\kappa$ split the lattice $\operatorname{Hom}(\mathbb{C}^\times, T)$ into the union of polyhedral regions to be called *chambers* (of one-parameter subgroups).

Suppose that ν is generic. Further, pick a generic element $\theta \in \tilde{\mathfrak{p}}_\mathbb{Z}$ with $X = X^\theta$ and $\lambda_0 \in \tilde{\mathfrak{p}}$. Let $\lambda := \lambda_0 + n\theta$ for $n \gg 0$. The following results were obtained in [BLPW] and [L7], see [L9, Proposition 4.1] for precise references.

Proposition 2.2. *The following is true:*

(1) *The category $\mathscr{O}_\nu(\mathscr{A}_\lambda)$ only depends on the chamber of ν.*
(2) *The natural functor $D^b(\mathscr{O}_\nu(\mathscr{A}_\lambda)) \to D^b(\mathscr{A}_\lambda\text{-mod})$ is a full embedding.*
(3) $\mathsf{C}_\nu(\mathscr{A}_\lambda) = \mathbb{C}[X^T]$.
(4) *More generally, we have $\mathsf{C}_{\nu_0}(\mathscr{A}_\lambda) = \bigoplus_Z \mathscr{A}^Z_{\iota_Z^*(\lambda)-\rho_Z}$, where the summation is taken over the irreducible components Z of $X^{\nu_0(\mathbb{C}^\times)}$, ι_Z is the embedding $Z \hookrightarrow X$, $\iota_Z^*: H^2(X,\mathbb{C}) \to H^2(Z,\mathbb{C})$ is the corresponding pull-back map, ρ_Z equals to half of the 1st Chern class of the contracting bundle of Z, and $\mathscr{A}^Z_{\iota_Z^*(\lambda)-\rho_Z}$ stands for the global sections of the filtered quantization of Z with period $\iota_Z^*(\lambda) - \rho_Z$.*
(5) *The category $\mathscr{O}_\nu(\mathscr{A}_\lambda)$ is highest weight, where the standard objects are $\Delta_\nu(p)$, the costandard objects are $\nabla_\nu(p)$ for $p \in X^T$. For an order, which is a part of the definition of a highest weight structure, we take the contraction order on X^T given by ν.*
(6) *Suppose ν_0 lies in the face of a chamber containing ν. Then $\Delta_{\nu_0}, \nabla_{\nu_0}$ restrict to exact functors $\mathscr{O}_\nu(\mathsf{C}_{\nu_0}(\mathscr{A}_\lambda)) \to \mathscr{O}_\nu(\mathscr{A}_\lambda)$. Moreover, $\Delta_\nu = \Delta_{\nu_0} \circ \underline{\Delta}$ and $\nabla_\nu = \nabla_{\nu_0} \circ \underline{\nabla}$, where we write $\underline{\Delta}$ and $\underline{\nabla}$ for the standardization and costandardization functors for $\mathscr{O}_\nu(\mathsf{C}_{\nu_0}(\mathscr{A}_\lambda))$.*

Now we explain what is meant by the contracting bundle in (4). It is the subvariety in X consisting of all points x such that $\lim_{t\to 0} \nu_0(t)x$ exists and lies in Z. Note that Z is a vector bundle over $X^{\nu_0(\mathbb{C}^\times)}$. Highest weight categories mentioned in (5) will be recalled in Section 2.3.

Now let us mention the holonomicity property, see [L7, Section 4.4] and [L8, Theorems 1.2,1.3].

Lemma 2.3. *Every module from category $\mathscr{O}_\nu(\mathscr{A}_\lambda)$ is holonomic in the sense of* [L8]. *In particular, if M is a simple object in $\mathscr{O}_\nu(\mathscr{A}_\lambda)$, then* $\operatorname{GK-dim} M = \frac{1}{2}\operatorname{GK-dim}(\mathscr{A}_\lambda / \operatorname{Ann} M)$.

We will also need the following important property, [BLPW, Lemma 6.4].

Lemma 2.4. *The classes of standard and costandard objects in $K_0(\mathscr{O}_\nu(\mathscr{A}_\lambda^\theta))$ coincide.*

2.3 Standardly stratified structures and cross-walling functors

Let us start by recalling standardly stratified categories following [LW].

Let $\mathscr{C}$ be a $\mathbb{C}$-linear abelian category equivalent to the category of finite-dimensional representations of a finite-dimensional algebra. Let $\mathscr{T}$ be an indexing set of the simple objects in $\mathscr{C}$. We write $L(\tau), P(\tau)$ for the simple and indecomposable projective objects indexed by $\tau \in \mathscr{T}$. The additional structure of a standardly stratified category on $\mathscr{C}$ is a partial *pre-order* $\leqslant$ on $\mathscr{T}$ that should satisfy certain axioms to be explained below. We write Ξ for the set of equivalence classes of $\leqslant$; this is a poset (with partial order again denoted by $\leqslant$) that comes with a natural surjection $\rho\colon \mathscr{T} \twoheadrightarrow \Xi$. The pre-order $\leqslant$ defines a filtration on $\mathscr{C}$ by Serre subcategories indexed by Ξ. Namely, to $\xi \in \Xi$ we assign the subcategory $\mathscr{C}_{\leqslant \xi}$ which is the Serre span of the simples $L(\tau)$ with $\rho(\tau) \leqslant \xi$. Define $\mathscr{C}_{<\xi}$ analogously and let $\mathscr{C}_\xi$ denote the quotient $\mathscr{C}_{\leqslant\xi}/\mathscr{C}_{<\xi}$. Let π_ξ denote the quotient functor $\mathscr{C}_{\leqslant\xi} \twoheadrightarrow \mathscr{C}_\xi$. Let us write $\Delta_\xi : \mathscr{C}_\xi \to \mathscr{C}_{\leqslant\xi}$ for the left adjoint functor of π_ξ. Also we write $\operatorname{gr}\mathscr{C}$ for $\bigoplus_\xi \mathscr{C}_\xi, \Delta$ for $\bigoplus_\xi \Delta_\xi :\ \operatorname{gr}\mathscr{C} \to \mathscr{C}$. We call Δ the *standardization functor*. Finally, for $\tau \in \rho^{-1}(\xi)$, we write $L_\xi(\tau)$ for $\pi_\xi(L(\tau))$, $P_\xi(\tau)$ for the projective cover of $L_\xi(\tau)$ in $\mathscr{C}_\xi$, and $\Delta(\tau)$ for $\Delta_\xi(P_\xi(\tau))$. The object $\Delta(\tau)$ is called *standard*. The object $\overline{\Delta}(\tau) := \Delta_\xi(L_\xi(\tau))$ is called *proper standard*. Note that there is a natural epimorphism $P(\tau) \twoheadrightarrow \Delta(\tau)$.

The axioms to be satisfied by $(\mathscr{C}, \leqslant)$ in order to give a standardly stratified structure are as follows.

(SS1) The functor $\Delta\colon\ \operatorname{gr}\mathscr{C} \to \mathscr{C}$ is exact.
(SS2) The projective $P(\tau)$ admits an epimorphism onto $\Delta(\tau)$ whose kernel has a filtration with successive quotients $\Delta(\tau')$, where $\tau' > \tau$.

When $\mathscr{C}_\xi = \mathrm{Vect}$ for all ξ (in which case $\Xi = \mathscr{T}$), we recover the classical notion of a highest weight category.

Note that $\mathscr{C}^{opp}$ is also a standardly stratified category with the same poset Ξ, [LW, Section 2.2]. So we have the exact costandardization functor ∇_ξ, the right adjoint of π_ξ.

We describe a standardly stratified structure on $\mathcal{O}_\nu(\mathscr{A}_\lambda)$, where λ is as in Proposition 2.2, that comes from a one-parameter subgroup ν_0 lying in a face of the chamber containing ν. The one-parameter subgroup ν_0 defines the order on the set of irreducible components of $X^{\nu_0(\mathbb{C}^\times)}$ (by contraction, see [L7, Section 6.1] for details). So we get a pre-order $\leqslant_{\nu_0}$ on the set X^T. It is easy to see (and was checked in [L7, Section 6.1]) that the order $\leqslant_\nu$ refines $\leqslant_{\nu_0}$.

The following proposition is the main result of [L7, Section 6].

Proposition 2.5. *Let λ be as in Proposition 2.2. Then the pre-order $\leqslant_{\nu_0}$ defines a standardly stratified structure on $\mathcal{O}_\nu(\mathscr{A}_\lambda)$. The associated graded category is $\mathcal{O}_\nu(\mathsf{C}_{\nu_0}(\mathscr{A}_\lambda))$. The standardization functor is Δ_{ν_0} and the costandardization functor is ∇_{ν_0}.*

Let us now discuss cross-walling functors that are derived equivalences between categories $\mathcal{O}$ corresponding to different generic one-parameter subgroups. Namely, let ν, ν' be two generic one-parameter subgroups and let λ be as in Proposition 2.2. The following proposition is established in [L7, Section 7].

Proposition 2.6. *The following are true:*

(1) *There is an equivalence $\mathfrak{CW}_{\nu'\leftarrow\nu} \colon D^b(\mathcal{O}_\nu(\mathscr{A}_\lambda)) \to D^b(\mathcal{O}_{\nu'}(\mathscr{A}_\lambda))$ such that we have a bifunctorial isomorphism*

$$\operatorname{Hom}_{D^b(\mathcal{O}_{\nu'}(\mathscr{A}_\lambda))}(\mathfrak{CW}_{\nu'\leftarrow\nu}(M), N) \xrightarrow{\sim} \operatorname{Hom}_{D^b(\mathscr{A}_\lambda\text{-mod})}(M, N),$$

where $M \in D^b(\mathcal{O}_\nu(\mathscr{A}_\lambda))$, $N \in D^b(\mathcal{O}_{\nu'}(\mathscr{A}_\lambda))$.

(2) *Suppose that ν_0 lies in a common face of the chambers containing ν, ν'. Then we have an isomorphism $\mathfrak{CW}_{\nu'\leftarrow\nu} \circ \Delta_{\nu_0} \xrightarrow{\sim} \Delta_{\nu_0} \circ \underline{\mathfrak{CW}}_{\nu'\leftarrow\nu}$, where we write $\underline{\mathfrak{CW}}_{\nu'\leftarrow\nu}$ for the cross-walling functor $D^b(\mathcal{O}_\nu(\mathsf{C}_{\nu_0}(\mathscr{A}_\lambda))) \to D^b(\mathcal{O}_{\nu'}(\mathsf{C}_{\nu_0}(\mathscr{A}_\lambda)))$.*

(3) *The functor $\mathfrak{CW}_{-\nu\leftarrow\nu}[-\frac{1}{2}\dim X]$ is a Ringel duality functor, i.e., an equivalence that maps $\Delta_\nu(p)$ to $\nabla_{-\nu}(p)$ for all $p \in X^T$.*

2.4 Wall-crossing functors and bijections

Let θ, θ' be two generic elements of $\tilde{\mathfrak{p}}_{\mathbb{Z}}$ and $\lambda \in \tilde{\mathfrak{p}}$. Following [BPW, Section 6.4], we will produce a derived equivalence $\mathfrak{WC}_{\theta'\leftarrow\theta} \colon D^b(\operatorname{Coh}(\mathscr{A}_\lambda^\theta)) \xrightarrow{\sim} D^b(\operatorname{Coh}(\mathscr{A}_\lambda^{\theta'}))$ assuming that abelian localization holds for (λ, θ) and that derived localization holds for (λ, θ'). Then we set $\mathfrak{WC}_{\theta'\leftarrow\theta} := L\operatorname{Loc}_\lambda^{\theta'} \circ \Gamma_\lambda^\theta$. Note that this functor is right t-exact.

We can give a different realization of $\mathfrak{WC}_{\theta'\leftarrow\theta}$. Namely, pick $\lambda' \in \lambda + \tilde{\mathfrak{p}}_{\mathbb{Z}}$ such that abelian localization holds for (λ', θ'). We identify $\operatorname{Coh}(\mathscr{A}_\lambda^\theta)$ with $\mathscr{A}_\lambda$-mod using Γ_λ^θ and $\operatorname{Coh}(\mathscr{A}_\lambda^{\theta'})$ with $\mathscr{A}_{\lambda'}$-mod using $\Gamma_{\lambda'}^{\theta'}(\mathscr{A}_{\lambda,\lambda'-\lambda}^{\theta'} \otimes_{\mathscr{A}_\lambda^{\theta'}} \bullet)$, where $\mathscr{A}_{\lambda,\lambda'-\lambda}^{\theta'}$ is the $\mathscr{A}_{\lambda'}^{\theta'}$-$\mathscr{A}_\lambda^{\theta'}$-bimodule quantizing the line bundle corresponding to $\lambda' - \lambda$ on $X^{\theta'}$. Under these identifications, the functor $\mathfrak{WC}_{\theta'\leftarrow\theta}$ becomes $\mathfrak{WC}_{\lambda'\leftarrow\lambda} :=$

$\mathcal{A}^{(\theta)}_{\lambda,\lambda'-\lambda} \otimes^L_{\mathcal{A}_\lambda} \bullet$, see [BPW, Section 6.4]. Here we write $\mathcal{A}^{(\theta)}_{\lambda,\lambda'-\lambda}$ for the global sections of $\mathcal{A}^{\theta}_{\lambda,\lambda'-\lambda}$.

We note that the functor $\mathfrak{WC}_{\lambda'\leftarrow\lambda}$ restricts to an equivalence $D^b(\mathcal{O}_\nu(\mathcal{A}_\lambda)) \to D^b(\mathcal{O}_\nu(\mathcal{A}_{\lambda'}))$, see [BLPW, Section 8.1].

We are especially interested in the case when θ, θ' lie in opposite chambers. In this case the *long wall-crossing* functor $\mathfrak{WC}_{\lambda'\leftarrow\lambda}$ is perverse in the sense of Chuang and Rouquier, see [L9, Section 3].

Let us recall the general definition. Let $\mathcal{T}^1, \mathcal{T}^2$ be triangulated categories equipped with t-structures that are homologically finite. Let $\mathcal{C}^1, \mathcal{C}^2$ denote the hearts of $\mathcal{T}^1, \mathcal{T}^2$, respectively.

We now define a perverse equivalence with respect to filtrations

$$\mathcal{C}^i = \mathcal{C}^i_0 \supset \mathcal{C}^i_1 \supset \cdots \supset \mathcal{C}^i_k = \{0\}$$

by Serre subcategories. By definition, this is a triangulated equivalence $\mathcal{F}: \mathcal{T}^1 \to \mathcal{T}^2$ subject to the following conditions:

(P1) For any j, the equivalence $\mathcal{F}$ restricts to an equivalence $\mathcal{T}^1_{\mathcal{C}^1_j} \to \mathcal{T}^2_{\mathcal{C}^2_j}$, where we write $\mathcal{T}^i_{\mathcal{C}^i_j}, i = 1,2$, for the category of all objects in $\mathcal{T}^i$ with homology (computed with respect to the t-structures of interest) in $\mathcal{C}^i_j$.

(P2) For $M \in \mathcal{C}^1_j$, we have $H_\ell(\mathcal{F}M) = 0$ for $\ell < j$ and $H_\ell(\mathcal{F}M) \in \mathcal{C}^2_{j+1}$ for $\ell > j$.

(P3) The functor $M \mapsto H_j(\mathcal{F}M)$ induces an equivalence $\mathcal{C}^1_j/\mathcal{C}^1_{j+1} \xrightarrow{\sim} \mathcal{C}^2_j/\mathcal{C}^2_{j+1}$ of abelian categories.

Lemma 2.7. *The equivalence $\mathfrak{WC}_{\lambda'\leftarrow\lambda}: D^b(\mathcal{O}_\nu(\mathcal{A}_\lambda)) \to D^b(\mathcal{O}_\nu(\mathcal{A}_{\lambda'}))$ is perverse with $\mathcal{C}^1_j = \{M \in \mathcal{O}_\nu(\mathcal{A}_\lambda) \mid \text{GK-dim}\, M \leqslant \dim X/2 - j\}$ (and $\mathcal{C}^2_j$ defined similarly).*

Now let us discuss the wall-crossing bijections. (P3) gives rise to a bijection $\mathrm{Irr}(\mathcal{C}^1) \to \mathrm{Irr}(\mathcal{C}^2)$ that is called a *wall-crossing bijection* and denoted by $\mathfrak{wc}_{\lambda'\leftarrow\lambda}$ for $\mathcal{F} := \mathfrak{WC}_{\lambda'\leftarrow\lambda}$.

Finally, let us recall the following important property of $\mathfrak{WC}_{\lambda'\leftarrow\lambda}$ established in [L7, Section 7.3].

Lemma 2.8. *In the case when $\theta = -\theta'$, the functor*

$$\mathfrak{WC}_{\lambda'\leftarrow\lambda}: D^b(\mathcal{O}_\nu(\mathcal{A}_\lambda)) \to D^b(\mathcal{O}_\nu(\mathcal{A}_{\lambda'}))$$

is a Ringel duality functor.

3 Gieseker varieties and their quantizations

The definition of Gieseker varieties $\mathcal{M}(n,r)$, $\mathcal{M}^\theta(n,r)$ (for $\theta = \det$ or $\det^{-1}$) was discussed in the introduction. We note that $\mathcal{M}^\theta(n,r)$, $\mathcal{M}^{-\theta}(n,r)$ are symplectomorphic via the isomorphism induced by $(A,B,i,j) \mapsto (B^*, -A^*, j^*, -i^*)$.

We also consider the varieties $\bar{\mathcal{M}}(n,r), \bar{\mathcal{M}}^\theta(n,r)$ that are obtained similarly but with the space $R = \operatorname{End}(V) \oplus \operatorname{Hom}(V,W)$ replaced with $\mathfrak{sl}(V) \oplus \operatorname{Hom}(V,W)$. So $\mathcal{M}(n,r) = \mathbb{C}^2 \times \bar{\mathcal{M}}(n,r)$ and $\mathcal{M}^\theta(n,r) = \mathbb{C}^2 \times \bar{\mathcal{M}}^\theta(n,r)$.

3.1 Quantizations

Let us discuss the quantizations $\bar{\mathcal{A}}^\theta_\lambda(n,r)$ of $\bar{\mathcal{M}}^\theta(n,r)$ defined via

$$\bar{\mathcal{A}}^\theta_\lambda(n,r) = \left[D(\bar{R})/D(\bar{R})\{x_{\bar{R}} - \lambda \operatorname{tr}(x) \mid x \in \mathfrak{g}\}|_{(T^*\bar{R})^{\theta-ss}}\right]^G,$$

so that $\bar{\mathcal{A}}_\lambda(n,r)$ is the algebra of global sections of $\bar{\mathcal{A}}^\theta_\lambda(n,r)$.

Recall that $\mathcal{A}_\lambda(n,r) = D(\mathbb{C}) \otimes \bar{\mathcal{A}}_\lambda(n,r)$. Let t be the coordinate on $\mathbb{C}$. Then the categories $\mathcal{O}_\nu(\mathcal{A}_\lambda(n,r))$ and $\mathcal{O}_\nu(\bar{\mathcal{A}}_\lambda(n,r))$ are equivalent: an equivalence $\mathcal{O}_\nu(\bar{\mathcal{A}}_\lambda(n,r)) \to \mathcal{O}_\nu(\mathcal{A}_\lambda(n,r))$ is given by $M \mapsto \mathbb{C}[t] \otimes M$ and a quasi-inverse functor sends N to the annihilator of ∂_t.

Now let us discuss periods. By [L2, Section 5], the period of $\bar{\mathcal{A}}^\theta_\lambda(n,r)$ is $\lambda - \rho$, where ρ is half the character of the action of $\mathrm{GL}(n)$ on $\Lambda^{top} R$. So $\rho = -r/2$ and the period of $\bar{\mathcal{A}}^\theta_\lambda(n,r)$ is $\lambda + r/2$.

So we have

$$\bar{\mathcal{A}}_\lambda(n,r)^{opp} \cong \bar{\mathcal{A}}_{-\lambda-r}(n,r). \tag{3.1}$$

Lemma 3.1. *We have $\bar{\mathcal{A}}_\lambda(n,r) \cong \bar{\mathcal{A}}_{-\lambda-r}(n,r)$.*

Proof. Recall from Section 2.1 that $\bar{\mathcal{A}}_\lambda(n,r) \xrightarrow{\sim} \Gamma(\bar{\mathcal{A}}^{\pm\theta}_\lambda(n,r))$. We have produced a symplectomorphism $\iota: \bar{\mathcal{M}}^\theta(n,r) \xrightarrow{\sim} \bar{\mathcal{M}}^{-\theta}(n,r)$ via $(A,B,i,j) \to (B^*, -A^*, j^*, -i^*)$. On the second cohomology this isomorphism induces multiplication by -1. It follows that $\Gamma(\bar{\mathcal{A}}^\theta_\lambda(n,r)) \cong \Gamma(\bar{\mathcal{A}}^{-\theta}_{-\lambda-r}(n,r))$ and our claim follows. □

We conclude that $\bar{\mathcal{A}}_\lambda(n,r)^{opp} \cong \bar{\mathcal{A}}_\lambda(n,r)$.

Let us finish by discussing the derived localization. The following is a special case of the main result of [MN1].

Lemma 3.2. *Derived localization holds for (λ,θ) if and only if $\bar{\mathcal{A}}_\lambda(n,r)$ has finite homological dimension.*

3.2 Fixed point subvarieties and their quantizations

Recall that the torus $T = \mathbb{C}^\times \times T_0$, where T_0 is a maximal torus in $\mathrm{GL}(r)$, acts on $\mathcal{M}^\theta(n,r)$.

Now we want to understand the structure of $\mathcal{M}^\theta(n,r)^{\nu_0(\mathbb{C}^\times)}$ for a non-generic ν_0 that lies in the interior of a codimension-1 face in a chamber. Let $\nu_0(t) =$

$(\mathrm{diag}(t^{d_1},\dots,t^{d_r}),t^k)$. The following claim follows directly from the description of the tangent space in the fixed points given in [NY, Theorem 3.2].

Lemma 3.3. *The walls in the lattice of one-parameter subgroups are given by $k=0$ and by $d_i - d_j = sk$ with $s \in \mathbb{Z}$ such that $|s| < n$.*

The description of $\mathscr{M}^\theta(n,r)^{\nu_0(\mathbb{C}^\times)}$ for ν_0 with $k=0$ is classical.

Lemma 3.4. *For $k=0$ (and pairwise different $d_1,\dots,d_r$), we have*

$$\mathscr{M}^\theta(n,r)^{\nu_0(\mathbb{C}^\times)} = \bigsqcup_{n_1+\dots+n_r=n} \prod_{i=1}^r \mathscr{M}^\theta(n_i,1).$$

Recall that $\mathscr{M}^\theta(n_i,1) = \mathrm{Hilb}_{n_i}(\mathbb{C}^2)$; when $n_i=0$, we assume that this variety is a single point. Lemma 3.4 gives an identification of $\mathscr{M}^\theta(n,r)^T$ with the set of r-multipartitions of n. Indeed, the fixed points of $\mathbb{C}^\times$ in $\mathscr{M}^\theta(n_i,1)$ are in a one-to-one correspondence with the partitions of n_i.

The description for $s \neq 0$ is less classical but is also known. Using an analog of the previous lemma, this description easily reduces to the case when $r=2$. Here the components of the fixed point subvariety are quiver varieties for finite type A Dynkin diagrams. We will not need a more precise description in this paper.

Now, for $\nu_0(t) = (\mathrm{diag}(t^{d_1},\dots,t^{d_r}),1)$ with $d_1 \gg d_2 \gg \cdots\cdots \gg d_r$, we want to describe the sheaves $\mathsf{C}_{\nu_0}(\mathscr{A}^\theta_\lambda(n,r))$.

Proposition 3.5. *For the irreducible component Z of $\mathscr{M}^\theta(n,r)^{\nu_0(\mathbb{C}^\times)}$ corresponding to the r-multipartition $(n_1,\dots,n_r)$ of n, we have*

$$\mathsf{C}_{\nu_0}(\mathscr{A}^\theta_\lambda(n,r))|_Z = \bigotimes_{i=1}^r \mathscr{A}^\theta_{\lambda+i-1}(n_i,1)$$

(the ith factor is absent if $n_i=0$). Moreover, for a Zariski generic $\lambda \in \mathbb{C}$, we have

$$\mathsf{C}_{\nu_0}(\mathscr{A}_\lambda(n,r)) = \bigoplus \bigotimes_{i=1}^r \mathscr{A}_{\lambda+i-1}(n_i,r),$$

where the summation is taken over all r-multipartitions $(n_1,\dots,n_r)$.

Proof. For $\mu = (n_1,\dots,n_r)$, let Z_μ denote the corresponding component of $X^{\nu_0(\mathbb{C}^\times)}$. Let Y_μ denote the contracting vector bundle for Z_μ. According to (4) of Proposition 2.2, we need to compute the 1st Chern class of Y_μ.

First, consider the following situation. Set $V := \mathbb{C}^n, W = \mathbb{C}^r$. Choose a decomposition $W = W^1 \oplus W^2$ with $\dim W^i = r_i$ and consider the one-parameter subgroup α of $\mathrm{GL}(W)$ acting trivially on W^2 and by $t \mapsto t$ on W^1. The components of the fixed points in $\mathscr{M}^\theta(n,r)$ are in one-to-one correspondence with decompositions of n into the sum of two parts. Pick such a decomposition $n = n_1 + n_2$ and consider a decomposition $V = V^1 \oplus V^2$ into the sum of two spaces of the corresponding dimensions. Then the corresponding component $Z' \subset \mathcal{M}^\theta(n,r)^{\alpha(\mathbb{C}^\times)}$ is $\mathcal{M}^\theta(n_1,r_1)\times\mathcal{M}^\theta(n_2,r_2)$.

Nakajima described the contracting bundle $Y \to Z'$, [Nak, Proposition 3.13]. This is the bundle on $Z' = \mathcal{M}^\theta(n_1,r_1)\times\mathcal{M}^\theta(n_2,r_2)$ that descends from the $\mathrm{GL}(n_1)\times\mathrm{GL}(n_2)$-module $\ker\beta^{12}/\operatorname{im}\alpha^{12}$, where α^{12},β^{12} are certain $\mathrm{GL}(n_1)\times\mathrm{GL}(n_2)$-equivariant linear maps

$$\mathrm{Hom}(V^2,V^1) \xrightarrow{\alpha^{12}} \mathrm{Hom}(V^2,V^1)^{\oplus 2}\oplus\mathrm{Hom}(W^2,V^1)\oplus\mathrm{Hom}(V^2,W^1) \xrightarrow{\beta^{12}} \mathrm{Hom}(V^2,V^1).$$

We do not need to know the precise form of the maps α^{12},β^{12}; what we need is that α^{12} is injective while β^{12} is surjective. So $\ker\beta^{12}/\operatorname{im}\alpha^{12}\cong\mathrm{Hom}(W^2,V^1)\oplus\mathrm{Hom}(V^2,W^1)$, an isomorphism of $\mathrm{GL}(n_1)\times\mathrm{GL}(n_2)$-modules.

It is easy to see that if $\alpha'\colon\mathbb{C}^\times\to\mathrm{GL}(r_1)$ is a homomorphism of the form $t\mapsto\mathrm{diag}(t^{d_1},\ldots,t^{d_k})$ with $d_1,\ldots,d_k\gg 0$, then the contracting bundle for the one-parameter subgroup $(\alpha',1)\colon\mathbb{C}^\times\to\mathrm{GL}(W^1)\times\mathrm{GL}(W^2)$ coincides with the sum of the contracting bundles for α' and α. So we see inductively that the vector bundle Y_μ on Z_μ descends from the $\prod_{i=1}^r\mathrm{GL}(n_i)$-module $\sum_{i=1}^r\left((\mathbb{C}^{n_i})^{\oplus r-i}\oplus(\mathbb{C}^{n_i*})^{\oplus i-1}\right)$. Therefore the 1st Chern class of Y_μ is $\sum_{i=1}^r(r+1-2i)c_i$, where we write c_i for the generator of $H^2(\mathcal{M}^\theta(n_i,1))$ (that is the first Chern class of $\mathcal{O}(1)$).

Now to prove the claim of this proposition we will use (4) of Proposition 2.2. The map ι_Z^* sends the generator c of $H^2(\mathcal{M}^\theta(n,r))$ to $\sum_{i=1}^r c_i$. The period of $\mathcal{A}_\lambda^\theta(n,r)$ is $(\lambda+r/2)c$. So the period of $\mathcal{A}^\theta_{\lambda+i-1}(n_i,1)$ is $(\lambda+1/2+i-1)c_i$. Now the statement of this proposition is a direct corollary of (4) of Proposition 2.2. □

3.3 Symplectic leaves and slices

Here we want to describe the symplectic leaves of $\bar{\mathcal{M}}(n,r)$ and study the structure of the variety $\bar{\mathcal{M}}(n,r)$ near a symplectic leaf.

Lemma 3.6. *The point* 0 *is a single leaf of* $\bar{\mathcal{M}}(n,r)$.

Proof. It is enough to show that the maximal ideal of 0 in $\mathbb{C}[T^*\bar{R}]^G$ is Poisson. Since $\bar{R}$ does not include the trivial G-module as a direct summand, we see that all homogeneous elements in $\mathbb{C}[T^*\bar{R}]^G$ have degree 2 or higher. It follows that the bracket of any two homogeneous elements also has degree 2 or higher, and our claim is proved. □

Now let us describe the slices to symplectic leaves in $\mathcal{M}(n,r)$.

Proposition 3.7. *The following is true:*

(1) *The symplectic leaves of* $\bar{\mathscr{M}}(n,r)$ *are parameterized by the unordered collections of numbers* $(n_1,\ldots,n_k)$ *with* $\sum_{i=1}^k n_i \leqslant n$ *(for* $r>1$*) and* $\sum_{i=1}^k n_i = n$ *(for* $r=1$*).*
(2) *There is a transversal slice to the leaf as above that is isomorphic to the formal neighborhood of* 0 *in* $\prod_{i=1}^k \bar{\mathscr{M}}(n_i,r)$.

Proof. Pick $x\in\bar{\mathscr{M}}(n,r)$. Consider the quiver $\overline{Q}^r$ with two vertices, $0,\infty$, two loops at 0, r arrows from 0 to ∞ and r arrow from ∞ to 0. We can view T^*R as the representation space of $\overline{Q}^r$ with dimension $(n,1)$. Pick a semisimple representation of $\overline{Q}^r$ lying over x. This representation decomposes as $r^0\oplus r^1\otimes U_1\oplus\cdots\oplus r^k\otimes U_k$, where r^0 is an irreducible representation of DQ^w with dimension $(n_0,1)$ and $r^1,\ldots,r^k$ are pairwise nonisomorphic irreducible representations of $\bar{Q}$; all $r^i, i=1,\ldots,k$, have dimension 1. Let $n_i=\dim U_i$.

According to [BL, Section 2.1.6], we have a decomposition $\bar{\mathscr{M}}(n,r)^{\wedge_x}\cong D\times\prod_{i=1}^k\bar{\mathscr{M}}(n_i,r)^{\wedge_0}$ of Poisson formal schemes, where D stands for the symplectic formal disk and $\bullet^{\wedge_x}$ indicates the formal neighborhood of x. From Lemma 3.6 it now follows that the locus described in the previous paragraph is a union of leaves. So in order to prove the entire proposition, it remains to show that the locus is irreducible. This is a consequence of [CB, Theorem 1.2] which says, in particular, that the locus of irreducible representations in $\mu^{-1}(0)$ is irreducible for any quiver representation. □

An isomorphism $\mathscr{M}(n,r)^{\wedge_x}\cong D\times\prod_{i=1}^k\bar{\mathscr{M}}(n_i,r)^{\wedge_0}$ can be quantized (see [BL, Section 5.4]) to

$$\bar{\mathscr{A}}_\lambda(n,r)_\hbar^{\wedge_x}=\mathbf{A}_\hbar^{\wedge_0}\widehat{\otimes}_{\mathbb{C}[[\hbar]]}\bar{\mathscr{A}}_\lambda(n_1,r)_\hbar^{\wedge_0}\widehat{\otimes}_{\mathbb{C}[[\hbar]]}\cdots\widehat{\otimes}_{\mathbb{C}[[\hbar]]}\bar{\mathscr{A}}_\lambda(n_k,r)_\hbar^{\wedge_0}. \tag{3.2}$$

Here the notation is as follows. By $\mathbf{A}_\hbar^{\wedge_0}$ we mean the formal Weyl algebra quantizing D. The subscript $\hbar$ means the Rees algebra, with respect to the filtration by order of differential operators. We note that the map $\hat{r}$ from [BL, Section 5.4] sends λ to $(\lambda,\lambda,\ldots,\lambda)$; hence all parameters on the right-hand side of (3.2) are all equal to λ.

3.4 Harish-Chandra bimodules and restriction functors

Let $\mathscr{A},\mathscr{A}'$ be two quantizations of the same positively graded Poisson algebra A, where, for simplicity, the degree of the Poisson bracket is -1. For example, we can take $A=\mathbb{C}[\bar{\mathscr{M}}(n,r)]$, $\mathscr{A}=\bar{\mathscr{A}}_\lambda(n,r),\mathscr{A}'=\bar{\mathscr{A}}_{\lambda'}(n,r)$, with filtrations coming from the filtration on $D(\bar{R})$ by order of the differential operator.

Let us recall the definition of a HC $\mathscr{A}$-$\mathscr{A}'$-bimodule. This is an $\mathscr{A}$-$\mathscr{A}'$-bimodule $\mathscr{B}$ that can be equipped with a *good filtration*, i.e., an $\mathscr{A}$-$\mathscr{A}'$-bimodule filtration subject to the following two properties:

- the induced left and right A-actions on $\operatorname{gr}\mathscr{B}$ coincide,
- $\operatorname{gr}\mathscr{B}$ is a finitely generated A-module.

By a homomorphism of Harish-Chandra bimodules, we mean a bimodule homomorphism. The category of HC $\mathscr{A}$-$\mathscr{A}'$-bimodules is denoted by $\mathrm{HC}(\mathscr{A}\text{-}\mathscr{A}')$. We also consider the full subcategory $D^b_{HC}(\mathscr{A}\text{-}\mathscr{A}')$ of the derived category of $\mathscr{A}$-$\mathscr{A}'$-bimodules with Harish-Chandra homology.

By the associated variety of a HC bimodule $\mathscr{B}$ (denoted by $\mathrm{V}(\mathscr{B})$) we mean the support in $\mathrm{Spec}(A)$ of the coherent sheaf $\mathrm{gr}\,\mathscr{B}$, where the associated graded bimodule is taken with respect to a good filtration. It is easy to see that $\mathrm{gr}\,\mathscr{B}$ is a Poisson A-module, so $\mathrm{V}(\mathscr{B})$ is the union of symplectic leaves, assuming $\mathrm{Spec}(A)$ has finitely many leaves.

Let $\mathscr{A}''$ be another filtered quantization of A. For $\mathscr{B}^1 \in \mathrm{HC}(\mathscr{A}'\text{-}\mathscr{A})$ and $\mathscr{B}^2 \in \mathrm{HC}(\mathscr{A}''\text{-}\mathscr{A}')$ we can take their tensor product $\mathscr{B}^2 \otimes_{\mathscr{A}'} \mathscr{B}^1$. This is easily seen to be a HC $\mathscr{A}''$-$\mathscr{A}$-bimodule. Also the derived tensor product of the objects from $D^b_{HC}(\mathscr{A}''\text{-}\mathscr{A}'), D^b_{HC}(\mathscr{A}'\text{-}\mathscr{A})$ lies in $D^-_{HC}(\mathscr{A}''\text{-}\mathscr{A})$ (and in $D^b_{HC}(\mathscr{A}''\text{-}\mathscr{A})$ provided $\mathscr{A}'$ has finite homological dimension). An analogous statement is true for one-sided derived Hom's.

We will need the following result from [L8] that is a part of [L8, Theorem 1.3].

Lemma 3.8. *Let $\mathscr{A}$ be the algebra of global sections of a quantization of a symplectic resolution. Then $\mathscr{A}$ has finite length as an $\mathscr{A}$-bimodule.*

Now let us proceed to restriction functors defined in [BL, Section 5.4] for HC bimodules over the algebras $\bar{\mathscr{A}}_\lambda(n,r)$. Let us write $\bar{\mathscr{A}}_\lambda(n)$ for $\bar{\mathscr{A}}_\lambda(n,r)$. Pick a point x in the leaf corresponding to the unordered collection $\mu = (n_1,\ldots,n_k)$ and set $\bar{\mathscr{A}}_\lambda(\mu) = \bigotimes_{i=1}^k \bar{\mathscr{A}}_\lambda(n_i)$. Then we have a functor

$$\bullet_{\dagger,\mu}\colon \mathrm{HC}(\bar{\mathscr{A}}_\lambda(n)\text{-}\bar{\mathscr{A}}_{\lambda'}(n)) \to \mathrm{HC}(\bar{\mathscr{A}}_\lambda(\mu)\text{-}\bar{\mathscr{A}}_{\lambda'}(\mu)).$$

We will need several facts about the restriction functors established in [BL, Section 5.5].

Proposition 3.9. *The following is true.*

(1) *The functor $\bullet_{\dagger,x}$ is exact and intertwines tensor products (as well as* Tor*'s). It also intertwines one-sided Hom's (as well as* Ext*'s).*
(2) *The associated variety $\mathrm{V}(\mathscr{B}_{\dagger,x})$ is uniquely characterized by $D \times \mathrm{V}(\mathscr{B}_{\dagger,\mu})^{\wedge_0} = \mathrm{V}(\mathscr{B})^{\wedge_x}$.*

Now let $\mathscr{L}$ denote the leaf of x. Consider the subcategory $\mathrm{HC}_{\overline{\mathscr{L}}}(\bar{\mathscr{A}}_\lambda(n)\text{-}\bar{\mathscr{A}}_{\lambda'}(n))$ of all objects whose associated variety is contained in $\overline{\mathscr{L}}$. Now suppose that $\mathscr{L}$ corresponds to μ. It follows that $\bullet_{\dagger,x}$ maps $\mathrm{HC}_{\overline{\mathscr{L}}}(\bar{\mathscr{A}}_\lambda(n)\text{-}\bar{\mathscr{A}}_{\lambda'}(n))$ to the category $\mathrm{HC}_{fin}(\bar{\mathscr{A}}_\lambda(\mu)\text{-}\bar{\mathscr{A}}_{\lambda'}(\mu))$ of finite-dimensional $\bar{\mathscr{A}}_\lambda(\mu)$-$\bar{\mathscr{A}}_{\lambda'}(\mu)$-bimodules. As was shown in [L9, Section 3.3], this functor admits a right adjoint

$$\bullet^{\dagger,x}\colon \mathrm{HC}_{fin}(\bar{\mathscr{A}}_\lambda(\mu)\text{-}\bar{\mathscr{A}}_{\lambda'}(\mu)) \to \mathrm{HC}_{\overline{\mathscr{L}}}(\bar{\mathscr{A}}_\lambda(n)\text{-}\bar{\mathscr{A}}_{\lambda'}(n)).$$

We would like to point out that the results and constructions explained above in this section generalize to products of the algebras $\bar{\mathscr{A}}_\lambda(?,r)$ in a straightforward way.

Let us finish this section with a further discussion of wall-crossing bimodules. Namely, for $\lambda \in \mathbb{C}, \chi \in \mathbb{Z}$, we have the $\bar{\mathscr{A}}_{\lambda+\chi}(n,r)$-$\bar{\mathscr{A}}_\lambda(n,r)$-bimodule

$$\bar{\mathscr{A}}^0_{\lambda,\chi}(n,r) := [D(\bar{R})/D(\bar{R})\{\xi_R - \lambda \operatorname{tr}\xi \,|\, \xi \in \mathfrak{g}\}]^{G,\chi}.$$

This gives an example of a HC $\bar{\mathscr{A}}_{\lambda+\chi}(n,r)$-$\bar{\mathscr{A}}_{\lambda}(n,r)$-bimodule.

Lemma 3.10. *The following is true.*

(1) *If $(\lambda+\chi,\theta)$ satisfies abelian localization, then $\bar{\mathscr{A}}^0_{\lambda,\chi}(n,r) \cong \bar{\mathscr{A}}^{(\theta)}_{\lambda,\chi}(n,r)$.*
(2) *(λ,θ) satisfies abelian localization if and only if, for all sufficiently large $n \in \mathbb{Z}$, the bimodules $\bar{\mathscr{A}}^0_{\lambda+mn\theta,n\theta}(n,r)$ and $\bar{\mathscr{A}}^0_{\lambda+(m+1)n\theta,-n\theta}(n,r)$ are mutually inverse Morita equivalences.*
(3) *If $x \in \overline{\mathscr{M}}(n,r)$ corresponds to a collection $\mu = (n_1,\ldots,n_k)$, then $\bar{\mathscr{A}}^0_{\lambda,\chi}(n,r)_{\dagger,x} = \bigotimes_{i=1}^k \bar{\mathscr{A}}^0_{\lambda,\chi}(n_i,r)$.*

Proof. (1) is [BL, Lemma 5.25]. (2) follows from [BL, Lemma 5.26] and [BPW, Proposition 5.20]. (3) is a special case of [BL, (5.11)]. □

4 Finite-dimensional representations and structure of category $\mathscr{O}$

4.1 Finite-dimensional representations

The goal of this section is to prove the following statement.

Proposition 4.1. *The sheaf $\bar{\mathscr{A}}^\theta_\lambda(n,r)$ has a coherent module supported on $\bar{\rho}^{-1}(0)$ if and only if $\lambda = \frac{s}{n}$ with s and n coprime. If this is the case, then the category $\operatorname{Coh}_{\bar{\rho}^{-1}(0)}(\bar{\mathscr{A}}^\theta_\lambda(n,r))$ of all coherent $\bar{\mathscr{A}}^\theta_\lambda(n,r)$-modules supported on $\bar{\rho}^{-1}(0)$ is equivalent to* Vect.

In the proof, and also below, we will need the following lemma.

Lemma 4.2. *Let $\mathscr{C}$ be a highest weight category, where the classes of standard and costandard objects coincide. Let $R\colon D^b(\mathscr{C}) \xrightarrow{\sim} D^b(\mathscr{C}^\vee)$ be a Ringel duality functor. The following conditions are equivalent:*

(1) *$\mathscr{C}$ is semisimple.*
(2) *We have $H_0(R(L)) \neq 0$ for every simple object L.*
(3) *every simple lies in the head of a costandard object.*

Recall that by a tilting object in a highest weight category one means an object that admits both a filtration by standards and a filtration by costandards. The labelling set for the indecomposable tilting objects is the same as for standard objects: namely, there is a unique indecomposable tilting object $T(p)$ that admits an embedding $\Delta(p) \hookrightarrow T(p)$ whose cokernel is standardly filtered.

Proof. Recall that R can be realized as $R\operatorname{Hom}_{\mathscr{C}}(T,\bullet)$, where T stands for the sum of all indecomposable tiltings in $\mathscr{C}$. The implication (1)⇒(2) is clear. The implication

(2)⇒(3) follows from the fact that since every tilting is filtered by costandards, every simple contained in the head of a tilting is contained in the head of some costandard.

Let us prove (3)⇒(1). Let τ be a maximal (with respect to the coarsest highest weight ordering) label. Then the simple $L(\tau)$ lies in the head of some costandard, say $\nabla(\sigma)$. But all simple constituents of $\nabla(\sigma)$ are $L(\xi)$ with $\xi \leqslant \sigma$. It follows that $\sigma = \tau$. Since $L(\tau)$ lies in the head of $\nabla(\tau)$ and also coincides with the socle, we see that $\nabla(\tau) = L(\tau)$. So $L(\tau)$ is injective. Since the classes of standards and costandard coincide, we see that $L(\tau) = \Delta(\tau)$. Since τ is maximal, $\Delta(\tau) = L(\tau)$ is projective. So $L(\tau)$ itself forms a block.

Since this holds for any maximal τ, we deduce that the category $\mathscr{C}$ is semisimple. □

Proof of Proposition 4.1. The proof is in several steps.

Step 1. Let us take dominant ν, i.e.,

$$\nu(t) = (\alpha(t), t),$$

where $\alpha(t) = \operatorname{diag}(t^{d_1}, \dots, t^{d_r})$ with $d_1 \gg d_2 \gg \cdots \gg d_r$. We can assume that λ is as in Proposition 2.2. Let us show that for the category $\mathscr{O}(\bar{\mathscr{A}}_\lambda(n,r))$, condition (2) of Lemma 4.2 is equivalent to all simples having support of dimension $rn-1$. By Lemmas 2.7 and 2.8, the functor R is a perverse equivalence with respect to the filtration by dimension of support. So the claim that all simples have support of dimension equal to $rn-1$ is equivalent to $H_0(RL) \neq 0$ for all simple L. By Lemma 2.4, the classes of standards and costandards in $K_0(\mathscr{O}_\nu(\bar{\mathscr{A}}_\lambda(n,r)))$ coincide and we are done by Lemma 4.2.

Step 2. Suppose that λ is Zariski generic and $\bar{\mathscr{A}}_\lambda(n,r)$ has a finite-dimensional representation. Let us prove that the denominator of λ is precisely n. Indeed, a finite-dimensional module lies in the category $\mathscr{O}_\nu$. It follows that the algebra $\mathsf{C}_{\nu_0}(\bar{\mathscr{A}}_\lambda(n,r))$ has a finite-dimensional representation; equivalently, there is a simple of GK dimension 1 in $\mathscr{O}_\nu(\mathsf{C}_{\nu_0}(\bar{\mathscr{A}}_\lambda(n,r)))$. Recall that

$$\mathsf{C}_{\nu_0}(\mathscr{A}_\lambda(n,r)) = \bigoplus \bigotimes_{i=1}^{r} \mathscr{A}_{\lambda+1-i}(n_i, 1),$$

where the sum is taken over all r-multipartitions $(n_1, \dots, n_r)$ of n. The minimal GK dimension of a module in the category $\mathscr{O}$ over $\bigotimes_{i=1}^r \mathscr{A}_{\lambda+1-i}(n_i,1)$ is bounded from below by the number of positive integers among $n_1, \dots, n_r$. So if there is a finite-dimensional $\bar{\mathscr{A}}_{\lambda+1-i}(n,r)$-module, then there is a GK dimension 1 module in the category $\mathscr{O}_\nu(\mathscr{A}_\lambda(n,1))$, and equivalently, a finite-dimensional $\bar{\mathscr{A}}_{\lambda+1-i}(n,1)$-module. The latter is a spherical rational Cherednik algebra for S_n. As was mentioned in Section 1.2, it has a finite-dimensional representation only when the denominator of λ is n.

Step 3. Let us prove that the algebra $\bar{\mathscr{A}}_\lambda(n,r)$ has no proper ideals of infinite codimension when $\lambda \notin \mathbb{Q}$ or the denominator of λ is bigger than or equal to n. Indeed, let $\mathscr{I}$ be a proper ideal. Pick x in an open symplectic leaf in $\mathrm{V}(\bar{\mathscr{A}}_\lambda(n,r)/\mathscr{I})$ and consider the ideal $\mathscr{I}_{\dagger,x}$ in the slice algebra for x. The latter ideal has finite codimension by (2) of Proposition 3.9. By Section 3.3, this slice algebra is $\bigotimes_{i=1}^k \bar{\mathscr{A}_\lambda}(n_i,r)$, where $n_1,\ldots,n_k$ are positive integers with $\sum n_i \leqslant n$. So, by our assumption on λ and Step 2, the slice algebra has no finite-dimensional representations. We arrive at a contradiction, which proves the statement in the beginning of the step. In particular, we see that the algebra $\bar{\mathscr{A}_\lambda}(n,r)$ is simple when λ is irrational or the denominator of λ is larger than n. We also see that in this case category $\mathscr{O}$ is semisimple; this follows from Step 1 and Lemma 2.3.

Step 4. Now consider the case when λ has denominator n. We claim that there is a finite-dimensional module in $\mathscr{O}_\nu(\bar{\mathscr{A}_\lambda}(n,r))$. The case when $n=1$ is easy, here $\bar{\mathscr{A}_\lambda}(1,r)=D^\lambda(\mathbb{P}^{r-1})$ and the finite-dimensional module is $H^0(\mathbb{P}^{r-1},\mathscr{O}(\lambda))$ for $\lambda \geqslant 0$ (recall that $\bar{\mathscr{A}_\lambda}(1,r)\cong \bar{\mathscr{A}}_{-\lambda-r}(1,r)$). So we assume that $n>1$.

By Lemma 2.3, for any $M \in \mathscr{O}_\nu(\bar{\mathscr{A}_\lambda}(n,r))$, we have

$$\text{GK-dim} M = \frac{1}{2}\dim \mathrm{V}(\bar{\mathscr{A}_\lambda}(n,r)/\operatorname{Ann} M).$$

Now Step 3 implies that all simples in $\mathscr{O}_\nu$ are either finite-dimensional or have support of dimension $rn-1$. If there are no finite-dimensional modules, then thanks to Step 1, $\mathscr{O}(\bar{\mathscr{A}_\lambda}(n,r))$ is semisimple.

Consider a one-parameter subgroup $\nu_0: t \mapsto (\alpha(t),1)$; it is in a face of the chamber of ν. The category $\mathscr{O}_\nu(\mathsf{C}_{\nu_0}(\bar{\mathscr{A}_\lambda}(n,r)))$ is not semisimple; it has category $\mathscr{O}$ for the Rational Cherednik algebra with parameter λ as a summand. By Proposition 2.5, $\mathscr{O}_\nu(\bar{\mathscr{A}_\lambda}(n,r))$ is not semisimple. Consequently, $\bar{\mathscr{A}_\lambda}(n,r)$ has a finite-dimensional representation.

Step 5. Let us show that the number of irreducible coherent $\bar{\mathscr{A}}^\theta_\lambda(n,r)$-modules supported on $\rho^{-1}(0)$ cannot be larger than 1. Recall that to a module in category $\mathscr{O}_\nu$ we can assign its characteristic cycle which is a formal linear combination of the irreducible components of the contracting locus of ν. This map from $K_0(\mathscr{O}_\nu)$ to the space of formal linear combinations of the irreducible components is injective, see [BLPW, Theorem 6.5]. So it is enough to show that there is only one lagrangian irreducible component in $\bar{\rho}^{-1}(0)$. Note that the lagrangian irreducible components of $\rho^{-1}(0)$ give a basis in $H^{2nr-2}(\mathscr{M}^\theta(n,r))$. According to [NY, Theorem 3.8], we have

$$\sum_i \dim H^{2i}(\mathscr{M}^\theta(n,r))t^i = \sum_\lambda t^{\sum_{i=1}^r (r|\lambda^{(i)}|-i(\lambda^{(i)t})_1)},$$

where the summation is over the set of the r-multipartitions $\lambda=(\lambda^{(1)},\ldots,\lambda^{(r)})$. The highest power of t on the right-hand side is $rn-1$; it occurs for a single λ, namely, for $\lambda=((n),\varnothing,\ldots,\varnothing)$. This shows that $\dim H^{2nr-2}(\mathscr{M}^\theta(n,r))=1$ and completes the proof of the claim at the beginning of this step.

Step 6. The previous step completes the proof of all claims of the theorem, except for the claim that the category of modules supported on $\rho^{-1}(0)$ is semisimple. The latter is an easy consequence of the observation that, in a highest weight category, we have $\operatorname{Ext}^1(L,L)=0$, for every simple L. □

We would like to point out that the argument of Step 4 generalizes to denominators less than n. So in those cases, there are also simple $\bar{\mathscr{A}}_\lambda(n,r)$-modules of support with dimension $< rn-1$.

4.2 Proof of Theorem 1.4

In this section we will prove Theorem 1.4. We have already seen (Step 3 of the proof of Proposition 4.1) that if the denominator is larger than n, then the category $\mathscr{O}$ is semisimple. The case of denominator n will follow from a more precise statement, see Theorem 4.3.

We introduce a certain model category. Let $\mathscr{C}_n$ denote the nontrivial block of category $\mathscr{O}$ for the Rational Cherednik algebra $H_{1/n}(n)$ associated to the symmetric group $\mathfrak{S}_n$. This is also the category of B-equivariant perverse sheaves on $\mathbb{P}^{n-1}$, where B is a Borel subgroup of PGL_n. We summarize some properties of this category.

(i) Its coarsest highest weight poset is linearly ordered: $p_n < p_{n-1} < \cdots < p_1$.
(ii) The projective objects $P(p_i)$ for $i > 1$ are the universal extensions

$$0 \to \Delta(p_{i-1}) \to P(p_i) \to \Delta(p_i) \to 0.$$

(iii) The indecomposable tilting object $T(p_{i-1})$ for $i > 1$ coincides with $P(p_i)$.
(iv) The simple objects $L(p_i)$ with $i > 1$ appear in the heads of tiltings, while $\mathrm{RHom}_{\mathscr{C}_n}(T, L(p_1))$ is concentrated in homological degree n.
(v) There is a unique simple in the Ringel dual category $\mathscr{C}_n^\vee$ that appears in the higher cohomology of $\mathrm{RHom}_{\mathscr{C}_n}(T, \bullet)$.

(i) and (ii) were obtained in [BEG]. (iii) and (v) follow from the observation that the Ringel dual of $\mathscr{C}_n$ is equivalent to $\mathscr{C}_n$. This is a consequence of [GGOR, Section 4.2]. (iv) follows, for example, from Proposition 4.1 and Lemma 4.2.

Theorem 4.3. *Consider a parameter of the form $\lambda = \frac{q}{n}$ with coprime q, n. Then the following is true.*

(1) *The category $\mathscr{O}_\nu(\bar{\mathscr{A}}_\lambda^\theta(n,r))$ has only one nontrivial block. This block is equivalent to $\mathscr{C}_{rn}$. This block contains an irreducible representation supported on $\bar{\rho}^{-1}(0)$.*
(2) *Suppose that ν is dominant. Then the labels in the non-trivial block of $\mathscr{O}(\bar{\mathscr{A}}_\lambda^\theta(n,r))$ are the hooks $h_{i,d} = (\varnothing, \dots, (n+1-d, 1^{d-1}), \dots, \varnothing)$ (where i is the number of the diagram where the hook appears) ordered by $h_{1,n} > h_{1,n-1} > \cdots > h_{1,1} > h_{2,n} > \cdots > h_{2,1} > \cdots > h_{r,1}$.*

Proof. The proof is in several steps. We again deal with the realization of our category as $\mathscr{O}(\bar{\mathscr{A}}_\lambda(n,r))$, where λ is as in Proposition 2.2. First, note that by Lemma 2.4, category $\mathscr{O}(\bar{\mathscr{A}}_\lambda(n,r))$ satisfies the conditions of Lemma 4.2.

Step 1. As we have seen in the proof of Proposition 4.1, all simples have maximal dimension of support, except for one, which we denote by L, and which is finite-dimensional. So all blocks but one consist of modules with support of dimension $rn-1$. So (2) of Lemma 4.2 holds for these blocks and they are equivalent to Vect. Let $\mathscr{C}$ denote the block of L. The label of L, denoted by p_{max}, is the largest in any highest weight ordering; this follows from the proof of (3)⇒(1) of Lemma 4.2. For all other labels p, the simple $L(p)$ lies in the head of the tilting generator T. In other words an analog of (iv) above holds for $\mathscr{C}$ with rn instead of n. In subsequent steps, we will show that $\mathscr{C} \cong \mathscr{C}_{rn}$.

Step 2. We show that an analog of (v) holds for $\mathscr{C}$. By Lemma 2.7, the higher homology of $\mathfrak{WC}_{\theta\to-\theta}(M)$ for $M \in \mathscr{O}(\bar{\mathscr{A}_\lambda}(n,r))$ cannot have support of maximal dimension. It follows that the higher homology is finite-dimensional, and so is the direct sums of several copies of a single simple in $\mathscr{O}(\bar{\mathscr{A}_{\lambda^-}}(n,r))$. Since $\mathfrak{WC}_{\theta\to-\theta}$ is a Ringel duality functor (Lemma 2.8), (v) follows.

Step 3. We show that in the coarsest highest weight order, there is a unique minimal label for $\mathscr{C}$, say p_{min}. This is equivalent to $\mathscr{C}^\vee$ having a unique maximal label because the orders on $\mathscr{C}$ and $\mathscr{C}^\vee$ are opposite. But $\mathscr{C}^\vee$ is equivalent to the nontrivial block in $\mathscr{O}(\bar{\mathscr{A}_{\lambda^-}}(n,r))$. So we are done by Step 1 (applied to $(\lambda^-,-\theta)$ instead of (λ,θ)) of this proof.

Step 4. We show that (v) implies that any tilting in $\mathscr{C}$ but one is projective. Let $R^\vee$ denote the Ringel duality equivalence $D^b(\mathscr{C}^\vee) \to D^b(\mathscr{C})^{opp}$ (recall that $\mathscr{C}^{\vee\vee}$ is naturally identified with $\mathscr{C}^{opp}$). We have $\operatorname{Ext}^i_{\mathscr{C}}(T(p),L(p')) = \operatorname{Hom}_{\mathscr{C}}(T(p),L(p')[i]) = \operatorname{Hom}_{\mathscr{C}^\vee}\big((R^\vee)^{-1}L(p')[i],(R^\vee)^{-1}T(p)\big)$. The objects $(R^\vee)^{-1}T(p)$ are injective, so $\operatorname{Ext}^i(T(p),L(p')) = \operatorname{Hom}\big(H^i((R^\vee)^{-1}L(p')),(R^\vee)^{-1}T(p)\big)$. Similar to the previous step (applied to $\mathscr{C}^\vee$ instead of $\mathscr{C}$ and $(R^\vee)^{-1}$ instead of R), there is a unique indecomposable injective $I^\vee(p^\vee)$ in $\mathscr{C}^\vee$ that admits nonzero maps from a higher cohomology of $(R^\vee)^{-1}L(p)$. So if $(R^\vee)^{-1}T(p) \neq I^\vee(p^\vee)$, then $T(p)$ is projective.

Step 5. We remark that $\Delta(p_{max})$ is projective but not tilting, while $\Delta(p_{min})$ is tilting but not projective. So the projectives in $\mathscr{C}$ are $\Delta(p_{max})$ and $T(p)$ for $p \neq p_{min}$. Similarly, the tiltings are $P(p), p \neq p_{max}$, and $\Delta(p_{min})$.

Step 6. Let Λ denote the highest weight poset for $\mathscr{C}$. We define a map ν: $\Lambda \setminus \{p_{min}\} \to \Lambda \setminus \{p_{max}\}$. It follows from Step 5 that the head of any tilting in $\mathscr{C}$ is simple. By definition, $\nu(p)$ is such that $L(\nu(p))$ is the head of $T(p)$. We remark that $\nu(p) < p$ for any highest weight order; this follows from Step 4.

Step 7. We show that any element $p \in \Lambda$ has the form $\nu^i(p_{max})$ with $i \geqslant 0$. Assume the converse and let us pick the maximal element not of this form, say p'. Since $p' \neq p_{max}$, we see that $L(p')$ lies in the head of some tilting. But the head of any indecomposable tilting is simple by Step 5. So $\Delta(p')$ is a top term of a filtration with standard subsequent quotients. By the definition of ν and the choice of p', $\Delta(p')$ is tilting itself. Step 5 implies $p' = p_{min}$. We now pick a minimal element p'' in $\Lambda \setminus \{p_{min}\}$. By Step 6, $\nu(p'') < p''$. So $\nu(p'') = p_{min}$. The claim in the beginning of the step is established. This proves (i) for $\mathscr{C}$.

Step 8. (ii) for $\mathscr{C}$ follows from Step 7, and (iii) follows from (ii) and Step 5.

Step 9. We show that $\#\Lambda = rn$. The minimal projective resolution for $\Delta(p_{min})$ has length $\#\Lambda$, all projectives there are different, and the last term is $\Delta(p_{max})$. It follows that $\mathrm{RHom}(\Delta(p_{min}), L(p_{max}))$ is concentrated in homological degree $\#\Lambda - 1$. The other tiltings are projectives and RHom's with them amount to Hom's. Since $\mathrm{RHom}(T, L(p_{max}))$ is concentrated in homological degree $rn-1$ (this follows from Lemmas 2.7 and 2.8), we are done.

Step 10. Let us complete the proof of (1). Let us order the labels in Λ decreasingly, $p_1 > \cdots > p_{rn}$. Using (ii) we get the following claims:

- $\mathrm{End}(P(p_i)) = \mathbb{C}[x]/(x^2)$ for $i > 1$ and $\mathrm{End}(P(p_1)) = \mathbb{C}$.
- $\mathrm{Hom}(P(p_i), P(p_j))$ is 1-dimensional if $|i-j| = 1$ and is 0 if $|i-j| > 1$.

Choose some basis elements $a_{i,i+1}, i = 1,\ldots,rn-1$ in $\mathrm{Hom}(P(p_{i+1}), P(p_i))$ and also basis elements $a_{i+1,i} \in \mathrm{Hom}(P(p_i), P(p_{i+1}))$. We remark that the image of the composition map $\mathrm{Hom}(P(p_{i+1}), P(p_i)) \times \mathrm{Hom}(P(p_i), P(p_{i+1})) \to \mathrm{End}(P(p_i))$ spans the maximal ideal. Choose generators a_{ii} in the maximal ideals of $\mathrm{End}(P(p_i))$, $i = 2,\ldots,rn$. We fix elements $a_{i,i+1}, 1 \leqslant i \leqslant rn-1$, and a_{22}. Normalize a_{21} by requiring that $a_{21}a_{12} = a_{22}$, automatically, $a_{12}a_{21} = 0$. Normalize a_{32} by $a_{23}a_{32} = a_{22}$ and then normalize a_{33} by $a_{33} = a_{32}a_{23}$. We continue normalizing $a_{i+1,i}$ and $a_{i+1,i+1}$ in this way. We then recover the multiplication table in $\mathrm{End}(\bigoplus P(p_i))$ in a unique way. This completes the proof of (1).

Step 11. We prove (2). Let us check that the labeling set Λ for the nontrivial block of $\mathscr{O}(\mathscr{A}_\lambda^{\bar{\theta}}(n,r))$ consists of hooks. For this, it is enough to check that $\Delta(h_{i,d})$ does not form a block on itself. This is done in a similar way to Step 4 in the proof of Theorem 1.2. Now, according to [Ko], the hooks are ordered as specified in (2) with respect to the geometric order on the torus fixed points in $\mathscr{M}^\theta(n,r)$ (note that the sign conventions here and in [Ko] are different). □

Remark 4.4. We can determine the label of the simple supported on $\bar{\rho}^{-1}(0)$ in the category $\mathscr{O}$ corresponding to an arbitrary generic ν. Namely, note that $\bar{\rho}^{-1}(0)$ coincides with the closure of a single contracting component and that contracting component corresponds to the maximal point. Now we can use results of [Ko] to find a label of the point: it always has only one nontrivial partition and this partition is either (n) or (1^n).

5 Localization theorems

In this section we prove Theorem 1.1. The proof is in the following steps.

- We apply results of McGerty and Nevins [MN2], and show that, first, if the abelian localization fails for (λ, θ), then λ is a rational number with denominator not exceeding n; second, the parameters $\lambda = \frac{q}{m}$ with $m \leqslant n$ and $-r < \lambda < 0$ are indeed singular; and, third, the functor Γ_λ^θ is exact when $\lambda > -r$, $\theta > 0$ or $\lambda < 0$, $\theta < 0$. Thanks to an isomorphism $\mathscr{A}_\lambda^\theta(n,r) \cong \mathscr{A}_{-\lambda-r}^{-\theta}(n,r)$ (see Sections 2.1 and 3.1), this reduces the proof of Theorem 1.1 to checking that the abelian localization holds for $\lambda = \frac{q}{m}$ with $q \geqslant 0$, $m \leqslant n$ and $\theta > 0$.

- Then we reduce the proof to the case when the denominator is precisely n and $\lambda, \theta > 0$.
- Then we study a connection between the algebras $\mathsf{C}_{v_0}(\bar{\mathscr{A}}_\lambda(n,r))$, $\Gamma(\mathsf{C}_{v_0}(\bar{\mathscr{A}}_\lambda^\theta(n,r)))$. We show that the numbers of simples in the categories $\mathscr{O}$ for these algebras coincide. We deduce the localization theorem from there.

The last step is a crucial one and it does not generalize to other quiver varieties.

5.1 Results of McGerty and Nevins and consequences

In [MN2], McGerty and Nevins found a sufficient condition for the functor

$$\Gamma_\lambda^\theta : \mathscr{A}_\lambda^\theta(n,r)\text{-mod} \to \mathscr{A}_\lambda(n,r)\text{-mod}$$

to be exact. They were dealing with more general Hamiltonian reductions but we will only need the Gieseker case. We explain what their result gives in the case of interest for us. Consider the quotient functors $\pi_\lambda : D_R\text{-mod}^{G,\lambda} \twoheadrightarrow \mathscr{A}_\lambda(n,r)\text{-mod}$ and $\pi_\lambda^\theta : D_R\text{-mod}^{G,\lambda} \twoheadrightarrow \operatorname{Coh}(\mathscr{A}_\lambda^\theta(n,r))$. Here D_R stands for the sheaf of differential operators viewed as a microlocal sheaf on T^*R, and $D_R\text{-mod}^{G,\lambda}$ is the category of (G,λ)-twisted equivariant D-modules on R. The functors $\pi_\lambda, \pi_\lambda^\theta$ are discussed in [BL, Section 2.3] or [BPW, Section 5.5].

Proposition 5.1. *The inclusion* $\ker \pi_\lambda^{\det} \subset \ker \pi_\lambda$ *holds provided* $\lambda > -r$. *Similarly,* $\ker \pi_\lambda^{\det^{-1}} \subset \pi_\lambda$ *provided* $\lambda < 0$.

I would like to thank Dmitry Korb for explaining to me the required modifications to [MN2, Section 8].

Proof. We consider the case $\theta = \det$; the opposite case follows from $\mathscr{A}_\lambda^\theta(n,r) \cong \mathscr{A}_{-r-\lambda}^{-\theta}(n,r)$ (established in Section 3.1). The proof closely follows [MN2, Section 8], where the case of $r = 1$ is considered. Instead of $R = \operatorname{End}(V) \oplus \operatorname{Hom}(V,W)$ they use $R' = \operatorname{End}(V) \oplus \operatorname{Hom}(W,V)$. Then, thanks to the partial Fourier transform, we have $D(R)\text{-mod}^{G,\lambda} \cong D(R')\text{-mod}^{G,\lambda+r}$. The set of weights in R' for a maximal torus $H \subset \operatorname{GL}(V)$ is independent of r, so we have the same Kempf-Ness subgroups as in the case $r = 1$: it is enough to consider the subgroups β with tangent vectors (in the notation of [MN2, Section 8]) $e_1 + \cdots + e_k$. The shift in [MN2, Section 8] becomes $\frac{rk}{2}$ (in the computation of *loc.cit.* we need to take the second summand r times, which is all that changes). So we get that $\ker \pi_\lambda^{\det} \subset \ker \pi_\lambda$, provided $k(-\frac{r}{2} - \lambda) \notin \frac{rk}{2} + \mathbb{Z}_{\geqslant 0}$ for all possible k, meaning $1 \leqslant k \leqslant n$ (the number $-\frac{r}{2} - \lambda$ is c' in [MN2, Section 8]). The condition simplifies to $\lambda \notin -r - \frac{1}{k}\mathbb{Z}_{\geqslant 0}$. This implies the claim of the proposition. $\square$

5.2 Reduction to denominator n and singular parameters

Proposition 5.1 allows us to show that certain parameters are singular (meaning that the homological dimension of $\mathscr{A}_\lambda(n,r)$ is infinite).

Corollary 5.2. *The parameters λ with denominator $\leqslant n$ and $-r<\lambda<0$ are singular.*

Proof. Assume the contrary. By Lemma 3.2, the functors $R\Gamma_\lambda^{\pm\theta}$ are equivalences. Since the functors $\Gamma_\lambda^{\pm\theta}$ are exact, we see that $\Gamma_\lambda^{\pm\theta}$ are equivalences of abelian categories. From the inclusions $\ker\pi_\lambda^{\pm\theta}\subset\ker\pi_\lambda$, we deduce that the functors $\pi_\lambda^{\pm\theta}$ are isomorphic. So the wall-crossing functor $\mathfrak{WC}_{-\theta\leftarrow\theta}=\pi_{\lambda^-}^{-\theta}\circ(\mathbb{C}_{\lambda^--\lambda}\otimes\bullet)\circ L\pi_\lambda^{\theta!}$ (see [BL, (2.10)] for the equality) is an equivalence of abelian categories. However, we have seen at the end of Section 4.1, the category $\mathscr{O}_\nu(\mathscr{A}_\lambda^\theta)$ is not semisimple. Combining Lemma 4.2 with Lemma 2.8, we see that $\mathfrak{WC}_{-\theta\leftarrow\theta}$ cannot be an abelian equivalence. □

Now let us observe that it is enough to check that abelian localization holds for $\lambda\geqslant 0$ and $\theta>0$. This follows from an isomorphism $\mathscr{A}_\lambda^\theta(n,r)\cong\mathscr{A}_{-\lambda-r}^{-\theta}(n,r)$. This an isomorphism of sheaves on $\mathscr{M}^\theta(n,r)\cong\mathscr{M}^{-\theta}(n,r)$ (see the proof of Lemma 3.1).

Lemma 5.3. *Suppose that for all n, abelian localization holds for $\mathscr{A}_\lambda^\theta(n,r)$ if $\lambda\geqslant 0$ and the denominator of λ is exactly n. Then abelian localization holds for all λ.*

Proof. By Lemma 3.10, abelian localization holds for (λ,θ) with $\theta>0$ if and only if the bimodules $\mathscr{A}^0_{\lambda+m\chi,\chi}(n,r),\mathscr{A}^0_{\lambda+(m+1)\chi,-\chi}(n,r)$ with some $\chi\in\mathbb{Z}_{>0}$ define mutually dual Morita equivalences; equivalently, the natural homomorphisms

$$\begin{aligned}&\mathscr{A}^0_{\lambda+m\chi,\chi}(n,r)\otimes_{\mathscr{A}_{\lambda+m\chi}(n,r)}\mathscr{A}^0_{\lambda+(m+1)\chi,-\chi}(n,r)\to\mathscr{A}_{\lambda+(m+1)\chi}(n,r),\\&\mathscr{A}^0_{\lambda+(m+1)\chi,-\chi}(n,r)\otimes_{\mathscr{A}_{\lambda+(m+1)\chi}(n,r)}\mathscr{A}^0_{\lambda+m\chi,\chi}(n,r)\to\mathscr{A}_{\lambda+m\chi}(n,r)\end{aligned}\tag{5.1}$$

are isomorphisms.

Assume the contrary: there is $\lambda>0$ with denominator $n'<n$ such that abelian localization does not hold for $\mathscr{A}_\lambda^\theta(n,r)$. Let K^1,C^1,K^2,C^2 denote the kernel and the cokernel of the first and second homomorphism in (5.1) (for some m), respectively. If one of these bimodules is nontrivial, then we can find $x\in\mathscr{M}(n,r)$ such that $K^i_{\dagger,x},C^i_{\dagger,x}$ are finite-dimensional and at least one of these bimodules is nonzero. From the classification of finite-dimensional irreducibles, see Proposition 4.1, we see that the slice algebras must be of the form $\bar{\mathscr{A}}_\lambda(n',r)^{\otimes k}$. But, by Lemma 3.10, $\mathscr{A}^0_{\lambda+(m+1)\chi,-\chi}(n,r)_{\dagger,x}=\bar{\mathscr{A}}^0_{\lambda+(m+1)\chi,-\chi}(n',r)^{\otimes k}$, $\mathscr{A}^0_{\lambda+m\chi,\chi}(n,r)_{\dagger,x}=\bar{\mathscr{A}}^0_{\lambda+m\chi,\chi}(n',r)^{\otimes k}$. Further, applying $\bullet_{\dagger,x}$ to (5.1), we again get natural homomorphisms. But the localization theorem holds for the algebra $\bar{\mathscr{A}}_\lambda(n',r)$ thanks to our inductive assumption, so the homomorphisms of the $\bar{\mathscr{A}}_\lambda(n',r)^{\otimes k}$-bimodules are isomorphisms. This contradiction finishes the proof of the lemma. □

5.3 Number of simples in $\mathscr{O}(\mathscr{A}_\lambda(n,r))$

So we need to prove that the localization theorem holds for positive parameters λ with denominator n (the case $\lambda=0$ occurs only if $n=1$, and in that case this is a classical localization theorem for differential operators on projective spaces). We will derive a proof from the claim that the number of simple objects in the categories $\mathscr{O}_\nu(\bar{\mathscr{A}_\lambda}(n,r))$ and $\mathscr{O}_\nu(\bar{\mathscr{A}}_\lambda^\theta(n,r))$ is the same. For this we will need to study the natural homomorphism

$$\varphi\colon \mathsf{C}_{\nu_0}(\bar{\mathscr{A}_\lambda}(n,r)) \to \Gamma(\mathsf{C}_{\nu_0}(\bar{\mathscr{A}}_\lambda^\theta(n,r))).$$

Here, as before, $\nu(t)=(\alpha(t),t), \nu_0(t)=(\alpha(t),1)$, where $\alpha\colon \mathbb{C}^\times \to \mathrm{GL}(r)$ is of the form $t\mapsto (t^{d_1},\dots,t^{d_r})$, and $d_1\gg d_2\gg\cdots\gg d_r$.

Recall, see Proposition 3.5, that

$$\Gamma(\mathsf{C}_{\nu_0}(\bar{\mathscr{A}}_\lambda^\theta(n,r))) = \bigoplus \bar{\mathscr{A}_\lambda}(n_1,\dots,n_r;r),$$

where the summation is taken over all compositions

$$n=n_1+\cdots+n_r \qquad \text{and} \qquad \bar{\mathscr{A}_\lambda}(n_1,\dots,n_r;r)\otimes D(\mathbb{C}) = \bigotimes_{i=1}^r \mathscr{A}_{\lambda+i-1}(n_i,1)$$

(the factor $D(\mathbb{C})$ is embedded into the right-hand side "diagonally"). Let $\mathscr{B}$ denote the maximal finite-dimensional quotient of $\Gamma(\mathsf{C}_{\nu_0}(\bar{\mathscr{A}}_\lambda^\theta(n,r)))$.

Proposition 5.4. *The composition of φ with the projection $\Gamma(\mathsf{C}_{\nu_0}(\bar{\mathscr{A}}_\lambda^\theta(n,r))) \twoheadrightarrow \mathscr{B}$ is surjective.*

Proof. The proof is in several steps. Recall that $\bar{\mathscr{A}}_{\lambda+i}(n,1)$ is the spherical subalgebra in the Cherednik algebra $H_c(n)$ for the reflection representation $\mathfrak{h}$ of $\mathfrak{S}_n$ with $c=\lambda+i$.

Step 1. We claim that it is sufficient to prove that the composition φ_i of φ with the projection $\Gamma(\mathsf{C}_{\nu_0}(\bar{\mathscr{A}}_\lambda^\theta(n,r))) \to \bar{\mathscr{A}}_{\lambda+i}(n,1)$ is surjective. Indeed, each $\bar{\mathscr{A}}_{\lambda+i}(n,1)$, $i=0,\dots,r-1$, has a unique finite-dimensional representation. The dimensions of these representations are pairwise different, see [BEG]. Namely, if $\lambda=\frac{q}{n}$, then the dimension is $\frac{(q+n-1)!}{q!n!}$. So $\mathscr{B}$ is the sum of r pairwise non-isomorphic matrix algebras. Therefore the surjectivity of the homomorphism $\mathsf{C}_{\nu_0}(\bar{\mathscr{A}_\lambda}(n,r)) \to \mathscr{B}$ follows from the surjectivity of all its r components. We remark that the other summands of $\mathsf{C}_{\nu_0}(\bar{\mathscr{A}_\lambda}(n,r))$ have no finite-dimensional representations.

Step 2. Generators of $\bar{\mathscr{A}}_{\lambda+i}(n,1)$ are known. The algebra $eH_c(n)e$ is generated by $S(\mathfrak{h})^W, S(\mathfrak{h}^*)^W$, see, e.g., the proof [EG, Proposition 4.9]. On the level of quantum Hamiltonian reduction, $S(\mathfrak{h})^W$ coincides with the image of $S(\bar{\mathfrak{g}})^G \subset D(\bar{\mathfrak{g}}\oplus\mathbb{C}^{n*})^G$, while $S(\mathfrak{h}^*)^W$ coincides with the image of $S(\bar{\mathfrak{g}}^*)^G$. Here and below we write $\bar{\mathfrak{g}}$ for $\mathfrak{sl}_n$. We will show that the images of $S(\bar{\mathfrak{g}})^G, S(\bar{\mathfrak{g}}^*)^G$ lie in the image of $\varphi_i\colon \mathsf{C}_{\nu_0}(\bar{\mathscr{A}_\lambda}(n,r)) \to \bar{\mathscr{A}}_{\lambda+i}(n,1)$. This will establish the surjectivity in Step 1.

Step 3. We produce a natural homomorphism $S(\bar{\mathfrak{g}}^*)^G \to \mathsf{C}_{\nu_0}(\bar{\mathscr{A}}_\lambda(n,r))$. First, recall that $\bar{\mathscr{A}}_\lambda(n,r)$ is a quotient of $D(\bar{\mathfrak{g}} \oplus (\mathbb{C}^{*n})^r)^G$. The algebra $S(\bar{\mathfrak{g}}^*)^G$ is included into $D(\bar{\mathfrak{g}} \oplus (\mathbb{C}^{*n})^{\oplus r})^G$ as the algebra of invariant functions on $\bar{\mathfrak{g}}$. So we get a homomorphism $S(\bar{\mathfrak{g}}^*)^G \to \bar{\mathscr{A}}_\lambda(n,r)$. Since the $\mathbb{C}^\times$-action ν_0 used to form $\mathsf{C}_{\nu_0}(\bar{\mathscr{A}}_\lambda(n,r))$ is nontrivial only on $(\mathbb{C}^{*n})^{\oplus r}$, we see that the image of $S(\bar{\mathfrak{g}}^*)^G$ lies in $\bar{\mathscr{A}}_\lambda(n,r)^{\nu_0(\mathbb{C}^\times)}$. So we get a homomorphism $\iota: S(\bar{\mathfrak{g}}^*)^G \to \mathsf{C}_{\nu_0}(\bar{\mathscr{A}}_\lambda(n,r))$.

Step 4. We claim that $\varphi_i \circ \iota$ coincides with the inclusion $S(\bar{\mathfrak{g}}^*)^G \to \bar{\mathscr{A}}_{\lambda+i}(n,1)$. We can filter the algebra $D(\bar{\mathfrak{g}} \oplus (\mathbb{C}^{*n})^{\oplus r})$ by order of the differential operator. This induces filtrations on $\bar{\mathscr{A}}_\lambda(n,r), \bar{\mathscr{A}}_\lambda^\theta(n,r)$. We have similar filtrations on the algebras $\bar{\mathscr{A}}_{\lambda+i}(n,1)$. The filtrations on $\bar{\mathscr{A}}_\lambda(n,r), \bar{\mathscr{A}}_\lambda^\theta(n,r)$ are preserved by ν_0, and hence we have filtrations on $\mathsf{C}_{\nu_0}(\bar{\mathscr{A}}_\lambda(n,r)), \Gamma(\mathsf{C}_{\nu_0}(\bar{\mathscr{A}}_\lambda^\theta(n,r)))$. It is clear from the construction of the projection $\Gamma(\mathsf{C}_{\nu_0}(\bar{\mathscr{A}}_\lambda^\theta(n,r))) \to \bar{\mathscr{A}}_{\lambda+i}(n,1)$ that it is compatible with the filtrations. On the other hand, the images of $S(\bar{\mathfrak{g}}^*)^G$ in both $\mathsf{C}_{\nu_0}(\bar{\mathscr{A}}_\lambda(n,r)), \bar{\mathscr{A}}_{\lambda+i}(n,1)$ lie in the filtration degree 0. So it is enough to prove that the associated graded homomorphism of $\varphi_i \circ \iota$ coincides with the associated graded homomorphism of the inclusion $S(\bar{\mathfrak{g}}^*)^G \to \bar{\mathscr{A}}_{\lambda+i}(n,1)$.

Step 5. The associated graded homomorphisms coincide with analogous homomorphisms defined on the classical level. The components of $\mathscr{M}^\theta(n,r)^{\nu_0(\mathbb{C}^\times)}$ that are Hilbert schemes are realized as follows. Pick an eigenbasis $w_1, \ldots, w_r \in \mathbb{C}^r$ for T_0. Then the Hilbert scheme component of $\mathscr{M}^\theta(n,r)^{T_0}$ corresponding to the composition $(0^{k-1}, n, 0^{r-k-1})$ (let us denote it by Z_k) consists of the G-orbits of $(A,B,0,j)$, where $j: \mathbb{C}^n \to \mathbb{C}^r$ is a map with image in $\mathbb{C}w_k$. In particular, the homomorphism $S(\bar{\mathfrak{g}}^*)^G \to \operatorname{gr}\mathscr{A}_{\lambda+k-1}(n,1)$ is dual to the morphism given by $(A,B,0,j) \to A$.

On the other hand, Z_k maps to $\mathscr{M}(r,n)//\nu_0(\mathbb{C}^\times)$ via $Z_k \hookrightarrow \mathscr{M}^\theta(n,r) \twoheadrightarrow \mathscr{M}(n,r) \twoheadrightarrow \mathscr{M}(n,r)//\nu_0(\mathbb{C}^\times)$. The corresponding homomorphism of algebras is the associated graded homomorphism of $\bar{\mathscr{A}}_\lambda(n,r)^{\nu_0(\mathbb{C}^\times)} \to \bar{\mathscr{A}}_{\lambda+k-1}(n,1)$. We have the morphism $\mathscr{M}(r,n)//\nu_0(\mathbb{C}^\times) \to \bar{\mathfrak{g}}//G$ induced by $(A,B,i,j) \mapsto A$. The corresponding homomorphism of algebras is the associated graded homomorphism of $S(\bar{\mathfrak{g}}^*)^G \to \bar{\mathscr{A}}_\lambda(n,r)^{\nu_0(\mathbb{C}^\times)}$. So we have checked that the associated graded homomorphism of $\varphi_k \circ \iota: S(\bar{\mathfrak{g}}^*)^G \to \bar{\mathscr{A}}_{\lambda+k-1}(n,1)$ coincides with that of the embedding $S(\bar{\mathfrak{g}}^*)^G \to \bar{\mathscr{A}}_{\lambda+k-1}(n,1)$. This proves the claim of Step 4.

Step 6. The coincidence of similar homomorphisms $S(\bar{\mathfrak{g}})^G \to \bar{\mathscr{A}}_{\lambda+i}(n,1)$ is established analogously. The proof of the surjectivity of $\mathsf{C}_{\nu_0}(\bar{\mathscr{A}}_\lambda(n,r)) \to \bar{\mathscr{A}}_{\lambda+i}(n,1)$ is now complete. □

We still have a Hamiltonian action of $\mathbb{C}^\times$ on $\mathsf{C}_{\nu_0}(\bar{\mathscr{A}}_\lambda(n,r))$ (via ν) that makes the homomorphism $\mathsf{C}_{\nu_0}(\bar{\mathscr{A}}_\lambda(n,r)) \to \Gamma(\mathsf{C}_{\nu_0}(\bar{\mathscr{A}}_\lambda^\theta(n,r)))$ equivariant. So we can form the category $\mathscr{O}(\mathsf{C}_{\nu_0}(\bar{\mathscr{A}}_\lambda(n,r)))$ for this action. By [L7, Section 5.5], we have an isomorphism $\mathsf{C}_\nu(\mathsf{C}_{\nu_0}(\bar{\mathscr{A}}_\lambda(n,r))) \cong \mathsf{C}_\nu(\bar{\mathscr{A}}_\lambda(n,r))$. So there is a natural bijection between the sets of simples in $\mathscr{O}_\nu(\mathsf{C}_{\nu_0}(\bar{\mathscr{A}}_\lambda(n,r)))$ and in $\mathscr{O}_\nu(\bar{\mathscr{A}}_\lambda(n,r))$.

Proposition 5.5. *The number of simples in $\mathscr{O}(\mathsf{C}_{\nu_0}(\bar{\mathscr{A}}_\lambda(n,r)))$ is bigger than or equal to the number of simples in $\mathscr{O}(\Gamma(\mathsf{C}_{\nu_0}(\bar{\mathscr{A}}_\lambda^\theta(n,r))))$.*

Proof. The proof is again in several steps.

Step 1. We have a natural homomorphism $\mathbb{C}[\bar{\mathfrak{g}}]^G \to \bigoplus \bar{\mathscr{A}}_\lambda(n_1,\ldots,n_r;r)$. It can be described as follows. We have an identification $\mathbb{C}[\bar{\mathfrak{g}}]^G \cong \mathbb{C}[\mathfrak{h}]^{\mathfrak{S}_n}$, where we write $\mathfrak{h}$ for a Cartan subalgebra of $\bar{\mathfrak{g}}$. The algebra $\mathbb{C}[\bar{\mathfrak{g}}]^G$ embeds into $\bar{\mathscr{A}}_\lambda(n_1,\ldots,n_r;r)$, which is a spherical Cherednik algebra for the group $\prod_{i=1}^r \mathfrak{S}_{n_i}$ acting on $\mathfrak{h}$, via the inclusion $\mathbb{C}[\mathfrak{h}]^{\mathfrak{S}_n} \subset \mathbb{C}[\mathfrak{h}]^{\mathfrak{S}_{n_1}\times\cdots\times\mathfrak{S}_{n_r}}$. For the homomorphism $\mathbb{C}[\bar{\mathfrak{g}}]^G \to \bigoplus \bar{\mathscr{A}}_\lambda(n_1,\ldots,n_r;r)$ we take the direct sum of these embeddings. In a similar way to Steps 4 and 5 in the proof of Proposition 5.4, one shows that the maps $\mathbb{C}[\bar{\mathfrak{g}}]^G \to \mathsf{C}_{\nu_0}(\bar{\mathscr{A}}_\lambda(n,r))$, $\Gamma(\mathsf{C}_{\nu_0}(\bar{\mathscr{A}}^\theta_\lambda(n,r)))$ are intertwined by the homomorphism $\mathsf{C}_{\nu_0}(\bar{\mathscr{A}}_\lambda(n,r)) \to \Gamma(\mathsf{C}_{\nu_0}(\bar{\mathscr{A}}^\theta_\lambda(n,r)))$.

Step 2. Let $\delta \in \mathbb{C}[\bar{\mathfrak{g}}]^G$ be the discriminant. We claim that $\mathsf{C}_{\nu_0}(\bar{\mathscr{A}}_\lambda(n,r))[\delta^{-1}] \xrightarrow{\sim} \Gamma(\mathsf{C}_{\nu_0}(\bar{\mathscr{A}}^\theta_\lambda(n,r)))[\delta^{-1}]$. Since δ is $\nu_0(\mathbb{C}^\times)$-stable, we have $\mathsf{C}_{\nu_0}(\bar{\mathscr{A}}_\lambda(n,r))[\delta^{-1}] = \mathsf{C}_{\nu_0}(\bar{\mathscr{A}}_\lambda(n,r)[\delta^{-1}])$. We will describe the algebra $\mathsf{C}_{\nu_0}\left(\bar{\mathscr{A}}_\lambda(n,r)[\delta^{-1}]\right)$ explicitly and see that $\mathsf{C}_{\nu_0}(\bar{\mathscr{A}}_\lambda(n,r)[\delta^{-1}]) \xrightarrow{\sim} \Gamma(\mathsf{C}_{\nu_0}(\bar{\mathscr{A}}^\theta_\lambda(n,r)))[\delta^{-1}]$.

Step 3. We now start describing $\bar{\mathscr{A}}_\lambda(n,r)[\delta^{-1}]$. Let $\bar{\mathfrak{g}}^{reg}$ denote the locus of the regular semisimple elements in $\bar{\mathfrak{g}}$. Then

$$\bar{\mathscr{A}}_\lambda(n,r)[\delta^{-1}] = D(\bar{\mathfrak{g}}^{reg} \times \operatorname{Hom}(\mathbb{C}^n,\mathbb{C}^r))///_\lambda G.$$

Here $///_\lambda$ denotes the quantum Hamiltonian reduction with parameter λ.

Recall that $\bar{\mathfrak{g}}^{reg} = G \times_{N_G(\mathfrak{h})} \mathfrak{h}^{reg}$, and so $\bar{\mathfrak{g}}^{reg} \times \operatorname{Hom}(\mathbb{C}^n,\mathbb{C}^r) = G \times_{N_G(\mathfrak{h})} (\mathfrak{h}^{reg} \times \operatorname{Hom}(\mathbb{C}^n,\mathbb{C}^r))$. It follows that

$$\begin{aligned} D(\bar{\mathfrak{g}}^{reg} \times \operatorname{Hom}(\mathbb{C}^n,\mathbb{C}^r))///_\lambda G &= D(\mathfrak{h}^{reg} \times \operatorname{Hom}(\mathbb{C}^n,\mathbb{C}^r))///_\lambda N_G(\mathfrak{h}) \\ &= (D(\mathfrak{h}^{reg}) \otimes D(\operatorname{Hom}(\mathbb{C}^n,\mathbb{C}^r))///_\lambda H)^{\mathfrak{S}_n} = \left(D(\mathfrak{h}^{reg}) \otimes D^\lambda(\mathbb{P}^{r-1})^{\otimes n}\right)^{\mathfrak{S}_n}. \end{aligned}$$

Here, in the second line, we write H for the Cartan subgroup of $G = \mathrm{GL}_n(\mathbb{C})$ and take the diagonal action of $\mathfrak{S}_n$. In the last expression, $\mathfrak{S}_n$ permutes the tensor factors. A similar argument shows that $\bar{\mathscr{M}}^\theta(n,r)_\delta = (T^*(\mathfrak{h}^{reg}) \times T^*(\mathbb{P}^{r-1})^n)/\mathfrak{S}_n$, and the restriction of $\bar{\mathscr{A}}^\theta_\lambda(n,r)$ to this open subset is $\left(D_{\mathfrak{h}^{reg}} \otimes (D^\lambda_{\mathbb{P}^{r-1}})^{\otimes n}\right)^{\mathfrak{S}_n}$.

Step 4. Now we are going to describe the algebra

$$\mathsf{C}_{\nu_0}\left(\left(D(\mathfrak{h}^{reg}) \otimes D^\lambda(\mathbb{P}^{r-1})^{\otimes n}\right)^{\mathfrak{S}_n}\right).$$

First, we claim that

$$\mathsf{C}_{\nu_0}\left(\left(D(\mathfrak{h}^{reg}) \otimes D^\lambda(\mathbb{P}^{r-1})^{\otimes n}\right)^{\mathfrak{S}_n}\right) = \left(\mathsf{C}_{\nu_0}\left(D(\mathfrak{h}^{reg}) \otimes D^\lambda(\mathbb{P}^{r-1})^{\otimes n}\right)\right)^{\mathfrak{S}_n}. \tag{5.2}$$

There is a natural homomorphism from the left-hand side to the right-hand side. To prove that it is an isomorphism, one can argue as follows. First, it is enough to show that its associated graded homomorphism is an isomorphism. Now, note that since the $\mathfrak{S}_n$-action on $\mathfrak{h}^{reg}$ is free, we have

$$\mathbb{C}[T^*\mathfrak{h}^{reg}] \otimes \operatorname{gr} \mathsf{C}_{\nu_0}(D^\lambda(\mathbb{P}^{r-1})^{\otimes n})$$
$$= \mathbb{C}[T^*\mathfrak{h}^{reg}] \otimes_{\mathbb{C}[T^*\mathfrak{h}^{reg}]^{\mathfrak{S}_n}} \left(\mathbb{C}[T^*\mathfrak{h}^{reg}] \otimes \operatorname{gr} \mathsf{C}_{\nu_0}(D^\lambda(\mathbb{P}^{r-1})^{\otimes n})\right)^{\mathfrak{S}_n}.$$

Since $\mathbb{C}[T^*\mathfrak{h}^{reg}]$ is $\nu_0(\mathbb{C}^\times)$-invariant, the previous equality implies (5.2).

Step 5. Now let us describe

$$\mathsf{C}_{\nu_0}\left(D(\mathfrak{h}^{reg}) \otimes D^\lambda(\mathbb{P}^{r-1})^{\otimes n}\right) = D(\mathfrak{h}^{reg}) \otimes \mathsf{C}_{\nu_0}\left((D^\lambda(\mathbb{P}^{r-1}))^{\otimes n}\right).$$

The $\mathbb{C}^\times$-action on the tensor product $(D^\lambda(\mathbb{P}^{r-1}))^{\otimes n}$ is diagonal, and it is easy to see that $\mathsf{C}_{\nu_0}\left((D^\lambda(\mathbb{P}^{r-1}))^{\otimes n}\right) = \left(\mathsf{C}_{\nu_0}(D^\lambda(\mathbb{P}^{r-1}))\right)^{\otimes n}$. So we need to compute $\mathsf{C}_{\nu_0}(D^\lambda(\mathbb{P}^{r-1}))$. We claim that this algebra is isomorphic to $\mathbb{C}^{\oplus r}$. Indeed, $D^\lambda(\mathbb{P}^{r-1})$ is a quotient of the central reduction $U_{\tilde{\lambda}}(\mathfrak{sl}_r)$ of $U(\mathfrak{sl}_r)$ at the central character $\tilde{\lambda} := \lambda\omega_r$. We remark that $\lambda\omega_r + \rho$ is regular because $\lambda \geqslant 0$. We have $\mathsf{C}_{\nu_0}(U_{\tilde{\lambda}}(\mathfrak{sl}_r)) = \mathbb{C}^{\oplus r!}$ and $\mathsf{C}_{\nu_0}(D^\lambda(\mathbb{P}^{r-1}))$ is a quotient of $\mathsf{C}_{\nu_0}(U_{\tilde{\lambda}}(\mathfrak{sl}_r))$. The number of irreducible representations of $\mathsf{C}_{\nu_0}(D^\lambda(\mathbb{P}^{r-1}))$ is equal to the number of simples in the category $\mathcal{O}$ for $D^\lambda(\mathbb{P}^{r-1})$ that coincides with r since abelian localization holds. An isomorphism $\mathsf{C}_{\nu_0}(D^\lambda(\mathbb{P}^{r-1})) = \mathbb{C}^{\oplus r}$ follows.

Step 6. So we see that

$$\mathsf{C}_{\nu_0}(\bar{\mathscr{A}}_\lambda(n,r)[\delta^{-1}]) = \left(D(\mathfrak{h}^{reg}) \otimes (\mathbb{C}^{\oplus r})^{\otimes n}\right)^{\mathfrak{S}_n}.$$

Using similar reasoning, we have

$$\Gamma([\bar{\mathscr{M}}^\theta(n,r)_\delta]^{\nu_0(\mathbb{C}^\times)}, \mathsf{C}_{\nu_0}(\bar{\mathscr{A}}^\theta_\lambda(n,r))) = \left(D(\mathfrak{h}^{reg}) \otimes (\mathbb{C}^{\oplus r})^{\otimes n}\right)^{\mathfrak{S}_n}.$$

The natural homomorphism

$$\mathsf{C}_{\nu_0}(\bar{\mathscr{A}}_\lambda(n,r)[\delta^{-1}]) \to \Gamma((\bar{\mathscr{M}}^\theta(n,r)_\delta)^{\nu_0(\mathbb{C}^\times)}, \mathsf{C}_{\nu_0}(\bar{\mathscr{A}}^\theta_\lambda(n,r))) \tag{5.3}$$

is an isomorphism by the previous two steps. Also we have a natural homomorphism

$$\Gamma(\mathsf{C}_{\nu_0}(\bar{\mathscr{A}}^\theta_\lambda(n,r)))[\delta^{-1}] \to \Gamma\left([\bar{\mathscr{M}}^\theta(n,r)_\delta]^{\nu_0(\mathbb{C}^\times)}, \mathsf{C}_{\nu_0}(\bar{\mathscr{A}}^\theta_\lambda(n,r))\right). \tag{5.4}$$

The latter homomorphism is an isomorphism. Indeed, $\mathsf{C}_{\nu_0}(\bar{\mathscr{A}}^\theta_\lambda(n,r))$ is the direct sum of quantizations of products of Hilbert schemes. The morphism $\prod \operatorname{Hilb}_{n_i}(\mathbb{C}^2) \to \prod\left(\mathbb{C}^{2n_i}/\mathfrak{S}_n\right)$ is an isomorphism over the non-vanishing locus of δ. This implies that (5.4) is an isomorphism.

By the construction of (5.3), it is the composition of

$$\mathsf{C}_{\nu_0}(\bar{\mathscr{A}}_\lambda(n,r)[\delta^{-1}]) \to \Gamma(\mathsf{C}_{\nu_0}(\bar{\mathscr{A}}^\theta_\lambda(n,r)))[\delta^{-1}]$$

and (5.4). So we have proved that $\mathsf{C}_{\nu_0}(\bar{\mathscr{A}}_\lambda(n,r))[\delta^{-1}] \to \Gamma(\mathsf{C}_{\nu_0}(\bar{\mathscr{A}}^\theta_\lambda(n,r)))[\delta^{-1}]$ is an isomorphism.

Step 7. For $p \in \mathcal{M}^\theta(n,r)^T$, let $L^0(p)$ be the corresponding irreducible $\Gamma(\mathsf{C}_{v_0}(\bar{\mathscr{A}}^\theta_\lambda(n,r)))$-module from category $\mathcal{O}$. These modules are either finite-dimensional (those are parameterized by the multi-partitions with one part equal to (n) and others empty) or has support of maximal dimension. It follows from Proposition 5.4 that all finite-dimensional $L^0(p)$'s restrict to pairwise non-isomorphic $\mathsf{C}_{v_0}(\bar{\mathscr{A}}_\lambda(n,r))$-modules. Now consider $L^0(p)$ with support of maximal dimension. We claim that the localizations $L^0(p)[\delta^{-1}]$ are pairwise non-isomorphic simple $\Gamma(\mathsf{C}_{v_0}(\bar{\mathscr{A}}^\theta_\lambda(n,r)))[\delta^{-1}]$-modules. We consider $p = (p^1,\dots,p^r)$ and $p' = (p'^1,\dots,p'^r)$ with $|p^i| = |p'^i|$ for all i and show that

- the localizations of $L^0(p), L^0(p')$ are simple
- and, moreover, are isomorphic only if $p = p'$.

The analogous claims hold if we localize to the regular locus for $\prod_{i=1}^r \mathfrak{S}_{|p^i|}$. Indeed, this localization realizes the KZ functor from [GGOR] that is a quotient onto its image. So the images of $L^0(p)$, $L^0(p')$ under this localization are simple and non-isomorphic. The modules $L^0(p)[\delta^{-1}]$, $L^0(p')[\delta^{-1}]$ are further restrictions of these localizations to the locus where $x_i \neq x_j$ for all i,j. But there is no monodromy of the D-modules in the image of the KZ functor along those additional hyperplanes and these D-modules have regular singularities everywhere. It follows that they remain simple and nonisomorphic (if $p \neq p'$).

Step 8. So we see that the $\mathsf{C}_{v_0}(\bar{\mathscr{A}}_\lambda(n,r))[\delta^{-1}]$-modules $L^0(p)[\delta^{-1}]$ are simple and pair-wise non-isomorphic. The $\mathsf{C}_{v_0}(\bar{\mathscr{A}}_\lambda(n,r))$-module $L^0(p)$ is finitely generated, because the algebra $\Gamma(\mathsf{C}_{v_0}(\bar{\mathscr{A}}^\theta_\lambda(n,r)))$ is a finitely generated $\mathsf{C}_{v_0}(\bar{\mathscr{A}}_\lambda(n,r))$-module. The latter follows from [L7, Lemma 5.4]. So the $\mathsf{C}_{v_0}(\bar{\mathscr{A}}_\lambda(n,r))$-module $L^0(p)$ lies in $\mathcal{O}_v(\mathsf{C}_{v_0}(\bar{\mathscr{A}}_\lambda(n,r)))$ thanks to the weight decomposition. There is a simple constituent $\underline{L}^0(p)$ of the $\mathsf{C}_{v_0}(\bar{\mathscr{A}}_\lambda(n,r))$-module $L^0(p)$ with $\underline{L}^0(p)[\delta^{-1}] = L^0(p)[\delta^{-1}]$ because the right-hand side is simple. The finite-dimensional modules $L^0(p)$, together with the modules of the form $\underline{L}^0(p)$, give a required number of pairwise nonisomorphic simple $\mathsf{C}_{v_0}(\bar{\mathscr{A}}_\lambda(n,r))$-modules. □

5.4 Completion of proofs

The following proposition completes the proof of Theorem 1.1.

Proposition 5.6. *Let λ be a positive parameter with denominator n. Then abelian localization holds for (λ,θ), with $\theta > 0$.*

Proof. The proof is in several steps.

Step 1. Since $\Gamma^\theta_\lambda : \mathcal{O}_v(\bar{\mathscr{A}}^\theta_\lambda(n,r)) \to \mathcal{O}_v(\bar{\mathscr{A}}_\lambda(n,r))$ is a quotient functor (see Section 5.1), to prove that it is an equivalence, it is enough to verify that the number of simples in these two categories is the same. The number of simples in $\mathcal{O}_v(\bar{\mathscr{A}}_\lambda(n,r))$ coincides with the number of simples in $\mathcal{O}(\mathsf{C}_{v_0}(\bar{\mathscr{A}}_v(n,r)))$ by the paragraph before Proposition 5.5. The latter is larger than or equal to the number of simples for $\mathcal{O}(\bigoplus \bar{\mathscr{A}}_\lambda(n_1,\ldots,n_r;r))$ which, in its turn, coincides with the number of r-multipartitions of n because abelian localization holds for all summands $\bar{\mathscr{A}}_\lambda(n_1,\ldots,n_r;r)$. We deduce that the number of simples in $\mathcal{O}_\nu(\bar{\mathscr{A}}_\lambda^\theta(n,r))$ and in $\mathcal{O}_\nu(\bar{\mathscr{A}}_\lambda(n,r))$ coincide. So we see that $\Gamma_\lambda^\theta : \mathcal{O}_\nu(\bar{\mathscr{A}}_\lambda^\theta(n,r)) \twoheadrightarrow \mathcal{O}_\nu(\bar{\mathscr{A}}_\lambda(n,r))$ is an equivalence. Now we are going to show that this implies that $\Gamma_\lambda^\theta : \bar{\mathscr{A}}_\lambda^\theta(n,r)\text{-mod} \to \bar{\mathscr{A}}_\lambda(n,r)\text{-mod}$ is an equivalence. Below we write $\mathcal{O}$ instead of $\mathcal{O}_\nu$. Our argument is similar to the proof of [L6, Theorem 2.1].

Step 2. Since Γ_λ^θ is an equivalence between the categories $\mathcal{O}$, we see that $\bar{\mathscr{A}}^0_{\lambda,\chi}(n,r)\otimes_{\bar{\mathscr{A}}_\lambda(n,r)}\bullet$ and $\bar{\mathscr{A}}^0_{\lambda+\chi,-\chi}(n,r)\otimes_{\bar{\mathscr{A}}_{\lambda+\chi}(n,r)}\bullet$ are mutually inverse equivalences between $\mathcal{O}(\bar{\mathscr{A}}_\lambda(n,r))$ and $\mathcal{O}(\bar{\mathscr{A}}_{\lambda+\chi}(n,r))$ for all $\lambda>0$ and $\chi\in\mathbb{Z}$ such that $(\lambda+\chi,\det)$ satisfies abelian localization (which happens as long as χ is sufficiently large). Set $\mathscr{B} := \bar{\mathscr{A}}^0_{\lambda+\chi,-\chi}(n,r)\otimes_{\bar{\mathscr{A}}_{\lambda+\chi}(n,r)}\bar{\mathscr{A}}^0_{\lambda,\chi}(n,r)$. This is an HC $\bar{\mathscr{A}}_\lambda(n,r)$-bimodule with a natural homomorphism to $\bar{\mathscr{A}}_\lambda(n,r)$ (compare to (5.1)) such that the induced homomorphism $\mathscr{B}\otimes_{\bar{\mathscr{A}}_\lambda(n,r)} M\to M$ is an isomorphism for any $M\in\mathcal{O}(\bar{\mathscr{A}}_\lambda(n,r))$. On the other hand, for any nonzero $x\in\mathscr{M}(n,r)$, the bimodules $\bar{\mathscr{A}}^0_{\lambda+\chi,-\chi}(n,r)_{\dagger,x}$, $\bar{\mathscr{A}}^0_{\lambda,\chi}(n,r)_{\dagger,x}$ are mutually inverse Morita equivalences. This follows from Lemma 3.10 since abelian localization holds for all slice algebras with parameters whose denominator is n. It follows that both kernel and cokernel of $\mathscr{B}\to\bar{\mathscr{A}}_\lambda(n,r)$ are finite-dimensional.

Step 3. Let L denote an irreducible finite-dimensional $\bar{\mathscr{A}}_\lambda(n,r)$-module. It is unique because of the equivalence $\mathcal{O}(\bar{\mathscr{A}}_\lambda(n,r))\cong\mathcal{O}(\bar{\mathscr{A}}_{\lambda+\chi}(n,r))$ and Proposition 4.1. Since the homomorphism $\mathscr{B}\otimes_{\bar{\mathscr{A}}_\lambda(n,r)}L\to L$ is an isomorphism and the cokernel of $\mathscr{B}\to\bar{\mathscr{A}}_\lambda(n,r)$ is finite-dimensional, we see that the cokernel is zero. Let K denote the kernel of $\mathscr{B}\twoheadrightarrow\bar{\mathscr{A}}_\lambda(n,r)$. We have an exact sequence

$$\operatorname{Tor}^1_{\bar{\mathscr{A}}_\lambda(n,r)}(\bar{\mathscr{A}}_\lambda(n,r),L)\to K\otimes_{\bar{\mathscr{A}}_\lambda(n,r)}L\to\mathscr{B}\otimes_{\bar{\mathscr{A}}_\lambda(n,r)}L\to L\to 0.$$

Clearly, the first term is zero, while the last homomorphism is an isomorphism. We deduce that $K\otimes_{\bar{\mathscr{A}}_\lambda(n,r)}L=0$. But K is a finite-dimensional $\bar{\mathscr{A}}_\lambda(n,r)$-bimodule and hence a $\bar{\mathscr{A}}_\lambda(n,r)/\operatorname{Ann}L$-bimodule. So its tensor product with L can only be zero if $K=0$.

Step 4. So we see that $\bar{\mathscr{A}}^0_{\lambda+\chi,-\chi}(n,r)\otimes_{\bar{\mathscr{A}}_{\lambda+\chi}(n,r)}\bar{\mathscr{A}}^0_{\lambda,\chi}(n,r)\cong\bar{\mathscr{A}}_\lambda(n,r)$. Similarly, $\bar{\mathscr{A}}^0_{\lambda,\chi}(n,r)\otimes_{\bar{\mathscr{A}}_\lambda(n,r)}\bar{\mathscr{A}}^0_{\lambda+\chi,-\chi}(n,r)\cong\bar{\mathscr{A}}_{\lambda+\chi}(n,r)$. It follows that Γ_λ^θ is an equivalence between $\operatorname{Coh}(\bar{\mathscr{A}}_\lambda^\theta(n,r))$ and $\bar{\mathscr{A}}_\lambda(n,r)\text{-mod}$. □

So Theorem 1.1 is proved.

Proof of Theorem 1.2. Let L denote a finite-dimensional irreducible representation of $\bar{\mathscr{A}}_\lambda(n,r)$. Since $L\operatorname{Loc}_\lambda^\theta(\bar{\mathscr{A}}_\lambda(n,r))=\bar{\mathscr{A}}_\lambda^\theta(n,r)$ and $R\Gamma_\lambda^\theta(\bar{\mathscr{A}}_\lambda^\theta(n,r))=\bar{\mathscr{A}}_\lambda(n,r)$, we see that $R\Gamma_\lambda^\theta\circ L\operatorname{Loc}_\lambda^\theta$ is the identity endofunctor of $D^-(\bar{\mathscr{A}}_\lambda(n,r)\text{-mod})$. The homology of $L\operatorname{Loc}_\lambda^\theta(L)$ is supported on $\bar{\rho}^{-1}(0)$. From Proposition 4.1 it follows that the denominator of λ is n. Thanks to Proposition 4.1, combined with Theorem

1.1, the present proof reduces to showing that for λ between $-r$ and 0 and with denominator n, the algebra $\bar{\mathscr{A}}_\lambda(n,r)$ has no finite-dimensional representations.

Recall that Γ_λ^θ is an exact functor. Since $R\Gamma_\lambda^\theta \circ L\mathrm{Loc}_\lambda^\theta$ is the identity, the functor Γ_λ^θ does not kill the simple $\bar{\mathscr{A}}_\lambda^\theta(n,r)$-module $\tilde{L}$ supported on $\bar{\rho}^{-1}(0)$. On the other hand, Γ_λ^θ does not kill modules whose support intersects $\bar{\mathscr{M}}^\theta(n,r)^{reg}$, the open subvariety in $\bar{\mathscr{M}}^\theta(n,r)$, where $\bar{\rho}$ is an isomorphism. In fact, every simple in $\mathscr{O}(\bar{\mathscr{A}}_\lambda^\theta(n,r))$ is either supported on $\bar{\rho}^{-1}(0)$ (if it is homologically shifted under the wall-crossing functor so that the global sections are finite dimensional) or its support intersects $\bar{\mathscr{M}}^\theta(n,r)^{reg}$ (if it is not). Indeed, the support of a simple object in $\mathscr{O}(\bar{\mathscr{A}}_{\lambda+\chi}(n,r))$ has dimension 0 or $\frac{1}{2}\dim \mathscr{M}(n,r)$. In the latter case it intersects the open leaf in $\mathscr{M}(n,r)$ by Lemma 2.3.

So we see that Γ_λ^θ does not kill any irreducible module in $\mathscr{O}(\bar{\mathscr{A}}_\lambda^\theta(n,r))$. So it is an equivalence. By the proof of Proposition 5.6, (λ,θ) satisfies abelian localization, which is impossible by Corollary 5.2. □

6 Two-sided ideals and dimensions of supports

6.1 Two-sided ideals

The goal of this section is to prove Theorem 1.3. We use the following notation. Let $\lambda \in \mathbb{C}$. Assume that λ has a denominator larger than n (or is irrational) or $\lambda \notin (-r,0)$. We write $\mathscr{A}$ for $\bar{\mathscr{A}}_\lambda(n,r)$. In the case when the denominator m of λ satisfies $m \leqslant n$, we write $\underline{\mathscr{A}}$ for $\bar{\mathscr{A}}_\lambda(m,r)$.

We start with classifying the two-sided ideals in $\underline{\mathscr{A}}$.

Lemma 6.1. *Let $m \leqslant n$. Then there is a unique proper ideal in $\underline{\mathscr{A}}$.*

Proof. The proper slice algebras for $\underline{\mathscr{A}}$ have no finite-dimensional representations, compare to the proof of Proposition 4.1. So every ideal $\mathscr{J} \subset \underline{\mathscr{A}}$ is either of finite codimension or $\mathrm{V}(\underline{\mathscr{A}}/\mathscr{J}) = \bar{\mathscr{M}}(m,r)$. The algebra $\underline{\mathscr{A}}$ has no zero divisors, so the second option is only possible when $\mathscr{J} = \{0\}$. Now suppose that $\mathscr{J}$ is of finite codimension. Then $\underline{\mathscr{A}}/\mathscr{J}$ (viewed as a left $\underline{\mathscr{A}}$-module) is the sum of several copies of the finite-dimensional irreducible $\underline{\mathscr{A}}$-module. So $\mathscr{J}$ coincides with the annihilator of the finite-dimensional irreducible module, and we are done. □

Let $\underline{\mathscr{J}}$ denote the unique proper two-sided ideal of $\underline{\mathscr{A}}$.

Now we are going to classify two-sided ideals in $\underline{\mathscr{A}}^{\otimes k}$. For this we need some notation. Set $\underline{\mathscr{I}}_i := \underline{\mathscr{A}}^{\otimes i-1} \otimes \underline{\mathscr{J}} \otimes \underline{\mathscr{A}}^{\otimes k-i-1}$. For a subset $\Lambda \subset \{1,\ldots,k\}$, define the ideals $\underline{\mathscr{I}}_\Lambda := \sum_{i\in\Lambda} \underline{\mathscr{I}}_i$, $\underline{\mathscr{I}}^\Lambda := \prod_{i\in\Lambda} \underline{\mathscr{I}}_i$.

Recall that a collection of subsets in $\{1,\ldots,k\}$ is called an *anti-chain* if none of these subsets is contained in another. Also recall that an ideal I in an associative algebra A is called *semiprime* if it is the intersection of prime ideals.

Lemma 6.2. *The following is true.*

(1) *Prime ideals in* $\underline{\mathscr{A}}^{\otimes k}$ *are precisely the ideals* $\underline{\mathscr{I}}_\Lambda$.
(2) *For every ideal* $\mathscr{I} \subset \underline{\mathscr{A}}^{\otimes k}$, *there is a unique anti-chain* $\Lambda_1,\ldots,\Lambda_q$ *of subsets in* $\{1,\ldots,k\}$ *such that* $\mathscr{I} = \bigcap_{i=1}^{p} \mathscr{I}_{\Lambda_i}$. *In particular, every ideal is semiprime.*
(3) *For every ideal* $\mathscr{I} \subset \underline{\mathscr{A}}^{\otimes k}$, *there is a unique anti-chain* $\Lambda'_1,\ldots,\Lambda'_q$ *of subsets of* $\{1,\ldots,k\}$ *such that* $\mathscr{I} = \sum_{i=1}^{q} \mathscr{I}^{\Lambda'_i}$.
(4) *The anti-chains in* (2) *and* (3) *are related as follows: from an antichain in* (2), *we form all possible subsets containing an element from each of* $\Lambda_1,\ldots,\Lambda_p$. *The minimal such subsets form an anti-chain in* (3).

The proof is very similar to [L1, Section 5.8]; we provide it for reader's convenience.

Proof. We prove (1). Let $\mathscr{I}$ be a prime ideal. Let x be a generic point in an open leaf $\mathscr{L} \subset \mathrm{V}(\underline{\mathscr{A}}^{\otimes k}/\mathscr{I})$ of maximal dimension. The corresponding slice algebra $\underline{\mathscr{A}}'$ has a finite-dimensional representation, and so is again the product of several copies of $\underline{\mathscr{A}}$. The leaf $\mathscr{L}$ is therefore the product of one-point leaves and open leaves in $\bar{\mathscr{M}}(m,r)$. An irreducible finite-dimensional representation of $\underline{\mathscr{A}}'$ is unique; let $\mathscr{I}'$ be its annihilator.

Consider the categories $\mathrm{HC}_{\overline{\mathscr{L}}}(\underline{\mathscr{A}}^{\otimes k})$ of all HC $\underline{\mathscr{A}}^{\otimes k}$-bimodules whose associated variety is contained in $\overline{\mathscr{L}}$ and $\mathrm{HC}_{fin}(\underline{\mathscr{A}}')$ of finite-dimensional $\underline{\mathscr{A}}'$-bimodules (that are automatically HC). So the functor $\bullet_{\dagger,x}$ restricts to $\mathrm{HC}_{\overline{\mathscr{L}}}(\underline{\mathscr{A}}^{\otimes k}) \to \mathrm{HC}_{fin}(\underline{\mathscr{A}}')$. Note that $\mathscr{I}_\dagger$ is an ideal of finite codimension in $\underline{\mathscr{A}}'$.

As we have mentioned in Section 3.4, the functor $\bullet_{\dagger,x}$ admits a right adjoint

$$\bullet^{\dagger,x}\colon \mathrm{HC}_{fin}(\underline{\mathscr{A}}') \to \mathrm{HC}_{\overline{\mathscr{L}}}(\underline{\mathscr{A}}^{\otimes k}).$$

Let $\mathscr{I}^1$ denote the kernel of the natural homomorphism $\underline{\mathscr{A}}^{\otimes k} \to (\underline{\mathscr{A}}'/\mathscr{I}')^{\dagger,x}$. It follows that $\mathscr{I} \subset \mathscr{I}^1$. By the construction of $\mathscr{I}^1$, we see that $\mathrm{V}(\underline{\mathscr{A}}^{\otimes k}/\mathscr{I}^1) = \overline{\mathscr{L}} = \mathrm{V}(\underline{\mathscr{A}}^{\otimes k}/\mathscr{I})$. It follows from [BoKr, Corollary 3.6] that $\mathscr{I} = \mathscr{I}^1$. So the number of prime ideals coincides with that of the non-empty subsets $\{1,\ldots,k\}$. On the other hand, the ideals $\mathscr{I}^\Lambda$ are all different (they have different associated varieties) and are all prime (the quotient $\underline{\mathscr{A}}^{\otimes k}/\mathscr{I}^\Lambda$ is the product of a matrix algebra and the algebra $\underline{\mathscr{A}}^{\otimes k-|\Lambda|}$ that has no zero divisors). This proves (1).

Let us prove (2) (and simultaneously (3)). We write $\mathscr{I}_{\Lambda_1,\ldots,\Lambda_p}$ for $\bigcap_{j=1}^{s} \mathscr{I}_{\Lambda_j}$. For ideals in $\underline{\mathscr{A}}^{\otimes k-1}$, we use a notation like $\underline{\mathscr{I}}_{\Lambda'_1,\ldots,\Lambda'_q}$. Reordering the indexes, we may assume that $k \in \Lambda_1,\ldots,\Lambda_s$ and $k \notin \Lambda_{s+1},\ldots,\Lambda_p$. Set $\Lambda'_j := \Lambda_j \setminus \{k\}$ for $j \leqslant s$. Then

$$\mathscr{I}_{\Lambda_1,\ldots,\Lambda_p} = (\underline{\mathscr{A}}^{\otimes k-1} \otimes \underline{\mathscr{J}} + \underline{\mathscr{I}}_{\Lambda'_1,\ldots,\Lambda'_s} \otimes \underline{\mathscr{A}}) \cap (\underline{\mathscr{I}}_{\Lambda_{s+1},\ldots,\Lambda_p} \otimes \underline{\mathscr{A}}). \tag{6.1}$$

We claim that the right-hand side of (6.1) coincides with

$$\underline{\mathscr{I}}_{\Lambda_{s+1},\ldots,\Lambda_p} \otimes \underline{\mathscr{J}} + \underline{\mathscr{I}}_{\Lambda'_1,\ldots,\Lambda'_s,\Lambda_{s+1},\ldots,\Lambda_p} \otimes \underline{\mathscr{A}}. \tag{6.2}$$

First, we notice that (6.2) is contained in (6.1). So we only need to prove the opposite inclusion. The projection of (6.1) to $\underline{\mathscr{A}}^{\otimes k-1} \otimes (\underline{\mathscr{A}}/\underline{\mathscr{J}})$ is contained in

$\underline{\mathscr{I}}_{\Lambda'_1,\ldots,\Lambda'_s,\Lambda_{s+1},\ldots,\Lambda_p}$ and hence also in the projection of (6.2). Also the intersection of (6.1) with $\underline{\mathscr{A}}^{\otimes k-1}\otimes\underline{\mathscr{J}}$ is contained in $\underline{\mathscr{I}}_{\Lambda_{s+1},\ldots,\Lambda_p}\otimes\underline{\mathscr{J}}$. So (6.1) is included into (6.2).

Repeating this argument with the sum similar to (6.2), but for the other $k-1$ factors of $\underline{\mathscr{A}}^{\otimes k}$, we conclude that $\mathscr{I}_{\Lambda_1,\ldots,\Lambda_p}=\sum_j \mathscr{I}^{\Lambda'_j}$, where the subsets $\Lambda'_j\subset\{1,\ldots,k\}$ are formed as described in (4). So we see that the ideals (2) are the same as the ideals in (3) and that (4) holds. What remains to do is to prove that every ideal has the form described in (2). To start with, we notice that every semiprime ideal has the form, as in (2), because of (1). In particular, the radical of any ideal has such a form.

Clearly, $\mathscr{I}^{\Lambda'_1}\mathscr{I}^{\Lambda'_2}=\mathscr{I}^{\Lambda'_1\cup\Lambda'_2}$. So it follows that any sum of the ideals $\mathscr{I}^{\Lambda'_j}$ coincides with its square. So if $\mathscr{I}$ is an ideal whose radical is $\mathscr{I}_{\Lambda_1,\ldots,\Lambda_p}$, then $\mathscr{I}$ coincides with its radical. This completes the proof. □

Now we are ready to establish a result that will imply Theorem 1.3 for non-singular parameters λ. Let $x_i\in\bar{\mathscr{M}}(n,r)$ be a point corresponding to the leaf with slice $\bar{\mathscr{M}}(m,r)^i$ (i.e., to the semisimple representations of the form $r^0\oplus(r^1)^m\oplus\cdots\oplus(r^i)^m$). We set $\mathscr{J}_i=\ker[\mathscr{A}\to\left((\underline{\mathscr{A}}/\underline{\mathscr{I}})^{\otimes i}\right)^{\dagger,x_i}]$.

Proposition 6.3. *The ideals $\mathscr{J}_i$, $i=1,\ldots,q$, have the following properties:*

(1) *The ideal $\mathscr{J}_i$ is prime for any i.*
(2) $\mathrm{V}(\mathscr{A}/\mathscr{J}_i)=\overline{\mathscr{L}}_i$, *where $\mathscr{L}_i$ is the symplectic leaf containing x_i.*
(3) $\mathscr{J}_1\subsetneq\mathscr{J}_2\subsetneq\cdots\subsetneq\mathscr{J}_q$.
(4) *Any proper two-sided ideal in $\mathscr{A}$ is one of $\mathscr{J}_i$.*

Proof. (2) follows from the construction of the ideals $\mathscr{J}_i$. Also from the construction it follows that $\mathscr{J}_i$ is the maximal among ideals with given associated variety. So it is prime; this is shown in a way similar to the proof of (1) of Lemma 6.2. This proves (1) of the proposition.

Let us prove (3). Since $(\mathscr{J}_i)_{\dagger,x_i}$ has finite codimension, we see that it coincides with the maximal ideal in $\underline{\mathscr{A}}^{\otimes i}$. So $(\mathscr{J}_j)_{\dagger,x_i}\subset(\mathscr{J}_i)_{\dagger,x_i}$ for $j<i$. Again from the construction of $\mathscr{J}_i$, it follows that $\mathscr{J}_j\subsetneq\mathscr{J}_i$.

Let us prove (4). The functor $\bullet_{\dagger,x_q}$ is faithful. Indeed, otherwise we have an HC bimodule $\mathscr{M}$ with $\mathrm{V}(\mathscr{M})\cap\mathscr{L}_q=\varnothing$. But $\mathscr{M}_{\dagger,x}$ has to be nonzero finite dimensional for some x and this is only possible when $x\in\mathscr{L}_i$ for some i. Since $\mathscr{L}_q\subset\overline{\mathscr{L}_i}$ for all i, this shows that $\bullet_{\dagger,x_q}$ is faithful. But this functor is also exact, so it embeds the lattice of the ideals in $\mathscr{A}$ into the lattice of ideals in $\underline{\mathscr{A}}^{\otimes q}$. We claim that this implies that every ideal in $\mathscr{A}$ is semiprime. Indeed, the functor $\bullet_{\dagger,x_q}$ is, in addition, tensor, and so preserves products of ideals. By (2) of Lemma 6.2, every two-sided ideal in $\underline{\mathscr{A}}^{\otimes q}$ coincides with its square. It follows that the same is true for two-sided ideals in $\mathscr{A}$. So every two-sided ideal in $\mathscr{A}$ is semiprime. But every prime ideal in $\mathscr{A}$ is some $\mathscr{J}_i$; this is proved analogously to (1) of Lemma 6.2. Since the ideals $\mathscr{J}_i$ form a chain, any semiprime ideal is prime and so coincides with some $\mathscr{J}_i$. □

Proof of Theorem 1.3. The case of finite homological dimension follows from Proposition 6.3. Let us now consider the case when $\lambda = \frac{s}{m}$ with $-rm < s < 0$. The algebra $\bar{\mathscr{A}}_\lambda(n,r)$ has no finite-dimensional representations and neither does any of the slice algebras $\bigotimes_{i=1}^k \bar{\mathscr{A}}_\lambda(n_i,r)$. By using the restriction functors (as in Step 3 in the proof of Proposition 4.1) we see that the algebra $\bar{\mathscr{A}}_\lambda(n,r)$ is simple. □

6.2 Restriction functors for asymptotic chamber

In this section we assume that ν is dominant, which means that

$$\nu(t) = (\mathrm{diag}(t^{d_1},\ldots,t^{d_r}),t)$$

with $d_1 \gg d_2 \gg \cdots \gg d_r$. We also assume that $\lambda > 0$ and is Zariski generic.

Let $\tau = (\tau_1,\ldots,\tau_k)$ be a partition of n. Set $\mathscr{O}_{\nu,\lambda}(\tau,r) := \boxtimes_{i=1}^k \mathscr{O}_\nu(\mathscr{A}_\lambda(\tau_i,r))$. We will produce exact functors $\mathrm{Res}_\tau\colon \mathscr{O}_{\nu,\lambda}(n,r) \to \mathscr{O}_{\nu,\lambda}(\tau,r)$ generalizing the Bezrukavnikov-Etingof functors, [BE], defined for $r=1$.

The filtration by order of the differential operators on $D(R)$ induces a filtration on $\mathscr{A}_\lambda(n,r)$. The degree zero part is $\mathbb{C}[R]^G = \mathbb{C}[\mathfrak{g}]^G$. Pick a point $b \in \mathfrak{g}/\!/G$ such that the stabilizer of the corresponding closed orbit is $\prod_{i=1}^k \mathrm{GL}(\tau_i)$ and consider the tensor product $\mathscr{A}_\lambda(n,r)^{\wedge_b} := \mathbb{C}[\mathfrak{g}/\!/G]^{\wedge_b} \otimes_{\mathbb{C}[\mathfrak{g}]^G} \mathscr{A}_\lambda(n,r)$. Since the adjoint action of $\mathbb{C}[\mathfrak{g}]^G$ is locally nilpotent, $\mathscr{A}_\lambda(n,r)^{\wedge_b}$ is naturally an algebra. Set $\mathscr{A}_\lambda(\tau,r)^{\wedge_0} := \boxtimes_{i=1}^k \mathscr{A}_\lambda(\tau_i,r)^{\wedge_0}$.

Lemma 6.4. *We have a* $\mathrm{GL}(r)$*-equivariant isomorphism of filtered algebras*

$$\vartheta\colon \mathscr{A}_\lambda(n,r)^{\wedge_b} \xrightarrow{\sim} \boxtimes_{i=1}^k \mathscr{A}_\lambda(\tau_i,r)^{\wedge_0}$$

(*the action on the right-hand side is diagonal*).

Proof. Consider the Rees algebras $\mathscr{A}_{\lambda\hbar}(n,r), \mathscr{A}_{\lambda\hbar}(\tau,r)$ and their full completions $\mathscr{A}_{\lambda\hbar}(n,r)^{\wedge_b}$ (at the point with closed G-orbit given by a diagonal matrix in $\mathfrak{g} \subset \mu^{-1}(0)$ corresponding to b) and $\mathscr{A}_{\lambda\hbar}(\tau,r)^{\wedge_0}$. As was mentioned in Section 3.3, see (3.2), we have a $\mathbb{C}[\hbar]$-linear isomorphism $\mathscr{A}_{\lambda\hbar}(n,r)^{\wedge_b} \cong \mathscr{A}_{\lambda\hbar}(\tau,r)^{\wedge_0}$ that can be made $\mathrm{GL}(r) \times \mathbb{C}^\times$-equivariant (here $\mathbb{C}^\times$ is the contracting action) because we take completions at $\mathrm{GL}(r) \times \mathbb{C}^\times$-stable points. By taking $\mathbb{C}^\times$-finite parts and taking the quotients by $\hbar - 1$, we get an isomorphism in the lemma. □

We will use an isomorphism from Lemma 6.4 to produce a functor Res_τ. First, we need to establish an equivalence of the category $\mathscr{O}_{\nu,\lambda}(\tau,r)$ with a certain category of $\mathscr{A}_\lambda(\tau,r)^{\wedge_0}$-modules. As usual, set $\nu_0(t) = (\mathrm{diag}(t^{d_1},\ldots,t^{d_r}),1)$. We consider the category $\mathscr{O}_{\nu,\lambda}(\tau,r)^{\wedge_0}$ consisting of all finitely generated $\mathscr{A}_\lambda(\tau,r)^{\wedge_0}$-modules such that $h_0 = d_1\nu_0$ acts locally finitely with eigenvalues bounded from above and generalized eigenspaces that are finitely generated over $\mathbb{C}[\mathfrak{g}_\tau/\!/G_\tau]^{\wedge_0}$, where $\mathfrak{g}_\tau$ is the standard Levi subalgebra of $\mathfrak{g}$ corresponding to τ. Note that all generalized

h_0-eigenspaces in a module from $\mathscr{O}_{\nu,\lambda}(n,r)$ are finitely generated over $\mathbb{C}[\mathfrak{g}_\tau]^{G_\tau}$. So we get an exact functor

$$N \mapsto N^{\wedge_0} := \mathbb{C}[\mathfrak{g}_\tau//G_\tau]^{\wedge_0} \otimes_{\mathbb{C}[\mathfrak{g}_\tau]^{G_\tau}} N \colon \mathscr{O}_{\nu,\lambda}(\tau,r) \to \mathscr{O}_{\nu,\lambda}(\tau,r)^{\wedge_0}.$$

Lemma 6.5. *The functor* $\bullet^{\wedge_0}$ *is a category equivalence. A quasi-inverse functor is given by taking the h-finite elements, where h is the image of* 1 *under the quantum comoment map for* ν.

Proof. Let N'_{fin} stand for the space of h-finite elements. It is easy to see that N'_{fin} is the sum of modules from $\mathscr{O}_{\nu,\lambda}(\tau,r)$. Note that all simultaneous generalized eigenspaces for (h,h_0) are finite-dimensional. This is because the generalized eigenspaces for h_0 are finitely generated modules over $\mathbb{C}[\mathfrak{g}_\tau//G_\tau]^{\wedge_0}$ and the generalized eigenspaces for h in such modules are finite-dimensional. By (4) of [L7, Lemma 4.1], N'_{fin} actually lies in $\mathscr{O}$. Also N'_{fin} is dense in N'. Now the proof is easy. □

We note that for $M \in \mathscr{O}_\nu(\mathscr{A}_\lambda(n,r))$, the $\mathscr{A}_\lambda(\tau,r)^{\wedge_0}$-module

$$\vartheta_*\left(\mathbb{C}[\mathfrak{g}//G]^{\wedge_b} \otimes_{\mathbb{C}[\mathfrak{g}]^G} M\right)$$

lies in $\mathscr{O}_{\nu,\lambda}(\tau,r)^{\wedge_0}$. We define $\mathrm{Res}_\tau(M)$ by $\left[\vartheta_*\left(\mathbb{C}[\mathfrak{g}//G]^{\wedge_b} \otimes_{\mathbb{C}[\mathfrak{g}]^G} M\right)\right]_{fin}$. This is an exact functor by construction.

Now let us study the properties of Res_τ.

First, the restriction functor behaves nicely on the level of associated varieties. The following result is a straightforward consequence of the construction; compare to (2) of Proposition 3.9.

Lemma 6.6. *The support* $\mathrm{Supp}(\mathrm{Res}_\tau(M))$ *is a unique conical (with respect to the contracting* $\mathbb{C}^\times$*-action) subvariety in* $\mathscr{M}(\tau,r)$ *such that*

$$\mathrm{Supp}(\mathrm{Res}_\tau(M)) \cap \mathscr{M}(\tau,r)^{\wedge_0} = \mathrm{Supp}(M) \cap \mathscr{M}(n,r)^{\wedge_b}$$

(where we consider the full completions).

Now let us show that in a proper sense, the restriction commutes with the parabolic induction. We can consider the functor

$$\underline{\mathrm{Res}}_\tau \colon \mathscr{O}_\nu(\mathsf{C}_{\nu_0}(\mathscr{A}_\lambda(n,r))) \to \mathscr{O}_\nu(\mathsf{C}_{\nu_0}(\mathscr{A}_\lambda(\tau,r))).$$

It is defined in the same way as Res_τ. On the summand corresponding to a composition $\mu = (n_1,\ldots,n_r)$, the functor $\underline{\mathrm{Res}}_\tau$ coincides with the direct sum of suitable Bezrukavnikov-Etingof restriction functors. More precisely, the corresponding summand of $\mathsf{C}_{\nu_0}(\mathscr{A}_\lambda(n,r))$ is $\mathscr{A}_\lambda(n_1,1) \otimes \mathscr{A}_{\lambda+1}(n_2,1) \otimes \cdots \otimes \mathscr{A}_{\lambda+r-1}(n_r,1)$. Then

$$\underline{\mathrm{Res}}_\tau(M) = \bigoplus_{S_\mu/W'} \mathrm{Res}_{W'}^{S_\mu} M, \tag{6.3}$$

where the summation is taken over all S_μ-orbits on S/S_τ; here W' stands for the standard parabolic stabilizer of the orbit. We write $\mathrm{Res}^{S_\mu}_{W'}$ for the Bezrukavnikov-Etingof restriction functor to the parabolic subgroup W' (or, more precisely, a version, where we do not change the space $\mathfrak{h}$).

We still have the Verma module functor

$$\Delta_{\nu_0}: \mathscr{O}_\nu(\mathsf{C}_{\nu_0}(\mathscr{A}_\lambda(\tau,r))) \to \mathscr{O}_\nu(\mathscr{A}_\lambda(\tau,r)).$$

Lemma 6.7. *We have isomorphisms of functors*

$$\Delta_{\nu_0} \circ \underline{\mathrm{Res}}_\tau \cong \mathrm{Res}_\tau \circ \Delta_{\nu_0}, \tag{6.4}$$

$$\nabla_{\nu_0} \circ \underline{\mathrm{Res}}_\tau \cong \mathrm{Res}_\tau \circ \nabla_{\nu_0}. \tag{6.5}$$

Proof. Recall that by the construction, Res_τ is isomorphic to the completion functor $\bullet^{\wedge_b}$.

Let us prove (6.4). Note first that

$$\begin{aligned}&\mathscr{A}_\lambda(n,r)^{\wedge_b}/\mathscr{A}_\lambda(n,r)^{\wedge_b}\left(\mathscr{A}_\lambda(n,r)^{\wedge_b}\right)^{>0,\nu_0}\\ &\qquad\cong \mathbb{C}[\mathfrak{g}/\!/G]^{\wedge_b}\otimes_{\mathbb{C}[\mathfrak{g}]^G}\left(\mathscr{A}_\lambda(n,r)/\mathscr{A}_\lambda(n,r)\mathscr{A}_\lambda(n,r)^{>0,\nu_0}\right).\end{aligned}$$

We can consider both sides of (6.4) as functors

$$\mathscr{O}_\nu(\mathsf{C}_{\nu_0}(\mathscr{A}_\lambda(n,r))) \to \mathscr{A}_\lambda(n,r)^{\wedge_b}\text{-mod}.$$

Then the right-hand side is given by taking the tensor product over $\mathsf{C}_{\nu_0}(\mathscr{A}_\lambda(n,r))$ with $\mathscr{A}_\lambda(n,r)^{\wedge_b}/\mathscr{A}_\lambda(n,r)^{\wedge_b}(\mathscr{A}_\lambda(n,r)^{\wedge_b})^{>0,\nu_0}$. Since the isomorphism $\mathscr{A}_\lambda(n,r)^{\wedge_b}\cong \mathscr{A}_\lambda(\tau,r)^{\wedge_0}$ is $\nu_0(\mathbb{C}^\times)$-equivariant, we get

$$\begin{aligned}&\mathscr{A}_\lambda(n,r)^{\wedge_b}/\mathscr{A}_\lambda(n,r)^{\wedge_b}\left(\mathscr{A}_\lambda(n,r)^{\wedge_b}\right)^{>0,\nu_0}\\ &\qquad\cong \mathscr{A}_\lambda(\tau,r)^{\wedge_0}/\mathscr{A}_\lambda(\tau,r)^{\wedge_0}\left(\mathscr{A}_\lambda(\tau,r)^{\wedge_0}\right)^{>0,\nu_0}.\end{aligned}$$

The bimodule on the right-hand side coincides with

$$\left(\mathscr{A}_\lambda(\tau,r)/\mathscr{A}_\lambda(\tau,r)\mathscr{A}_\lambda(\tau,r)^{>0,\nu_0}\right)\otimes_{\mathbb{C}[\mathfrak{g}_\tau]^{G_\tau}}\mathbb{C}[\mathfrak{g}_\tau/\!/G_\tau]^{\wedge_0}.$$

It follows that the functor $\Delta_{\nu_0}\circ\underline{\mathrm{Res}}_\tau$ is given by taking the tensor product over $\mathsf{C}_{\nu_0}(\mathscr{A}_\lambda(n,r))$ with $\mathscr{A}_\lambda(\tau,r)^{\wedge_0}/\mathscr{A}_\lambda(\tau,r)^{\wedge_0}(\mathscr{A}_\lambda(\tau,r)^{\wedge_0})^{>0,\nu_0}$. An isomorphism $\mathrm{Res}_\tau\circ\Delta_{\nu_0}\cong\Delta_{\nu_0}\circ\underline{\mathrm{Res}}_\tau$ follows.

Let us proceed to proving (6.5). As in the proof of (6.4), both functors are isomorphic to

$$\begin{aligned}&\mathrm{Hom}_{\mathsf{C}_{\nu_0}(\mathscr{A}_\lambda(n,r)^{\wedge_b})}\Big(\mathscr{A}_\lambda(n,r)^{\wedge_b}/\left(\mathscr{A}_\lambda(n,r)^{\wedge_b}\right)^{<0,\nu_0}\mathscr{A}_\lambda(n,r)^{\wedge_b},\\ &\qquad\qquad \mathsf{C}_{\nu_0}(\mathscr{A}_\lambda(n,r)^{\wedge_b})\otimes_{\mathsf{C}_{\nu_0}(\mathscr{A}_\lambda(n,r))}\bullet\Big),\end{aligned}$$

where we take the restricted Hom (with respect to the natural grading on the first argument). □

For $M \in \mathscr{O}_v(\mathsf{C}_{v_0}(\mathscr{A}_\lambda(\tau,r)))$, we write $L_{v_0}(M)$ for the maximal quotient of $\Delta_{v_0}(M)$ which does not intersect the highest weight subspace M. Equivalently, $L_{v_0}(M)$ is the image of the natural homomorphism $\Delta_{v_0}(M) \to \nabla_{v_0}(M)$ (induced by the identity map $M \to M$).

The following corollary of Lemma 6.7 will play a crucial role in computing the annihilators of simple objects in $\mathscr{O}_{v,\lambda}(n,r)$.

Corollary 6.8. *For* $M \in \mathscr{O}_v(\mathsf{C}_{v_0}(\mathscr{A}_\lambda(n,r)))$, *we have*

$$\mathrm{Res}_\tau(L_{v_0}(M)) \cong L_{v_0}(\underline{\mathrm{Res}}_\tau M). \tag{6.6}$$

Proof. The proof reduces to the case when $M \in \mathscr{O}_v(\mathsf{C}_{v_0}(\mathscr{A}_\lambda(n,r))|_\mu)$, where $\mathsf{C}_{v_0}(\mathscr{A}_\lambda(n,r))|_\mu$ denotes the summand of $\mathsf{C}_{v_0}(\mathscr{A}_\lambda(n,r))$ corresponding to a composition μ. The natural map $\Delta_{v_0}(M) \to \nabla_{v_0}(M)$ gives rise to a map $\mathrm{Res}_\tau \circ \Delta_{v_0}(M) \to \mathrm{Res}_\tau \circ \nabla_{v_0}(M)$. This map is the identity on the highest weight spaces for h_0 because of the construction of Res_τ. The identifications

$$\mathrm{Res}_\tau \circ \Delta_{v_0}(M) \cong \Delta_{v_0} \circ \underline{\mathrm{Res}}_\tau(M), \quad \mathrm{Res}_\tau \circ \nabla_{v_0}(M) \cong \nabla_{v_0} \circ \underline{\mathrm{Res}}_\tau(M)$$

are the identity on the highest weight spaces, by the proof of Lemma 6.7. So applying the functor Res_τ to the morphism $\Delta_{v_0}(M) \to \nabla_{v_0}(M)$, we get the natural morphism $\Delta_{v_0} \circ \underline{\mathrm{Res}}_\tau(M) \to \nabla_{v_0} \circ \underline{\mathrm{Res}}_\tau(M)$. Since Res_τ is exact, (6.6) follows. □

To finish this section let us study the interaction of Res_τ with highest weight structures.

Proposition 6.9. *The functor* Res_τ *maps standard objects to standardly filtered ones, costandard objects to costandardly filtered ones, tiltings to tiltings, projectives to projectives, injectives to injectives.*

Proof. The proof is in several steps.

Step 1. Let us show that Res_τ maps standard objects to standardly filtered ones. Lemma 6.7 reduces this statement to showing that $\underline{\mathrm{Res}}_\tau$ maps standard objects to standardly filtered ones. This property of $\underline{\mathrm{Res}}_\tau$ follows from [S, Section 2].

Step 2. Similarly, we see that Res_τ maps costandard objects into costandardly filtered ones. This is based on the similar claim for $\underline{\mathrm{Res}}_\tau$, which follows from [L3]. Because of this, Res_τ sends tiltings to tiltings.

Step 3. Let us show that Res_τ maps injectives to injectives. The functor

$$\mathfrak{WC}_{\lambda\leftarrow\lambda^-} = \mathscr{A}^0_{\lambda^-,\lambda-\lambda^-}(n,r) \otimes^L_{\mathscr{A}_{\lambda^-}(n,r)} \bullet$$

is a Ringel duality functor, and hence induces equivalences $\mathscr{O}_v(\mathscr{A}_{\lambda^-}(n,r))\text{-tilt} \xrightarrow{\sim} \mathscr{O}_v(\mathscr{A}_\lambda(n,r))\text{-inj}$. It follows that

$$\mathscr{A}^0_{\lambda^-,\lambda-\lambda^-}(n,r) \otimes_{\mathscr{A}^-_\lambda(n,r)} \bullet$$

gives this equivalence. On the other hand, from the definition of Res_τ we see that

(6.7) $\mathrm{Res}_\tau(\mathscr{A}^0_{\lambda^-,\lambda-\lambda^-}(n,r)\otimes_{\mathscr{A}_{\lambda^-}(n,r)}\bullet)\cong\mathscr{A}^0_{\lambda^-,\lambda-\lambda^-}(n,r)_{\dagger,\tau}\otimes_{\mathscr{A}_{\lambda^-}(\tau;r)}\mathrm{Res}_\tau(\bullet).$

But the bimodule $\mathscr{A}^0_{\lambda^-,\lambda-\lambda^-}(n,r)_{\dagger,\tau}$ is a wall-crossing bimodule for the algebras $\mathscr{A}_?(\tau;r)$, by Lemma 3.10. So tensoring with this bimodule maps tiltings to injectives. Hence the right-hand side of (6.7) maps tiltings to injectives, and so does the left-hand side. Since $\mathscr{A}^0_{\lambda^-,\lambda-\lambda^-}(n,r)\otimes_{\mathscr{A}_{\lambda^-}(n,r)}\bullet$ is an equivalence $\mathscr{O}_\nu(\mathscr{A}_\lambda^-(n,r))\text{-tilt}\xrightarrow{\sim}\mathscr{O}_\nu(\mathscr{A}_\lambda(n,r))\text{-inj}$, we see that Res_τ maps injectives to injectives.

Step 4. To see that Res_τ maps projectives to projectives, one can argue similarly by using the functor $=\mathfrak{WC}^{-1}_{\lambda^-\leftarrow\lambda}=R\mathrm{Hom}_{\mathscr{A}_{\lambda^-}(n,r)}(\mathscr{A}^0_{\lambda,\lambda^--\lambda}(n,r),\bullet)$ instead of $\mathfrak{WC}_{\lambda\leftarrow\lambda^-}$. □

6.3 Supports for asymptotic chambers

Recall that the simples of $\mathscr{O}_{\nu,\lambda}(n,r)$ are indexed by the r-multipartitions

$$\sigma=(\sigma^{(1)},\dots,\sigma^{(r)})$$

of n; moreover, $L_\nu(\sigma)=L_{\nu_0}(\underline{L}(\sigma))$, where $\underline{L}(\sigma)=L^A(\sigma^{(1)})\boxtimes\cdots\boxtimes L^A(\sigma^{(r)})$. We write $L^A(\sigma')$ for the irreducible module in the category $\mathscr{O}$ for the Rational Cherednik algebra $eH_{1,\lambda}(|\sigma'|)e$. Here σ' is a partition.

Here is our main result of this section; it describes the dimensions of the supports of the simples $L_\nu(\sigma)$ in terms of σ.

Theorem 6.10. *Let m denote the denominator of λ (equal to $+\infty$ if $\lambda\notin\mathbb{Q}$). Assume $m>1$. Divide $\sigma^{(1)}$ by m with remainder: $\sigma^{(1)}=m\sigma^{(1)q}+\sigma^{(1)r}$ (componentwise operations, "q" and "r" stand for the quotient and the remainder). Then* $\dim\mathrm{Supp}\,L_\nu(\sigma)=rn-|\sigma^{(1)q}|(rm-1)$.

Proof. Our proof is by induction on n. For $n<m$, the category $\mathscr{O}$ is semisimple, and all simples have support of dimension rn, while for $n=m$, the result follows from Theorem 4.3. So we will assume that the claim of the theorem is proved for all dimensions less than n.

Let us write n_i for $|\sigma^{(i)}|$. By [L8, Theorem 1.2], all irreducible components of $\mathrm{Supp}\,L_\nu(\sigma)$ have the same dimension. So if $\mathrm{Res}_\tau(L_\nu(\sigma))\neq 0$, then

$$\dim\mathrm{Supp}\,L_\nu(\sigma)=\dim\mathrm{Res}_\tau(L_\nu(\sigma)).$$

By Corollary 6.8, we have

$$\mathrm{Res}_\tau(L_\nu(\sigma))=L_\nu\left(\underline{\mathrm{Res}}_\tau\left(L^A(\sigma^{(1)})\boxtimes L^A(\sigma^{(2)})\boxtimes\cdots\boxtimes L^A(\sigma^{(r)})\right)\right).$$

Let us first take $\tau = (n-1,1)$. Let us compute

$$\underline{\mathrm{Res}}_\tau \left(L^A(\sigma^{(1)}) \boxtimes L^A(\sigma^{(2)}) \boxtimes \cdots \boxtimes L^A(\sigma^{(r)})\right).$$

According to (6.3) it equals

$$\bigoplus_{i=1}^{r} L^{(A)}(\sigma^{(1)}) \boxtimes \cdots \boxtimes EL^{(A)}(\sigma^{(i)}) \boxtimes \cdots \boxtimes L^{(A)}(\sigma^{(r)}),$$

where we write E for the Bezrukavnikov-Etingof functor restricting from S_{n_i} to S_{n_i-1}. According to [S], $E = \bigoplus_{i=0}^{m-1} E_i$, where the E_i's are the categorification functors for a categorical action of $\hat{\mathfrak{sl}}_m$ on $\bigoplus_{k=0}^{+\infty} \mathscr{O}(H_{\lambda+i-1}(k))$. Let us divide $\sigma^{(i)}$ by m with remainder: $\sigma^{(i)} = m\sigma^{(i)q} + \sigma^{(i)r}$. We have $EL^A(\sigma^{(i)}) = 0$ if and only if $\sigma^{(i)r} = 0$, [Wi]. Moreover, if $\sigma^{(i)r} \neq 0$, then $EL(\sigma^{(i)})$ surjects onto $L(\underline{\sigma}^{(i)})$, where $\underline{\sigma}^{(i)}$ is obtained from $\sigma^{(i)}$ by removing a box from $\sigma^{(i)r}$. This follows from results of [L4].

So assume that $\sigma^{(i)r} \neq \varnothing$ for some i. Let $\underline{\sigma}$ be the r-partition obtained from σ by replacing $\sigma^{(i)}$ with $\underline{\sigma}^{(i)}$. Then $\mathrm{Res}_\tau(L(\sigma)) \twoheadrightarrow L(\underline{\sigma}) \boxtimes L$, where L is some simple object in $\mathscr{O}_\nu(\mathscr{A}_\lambda(1,r))$. By our inductive assumption,

$$\dim \mathrm{Supp} L(\underline{\sigma}) \boxtimes L = rn - (rm-1)|\sigma^{(1)q}|.$$

It follows that $\dim \mathrm{Supp} L(\sigma) = \dim \mathrm{Supp}\,\mathrm{Res}_\tau(L(\sigma)) \geqslant rn - (rm-1)|\sigma^{(1)q}|$. On the other hand, if $L^A(\hat{\sigma}^{(i)})$ is a simple occurring in $EL^A(\sigma^{(i)})$, then $|\hat{\sigma}^{(i)q}| \geqslant |\sigma^{(i)q}|$. From here we deduce that $\dim \mathrm{Supp}\,\mathrm{Res}_\tau(L(\sigma)) \leqslant rn - (rm-1)|\sigma^{(1)q}|$. So if $\mathrm{Res}_\tau(L(\sigma)) \neq 0$, then $\dim \mathrm{Supp} L(\sigma) = rn - (rm-1)|\sigma^{(1)q}|$.

So we only need to consider the case when $\mathrm{Res}_\tau(L(\sigma)) = 0$ for $\tau = (n-1,1)$. By the previous paragraph, this means that all $\sigma^{(i)}$ are divisible by m. Take $\tau' = (m^{n/m})$. Then, since the support of $L^A(\sigma^{(i)})$ corresponds to the partition $m^{n_i/m}$ of n_i, we can use (6.3) to get

$$\underline{\mathrm{Res}}_{\tau'} \left(L^A(\sigma^{(1)}) \boxtimes \cdots \boxtimes L^A(\sigma^{(r)})\right) = \mathrm{Res}\, L^A(\sigma^{(1)}) \boxtimes \cdots \boxtimes \mathrm{Res}\, L^A(\sigma^{(r)}),$$

where on the right-hand side, we write Res for the Bezrukavnikov-Etingof restriction functor from S_{mn_i} to $S_m^{n_i/m}$. According to [Wi], we have $\mathrm{Res}\, L^A(\sigma^{(i)}) = L^A(m)^{\boxtimes n_i/m} \boxtimes V_{\sigma^{(i)q}}$, where $V_{\sigma^{(i)q}}$ is a nonzero multiplicity space (that is the irreducible $S_{|\sigma^{(i)q}|}$-module corresponding to $\sigma^{(i)q}$).

It follows that $\mathrm{Res}_{\tau'} L(\sigma)$ is the sum of several copies of

$$L_1^{\boxtimes n_1/m} \boxtimes L_2^{\boxtimes n_2/m} \boxtimes \cdots \boxtimes L_r^{\boxtimes n_r/m}.$$

Here we write L_i for the irreducible in $\mathscr{O}_\nu(\mathscr{A}_\lambda(m,r))$ corresponding to the r-partition with ith part (m). By Theorem 4.3, we have $\dim \mathrm{Supp} L_1 = 1$ and $\dim \mathrm{Supp} L_i = rm$

for $i > 1$. Since $\dim \operatorname{Supp} L(\sigma) = \dim \operatorname{Supp} \operatorname{Res}_{\tau'} L(\sigma)$, this implies the claim of the theorem when $\operatorname{Res}_\tau(L(\sigma)) = 0$ and finishes the proof. □

Proof of Theorem 1.5. In the proof we can assume that λ is sufficiently large. By [L8, Theorem 1.2], we have $\dim \operatorname{Supp} L_\nu(\sigma) = \frac{1}{2} \dim \mathrm{V}(\mathscr{A}_\lambda(n,r)/\operatorname{Ann} L_\nu(\sigma))$. Theorem 1.3 implies that the two-sided ideals in $\mathscr{A}_\lambda(n,r)$ are determined by the dimensions of their associated varieties. This finishes the proof of this theorem. □

6.4 Cross-walling bijections

We have computed the supports in the case when ν is dominant. For other chambers of one-parameter subgroups, the supports of simples in the categories $\mathscr{O}$ can be computed using the cross-walling bijections defined in a more general situation below. We plan to compute these bijections combinatorially in a subsequent paper.

Let $X = X^\theta$ be a conical symplectic resolution equipped with a Hamiltonian action of a torus T with finitely many fixed points. Let $\mathscr{A}_\lambda^\theta$ be a quantization of X and $\mathscr{A}_\lambda := \Gamma(\mathscr{A}_\lambda^\theta)$. We suppose that λ is sufficiently ample so that Propositions 2.2 and 2.5 hold.

Let $\nu, \nu' : \mathbb{C}^\times \to T$ be two generic one-parameter subgroups lying in two chambers which are opposite with respect to a common face. We are going to define a bijection $\mathfrak{cw}_{\nu' \leftarrow \nu} : X^T \to X^T$. Let ν_0 be a one-parameter subgroup that is generic in the common face of the chambers of ν, ν' such that these chambers are opposite with respect to the face. Consider the cross-walling functor $\underline{\mathfrak{CW}}_{\nu' \leftarrow \nu} : D^b(\mathscr{O}_\nu(\mathsf{C}_{\nu_0}(\mathscr{A}_\lambda^\theta))) \to D^b(\mathscr{O}_{\nu'}(\mathsf{C}_{\nu_0}(\mathscr{A}_\lambda^\theta)))$. For $X^{\nu_0(\mathbb{C}^\times)}$, the one-parameter subgroups ν, ν' lie in opposite chambers. So $\underline{\mathfrak{CW}}_{\nu' \leftarrow \nu}^{-1}$ is the direct sum (over the irreducible components of $X^{\nu_0(\mathbb{C}^\times)}$) of Ringel duality functors with various homological shifts. Each Ringel duality functor is a perverse equivalence, Lemma 2.8, and hence gives rise to a bijection between the set of simples, see Section 2.4. We take the disjoint union of these bijections for $\mathfrak{cw}_{\nu' \leftarrow \nu}$.

The main result of this section is the following.

Proposition 6.11. *We have*

$$\operatorname{Ann}(L_\nu(p)) = \operatorname{Ann}(L_{\nu'}(\mathfrak{cw}_{\nu' \leftarrow \nu}(p)))$$

and

$$\dim \operatorname{Supp} L_\nu(p) = \dim \operatorname{Supp} L_{\nu'}(\mathfrak{cw}_{\nu' \leftarrow \nu}(p)).$$

Proof. Note that the former equality implies the latter by Lemma 2.3.

Let us prove the equality of the annihilators. Recall that $\mathfrak{CW}_{\nu' \leftarrow \nu} \circ \Delta_{\nu_0} \cong \Delta_{\nu_0} \circ \underline{\mathfrak{CW}}_{\nu' \leftarrow \nu}$, Proposition 2.6. In particular, $\mathfrak{CW}_{\nu' \to \nu}$ preserves the filtrations on $D^b(\mathscr{O}_?(\mathscr{A}_\lambda))$ (where $? = \nu$ or ν') defined by ν_0. Moreover, let Z be the component in $X^{\nu_0(\mathbb{C}^\times)}$ containing p. Let $D^b_{\leqslant Z}(\mathscr{O}_?(\mathscr{A}_\lambda)) \subset D^b(\mathscr{O}_\nu(\mathscr{A}_\lambda))$ denote the corresponding filtration subcategory. Let $\pi_Z\colon D^b_{\leqslant Z}(\mathcal{O}_?(\mathscr{A}_\lambda)) \twoheadrightarrow D^b(\mathcal{O}_?(\mathsf{C}_{v_0}(\mathscr{A}_\lambda)_Z))$ be the quotient functor so that

$$\pi_Z \circ \mathfrak{CW}_{v' \leftarrow v} = \underline{\mathfrak{CW}}_{v' \leftarrow v} \circ \pi_Z. \tag{6.8}$$

By the construction of the bijection $\mathfrak{cw}_{v' \leftarrow v}$, the simple $\underline{L}_{v'}(\mathfrak{cw}_{v' \leftarrow v}(p))$ is a constituent in $H_\bullet(\underline{\mathfrak{CW}}_{v' \leftarrow v}\underline{L}_v(p))$. But $\underline{L}_{v'}(\mathfrak{cw}_{v' \leftarrow v}(p)) = \pi_Z(L_{v'}(\mathfrak{cw}_{v' \leftarrow v}(p)))$ and, by (6.8),

$$H_\bullet(\underline{\mathfrak{CW}}_{v' \leftarrow v}\underline{L}_v(p)) = \pi_Z(H_\bullet(\mathfrak{CW}_{v' \leftarrow v}L_v(p))).$$

It follows that $L_{v'}(\mathfrak{cw}_{v' \leftarrow v}(p))$ is a constituent in $H_\bullet(\mathfrak{CW}_{v' \leftarrow v}L_v(p))$.

Set $\mathscr{I} := \bigcap_n (\operatorname{Ann} L_v(p))^n$ so that $\mathscr{I}^2 = \mathscr{I}$. Since the regular $\mathscr{A}_\lambda$-bimodule has finite length, see Lemma 3.8, we see that $\mathscr{I}$ coincides with some power of $\operatorname{Ann} L_v(p)$. It is enough to show that $H_\bullet(\mathfrak{CW}_{v' \leftarrow v}L_v(p))$ is annihilated by $\mathscr{I}$.

For any HC $\mathscr{A}_\lambda$-bimodule $\mathscr{B}$, we have

$$\mathscr{B} \otimes^L_{\mathscr{A}_\lambda} \mathfrak{CW}_{v' \leftarrow v}(\bullet) \cong \mathfrak{CW}_{v' \leftarrow v}(\mathscr{B} \otimes^L_{\mathscr{A}_\lambda} \bullet). \tag{6.9}$$

This follows from the construction of the functors $\mathfrak{CW}_{v' \leftarrow v}$, see [BLPW, Section 8.2] together with the observation that the functors $\mathscr{B} \otimes^L_{\mathscr{A}_\lambda} \bullet, R\operatorname{Hom}_{\mathscr{A}_\lambda}(\mathscr{B}, \bullet)$ preserve the categories $D^b(\mathcal{O}_{v''})$ for all generic v''.

We will apply (6.9) to $\mathscr{B} = \mathscr{I}$. We have

$$\mathscr{I}^{\otimes^L k} \otimes^L \mathfrak{CW}_{v' \leftarrow v}(L_v(p)) \cong \mathfrak{CW}_{v' \leftarrow v}(\mathscr{I}^{\otimes^L, k} \otimes^L L_v(p)), \tag{6.10}$$

where all derived tensor products are taken over $\mathscr{A}_\lambda$ and we write $\mathscr{I}^{\otimes^L k}$ for the k-th derived tensor power of $\mathscr{I}$. Note that thanks to $\mathscr{I}^2 = \mathscr{I}$, we get $\mathscr{I} \otimes_{\mathscr{A}_\lambda} \mathscr{A}_\lambda / \mathscr{I} = 0$. It follows that $\mathscr{I} \otimes_{\mathscr{A}_\lambda} M = 0$ for all $\mathscr{A}_\lambda$-modules M annihilated by $\mathscr{I}$. Moreover, thanks to the exact sequence $0 \to \mathscr{I} \to \mathscr{A}_\lambda \to \mathscr{A}_\lambda/\mathscr{I} \to 0$, all homology of $\mathscr{I} \otimes^L_{\mathscr{A}_\lambda} M$ are annihilated by $\mathscr{I}$ provided M is annihilated by $\mathscr{I}$. We deduce that $H_i(\mathscr{I}^{\otimes^L, k} \otimes^L L_v(p)) = 0$ for $i < k$. We deduce that the homology of the right-hand side of (6.10) vanishes in degrees $\leqslant k - k_0$, where k_0 is the maximal number such that $H_{-k_0}(\mathfrak{CW}_{v' \leftarrow v}L) \neq 0$, here L runs over the simples in $\mathcal{O}_v(\mathscr{A}_\lambda)$. On the other hand, let ℓ be the minimal number such that $H_\ell(\mathfrak{CW}_{v' \leftarrow v}(L_v(p)))$ is not annihilated by any power of $\mathscr{I}$. Using the spectral sequence for the composition of derived functors, we see that H_ℓ of the left hand side of (6.10) is nonzero for any k. This gives a contradiction which completes the proof. □

References

[BEG] Yu. Berest, P. Etingof, V. Ginzburg, *Finite-dimensional representations of rational Cherednik algebras*. Int. Math. Res. Not. 2003, no. 19, 1053–1088.

[BE] R. Bezrukavnikov, P. Etingof, *Parabolic induction and restriction functors for rational Cherednik algebras*. Selecta Math., 14(2009), 397–425.

[BeKa] R. Bezrukavnikov, D. Kaledin, *Fedosov quantization in the algebraic context*. Moscow Math. J. 4 (2004), 559–592.

[BL] R. Bezrukavnikov, I. Losev, *Etingof conjecture for quantized quiver varieties*. arXiv:1309.1716.

[BoKr] W. Borho, H. Kraft, *Über die Gelfand-Kirillov-Dimension*. Math. Ann. 220(1976), 1–24.

[BLPW] T. Braden, A. Licata, N. Proudfoot, B. Webster, *Quantizations of conical symplectic resolutions II: category O and symplectic duality*. Astérisque 384 (2016), 75–179.

[BPW] T. Braden, N. Proudfoot, B. Webster, *Quantizations of conical symplectic resolutions I: local and global structure*. Astérisque 384 (2016), 1–73.

[CB] W. Crawley-Boevey, *Geometry of the moment map for representations of quivers,* Comp. Math. **126** (2001), 257–293.

[EG] P. Etingof and V. Ginzburg, *Symplectic reflection algebras, Calogero-Moser space, and deformed Harish-Chandra homomorphism*, Invent. Math. **147** (2002), 243–348.

[GG] W.L. Gan, V. Ginzburg, *Almost commuting variety, $\mathcal{D}$-modules and Cherednik algebras*. IMRP, 2006, doi: 10.1155/IMRP/2006/26439.

[GGOR] V. Ginzburg, N. Guay, E. Opdam and R. Rouquier, *On the category $\mathcal{O}$ for rational Cherednik algebras*, Invent. Math., **154** (2003), 617–651.

[GS1] I. Gordon, T. Stafford, *Rational Cherednik algebras and Hilbert schemes*, Adv. Math. 198 (2005), no. 1, 222–274.

[GS2] I. Gordon, T. Stafford, *Rational Cherednik algebras and Hilbert schemes. II. Representations and sheaves*, Duke Math. J. 132 (2006), no. 1, 73–135.

[KR] M. Kashiwara and R. Rouquier, *Microlocalization of rational Cherednik algebras*, Duke Math. J. **144** (2008) 525–573.

[Ko] D. Korb, *Order on the Fixed Points of the Gieseker Variety With Respect to the Torus Action*. arXiv:1312.3025.

[L1] I. Losev, *Completions of symplectic reflection algebras*. Selecta Math., 18(2012), N1, 179–251.

[L2] I. Losev, *Isomorphisms of quantizations via quantization of resolutions*. Adv. Math. 231(2012), 1216–1270.

[L3] I. Losev. *On isomorphisms of certain functors for Cherednik algebras*. Repres. Theory, 17 (2013), 247–262.

[L4] I. Losev, *Highest weight $\mathfrak{sl}_2$-categorifications I: crystals*. Math. Z. 274(2013), 1231–1247.

[L5] I. Losev, *Proof of Varagnolo-Vasserot conjecture on cyclotomic categories $\mathcal{O}$*. Selecta Math. 22(2016), 631–668.

[L6] I. Losev, *Abelian localization for cyclotomic Cherednik algebras*. Int Math Res Notices (2015) vol. 2015, 8860–8873.

[L7] I. Losev, *On categories $\mathcal{O}$ for quantized symplectic resolutions*. Compos. Math. 153 (2017), N12, 2445–2481.

[L8] I. Losev, *Bernstein inequality and holonomic modules* (with a joint appendix by I. Losev and P. Etingof). Adv. Math. 308 (2017), 941–963.

[L9] I. Losev, *Wall-crossing functors for quantized symplectic resolutions: perversity and partial Ringel dualities*. arXiv:1604.06678.

[LW] I. Losev, B. Webster, *On uniqueness of tensor products of irreducible categorifications*. Selecta Math. 21(2015), N2, 345–377.

[MN1] K. McGerty, T. Nevins, *Derived equivalence for quantum symplectic resolutions*, Selecta Math. (N.S.) 20 (2014), no. 2, 675–717.

[MN2] K. McGerty, T. Nevins, *Compatability of t-structures for quantum symplectic resolutions*. Duke Math J. 165(2016), 2529–2585.

[Nak] H. Nakajima, *Quiver varieties and tensor products*. Invent. Math. 146 (2001), no. 2, 399–449.

[NY] H. Nakajima, K. Yoshioka, *Lectures on Instanton Counting*. Algebraic structures and moduli spaces, 31–101, CRM Proc. Lecture Notes, 38, Amer. Math. Soc., Providence, RI, 2004.

[Nam] Y. Namikawa, *Poisson deformations and birational geometry*, J. Math. Sci. Univ. Tokyo 22 (2015), no. 1, 339–359.

[R] R. Rouquier, *q-Schur algebras for complex reflection groups*. Mosc. Math. J. 8 (2008), 119–158.

[S] P. Shan, *Crystals of Fock spaces and cyclotomic rational double affine Hecke algebras*. Ann. Sci. Ecole Norm. Sup. **44** (2011), 147–182.

[We] B. Webster, *On generalized category $\mathcal{O}$ for a quiver variety*. Math. Ann. 368 (2017), no. 1–2, 483–536.

[Wi] S. Wilcox, *Supports of representations of the rational Cherednik algebra of type A*. Adv. Math. 314(2017), 426-492.

Generalized Bruhat Cells and Completeness of Hamiltonian Flows of Kogan-Zelevinsky Integrable Systems

Jiang-Hua Lu and Yipeng Mi

In memory of Bertram Kostant

Abstract Let G be any connected and simply connected complex semisimple Lie group, equipped with a standard holomorphic multiplicative Poisson structure. We show that the Hamiltonian flows of all the Fomin-Zelevinsky twisted generalized minors on every double Bruhat cell of G are complete in the sense that all the integral curves of their Hamiltonian vector fields are defined on $\mathbb{C}$. It follows that the Kogan-Zelevinsky integrable systems on G have complete Hamiltonian flows, generalizing the result of Gekhtman and Yakimov for the case $SL(n,\mathbb{C})$. We in fact construct a class of integrable systems with complete Hamiltonian flows associated to *generalized Bruhat cells* which are defined using arbitrary sequences of elements in the Weyl group of G, and we obtain the results for double Bruhat cells through the so-called open *Fomin-Zelevinsky embeddings* of (reduced) double Bruhat cells in generalized Bruhat cells. The Fomin-Zelevinsky embeddings are proved to be Poisson, and they provide global coordinates on double Bruhat cells, called *Bott-Samelson coordinates*, in which all the Fomin-Zelevinsky minors become polynomials and the Poisson structure can be computed explicitly.

Keywords: Poisson structures • integrable systems • double Bruhat cells • generalized Bruhat cells

Mathematics Subject Classification (2010): Primary: 17B80; Secondary: 17B63, 20G20

Research in this paper was partially supported by the Research Grants Council of the Hong Kong SAR, China (GRF HKU 703712 and 17304415).

Jiang-Hua Lu
Department of Mathematics, The University of Hong Kong, Pokfulam Road, Hong Kong,
e-mail: jhlu@maths.hku.hk

Yipeng Mi
Department of Mathematics, The University of Hong Kong, Pokfulam Road, Hong Kong,
e-mail: littlemi@connect.hku.hk

V. G. Kac, V. L. Popov (eds.), *Lie Groups, Geometry, and Representation Theory*,
Progress in Mathematics 326, https://doi.org/10.1007/978-3-030-02191-7_12

1 Introduction

1.1 Introduction

Complex Poisson manifolds provide a natural setting for complex integrable systems. Here recall that a complex Poisson manifold is a pair (X, π), where X is a complex manifold and π, the *Poisson structure*, is a holomorphic section of $\wedge^2 T_X$ such that the bilinear map $\mathscr{H}_X \otimes_{\mathbb{C}} \mathscr{H}_X \to \mathscr{H}_X$ given by

$$(1) \qquad \{f, g\} \stackrel{\text{def}}{=} \pi(df, dg), \qquad f, g \in \mathscr{H}_X,$$

makes the sheaf $\mathscr{H}_X$ of local holomorphic functions on X into a sheaf of Poisson algebras. In such a case, the operation $\{\, , \}$ is called a *Poisson bracket* on $\mathscr{H}_X$, and $f, g \in \mathscr{H}_X$ are said to Poisson commute if $\{f, g\} = 0$. Given a complex Poisson manifold (X, π) and a (local) holomorphic function f on X, the *Hamiltonian vector field* H_f of f is the holomorphic vector field on X defined by

$$H_f(g) = \{f, g\}, \qquad g \in \mathscr{H}_X,$$

and f is said to have *complete Hamiltonian flow* if all the integral curves of H_f on X are defined on the whole of $\mathbb{C}$. If the generic symplectic leaves of π in X have dimension $2m$, a set $(y_1, \ldots, y_m)$ of m holomorphic functions on X is said to define an *integrable system on X with complete Hamiltonian flows* if they are functionally independent when restricted to generic symplectic leaves, pairwise Poisson commute, and each y_j has complete Hamiltonian flow on X. An integrable system on X with complete Hamiltonian flows thus defines an analytic action of $\mathbb{C}^m$ on X.

In this paper, we are mainly concerned with Hamiltonian flows and integrable systems defined by regular functions on a smooth complex affine variety, equipped with a Poisson structure that is algebraic. We refer to [32] for a systematic study of complex Poisson manifolds, algebraic Poisson varieties, as well as their relations to integrable systems.

For an integer $m \geq 1$, let $M_m(\mathbb{C})$ be the vector space of all $m \times m$ matrices with complex entries. In [30, 31], B. Kostant and N. Wallach proved the remarkable result that the complex Gelfand-Zeitlin integrable system on $M_m(\mathbb{C})$ has complete Hamiltonian flows, thus giving rise to an analytic action of $\mathbb{C}^{m(m-1)/2}$ on $M_m(\mathbb{C})$. Here the Hamiltonian vector fields are defined with respect to the linear Kostant-Kirillov-Souriau Poisson structure on $M_m(\mathbb{C})$, by identifying $M_m(\mathbb{C})$ with its dual vector space via the trace form and by using the natural Lie algebra structure on $M_m(\mathbb{C})$. Various aspects of this (complex) Gelfand-Zeitlin system on $M_m(\mathbb{C})$, such as its relation to representation theory and the geometry of the orbits of the action by $\mathbb{C}^{m(m-1)/2}$, have been studied in detail by Kostant and Wallach [30, 31] and by M. Colarusso and S. Evens [6, 7, 8, 9, 10].

The vector space $M_m(\mathbb{C})$ also has a well-known quadratic Poisson structure with its origin in quantum groups [19]. In [19], M. Gekhtman and M. Yakimov gave an analog of the Gelfand-Zeitlin integrable system on $M_m(\mathbb{C})$ with respect to the quadratic Poisson structure and proved its completeness in a certain sense. In the same paper [19], they also showed that the Hamiltonian flows (with respect to the quadratic Poisson structure) of all minors on $M_m(\mathbb{C})$ are complete, and that the Kogan-Zelevinsky integrable systems, originally introduced by M. Kogan and A. Zelevinsky [29] for any connected and simply connected complex semisimple Lie group (see Example 3.10 for more detail), have complete Hamiltonian flows for the case of $SL_m(\mathbb{C})$.

Let $\mathcal{Q}$ be the algebra of all quasi-polynomials in one complex variable [1, §26], i.e., $\mathcal{Q}$ consists of all holomorphic functions on $\mathbb{C}$ of the form

$$(2) \qquad \gamma(c) = \sum_{k=1}^{N} q_k(c) e^{a_k c}, \qquad c \in \mathbb{C},$$

with each $q_k(c) \in \mathbb{C}[c]$ and the a_k's being pairwise distinct complex numbers. It is in fact shown in [19] that all the integral curves of the Hamiltonian vector field (with respect to the quadratic Poisson structure on $M_n(\mathbb{C})$) of every minor on $M_m(\mathbb{C})$ are of the form

$$c \longmapsto (\gamma_{jk}(c))_{j,k=1,\ldots,m}, \qquad c \in \mathbb{C},$$

where $\gamma_{jk} \in \mathcal{Q}$ for all $j,k = 1,\ldots,m$. Motivated by the results in [19], we make the following definition.

Definition 1.1. Let X be a smooth affine variety with an algebraic Poisson structure, and assume that $X \subset \mathbb{C}^n$. A regular function y on X is said to have *complete Hamiltonian flow with property* $\mathcal{Q}$ (with respect to the given embedding of X in $\mathbb{C}^n$) if all the integral curves of the Hamiltonian vector field of y are defined on $\mathbb{C}$ and are of the form

$$c \longmapsto (\gamma_1(c),\ldots,\gamma_n(c)), \qquad c \in \mathbb{C},$$

with $\gamma_j \in \mathcal{Q}$ for every $j = 1,2,\ldots,n$. An integrable system on X defined by a set $(y_1,\ldots,y_m)$ of regular functions on X is said to have *complete Hamiltonian flows with property* $\mathcal{Q}$ if each y_j has complete Hamiltonian flow with property $\mathcal{Q}$. ◇

As $\mathcal{Q}$ is an algebra, the property for a regular function, and thus also for an integrable system, to have complete Hamiltonian flow with property $\mathcal{Q}$ is independent of the embedding of X into affine spaces, and is thus also invariant under biregular Poisson isomorphisms between affine Poisson varieties (see Lemma 1.5). In view of Definition 1.1, the results in [19] on the minors on $M_m(\mathbb{C})$ can be rephrased as saying that, with respect to the quadratic Poisson structure on $M_m(\mathbb{C})$, all the minors on $M_m(\mathbb{C})$ have complete Hamiltonian flows with property $\mathcal{Q}$, and that the Kogan-Zelevinsky integrable systems for $SL(m,\mathbb{C}) \subset M_m(\mathbb{C})$ have complete Hamiltonian flows with property $\mathcal{Q}$.

In this paper, we generalize the above result of [19] for $SL(m,\mathbb{C})$ to any connected and simply connected complex semisimple Lie group G. More precisely, recall that double Bruhat cells in G are defined as

$$G^{u,v} = (BuB) \cap (B_-vB_-),$$

where (B,B_-) is a pair of opposite Borel subgroups of G, and $u,v \in W$, the Weyl group of G with respect to $T = B \cap B_-$. By [3, Proposition 2.8], each double Bruhat cell $G^{u,v}$ is an affine variety (see also [17, Proposition 3.1] and [17, Corollary 2.5], which imply that $G^{u,v}$ is biregularly isomorphic to a principal Zariski open subset of an affine space). We show that all the *Fomin-Zelevinsky twisted generalized minors* (see Definition 3.6) on every double Bruhat cell $G^{u,v}$ in G have complete Hamiltonian flows with property $\mathcal{Q}$. Consequently, the Kogan-Zelevinsky integrable systems on every $G^{u,u}$, defined by certain Fomin-Zelevinsky twisted generalized minors and using reduced words of u, have complete Hamiltonian flows with property $\mathcal{Q}$. Here the group G is equipped with the so-called *standard multiplicative Poisson structure* $\pi_{\rm st}$, the definition of which depends on the choice of (B,B_-) and the additional choice of a non-degenerate symmetric bilinear form $\langle\, ,\, \rangle_{\mathfrak{g}}$ on the Lie algebra $\mathfrak{g}$ of G (see §2.1 for its precise definition). For $G = SL(m,\mathbb{C})$ and the standard choices of B, B_-, and $\langle\, ,\, \rangle_{\mathfrak{g}}$, $\pi_{\rm st}$ extends to the quadratic Poisson structure on $M_m(\mathbb{C})$ used in [19].

We remark that the Fomin-Zelevinsky twisted generalized minors on double Bruhat cells have been studied extensively in the literature (see, for example, [2, 3, 17, 20, 21, 22, 23, 24] and references therein), especially because of their important roles in the theories of total positivity, crystal bases, and (classical and quantum) cluster algebras. In particular, certain Fomin-Zelevinsky minors on $G^{u,v}$ form an initial cluster of the Berenstein-Fomin-Zelevinsky upper cluster algebra structure on the coordinate ring of $G^{u,v}$ (see [3]), which is compatible with the Poisson structure $\pi_{\rm st}$ on $G^{u,v}$ in the sense [18] that any two functions f_1, f_2 in the same cluster have *log-canonical* Poisson bracket, i.e.,

$$\{f_1, f_2\} = a f_1 f_2$$

for some constant a. We also refer to [24] for a remarkable proof, *using the Poisson structure* $\pi_{\rm st}$, that the Berenstein-Fomin-Zelevinsky upper cluster algebra structure on the coordinate ring of $G^{u,v}$ coincides with the corresponding cluster algebra structure.

Given a double Bruhat cell $G^{u,v}$, our main tool for studying the Poisson manifold $(G^{u,v}, \pi_{\rm st})$ is a modification of the map defined by Fomin and Zelevinsky in [17, Proposition 3.1], which we call the (open) *Fomin-Zelevinsky embedding* (see Definition 3.4)

$$F^{u,v}:\ G^{u,v} \longrightarrow T \times \mathcal{O}^{(v^{-1},u)},$$

and the resulting global *Bott-Samelson coordinates* on $G^{u,v}$. Here $\mathcal{O}^{(v^{-1},u)}$ is the so-called *generalized Bruhat cell* associated to the elements v^{-1} and u of the Weyl group W. The generalized Bruhat cell $\mathcal{O}^{\mathbf{u}}$, where $\mathbf{u} = (u_1, \ldots, u_n)$ is *any* sequence of elements in W, is defined in [35, §1.3] as

$$\mathcal{O}^{\mathbf{u}} = Bu_1B \times_B \cdots \times_B Bu_nB/B \subset F_n \overset{\rm def}{=} G \times_B \cdots \times_B G/B,$$

where F_n is the quotient space of G^n by a certain right action of the product group B^n (see §2.2 for detail). The product Poisson structure $\pi_{\rm st}^n$ on G^n projects to a well-defined Poisson structure π_n on F_n, with respect to which every generalized Bruhat cell $\mathcal{O}^{\mathbf{u}}$ is a Poisson submanifold. Fixing a pinning for (G,T) (see §1.4), the choice of a reduced word for each u_j then gives rise to a parametrization of $\mathcal{O}^{(u_1,\ldots,u_n)}$ by $\mathbb{C}^{l(u_1)+\cdots+l(u_n)}$, where l is the length function on W, and we refer to the resulting coordinates on $\mathcal{O}^{(u_1,\ldots,u_n)}$ as *Bott-Samelson coordinates* (see Definition 2.7). Explicit formulas for the Poisson structure π_n on $\mathcal{O}^{(u_1,\ldots,u_n)}$ in Bott-Samelson coordinates have been studied in [14, 15]. By *Bott-Samelson coordinates on a double Bruhat cell* $G^{u,v}$, we mean the coordinates on $G^{u,v}$ obtained through the open Fomin-Zelevinsky embedding $F^{u,v}$ and the combination of coordinates on T and Bott-Samelson coordinates on $\mathcal{O}^{(v^{-1},u)}$. We prove in Theorem 3.11 that the open Fomin-Zelevinsky embedding

$$F^{u,v}\colon\ (G^{u,v},\ \pi_{\mathrm{st}}) \longrightarrow (T \times \mathcal{O}^{(v^{-1},u)},\ 0 \bowtie \pi_2)$$

is Poisson, its image being a single T-orbit of symplectic leaves of $(T \times \mathcal{O}^{(v^{-1},u)}$, $0 \bowtie \pi_2)$. Here $0 \bowtie \pi_2$ is the sum of $0 \times \pi_2$, the product of the zero Poisson structure on T with the Poisson structure π_2 on $\mathcal{O}^{(v^{-1},u)}$, and a certain *mixed term* defined using the T-action on $\mathcal{O}^{(v^{-1},u)}$ (see §2.7 for detail). We then show that in Bott-Samelson coordinates on $G^{u,v}$, all the Fomin-Zelevinsky twisted generalized minors on $G^{u,v}$ belong to a special class of polynomials whose Hamiltonian flows are, *manifestly*, complete with property $\mathcal{Q}$.

We remark that for any sequence $(u_1,\ldots,u_n) \in W^n$, it is shown in [15, §5] that through the identification of the algebra of regular functions on $\mathcal{O}^{(u_1,\ldots,u_n)}$ with the polynomial algebra $\mathbb{C}[x_1,\ldots,x_l]$ via Bott-Samelson coordinates, where $l = l(u_1)+\cdots+l(u_n)$, the Poisson structure π_n on $\mathcal{O}^{(u_1,\ldots,u_n)}$ makes $\mathbb{C}[x_1,\ldots,x_l]$ into a so-called *symmetric nilpotent semi-quadratic Poisson algebra* as defined in [21, Definition 4], also called a *Poisson CGL extension* in [23]. The Fomin-Zelevinsky embedding, being an open Poisson embedding, thus gives an explicit identification of the coordinate ring of $G^{u,v}$, as a Poisson algebra, with a certain explicit localization of a Poisson CGL extension. Identifications of this kind, and their quantum analogs, have been used and play a crucial role in the work of K. Goodearl and M. Yakimov in their proof that the Berenstein-Fomin-Zelevinsky upper cluster algebra structure on the coordinate ring of $G^{u,v}$ coincides with the corresponding cluster algebra structure (both at the classical and quantum levels). See [20, 21, 22, 23, 24]. We thank M. Yakimov for pointing out to us that their identification in [24] of the coordinate ring of $G^{u,v}$ with the localization of a Poisson CGL extension is through a Poisson analog of the method in [20, Proposition 4.4] to prove [20, Theorem 4.1].

We also point out that based on the results in [14, 15], B. Elek has written a computer program in the language of GAP that computes the Poisson structure π_n on $\mathcal{O}^{(u_1,\ldots,u_n)}$ in the Bott-Samelson coordinates for any G and any sequence $(u_1,\ldots,u_n) \in W^n$. This, in principle, allows one to compute explicitly the Hamiltonian flows of the Fomin-Zelevinsky minors in Bott-Samelson coordinates. We give examples for $G = SL(2,\mathbb{C})$, $G = SL(3,\mathbb{C})$, and $G = G_2$ in Example 2.32, Example 3.16 and Example 3.17, respectively.

In the first part of the paper, §2, we develop a general theory on complete Hamiltonian flows associated to an *arbitrary* generalized Bruhat cell $\mathcal{O}^{\mathbf{u}}$ and the

Poisson structure π_n, where $\mathbf{u} = (u_1, \ldots, u_n) \in W^n$. We introduce a collection $\mathscr{Y}^{\mathbf{u}}$ (see Definition 2.17) of regular functions on $\mathscr{O}^{\mathbf{u}}$ and prove in Theorem 2.19, using Bott-Samelson coordinates on $\mathscr{O}^{\mathbf{u}}$, that every $y \in \mathscr{Y}^{\mathbf{u}}$ has complete Hamiltonian flow with property $\mathscr{Q}$. We also study a Poisson structure $0 \bowtie \pi_n$ on $T \times \mathscr{O}^{\mathbf{u}}$ and introduce a collection $\widetilde{\mathscr{Y}}^{\mathbf{u}}$ (see Definition 2.25) of regular functions on $T \times \mathscr{O}^{\mathbf{u}}$, which are shown in Theorem 2.26 to all have complete Hamiltonian flows with property $\mathscr{Q}$. Moreover, in §2.8 we construct on $(\mathscr{O}^{(\mathbf{u}^{-1}, \mathbf{u})}, \pi_{2n})$ an integrable system that has complete Hamiltonian flows with property $\mathscr{Q}$, where $\mathbf{u}^{-1} = (u_n^{-1}, \ldots, u_1^{-1})$, and $(\mathscr{O}^{(\mathbf{u}^{-1}, \mathbf{u})}, \pi_{2n})$ can be regarded as a *double* of $(\mathscr{O}^{\mathbf{u}}, \pi_n)$.

In §3, the second part of the paper, we relate double Bruhat cells and generalized Bruhat cells through the Fomin-Zelevinsky embeddings. Given a double Bruhat cell $G^{u,v}$ and respective reduced words $\mathbf{u}$ and $\mathbf{v}$ of u and v, we introduce in §3.1 the Fomin-Zelevinsky embeddings

$$\hat{F}^{\mathbf{u},\mathbf{v}}: \; G^{u,v}/T \longrightarrow \mathscr{O}^{(\mathbf{v}^{-1}, \mathbf{u})} \cong \mathscr{O}^{(v^{-1}, u)},$$
$$F^{\mathbf{u},\mathbf{v}}: \; G^{u,v} \longrightarrow T \times \mathscr{O}^{(\mathbf{v}^{-1}, \mathbf{u})} \cong T \times \mathscr{O}^{(v^{-1}, u)},$$

where recall that $G^{u,v}/T$ is the *reduced double Bruhat cell* corresponding to $G^{u,v}$. We prove in Proposition 3.9 that all the Fomin-Zelevinsky twisted generalized minors on $G^{u,v}$ are pullbacks by $F^{\mathbf{u},\mathbf{v}}$ of regular functions on $T \times \mathscr{O}^{(\mathbf{v}^{-1}, \mathbf{u})}$ from the collection $\widetilde{\mathscr{Y}}^{(\mathbf{v}^{-1}, \mathbf{u})}$. After proving in Theorem 3.12 that the Fomin-Zelevinsky embeddings are Poisson, we establish in Corollary 3.14 and Corollary 3.15 that all the Fomin-Zelevinsky minors on every $G^{u,v}$ and all the Kogan-Zelevinsky integrable systems on every $G^{u,u}$ have complete Hamiltonian flows with property $\mathscr{Q}$. For each reduced word $\mathbf{u}$ of u, by pulling back the integrable system on $\mathscr{O}^{(\mathbf{u}^{-1}, \mathbf{u})}$ constructed in §2.8 using $\hat{F}^{\mathbf{u},\mathbf{u}}$, we also obtain an integrable system on $G^{u,u}/T$ that has complete Hamiltonian flows with property $\mathscr{Q}$. See §3.5.

The proof of the fact that the Fomin-Zelevinsky embeddings are Poisson, as given in §3.3, uses certain formulas for Poisson brackets between Fomin-Zelevinsky minors by Kogan and Zelevinsky [29] and their explicit expressions in Bott-Samelson coordinates. As such a proof depends on many computations in different coordinates, we give a more conceptual proof in the Appendix, which involves some general theory on Poisson Lie groups and certain other interesting facts about the Poisson Lie group (G, π_{st}).

1.2 Further studies

We expect most of the results in §2 on the standard Poisson structure on generalized Bruhat cells to hold for arbitrary symmetric Poisson CGL extensions. In particular, it would be very interesting to construct a *double* of a symmetric Poisson CGL extension, as we have done from $(\mathscr{O}^{\mathbf{u}}, \pi_n)$ to $(\mathscr{O}^{(\mathbf{u}^{-1}, \mathbf{u})}, \pi_{2n})$, and produce an integrable system on the double that has complete Hamiltonian flows with property $\mathscr{Q}$.

Consider now a double Bruhat cell $G^{u,u}$, where $u \in W$. In addition to carrying Kogan-Zelevinsky integrable systems, which we have now proved to have complete Hamiltonian flows with property $\mathcal{Q}$, it is shown in [36] that $G^{u,u}$ also has the natural structure of a *Poisson groupoid* over the Bruhat cell $BuB/B \subset G/B$. In fact, each symplectic leaf of $\pi_{\rm st}$ in $G^{u,u}$ is shown in [36] to be a symplectic groupoid over BuB/B. It is very natural, then, to formulate any compatibility between the Kogan-Zelevinsky integrable systems and the groupoid structure on $G^{u,u}$, and to ask whether there is any groupoid-theoretical interpretation of the $\mathbb{C}^{l(u)}$-action on $G^{u,u}$ defined by the Kogan-Zelevinsky systems.

Finally, as with the complex Gelfand-Zeitlin integrable system on the matrix space $M_m(\mathbb{C})$ pioneered by Kostant and Wallach, it would be very interesting to see when the analytic action of $\mathbb{C}^{l(u)}$ on $G^{u,u}$ descends to an algebraic action of $(\mathbb{C}^\times)^{l(u)}$ and what roles the property $\mathcal{Q}$ of the Hamiltonian flows plays in understanding the geometry of the orbits of the action.

1.3 Why property $\mathcal{Q}$?

Recall Definition 1.1 of property $\mathcal{Q}$ of Hamiltonian flows. We first make a simple observation which explains the appearance of property $\mathcal{Q}$ in this paper. Let $(x_1,\ldots,x_n)$ be the standard coordinates on $\mathbb{C}^n$ and let $A = \mathbb{C}[x_1,\ldots,x_n]$. We identify an algebraic Poisson structure on $\mathbb{C}^n$ with the corresponding Poisson bracket $\{\,,\}$ on the polynomial algebra A.

Lemma 1.2. *Let $\{\,,\}$ be any algebraic Poisson structure on $\mathbb{C}^n$ and let $y \in A$. Assume that there is a set $\mathcal{G} = (\tilde{x}_1,\ldots,\tilde{x}_n)$ of generators for A such that for each $j = 1,2,\ldots,n$, there exist $\kappa_j \in \mathbb{C}$ and $f_j \in \mathbb{C}[\tilde{x}_1,\ldots,\tilde{x}_{j-1}]$ (f_j is a constant when $j = 1$) such that*

$$\{y, \tilde{x}_j\} = \kappa_j \tilde{x}_j y + f_j. \tag{3}$$

Then y has complete Hamiltonian flow with property $\mathcal{Q}$.

Proof. As the change of coordinates $(x_1,\ldots,x_n) \to (\tilde{x}_1,\ldots,\tilde{x}_n)$ is biregular, we may assume that $(\tilde{x}_1,\ldots,\tilde{x}_n) = (x_1,\ldots,x_n)$. The Hamiltonian vector field of y with respect to $\{\,,\}$ is then given by

$$H_y = \{y,x_1\}\frac{\partial}{\partial x_1} + \{y,x_2\}\frac{\partial}{\partial x_2} + \cdots + \{y,x_n\}\frac{\partial}{\partial x_n}.$$

As y is constant on any integral curve of H_y, solving for an integral curve of H_y is the same as solving for the functions $(x_1(c),\ldots,x_n(c))$ from the system of ODEs

$$\frac{dx_1}{dc} = \kappa_1 x_1 y_0 + f_1,$$
$$\frac{dx_2}{dc} = \kappa_2 x_2 y_0 + f_2(x_1), \qquad \cdots,$$
$$\frac{dx_n}{dc} = \kappa_n x_n y_0 + f_n(x_1, \ldots, x_{n-1}),$$

where y_0 and f_1 are constants. Solving for the x_j's successively, one sees that each x_j is defined on $\mathbb{C}$ and is in $\mathscr{Q}$. □

Remark 1.3. In our applications of Lemma 1.2, for example, in the proofs of Theorem 2.19 and Theorem 2.26, the set $(\tilde{x}_1, \ldots, \tilde{x}_n)$ of generators of $\mathbb{C}[x_1, \ldots, x_n]$ is often a shuffle of $(x_1, \ldots, x_n)$. ◇

Lemma 1.4. *With the same assumptions as in Lemma 1.2, suppose that X is a smooth affine Poisson subvariety of $\mathbb{C}^n$ and $g \in A$ is such that $g|_X \neq 0$. Consider $X' = \{x \in X : g(x) \neq 0\} \subset X$, regarded as an affine variety in $\mathbb{C}^{n+1}$ via the embedding*

$$X' \longrightarrow \mathbb{C}^{n+1}, \; x \longmapsto (x, \; 1/g(x)), \qquad x \in X'. \tag{4}$$

Equip X' with the (algebraic) Poisson structure from X by restriction. If $\{y, g\} = \kappa y g$ for some $\kappa \in \mathbb{C}$, then $y|_{X'}$ has complete Hamiltonian flow in X' with property $\mathscr{Q}$ (with respect to the embedding in (4)*).*

Proof. Let $I = \langle a_1, \ldots, a_m \rangle$ be the radical ideal of A defining X. Then there exist $b_{j,k} \in A$, $j,k = 1, \ldots, m$, such that $\{y, a_k\} = \sum_{j=1}^m b_{j,k} a_j$ for $k = 1, \ldots, m$. Let γ: $\mathbb{C} \to \mathbb{C}^n$ be an integral curve of H_y with $\gamma(0) \in X$. For $j,k = 1, \ldots, m$, let $f_k = a_k \circ \gamma$ and $g_{j,k} = b_{j,k} \circ \gamma$. Then the functions $f_1, \ldots, f_m$ on $\mathbb{C}$ satisfy the system of linear ODEs

$$\frac{df_k}{dc} = \sum_{j=1}^m g_{j,k}(c) f_j, \qquad k = 1, \ldots, m,$$

with the initial conditions $f_1(0) = \cdots = f_m(0) = 0$. By the uniqueness of solutions of ODEs, $f_k = 0$ for $k = 1, \ldots, m$. Thus $\gamma(\mathbb{C}) \subset X$.

Identifying the algebra $\mathscr{O}_{X'}$ of regular functions on X' with the localization $\mathscr{O}_X[(g|_X)^{-1}]$, the extension of $\{\, , \}$ from $\mathscr{O}_X$ to $\mathscr{O}_{X'}$ indeed makes X' into a (smooth) affine Poisson variety. Let $p \in X'$ and let γ: $\mathbb{C} \to X$ be the integral curve of H_y through p in X. Consider the function $g(\gamma(c))$, $c \in \mathbb{C}$. It follows from $\{y, g\} = \kappa y g$ that

$$\frac{d}{dc}(g(\gamma(c))) = \{y, g\}(\gamma(c)) = \kappa y(p) g(\gamma(c)), \qquad c \in \mathbb{C}.$$

Thus $g(\gamma(c)) = g(p)e^{\kappa y(p)c}$ for all $c \in \mathbb{C}$. In particular, $g(\gamma(c)) \neq 0$ for all $c \in \mathbb{C}$ and thus $\gamma(\mathbb{C}) \subset X'$. Moreover, as $1/g(\gamma(c)) = e^{-\kappa y(p)c}/g(p)$ for all $c \in \mathbb{C}$, it follows by definition that $y|_{X'} \in \mathscr{O}_{X'}$ has complete Hamiltonian flow in X' with property $\mathscr{Q}$. □

We now show that property $\mathscr{Q}$ for a regular function on an affine Poisson variety X is independent of the embedding of X into an affine space.

Lemma 1.5. *Let X be a smooth affine variety with an algebraic Poisson structure π, and let $\phi: X \to \mathbb{C}^n$ and $\psi: X \to \mathbb{C}^m$ be embeddings of X as affine varieties. Suppose that a regular function y on X has complete Hamiltonian flow with property $\mathscr{Q}$ with respect to ϕ. Then y also has complete Hamiltonian flow with property $\mathscr{Q}$ with respect to ψ.*

Proof. Let $\gamma: \mathbb{C} \to X, c \mapsto \gamma(c)$, be an integral curve of the Hamiltonian vector field of y. Writing $\phi \circ \gamma: \mathbb{C} \to \mathbb{C}^n$ as

$$\phi(\gamma(c)) = (\gamma_1(c), \ldots, \gamma_n(c)), \qquad c \in \mathbb{C},$$

we know by assumption that $\gamma_j \in \mathscr{Q}$ for $j = 1, \ldots, n$. Let $z_1, \ldots, z_m$ be the coordinates on $\mathbb{C}^m$. Then for $k = 1, \ldots, m$, $\psi^*(z_k)$ is a regular function on X, so there exists $f_k \in \mathbb{C}[x_1, \ldots, x_n]$ such that $\psi^*(z_k) = \phi^*(f_k)$ as regular functions on X. Define

$$F: \quad \mathbb{C}^n \longrightarrow \mathbb{C}^m, \; F(x) = (f_1(x), \ldots, f_m(x)), \qquad x \in \mathbb{C}^n.$$

Then $F \circ \phi = \psi: X \to \mathbb{C}^m$, and thus

$$\psi(\gamma(c)) = F(\phi(\gamma(c)) = F(\gamma_1(c), \ldots, \gamma_j(c)), \qquad c \in \mathbb{C}.$$

As $\mathscr{Q}$ is a subalgebra of the algebra of all $\mathbb{C}$-valued functions on $\mathbb{C}$, one has $\psi \circ \gamma \in \mathscr{Q}^m$. □

1.4 Notation

We now fix some notation that will be used throughout the paper.

We will fix the connected and simply connected complex semisimple Lie group G and also fix a pair (B, B_-) of opposite Borel subgroups of G. Let N and N_- be the respective uniradicals of B and B_-. Let $T = B \cap B_-$, and let $\mathfrak{t}$ be the Lie algebra of T. Let $\Delta^+ \subset \mathfrak{t}^*$ be the system of positive roots determined by B, and let

$$\mathfrak{g} = \mathfrak{t} + \sum_{\alpha \in \Delta^+} (\mathfrak{g}_\alpha + \mathfrak{g}_{-\alpha})$$

be the root decomposition of $\mathfrak{g}$. Let $N_G(T)$ be the normalizer subgroup of T in G, and let $W = N_G(T)/T$ be the Weyl group. Define the right action of W on T by

$$t^w = \dot{w}^{-1} t \dot{w}, \qquad t \in T,$$

where for $w \in W$, $\dot{w}$ is any representative of w in $N_G(T)$.

Let $\Gamma \subset \Delta^+$ be the set of all simple roots in Δ^+. For $\alpha \in \Gamma$, we will fix root vectors $e_\alpha \in \mathfrak{g}_\alpha$ and $e_{-\alpha} \in \mathfrak{g}_{-\alpha}$ such that $h_\alpha \stackrel{\text{def}}{=} [e_\alpha, e_{-\alpha}]$ satisfies $\alpha(h_\alpha) = 2$. Recall that the collection $(T, B, \{e_\alpha, e_{-\alpha} : \alpha \in \Gamma\})$ is called a *pinning* of G. For $\alpha \in \Gamma$, let

$$u_\alpha, u_{-\alpha}\colon\ \mathbb{C} \longrightarrow G,\ u_\alpha(c) = \exp(ce_\alpha) \ \text{ and } \ u_{-\alpha}(c) = \exp(ce_{-\alpha}), \qquad c \in \mathbb{C},$$

be the corresponding one-parameter unipotent subgroups of G, and define the representative of the simple reflection s_α in $N_G(T)$ by

$$\overline{s_\alpha} = u_\alpha(-1)u_{-\alpha}(1)u_\alpha(-1).$$

By [17], if $w \in W$ and $w = s_{\alpha_1}s_{\alpha_2}\cdots s_{\alpha_l}$ is a reduced decomposition of w, where $\alpha_j \in \Gamma$ for $j = 1, \ldots, l$, the representatives

$$\overline{w} \stackrel{\text{def}}{=} \overline{s_{\alpha_1}}\,\overline{s_{\alpha_2}} \cdots \overline{s_{\alpha_l}} \quad \text{ and } \quad \overline{\overline{w}} \stackrel{\text{def}}{=} \left(\overline{w^{-1}}\right)^{-1}$$

of w in $N_G(T)$ are independent of the choices of the reduced decompositions of w, and $\overline{w_1w_2} = \overline{w_1}\,\overline{w_2}$ and $\overline{\overline{w_1w_2}} = \overline{\overline{w_1}}\,\overline{\overline{w_2}}$ if $l(w_1w_2) = l(w_1) + l(w_2)$.

Let $X(T)$ be the character group of T. For $\lambda \in X(T)$, denote the corresponding group homomorphism $T \to \mathbb{C}^\times$ by $t \mapsto t^\lambda$ for $t \in T$. Define the left action of W on $X(T)$ by

$$(t^w)^\lambda = t^{w(\lambda)}, \qquad t \in T, w \in W, \lambda \in X(T).$$

For $\lambda \in X(T)$, we will also use λ to denote the corresponding element in $\mathfrak{t}^* = \mathbb{C} \otimes_{\mathbb{Z}} X(T)$. The linear pairing between $\mathfrak{t}$ and $\mathfrak{t}^*$ is denoted by $(\,,\,)$.

Let $\mathscr{P}^+ \subset X(T)$ be the set of all dominant weights, and for $\alpha \in \Gamma$, let $\omega_\alpha \subset \mathscr{P}^+$ be the corresponding fundamental weight. Recall from [17] that corresponding to each $\alpha \in \Gamma$ one has the *principal minor* Δ^{ω_α}, which is the regular function on G whose restriction to B_-B is given by

$$\Delta^{\omega_\alpha}(g) = [g]_0^{\omega_\alpha},$$

where for $g \in B_-B = N_-TN$, we write

$$g = [g]_-[g]_0[g]_+, \qquad \text{where } \ [g]_- \in N_-,\ [g]_0 \in T,\ [g]_+ \in N. \tag{5}$$

Recall also from [17] that a *generalized minor* on G is a regular function on G of the form

$$g \longmapsto \Delta_{w_1\omega_\alpha, w_2\omega_\alpha}(g) \stackrel{\text{def}}{=} \Delta^{\omega_\alpha}(\overline{w_1}^{-1}g\overline{w_2}), \qquad g \in G,$$

where $w_1, w_2 \in W$ and $\alpha \in \Gamma$. Let $\lambda \in \mathscr{P}^+$, and write λ as $\lambda = \sum_\alpha n_\alpha \omega_\alpha$ with $n_\alpha \in \mathbb{N}$ for each $\alpha \in \Gamma$. One then has the regular function Δ^λ on G defined by

$$\Delta^\lambda = \prod_{\alpha \in \Gamma} (\Delta^{\omega_\alpha})^{n_\alpha}. \tag{6}$$

For $w_1, w_2 \in W$, define the regular function $\Delta_{w_1\lambda, w_2\lambda}$ on G by

$$\Delta_{w_1\lambda, w_2\lambda}(g) = \Delta^\lambda(\overline{w_1}^{-1}g\overline{w_2}), \qquad g \in G. \tag{7}$$

It is easy to see that $\Delta_{w_1\lambda, w_2\lambda}$ is a matrix coefficient of the highest weight representation of G with highest weight λ (see [37, Definition 6.2]), and

(8) $$\Delta_{w_1\lambda, w_2\lambda} = \prod_{\alpha\in\Gamma} (\Delta_{w_1\omega_\alpha, w_2\omega_\alpha})^{n_\alpha}.$$

Thus $\Delta_{w_1\lambda, w_2\lambda}$ is a monomial, with non-negative exponents, of generalized minors.

1.5 Acknowledgments

We are grateful to M. Yakimov for answering our questions on double Bruhat cells and for pointing out to us certain references. We would also like to thank Jun Peng and Shizhuo Yu for helpful discussions, and we thank the referee for helpful comments.

2 Complete Hamiltonian flows and integrable systems on generalized Bruhat cells

2.1 The complex Poisson Lie group $(G, \pi_{\rm st})$

Let G be any connected and simply connected complex semisimple Lie group with Lie algebra $\mathfrak{g}$, and let the notation be as in §1.4. We will also fix a non-degenerate symmetric invariant bilinear form $\langle\, ,\rangle_{\mathfrak{g}}$ on $\mathfrak{g}$. The restriction of $\langle\, ,\rangle_{\mathfrak{g}}$ to $\mathfrak{t}$ will be denoted by $\langle\, ,\rangle$. As $\langle\, ,\rangle$ is non-degenerate, one has the isomorphism $\#\colon \mathfrak{t}^* \to \mathfrak{t}$, $\lambda \mapsto \lambda^{\#}$, given by

(9) $$\langle \lambda^{\#},\, x\rangle = \lambda(x) = (\lambda, x), \qquad \lambda \in \mathfrak{t}^*,\, x \in \mathfrak{t}.$$

Let $\langle\, ,\rangle$ also denote the bilinear form on $\mathfrak{t}^*$ given by

$$\langle \lambda_1, \lambda_2\rangle = \langle \lambda_1^{\#},\, \lambda_2^{\#}\rangle, \qquad \lambda_1, \lambda_2 \in \mathfrak{t}^*.$$

The choice of the triple $(T, B, \langle\, ,\rangle_{\mathfrak{g}})$ gives rise to the *standard quasi-triangular r-matrix* $r_{\rm st}$ on $\mathfrak{g}$ defined by (see [11, 16])

$$r_{\rm st} = \sum_{i=1}^{r} h_i \otimes h_i + \sum_{\alpha\in\Delta^+} \langle \alpha,\, \alpha\rangle e_{-\alpha} \otimes e_\alpha \in \mathfrak{g}\otimes\mathfrak{g},$$

where $\{h_i\}_{i=1}^r$ is any orthonormal basis of $\mathfrak{h}$ with respect to $\langle\, ,\rangle$, and where, for $\alpha \in \Delta^+$, the root vectors $e_\alpha \in \mathfrak{g}_\alpha$ and $e_{-\alpha} \in \mathfrak{g}_{-\alpha}$ are such that $\alpha([e_\alpha, e_{-\alpha}]) = 2$. Correspondingly, one has the *standard multiplicative* holomorphic Poisson bi-vector field $\pi_{\rm st}$ on G given by

(10) $$\pi_{\rm st}(g) = l_g r_{\rm st} - r_g r_{\rm st} = l_g \Lambda_{\rm st} - r_g \Lambda_{\rm st}, \qquad g \in G,$$

where l_g (resp. r_g) for $g \in G$ is the left (resp. right) translation on G by g, and

$$\Lambda_{\mathrm{st}} = \sum_{\alpha \in \Delta^+} \frac{\langle \alpha, \alpha \rangle}{2} (e_{-\alpha} \otimes e_{\alpha} - e_{\alpha} \otimes e_{-\alpha}) \in \wedge^2 \mathfrak{g}$$

is the skew-symmetric part of r_{st}. We refer to [5, 11, 16] for the relation between the Poisson Lie group (G, π_{st}) and the corresponding quantum group of G. By, for example, [25, 26, 29], the double Bruhat cells $G^{u,v} = (BuB) \cap (B_- vB_-)$, where $u, v \in W$, are precisely the T-orbits (under left or right translation) of symplectic leaves of π_{st} in G. In particular, BuB and $B_- uB_-$, for any $u \in W$, are Poisson submanifolds of (G, π_{st}). The restriction of π_{st} to $G^{u,v}$ will still be denoted by π_{st}.

Remark 2.1. The Poisson structure π_{st} in (10) is the negative of the one defined in [29, §2.5] by Kogan and Zelevinsky, where the bilinear form $\langle \, , \rangle$ on $\mathfrak{t}^*$ is denoted as $(\, ,)$. ◇

2.2 The Poisson structure π_n on the generalized Bruhat cell $\mathcal{O}^{\mathbf{u}}$

For an integer $n \geq 1$, let the product group B^n act on G^n by

$$(g_1, g_2, \ldots, g_n) \cdot (b_1, b_2, \ldots, b_n) = (g_1 b_1, b_1^{-1} g_2 b_2, \ldots, b_{n-1}^{-1} g_n b_n), \quad g_j \in G, b_j \in B,$$

and denote the corresponding quotient space by

$$F_n = G \times_B G \times_B \cdots \times_B G/B. \tag{11}$$

Let $\varpi_n \colon G^n \to F_n$ be the natural projection. For a sequence $\mathbf{u} = (u_1, \ldots, u_n) \in W^n$, let

$$\mathcal{O}^{\mathbf{u}} = Bu_1 B \times_B \cdots \times_B Bu_n B/B \stackrel{\mathrm{def}}{=} \varpi_n((Bu_1 B) \times \cdots \times (Bu_n B)) \subset F_n. \tag{12}$$

The Bruhat decomposition $G = \bigsqcup_{u \in W} BuB$ of G then gives the decomposition

$$F_n = \bigsqcup_{\mathbf{u} \in W^n} \mathcal{O}^{\mathbf{u}} \quad \text{(disjoint union)}. \tag{13}$$

Following [35], $\mathcal{O}^{\mathbf{u}} \subset F_n$ is called a *generalized Bruhat cell*, and when $\mathbf{u}$ is a sequence of *simple reflections*, the generalized Bruhat cell $\mathcal{O}^{\mathbf{u}}$ is said to be *of Bott-Samelson type*. Let $l \colon W \to \mathbb{N}$ be again the length function on W. It is clear that

$$\dim \mathcal{O}^{\mathbf{u}} = l(\mathbf{u}) = l(u_1) + \cdots + l(u_n).$$

It is shown in [34, §7.1] (see also [35, Theorem 1.1]) that

$$\pi_n \stackrel{\mathrm{def}}{=} \varpi_n(\pi_{\mathrm{st}}^n),$$

where $\pi_{\rm st}^n$ is the product Poisson structure on G^n, is a well-defined Poisson structure on F_n. As BuB is a Poisson submanifold of G with respect to $\pi_{\rm st}$ for every $u \in W$, all the generalized Bruhat cells $\mathcal{O}^{\mathbf{u}}$ in F_n are Poisson submanifolds with respect to π_n. The restriction of π_n to each $\mathcal{O}^{\mathbf{u}}$, still denoted by π_n, will be referred to as the *standard Poisson structure* on the generalized Bruhat cell $\mathcal{O}^{\mathbf{u}}$.

Note that the action of T on F_n given by

$$(14)\qquad t\cdot[g_1,g_2,\ldots,g_n] = [tg_1,g_2,\ldots,g_n], \qquad t\in T,\ g_1,g_2,\ldots,g_n \in G,$$

preserves the Poisson structure π_n on F_n, where for $(g_1,\ldots,g_n) \in G^n$,

$$[g_1,\ldots,g_n] = \varpi_n(g_1,\ldots,g_n) \in F_n.$$

Define a *T-leaf* of (F_n,π_n) to be the union $\bigcup_{t\in T}(t\cdot\Sigma)$, where Σ is a symplectic leaf of (F_n,π_n). For $\mathbf{u}\in W^n$ and $w\in W$, define

$$(15)\qquad \mathcal{O}^{\mathbf{u}}_w = \{[g_1,g_2,\ldots,g_n] \in \mathcal{O}^{\mathbf{u}} : g_1g_2\cdots g_n \in B_-wB\} \subset \mathcal{O}^{\mathbf{u}}.$$

It is shown in [35, Theorem 1.1] that the decomposition

$$(16)\qquad F_n = \bigsqcup_{\mathbf{u}\in W^n,\, w\in W} \mathcal{O}^{\mathbf{u}}_w \qquad \text{(disjoint union)}$$

is that of F_n into the T-leaves of π_n (see [35, Theorem 1.1] for a criterion for $\mathcal{O}^{\mathbf{u}}_w \neq \emptyset$). In particular, for $\mathbf{u}\in W^n$, one has the (unique) open T-leaf

$$(17)\qquad \mathcal{O}^{\mathbf{u}}_e = \{[g_1,g_2,\ldots,g_n] \in \mathcal{O}^{\mathbf{u}} : g_1g_2\cdots g_n \in B_-B\}$$

of π_n in $\mathcal{O}^{\mathbf{u}}$, which will play an important role in this paper.

Remark 2.2. We remark that every generalized Bruhat cell $\mathcal{O}^{\mathbf{u}} \subset F_n$ with the Poisson structure π_n is Poisson isomorphic to a generalized Bruhat cell of Bott-Samelson type with the Poisson structure $\pi_{l(\mathbf{u})}$, where $l(\mathbf{u}) = l(u_1)+\cdots l(u_n)$. Indeed, if $u_j = s_{j,1}s_{j,2}\cdots s_{j,l(u_j)}$ is a reduced decomposition of u_j, one then has the sequence

$$\tilde{\mathbf{u}} = (s_{1,1},\, s_{1,2},\, \ldots,\, s_{1,l(u_1)},\, \ldots,\, s_{n,1},\, s_{n,2},\, \ldots s_{n,l(u_n)})$$

of simple reflections of length $l(\mathbf{u})$, and the map $G^{l(u_1)}\times\cdots\times G^{l(u_n)} \to G^n$ given by

$$(g_{1,1},\ldots,g_{1,l(u_1)},\, \ldots,\, g_{n,1},\ldots,g_{n,l(u_n)}) \mapsto (g_{1,1}g_{1,2}\cdots g_{1,l(u_1)},\, \ldots,\, g_{n,1}g_{n,2}\cdots g_{n,l(u_n)})$$

induces a T-equivariant Poisson isomorphism (see [35, §1.3])

$$(18)\qquad (F_{l(\mathbf{u})},\, \pi_{l(\mathbf{u})}) \supset (\mathcal{O}^{\tilde{\mathbf{u}}},\, \pi_{l(\mathbf{u})}) \longrightarrow (\mathcal{O}^{\mathbf{u}},\, \pi_n) \subset (F_n,\, \pi_n).$$

Consequently, to study the Poisson manifold $(\mathcal{O}^{\mathbf{u}},\pi_n)$ for an arbitrary generalized Bruhat cell $\mathcal{O}^{\mathbf{u}}$, it is enough to study the case when $\mathcal{O}^{\mathbf{u}}$ is of Bott-Samelson type. ◇

2.3 The Poisson structure π_n in Bott-Samelson coordinates on $\mathcal{O}^{\mathbf{u}}$

Assume first that $\mathbf{u} = (s_{\alpha_1}, s_{\alpha_2}, \ldots, s_{\alpha_n})$ is any sequence of simple reflections in W, and for notational simplicity, we will also write

(19) $$\mathbf{u} = (s_{\alpha_1}, s_{\alpha_2}, \ldots, s_{\alpha_n}) = (s_1, s_2, \ldots, s_n).$$

Corresponding to $\mathbf{u}$, one has the Bott-Samelson variety

$$Z_{\mathbf{u}} = \varpi_n(P_1 \times P_2 \times \cdots \times P_n) \subset F_n,$$

where $P_j = B \cup Bs_jB$ for each j. It is easy to see that $Z_{\mathbf{u}}$ is a (smooth and projective) Poisson submanifold of F_n with respect to the Poisson structure π_n, and $(\mathcal{O}^{\mathbf{u}}, \pi_n)$ is embedded in $(Z_{\mathbf{u}}, \pi_n)$ as an open Poisson submanifold.

The choice of the pinning in §1.4 gives rise [13, 15] to the atlas

$$\mathscr{A} = \{(\phi^\gamma \colon \mathbb{C}^n \longrightarrow \phi^\gamma(\mathbb{C}^n)) : \gamma \in \Upsilon_{\mathbf{u}}\}$$

on $Z_{\mathbf{u}}$, where $\Upsilon_{\mathbf{u}}$ is the set of all *subexpressions of* $\mathbf{u}$, i.e., the set of all sequences

$$\gamma = (\gamma_1, \ldots, \gamma_n) \in W^n,$$

where $\gamma_j = e$ or $\gamma_j = s_{\alpha_j}$ for each $j = 1, \ldots, n$, with e being the identity element of W. More specifically, for $\gamma = (\gamma_1, \ldots, \gamma_n) \in \Upsilon_{\mathbf{u}}$, one has the embedding $\phi^\gamma \colon \mathbb{C}^n \to Z_{\mathbf{u}}$,

$$\phi^\gamma(x_1, \ldots, x_n) = [u_{-\gamma_1(\alpha_1)}(x_1)\overline{\gamma_1}, u_{-\gamma_2(\alpha_2)}(x_2)\overline{\gamma_2}, \ldots, u_{-\gamma_n(\alpha_n)}(x_n)\overline{\gamma_n}] \in Z_{\mathbf{u}},$$

where $\bar{e} = e \in G$. For each $\gamma \in \Upsilon_{\mathbf{u}}$, it is shown in [14, 15] that the Poisson structure π_n is algebraic in the coordinate chart $\phi^\gamma \colon \mathbb{C}^n \to \phi^\gamma(\mathbb{C}^n)$, and the Poisson brackets among the coordinate functions are expressed using root strings and structure constants of the Lie algebra $\mathfrak{g}$. Note, in particular, that when $\gamma = \mathbf{u}$, $\phi^{\mathbf{u}}(\mathbb{C}^n) = \mathcal{O}^{\mathbf{u}}$.

Definition 2.3. For a sequence $\mathbf{u} = (s_1, s_2, \ldots, s_n)$ of simple reflections, we call the coordinates $(x_1, \ldots, x_n)$ on $\mathcal{O}^{\mathbf{u}}$ via $\phi^{\mathbf{u}} \colon \mathbb{C}^n \to \mathcal{O}^{\mathbf{u}}$ given by

(20) $$\phi^{\mathbf{u}}(x_1, \ldots, x_n) = [u_{\alpha_1}(x_1)\overline{s}_1, u_{\alpha_2}(x_2)\overline{s}_2, \ldots, u_{\alpha_n}(x_n)\overline{s}_n], \qquad (x_1, \ldots, x_n) \in \mathbb{C}^n,$$

Bott-Samelson coordinates on $\mathcal{O}^{\mathbf{u}}$, and we call $\pi_{(G,\mathbf{u})} = (\phi^{\mathbf{u}})^{-1}(\pi_n)$ the *Bott-Samelson Poisson structure on* $\mathbb{C}^n$ *associated to* $(G, \mathbf{u})$. When necessary, we will denote the induced Poisson polynomial algebra by $(\mathbb{C}[x_1, \ldots, x_n], \pi_{(G,\mathbf{u})})$ or $(\mathbb{C}[x_1, \ldots, x_n], \{\,,\}_{(G,\mathbf{u})})$. ◇

Note that in the Bott-Samelson coordinates, the T-action on $\mathcal{O}^{\mathbf{u}}$ in (14) is given by

(21) $$t \cdot \phi^{\mathbf{u}}(x_1, x_2, \ldots, x_n) = \phi^{\mathbf{u}}(t^{\alpha_1}x_1, t^{s_1(\alpha_2)}x_2, \ldots, t^{s_1 s_2 \cdots s_{n-1}(\alpha_n)}x_n), \qquad t \in T.$$

Remark 2.4. By a *T-Poisson algebra* we mean a Poisson algebra with a T-action by Poisson automorphisms. By (21), the Poisson polynomial algebra $(\mathbb{C}[x_1,\ldots,x_n], \pi_{(G,\mathbf{u})})$ is a T-Poisson algebra with the T-action

$$t \cdot_{\mathbf{u}} x_j = t^{s_1 s_2 \cdots s_{j-1}(\alpha_j)} x_j, \qquad t \in T,\ j = 1,\ldots,n. \tag{22}$$

For $1 \le i \le k \le n$, let $\mathbf{u}_{[i,k]} = (s_i, s_{i+1}, \ldots, s_k)$, so one also has the T-Poisson algebra $(\mathbb{C}[x_i, x_{i+1}, \ldots, x_k],\ \pi_{\mathbf{u}_{[i,k]}})$ with the T-action

$$t \cdot_{\mathbf{u}_{[i,k]}} x_j = t^{s_i s_{i+1} \cdots s_{j-1}(\alpha_j)} x_j, \qquad t \in T,\ j = i,\ldots,k. \tag{23}$$

By [15, Theorem 4.14], the inclusion

$$(\mathbb{C}[x_i,\ldots,x_k],\ \pi_{(G,\mathbf{u}_{[i,k]})}) \hookrightarrow (\mathbb{C}[x_1,\ldots,x_n],\ \pi_{(G,\mathbf{u})})$$

is a Poisson algebra embedding but not T-equivariant due to (22) and (23). $\diamond$

We do not need the full strength of the explicit formulas for $\{\,,\}_{(G,\mathbf{u})}$ given in [15, Theorem 4.14], but we will need the following property.

Lemma 2.5. *Assume that $f \in \mathbb{C}[x_1,\ldots,x_n]$ is a T-weight vector with weight $\lambda_f \in X(T)$.*

1) *If $f \in \mathbb{C}[x_2,\ldots,x_n]$, then*

$$\{x_1, f\}_{(G,\mathbf{u})} + \langle \lambda_f,\ \alpha_1\rangle x_1 f \in \mathbb{C}[x_2,\ldots,x_n];$$

2) *If $f \in \mathbb{C}[x_1,\ldots,x_{n-1}]$, then*

$$\{f, x_n\}_{(G,\mathbf{u})} + \langle \lambda_f,\ s_1 s_2 \cdots s_{n-1}(\alpha_n)\rangle x_n f \in \mathbb{C}[x_1,\ldots,x_{n-1}].$$

Proof. For $\zeta \in \mathfrak{t}$, let ∂_ζ be the derivation of $\mathbb{C}[x_1,\ldots,x_n]$ given by

$$\partial_\zeta(x_j) = (s_1 s_2 \cdots s_{j-1}(\alpha_j),\ \zeta) x_j, \qquad j = 1,2,\ldots,n.$$

By [15, Theorem 5.12], there exist a derivation δ of $\mathbb{C}[x_2,\ldots,x_n]$ and a derivation δ' on $\mathbb{C}[x_1,\ldots,x_{n-1}]$ such that

$$\{x_1,\ f\}_{(G,\mathbf{u})} = -\frac{\langle \alpha_1, \alpha_1\rangle}{2} x_1 \partial_{h_{\alpha_1}}(f) + \delta(f), \qquad f \in \mathbb{C}[x_2,\ldots,x_n], \tag{24}$$

$$\{f,\ x_n\}_{(G,\mathbf{u})} = -\frac{\langle \alpha_n, \alpha_n\rangle}{2} x_n \partial_{s_1 s_2 \cdots s_{n-1}(h_{\alpha_n})}(f) + \delta'(f), \qquad f \in \mathbb{C}[x_1,\ldots,x_{n-1}], \tag{25}$$

from which both assertions of Lemma 2.5 now follow. $\square$

One advantage of embedding $(\mathcal{O}^{\mathbf{u}}, \pi_n)$ into $(Z_{\mathbf{u}}, \pi_n)$ is that we can make use of other coordinate charts on $Z_{\mathbf{u}}$ in which the Poisson structure π_n may be simple. Indeed, let

$$(e) \stackrel{\text{def}}{=} (e, e, \ldots, e)$$

be the subexpression of $\mathbf{u}$ in which all the entries of (e) are the identity element of W, and consider the coordinate chart $\phi^{(e)}: \mathbb{C}^n \to \phi^{(e)}(\mathbb{C}^n) \subset Z_{\mathbf{u}}$,

$$\phi^{(e)}(\varepsilon_1,\ldots,\varepsilon_n) = [u_{-\alpha_1}(\varepsilon_1),\, u_{-\alpha_2}(\varepsilon_2),\, \ldots,\, u_{-\alpha_n}(\varepsilon_n)], \qquad (\varepsilon_1,\ldots,\varepsilon_n) \in \mathbb{C}^n. \tag{26}$$

Lemma 2.6. [15, Theorem 4.14] *In the coordinates $(\varepsilon_1,\varepsilon_2,\ldots,\varepsilon_n)$ on $\phi^{(e)}(\mathbb{C}^n)$, the Poisson structure π_n is log-canonical. More precisely,*

$$\{\varepsilon_i,\, \varepsilon_j\} = \langle \alpha_i,\, \alpha_j \rangle \varepsilon_i \varepsilon_j, \qquad 1 \le i < j \le n.$$

Lemma 2.6 will be used in the proof of Proposition 2.13 to compute the log-Hamiltonian vector fields for certain regular functions on $\mathcal{O}^{\mathbf{u}}$.

Definition 2.7. For an arbitrary sequence $\mathbf{u} = (u_1,\ldots,u_n) \in W^n$, choose any reduced word for each u_j and form the sequence $\tilde{\mathbf{u}}$ of simple reflections as in Remark 2.2. Through the Poisson isomorphism $(\mathcal{O}^{\tilde{\mathbf{u}}}, \pi_{l(\mathbf{u})}) \to (\mathcal{O}^{\mathbf{u}}, \pi_n)$ in (18) in Remark 2.2, the Bott-Samelson coordinates on $\mathcal{O}^{\tilde{\mathbf{u}}}$ will also be called Bott-Samelson coordinates on $\mathcal{O}^{\mathbf{u}}$. ◇

Remark 2.8. For the remainder of §2, we will work with generalized Bruhat cells $\mathcal{O}^{\mathbf{u}}$ of Bott-Samelson type, i.e., we will assume that $\mathbf{u}$ is a sequence of *simple reflections*. The results are trivially extended to arbitrary generalized Bruhat cells by the Poisson isomorphism (18) in Remark 2.2. See also Remark 2.31. ◇

2.4 Homogeneous Poisson regular functions on $(\mathcal{O}^{\mathbf{u}}, \pi_n)$

Let (X,π) be a smooth affine Poisson variety, and let $(\mathcal{O}_X, \{\,,\,\})$ be the corresponding Poisson algebra of regular functions on X. Recall that $y \in \mathcal{O}_X$ is said to be *Poisson* if the principal ideal of $\mathcal{O}_X$ generated by y is a Poisson ideal. For such a Poisson element y, define the *log-Hamiltonian vector field $H_{\log(y)}$* on X by

$$H_{\log(y)}(f) = \frac{1}{y}\{y,\, f\}, \qquad f \in \mathcal{O}_X. \tag{27}$$

If a complex algebraic torus T acts on (X,π) preserving the Poisson structure, let T act on $\mathcal{O}_X$ by

$$(t \cdot f)(x) = f(t \cdot x), \qquad t \in T,\, x \in X,$$

so that $\mathcal{O}_X$ becomes a T-Poisson algebra. An element $y \in \mathcal{O}_X$ that is a T-weight vector will also be said to be *T-homogeneous*.

Lemma 2.9. *Let $(\mathbb{C}[x_1,\ldots,x_n], \{\,,\,\})$ be a polynomial T-Poisson algebra for which each x_j is T-homogeneous. If $y \in \mathbb{C}[x_1,\ldots,x_n]$ is T-homogeneous and Poisson, so is every one of its prime factors.*

Proof. Let $y_1 \in \mathbb{C}[x_1,\ldots,x_n]$ be a prime factor of y and write $y = y_1^k y_2$, where k is a positive integer, $y_2 \in \mathbb{C}[x_1,\ldots,x_n]$, and y_1 does not divide y_2. For any

$f \in \mathbb{C}[x_1, \ldots, x_n]$, it follows from

$$y_1^k y_2 H_{\log(y)}(f) = \{y_1^k y_2, f\} = y_1^k \{y_2, f\} + k y_1^{k-1} y_2 \{y_1, f\}$$

that $y_1 | \{y_1, f\}$. Thus y_1 is Poisson. Equip the (algebraic) character group $X(T)$ of T with a total ordering $\preceq$ such that

$$\lambda_1 \preceq \lambda_2 \text{ and } \lambda_3 \preceq \lambda_4 \implies \lambda_1 + \lambda_3 \preceq \lambda_2 + \lambda_4$$

for any $\lambda_j \in X(T)$, $j = 1, 2, 3, 4$. For each non-zero $f \in \mathbb{C}[x_1, \ldots, x_n]$, let $\lambda_{f,\min}$ and $\lambda_{f,\max}$ be the respective minimal and maximal weights appearing in the decomposition of f into the sum of T-weight vectors. It follows from

$$\lambda_{y_1,\min} \lambda_{y_1^{k-1} y_2,\min} = \lambda_{y,\min} = \lambda_{y,\max} = \lambda_{y_1,\max} \lambda_{y_1^{k-1} y_2,\max}$$

that both y_1 and $y_1^{k-1} y_2$ are T-weight vectors. □

Let again $\mathbf{u} = (s_{\alpha_1}, \ldots, s_{\alpha_n}) = (s_1, \ldots, s_n)$. We will identify all the regular functions on $(\mathscr{O}^{\mathbf{u}}, \pi_n)$ that are T-homogeneous and Poisson, and we will compute their log-Hamiltonian vector fields. For $\alpha \in \Gamma$ and $c \in \mathbb{C}$, set

(28) $$p_\alpha(c) = u_\alpha(c) \overline{s}_\alpha \in G.$$

Then the parametrization $\phi^{\mathbf{u}} : \mathbb{C}^n \to \mathscr{O}^{\mathbf{u}}$ can be written as

(29) $$\phi^{\mathbf{u}}(x_1, x_2, \cdots, x_n) = [p_{\alpha_1}(x_1), \, p_{\alpha_2}(x_2), \, \ldots, \, p_{\alpha_n}(x_n)], \qquad (x_1, \ldots, x_n) \in \mathbb{C}^n.$$

Recall from §1.4 that $\mathscr{P}^+ \subset X(T)$ is the set of all dominant weights on T, and that for $\lambda \in \mathscr{P}^+$, Δ^λ is the regular function on G defined in (6).

Definition 2.10. For $\lambda \in \mathscr{P}^+$, let y^λ be the regular function on $\mathscr{O}^{\mathbf{u}}$ given by

(30) $$y^\lambda([p_{\alpha_1}(x_1), \, p_{\alpha_2}(x_2), \, \ldots, \, p_{\alpha_n}(x_n)]) = \Delta^\lambda(p_{\alpha_1}(x_1) p_{\alpha_2}(x_2) \cdots p_{\alpha_n}(x_n)),$$

where $(x_1, \ldots, x_n) \in \mathbb{C}^n$. ◇

Lemma 2.11. *For any $\lambda \in \mathscr{P}^+$, the regular function y^λ on $\mathscr{O}^{\mathbf{u}}$ is T-homogeneous with weight $\lambda - s_1 s_2 \cdots s_n(\lambda)$.*

Proof. Let $x = (x_1, \ldots, x_n) \in \mathbb{C}^n$, $q(x) = [p_{\alpha_1}(x_1), \, \ldots, \, p_{\alpha_n}(x_n)] \in \mathscr{O}^{\mathbf{u}}$, and $t \in T$. It follows from

(31) $$t p_{\alpha_1}(x_1) \cdots p_{\alpha_n}(x_n) = p_{\alpha_1}(t^{\alpha_1} x_1) \cdots p_{\alpha_n}(t^{s_1 \cdots s_{n-1}(\alpha_n)} x_n) t^{s_1 s_2 \cdots s_n} \in G$$

and (21) that

$$\begin{aligned}(t\cdot y^{\lambda})(q(x)) &= \Delta^{\lambda}(p_{\alpha_1}(t^{\alpha_1}x_1)\cdots p_{\alpha_n}(t^{s_1\cdots s_{n-1}(\alpha_n)}x_n))\\ &= \Delta^{\lambda}\left(tp_{\alpha_1}(x_1)\cdots p_{\alpha_n}(x_n)(t^{s_1s_2\cdots s_n})^{-1}\right)\\ &= t^{\lambda-s_1s_2\cdots s_n(\lambda)}y^{\lambda}(q(x)).\end{aligned}$$

Thus y^{λ} is T-homogeneous with weight $\lambda - s_1s_2\cdots s_n(\lambda)$. □

Let $\alpha\in\Gamma$. We make some remarks on the function y^{ω_α} on $\mathcal{O}^{\mathbf{u}}$ that will be used in the next Proposition 2.13 and in §2.8. For the given $\mathbf{u}=(s_1,\ldots,s_n)=(s_{\alpha_1},\ldots,s_{\alpha_n})$, if $\alpha\in\{\alpha_1,\ldots,\alpha_n\}$, let

$$i_\alpha=\min\{1\le k\le n:\alpha_k=\alpha\}\qquad\text{and}\qquad j_\alpha=\max\{1\le k\le n:\alpha_k=\alpha\}.$$

We will regard y^{ω_α} as an element in $\mathbb{C}[x_1,\ldots,x_n]$ via the parametrization $\phi^{\mathbf{u}}$ of $\mathcal{O}^{\mathbf{u}}$.

Lemma 2.12. *Let $\alpha\in\Gamma$. Then $y^{\omega_\alpha}=1$ if $\alpha\notin\{\alpha_1,\ldots,\alpha_n\}$. Otherwise, there exist $A_\alpha,B_\alpha,C_\alpha,D_\alpha\in\mathbb{C}[x_{i_\alpha+1},\ldots,x_{j_\alpha-1}]$ and $A_\alpha\neq0$, such that*

$$y^{\omega_\alpha}(x)=A_\alpha x_{i_\alpha}x_{j_\alpha}+B_\alpha x_{i_\alpha}+C_\alpha x_{j_\alpha}+D_\alpha\in\mathbb{C}[x_{i_\alpha},\ldots,x_{j_\alpha}].$$

Proof. Let V_{ω_α} be the irreducible representation of G with highest weight ω_α and let $v_0\in V_{\omega_\alpha}$ be a highest weight vector. Let $\mathscr{P}'(\omega_\alpha)$ be the set of all weights in V_{ω_α} that are not equal to ω_α, and for each $\mu\in\mathscr{P}'(\omega_\alpha)$, let $V_{\omega_\alpha}(\mu)\subset V_{\omega_\alpha}$ be the corresponding weight space. Then the function Δ^{ω_α} on G is given by

$$gv_0-\Delta^{\omega_\alpha}(g)v_0\in\sum_{\mu\in\mathscr{P}'_{\omega_\alpha}}V_{\omega_\alpha}(\mu),\qquad g\in G. \tag{32}$$

Let $\mathscr{P}''(\omega_\alpha)=\{\mu\in\mathscr{P}'(\omega_\alpha):\mu\neq\omega_\alpha-\alpha\}$. Then one has the direct sum decomposition

$$V_{\omega_\alpha}=\mathbb{C}v_0+\mathbb{C}\overline{s_\alpha}v_0+\sum_{\mu\in\mathscr{P}''(\omega_\alpha)}V_{\omega_\alpha}(\mu).$$

Let $SL(2,\mathbb{C})$ act on V_{ω_α} via the group homomorphism $SL(2,\mathbb{C})\to G$ determined by the choices of $e_\alpha\in\mathfrak{g}_\alpha$ and $e_{-\alpha}\in\mathfrak{g}_{-\alpha}$ in §1.4. Then both $\mathbb{C}v_0+\mathbb{C}\overline{s_\alpha}v_0$ and $\sum_{\mu\in\mathscr{P}''(\omega_\alpha)}V_{\omega_\alpha}(\mu)$ are $SL(2,\mathbb{C})$-invariant, and the resulting representation of $SL(2,\mathbb{C})$ on $\mathbb{C}v_0+\mathbb{C}\overline{s_\alpha}v_0$ is isomorphic to the standard one on $\mathbb{C}^2$. In particular, for any $c\in\mathbb{C}$, one has

$$u_\alpha(c)\overline{s_\alpha}v_0=cv_0+\overline{s_\alpha}v_0,$$

from which it follows (see also [28, Lemma 7]) that for any $c\in\mathbb{C}$,

$$\begin{aligned}\Delta^{\omega_\alpha}(gu_\alpha(c)\overline{s_\alpha})&=c\Delta^{\omega_\alpha}(g)+\Delta^{\omega_\alpha}(g\overline{s_\alpha}),\\ \Delta^{\omega_\alpha}(u_\alpha(c)\overline{s_\alpha}g)&=c\Delta^{\omega_\alpha}(g)+\Delta^{\omega_\alpha}(\overline{s_\alpha}g).\end{aligned} \tag{33}$$

Similarly, one shows (see [17, §2.3]) that if $\alpha'\in\Gamma$ and $\alpha'\neq\alpha$, then

$$\Delta^{\omega_\alpha}(gu_{\alpha'}(c)\overline{s_{\alpha'}})=\Delta^{\omega_\alpha}(u_{\alpha'}(c)\overline{s_{\alpha'}}g)=\Delta^{\omega_\alpha}(g),\qquad g\in G,c\in\mathbb{C}. \tag{34}$$

Lemma 2.12 now follows by applying (33) and (34). □

Recall from (17) the Zariski open subset $\mathcal{O}_e^{\mathbf{u}}$ of $\mathcal{O}^{\mathbf{u}}$ defined by

$$(35)\quad \mathcal{O}_e^{\mathbf{u}} = \{[p_{\alpha_1}(x_1),\, p_{\alpha_2}(x_2),\, \ldots,\, p_{\alpha_n}(x_n)] : p_{\alpha_1}(x_1)p_{\alpha_2}(x_2)\cdots p_{\alpha_n}(x_n) \in B_-B\}.$$

Define $\tau\colon \mathcal{O}_e^{\mathbf{u}} \to T$ by (recall notation in (5))

$$(36)\qquad \tau(\phi^{\mathbf{u}}(x)) = [p_{\alpha_1}(x_1)p_{\alpha_2}(x_2)\cdots p_{\alpha_n}(x_n)]_0, \qquad \phi^{\mathbf{u}}(x) \in \mathcal{O}_e^{\mathbf{u}}.$$

Note that, by definition, for each $\lambda \in \mathcal{P}^+$, one has

$$(37)\qquad y^\lambda(\phi^{\mathbf{u}}(x)) = (\tau(\phi^{\mathbf{u}}(x)))^\lambda, \qquad \text{for } \phi^{\mathbf{u}}(x) \in \mathcal{O}_e^{\mathbf{u}}.$$

Let $\sigma\colon \mathfrak{t} \to \mathfrak{X}^1(Z_{\mathbf{u}})$ be the action of $\mathfrak{t}$ on $Z_{\mathbf{u}}$ induced by the T-action in (14), i.e.,

$$(38)\qquad \sigma(\zeta)(q) = \frac{d}{ds}\Big|_{s=0} \exp(s\zeta)\cdot q, \qquad \zeta \in \mathfrak{t},\ q \in Z_{\mathbf{u}}.$$

Recall the map $\#\colon \mathfrak{t}^* \to \mathfrak{t}, \lambda \mapsto \lambda^\#$, given in (9).

Proposition 2.13. *The set $\{cy^\lambda : c \in \mathbb{C},\, \lambda \in \mathcal{P}^+\}$ is the set of all regular functions on $\mathcal{O}^{\mathbf{u}}$ that are T-homogeneous and Poisson with respect to π_n, and for $\lambda \in \mathcal{P}^+$,*

$$(39)\qquad H_{\log(y^\lambda)} = -\sigma(\lambda^\# + s_1s_2\cdots s_n(\lambda^\#)).$$

Proof. Recall from (26) the coordinate chart $\phi^{(e)}\colon \mathbb{C}^n \to \mathcal{O}^{(e)}$, and note that

$$\mathcal{O}^{(e)} \cap \mathcal{O}^{\mathbf{u}} = \{[u_{-\alpha_1}(\varepsilon_1), u_{-\alpha_2}(\varepsilon_2), \ldots, u_{-\alpha_n}(\varepsilon_n)] : \varepsilon_i \neq 0, i = 1, \ldots, n\} \subset \mathcal{O}_e^{\mathbf{u}}.$$

For $\alpha \in \Gamma$, let $\alpha^\vee\colon \mathbb{C}^\times \to T$ be the co-character of T corresponding to $h_\alpha \in \mathfrak{t}$, so that

$$(\alpha^\vee(c))^\mu = c^{(\mu, h_\alpha)}, \qquad \forall\, c \in \mathbb{C}^\times,\ \mu \in X(T),$$

where recall that $h_\alpha \in [\mathfrak{g}_\alpha, \mathfrak{g}_{-\alpha}]$ is such that $\alpha(h_\alpha) = 2$. We first prove that

$$(40)\qquad \tau(\phi^{(e)}(\varepsilon)) = \prod_{i=1}^n (\alpha_i^\vee(\varepsilon_i^{-1}))^{s_{i+1}s_{i+2}\cdots s_n}, \qquad \varepsilon = (\varepsilon_1, \ldots, \varepsilon_n) \in (\mathbb{C}^\times)^n.$$

To this end, let $x = (x_1, \ldots, x_n) \in \mathbb{C}^n$ be such that $\phi^{(e)}(\varepsilon) = \phi^{\mathbf{u}}(x) \in \mathcal{O}^{(e)} \cap \mathcal{O}^{\mathbf{u}}$. Then for $j = 1, 2, \ldots, n$, there exist $t_j \in T$ and $u_j \in N$ such that

$$u_{-\alpha_1}(\varepsilon_1)\cdots u_{-\alpha_{j-1}}(\varepsilon_{j-1})u_{-\alpha_j}(\varepsilon_j) = p_{\alpha_1}(x_1)\cdots p_{\alpha_{j-1}}(x_{j-1})p_{\alpha_j}(x_j)t_ju_j,$$

and $\tau(\phi^{(e)}(\varepsilon)) = t_n^{-1} \in T$ by definition. A calculation in $SL(2, \mathbb{C})$ shows that

$$(41)\qquad u_{-\alpha}(c) = u_\alpha(c^{-1})\overline{s_\alpha}\alpha^\vee(c)u_\alpha(c^{-1}), \qquad \forall\, \alpha \in \Gamma,\ c \in \mathbb{C}^\times.$$

It follows that $t_1 = \alpha_1^\vee(\varepsilon_1)$, and

$$t_j = (t_{j-1})^{s_j} \alpha_j^\vee(\varepsilon_j), \qquad j = 2, \ldots, n.$$

Thus (40) holds. It now follows from (37) and (40) that on $\mathscr{O}^{(e)} \cap \mathscr{O}^{\mathbf{u}} \subset \mathscr{O}^{(e)}$ and in the coordinates $(\varepsilon_1, \ldots, \varepsilon_n)$ on $\mathscr{O}^{(e)}$, one has

$$y^\lambda = \prod_{i=1}^n \varepsilon_i^{-(s_{i+1}s_{i+2}\cdots s_n(\lambda),\, h_{\alpha_i})}. \tag{42}$$

To prove (39), it is enough to show that it holds on the Zariski open subset $\mathscr{O}^{(e)} \cap \mathscr{O}^{\mathbf{u}}$ of $\mathscr{O}^{\mathbf{u}}$. It is easy to see that the coordinate function ε_j on $\mathscr{O}^{(e)}$ satisfies

$$\sigma(\zeta)(\varepsilon_j) = -\alpha_j(\zeta)\varepsilon_j, \qquad \zeta \in \mathfrak{t},\ j = 1, 2, \ldots, n.$$

It is thus enough to prove that, for any $\lambda \in \mathscr{P}^+$,

$$\{y^\lambda, \varepsilon_j\} = \langle \lambda + s_1 \cdots s_{n-1}s_n(\lambda),\ \alpha_j \rangle y^\lambda \varepsilon_j, \qquad j = 1, 2, \ldots, n. \tag{43}$$

For $i = 1, 2, \ldots, n$, let $m_i = (s_{i+1} \cdots s_{n-1}s_n(\lambda),\ h_{\alpha_i})$. Let $j = 1, 2, \ldots, n$. By (42) and Lemma 2.6, one has

$$\{y^\lambda, \varepsilon_j\} = \langle \beta_j, \alpha_j \rangle y^\lambda \varepsilon_j,$$

where $\beta_j = -\sum_{i=1}^{j-1} m_i\alpha_i + \sum_{i=j+1}^n m_i\alpha_i$. It is easy to see that

$$\beta_j = \lambda + s_1s_2\cdots s_n(\lambda) - s_js_{j+1}\cdots s_n(\lambda) - s_{j+1}\cdots s_n(\lambda),$$

and as $\langle s_js_{j+1}\cdots s_n(\lambda) + s_{j+1}\cdots s_n(\lambda),\ \alpha_j \rangle = 0$, (43) holds. It follows that y^λ is Poisson with respect to π_n, and that its log-Hamiltonian vector field on $\mathscr{O}^{\mathbf{u}}$ is given as in (39).

Assume now that y is any non-zero regular function on $\mathscr{O}^{\mathbf{u}}$ that is T-homogeneous and Poisson with respect to π_n. It remains to show that $y = cy^\lambda$ for some $c \in \mathbb{C}^\times$ and $\lambda \in \mathscr{P}^+$. Let y_1 be a prime factor of y. By Lemma 2.9, y_1 is T-homogeneous and Poisson. Let $X_1 = \{q \in \mathscr{O}^{\mathbf{u}} : y_1(q) = 0\}$. Then X_1 is an irreducible T-invariant Poisson divisor of $(\mathscr{O}^{\mathbf{u}}, \pi_n)$. On the other hand, consider the T-leaf decomposition (see §2.2)

$$\mathscr{O}^{\mathbf{u}} = \bigsqcup_{w \in W} \mathscr{O}^{\mathbf{u}}_w$$

of $(\mathscr{O}^{\mathbf{u}}, \pi_n)$, where $\mathscr{O}^{\mathbf{u}}_w$ is defined in (15), and note that

$$\mathscr{O}^{\mathbf{u}} \backslash \mathscr{O}^{\mathbf{u}}_e = \bigcup_{\alpha \in \{\alpha_1, \alpha_2, \ldots, \alpha_n\}} \overline{\mathscr{O}^{\mathbf{u}}_{s_\alpha}}$$

is the decomposition of the divisor $\mathscr{O}^{\mathbf{u}} \backslash \mathscr{O}^{\mathbf{u}}_e$ into its irreducible components, where $\overline{\mathscr{O}^{\mathbf{u}}_{s_\alpha}}$ is the Zariski closure of $\mathscr{O}^{\mathbf{u}}_{s_\alpha}$ in $\mathscr{O}^{\mathbf{u}}$. If $X_1 \cap \mathscr{O}^{\mathbf{u}}_e \neq \emptyset$, as $\mathscr{O}^{\mathbf{u}}_e$ is a single T-leaf, X_1 would contain an open subset of $\mathscr{O}^{\mathbf{u}}$, which is not possible. Thus X_1 is an irreducible component of the divisor $\mathscr{O}^{\mathbf{u}} \backslash \mathscr{O}^{\mathbf{u}}_e$, and hence $X_1 = \overline{\mathscr{O}^{\mathbf{u}}_{s_{\alpha_i}}}$ for some $i = 1, 2, \ldots, n$. On the other hand, by [17, Proposition 2.4], the Zariski closure $\overline{B_-s_{\alpha_i}B}$ of $B_-s_{\alpha_i}B$ on G

is precisely the zero set of $\Delta^{\omega_{\alpha_i}}$. It follows that $\overline{\mathscr{O}^{\mathbf{u}}_{s_{\alpha_i}}}$ is the zero set of the function $y^{\omega_{\alpha_i}}$. As $\overline{\mathscr{O}^{\mathbf{u}}_{s_{\alpha_i}}}$ is irreducible, $y^{\omega_{\alpha_i}}$ cannot have more than one prime factor. By Lemma 2.12, there exist $1 \leq j \leq n$, $E, F \in \mathbb{C}[x_1, \ldots, x_{j-1}]$, $E \neq 0$, such that $y^{\omega_{\alpha_i}} = x_j E + F$, which implies that $y^{\omega_{\alpha_i}}$ must be prime itself. It follows that $y_1 = c y^{\omega_{\alpha_i}}$ for some $c \in \mathbb{C}^\times$. This finishes the proof of Proposition 2.13. □

Remark 2.14. By Proposition 2.13 and its proof, the regular functions on $\mathscr{O}^{\mathbf{u}}$ that are T-homogeneous, Poisson with respect to π_n and prime, are, up to scalar multiples, precisely the elements y^{ω_α} with $\alpha \in \{\alpha_1, \alpha_2, \ldots, \alpha_n\}$. Moreover, one has

$$\mathscr{O}^{\mathbf{u}}_e = \{q \in \mathscr{O}^{\mathbf{u}} : y^{\omega_\alpha}(q) \neq 0 \, \forall \, \alpha \in \{\alpha_1, \ldots, \alpha_n\}\}. \tag{44}$$

In particular, $\mathscr{O}^{\mathbf{u}}_e$ is an affine variety with an embedding

$$\mathscr{O}^{\mathbf{u}}_e \longrightarrow \mathbb{C}^{n+1}, \quad q \longmapsto ((\phi^{\mathbf{u}})^{-1}(q), \, 1/\varphi(q)), \tag{45}$$

where $\varphi = \prod_{\alpha \in \{\alpha_1, \ldots, \alpha_n\}} y^{\omega_\alpha}$. The restriction of π_n to $\mathscr{O}^{\mathbf{u}}_e$ then makes $\mathscr{O}^{\mathbf{u}}_e$ into a smooth affine Poisson variety. ◇

Corollary 2.15. *For any $\lambda \in \mathscr{P}^+$ and for any regular function f on $\mathscr{O}^{\mathbf{u}}$ that is T-homogeneous with weight λ_f, one has*

$$\{y^\lambda, f\} = -\langle \lambda + s_1 s_2 \cdots s_n(\lambda), \, \lambda_f \rangle y^\lambda f.$$

In particular, for $\lambda, \lambda' \in \mathscr{P}^+$, one has

$$\{y^\lambda, y^{\lambda'}\} = (\langle \lambda, \, s_1 s_2 \cdots s_n(\lambda') \rangle - \langle \lambda', \, s_1 s_2 \cdots s_n(\lambda) \rangle) \, y^\lambda y^{\lambda'}. \tag{46}$$

Remark 2.16. For $w \in W$, equip T with the Poisson structure $\pi_{(T,w)}$ given by

$$\{t^\lambda, t^{\lambda'}\}_{(T,w)} = (\langle \lambda, w(\lambda') \rangle - \langle \lambda', w(\lambda) \rangle) t^{\lambda+\lambda'}, \qquad \lambda, \lambda' \in X(T).$$

It follows from (46) that $\tau \colon (\mathscr{O}^{\mathbf{u}}_e, \pi_n) \to (T, \pi_{(T, s_1 s_2 \cdots s_n)})$ is a Poisson morphism, where $\tau \colon \mathscr{O}^{\mathbf{u}}_e \to T$ is given in (36). ◇

2.5 *The collection $\mathscr{Y}^{\mathbf{u}}$ of regular functions on $(\mathscr{O}^{\mathbf{u}}, \pi_n)$*

Let again

$$\mathbf{u} = (s_{\alpha_1}, \ldots, s_{\alpha_n}) = (s_1, \ldots, s_n)$$

be any sequence of simple reflections for G, and recall the parametrization

$$\phi^{\mathbf{u}} \colon \; \mathbb{C}^n \longrightarrow \mathscr{O}^{\mathbf{u}}, \;\; \phi^{\mathbf{u}}(x_1, x_2, \ldots, x_n) = [p_{\alpha_1}(x_1), \, p_{\alpha_2}(x_2), \, \ldots, \, p_{\alpha_n}(x_n)],$$

where recall that $p_\alpha(c) = u_\alpha(c) \overline{s}_\alpha$ for $\alpha \in \Gamma$ and $c \in \mathbb{C}$.

Definition 2.17. Define, for $1 \le i \le n$, $0 \le i-1 \le j \le n$, and $\lambda \in \mathscr{P}^+$, the regular function $y^{\lambda}_{[i,j]}$ on $\mathscr{O}^{\mathbf{u}}$ by $y^{\lambda}_{[i,j]} = 1$ when $j = i-1$, and

$$y^{\lambda}_{[i,j]}([p_{\alpha_1}(x_1), \ldots, p_{\alpha_n}(x_n)]) = \Delta^{\lambda}(p_{\alpha_i}(x_i)p_{\alpha_{i+1}}(x_{i+1})\cdots p_{\alpha_j}(x_j)), \quad i \le j. \tag{47}$$

Moreover, set

$$\mathscr{Y}^{\mathbf{u}} = \left\{ c y^{\lambda}_{[i,j]} : c \in \mathbb{C},\ 1 \le i \le n, 0 \le i-1 \le j \le n,\ \lambda \in \mathscr{P}^+ \right\}. \tag{48}$$

Note that the constant functions and the coordinate functions $x_1, \ldots, x_n$ are all in $\mathscr{Y}^{\mathbf{u}}$. ◇

Lemma 2.18. *For $1 \le i \le n$, $0 \le i-1 \le j \le n$, and $\lambda \in \mathscr{P}^+$, $y^{\lambda}_{[i,j]}$ is T-homogeneous with weight $s_1 s_2 \cdots s_{i-1}(\lambda) - s_1 s_2 \cdots s_j(\lambda) \in X(T)$.*

Proof. The statement clearly holds when $j = i-1$. Let $1 \le i \le j \le n$. For $t \in T$ and $x = (x_1, \ldots, x_n) \in \mathbb{C}^n$, similar to (31), one has

$$t p_{\alpha_1}(x_1) \cdots p_{\alpha_{i-1}}(x_{i-1}) = p_{\alpha_1}(t^{\alpha_1}x_1) \cdots p_{\alpha_{i-1}}(t^{s_1 \cdots s_{i-2}(\alpha_{i-1})}x_{i-1}) t^{s_1 \cdots s_{i-1}},$$
$$t p_{\alpha_1}(x_1) \cdots p_{\alpha_j}(x_j) = p_{\alpha_1}(t^{\alpha_1}x_1) \cdots p_{\alpha_j}(t^{s_1 \cdots s_{j-1}(\alpha_j)}x_j) t^{s_1 s_2 \cdots s_j}.$$

It follows that

$$p_{\alpha_i}(t^{s_1 \cdots s_{i-1}(\alpha_i)}x_i) \cdots p_{\alpha_j}(t^{s_1 \cdots s_{j-1}(\alpha_j)}x_j) = t^{s_1 \cdots s_{i-1}} p_{\alpha_i}(x_i) \cdots p_{\alpha_j}(x_j)(t^{-1})^{s_1 \cdots s_j}.$$

Similarly to the proof of Lemma 2.11, one sees that $y^{\lambda}_{[i,j]}$ is T-homogeneous with weight $s_1 s_2 \cdots s_{i-1}(\lambda) - s_1 s_2 \cdots s_j(\lambda)$. □

Consider now the Poisson structure π_n on $\mathscr{O}^{\mathbf{u}}$. By Remark 2.14, we also have the affine Poisson variety $(\mathscr{O}^{\mathbf{u}}_e, \pi_n|_{\mathscr{O}^{\mathbf{u}}_e})$. Regard every $y \in \mathscr{Y}^{\mathbf{u}}$ also as a regular function on $\mathscr{O}^{\mathbf{u}}_e$ by restriction, we now prove the first main result of this paper.

Theorem 2.19. *Every $y \in \mathscr{Y}^{\mathbf{u}}$ has complete Hamiltonian flows with property $\mathscr{Q}$ in both $\mathscr{O}^{\mathbf{u}}$ and $\mathscr{O}^{\mathbf{u}}_e$.*

Proof. Identifying $\mathscr{O}^{\mathbf{u}} \cong \mathbb{C}^n$ by $\phi^{\mathbf{u}}$, we now apply Lemma 1.2. Let $\lambda \in \mathscr{P}^+$, $1 \le i \le j \le n$, and $y = y^{\lambda}_{[i,j]} \in \mathbb{C}[x_1, \ldots, x_n]$. Consider the re-ordering of the coordinates

$$(\tilde{x}_1, \ldots, \tilde{x}_n) = (x_i, \ldots, x_j, x_{j+1}, \ldots, x_n, x_{i-1}, \ldots, x_1).$$

Case 1: $i \le k \le j$. Applying Corollary 2.15 and Remark 2.4 to the Poisson subalgebra $\mathbb{C}[x_i, \ldots, x_j]$, one sees that $\{y, x_k\}$ is log-canonical;

Case 2: $j+1 \le k \le n$. Applying Lemma 2.18 and 2) of Lemma 2.5 to the Poisson subalgebra $\mathbb{C}[x_i, \ldots, x_j, x_{j+1}, \ldots, x_k]$, one sees that there exists $c \in \mathbb{C}$ such that

$$\{y, x_k\} - c y x_k \in \mathbb{C}[x_i, \ldots, x_j, x_{j+1}, \ldots, x_{k-1}];$$

Case 3: $1 \le k \le i-1$. Applying Lemma 2.18 and 1) of Lemma 2.5 to the Poisson subalgebra $\mathbb{C}[x_k, \ldots, x_{i-1}, x_i, \ldots, x_j]$, one sees that there exists $c' \in \mathbb{C}$ such that

$$\{y, x_k\} - c' y x_k \in \mathbb{C}[x_{k+1}, \ldots, x_i, \ldots, x_j] \subset \mathbb{C}[x_i, \ldots, x_j, x_{j+1}, \ldots, x_n, x_{i-1}, \ldots, x_{k+1}].$$

In all three cases, the conditions in Lemma 1.2 on the Poisson brackets between y and the coordinate functions $\tilde{x}_1, \ldots, \tilde{x}_n$ are satisfied. Thus y has complete Hamiltonian flow in $\mathbb{C}^n \cong \mathscr{O}^{\mathbf{u}}$ with property $\mathscr{Q}$.

Let $\varphi = \prod_{\alpha \in \{\alpha_1, \ldots, \alpha_n\}} y^{\omega_\alpha}$. By Corollary 2.15, the Poisson bracket $\{y, \varphi\}$ is log-canonical. It now follows from Lemma 1.4 that y, as a regular function on the affine Poisson variety $(\mathscr{O}_e^{\mathbf{u}}, \pi_n|_{\mathscr{O}_e^{\mathbf{u}}})$, has complete Hamiltonian flow in $\mathscr{O}_e^{\mathbf{u}}$ with property $\mathscr{Q}$. □

Notation 2.20. Set $s_{[k,l]} = s_k s_{k+1} \cdots s_l$ for $1 \le k \le l \le n$, and $s_{[k,l]} = e$ when $k > l$. ◇

Lemma 2.21. *For $1 \le i' \le i \le n$, $0 \le i-1 \le j \le j' \le n$, and for any $\lambda \in \mathscr{P}^+$, one has*

$$\left\{ y^{\lambda}_{[i,j]},\ y^{\lambda'}_{[i',j']} \right\} = \langle s_{[1,i-1]}(\lambda) - s_{[1,j]}(\lambda),\ s_{[1,i'-1]}(\lambda') + s_{[1,j']}(\lambda') \rangle\ y^{\lambda}_{[i,j]}\, y^{\lambda'}_{[i',j']}.$$

Proof. The statement clearly holds when $j = i-1$. Assume that $1 \le i' \le i \le j \le j' \le n$. By Corollary 2.15 and Lemma 2.18 applied to $(\mathscr{O}^{(s_{i'}, s_{i'+1}, \ldots, s_{j'})}, \pi_{j'-i'+1})$, one has

$$\begin{aligned}\left\{ y^{\lambda}_{[i,j]},\ y^{\lambda'}_{[i',j']} \right\} &= \langle \lambda' + s_{[i',j']}(\lambda'),\ s_{[i',i-1]}(\lambda) - s_{[i',j]}(\lambda) \rangle\ y^{\lambda}_{[i,j]}\, y^{\lambda'}_{[i',j']} \\ &= \langle s_{[1,i-1]}(\lambda) - s_{[1,j]}(\lambda),\ s_{[1,i'-1]}(\lambda') + s_{[1,j']}(\lambda') \rangle\ y^{\lambda}_{[i,j]}\, y^{\lambda'}_{[i',j']}.\end{aligned}$$ □

2.6 The Poisson structure $0 \bowtie \pi_X$ on $T \times X$

Let T be a complex algebraic torus, and suppose that (X, π_X) is a smooth affine Poisson variety with a T-action by Poisson isomorphisms. Denote again the Lie algebra of T by $\mathfrak{t}$, and let $\theta\colon \mathfrak{t} \to \mathfrak{X}^1(X)$ be the induced Lie algebra action of $\mathfrak{t}$ on X, i.e.,

$$\theta(\zeta)(x) = \left.\frac{d}{ds}\right|_{s=0} \exp(s\zeta)x, \qquad \zeta \in \mathfrak{t},\ x \in X.$$

For any linear map $M\colon \mathfrak{t}^* \to \mathfrak{t}$, we define the Poisson structure $0 \bowtie_M \pi_X$ on $T \times X$ by

$$0 \bowtie_M \pi_X = (0, \pi_X) + \sum_{i=1}^{r} ((M^*(\zeta_i^*))^L, 0) \wedge (0, \theta(\zeta_i)), \tag{49}$$

where $\{\zeta_i\}_{i=1}^r$ is any basis of $\mathfrak{t}$, $\{\zeta_i^*\}_{i=1}^r$ its dual basis of $\mathfrak{t}^*$, and for $\zeta \in \mathfrak{t}$, ζ^L denotes the left (also right) invariant vector field on T with value ζ at the identity element. In particular, for $\lambda \in X(T)$ and $f \in \mathscr{O}_X$,

$$\{t^\lambda, f\} = t^\lambda\, \theta(M(\lambda))(f).$$

Notation 2.22. Assume again that $\langle\, ,\rangle$ is a symmetric non-degenerate bilinear form on $\mathfrak{t}$, and let $\#\colon \mathfrak{t}^* \to \mathfrak{t}$ be as given in (9). For $M = -\#\colon \mathfrak{t}^* \to \mathfrak{t}$, the corresponding Poisson structure on $T \times X$ will be denoted by $0 \bowtie \pi_X$. Then

$$0 \bowtie \pi_X = (0,\, \pi_X) - \sum_{i=1}^{r} (h_i^L,\, 0) \wedge (0,\, \theta(h_i)),$$

where $\{h_i\}_{i=1}^r$ is any orthonormal basis of $\mathfrak{t}$ with respect to $\langle\, ,\rangle$. ◇

Recall again that for $x \in X$, the *T-leaf* of π_X through x is the union $L = \bigcup_{t\in T} t\Sigma_x$, where Σ_x is the symplectic leaf of π_X through x, and the *leaf stabilizer* (see [35, §2]) of $\mathfrak{t}$ at x is the subspace $\mathfrak{t}_x$ of $\mathfrak{t}$ given by

$$\mathfrak{t}_x = \{\zeta \in \mathfrak{t} : \theta(\xi)(x) \in T_x\Sigma_x\}.$$

(Here and for the rest of §2.6, the T in the tangent space T_xX should not be confused with the torus T.) It is shown in [35, §2] that for a given T-leaf of π_X, the leaf stabilizer $\mathfrak{t}_x$ is independent of $x \in L$ and that the codimension of any symplectic leaf of π_X in L is equal to the codimension of $\mathfrak{t}_x$ in $\mathfrak{t}$. Equip $T \times X$ with the diagonal action

$$t_1 \cdot (t,x) = (t_1 t,\, t_1 x), \qquad t_1, t \in T,\, x \in X.$$

It is clear that $0 \bowtie \pi_X$ is T-invariant. In the rest of §2.6, we determine the T-leaves and the leaf stabilizers of $0 \bowtie \pi_X$ in $T \times X$. We first recall from [33, 35] the construction of the *Drinfeld Lagrangian subspace* $\mathfrak{l}_x$ of $\mathfrak{t} \oplus \mathfrak{t}$ associated to the T-action on (X, π_X).

For any Poisson manifold (Z, π) and $z \in Z$, let $\pi_z^\flat\colon T_z^*Z \to T_zZ$ be given by

$$\pi_z^\flat(\alpha_z)(\beta_z) = \pi(z)(\alpha_z, \beta_z), \qquad \alpha_z, \beta_z \in T_z^*Z,$$

so that $\mathrm{Im}(\pi_z^\flat)$ is the tangent space at z to the symplectic leaf of π through z. Let now $L \subset X$ be a T-leaf of π_X, and let $x \in X$. Consider the linear map

$$\theta_x\colon\ \mathfrak{t} \longrightarrow T_xX,\ \ \theta_x(\zeta) = \theta(\zeta)(x), \qquad \zeta \in \mathfrak{t},$$

and its dual map $\theta_x^*\colon T_x^*X \to \mathfrak{t}^*$. Recall from [33] and [35, §2] that associated to the T-Poisson manifold (L, π_X) one has the Lie algebroid $A = (L \times \mathfrak{t}) \bowtie T^*L$ over L with the surjective anchor map $\rho = -\theta + \pi_X^\flat$ from A to the tangent bundle of L. At any $x \in L$, the kernel of $\rho_x\colon \mathfrak{t} \oplus T_x^*L \to T_xL$ is then

$$\ker \rho_x = \{(\zeta,\, \alpha_x) \in \mathfrak{t} \oplus T_x^*L : \theta_x(\zeta) = \pi_{X,x}^\flat(\alpha_x)\}.$$

Consider the vector space $\mathfrak{t} \oplus \mathfrak{t}^* \cong \mathfrak{t} \oplus \mathfrak{t}$ with the symmetric bilinear form

$$\langle(\zeta_1,\, \zeta_2),\, (\zeta_1',\, \zeta_2')\rangle = \langle \zeta_1, \zeta_2'\rangle + \langle \zeta_1', \zeta_2\rangle, \qquad \zeta_1, \zeta_2, \zeta_1', \zeta_2' \in \mathfrak{t}. \tag{50}$$

For each $x \in L$, the linear map

$$\ker \rho_x \longrightarrow \mathfrak{t} \oplus \mathfrak{t}, \ (\zeta, \alpha_x) \longmapsto (\zeta, (\theta_x^*(\alpha_x))^\#),$$

is then injective [12, 33], and its image, denoted by

$$\mathfrak{l}_x \stackrel{\text{def}}{=} \{(\zeta, (\theta_x^*(\alpha_x))^\#) : (\zeta, \alpha_x) \in \ker \rho_x\},$$

is Lagrangian with respect to the bilinear form $\langle , \rangle$ on $\mathfrak{t} \oplus \mathfrak{t}$ given in (50), and it is called the *Drinfeld Lagrangian subspace* of $\mathfrak{t} \oplus \mathfrak{t}$ associated to $x \in L$ and the T-action on (L, π_X). By definition, the leaf stabilizer $\mathfrak{t}_x$ of $\mathfrak{t}$ at x is $p_1(\mathfrak{l}_x)$, where $p_1: \mathfrak{t} \oplus \mathfrak{t} \to \mathfrak{t}$ is the projection to the first factor. Let $p_2: \mathfrak{t} \oplus \mathfrak{t} \to \mathfrak{t}$ be the projection to the second factor, and introduce

(51) $\text{Flip}(\mathfrak{l}_x) = \{(\zeta_1, \zeta_2) : (\zeta_2, \zeta_1) \in \mathfrak{l}_x\} \subset \mathfrak{t} \oplus \mathfrak{t}$,

(52) $\tilde{\mathfrak{t}}_x = p_2(\mathfrak{l}_x) = \{(\theta_x^*(\alpha_x))^\# : \alpha_x \in T_x^*L, \ \pi_{X,x}^\flat(\alpha_x) = \theta_x(\zeta) \text{ for some } \zeta \in \mathfrak{t}\} \subset \mathfrak{t}$.

Lemma 2.23. *With respect to the diagonal action of T on $T \times X$, the T-leaves of the Poisson structure $0 \bowtie \pi_X$ in $T \times X$ are precisely all the $T \times L$, where L is a T-leaf of π_X in X. Moreover, for a T-leaf L in X and $(t,x) \in T \times L$, the Drinfeld Lagrangian subspace at (t,x) for the diagonal T-action on $(T \times L, 0 \bowtie \pi_X)$ is $\mathfrak{l}_{(t,x)} = \text{Flip}(\mathfrak{l}_x)$, the leaf stabilizer of $\mathfrak{t}$ at (t,x) is $\tilde{\mathfrak{t}}_x$, and the rank of the Poisson structure $0 \bowtie \pi_X$ at (t,x) is equal to* $\dim L + \dim \tilde{\mathfrak{t}}_x$.

Proof. Let $\pi = 0 \bowtie \pi_X$ and $(t,x) \in T \times X$. For $a \in \mathfrak{t}^*$, let a^L be the left invariant 1-form on T with value a at the identity element. It follows from the definition of π that

$$\pi_{(t,x)}^\flat(a^L(t), \alpha_x) = \left(((\theta_x^*(\alpha_x))^\#)^L(t), \ \pi_{X,x}^\flat(\alpha_x) - \theta_x(a^\#)\right), \qquad a \in \mathfrak{t}^*, \ \alpha_x \in T_x^*X.$$

Let $\tilde{\theta}_{(t,x)}: \mathfrak{t} \to T_{(t,x)}(T \times X), \ \zeta \mapsto (\zeta^L(t), \theta_x(\zeta))$, and let L be the T-leaf of π_X through x. It is then clear that

$$\text{Im}(\tilde{\theta}_{(t,x)}) + \text{Im}(\pi_{X,x}^\flat) = T_{(t,x)}(T \times L),$$

from which it follows that $T \times L$ is the T-leaf of $\pi = 0 \bowtie \pi_X$ through (t,x).

Let $\mathfrak{l}_{(t,x)} \subset \mathfrak{t} \oplus \mathfrak{t}$ be the Drinfeld Lagrangian subspace for the diagonal T-action on $T \times X$. By definition,

$$\begin{aligned}\mathfrak{l}_{(t,x)} = \Big\{(\zeta, (a + \theta_x^*(\alpha_x))^\#) : \zeta \in \mathfrak{t}, \ a \in \mathfrak{t}^*, \ \alpha_x \in T_x^*L, \ \zeta = (\theta_x^*(\alpha_x))^\#, \\ \theta_x(\zeta) = \pi_{X,x}^\flat(\alpha_x) - \theta_x(a^\#)\Big\} \\ = \{(\zeta, a^\# + \zeta) : \zeta \in \mathfrak{t}, \ a \in \mathfrak{t}^*, \ (a^\# + \zeta, \zeta) \in \mathfrak{l}_x\} = \text{Flip}(\mathfrak{l}_x).\end{aligned}$$

The remaining statements in Lemma 2.23 follow from the fact that $\mathfrak{l}_{(t,x)} = \text{Flip}(\mathfrak{l}_x)$. □

Let now $\mathbf{u} = (u_1, u_2, \ldots, u_n) \in W^n$ be arbitrary, and consider the generalized Bruhat cell $\mathcal{O}^{\mathbf{u}} \subset F_n$, equipped with the Poisson structure π_n and the T-action given in (14), where $T = B \cap B_-$. One then has the Poisson structure $0 \bowtie \pi_n$ on $T \times \mathcal{O}^{\mathbf{u}}$. By definition, the projections to the two factors

$$(T \times \mathcal{O}^{\mathbf{u}},\, 0 \bowtie \pi_n) \longrightarrow (T,\, 0) \quad \text{and} \quad (T \times \mathcal{O}^{\mathbf{u}},\, 0 \bowtie \pi_n) \longrightarrow (\mathcal{O}^{\mathbf{u}},\, \pi_n)$$

are Poisson, where $(T, 0)$ denotes T with the zero Poisson structure, and for any $\lambda \in X(T)$ and any T-homogeneous regular function f on $\mathcal{O}^{\mathbf{u}}$ with T-weight λ_f,

$$\{t^\lambda,\, f\} = -\langle \lambda,\, \lambda_f \rangle t^\lambda f. \tag{53}$$

Let $1 - u_1 u_2 \cdots u_n \colon \mathfrak{t} \to \mathfrak{t}$ be given by $\zeta \mapsto \zeta - u_1 u_2 \cdots u_n(\zeta)$, $\zeta \in \mathfrak{t}$. Recall that $l(\mathbf{u}) = l(u_1) + l(u_2) + \cdots + l(u_n)$.

Proposition 2.24. *Equip $T \times \mathcal{O}^{\mathbf{u}}$ with the diagonal T-action. Then $T \times \mathcal{O}^{\mathbf{u}}_e$ is a single T-leaf of $(T \times \mathcal{O}^{\mathbf{u}}, 0 \bowtie \pi_n)$. The leaf stabilizer of $\mathfrak{t}$ in $T \times \mathcal{O}^{\mathbf{u}}_e$ is* $\mathrm{Im}(1 - u_1 u_2 \cdots u_n)$*, and the symplectic leaves of $0 \bowtie \pi_n$ in $T \times \mathcal{O}^{\mathbf{u}}$ have dimension equal to*

$$l(\mathbf{u}) + \dim \mathrm{Im}(1 - u_1 u_2 \cdots u_n).$$

Proof. By [35, Theorem 1.1], $\mathcal{O}^{\mathbf{u}}_e$ is a single T-leaf of π_n for the T-action on $\mathcal{O}^{\mathbf{u}}$. By Lemma 2.23, $T \times \mathcal{O}^{\mathbf{u}}_e$ is a single T-leaf of $0 \bowtie \pi_n$ for the diagonal T-action on $T \times \mathcal{O}^{\mathbf{u}}$. Let $x \in \mathcal{O}^{\mathbf{u}}_e$. The rest of Proposition 2.24 would follow from Lemma 2.23 once we prove that the Drinfeld Lagrangian subspace $\mathfrak{l}_x \subset \mathfrak{t} \oplus \mathfrak{t}$ for the T-action on $(\mathcal{O}^{\mathbf{u}}_e, \pi_n)$ is given by

$$\mathfrak{l}_x = \{(\zeta + u_1 u_2 \cdots u_n(\zeta),\ -\zeta + u_1 u_2 \cdots u_n(\zeta)) : \ \zeta \in \mathfrak{t}\}. \tag{54}$$

To prove (54), note that by Remark 2.2, we may assume that $\mathbf{u} = (s_1, s_2, \ldots, s_n)$ is a sequence of simple reflections. By [35, Theorem 1.1], the leaf stabilizer of $\mathfrak{t}$ in $\mathcal{O}^{\mathbf{u}}_e$ is $\mathrm{Im}(1 + s_1 s_2 \cdots s_n)$. Consider now the regular functions y^λ on $\mathcal{O}^{\mathbf{u}}$ defined in §2.4, where $\lambda \in \mathcal{P}^+$. By Proposition 2.13, for every $\lambda \in \mathcal{P}^+$, one has

$$\sigma(\lambda^\# + s_1 s_2 \cdots s_n(\lambda^\#)) = -\pi_n^\flat \left(\frac{dy^\lambda}{y^\lambda} \right).$$

As the function y^λ has T-weight $\lambda - s_1 s_2 \cdots s_n(\lambda)$ by Lemma 2.11, it follows from definition that $\mathfrak{l}_x$ is as described. □

2.7 The collection $\widetilde{\mathcal{Y}}^{\mathbf{u}}$ of regular functions on $(T \times \mathcal{O}^{\mathbf{u}}, 0 \bowtie \pi_n)$

Let again $\mathbf{u} = (s_1, \ldots, s_n)$ be any sequence of simple reflections in W, and recall from Definition 2.17 the collection $\mathcal{Y}^{\mathbf{u}}$ of regular functions on $\mathcal{O}^{\mathbf{u}}$.

Definition 2.25. Let $\widetilde{\mathscr{Y}^{\mathbf{u}}}$ be the set of all regular functions on $T \times \mathscr{O}^{\mathbf{u}}$ of the form $t^{\lambda} y$ with $\lambda \in X(T)$ and $y \in \mathscr{Y}^{\mathbf{u}}$, i.e,

$$\widetilde{\mathscr{Y}^{\mathbf{u}}} = \left\{ ct^{\lambda} y_{[i,j]}^{\lambda'} : c \in \mathbb{C},\ 1 \le i \le n, 0 \le i-1 \le j \le n,\ \lambda \in X(T),\ \lambda' \in \mathscr{P}^{+} \right\}.$$

◇

Note that both $T \times \mathscr{O}^{\mathbf{u}}$ and $T \times \mathscr{O}_e^{\mathbf{u}}$ are principal Zariski open subsets in affine spaces and are thus affine varieties. Denote the restriction of the Poisson structure $0 \bowtie \pi_n$ to $T \times \mathscr{O}_e^{\mathbf{u}}$ also by $0 \bowtie \pi_n$. One then has the (smooth) Poisson affine varieties $(T \times \mathscr{O}^{\mathbf{u}}, 0 \bowtie \pi_n)$ and $(T \times \mathscr{O}_e^{\mathbf{u}}, 0 \bowtie \pi_n)$. We also regard $\tilde{y} \in \widetilde{\mathscr{Y}^{\mathbf{u}}}$ as a regular function on $T \times \mathscr{O}_e^{\mathbf{u}}$ by restriction.

Theorem 2.26. *Every $\tilde{y} \in \widetilde{\mathscr{Y}^{\mathbf{u}}}$ has complete Hamiltonian flows with property $\mathscr{Q}$ in both $T \times \mathscr{O}^{\mathbf{u}}$ and $T \times \mathscr{O}_e^{\mathbf{u}}$.*

Proof. Let $\beta_1, \ldots, \beta_r$ be all the simple roots. Let $(\xi_1, \ldots, \xi_r)$ be the coordinates on $\mathbb{C}^r$, and identify T with $(\mathbb{C}^{\times})^r \subset \mathbb{C}^r$ by

$$T \ni t \longmapsto (\xi_1(t), \ldots, \xi_r(t)) \in \mathbb{C}^r,$$

where $\xi_j(t) = t^{\omega_{\beta_j}}$. Let again $(x_1, \ldots, x_n)$ be the coordinates on $\mathscr{O}^{\mathbf{u}}$ via the parametrization $\phi^{\mathbf{u}} : \mathbb{C}^n \to \mathscr{O}^{\mathbf{u}}$. Note then that by (53), the Poisson structure $0 \bowtie \pi_n$ on $T \times \mathscr{O}^{\mathbf{u}} \cong (\mathbb{C}^{\times})^r \times \mathbb{C}^n$ extends to a unique algebraic Poisson structure on $\mathbb{C}^r \times \mathbb{C}^n$, still denoted as $0 \bowtie \pi_n$, such that for any T-homogeneous regular function f on $\mathscr{O}^{\mathbf{u}}$ with T-weight λ_f,

$$\{\xi_k, f\} = -\langle \omega_{\beta_k}, \lambda_f \rangle \xi_k f, \qquad k = 1, \ldots, r. \tag{55}$$

Consider $\mathbb{C}^{1+r+n}$ with coordinates $(\xi_0, \xi_1, \ldots, \xi_r, x_1, \ldots, x_n)$ and extend the Poisson structure $0 \bowtie \pi_n$ on $\mathbb{C}^{r+n}$ to a Poisson structure π on $\mathbb{C}^{1+r+n}$ such that

$$\{\xi_0, \xi_k\} = 0 \;\text{ and }\; \{\xi_0, x_j\} = \langle \omega_{\beta_1} + \cdots + \omega_{\beta_r}, s_1 s_2 \cdots s_{j-1}(\alpha_j) \rangle \xi_0 x_j, \;\; j = 1, \ldots, n,$$

and such that the projection $(\mathbb{C}^{1+r+n}, \pi) \to (\mathbb{C}^{r+n}, 0 \bowtie \pi_n)$ to the last $(r+n)$ coordinates is Poisson. It then follows from the definition that the function $\xi_0 \xi_1 \cdots \xi_r$ on $\mathbb{C}^{1+r+n}$ is a Casimir function. Consider now the affine Poisson subvariety X of $(\mathbb{C}^{1+r+n}, \pi)$ given by

$$X = \{(\xi_0, \xi_1, \ldots, \xi_r, x_1, \ldots, x_n) : \xi_0 \xi_1 \cdots \xi_r = 1\}.$$

The embedding $\Phi : (T \times \mathscr{O}^{\mathbf{u}}, 0 \bowtie \pi_n) \to (\mathbb{C}^{1+r+n}, \pi)$ given by

$$(t, q) \longmapsto ((\xi_1(t) \cdots \xi_r(t))^{-1}, \xi_1(t), \ldots, \xi_r(t), (\phi^{\mathbf{u}})^{-1}(q)), \qquad (t, q) \in T \times \mathscr{O}^{\mathbf{u}},$$

is then Poisson with X as its image. Let $\tilde{y} = t^{\lambda} y_{[i,j]}^{\lambda'} \in \widetilde{\mathscr{Y}^{\mathbf{u}}}$ as in Definition 2.25. Since for each $k = 1 \ldots, r$, $\xi_k^{-1} = \Phi^*(\xi_0 \xi_1 \cdots \xi_{k-1} \xi_{k+1} \cdots \xi_r)$ as regular functions on $T \times \mathscr{O}^{\mathbf{u}}$, there exists a monomial $\mu \in \mathbb{C}[\xi_0, \xi_1, \ldots, \xi_r]$ such that $\Phi^*\left(\mu y_{[i,j]}^{\lambda'}\right) = \tilde{y}$. Consider now the polynomial $y = \mu y_{[i,j]}^{\lambda'} \in \mathbb{C}[\xi_0, \xi_1, \ldots, \xi_r, x_1, \ldots, x_n]$. By considering a re-ordering $(\tilde{x}_1, \ldots, \tilde{x}_n)$ of the coordinates $(x_1, \ldots, x_n)$ as in the proof of Theorem 2.19, and then considering the coordinates $(\xi_0, \xi_1, \ldots, \xi_r, \tilde{x}_1, \ldots, \tilde{x}_n)$ on $\mathbb{C}^{1+r+n}$, one proves as in Theorem 2.19 that y has complete Hamiltonian flow in $\mathbb{C}^{1+r+n}$ with property $\mathscr{Q}$. By Lemma 1.4, $\tilde{y} = \Phi^*(y)$ has complete Hamiltonian flow in $T \times \mathscr{O}^{\mathbf{u}}$ with property $\mathscr{Q}$. Note also that $\Phi(T \times \mathscr{O}_e^{\mathbf{u}}) = \{x \in X : g(x) \neq 0\}$, where

$$g(\xi_0, \xi_1, \ldots, \xi_r, x_1, \ldots, x_n) = \prod_{\alpha \in \{\alpha_1, \ldots, \alpha_n\}} y^{\omega_\alpha}(\phi^{\mathbf{u}}(x_1, \ldots, x_n)) \in \mathbb{C}[\xi_0, \xi_1, \ldots, \xi_r, x_1, \ldots, x_n].$$

By Corollary 2.15, $\{y, g\}$ is log-canonical. By Lemma 1.4, one sees that $\tilde{y}|_{T \times \mathscr{O}_e^{\mathbf{u}}}$ has complete Hamiltonian flow in $T \times \mathscr{O}_e^{\mathbf{u}}$ with property $\mathscr{Q}$. □

It follows from Lemma 2.21 that for $1 \leq i' \leq i \leq n+1$, $i-1 \leq j \leq j' \leq n$, and any $\lambda_1, \lambda_2, \lambda_1', \lambda_2' \in \mathscr{P}^+$, the Poisson bracket between $t^{\lambda_1} y_{[i,j]}^{\lambda_2}$ and $t^{\lambda_1'} y_{[i',j']}^{\lambda_2'}$ is log-canonical. We single out the following cases, which will appear in our application to double Bruhat cells in §3, and for which the Poisson brackets are especially simple.

Notation 2.27. For $1 \leq i \leq n+1$, $i-1 \leq j \leq n$, and $\lambda \in \mathscr{P}^+$, let

$$\tilde{y}_{[i,j]}^{\lambda} = t^{-s_{[1,i-1]}(\lambda)} y_{[i,j]}^{\lambda} \in \widetilde{\mathscr{Y}^{\mathbf{u}}},$$

where $y_{[i,j]}^{\lambda} = 1$ if $j = i-1$. Note then that with respect to the diagonal T-action on $T \times \mathscr{O}^{\mathbf{u}}$, $\tilde{y}_{[i,j]}^{\lambda}$ is T-homogeneous with weight $-s_{[1,j]}(\lambda)$ when $i \leq j$ (see Notation 2.20). ◇

Lemma 2.28. *For any $\lambda, \lambda' \in \mathscr{P}^+$ and $1 \leq i' \leq i \leq n+1$, $i-1 \leq j \leq j' \leq n$, and with respect to the Poisson structure $0 \bowtie \pi_n$ on $T \times \mathscr{O}^{\mathbf{u}}$, one has*

$$\left\{\tilde{y}_{[i,j]}^{\lambda}, \tilde{y}_{[i',j']}^{\lambda'}\right\} = (\langle s_{[1,i-1]}(\lambda), s_{[1,i'-1]}(\lambda')\rangle - \langle s_{[1,j]}(\lambda), s_{[1,j']}(\lambda')\rangle)\, \tilde{y}_{[i,j]}^{\lambda} \tilde{y}_{[i',j']}^{\lambda'}. \tag{56}$$

Proof. If $j' = i'-1$, then $j = i-1$, and both sides of (56) are zero. Assume that $j' \geq i'$. Set $a = s_{[1,i-1]}(\lambda)$, $b = s_{[1,j]}(\lambda)$, $a' = s_{[1,i'-1]}(\lambda')$, and $b' = s_{[1,j']}(\lambda')$. It follows from the definition of $0 \bowtie \pi_n$ and Lemma 2.21 that

$$\left\{\tilde{y}_{[i,j]}^{\lambda}, \tilde{y}_{[i',j']}^{\lambda'}\right\} = c\, \tilde{y}_{[i,j]}^{\lambda} \tilde{y}_{[i',j']}^{\lambda'},$$

with $c = -\langle a', a-b\rangle + \langle a-b, a'+b'\rangle + \langle a, a'-b'\rangle = \langle a, a'\rangle - \langle b, b'\rangle$. □

2.8 Integrable systems on $\mathcal{O}^{(\mathbf{u}^{-1},\mathbf{u})}$

Let again $\mathbf{u} = (s_{\alpha_1}, \ldots, s_{\alpha_n}) = (s_1, \ldots, s_n)$ be any sequence of simple reflections in W. Consider the sequence

$$(\mathbf{u}^{-1}, \mathbf{u}) = (s_n,\ \ldots,\ s_2,\ s_1,\ s_1,\ s_2,\ \ldots, s_n)$$

and the Poisson manifold $(\mathcal{O}^{(\mathbf{u}^{-1},\mathbf{u})}, \pi_{2n})$. Recall the parametrization $\phi^{(\mathbf{u}^{-1},\mathbf{u})} : \mathbb{C}^{2n} \to \mathcal{O}^{(\mathbf{u}^{-1},\mathbf{u})}$ given by

$$\phi^{(\mathbf{u}^{-1},\mathbf{u})}(x_1, \ldots, x_{2n}) = [p_{\alpha_n}(x_1),\ \ldots,\ p_{\alpha_1}(x_n),\ p_{\alpha_1}(x_{n+1}),\ \ldots,\ p_{\alpha_n}(x_{2n})].$$

For $k = 1, \ldots, n$, define the regular functions $y_k \in \mathcal{Y}^{(\mathbf{u}^{-1},\mathbf{u})}$ on $\mathcal{O}^{(\mathbf{u}^{-1},\mathbf{u})}$ (see notation in Definition 2.17):

$$(57)\quad y_k = \Delta^{\omega_{\alpha_k}}(p_{\alpha_k}(x_{n+1-k}) \cdots p_{\alpha_1}(x_n)\, p_{\alpha_1}(x_{n+1}) \cdots p_{\alpha_k}(x_{n+k})) = y^{\omega_{\alpha_k}}_{[n+1-k,\, n+k]}.$$

Theorem 2.29. *The functions $y_1, \ldots, y_n$ form an integrable system on $(\mathcal{O}^{(\mathbf{u}^{-1},\mathbf{u})}, \pi_{2n})$ that has complete Hamiltonian flows with property $\mathcal{Q}$.*

Proof. By Lemma 2.21, $\{y_k, y_{k'}\} = 0$ for $1 \le k < k' \le n$. By [35, Theorem 1.1], the Poisson structure π_{2n} is non-degenerate everywhere in $\mathcal{O}_e^{(\mathbf{u}^{-1},\mathbf{u})}$. It follows from Lemma 2.12 that the functions $y_1, \ldots, y_n$ are functionally independent on $\mathcal{O}^{(\mathbf{u}^{-1},\mathbf{u})}$. By Theorem 2.26, $y_1, \ldots, y_n$ form an integrable system on $(\mathcal{O}^{(\mathbf{u}^{-1},\mathbf{u})}, \pi_{2n})$ that has complete Hamiltonian flows with property $\mathcal{Q}$. □

Remark 2.30. Introduce the regular function $\tilde{y}_k$ on $T \times \mathcal{O}^{(\mathbf{u}^{-1},\mathbf{u})}$ by (see Notation 2.27)

$$(58)\quad \tilde{y}_k = t^{-s_n s_{n-1} \cdots s_k(\omega_{\alpha_k})} \Delta^{\omega_{\alpha_k}}(p_{\alpha_{k-1}}(x_{n+2-k}) \cdots p_{\alpha_1}(x_n) p_{\alpha_1}(x_{n+1}) \cdots p_{\alpha_{k-1}}(x_{n+k-1}))$$
$$= \tilde{y}^{\omega_{\alpha_k}}_{[n-k+2,\, n+k-1]}, \qquad k = 1, \ldots, n.$$

As we will see in Example 3.10, when $\mathbf{u}$ is a reduced word, the functions $\tilde{y}_1, \ldots, \tilde{y}_n$ form an integrable system on $T \times \mathcal{O}_e^{(\mathbf{u}^{-1},\mathbf{u})}$, which under the Fomin-Zelevinsky isomorphism (see Definition 3.4) is precisely the Kogan-Zelevinsky integrable system on the double Bruhat cell $G^{u,u}$ defined by $\mathbf{u}$, where $u = s_1 s_2 \cdots s_n$. For an arbitrary $\mathbf{u} = (s_1, \ldots, s_n)$, not necessarily reduced, we know by Proposition 2.24 that $T \times \mathcal{O}_e^{(\mathbf{u}^{-1},\mathbf{u})}$ is a single T-leaf of the Poisson structure $0 \bowtie \pi_{2n}$ (for the diagonal T-action), all the symplectic leaves have dimension $2n$, and by Lemma 2.28, the functions $\tilde{y}_1, \ldots, \tilde{y}_n$ pairwise Poisson commute. However, we are unable to show that the functions $\tilde{y}_1, \ldots, \tilde{y}_n$ are functionally independent when restricted to a symplectic leaf of $0 \bowtie \pi_{2n}$, as we need to better understand the symplectic leaves in $T \times \mathcal{O}_e^{(\mathbf{u}^{-1},\mathbf{u})}$. We will leave this to future research. ◇

Remark 2.31. For any arbitrary sequence $\mathbf{u} = (u_1, \ldots, u_n)$ in W, by choosing a reduced word for each u_j and by identifying $\mathcal{O}^{\mathbf{u}}$ with an $\mathcal{O}^{\tilde{\mathbf{u}}}$ of Bott-Samelson type as in Remark 2.2, one then has the collections of regular functions $\tilde{\mathscr{Y}}$ on $\mathcal{O}^{\mathbf{u}}$ and $\widetilde{\mathscr{Y}^{\mathbf{u}}}$ on $T \times \mathcal{O}^{\mathbf{u}}$, and Theorem 2.19 and Theorem 2.26 apply. Moreover, one has the integrable system on $\mathcal{O}^{(\mathbf{u}^{-1},\mathbf{u})}$ defined in §2.29, where $\mathbf{u}^{-1} = (u_n^{-1}, \ldots, u_1^{-1})$, which has complete Hamiltonian flows with property $\mathscr{Q}$. ◇

Example 2.32. Consider $G = SL(2,\mathbb{C})$ and $\mathbf{u}_n = (s,s,\ldots,s)$ of length n, where $s = s_\alpha$, and α is the only simple root of $SL(2,\mathbb{C})$. Take the bilinear form $\langle\, , \rangle_{\mathfrak{g}}$ on $\mathfrak{g} = \mathfrak{sl}(2,\mathbb{C})$ such that $\langle \alpha, \alpha \rangle = 1$. Then the Bott-Samelson Poisson structure associated to $(SL(2,\mathbb{C}), \mathbf{u}_n)$ on $\mathbb{C}^n$ is explicitly given by (see [15, §6.2])

$$\{x_i,\, x_{i+1}\} = x_i x_{i+1} - 1, \qquad 1 \le i \le n-1,$$
$$\{x_i,\, x_j\} = (-1)^{j-i+1} x_i x_j, \qquad 1 \le i < j \le n,\ j-i \ge 2.$$

Let ω be the only fundamental weight. To see what the polynomials $y^{\omega}_{[i,j]} \in \mathbb{C}[x_i, \ldots, x_j]$ are for $1 \le i \le j \le n$, for any integer $k \ge 1$, introduce $E_k \in \mathbb{Z}[x_1, \ldots, x_k]$ via the identity

$$\begin{pmatrix} x_1 & -1 \\ 1 & 0 \end{pmatrix} \begin{pmatrix} x_2 & -1 \\ 1 & 0 \end{pmatrix} \cdots \begin{pmatrix} x_k & -1 \\ 1 & 0 \end{pmatrix} = \begin{pmatrix} E_k(x_1, \ldots, x_k) & * \\ * & * \end{pmatrix} \in SL(2,\mathbb{C}). \tag{59}$$

The polynomials E_k are, with a variation of signs, the well-known *Euler continuants* [27], and are related to continued fractions via

$$\frac{E_j(x_1,\ldots,x_j)}{E_{j-1}(x_2,\ldots,x_j)} = x_1 - \cfrac{1}{x_2 - \cfrac{1}{\cdots - \cfrac{1}{x_{j-1} - \cfrac{1}{x_j}}}},$$

and can be recursively defined by $E_0 = 1$, $E_1(x_1) = x_1$, and

$$E_j(x_1, \ldots, x_j) = x_j E_{j-1}(x_1, \ldots, x_{j-1}) - E_{j-2}(x_1, \ldots, x_{j-2}), \qquad j \ge 2.$$

In our example, as $q_\alpha(c) = \begin{pmatrix} c & -1 \\ 1 & 0 \end{pmatrix}$ for $c \in \mathbb{C}$, it follows from the definition that

$$y^{\omega}_{[i,j]} = E_{j-i+1}(x_i, x_{i+1}, \ldots, x_j) \in \mathbb{Z}[x_i, x_{i+1}, \ldots, x_j], \qquad 1 \le i \le j \le n.$$

As $(\mathbf{u}_n^{-1}, \mathbf{u}_n) = \mathbf{u}_{2n}$, the polynomials $y_1, \ldots, y_n$ forming an integrable system on $\mathbb{C}^{2n}$ that has complete Hamiltonian flows with property $\mathscr{Q}$, as stated in Theorem 2.29, are given by

$$y_k = E_{2k}(x_{n-k+1}, x_{n-k+2}, \ldots, x_{n+k}), \qquad k = 1, 2, \ldots, n.$$

For example, for $n = 2$, we have $y_1 = x_2x_3 - 1$ and $y_2 = x_1x_2x_3x_4 - x_1x_2 - x_1x_4 - x_3x_4 + 1$. For $p = (x_1,x_2,x_3,x_4) \in \mathbb{C}^4$, the integral curve $\gamma_i\colon \mathbb{C} \to \mathbb{C}^4$ of the Hamiltonian vector field of y_i through p is given by

$$\gamma_1(c) = \begin{cases} (x_1 + cx_3,\ x_2,\ x_3,\ x_4 - cx_2), & y_1 = 0, \\ (x_1 + x_3(e^{cy_1} - 1)/y_1,\ e^{-cy_1}x_2,\ e^{cy_1}x_3,\ x_4 + x_2(e^{-cy_1} - 1)/y_1), & y_1 \neq 0, \end{cases}$$

$$\gamma_2(c) = (x_1e^{-cy_2},\ x_2e^{cy_2},\ x_3e^{-cy_2},\ x_4e^{cy_2}),$$

where for $i = 1,2$, y_i also denotes the value $y_i(p)$ of the function y_i at p. ◇

3 Complete Hamiltonian flows of generalized minors and the Kogan-Zelevinsky integrable systems

In this section we introduce Bott-Samelson coordinates on double Bruhat cells via the Fomin-Zelevinsky embeddings, and we apply results from §2 to establish the completeness of Hamiltonian flows of all Fomin-Zelevinsky minors and Kogan-Zelevinsky integrable systems on double Bruhat cells.

3.1 Fomin-Zelevinsky embeddings and Bott-Samelson coordinates on double Bruhat cells

For $w \in W$ and a representative $\dot{w} \in N_G(T)$ of w, let

$$C_{\dot{w}} = (N\dot{w}) \cap (\dot{w}N_-) \quad \text{and} \quad D_{\dot{w}} = (N_-\dot{w}) \cap (\dot{w}N).$$

The multiplication map of G then gives diffeomorphisms

$$C_{\dot{w}} \times B \longrightarrow BwB, \qquad B_- \times C_{\dot{w}} \longrightarrow B_-wB_-, \qquad \text{and} \qquad D_{\dot{w}} \times B_- \longrightarrow B_-wB_-.$$

With the representatives $\overline{w}$ and $\overline{\overline{w}}$ in $N_G(T)$ from §1.4 for $w \in W$, one then has

$$C_{\overline{\overline{w}}^{-1}} = C_{\overline{w^{-1}}} = \{g^{-1} : g \in D_{\overline{\overline{w}}}\}.$$

Notation 3.1. For $u, v \in W$, we will write each $g \in G^{u,v}$ uniquely as

$$g = n\overline{u}n't' = m\overline{\overline{v}}m't, \tag{60}$$

where $n\overline{u} \in C_{\overline{u}}$, $m\overline{\overline{v}} \in D_{\overline{\overline{v}}}$, $n' \in N$, $m' \in N_-$, $t,t' \in T$. With $g \in G^{u,v}$ thus written,

$$\overline{u}^{-1}g,\ g\overline{\overline{v}}^{-1},\ (m\overline{\overline{v}})^{-1}n\overline{u} \in N_-TN, \qquad [\overline{u}^{-1}g]_0 = t', \qquad [g\overline{\overline{v}}^{-1}]_0^v = t, \tag{61}$$

$$[(m\overline{\overline{v}})^{-1}n\overline{u}]_- = m', \qquad [(m\overline{\overline{v}})^{-1}n\overline{u}]_0 = t(t')^{-1}, \qquad [(m\overline{\overline{v}})^{-1}n\overline{u}]_+ = (n')^{-1}. \tag{62}$$

◇

Fix $u, v \in W$, and recall from §2.2 the generalized Bruhat cell

$$\mathscr{O}^{(v^{-1},u)} = Bv^{-1}B \times_B BuB/B \subset F_2$$

and its Zariski open subset $O_e^{(v^{-1},u)} = \{[g_1, g_2] \in \mathscr{O}^{(v^{-1},u)} : g_1 g_2 \in B_-B\}$. Note that

$$C_{\overline{v^{-1}}} \times C_{\overline{u}} \longrightarrow \mathscr{O}^{(v^{-1},u)}, \ (c_1, c_2) \longmapsto [c_1, c_2],$$

is an isomorphism. The following is a modification of [17, Proposition 3.1].

Proposition 3.2. *For any $u, v \in W$, the map*

$$F^{u,v}: \ G^{u,v} \longrightarrow T \times \mathscr{O}_e^{(v^{-1},u)}, \ \ g \longmapsto (t, \ [(m\overline{\overline{v}})^{-1}, \ n\overline{u}]), \tag{63}$$

where $g \in G^{u,v}$ is as in (60)*, is a biregular isomorphism, with its inverse given by*

$$\begin{aligned} T \times \mathscr{O}_e^{(v^{-1},u)} \longrightarrow G^{u,v}: \ (t, [(m\overline{\overline{v}})^{-1}, n\overline{u}]) &\longmapsto n\overline{u}[(m\overline{\overline{v}})^{-1}n\overline{u}]_+^{-1}[(m\overline{\overline{v}})^{-1}n\overline{u}]_0^{-1}t \\ &= m\overline{\overline{v}}\,[(m\overline{\overline{v}})^{-1}n\overline{u}]_- t, \end{aligned} \tag{64}$$

where $m\overline{\overline{v}} \in D_{\overline{\overline{v}}}$, and $n\overline{u} \in C_{\overline{u}}$.

Proof. Consider the Zariski open subset $\mathscr{U}^{u,v}$ of

$$(C_{\overline{u}}\overline{u}^{-1}) \times T \times (\overline{\overline{v}}C_{\overline{\overline{v}}^{-1}}) = (N \cap \overline{u}N_-\overline{u}^{-1}) \times T \times (N_- \cap \overline{\overline{v}}N\overline{\overline{v}}^{-1})$$

consisting of all $(n, t', m) \in (C_{\overline{u}}\overline{u}^{-1}) \times T \times (\overline{\overline{v}}C_{\overline{\overline{v}}^{-1}})$ such that $m^{-1}n \in \overline{\overline{v}}B_-B\overline{u}^{-1}$. Proposition 3.1 of [17] says that, with $g \in G^{u,v}$ as in (60),

$$G^{u,v} \longrightarrow \mathscr{U}^{u,v}, \ g \longmapsto (n, t', m^{-1}), \quad g \in G^{u,v},$$

is a biregular isomorphism from $G^{u,v}$ to $\mathscr{U}^{u,v}$. It follows that the map

$$G^{u,v} \longrightarrow T \times \mathscr{O}_e^{(v^{-1},u)}, \ g \longmapsto (t', \ [(m\overline{\overline{v}})^{-1}, \ n\overline{u}]), \quad g \in G^{u,v},$$

is a biregular isomorphism. As $t = t'[(m\overline{\overline{v}})^{-1}n\overline{u}]_0$, it follows that $F^{u,v}$ is a biregular isomorphism, and the formula for its inverse follows from (62). □

Remark 3.3. In the notation of Proposition 3.2, the isomorphism $F^{u,v}$ is T-equivariant if T acts on $G^{u,v}$ by left translation and on $T \times \mathscr{O}_e^{(v^{-1},u)}$ by

$$t_1 \cdot (t, q) = (t_1^v t, \ t_1^v q), \quad t_1, t \in T, \ q \in \mathscr{O}_e^{(v^{-1},u)}.$$

If T acts on $G^{u,v}$ by right translation, then $F^{u,v}$ is T-equivariant if T acts on $T \times \mathscr{O}_e^{(v^{-1},u)}$ by $(t,q) \cdot t_1 = (tt_1, q)$ for $t_1, t \in T$ and $q \in \mathscr{O}_e^{(v^{-1},u)}$. ◇

Recall from [2] that for $u,v \in W$, the quotient $G^{u,v}/T \subset G/T$ of the double Bruhat cell $G^{u,v}$ is called a *reduced double Bruhat cell.*

Definition 3.4. For $u,v \in W$, the biregular isomorphisms

$$\hat{F}^{u,v}:\ G^{u,v}/T \longrightarrow \mathcal{O}_e^{(v^{-1},u)},\quad g.T \longmapsto [(m\overline{\overline{v}})^{-1}, n\overline{u}], \tag{65}$$

$$F^{u,v}:\ G^{u,v} \longrightarrow T \times \mathcal{O}_e^{(v^{-1},u)},\quad g \longmapsto (t, [(m\overline{\overline{v}})^{-1}, n\overline{u}]), \tag{66}$$

where, again, $g \in G^{u,v}$ is decomposed as in (60), will be called, respectively, the *Fomin-Zelevinsky isomorphisms* for $G^{u,v}/T$ and for $G^{u,v}$. The corresponding open embeddings via the embedding $\mathcal{O}_e^{(v^{-1},u)} \hookrightarrow \mathcal{O}^{(v^{-1},u)}$ are also denoted by

$$\hat{F}^{u,v}:\ G^{u,v}/T \longrightarrow \mathcal{O}^{(v^{-1},u)} \quad \text{and} \quad F^{u,v}:\ G^{u,v} \longrightarrow T \times \mathcal{O}^{(v^{-1},u)}, \tag{67}$$

and will be respectively called the *Fomin-Zelevinsky embeddings* of $G^{u,v}/T$ and $G^{u,v}$. ◇

Let again $u,v \in W$, and let $a = l(v)$ and $b = l(u)$. Fix reduced words

$$\mathbf{v} = (s_a,\ s_{a-1},\ \cdots,\ s_1) \quad \text{and} \quad \mathbf{u} = (s_{a+1},\ s_{a+2},\ \cdots,\ s_{a+b}) \tag{68}$$

of v and u, and consider the generalized Bruhat cell $\mathcal{O}^{(\mathbf{v}^{-1},\mathbf{u})} \subset F_{a+b}$, where

$$(\mathbf{v}^{-1}, \mathbf{u}) = (s_1, s_2, \ldots, s_a, s_{a+1}, s_{a+2}, \ldots, s_{a+b}). \tag{69}$$

Consider the isomorphism $\mu: \mathcal{O}^{(\mathbf{v}^{-1},\mathbf{u})} \to \mathcal{O}^{(v^{-1},u)}$ given by

$$\mu([g_1, \ldots, g_a, g_{a+1}, \ldots, g_{a+b}]) = [g_1 g_2 \cdots g_a,\ g_{a+1} g_{a+2} \cdots g_{a+b}], \tag{70}$$

where $g_j \in Bs_jB$ for $j = 1, \ldots, a+b$, and define the open embeddings

$$\hat{F}^{\mathbf{u},\mathbf{v}} \overset{\text{def}}{=} \mu^{-1} \circ \hat{F}^{u,v}:\ G^{u,v}/T \longrightarrow \mathcal{O}^{(\mathbf{v}^{-1},\mathbf{u})}, \tag{71}$$

$$F^{\mathbf{u},\mathbf{v}} \overset{\text{def}}{=} \mu^{-1} \circ F^{u,v}:\ G^{u,v} \longrightarrow T \times \mathcal{O}^{(\mathbf{v}^{-1},\mathbf{u})}. \tag{72}$$

Note that the images of $\hat{F}^{\mathbf{u},\mathbf{v}}$ and $F^{\mathbf{u},\mathbf{v}}$ are respectively $\mathcal{O}_e^{(\mathbf{v}^{-1},\mathbf{u})}$ and $T \times \mathcal{O}_e^{(\mathbf{v}^{-1},\mathbf{u})}$. Recall also the coordinates $(x_1, \ldots, x_{a+b})$ on $\mathcal{O}^{(\mathbf{v}^{-1},\mathbf{u})}$ given by $\phi^{(\mathbf{v}^{-1},\mathbf{u})}: \mathbb{C}^{a+b} \to \mathcal{O}^{(\mathbf{v}^{-1},\mathbf{u})}$.

Definition 3.5. We call $\hat{F}^{\mathbf{u},\mathbf{v}}$ (resp. $F^{\mathbf{u},\mathbf{v}}$) the Fomin-Zelevinsky embedding of $G^{u,v}/T$ (resp. of $G^{u,v}$) associated to the reduced words $\mathbf{u}$ and $\mathbf{v}$ of u and v in (68), and we call $(x_1, \ldots, x_{a+b})$ (resp. $(t, x_1, \ldots, x_{a+b})$, where $t \in T$,) the *Bott-Samelson coordinates* on $G^{u,v}/T$ (resp. on $G^{u,v}$) associated to the reduced words $\mathbf{u}$ and $\mathbf{v}$ of u and v in (68). ◇

3.2 Fomin-Zelevinsky minors on $G^{u,v}$ in Bott-Samelson coordinates

Let again $u, v \in W$. Recall from [17] the Fomin-Zelevinsky twist map

$$(73) \qquad G^{u,v} \longrightarrow G^{u^{-1},v^{-1}}, \; g \longmapsto g' \stackrel{\text{def}}{=} \left([\overline{u}^{-1}g]_-^{-1}\overline{u}^{-1}g\overline{\overline{v}}^{-1}[g\overline{\overline{v}}^{-1}]_+^{-1}\right)^{\theta},$$

where θ is the involutive automorphism of G satisfying

$$\theta(t) = t^{-1}, \qquad \theta(u_\alpha(c)) = u_{-\alpha}(c), \qquad t \in T, \alpha \in \Gamma, c \in \mathbb{C}.$$

Recall from §1.4 the regular functions $\Delta_{w_1\lambda, w_2\lambda}$ on G, where $w_1, w_2 \in W$ and $\lambda \in \mathscr{P}^+$. By [17], a *twisted generalized minor* on $G^{u,v}$ is a regular function of the form

$$g \longmapsto \Delta_{w_1\omega_\alpha, w_2\omega_\alpha}(g'), \qquad g \in G^{u,v},$$

where $w_1, w_2 \in W$ and $\alpha \in \Gamma$.

Definition 3.6. By a *Fomin-Zelevinsky twisted generalized minor*, or simply a *Fomin-Zelevinsky minor*, on $G^{u,v}$, we mean a twisted generalized minor on $G^{u,v}$ of the form

$$g \longmapsto \Delta_{u_2^{-1}\omega_\alpha, \, v_1\omega_\alpha}(g'), \qquad g \in G^{u,v},$$

where $\alpha \in \Gamma$, and $u_2, v_1 \in W$ are such that $l(u) = l(u_1) + l(u_2)$, and $l(v) = l(v_1) + l(v_2)$ with $u_1, v_2 \in W$ determined by $u = u_1u_2$, $v = v_1v_2$. ◇

Fixing again the reduced words **u** and **v** of u and v as in (68), we now express the Fomin-Zelevinsky minors in the corresponding Bott-Samelson coordinates on $G^{u,v}$.

Notation 3.7. For $i = 0, 1, \ldots, a$, $j = 1, 2, \ldots, b+1$, and $\lambda \in \mathscr{P}^+$, let

$$(74) \qquad v_{[1,i]} = \begin{cases} e, & i = 0, \\ s_a s_{a-1} \cdots s_{a+1-i}, & i = 1, 2, \ldots, a, \end{cases}$$

$$v_{[i+1,a]} = \left(v_{[1,i]}\right)^{-1} v,$$

$$(75) \qquad u_{[j,b]} = \begin{cases} s_{a+j}s_{a+j+1} \cdots s_{a+b}, & j = 1, 2, \ldots, b, \\ e, & j = b+1, \end{cases}$$

and let $M_{i,j}^\lambda$ be the regular function on $G^{u,v}$ given by

$$(76) \qquad M_{i,j}^\lambda(g) \stackrel{\text{def}}{=} \Delta_{(u_{[j,b]})^{-1}\lambda, \, v_{[1,i]}\lambda}(g'), \qquad g \in G^{u,v}.$$

Note in particular that

$$M_{0,1}^\lambda(g) = \Delta_{u^{-1}\lambda, \lambda}(g'), \qquad M_{a,b+1}^\lambda(g) = \Delta_{\lambda, v\lambda}(g'), \qquad g \in G^{u,v}.$$

By (8) in §1.4, each $M_{i,j}^{\lambda}$ is a monomial, with non-negative powers, of Fomin-Zelevinsky minors on $G^{u,v}$. Recall also that for $\lambda \in \mathscr{P}^+$, $\Delta^{\lambda}(g) = \Delta_{\lambda,\lambda}(g)$ for $g \in G$. ◇

Lemma 3.8. *Let $0 \le i \le a$, $1 \le j \le b+1$, and $\lambda \in \mathscr{P}^+$. For $g \in G^{u,v}$ as in (60), one has*

$$M_{i,j}^{\lambda}(g) = t^{-(v_{[i+1,a]})^{-1}(\lambda)} \Delta^{\lambda}\left(\overline{\overline{v_{[i+1,a]}}}\,(m\overline{\overline{v}})^{-1} n\overline{u}\,\overline{u_{[j,b]}}^{-1}\right).$$

Proof. Introduce (another twist on $G^{u,v}$)

$$(77)\quad G^{u,v} \longrightarrow G^{v,u}, \quad g \longmapsto \hat{g} \stackrel{\text{def}}{=} ((g')^{-1})^{\theta} = \left([\overline{u}^{-1}g]_-^{-1}\overline{u}^{-1}\,g\overline{\overline{v}}^{-1}[g\overline{\overline{v}}^{-1}]_+^{-1}\right)^{-1}.$$

By [17, (2.15)], for every $\alpha \in \Gamma$, one has

$$\Delta^{\omega_\alpha}(g) = \Delta^{\omega_\alpha}((g^{-1})^{\theta}), \qquad \forall\, g \in G.$$

By [17, Proposition 2.1], for any $w_1, w_2 \in W$ and $\alpha \in \Gamma$, one has

$$(78)\qquad \Delta_{w_1\omega_\alpha, w_2\omega_\alpha}(g') = \Delta^{\omega_\alpha}((\overline{w_2}^{-1})^{\theta}\hat{g}\,\overline{w_1}^{\theta}) = \Delta^{\omega_\alpha}(\overline{\overline{w_2}}^{-1}\hat{g}\,\overline{\overline{w_1}}), \qquad g \in G^{u,v}.$$

Let $g \in G^{u,v}$ be as in (60) and further write

$$(79)\qquad m' = m_1 m_2, \qquad \text{where } m_1 \in N_- \cap (\overline{\overline{v}}^{-1}N_-\overline{\overline{v}}),\ m_2 \in N_- \cap (\overline{\overline{v}}^{-1}N\overline{\overline{v}}).$$

It is then easy to see that

$$\hat{g} = (t^{-1})^{v^{-1}}\overline{\overline{v}}m_1^{-1}(m\overline{\overline{v}})^{-1}n\overline{u}, \qquad g \in G^{u,v}.$$

Using (78), one has

$$\begin{aligned} M_{i,j}^{\lambda}(g) &= \Delta^{\lambda}\left(\overline{\overline{v_{[1,i]}}}^{-1}\,\hat{g}\,\overline{\overline{(u_{[j,b]})^{-1}}}\right) \\ &= t^{-(v_{[i+1,a]})^{-1}(\lambda)}\Delta^{\lambda}\left(\overline{\overline{v_{[i+1,a]}}}m_1^{-1}\,(m\overline{\overline{v}})^{-1}n\overline{u}\,\overline{u_{[j,b]}}^{-1}\right). \end{aligned}$$

As $l(v) = l(v_{[1,i]}) + l(v_{[i+1,a]})$ and $m_1^{-1} \in N_- \cap (\overline{\overline{v}}^{-1}N_-\overline{\overline{v}})$, $\overline{\overline{v_{[i+1,a]}}}m_1^{-1}\overline{\overline{v_{[i+1,a]}}}^{-1} \in N_-$ by [4, VI, 1.6], from which Lemma 3.8 follows. □

To express the functions $M_{i,j}^{\lambda}$ on $G^{u,v}$ in the Bott-Samelson coordinates $(t, x_1, \ldots, x_{a+b})$ defined by the given reduced words $\mathbf{u}$ and $\mathbf{v}$ of u and v, set for $1 \le k \le l \le a+b$,

$$(80)\quad p_{[k,l]}(x) = p_{\alpha_k}(x_k)p_{\alpha_{k+1}}(x_{k+1})\cdots p_{\alpha_l}(x_l) \in G, \qquad x = (x_1, \ldots, x_{a+b}) \in \mathbb{C}^{a+b}.$$

Here recall that $p_\alpha(c) = u_\alpha(x)\overline{s_\alpha}$ for $\alpha \in \Gamma$ and $c \in \mathbb{C}$. Define $p_{[k,l]}(x) = e$ if $k > l$.

Proposition 3.9. *Let $0 \le i \le a$, $1 \le j \le b+1$, and $\lambda \in \mathscr{P}^+$. In the Bott-Samelson coordinates $(t, x_1, \ldots, x_{a+b})$ of $g \in G^{u,v}$, one has*

$$M^{\lambda}_{i,j}(g) \stackrel{\text{def}}{=} \Delta_{(u_{[j,b]})^{-1}\lambda,\, v_{[1,i]}\lambda}(g') = t^{-s_1s_2\cdots s_{a-i}(\lambda)}\Delta^{\lambda}(p_{[a+1-i,a-1+j]}(x)).$$

Consequently, with the open Fomin-Zelevinsky embedding $F^{\mathbf{u},\mathbf{v}}: G^{u,v} \to T \times \mathscr{O}^{(\mathbf{v}^{-1},\mathbf{u})}$ *in* (72) *and in the notation of Notation* 2.27, *one has*

$$M^{\lambda}_{i,j} = (F^{\mathbf{u},\mathbf{v}})^* \left(\tilde{y}^{\lambda}_{[a+1-i,a-1+j]}\right), \tag{81}$$

where $\tilde{y}^{\lambda}_{[a+1-i,a-1+j]} = t^{-s_1s_2\cdots s_{a-i}(\lambda)} y^{\lambda}_{[a+1-i,a-1+j]} \in \widetilde{\mathscr{Y}}^{(\mathbf{v}^{-1},\mathbf{u})}$.

Proof. Let $g \in G^{u,v}$ be as in (60). Writing

$$(m\overline{\overline{v}})^{-1}n\overline{u} = p_{[1,a+b]}(x) = p_{[1,a-i]}(x)\,p_{[a-i+1,a+j-1]}(x)\,p_{[a+j,a+b]}(x),$$

and noting that $\overline{\overline{v_{[i+1,a]}}}\; p_{[1,a-i]}(x) \in N_-$ and $p_{[a+j,a+b]}(x)\; \overline{u_{[j,b]}}^{-1} \in N$, it follows from Lemma 3.8 that

$$\begin{aligned} M^{\lambda}_{i,j}(g) &= t^{-(v_{[i+1,a]})^{-1}(\lambda)}\Delta^{\lambda}\left(\overline{\overline{v_{[i+1,a]}}}\,(m\overline{\overline{v}})^{-1}n\overline{u}\,\overline{u_{[j,b]}}^{-1}\right) \\ &= t^{-s_1s_2\cdots s_{a-i}(\lambda)}\Delta^{\lambda}(p_{[a+1-i,a-1+j]}(x)). \end{aligned}$$

□

Example 3.10. Let $u \in W$ be arbitrary. In [29, §2.5], Kogan and Zelevinsky introduced an integrable system $M_1, M_3, \ldots, M_{2n-1}$ on $(G^{u,u}, \pi_{\mathrm{st}})$ corresponding to each reduced word $\mathbf{u} = (s_1, s_2, \ldots, s_n)$ of u. In terms of our notation here,

$$M_{2k-1}(g) = M^{\omega_{\alpha_k}}_{k-1,k}(g) = \Delta_{s_ns_{n-1}\cdots s_k\omega_{\alpha_k},\, s_1s_2\cdots s_{k-1}\omega_{\alpha_k}}(g'), \qquad g \in G^{u,u},\ k = 1, \ldots, n.$$

Using the open Fomin-Zelevinsky embedding $F^{\mathbf{u},\mathbf{u}}: G^{u,u} \to T \times O^{(\mathbf{u}^{-1},\mathbf{u})}$, one thus has

$$M_{2k-1} = (F^{\mathbf{u},\mathbf{u}})^* \left(\tilde{y}^{\omega_{\alpha_k}}_{[n+2-k,n+k-1]}\right), \qquad k = 1, \ldots, n.$$

A concrete example for $G = SL(3,\mathbb{C})$ will be given in Example 3.16. ◇

3.3 The Fomin-Zelevinsky embeddings are Poisson

Turning to Poisson structures, note that since the multiplicative Poisson structure π_{st} on G is invariant under (both left and right) translations by elements in T,

$$\hat{\pi}_{\mathrm{st}} \stackrel{\text{def}}{=} \varpi(\pi_{\mathrm{st}})$$

is a well-defined Poisson structure on G/T, where $\varpi: G \to G/T$ is the natural projection. It is evident that each reduced double Bruhat cell $G^{u,v}/T$ is a Poisson submanifold of G/T with respect to $\hat{\pi}_{\mathrm{st}}$. The restriction of $\hat{\pi}_{\mathrm{st}}$ to $G^{u,v}/T$ will also be denoted as $\hat{\pi}_{\mathrm{st}}$.

Theorem 3.11. *For any $u, v \in W$, the open Fomin-Zelevinsky embeddings*

$$\hat{F}^{u,v}\colon\quad (G^{u,v}/T,\, \hat{\pi}_{\rm st}) \longrightarrow \left(\mathscr{O}^{(v^{-1},u)},\, \pi_2\right),$$

$$F^{u,v}\colon\quad (G^{u,v},\, \pi_{\rm st}) \longrightarrow \left(T \times \mathscr{O}^{(v^{-1},u)},\, 0 \bowtie \pi_2\right)$$

are Poisson, inducing biregular Poisson isomorphisms

$$\hat{F}^{u,v}\colon\quad (G^{u,v}/T,\, \hat{\pi}_{\rm st}) \xrightarrow{\sim} \left(\mathscr{O}_e^{(v^{-1},u)},\, \pi_2\right),$$

$$F^{u,v}\colon\quad (G^{u,v},\, \pi_{\rm st}) \xrightarrow{\sim} \left(T \times \mathscr{O}_e^{(v^{-1},u)},\, 0 \bowtie \pi_2\right).$$

Given any reduced words $\mathbf{u}$ and $\mathbf{v}$ of u and v as in (68), recall the isomorphism $\mu\colon \mathscr{O}^{(\mathbf{v}^{-1},\mathbf{u})} \to \mathscr{O}^{(v^{-1},u)}$ defined in (70). By Remark 2.2, $\mu\colon (\mathscr{O}^{(\mathbf{v}^{-1},\mathbf{u})}, \pi_{a+b}) \to (\mathscr{O}^{(v^{-1},u)}, \pi_2)$ is Poisson. Thus Theorem 3.11 is equivalent to the following theorem.

Theorem 3.12. *For any $u, v \in W$ and reduced words $\mathbf{u}$ and $\mathbf{v}$ of u and v, the open Fomin-Zelevinsky embeddings*

$$\hat{F}^{\mathbf{u},\mathbf{v}}\colon\quad (G^{u,v}/T,\, \hat{\pi}_{\rm st}) \longrightarrow \left(\mathscr{O}^{(\mathbf{v}^{-1},\mathbf{u})},\, \pi_{a+b}\right), \tag{82}$$

$$F^{\mathbf{u},\mathbf{v}}\colon\quad (G^{u,v},\, \pi_{\rm st}) \longrightarrow \left(T \times \mathscr{O}^{(\mathbf{v}^{-1},\mathbf{u})},\, 0 \bowtie \pi_{a+b}\right) \tag{83}$$

are Poisson, inducing biregular Poisson isomorphisms

$$\hat{F}^{\mathbf{u},\mathbf{v}}\colon\quad (G^{u,v}/T,\, \hat{\pi}_{\rm st}) \xrightarrow{\sim} \left(\mathscr{O}_e^{(\mathbf{v}^{-1},\mathbf{u})},\, \pi_{a+b}\right),$$

$$F^{\mathbf{u},\mathbf{v}}\colon\quad (G^{u,v},\, \pi_{\rm st}) \xrightarrow{\sim} \left(T \times \mathscr{O}_e^{(\mathbf{v}^{-1},\mathbf{u})},\, 0 \bowtie \pi_{a+b}\right).$$

We now give a proof of Theorem 3.12 using a result of Kogan and Zelevinsky in [29] on the Poisson brackets between certain Fomin-Zelevinsky minors on $G^{u,v}$ and our expressions of the minors in Bott-Samelson coordinates. A more conceptual proof of Theorem 3.11 without computations in coordinates is given in the Appendix.

We first compute the Poisson brackets with respect to $0 \bowtie \pi_{a+b}$ of the functions $\tilde{y}^{\lambda}_{[a+1-i,a-1+j]}$ on $T \times \mathscr{O}^{(\mathbf{v}^{-1},\mathbf{u})}$ appearing in (81).

Lemma 3.13. *For $0 \leq i \leq i' \leq a$, $1 \leq j \leq j' \leq b+1$, and $\lambda, \lambda' \in \mathscr{P}^+$,*

$$\left\{\tilde{y}^{\lambda}_{[a+1-i,a-1+j]},\, \tilde{y}^{\lambda'}_{[a+1-i',a-1+j']}\right\} = c\, \tilde{y}^{\lambda}_{[a+1-i,a-1+j]}\, \tilde{y}^{\lambda'}_{[a+1-i',a-1+j']},$$

where (*see Notation* 2.20)

$$c = \langle s_{[1,a-i]}(\lambda),\, s_{[1,a-i']}(\lambda')\rangle - \langle s_{[1,a-1+j]}(\lambda),\, s_{[1,a-1+j']}(\lambda')\rangle \tag{84}$$

$$= \langle v_{[1,i]}(\lambda),\, v_{[1,i']}(\lambda')\rangle - \langle (u_{[j,b]})^{-1}(\lambda),\, (u_{[j',b]})^{-1}(\lambda')\rangle. \tag{85}$$

Proof. By Lemma 2.28, the Poisson bracket between the two functions is log-canonical with the coefficient c given as in (84). For (85), we only need to note that $s_{[1,a-i]} = v^{-1}v_{[1,i]}$, $s_{[1,a-i']} = v^{-1}v_{[1,i']}$, $s_{[1,a-1+j]} = v^{-1}u(u_{[j,b]})^{-1}$, and $s_{[1,a-1+j']} = v^{-1}u(u_{[j',b]})^{-1}$. □

Proof of Theorem 3.12 *by computation in coordinates.* Associated to any double reduced word $(i_1,\ldots,i_{a+b})$ of (u,v), appended with r entries $i_{a+b+1},\ldots,i_{a+b+r}$ representing the r simple reflections in W, Fomin and Zelevinsky defined in [17] a sequence M_k, $k=1,\ldots,a+b+r$, of twisted generalized minors on $G^{u,v}$, which are of the form

$$M_k(g) = \Delta_{\gamma^k,\delta^k}(g'), \qquad g \in G^{u,v},\ k=1,\ldots,a+b+r,$$

where $\gamma^k = u_{\geq k}(\omega_{|i_k|})$ and $\delta^k = v_{<k}(\omega_{|i_k|})$ (we refer to [17, 29] for the detail on the notation). Moreover, $\{M_k : k=1,\ldots,a+b+r\}$ generates the field of rational functions on $G^{u,v}$. It is shown in [29, Theorem 2.6] that with respect to the Poisson structure $\pi_{\rm st}$ on $G^{u,v}$, one has

$$\{M_k, M_{k'}\} = (\langle\delta^k,\delta^{k'}\rangle - \langle\gamma^k,\gamma^{k'}\rangle)M_kM_{k'}, \qquad k \leq k'. \tag{86}$$

Here recall that our Poisson structure $\pi_{\rm st}$ on G is the negative of the one used in [29].

Given reduced words $\mathbf{u}$ and $\mathbf{v}$ of u and v, use any shuffle of $\mathbf{u}$ and $\mathbf{v}$ to form a double reduced word $(i_1,\ldots,i_{a+b})$ of (u,v), and consider the sequence of functions M_k, $k=1,\ldots,a+b+r$ described above. In our notation in (76), it is easy to see that for each $1\leq k\leq a+b+r$, there exist $0\leq i(k)\leq a$ and $1\leq j(k)\leq b+1$ such that

$$M_k(g) = M^{\omega_{|i_k|}}_{i(k),j(k)}(g) = \Delta_{(u_{[j(k),b]})^{-1}\omega_{|i_k|},\, v_{[1,i(k)]}\omega_{|i_k|}}(g'), \qquad g\in G^{u,v},$$

and $i(k)\leq i(k')$ and $j(k)\leq j(k')$ if $k\leq k'$. By (81), one has

$$M_k = (F^{\mathbf{u},\mathbf{v}})^*\left(\tilde{y}^{\omega_{|i_k|}}_{[a+1-i(k),a-1+j(k)]}\right), \qquad k=1,\ldots,a+b+r.$$

By Lemma 3.13 and comparing with (86), one sees immediately that $F^{\mathbf{u},\mathbf{v}}$: $(G^{u,v},\pi_{\rm st}) \to (T\times\mathcal{O}^{(\mathbf{v}^{-1},\mathbf{u})}, 0\bowtie\pi_{a+b})$ is Poisson. The statement about $\hat{F}^{\mathbf{u},\mathbf{v}}$ follows from Remark 3.3 and the definitions of the Poisson structures $\hat{\pi}_{\rm st}$ and $0\bowtie\pi_{a+b}$.

3.4 Completeness of Hamiltonian flows of Fomin-Zelevinsky minors

For $u,v\in W$, regarding $(G^{u,v},\pi_{\rm st})$ as an affine Poisson variety, we now have one of the main results of the paper.

Corollary 3.14. *For any $u,v\in W$ and with respect to the Poisson structure $\pi_{\rm st}$, every Fomin-Zelevinsky minor on $G^{u,v}$ has complete Hamiltonian flow with property $\mathcal{Q}$.*

Proof. Choose any reduced words $\mathbf{u}$ and $\mathbf{v}$ of u and v and consider the corresponding Fomin-Zelevinsky Poisson isomorphism $F^{\mathbf{u},\mathbf{v}}\colon (G^{u,v}, \pi_{\rm st}) \to (T \times \mathcal{O}_e^{(\mathbf{v}^{-1},\mathbf{u})}, 0 \bowtie \pi_{a+b})$. By Proposition 3.9, every Fomin-Zelevinsky minor on $G^{u,v}$ is identified, via $F^{\mathbf{u},\mathbf{v}}$, with a regular function $\tilde{y} \in \widetilde{\mathscr{Y}}^{(\mathbf{v}^{-1},\mathbf{u})}$, which, by Theorem 2.26, has complete Hamiltonian flow in $T \times \mathcal{O}_e^{(\mathbf{v}^{-1},\mathbf{u})}$ with property $\mathcal{Q}$. □

For $u \in W$, recall from Example 3.10 the Kogan-Zelevinsky integrable systems on $G^{u,u}$ defined using reduced words of u.

Corollary 3.15. *For every $u \in W$, all the Kogan-Zelevinsky integrable systems on $G^{u,u}$ have complete Hamiltonian flows with property $\mathcal{Q}$.*

3.5 Integrable systems on $(G^{u,u}/T, \hat{\pi}_{\rm st})$

Let again $u \in W$ be arbitrary, and let $\mathbf{u} = (s_1, s_2, \ldots, s_n)$ be any reduced word of u. Consider the open Fomin-Zelevinsky embedding

$$\hat{F}^{(\mathbf{u},\mathbf{u})}\colon\ (G^{u,u}/T,\ \hat{\pi}_{\rm st}) \longrightarrow (\mathcal{O}^{(\mathbf{u}^{-1},\mathbf{u})},\ \pi_{2n}).$$

Recall from §2.8 the integrable system $(y_1, \ldots, y_n)$ on $(\mathcal{O}^{(\mathbf{u}^{-1},\mathbf{u})},\ \pi_{2n})$. Define

$$\hat{M}_k = (\hat{F}^{(\mathbf{u},\mathbf{u})})^* y_k, \qquad k = 1, \ldots, n,$$

Then $(\hat{M}_1, \ldots, \hat{M}_n)$ is an integrable system on $G^{u,u}/T$ that has complete Hamiltonian flows with property $\mathcal{Q}$. To express the regular function $\hat{M}_k$ on $G^{u,u}/T$ using the Fomin-Zelevinsky twist map $G^{u,u} \to G^{u^{-1},u^{-1}}, g \mapsto g'$, note by Proposition 3.9 that for $g \in G^{u,u}$,

$$\begin{aligned}\hat{M}_k(gT) &= t^{s_n s_{n-1}\cdots s_{k+1}(\omega_{\alpha_k})} \Delta_{s_n s_{n-1}\cdots s_{k+1}\omega_{\alpha_k},\, s_1 s_2\cdots s_k \omega_{\alpha_k}}(g') \\ &= t^{s_n s_{n-1}\cdots s_{k+1}(\omega_{\alpha_k})} M_{k,k+1}^{\omega_{\alpha_k}}(g),\end{aligned}$$

where recall from (61) that $t = [g\overline{\overline{u}}^{-1}]_0^u$, and the function $M_{k,k+1}^{\omega_{\alpha_k}}$ on $G^{u,u}$ is defined in Notation 3.7. It is easy to see that $t = ([\overline{\overline{u}}g']_0^u)^{-1}$. One thus has, for $k = 1, \ldots, n$,

$$\hat{M}_k(gT) = \left(\left[\overline{\overline{u}}g'\right]_0^{-1} \right)^{s_1 s_2\cdots s_k(\omega_{\alpha_k})} \Delta_{s_n s_{n-1}\cdots s_{k+1}\omega_{\alpha_k}, s_1 s_2\cdots s_k \omega_{\alpha_k}}(g'), \qquad g \in G^{u,u}.$$

3.6 Examples

We give two examples to illustrate the Fomin-Zelevinsky embeddings and how Fomin-Zelevinsky minors and their Hamiltonian flows can be expressed in

Bott-Samelson coordinates. The Poisson brackets in Example 3.16 and Example 3.17 are computed using the GAP computer program written by B. Elek.

Example 3.16. Consider $G = SL(3,\mathbb{C})$ with the standard choices of maximal torus T and the simple roots α_1 and α_2, and choose the bilinear form on $\mathfrak{g}$ such that the induced bilinear form $\langle\, , \rangle$ on $\mathfrak{t}^*$ satisfies $\langle \alpha_2, \alpha_2\rangle = \langle \alpha_1, \alpha_1\rangle = 2$. Let w_0 be the longest element in the Weyl group, and take the reduced word $\mathbf{u} = (s_{\alpha_1}, s_{\alpha_2}, s_{\alpha_1})$ of w_0. Let ω_1 and ω_2 be the fundamental weights corresponding to α_1 and α_2, and consider Bott-Samelson coordinates $(a,b,x_1,\ldots,x_6)$ on $T \times \mathcal{O}^{(\mathbf{u}^{-1},\mathbf{u})}$, where

$$a(t) = t^{\omega_1} \quad \text{and} \quad b(t) = t^{\omega_2}, \quad t \in T.$$

For $g = (a_{ij})_{i,j=1,2,3}$ and $1 \le i < k \le 3$ and $1 \le j < l \le 3$, let $\Delta_{ik,jl}(g)$ be the determinant of the submatrix of g formed by the i'th and the k'th row and the j'th and the l'th column. Recall that $g \in G^{w_0,w_0}$ if and only if $a_{13}a_{31}\Delta_{12,23}\Delta_{23,12} \neq 0$. A direct calculation shows that the open Fomin-Zelevinsky embedding

$$F^{\mathbf{u},\mathbf{u}}\colon\ G^{w_0,w_0} \longrightarrow T \times \mathcal{O}^{(\mathbf{u}^{-1},\mathbf{u})} \cong (\mathbb{C}^\times)^2 \times \mathbb{C}^6$$

is explicitly given by $SL(3,\mathbb{C}) \ni g = (a_{ij}) \mapsto (a,b,x_1,\ldots,x_6)$, with

$$\begin{aligned} &a = 1/\Delta_{12,23}, \quad b = 1/a_{13}, \quad x_1 = \Delta_{13,23}/\Delta_{12,23}, \quad x_2 = a_{33}/a_{13}, \\ &x_3 = a_{23}/a_{13}, \quad x_4 = \Delta_{13,12}/\Delta_{23,12}, \quad x_5 = \Delta_{12,12}/\Delta_{23,12}, \quad x_6 = a_{21}/a_{31}. \end{aligned}$$

The Zariski open subset $T \times \mathcal{O}_e^{(\mathbf{u}^{-1},\mathbf{u})}$ of $T \times \mathcal{O}^{(\mathbf{u}^{-1},\mathbf{u})}$ is defined in the coordinates $(a,b,x_1,\ldots,x_6)$ by $y_1 \neq 0$ and $y_2 \neq 0$, where

$$y_1 = x_1x_3x_4x_6 - x_2x_4x_6 - x_1x_6 - x_1x_3x_5 + x_2x_5 + 1 \quad \text{and} \quad y_2 = x_2x_5 - x_3x_4 + 1.$$

One can see that $F^{\mathbf{u},\mathbf{u}}(G^{w_0,w_0}) \subset T \times \mathcal{O}_e^{(\mathbf{u}^{-1},\mathbf{u})}$ directly by checking that

$$y_1(F^{\mathbf{u},\mathbf{u}}(g)) = \frac{1}{a_{31}\Delta_{12,23}} \quad \text{and} \quad y_2(F^{\mathbf{u},\mathbf{u}}(g)) = \frac{1}{a_{13}\Delta_{23,12}}, \quad g \in G^{w_0,w_0},$$

and one can also solve for the inverse of $F^{\mathbf{u},\mathbf{u}}\colon G^{w_0,w_0} \to T \times \mathcal{O}_e^{(\mathbf{u}^{-1},\mathbf{u})}$ directly.

In the coordinates $(a,b,x_1,\ldots,x_6)$, the Poisson structure $0 \bowtie \pi_6$ on $T \times \mathcal{O}^{(\mathbf{u}^{-1},\mathbf{u})}$ has

$$\begin{aligned} &\{x_1,x_2\} = -x_1x_2, \quad \{x_1,x_3\} = x_1x_3 - 2x_2, \quad \{x_1,x_4\} = -x_1x_4, \\ &\{x_1,x_5\} = x_1x_5 - 2x_4, \quad \{x_1,x_6\} = 2x_1x_6 - 2, \\ &\{x_2,x_3\} = -x_2x_3, \quad \{x_2,x_4\} = x_2x_4, \quad \{x_2,x_5\} = 2x_2x_5 - 2x_3x_4 + 2, \\ &\{x_2,x_6\} = x_2x_6 - 2x_3, \quad \{x_3,x_4\} = 2x_3x_4 - 2, \quad \{x_3,x_5\} = x_3x_5, \\ &\{x_3,x_6\} = -x_3x_6, \quad \{x_4,x_5\} = -x_4x_5, \quad \{x_4,x_6\} = x_4x_6 - 2x_5, \quad \{x_5,x_6\} = -x_5x_6, \end{aligned}$$

$$\{a, x_j\} = -\langle \omega_1, \lambda_j \rangle a x_j, \quad \{b, x_j\} = -\langle \omega_2, \lambda_j \rangle b x_j, \quad j = 1, \ldots, 6, \quad \text{where}$$
$$\lambda_1 = \alpha_1,\ \lambda_2 = \alpha_1 + \alpha_2,\ \lambda_3 = \alpha_2,\ \lambda_4 = -\alpha_2,\ \lambda_5 = -\alpha_1 - \alpha_2,\ \lambda_6 = -\alpha_1.$$

One computes directly that the regular functions $M_1, \ldots, M_6$ on G^{w_0,w_0} defined by Kogan and Zelevinsky in [29, Corollary 2.7] are given by

$$M_1 = b, \quad M_2 = bx_4, \quad M_3 = a, \quad M_4 = ax_5,$$
$$M_5 = ab^{-1}(x_3x_4 - 1), \quad M_6 = ab^{-1}(x_3x_4x_6 - x_6 - x_3x_5),$$

and that M_1, M_3, M_5 form an integrable system on G^{w_0,w_0}. For $i = 1,3,5$, denote by $c \mapsto \phi_{i,c} \in \mathrm{Diff}((\mathbb{C}^\times)^2 \times \mathbb{C}^6)$, $c \in \mathbb{C}$, the Hamiltonian flow of M_i. A direct calculation gives that for $p = (a,b,x_1,x_2,x_3,x_4,x_5,x_6) \in (\mathbb{C}^\times)^2 \times \mathbb{C}^6$,

$$\phi_{1,c}(p) = (a,\ b,\ x_1,\ x_2e^{-cM_1},\ x_3e^{-cM_1},\ x_4e^{cM_1},\ x_5e^{cM_1},\ x_6),$$
$$\phi_{3,c}(p) = (a,\ b,\ x_1e^{-cM_3},\ x_2e^{-cM_3},\ x_3,\ x_4,\ x_5e^{cM_3},\ x_6e^{cM_3}),$$
$$\phi_{5,c}(p) = (a,\ b,\ x_1e^{-cM_5} + ab^{-1}x_2x_4(e^{cM_5} - e^{-cM_5})/M_5,\ x_2,\ x_3e^{-cM_5},\ x_4e^{cM_5},$$
$$x_5,\ x_6e^{cM_5} + ab^{-1}x_3x_5(e^{-cM_5} - e^{cM_5})/M_5), \quad \text{if } M_5 \neq 0,$$
$$\phi_{5,c}(p) = (a,\ b,\ x_1 + 2ab^{-1}x_2x_4c,\ x_2,\ x_3,\ x_4,\ x_5,\ x_6 - 2ab^{-1}x_3x_5c), \quad \text{if } M_5 = 0,$$

where for $i = 1,3,5$, M_i also denotes the value $M_i(p)$ of the function M_i at p. ◇

Example 3.17. Consider G_2 with the simple roots α_1 and α_2 and $\langle \alpha_2, \alpha_2 \rangle = 3\langle \alpha_1, \alpha_1 \rangle = 6$. Take $\mathbf{u} = (s_{\alpha_2}, s_{\alpha_1}, s_{\alpha_2})$ so that $(\mathbf{u}^{-1}, \mathbf{u}) = (s_{\alpha_2}, s_{\alpha_1}, s_{\alpha_2}, s_{\alpha_2}, s_{\alpha_1}, s_{\alpha_2})$. The Poisson structure π_6 on $\mathcal{O}^{(\mathbf{u}^{-1},\mathbf{u})}$ in the Bott-Samelson coordinates $(x_1, \ldots, x_6)$ is given by

$$\{x_1,x_2\} = -3x_1x_2, \quad \{x_1,x_3\} = -6x_2^3 - 3x_1x_3, \quad \{x_1,x_4\} = 3x_1x_4,$$
$$\{x_1,x_5\} = -6x_2^2x_4 + 3x_1x_5, \quad \{x_1,x_6\} = -18x_2^2x_5^2 + 6x_1x_6 - 18x_2x_5 + 6x_3x_4 - 6,$$
$$\{x_2,x_3\} = -3x_2x_3, \quad \{x_2,x_4\} = 3x_2x_4,$$
$$\{x_2,x_5\} = 2x_2x_5 - 2x_3x_4 + 2, \quad \{x_2,x_6\} = -6x_3x_5^2 + 3x_2x_6,$$
$$\{x_3,x_4\} = 6x_3x_4 - 6, \quad \{x_3,x_5\} = 3x_3x_5, \quad \{x_3,x_6\} = 3x_3x_6,$$
$$\{x_4,x_5\} = -3x_4x_5, \quad \{x_4,x_6\} = -6x_5^3 - 3x_4x_6, \quad \{x_5,x_6\} = -3x_5x_6.$$

By Theorem 2.29, one has the integrable system $\{y_1, y_2, y_3\}$ on $\mathcal{O}^{(\mathbf{u}^{-1},\mathbf{u})}$, where $y_1 = x_3x_4 - 1$, $y_2 = x_2x_5 - x_3x_4 + 1$, and

$$y_3 = x_1x_3x_4x_6 - x_2^3x_4x_6 - x_1x_6 - x_1x_3x_5^3 + x_2^3x_5^3 + 3x_2^2x_5^2 - 3x_2x_3x_4x_5$$
$$+ 3x_2x_5 + x_3^2x_4^2 - 2x_3x_4 + 1.$$

For $i = 1,2,3$, denote by $c \mapsto \varphi_{i,c} \in \mathrm{Diff}(\mathbb{C}^6)$, $c \in \mathbb{C}$, the Hamiltonian flow of y_i. For $p = (x_1,x_2,x_3,x_4,x_5,x_6) \in \mathbb{C}^6$, a direct calculation gives

$$\varphi_{1,c}(p) = \left(x_1 + x_2^3 x_4 (e^{6cy_1} - 1)/y_1,\ x_2,\ x_3 e^{-6cy_1},\right.$$
$$\left. x_4 e^{6cy_1},\ x_5,\ x_6 + x_3 x_5^3 (e^{-6cy_1} - 1)/y_1\right), \text{ if } y_1 \neq 0,$$
$$\varphi_{1,c}(p) = (x_1 + 6x_2^3 x_4 c,\ x_2,\ x_3,\ x_4,\ x_5,\ x_6 - 6x_3 x_5^3 c), \text{ if } y_1 = 0,$$
$$\varphi_{2,c}(p) = \left(x_1,\ x_2 e^{-2cy_2},\ x_3 e^{-6cy_2},\ x_4 e^{6cy_2},\ x_5 e^{2cy_2},\ x_6\right),$$
$$\varphi_{3,c}(p) = \left(x_1 e^{-6cy_3},\ x_2 e^{-6cy_3},\ x_3 e^{-12cy_3},\ x_4 e^{12cy_3},\ x_5 e^{6cy_3},\ x_6 e^{6cy_3}\right),$$

where for $i = 1,2,3$, y_i also denotes the value $y_i(p)$ of the function y_i at p. $\diamond$

Appendix A The Fomin-Zelevinsky embeddings are Poisson

In this appendix we give a proof of Theorem 3.11 without computations in coordinates. We first recall a construction of mixed product Poisson structures given in [34] and express some Poisson structures related to the standard Poisson Lie group $(G, \pi_{\rm st})$ as mixed product Poisson structures. We refer to [34] for basic facts on Poisson Lie groups and Poisson Lie group actions needed in this appendix.

Throughout the appendix, if (X, π) is a Poisson manifold and if X_1 is a Poisson submanifold with respect to π, the restriction of π to X_1 will also be denoted by π. If A is a Lie group with Lie algebra $\mathfrak{a}$, for $\xi \in \mathfrak{a}$, ξ^L (resp. ξ^R) will denote the left (resp. right) invariant vector field on A with value ξ at the identity of A.

A.1 *Quotient and mixed product Poisson structures*

Assume first that (A, π_A) is a Poisson Lie group with a right and a left Poisson action

$$(Z, \pi_Z) \times (A, \pi_A) \longrightarrow (Z, \pi_Z) \quad \text{and} \quad (A, \pi_A) \times (Y, \pi_Y) \longrightarrow (Y, \pi_Y)$$

on Poisson manifolds (Z, π_Z) and (Y, π_Y). Suppose that the diagonal action

$$(Z \times Y) \times A \longrightarrow Z \times Y, \ (z, y) \cdot a = (za,\ a^{-1} y)$$

is free and that the quotient, denoted by $Z \times_A Y$, is a manifold. Then [34, §7] the projection of the product Poisson structure $\pi_Z \times \pi_Y = (\pi_Z, 0) + (0, \pi_Y)$ on $Z \times Y$ is a well-defined Poisson structure, called the *quotient* Poisson structure, on $Z \times_A Y$.

Assume now that a pair of dual Poisson Lie groups (A, π_A) and (A^*, π_{A^*}) act respectively on the Poisson manifolds (Y, π_Y) and (X, π_X) by left and right Poisson actions

$$\lambda:\ (A, \pi_A) \times (Y, \pi_Y) \longrightarrow (Y, \pi_Y) \quad \text{and} \quad \rho:\ (X, \pi_X) \times (A^*, \pi_{A^*}) \longrightarrow (X, \pi_X).$$

Let $\mathfrak{a}$ be the Lie algebra of A, so that the Lie algebra of A^* is identified with the dual vector space $\mathfrak{a}^*$ of $\mathfrak{a}$. Denote the corresponding Lie algebra actions by

$$\rho:\ \mathfrak{a}^* \longrightarrow \mathfrak{X}^1(X),\ \rho(\xi^*)(x) = \frac{d}{dt}\Big|_{t=0}(x\exp(t\xi^*)), \qquad x \in X, \xi^* \in \mathfrak{a}^*,$$
$$\lambda:\ \mathfrak{a} \longrightarrow \mathfrak{X}^1(Y),\ \lambda(\xi)(y) = \frac{d}{dt}\Big|_{t=0}(\exp(t\xi)y), \qquad y \in Y, \xi \in \mathfrak{a}.$$

Then [34, §2] the bivector field $\pi_X \times_{(\rho,\lambda)} \pi_Y$ on the product manifold $X \times Y$ given by

$$\pi_X \times_{(\rho,\lambda)} \pi_Y = (\pi_X, 0) + (0, \pi_Y) - \sum_{i=1}^{m} (\rho(\xi_i^*), 0) \wedge (0,\ \lambda(\xi_i)) \tag{87}$$

is a Poisson structure on $X \times Y$, called the *mixed product of* π_X *and* π_Y *associated to* (ρ,λ), where $(\xi_i)_{i=1}^m$ is any basis for $\mathfrak{a}$ and $(\xi_i^*)_{i=1}^m$ the dual basis for $\mathfrak{a}^*$. Here, for $V \in \mathfrak{X}^1(X)$ and $U \in \mathfrak{X}^1(Y)$, $(V,0) \wedge (0,U) \in \mathfrak{X}^2(X \times Y)$ is such that

$$((V,0) \wedge (0,U))(f,g) = V(f)U(g)$$

for functions f on X and g on Y. In the same setting, let λ_A be the left action of A on itself by left multiplication, and equip $X \times A$ with the mixed product Poisson structure $\pi_{X\times A} = \pi_X \times_{(\rho,\lambda_A)} \pi_A$. One then has the (free) right Poisson action

$$(X \times A,\ \pi_{X\times A}) \times (A, \pi_A) \longrightarrow (X \times A,\ \pi_{X\times A}),\ \ ((x,a),a') \longmapsto (x, aa'),$$

of the Poisson Lie group (A, π_A), which, together with the left Poisson action λ of (A, π_A) on (Y, π_Y), gives rise to the quotient Poisson structure, denoted by π, on $(X \times A) \times_A Y$. Let $\varpi\colon X \times A \times Y \to (X \times A) \times_A Y$ be the quotient map, and set $[(x,a),y] = \varpi(x,a,y)$ for $(x,a,y) \in X \times A \times Y$. Note that one has the diffeomorphism

$$\Psi:\ (X \times A) \times_A Y \longrightarrow X \times Y,\ [(x,a),y] \longmapsto (x, ay), \qquad x \in X, a \in A, y \in Y.$$

Lemma A.1. *As Poisson structures on* $X \times Y$*, one has* $\Psi(\pi) = \pi_X \times_{(\rho,\lambda)} \pi_Y$.

Proof. Let e be the identity element of A, and let $(x,y) \in X \times Y$. By definition and a straightforward calculation, one has

$$\begin{aligned}
&\Psi(\pi([(x,e),y])) \\
&\quad = \Psi(\varpi(\pi_{X\times A}(x,e), 0) + \varpi(0, \pi_Y(y))) \\
&\quad = (\pi_X(x),\ 0) + (0,\ \pi_Y(y)) - \sum_{i=1}^{m} \Psi(\varpi(\rho(\xi_i^*),0,0)) \wedge \Psi(\varpi(0, \xi_i^R, 0)) \\
&\quad = (\pi_X \times_{(\rho,\lambda)} \pi_Y)(x,y).
\end{aligned}$$

□

A.2 Drinfeld double of the Poisson Lie group (G, π_{st})

We now turn to a complex semisimple Poisson Lie group (G, π_{st}) as defined in §2.1. As both B and B_- are Poisson submanifolds of G with respect to π_{st}, we have the Poisson Lie subgroups (B, π_{st}) and (B_-, π_{st}) of (G, π_{st}).

Denote the Lie algebras of B, B_-, N and N_- respectively by $\mathfrak{b}, \mathfrak{b}_-, \mathfrak{n}$ and $\mathfrak{n}_-$, and define a non-degenerate bilinear pairing between $\mathfrak{b}_-$ and $\mathfrak{b}$ by

$$\langle x_- + x_0,\ y_+ + y_0\rangle_{(\mathfrak{b}_-, \mathfrak{b})} = \frac{1}{2}\langle x_-,\ y_+\rangle_{\mathfrak{g}} + \langle x_0,\ y_0\rangle_{\mathfrak{g}}, \tag{88}$$

where $x_- \in \mathfrak{n}_-$, $x_0, y_0 \in \mathfrak{t}$, $y_+ \in \mathfrak{n}$. Then (B_-, π_{st}) and $(B, -\pi_{\text{st}})$ form a pair of dual Poisson Lie groups with respect to the pairing $\langle\, ,\, \rangle_{(\mathfrak{b}_-, \mathfrak{b})}$ (see [34, §2.3]).

The *Drinfeld double* of the Poisson Lie group (G, π_{st}) is the product Lie group $G \times G$ together with the multiplicative Poisson structure Π_{st} given by

$$\Pi_{\text{st}} = (\pi_{\text{st}}, 0) + (0, \pi_{\text{st}}) + \mu_1 + \mu_2, \tag{89}$$

where

$$\mu_1 = \sum_{i=1}^{m} (\xi_i^R, 0) \wedge (0, x_i^R), \qquad \mu_2 = -\sum_{i=1}^{m} (\xi_i^L, 0) \wedge (0, x_i^L), \tag{90}$$

and $\{x_i\}_{i=1}^m$ is any basis of $\mathfrak{b}_-$ and $\{\xi_i\}_{i=1}^m$ its dual basis of $\mathfrak{b}$ under the pairing $\langle\, ,\, \rangle_{(\mathfrak{b}_-, \mathfrak{b})}$ in (88). See [16] and [35, §6.1]. See also [34, §6.2] and [36, §4.2] for an interpretation of Π_{st} as a mixed product Poisson structure on $G \times G$ defined by Poisson Lie group actions.

As a Drinfeld double of (G, π_{st}), the Poisson Lie group $(G \times G, \Pi_{\text{st}})$ contains (G, π_{st}), identified with the diagonal of $G \times G$, as a Poisson Lie subgroup [34, §6.2]. Moreover, any left $(B \times B)$ and right $(B_- \times B_-)$-invariant submanifold of $G \times G$ is a Poisson submanifold with respect to Π_{st}. We will use these facts in §A.3 to decompose some Poisson structures related π_{st} as product Poisson structures.

A.3 Some mixed product Poisson structures associated to (G, π_{st})

Consider the flag manifold G/B with the Poisson structure π_1. Then each Bruhat cell BwB/B, $w \in W$, is a Poisson submanifold with respect to π_1. We will also consider the quotient manifold $B_-\backslash G$ with the quotient Poisson structure π'_{-1} of π_{st}. Then, again, for each $w \in W$, $B_-\backslash B_- wB_-$ is a Poisson submanifold of $(B_-\backslash G,\ \pi'_{-1})$.

Fix $w \in W$ and let $\dot{w} \in N_G(T)$ be any representative of w. Recall from §3.1 that

$$C_{\dot{w}} = (N\dot{w}) \cap (\dot{w}N_-),$$

and that the multiplication map of G gives diffeomorphisms

$$C_{\dot{w}} \times B \longrightarrow BwB \quad \text{and} \quad B_- \times C_{\dot{w}} \longrightarrow B_-wB_-.$$

By [36, Lemma 14], one has the anti-Poisson isomorphism

$$(91) \qquad I_{\dot{w}}\colon\ (B\dot{w}B/B,\, \pi_1) \longrightarrow (B_-\backslash B_-\dot{w}B_-,\, \pi'_{-1}),\ \ c.B \longmapsto B_-.c, \qquad c \in C_{\dot{w}}.$$

Notation A.2. Let $\rho_{\dot{w}}$ be the right action of B_- on $B\dot{w}B/B$ obtained via the right B_- action on $B_-\backslash B_-wB_-$ through $I_{\dot{w}}$. More precisely, $\rho_{\dot{w}}$ denotes the right Poisson action

$$(92) \qquad \begin{aligned} \rho_{\dot{w}}\colon\ & (BwB/B,\, \pi_1) \times (B_-,\, -\pi_{\rm st}) \longrightarrow (BwB/B,\, \pi_1),\\ & (c.B,\, b_-) \longmapsto c'.B, \qquad c, c' \in C_{\dot{w}},\, b_-, b'_- \in B_-,\, cb_- = b'_-c', \end{aligned}$$

of the Poisson Lie group $(B_-, -\pi_{\rm st})$. Let λ_B be the left action of the Poisson Lie group $(B, \pi_{\rm st})$ on itself by left translation. Note that $(B_-, -\pi_{\rm st})$ and $(B, \pi_{\rm st})$ form a pair of dual Poisson Lie groups under the pairing $-\langle\, ,\, \rangle_{(\mathfrak{b}_-,\mathfrak{b})}$ in (88). ◇

Consider now the isomorphism

$$(93) \qquad J_{\dot{w}}\colon\ BwB \longrightarrow (BwB/B) \times B,\ \ cb \longmapsto (c.B,\, b), \qquad c \in C_{\dot{w}},\, b \in B.$$

Lemma A.3. *As Poisson structures on $(BwB/B) \times B$, one has*

$$(94) \qquad J_{\dot{w}}(\pi_{\rm st}) = \pi_1 \times_{(\rho_{\dot{w}},\lambda_B)} \pi_{\rm st} = (\pi_1, 0) + (0, \pi_{\rm st}) + \sum_{i=1}^{m} (\rho_{\dot{w}}(x_i), 0) \wedge (0, \xi_i^R),$$

where $\{x_i\}_{i=1}^m$ is any basis of $\mathfrak{b}_-$ and $\{\xi_i\}_{i=1}^m$ its dual of $\mathfrak{b}$ under the pairing in (88).

Proof. We prove Lemma A.3 using a similar result for $(B_-wB_-, \pi_{\rm st})$ in [36, Remark 9]. Consider the map

$$J'_{\dot{w}}\colon\ B_-wB_- \longrightarrow B_- \times (BwB/B),\ \ b_-c \longmapsto (b_-,\, c.B), \qquad b_- \in B_-,\, c \in C_{\dot{w}}.$$

Let ρ_{B_-} be the right Poisson action of the Poisson Lie group $(B_-, \pi_{\rm st})$ on itself by right translation, and let λ_+ be the left Poisson action of the Poisson Lie group $(B, -\pi_{\rm st})$ on $(BwB/B, -\pi_1)$ by left translation. Recall that $(B_-, \pi_{\rm st})$ and $(B, -\pi_{\rm st})$ form a pair of dual Poisson Lie groups under the pairing $\langle\, ,\, \rangle_{(\mathfrak{b}_-,\mathfrak{b})}$ of their Lie algebras. Define the Poisson structure π' on $B_- \times (BwB/B)$ by

$$\pi' = \pi_{\rm st} \times_{(\rho_{B_-},\lambda_+)} (-\pi_1) = (\pi_{\rm st}, 0) + (0, -\pi_1) - \sum_{i=1}^{m} (x_i^L, 0) \wedge (0, \lambda_+(\xi_i)).$$

It is shown in [36, Remark 9] that $J'_{\dot{w}}(\pi_{\rm st}) = \pi'$ as Poisson structures on $B_- \times (BwB/B)$. Consider (see [17, §1.5]) the involutive anti-automorphism Θ of G determined by

$$\Theta(t) = t, \ \ t \in T, \qquad \Theta(u_\alpha(c)) = u_{-\alpha}(-c), \qquad \alpha \in \Gamma,\, c \in \mathbb{C}.$$

It is easy to see from the definition that $\Theta(\pi_{\rm st}) = \pi_{\rm st}$. Consider now the sequence of Poisson isomorphisms

$$\begin{aligned}(BwB,\,\pi_{\rm st}) &\stackrel{\Theta}{\longrightarrow} (B_-w^{-1}B_-,\,\pi_{\rm st}) \stackrel{J'_{\Theta(\dot{w})}}{\longrightarrow} (B_- \times (Bw^{-1}B/B),\,\pi')\\ &\stackrel{\Theta\times\Theta}{\longrightarrow} (B\times(B_-\backslash B_-wB_-),\,\pi^{(2)}) \stackrel{S}{\longrightarrow} ((B_-\backslash B_-wB_-)\times B,\,\pi^{(3)})\\ &\stackrel{I_{\dot{w}}^{-1}\times \mathrm{Id}_B}{\longrightarrow} ((BwB/B)\times B,\,\pi^{(4)}),\end{aligned}$$

where S is the map that switches the two factors, and $I_{\dot{w}}$ is given in (91). It is easy to see that $J_{\dot{w}}$ is the composition of the above five maps. A direct calculation and the facts that Θ is Poisson and $I_{\dot{w}}$ is anti-Poisson show that

$$\pi^{(4)} = (\pi_1,0)+(0,\pi_{\rm st})+\sum_{i=1}^m(\rho_{\dot{w}}(x_i),0)\wedge(0,\xi_i^R).$$

As $\{-x_i\}_{i=1}^m$ and $\{\xi_i\}_{i=1}^m$ are bases of $\mathfrak{b}_-$ and $\mathfrak{b}$ that are dual to each other under the pairing $-\langle\,,\,\rangle_{(\mathfrak{b}_-,\mathfrak{b})}$ between $\mathfrak{b}_-$ and $\mathfrak{b}$, we know that

$$J_{\dot{w}}(\pi_{\rm st}) = \pi^{(4)} = \pi_1 \times_{(\rho_{\dot{w}},\lambda_B)} \pi_{\rm st}. \qquad \square$$

Let now (Y,π_Y) be any Poisson manifold with a left Poisson action

$$\lambda:\quad (B,\pi_{\rm st})\times(Y,\pi_Y)\longrightarrow(Y,\pi_Y)$$

by the Poisson Lie group $(B,\pi_{\rm st})$. Consider the quotient manifold $BwB\times_B Y$ with the quotient Poisson structure $\pi \stackrel{\rm def}{=} \varpi(\pi_{\rm st}\times\pi_Y)$, where $\varpi: BwB\times Y\to BwB\times_B Y$ is the natural projection. On the other hand, note that one has the isomorphism

(95) $$J_{\dot{w},Y}:\ BwB\times_B Y\longrightarrow (BwB/B)\times Y,\quad [cb,y]\longmapsto (c.B,\,by),$$

where $c\in C_{\dot{w}}$, $b\in B$ and $y\in Y$.

Lemma A.4. *As Poisson structures on $(BwB/B)\times Y$, one has*

$$J_{\dot{w},Y}(\pi) = \pi_1\times_{(\rho_{\dot{w}},\lambda)}\pi_Y = (\pi_1,0)+(0,\pi_Y)+\sum_{i=1}^m(\rho_{\dot{w}}(x_i),0)\wedge(0,\lambda(\xi_i)),$$

where, again, $\{x_i\}_{i=1}^m$ is any basis of $\mathfrak{b}_-$ and $\{\xi_i\}_{i=1}^m$ its dual the basis of $\mathfrak{b}$ under the pairing $\langle\,,\,\rangle_{(\mathfrak{b}_-,\mathfrak{b})}$ in (88).

Proof. The assertion follows from Lemma A.3 and Lemma A.1. $\square$

A.4 Fomin-Zelevinsky embeddings for reduced double Bruhat cells

Let $u,v \in W$ and recall that the reduced double Bruhat cell $G^{u,v}/T \subset G/T$ has the Poisson structure $\hat{\pi}_{\rm st}$, which is the projection to G/T of the Poisson structure $\pi_{\rm st}$ on G. Recall from §2.2 and §3.1 the generalized Bruhat cell $\mathscr{O}^{(v^{-1},u)} \subset F_2$ and its Zariski open subset $\mathscr{O}_e^{(v^{-1},u)}$, and recall from §3.1 the Fomin-Zelevinsky embedding

$$\hat{F}^{u,v}:\; G^{u,v}/T \longrightarrow \mathscr{O}^{(v^{-1},u)},\; \hat{F}^{u,v}(gT) = [(m\overline{v})^{-1},\, n\overline{u}], \qquad g \in G^{u,v},$$

which induces a biregular isomorphism from $G^{u,v}/T$ to $\mathscr{O}_e^{(v^{-1},u)}$. Here again $g \in G^{u,v}$ is uniquely written as $g = n\overline{u}n't' = m\overline{\overline{v}}m't$ with $n\overline{u} \in C_{\overline{u}}$, $(m\overline{\overline{v}})^{-1} \in C_{\overline{\overline{v}}^{-1}}$, $n' \in N$, $m' \in N_-$ and $t,t' \in T$. The following proposition is the first part of Theorem 3.11.

Proposition A.5. *For any $u,v \in W$, the Fomin-Zelevinsky embedding*

$$\hat{F}^{u,v}:\; (G^{u,v}/T,\, \hat{\pi}_{\rm st}) \longrightarrow (\mathscr{O}^{(v^{-1},u)},\, \pi_2)$$

is Poisson.

Proof. Let $(G \times G, \Pi_{\rm st})$ be the Drinfeld double of the Poisson Lie group $(G, \pi_{\rm st})$, and let Π_1 be the quotient of $\Pi_{\rm st}$ to $(G \times G)/(B \times B_-) = G/B \times G/B_-$. As the diagonal embedding $(G, \pi_{\rm st}) \to (G \times G, \Pi_{\rm st})$ is Poisson, the map

$$\phi_1:\; (G/T, \pi_{\rm st}) \longrightarrow (G/B \times G/B_-,\, \Pi_1),\;\; g.T \longmapsto (g.B, g.B_-),$$

is Poisson. Let π_{-1} be the projection of $\pi_{\rm st}$ to G/B_-, and let λ_+ and λ_- be the respective left actions of G on G/B and G/B_- by left translation. In the notation of (89), one has

$$\Pi_1 = (\pi_1, 0) + (0, \pi_{-1}) + \sum_{i=1}^{m} (\lambda_+(\xi_i), 0) \wedge (0,\, \lambda_-(x_i)).$$

Let $\phi_2: G/B \times G/B_- \to (B_-\backslash G) \times (G/B), (g.B, h.B_-) \mapsto (B_-.h^{-1}, g.B)$. Then

$$\phi_2(\Pi_1) = (-\pi'_{-1}, 0) + (0, \pi_1) + \sum_{i=1}^{m} (\rho_-(x_i), 0) \wedge (0, \lambda_+(\xi_i)) = (-\pi'_{-1}) \times_{(\rho_-, \lambda_+)} \pi_1,$$

where ρ_- is the right Poisson Lie group action

$$\begin{aligned} \rho_-:\; (B_-\backslash G, -\pi'_{-1}) \times (B_-, -\pi_{\rm st}) &\longrightarrow (B_-\backslash G, -\pi'_{-1}), \\ ((B_-).g,\, b_-) &\longmapsto (B_-).gb_-. \end{aligned}$$

Restricting ϕ_1 to $(G^{u,v}/T, \hat{\pi}_{\rm st})$ and using $I_{\overline{\overline{v}}^{-1}}$ in (91), one has the Poisson morphism

$$(I_{\overline{\overline{v}}^{-1}})^{-1} \circ \phi_2 \circ (\phi_1|_{G^{u,v}/T}):$$
$$(G^{u,v}/T, \hat{\pi}_{\rm st}) \to \left((Bv^{-1}B/B) \times (BuB/B),\, \pi_1 \times_{(\rho_{\overline{\overline{v}}^{-1}}, \lambda_+)} \pi_1\right).$$

Taking now $w = v^{-1}$, $\dot{w} = \overline{\overline{v}}^{-1}$, and $(Y, \pi_Y) = (BuB/B, \pi_1)$ in Lemma A.4, one sees that

$$\hat{F}^{u,v} = (J_{\overline{\overline{v}}^{-1}})^{-1} \circ (I_{\overline{\overline{v}}^{-1}})^{-1} \circ \phi_2 \circ (\phi_1|_{G^{u,v}/T})$$

is Poisson from $(G^{u,v}/T, \hat{\pi}_{\rm st})$ to $(\mathscr{O}^{(v^{-1},u)}, \pi_2)$. □

A.5 Double Bruhat cells vs reduced double Bruhat cells

Recall that the projection map $\varpi\colon (G, \pi_{\rm st}) \to (G/T, \hat{\pi}_{\rm st})$ is Poisson by definition. For $u, v \in W$, we now give another relation between the two Poisson manifolds $(G^{u,v}, \pi_{\rm st})$ and $(G^{u,v}/T, \hat{\pi}_{\rm st})$.

Lemma A.6. *Equip T with the zero Poisson structure. For any $v \in W$, the map*

$$(B_-vB_-, \pi_{\rm st}) \longrightarrow (T, 0), \;\; g \longmapsto [g\overline{\overline{v}}^{-1}]_0^v, \qquad g \in B_-vB_-,$$

is Poisson.

Proof. By [36, Lemma 10], both N_- and $C_{\overline{\overline{v}}} = (N\overline{\overline{v}}) \cap (\overline{\overline{v}}N_-)$ are coisotropic submanifolds of G with respect to the Poisson structure $\pi_{\rm st}$. It follows by the multiplicativity of $\pi_{\rm st}$ that $N_-\overline{\overline{v}}N_- = N_-C_{\overline{\overline{v}}}$ is also coisotropic with respect to $\pi_{\rm st}$. Writing $g \in B_-vB_-$ as $g = g_1t$, where $g_1 \in N_-\overline{\overline{v}}N_-$ and $t \in T$, one has $\pi_{\rm st}(g) = r_t\pi_{\rm st}(g_1)$, so $\varpi(\pi_{\rm st}(g)) = \varpi(\pi_{\rm st}(g_1)) = 0$. □

It is clear that the T-action on G/T by left translation preserves the Poisson structure $\hat{\pi}_{\rm st}$. Recall from §2.7 that associated to any linear map $M\colon \mathfrak{t}^* \to \mathfrak{t}$ one has the Poisson structure $0 \bowtie_M \hat{\pi}_{\rm st}$ on $T \times (G/T)$. By slight abuse of notation, for $v \in W$, we set

$$0 \bowtie_v \hat{\pi}_{\rm st} = 0 \bowtie_M \hat{\pi}_{\rm st}$$

to be the Poisson structure on $T \times (G/T)$ associated to the linear map

$$M\colon \;\; \mathfrak{t}^* \longrightarrow \mathfrak{t}, \;\; M(\lambda) = -v(\lambda^\#), \qquad \lambda \in \mathfrak{t}^*.$$

It is clear that $T \times ((B_-vB_-)/T)$ and $T \times (G^{u,v}/T)$, $u \in W$, are Poisson submanifolds of $(T \times (G/T),\, 0 \bowtie_v \hat{\pi}_{\rm st})$. The following proposition is based on a similar statement by Shizhuo Yu for the case of BvB.

Proposition A.7. *The maps*

$$\nu': \ (B_-vB_-, \pi_{\rm st}) \longrightarrow (T \times (B_-vB_-/T),\, 0 \bowtie_v \hat{\pi}_{\rm st}), \ \ g \longmapsto \left([g\overline{\overline{v}}^{-1}]_0^v,\, g.T\right),$$
$$\nu: \ (G^{u,v},\, \pi_{\rm st}) \longrightarrow (T \times (G^{u,v}/T),\, 0 \bowtie_v \hat{\pi}_{\rm st}), \ \ g \longmapsto \left([g\overline{\overline{v}}^{-1}]_0^v,\, g.T\right),$$

are Poisson isomorphisms.

Proof. As ν is the restriction of ν' to $G^{u,v}/T$, it is enough to prove the assertion for ν'. Note that if $g \in B_-vB_-$ is written as $g = m\overline{\overline{v}}m't$, where $m, m' \in N_-$ and $t \in T$, then $[g\overline{\overline{v}}^{-1}]_0^v = t$. It follows that ν' is an isomorphism. Consider now the map

$$\kappa: \ G \times (B_-vB_-) \longrightarrow T \times (G/T), \ (g', g) \longmapsto \left(([g\overline{\overline{v}}^{-1}]_0^v)^{-1},\, g'.T\right),$$

and equip $G \times (B_-vB_-)$ with the Poisson structure $\Pi_{\rm st}$. By Lemma A.6, κ maps the bi-vector field $(\pi_{\rm st}, 0) + (0, \pi_{\rm st})$ on $G \times (B_-vB_-)$ to $(0, \hat{\pi}_{\rm st})$. Moreover, it is clear that for any $x \in \mathfrak{n}_-$, $\kappa(0, x^R) = \kappa(0, x^L) = 0$. Thus $\kappa(\mu_1) = \kappa(\mu_1')$ and $\kappa(\mu_2) = \kappa(\mu_2')$, where μ_1 and μ_2 are given in (90), and

$$\mu_1' = \sum_{i=1}^r (\xi_i^R, 0) \wedge (0, x_i^R) \quad \text{and} \quad \mu_2' = -\sum_{i=1}^r (\xi_i^L, 0) \wedge (0, x_i^L),$$

with $\{x_i\}_{i=1,\ldots,r}$ and $\{\xi_i\}_{i=1,\ldots,r}$ any bases of $\mathfrak{t}$ such that $\langle x_i,\, \xi_j\rangle = \delta_{i,j}$ for $i, j = 1, \ldots, r$. On the other hand, it is clear that $\kappa(x^L, 0) = 0$ for every $x \in \mathfrak{t}$. Thus $\kappa(\mu_2') = 0$, and we only need to compute $\kappa(\mu_1')$. It is easy to see that for any $x \in \mathfrak{t}$,

$$\kappa(x^R, 0) = (0, \theta(x)) \quad \text{and} \quad \kappa(0, x^R) = ((v^{-1}(x))^R, 0),$$

where $\theta(x) \in \mathfrak{X}^1(G/T)$ is given by

$$\theta(x)(g.T) = \frac{d}{ds}\Big|_{s=0} (\exp(sx)g.T), \qquad g \in G,$$

and recall that x^R is the right (also left) invariant vector field on T with value x at the identity element of T. Thus

$$\kappa(\mu_1') = -\sum_{i=1}^r ((v^{-1}(x_i))^L, 0) \wedge (0, \theta(\xi_i)).$$

Let $\{h_i\}$ be an orthonormal basis of $\mathfrak{t}$ with respect to $\langle\, ,\, \rangle$. Then

$$\kappa(\Pi_{\rm st}) = (0, \hat{\pi}_{\rm st}) - \sum_{i=1}^r (h_i^R, 0) \wedge (0, \theta(v(h_i))) = 0 \bowtie_v \hat{\pi}_{\rm st}.$$

It follows that $\kappa: (G \times (B_-vB_-), \Pi_{\rm st}) \to (T \times (G/T), 0 \bowtie_v \hat{\pi}_{\rm st})$ is a Poisson map. As

$$\iota: \ (B_-vB_-, \pi_{\rm st}) \longrightarrow (G \times (B_-vB_-), \Pi_{\rm st}), \ g \longmapsto (g, g), \qquad g \in B_-vB_-,$$

is Poisson, $\nu' = \kappa \circ \iota : (B_-\nu B_-, \pi_{\rm st}) \to (T \times (B_-\nu B_-/T), 0 \bowtie_\nu \hat{\pi}_{\rm st})$ is Poisson. □

A.6 Fomin-Zelevinsky embeddings for double Bruhat cells

Let again $u, v \in W$ and recall the Fomin-Zelevinsky embedding

$$F^{u,v} : \; G^{u,v} \longrightarrow T \times \mathcal{O}^{(v^{-1},u)}, \; F^{u,v}(g) = ([g\overline{\overline{v}}^{-1}]_0^v, \, \hat{F}^{u,v}(g.T)),$$

where $\hat{F}^{u,v} : G^{u,v}/T \to \mathcal{O}^{(v^{-1},u)}$ is the Fomin-Zelevinsky embedding for $G^{u,v}/T$. Recall from Notation 2.22 the definition of the Poisson structure $0 \bowtie \pi_2$ on $T \times \mathcal{O}^{(v^{-1},u)}$. The following proposition is the second part of Theorem 3.11.

Proposition A.8. *For any $u, v \in W$, the Fomin-Zelevinsky embedding*

$$F^{u,v} : \; (G^{u,v}, \, \pi_{\rm st}) \longrightarrow (\mathcal{O}^{(v^{-1},u)}, \, 0 \bowtie \pi_2)$$

is Poisson.

Proof. It is easy to see that the map $\hat{F}^{u,v} : G^{u,v}/T \to \mathcal{O}^{(v^{-1},u)}$ is T-equivariant, where $G^{u,v}/T$ has the T-action by

$$t \cdot (g.T) = (t^{v^{-1}} g).T, \qquad t \in T, \, g \in G^{u,v},$$

and $\mathcal{O}^{(v^{-1},u)} \subset F_2$ has the T-action given in (14). It now follows from Proposition A.7 and the definitions of the Poisson structures $0 \bowtie_v \hat{\pi}_{\rm st}$ and $0 \bowtie \pi_2$ that $F^{u,v}$ is Poisson. □

References

[1] V. Arnold, *Ordinary Differential Equations*, translation from Russian by Roger Cooke, Springer-Verlag, 1992.
[2] A. Berenstein and A. Zelevinsky, Tensor product multiplicities, canonical bases and totally positive varieties, *Invent. Math.* **143** (2001), 77–128.
[3] A. Berenstein, S. Fomin, and A. Zelevinsky, Cluster algebras III, Upper bounds and double Bruhat cells, *Duke Math. J.* **126** (2005), 1–52.
[4] N. Bourbaki, *Groupes et algèbras de Lie*, Ch. IV–VI, Hermann, Paris, 1968.
[5] V. Chari and A. Pressley, *A Guide to Quantum Groups*, Cambridge University Press, 1994.
[6] M. Colarusso, The orbit structure of the Gelfand-Zeitlin group on $n \times n$ matrices, *Pacific J. Math.* **250** (1) (2011), 109–138.
[7] M. Colarusso and S. Evens, On algebraic integrability of Gelfand-Zeitlin fields, *Trans. Groups* **15** (1) (2010), 46–71.
[8] M. Colarusso and S. Evens, K-orbits on the flag variety and strongly regular nilpotent matrices, *Selecta. Math. (N.S.)* **18** (1) (2012), 159–177.

[9] M. Colarusso and S. Evens, The Gelfand-Zeitlin integrable system and K-orbits on the flag variety, in *Symmetry: Representation Theory and Its Applications*, Prog. Math., **257**, 85–119, Birkhäuser, New York, 2014.
[10] M. Colarusso and S. Evens, Eigenvalue coincidences and K-orbits, I, *J. Algebra*, **422** (2015), 611–632.
[11] V. G. Drinfeld, Quantum groups, *Proceedings of the International Congress of Mathematicians*, **1** (2) (Berkeley, Calif., 1986), 798–820, Amer. Math. Soc., Providence, RI, 1987.
[12] V. G. Drinfeld, On Poisson homogeneous spaces of Poisson-Lie groups, *Theo. Math. Phys.* **95** (2) (1993), 226–227.
[13] O. Dudas, Note on the Deodhar decomposition of a double Schubert cell, arXiv:0807.2198.
[14] B. Elek, *Computing the standard Poisson structure on Bott-Samelson varieties in coordinates*, Mphil Thesis, the University of Hong Kong, 2012.
[15] B. Elek, and J.-H. Lu, Bott-Samelson varieties and Poisson Ore extensions, arXiv:1601.00047.
[16] P. Etingof and O. Schiffmann, *Lectures on Quantum Groups*, 2nd edition, International Press, 2002.
[17] S. Fomin and A. Zelevinsky, Double Bruhat cells and total positivity, *J. Amer. Math. Soc.* 12 (1999), 335–380.
[18] M. Gekhtman, M. Shapiro, and A. Vainshtein, *Cluster algebras and Poisson geometry*, Mathematical Surveys and Monographs, the AMS, **167** (2010).
[19] M. Gekhtman and M. Yakimov, Completeness of determinantal Hamiltonian flows on the matrix affine space, *Lett. Math. Phys.* **90** (2009), 161–173.
[20] K. Goodearl and M. Yakimov, The Berenstein-Zelevinsky quantum cluster algebra conjecture, arXiv:1602.00498.
[21] K. Goodearl and M. Yakimov, Quantum cluster algebras and quantum nilpotent algebras, *Proc. Natl. Acad. Sci.* USA **111** (27) (2014), 9696–9703.
[22] K. Goodearl and M. Yakimov, *Quantum cluster algebra structures on quantum nilpotent algebras*, Memoirs Amer. Math. Soc. **247** (1169) (2017).
[23] K. Goodearl and M. Yakimov, Cluster algebra structures on Poisson nilpotent algebras, arXiv:1801.01963.
[24] K. Goodearl and M. Yakimov, Cluster structures on double Bruhat cells, in preparation.
[25] T. J. Hodges and T. Levasseur, Primitive ideals of $C_q[SL(3)]$, *Comm. Math. Phys.* **156** (3) (1993), 581–605.
[26] T. Hoffmann, J. Kellendonk, N. Kutz, and N. Reshetikhin, Factorization dynamics and Coxeter-Toda lattices, *Comm. Math. Phys.* **212** (2) (2000), 297–321.
[27] D. Knuth, *The Art of Computer Programming,* Vol. 2, 3rd edition, Addison-Wesley, 1997.
[28] A. Knutson, Frobenius splitting, point-counting, and degeneration, arXiv:0911.4941.
[29] M. Kogan and A. Zelevinsky, On symplectic leaves and integrable systems in standard complex semisimple Poisson Lie groups, *Int. Math. Res. Not.* **32** (2002), 1685–1702.
[30] B. Kostant and N. Wallach, Gelfand-Zeitlin theory from the perspective of classical mechanics. I, in: *Studies in Lie Theory*, Prog. Math., Vol. 243, Birkhäuser, 2006, 319–364.
[31] B. Kostant and N. Wallach, Gelfand-Zeitlin theory from the perspective of classical mechanics. II, *The Unity of Mathematics*, Prog. Math., Vol. 244, Birkhäuser, 2006, 387–420.
[32] C. Laurent-Gengoux, A. Pichereau, and P. Vanhaecke, *Poisson structures*, Grundlehren der mathematischen Wissenschaften, Vol. 347, Springer, 2013.
[33] J.-H. Lu, Poisson homogeneous spaces and Lie algebroids associated to Poisson actions, *Duke Math. J.* **86** (2) (1997), 261–304.
[34] J.-H. Lu and V. Mouquin, Mixed product Poisson structures associated to Poisson Lie groups and Lie bialgebras, *Intern. Math. Res. Notices* **2016**, 1–58.
[35] J.-H. Lu and V. Mouquin, On the T-leaves of some Poisson structures related to products of flag varieties, *Adv. Math.* **306** (2017), 1209–1261.
[36] J.-H. Lu and V. Mouquin, Double Bruhat cells and symplectic groupoids, *Trans. Groups* (2017), doi.org/10.1007/s00031-017-9437-6.
[37] R. Marsh and K. Rietsch, Parametrizations of flag varieties, *Rep. Theory*, AMS, **8** (2004), 212–242.

On the Definition of Almost Characters

G. Lusztig

Dedicated to the memory of Bertram Kostant

Abstract We define almost characters of $\mathcal{G}(\mathbf{F}_q)$, where $\mathcal{G}$ is a reductive connected group over a finite field $\mathbf{F}_q$, as explicit linear combinations of irreducible characters. Previously, these were defined assuming that the centre of $\mathcal{G}$ is connected.

Keywords: almost character • Fourier transform • character sheaf

Mathematics Subject Classification (2010): 20G99

Introduction

0.1

Let $\mathcal{G}(\mathbf{F}_q)$ be the group of rational points of a connected reductive group $\mathcal{G}$ defined over a finite field $\mathbf{F}_q$. The vector space of $\mathbf{C}$-valued class functions on $\mathcal{G}(\mathbf{F}_q)$ has a basis β given by the characters of the irreducible representations of $\mathcal{G}(\mathbf{F}_q)$. It has a second basis β' (whose elements are defined only up to multiplication by roots of 1) given by the characteristic functions of the Frobenius-invariant character sheaves on $\mathcal{G}$. To compare these two bases it is convenient to introduce a third basis β'' (whose elements are again defined only up to multiplication by roots of 1). Now the functions in β'' are given by explicit linear combinations of elements of β; they are called almost characters. In this paper we give a definition of β''. (A definition of almost characters assuming that $\mathcal{G}$ has connected centre appeared in [L2].) The interest of the basis β'' is that it conjecturally coincides with β'.

This paper was written in response to a question that Meinolf Geck asked me; I wish to thank him for asking the question and for comments on an earlier version.

Supported in part by National Science Foundation grant 1566618.

Department of Mathematics, M.I.T., Cambridge, MA 02139

V. G. Kac, V. L. Popov (eds.), *Lie Groups, Geometry, and Representation Theory*, Progress in Mathematics 326, https://doi.org/10.1007/978-3-030-02191-7_13

0.2 Notation

$\mathbf{k}$ denotes a fixed algebraically closed field of characteristic $p \geq 0$. All algebraic groups are assumed to be affine and over $\mathbf{k}$. If H is an algebraic group, we denote by H^0 the identity component of H. For $g \in H$ we denote by g_s (resp. g_u) the semisimple (resp. unipotent) part of g, so that $g = g_s g_u = g_u g_s$.

If x is an element of a group Γ, we denote by $Z_\Gamma(x)$ the centralizer of x in Γ.

1 Nonabelian Fourier transform

1.1

In this paper a representation of a group Γ is always assumed to be in a finite dimensional $\mathbf{C}$-vector space.

We now assume that Γ is a finite group. Let $M(\Gamma)$ be the set consisting of all pairs (x,σ) where $x \in \Gamma$ and σ is an irreducible representation of $Z_\Gamma(x)$ (up to isomorphism); these pairs are taken modulo the equivalence relation $(x,\sigma) \mapsto (vxv^{-1}, {}^v\sigma)$ for any $v \in \Gamma$ (here ${}^v\sigma$ is the irreducible representation of $Z_\Gamma(vxv^{-1})$ defined by composing σ with $z \mapsto v^{-1}zv$). Following [L1] we define a pairing $\{,\}: M(\Gamma) \times M(\Gamma) \to \mathbf{C}$ by

(a)
$$\{(x,\sigma),(y,\tau)\} = \sum_{w \in \Gamma; xwyw^{-1} = wyw^{-1}x} |Z_\Gamma(x)|^{-1}|Z_\Gamma(y)|^{-1}\mathrm{tr}(w^{-1}x^{-1}w,\tau)\mathrm{tr}(wyw^{-1},\sigma).$$

1.2

Let c be an integer ≥ 1. Let $\mu_c = \{\alpha \in \mathbf{C}; \alpha^c = 1\}$.

A $\mathbf{Z}/c$-grading for a finite group Γ is by definition a surjective homomorphism $\zeta: \Gamma \to \mathbf{Z}/c$, such that for some $x \in \zeta^{-1}(1)$ we have $x^c = 1$. For such Γ, ζ and for $i = 0,1$ let $M^i_\zeta(\Gamma)$ be the subset of $M(\Gamma)$ consisting of all $(x,\sigma) \in M(\Gamma)$ such that $x \in \zeta^{-1}(i)$ and

(a) $Z_\Gamma(x) \cap \zeta^{-1}(1) \neq \emptyset$, $\sigma|_{Z_\Gamma(x)\cap\zeta^{-1}(0)}$ is irreducible.

(When $i = 1$ the conditions (a) are automatically satisfied.)

For $\alpha \in \mu_c$ let $\varepsilon_\alpha: \Gamma \to \mathbf{C}^*$ be the homomorphism which takes any $x \in \zeta^{-1}(1)$ to α; the restriction of ε_α to various subgroups of Γ is denoted again by ε_α.

Let $V(\Gamma)$ be the vector space of formal linear combinations $\sum_{(x,\sigma)\in M(\Gamma)} b_{x,\sigma}(x,\sigma)$ with $b_{x,\sigma} \in \mathbf{C}$. For $i = 0,1$ we define a subspace $V^i_\zeta(\Gamma)$ of $V(\Gamma)$ by the conditions

$b_{x,\sigma} = 0$ if $(x,\sigma) \notin M^i_\zeta(\Gamma)$,

$b_{x,\sigma\otimes\varepsilon_\alpha}=\alpha^{i-1}b_{x,\sigma}$ for any $\alpha\in\mu_c$.
Let $A\colon V(\Gamma)\to V(\Gamma)$ be the linear map given by

$$A(x,\sigma)=\sum_{(y,\tau)\in M(\Gamma)}\{(x,\sigma),(y,\tau)\}(y,\tau)$$

for any $(x,\sigma)\in M(\Gamma)$. For $i=0,1$, we show that
(b) $A(V^i_\zeta(\Gamma))\subset V^{1-i}_\zeta(\Gamma)$.
Since $V^i_\zeta(\Gamma)$ is generated by the vectors $\sum_{\alpha\in\mu_c}\alpha^{i-1}(x,\sigma\otimes\varepsilon_\alpha)$ for various $(x,\sigma)\in M^i_\zeta(\Gamma)$, it is enough to show that for $(x,\sigma)\in M^i_\zeta(\Gamma)$ we have
$A\big(\sum_{\alpha\in\mu_c}\alpha^{i-1}(x,\sigma\otimes\varepsilon_\alpha)\big)\in V^{1-i}_\zeta(\Gamma)$ or equivalently
(c) $\sum_{(z,\rho)\in M(\Gamma)}b'_{z,\rho}(z,\rho)\in V^{1-i}_\zeta(\Gamma)$
where

$$b'_{z,\rho}=\sum_{w\in\Gamma;xwzw^{-1}=wzw^{-1}x}|Z_\Gamma(x)|^{-1}|Z_\Gamma(z)|^{-1}$$
$$\times\,\mathrm{tr}(w^{-1}x^{-1}w,\rho)\sum_{\alpha\in\mu_c}\alpha^{i-1}\mathrm{tr}(wzw^{-1},\sigma\otimes\varepsilon_\alpha).$$

The last sum is equal to $c\delta_{1-i,\zeta(z)}\mathrm{tr}(wzw^{-1},\sigma)$ hence, if $b'_{z,\rho}\neq 0$, then $z\in\zeta^{-1}(1-i)$. If in addition we have $i=1$, then $Z_\Gamma(z)$ contains $w^{-1}xw\in\zeta^{-1}(1)$ (for some w) which belongs to $Z_\Gamma(z)$. Thus $Z_\Gamma(z)\cap\zeta^{-1}(1)\neq 0$; moreover in this case we have $\mathrm{tr}(w^{-1}x^{-1}w,\rho)\neq 0$ for some w and $w^{-1}x^{-1}w\in\zeta^{-1}(-1)$; this forces $\rho|_{Z_\Gamma(z)\cap\zeta^{-1}(0)}$ to be irreducible.

For $i=0,1$ we have $\mathrm{tr}(w^{-1}x^{-1}w,\rho\otimes\varepsilon_{\alpha'})=\alpha'^{-i}\mathrm{tr}(w^{-1}xw,\rho)$ for any $\alpha'\in\mu_c$, so that $b'_{z,\rho\otimes\varepsilon_{\alpha'}}=\alpha'^{-i}b'_{z,\rho}$ for any α', z, ρ. We see that (c) holds and (b) is proved. From [L1] it is known that $A^2=1$. Combining this with (b) we see that
(d) $A\colon V^1_\zeta(\Gamma)\to V^0_\zeta(\Gamma)$, $A\colon V^0_\zeta(\Gamma)\to V^1_\zeta(\Gamma)$ *are inverse isomorphisms.*
(This was stated without proof in [L2, (4.16.1)]).

1.3

A $\mathbf{Z}$-grading for a group Δ is by definition a surjective homomorphism $\eta\colon\Delta\to\mathbf{Z}$ such that $\eta^{-1}(0)$ is finite. We now fix such Δ,η. Let Δ^* be the set of central elements $\xi\in\Delta$ such that $\xi=y^c$ for some $y\in\eta^{-1}(1)$ and some $c\geq 1$. We have $\Delta^*\neq\emptyset$; indeed, if $y\in\eta^{-1}(1)$, then y^c is in the centre of Δ for some $c\geq 1$. Moreover,
(a) *we can find $d\geq 1$ such that $y\mapsto y^d$ is a constant function on $\eta^{-1}(1)$; for this d we have $y^d\in\Delta^*$ for all $y\in\eta^{-1}(1)$.*
If $\xi\in\Delta^*$, $c'\geq 1$ then $\xi^{c'}\in\Delta^*$. Moreover,
(b) *if $\xi\in\Delta^*$, $\xi'\in\Delta^*$, then $\xi^a=\xi'^{a'}$ for some $a\geq 1$, $a'\geq 1$.*
Indeed, we have $\xi=y^c$, $\xi'=y'^{c'}$ for some $c\geq 1$, $c'\geq 1$ and some y,y' in $\eta^{-1}(1)$.

Hence if d is as in (a), we have $\xi^{c'd} = y^{cc'd} = y'^{cc'd} = \xi'^{cd}$, so that (b) holds with $a = c'd$, $a' = cd$.

For $i = 0, 1$ let $M^i_\eta(\Delta)$ be the set of all pairs (x, σ) where $x \in \eta^{-1}(i)$ and σ (an irreducible representation up to isomorphism of $Z_\Delta(x) \cap \eta^{-1}(0)$) are such that

(c) $Z_\Delta(x) \cap \eta^{-1}(1) \neq \emptyset$, *$\sigma$ can be extended to a representation of* $Z_\Delta(x)$*;*

these pairs are taken modulo the equivalence relation $(x, \sigma) \sim (vxv^{-1}, {}^v\sigma)$ for any $v \in \Delta$ (here ${}^v\sigma$ is the irreducible representation of $Z_\Delta(vxv^{-1}) \cap \delta^{-1}(0)$ defined by composing σ with $z \mapsto v^{-1}zv$).

(When $i = 1$ the conditions (c) are automatically satisfied.)

For any $\xi \in \Delta^*$, we set $\Delta_\xi = \Delta/\langle\xi\rangle$ where $\langle\xi\rangle$ is the cyclic subgroup of the centre of Δ generated by ξ; this is a finite group. For any $x \in \Delta$ we denote by x_ξ the image of x in Δ_ξ. Now η induces a surjective homomorphism $\zeta\colon \Delta_\xi \to \mathbf{Z}/\eta(\xi)$ which is a $\mathbf{Z}/\eta(\xi)$-grading for Δ_ξ. (We can find $y \in \eta^{-1}(1)$ such that $y^{\eta(\xi)} = \xi$; then $y_\xi^{\eta(\xi)} = 1$ and $\zeta(y_\xi) = 1$.)

For $i = 0, 1$ let $U^i_\eta(\Delta)$ be the vector space of formal linear combinations $\sum_{(y,\tau)\in M^i_\eta(\Delta)} b_{y,\tau}(y, \tau)$ with $b_{x,\sigma} \in \mathbf{C}$. To any $(x, \sigma) \in M^0_\eta(\Delta)$ we associate an element $(x, \sigma)\hat{} \in U^1_\eta(\Delta)$ (defined up to multiplication by a root of 1) as follows.

We choose $\xi \in \Delta^*$ and an extension of σ to a representation $\tilde{\sigma}$ of $Z_\Gamma(x)$ on which ξ acts as 1; we set

$$(x,\sigma)\hat{}_{\xi,\tilde{\sigma}} = \sum_{(y,\tau)\in M^1_\eta(\Delta)} \sum_{w\in\Delta/\langle\xi\rangle; xwyw^{-1}=wyw^{-1}x} \eta(\xi)|Z_\Delta(x)/\langle\xi\rangle|^{-1}|Z_\Delta(y)/\langle\xi\rangle|^{-1} \times \operatorname{tr}(w^{-1}x^{-1}w, \tilde{\tau})\operatorname{tr}(wyw^{-1}, \tilde{\sigma})(y,\tau) \in U^1_\eta(\Delta) \quad \text{(d)}$$

where $\tilde{\tau}$ is the representation of $Z_\Delta(y)$ extending τ on which y acts as 1.

Assume now that $\xi, \tilde{\sigma}$ are replaced by another pair $\xi', \tilde{\sigma}'$ with the same properties as $\xi, \tilde{\sigma}$. We want to show that

(e) $(x,\sigma)\hat{}_{\xi',\tilde{\sigma}'}$ equals $(x,\sigma)\hat{}_{\xi,\tilde{\sigma}}$ times a root of 1.

Using (b) we can assume that $\xi' = \xi^m$ for some $m \geq 1$. From the definitions we see that

$(x,\sigma)\hat{}_{\xi^m,\tilde{\sigma}} = (x,\sigma)\hat{}_{\xi,\tilde{\sigma}}$

and

$(x,\sigma)\hat{}_{\xi^m,\tilde{\sigma}}$ is equal to $(x,\sigma)\hat{}_{\xi^m,\tilde{\sigma}'}$ times a root of 1

so that (e) follows. We shall write $(x,\sigma)\hat{}$ instead of $(x,\sigma)\hat{}_{\xi,\tilde{\sigma}} \in U^1_\eta(\Delta)$; this is well defined only up to multiplication by a root of 1.

1.4

For any $(x,\sigma)\in M^0_\eta(\Delta)$ we choose $\xi\in\Delta^*$ and an extension of σ to a representation $\tilde\sigma$ of $Z_\Gamma(x)$ on which ξ acts as 1; we can assume that ξ is independent of (x,σ). We show:

(a) *The elements* $(x,\sigma)\hat{}_{\xi,\tilde\sigma}$ *for various* $(x,\sigma)\in M^0_\eta(\Delta)$ *form a* $\mathbf{C}$*-basis of* $U^1_\eta(\Delta)$.

We set $c=\eta(\xi)$. Let $\zeta:\Delta_\xi\to\mathbf{Z}/c$ be as in 1.3. We define a linear map $f: U^0_\eta(\Delta)\to V^0_\zeta(\Delta_\xi)$ by

$(x,\sigma)\mapsto\sum_{\alpha\in\mu_c}\alpha^{-1}(x_\xi,\tilde\sigma\otimes\varepsilon_\alpha)$

(here we view $\tilde\sigma$ as an irreducible representation of $Z_{\Delta_\xi}(x_\xi)$); this is easily seen to be an isomorphism. We define a linear map $f': V^1_\zeta(\Delta_\xi)\to U^1_\eta(\Delta)$ by

$$\sum_{(y,\tau)\in M^1_\zeta(\Delta_x)} b_{y,\tau}(y,\tau)\mapsto\sum_{(y,\tau)\in M^1_\zeta(\Delta_\xi);y=1\text{ on }\tau} b_{y,\tau}(\tilde y,\tau|_{Z_{\Delta_\xi}(y)\cap\zeta^{-1}(0)}),$$

where for $(y,\tau)\in M^1_\zeta(\Delta_\xi)$, we define $\tilde y\in\eta^{-1}(1)$ by $\tilde y_\xi=y$. (Note that $Z_{\Delta_\xi}(y)\cap\zeta^{-1}(0)=Z_\Delta(\tilde y)\cap\eta^{-1}(0)$.) This is easily seen to be an isomorphism. The composition

$$V^0_\Delta\xrightarrow{f}V^0_{\Delta_\xi}\xrightarrow{A}V^1_{\Delta_\xi}\xrightarrow{f'}V^1_\Delta$$

is an isomorphism (see 1.2(d)). It takes $(x,\sigma)\in M^0_\eta(\Delta)$ to $(x,\sigma)\hat{}_{\xi,\tilde\sigma}$. This proves (a).

2 Special elements

2.1

In this section we assume that $p=0$. We fix an integer $q\neq 0$.

Let $\mathcal{H}$ be a connected reductive group. For any element $h\in\mathcal{H}$, let $\mathcal{B}_u$ be the variety of Borel subgroups of $\mathcal{H}$ that contain h.

Let $\mathcal{H}^{sp}_u$ be the set of special unipotent elements of $\mathcal{H}$, in the sense of [L2, (13.1.1)]; for $u\in\mathcal{H}^{sp}_u$ let $\bar A(u)$ be the canonical quotient group of $Z_{\mathcal{H}}(u)/Z^0_{\mathcal{H}}(u)$ defined in [L2, 13.1] (paragraph following (13.1.2)) and let $I_{\mathcal{H}}(u)$ be the kernel of the obvious homomorphism $Z_{\mathcal{H}}(u)/Z^0_{\mathcal{H}}(u)\to\bar A(u)$. (When we refer to [L2, 13.1], we assume that the ground field in *loc. cit.* is replaced by our $\mathbf{k}$.) For further information on $A(u)$ see [L4]. We show:

(a) *Assume that* $u\in\mathcal{H}^{sp}_u$. *Then* $u^q\in\mathcal{H}^{sp}_u$ *and we have canonically*

$Z_{\mathcal{H}}(u)/Z^0_{\mathcal{H}}(u)=Z_{\mathcal{H}}(u^q)/Z^0_{\mathcal{H}}(u^q)$, $I_{\mathcal{H}}(u)=I_{\mathcal{H}}(u^q)$.

We have clearly $Z_{\mathcal{H}}(u)\subset Z_{\mathcal{H}}(u^q)$. Since $q\neq 0$ and $p=0$, u is conjugate to u^q in H; it follows that $Z_{\mathcal{H}}(u)=Z_{\mathcal{H}}(u^q)$. This implies the first equality in (a). We have clearly $\mathcal{B}_u \subset \mathcal{B}_{u^q}$. Since $\mathcal{B}_u$, $\mathcal{B}_{u^q}$ are isomorphic, we must have $\mathcal{B}_u = \mathcal{B}_{u^q}$. Using this and the definition we see that the second equality in (a) holds.

2.2

In the remainder of this paper G denotes a connected reductive group over $\mathbf{k}$. Moreover we fix an automorphism $j\colon G \xrightarrow{\sim} G$ of finite order.

2.3

We say that $g \in G$ is special if

(i) $g_s^{q^r} = g_s$ for some $r \geq 1$ and

(ii) g_u is a special unipotent element of the connected reductive group $Z_G^0(g_s)$. Let G^{sp} be the set of special elements of G. For $g \in G^{sp}$, setting $\tilde{\mathcal{H}} = Z_G(g_s)$, $\mathcal{H} = \tilde{\mathcal{H}}^0$, we have obvious embeddings

$$I_{\mathcal{H}}(g_u) \subset Z_{\mathcal{H}}(g_u)/Z^0_{\mathcal{H}}(g_u) \subset Z_{\tilde{\mathcal{H}}}(g_u)/Z^0_{\tilde{\mathcal{H}}}(g_u) = Z_G(g)/Z^0_G(g);$$

we denote by $Z'_G(g)$ the inverse image of $I_{\mathcal{H}}(g_u)$ under the obvious surjective homomorphism $Z_G(g) \to Z_G(g)/Z^0_G(g)$ and by $Z'_{\tilde{\mathcal{H}}}(g_u)$ the inverse image of $I_{\mathcal{H}}(g_u)$ under the obvious surjective homomorphism $Z_{\tilde{\mathcal{H}}}(g_u) \to Z_{\tilde{\mathcal{H}}}(g_u)/Z^0_{\tilde{\mathcal{H}}}(g_u)$, We have $Z'_G(g) = Z'_{\tilde{\mathcal{H}}}(g_u)$. This is a normal subgroup of $Z_G(g) = Z_{\tilde{\mathcal{H}}}(g_u)$.

For $g \in G^{sp}$ we set

$$\bar{Z}_G(g) = Z_G(g)/Z'_G(g),$$

a finite group.

We show:

(a) *Assume that $g \in G^{sp}$. Then $g^q \in G^{sp}$ and $Z_G(g) = Z_G(g^q)$, $Z'_G(g) = Z'_G(g^q)$.*

Let $r \geq 1$ be such that $g_s^{q^r} = g_s$. We have clearly $Z_G(g_s) \subset Z_G(g_s^q) \subset Z_G(g_s^{q^r}) = Z_G(g_s^q)$, hence $Z_G(g_s) = Z_G(g_s^q)$. We have $Z_G(g_u) \subset Z_G(g_u^q)$; as in the proof of 2.1(a), g_u, g_u^q are conjugate in $Z_G^0(g_s)$, hence they are also conjugate in G and $Z_G(g_u) = Z_G(g_u^q)$. Since $Z_G(g) = Z_G(g_s) \cap Z_G(g_u)$ and $Z_G(g^q) = Z_G(g_s^q) \cap Z_G(g_u^q)$, we deduce that $Z_G(g) = Z_G(g^q)$, hence $Z^0_G(g) = Z^0_G(g^q)$ and $Z_G(g)/Z^0_G(g) = Z_G(g^q)/Z^0_G(g^q)$. Let $\mathcal{H} = Z^0_G(g_s) = Z^0_G(g_s^q)$. By 2.1(a), we have $I_{\mathcal{H}}(g_u) = I_{\mathcal{H}}(g_u^q)$. It follows that $Z'_G(g) = Z'_G(g^q)$, as required.

We show:

(b) *Let $g \in G^{sp}$. Then $g_s^q g_u \in G^{sp}$ and $Z_G(g) = Z_G(g_s^q g_u)$, $Z'_G(g) = Z'_G(g_s^q g_u)$.*

Let $\tilde{\mathcal{H}} = Z_G(g_s)$. As in the proof of (a) we have $Z_G(g_s^q) = \tilde{\mathcal{H}}$. We have $Z_G(g) = Z_{\tilde{\mathcal{H}}}(g_u)$ and $Z_G(g_s^q g_u) = Z_{\tilde{\mathcal{H}}}(g_u)$; moreover, $Z'_G(g) = Z'_{\tilde{\mathcal{H}}}(g_u)$ and $Z'_G(g_s^q g_u) = Z'_{\tilde{\mathcal{H}}}(g_u)$; (b) follows.

The following result is immediate.

(c) *If* $\mathfrak{s}: G \to G$ *is an automorphism and* $g \in G^{sp}$*, then* $g' \mapsto \mathfrak{s}(g')$ *defines an isomorphism* $Z'_G(g) \to Z'_G(\mathfrak{s}(g))$.

2.4

Let $g \in G$ and let $\tilde{\mathcal{H}} = Z_G(g_s), \mathcal{H} = \tilde{\mathcal{H}}^0$. We show:

(a) $j(g^q)$ is G-conjugate to $j(g_s^q g_u)$.

It is enough to show that g^q is G-conjugate to $g_s^q g_u$. It is also enough to show that g_u^q is $\mathcal{H}$-conjugate to g_u. This holds since $p = 0$.

Let G_*^{sp} be the set of all $g \in G^{sp}$ which satisfy:

(i) g is G-conjugate to $j(g^q)$; g is G-conjugate to $j(g_s^q g_u)$ (if $p = 0$);

(The two conditions in (i) are equivalent by (a).)

Clearly, G_*^{sp} is a union of G-conjugacy classes.

2.5

Let $g \in G_*^{sp}$. For any $e \in \mathbf{Z}$, we set

$$H_e(g) = \{x \in G; xgx^{-1} = j^e(g^{q^e})\}, \quad 'H_e(g) = \{x \in G; xgx^{-1} = j^e(g_s^{q^e} g_u)\}.$$

Note that $H_1(g) \neq \emptyset$, $'H_1(g) \neq \emptyset$. We set $H(g) = \sqcup_{e \in \mathbf{Z}} H_e(g)$, $'H(g) = \sqcup_{e \in \mathbf{Z}} {'H_e}(g)$. We regard $H(g)$ as a group with multiplication defined by $x * x' = j^{e'}(x)x'$ for $x \in H_e(g)$, $x' \in H_{e'}(g)$; we regard $'H(g)$ as a group with multiplication defined by $x * x' = j^{e'}(x)x'$ for $x \in {'H_e}(g)$, $x' \in {'H_{e'}}(g)$. Note that the map $H(g) \to \mathbf{Z}$ (resp. $'H_e(g)$) which takes $H_e(g)$ (resp. $'H_e(g)$) to e for any $e \in \mathbf{Z}$ is a (surjective) group homomorphism with kernel $H_0(g) = Z_G(g)$ (resp. $'H_0(g) = Z_G(g)$). Thus, $Z'_G(g)$ can be regarded as a subgroup of $H(g)$ and also as a subgroup of $'H(g)$. It is in fact a normal subgroup. (It is enough to show that for $x_0 \in Z'_G$ and x in $H_e(g)$ or on $'H_e(g)$, we have $x^{-1}j^e(x_0)x \in Z'_G$. We have $j^e(x_0) \in Z'_G(j^e(g))$, hence $x^{-1}j^e(x_0)x \in Z'_G(x^{-1}j^e(g)x)$; by 2.3(a),(b), the last group equals $Z'_G(x^{-1}j^e(g^q)x) = Z'_G(g)$ if $x \in H_e(g)$ and to $Z'_G(x^{-1}j^e(g_s^q g_u)x) = Z'_G(g)$ if $x \in {'H_e}(g)$.)

We set $\Delta(g) = H(g)/Z'_G(g)$, $'\Delta(g) = {'H}(g)/Z'_G(g)$. These have group structures induced from those of $H(h), 'H(g)$. The homomorphisms $H(g) \to \mathbf{Z}$, $'H(g) \to \mathbf{Z}$ (as above) induce (surjective) group homomorphisms $\eta : \Delta(g) \to \mathbf{Z}$, $'\eta : {'\Delta}(g) \to \mathbf{Z}$ with finite kernel $Z_G(g)/Z'_G(g) = \bar{Z}_G(g)$. Thus $\eta, '\eta$ are $\mathbf{Z}$-gradings for $\Delta(g), '\Delta(g)$.

2.6

Let $g \in G_*^{sp}$. Let $\mathcal{H} = Z_G^0(g_s)$ and let $\zeta : \lambda \mapsto \zeta_\lambda$ be a homomorphism of algebraic groups $\mathbf{k}^* \to \mathcal{H}$ (coming from the Jacobson-Morozov theorem applied to $u \in \mathcal{H}$) such that $\zeta_\lambda u \zeta_\lambda^{-1} = u^\lambda$ for any $\lambda \in \mathbf{k}^*$. We define a map $\Psi : {}'H(g) \to H(g)$ by $x \mapsto y$ where $x \in {}'H_e(g)$, $y = j^e(\zeta_{q^e})x$, $e \in \mathbf{Z}$. The computation

$$ygy^{-1} = j^e(\zeta_{q^e})xgx^{-1}j^e(\zeta_{q^e}^{-1}) = j^e(\zeta_{q^e})\,j^e(g_s^{q^e} g_u)\,j^e(\zeta_{q^e}^{-1})$$
$$= j^e(\zeta_{q^e} g_s^{q^e} g_u \zeta_{q^e}^{-1}) = j^e(g_s^{q^e} g_u^{q^e}) = j^e(g^{q^e})$$

shows that Ψ is well defined and it maps ${}'H_e(g)$ to $H_e(g)$. It is in fact a bijection with inverse given by $y \mapsto x$ where $y \in H_e(g)$, $x = j^e(\zeta_{q^{-e}})y$, $e \in \mathbf{Z}$. For $x \in {}'H_e(g)$, $x' \in {}'H_{e'}(g)$, we have

$\Psi(x * x') = j^{e+e'}(\zeta_{q^{e+e'}})\,j^{e'}(x)x'$,

$\Psi(x) * \Psi(x') = j^{e+e'}(\zeta_{q^e})\,j^{e'}(x\zeta_{q^{e'}})x'$;

hence

(a) $$\Psi(x * x') = \Psi(x) * \Psi(x')c$$

where

$$c = x'^{-1} j^{e'}(x\zeta_{q^{e'}})^{-1} j^{e+e'}(\zeta_{q^{e'}})\,j^{e'}(x)x' = x'^{-1} j^{e'}(\zeta_{q^{e'}}^{-1} x^{-1} j^e(\zeta_{q^{e'}})x)x'.$$

We show that

(b) $$c \in Z_G(g)^0.$$

We first show that for any $\lambda \in \mathbf{C}$

$$c_\lambda := x'^{-1} j^{e'}(\zeta_\lambda^{-1} x^{-1} j^e(\zeta_\lambda)x)x'$$

belongs to $Z_G(g)$, or equivalently that

$$x' c_\lambda x'^{-1} \in Z_G(x' g x'^{-1}) = Z_G(j^{e'}(g_s^{q^{e'}} g_u)) = j^{e'}(Z_G(g)).$$

It is enough to show that $\zeta_\lambda^{-1} x^{-1} j^e(\zeta_\lambda)x \in Z_G(g)$. We have

$$\zeta_\lambda^{-1} x^{-1} j^e(\zeta_\lambda)xg = \zeta_\lambda^{-1} x^{-1} j^e(\zeta_\lambda) j^e(g_s^{q^e} g_u)x$$
$$= \zeta_\lambda^{-1} x^{-1} j^e(\zeta_\lambda g_s^{q^e} g_u)x = \zeta_\lambda^{-1} x^{-1} j^e(g_s^{q^e} g_u^\lambda z_\lambda)x$$
$$= \zeta_\lambda^{-1} g_s x^{-1} j^e(g_u^\lambda z_\lambda)x = \zeta_\lambda^{-1} g_s g_u^\lambda x^{-1} j^e(z_\lambda)x = g z_\lambda^{-1} x^{-1} j^e(z_\lambda)x$$

so that we have indeed $\zeta_\lambda^{-1} x^{-1} j^e(\zeta_\lambda)x \in Z_G(g)$. Thus we have $c_\lambda \in Z_G(g)$ for any $\lambda \in \mathbf{C}$. Note that $c_0 = 1$. Since $\lambda \mapsto c_\lambda$ is a morphism $\mathbf{C} \to Z_G(g)$ whose image contains 1, its image must be contained in $Z_G^0(g)$. Since $c = c_{q^{e'}}$, we see that (b) holds. From (a),(b) we see that for x,x' in $'H(g)$, we have

(c) $$\Psi(x*x') \in \Psi(x)*\Psi(x')*Z'_G(g)$$

(recall that $Z^0_G(g) \subset Z'_G(g)$.) From the definitions, we have $\Psi(x) = x$ for any $x \in {}'H_0(g)$; in particular, $\Psi(Z'_G(g)) = Z'_G(g)$. Using this and (c) we see that for $x \in {}'H(g)$, we have $\Psi(x*Z'_G(g)) \subset \Psi(x)*Z'_G(g)$ so that Ψ induces a map $\bar{\Psi}\colon {}'\Delta(g) \xrightarrow{\sim} \Delta(g)$; here we view $'\Delta(g)$ (resp. $\Delta(g)$) as the set of orbits of the right translation action of $Z'_G(g)$ on $'H(g)$ (resp. $H(g)$). From (c) we see that $\bar{\Psi}$ is a group homomorphism (in fact a group isomorphism). Note also that $\bar{\Psi}$ induces the identity map $\bar{Z}_G(g) = {}'\Delta(g) \to \Delta(g) = \bar{Z}_G(g)$ and it is compatible with the $\mathbf{Z}$-gradings.

We now try to see to what extent does Ψ depend on the choice of ζ. By a result of Kostant [Ko], any other choice of ζ must be of the form $\zeta'\colon \lambda \mapsto h\zeta_\lambda h^{-1}$ for some $h \in Z^0_{\mathcal{H}}(g_u) = Z^0_G(g)$. If we define Ψ in terms of ζ', we obtain a bijection $\Psi'\colon {}'H(g) \to H(g)$ given on $'H_e(g)$ by

$$\Psi'\colon x \mapsto j^e(\zeta'_{q^e})x = j^e(h)\Psi(x)h^{-1}hx^{-1}j^e(h^{-1})x = h*\Psi(x)*h^{-1}*d$$

where $d = hx^{-1}j^e(h^{-1})x$. We show that

(d) $$hx^{-1}j^e(h^{-1})x \in Z^0_G(g) \text{ for any } h \in Z^0_G(g).$$

We have

$$\begin{aligned} hx^{-1}j^e(h^{-1})xg &= hx^{-1}j^e(h^{-1})j^e(g_s^{q^e}g_u)x = hx^{-1}j^e(h^{-1}g_s^{q^e}g_u)x \\ &= hx^{-1}j^e(g_s^{q^e}g_uh^{-1})x = hgx^{-1}j^e(h^{-1})x = ghx^{-1}j^e(h^{-1})x. \end{aligned}$$

Thus, $hx^{-1}j^e(h^{-1})x \in Z_G(g)$ for any $h \in Z^0_G(g)$. We see that $h \mapsto hx^{-1}j^e(h^{-1})x$ is a morphism $Z^0_G(g) \to Z_G(g)$. Its image is connected and its value at $h = 1$ is 1; hence its image is contained in $Z^0_G(g)$. This proves (d). From (d) we see that Ψ and Ψ' induce the same isomorphism $'\Delta(g) \to \Delta(g)$. Thus the isomorphism $'\Delta(g) \to \Delta(g)$ that we constructed is independent of choices, so it is canonical; we use it to identify these two groups.

2.7

If Δ is a group with a given $\mathbf{Z}$-grading $\eta\colon \Delta \to \mathbf{Z}$, we say that an automorphism $f\colon \Delta \to \Delta$ is inner if it is of the form $x \mapsto x_0xx_0^{-1}$ for some $x_0 \in \eta^{-1}(0)$. We consider an index set I and a family $(\Delta_\alpha, \mathcal{R}_{\alpha\alpha'})$ where for any $\alpha \in I$, Δ_α is a group with a given $\mathbf{Z}$-grading η and for any α, α' in I, $\mathcal{R}_{\alpha\alpha'}$ is a nonempty set of group isomorphisms $\Delta_{\alpha'} \to \Delta_\alpha$ compatible with the $\mathbf{Z}$-gradings. We say that this family is coherent if the following conditions are satisfied.

(i) For any $\alpha \in I$, $\mathcal{R}_{\alpha\alpha}$ consists of the inner automorphisms of Δ_α.

(ii) For any α, α' in I and any $f \in \mathcal{R}_{\alpha,\alpha'}$, we have $\mathcal{R}_{\alpha,\alpha'} = \mathcal{R}_{\alpha,\alpha} f = f \mathcal{R}_{\alpha',\alpha'}$.

(iii) For any $\alpha, \alpha', \alpha''$ in I, composition defines an inclusion $\mathcal{R}_{\alpha\alpha'}\mathcal{R}_{\alpha'\alpha''} \subset \mathcal{R}_{\alpha,\alpha''}$.

In this case, for $i = 0, 1$ and for any α, α' in I, the bijection $M^i_\eta(\Delta_{\alpha'}) \xrightarrow{\sim} M^i_\eta(\Delta_\alpha)$ (notation of 1.3) induced by $f : \Delta_{\alpha'} \to \Delta_\alpha$ (in $\mathcal{R}_{\alpha,\alpha'}$) is independent of the choice of f. (We use the fact that an inner automorphism of $\Delta_\alpha, \eta_\alpha$ induces the identity map on $M^i_\eta(\Delta_\alpha)$.) It follows that the sets $M^i_\eta(\Delta_\alpha)$ with $\alpha \in I$ can be canonically identified with a single set associated to the family.

2.8

Let $\underline{G}^{sp}$, $\underline{G}^{sp}_*$ be the set of G-conjugacy classes contained in G^{sp}, G^{sp}_*. Let $C \in \underline{G}^{sp}_*$. We consider the family $(\Delta(g), \mathcal{R}_{g,g'})$ indexed by C, where for $g \in C$, $\Delta(g)$ is the group with a $\mathbf{Z}$-grading defined in 2.5 and for g, g' in C, $\mathcal{R}_{g,g'}$ consists of the isomorphisms $\tau_k : \Delta(g') \to \Delta(g)$ (for various $k \in G$ such that $kg'k^{-1} = g$) where τ_k is induced by the isomorphism $H(g') \to H(g)$ whose restriction $H_e(g') \to H_e(g)$ is $x \mapsto j^e(k)xk^{-1}$. From the definitions we see that this family is coherent. Hence for $i = 0, 1$, the sets $M^i_\eta(\Delta(g))$ with $g \in C$ can be canonically identified (as in 2.7) with a single set $M^i(C)$ associated to C. We set

(a) $$AC(G) = \sqcup_{C \in \underline{G}^{sp}_*} M^0(C),$$

(b) $$\mathcal{C}(G) = \sqcup_{C \in \underline{G}^{sp}_*} M^1(C).$$

(Note that $AC(G), \mathcal{C}(G)$ depend on q and j.) The set $\mathcal{C}(G)$ appeared in [L3].

2.9

We define $\phi : G^{sp} \to G^{sp}$ by $g \mapsto j(g^q)$. This is well defined by 2.3(a). Clearly, ϕ induces a map $\underline{G}^{sp} \to \underline{G}^{sp}$ denoted again by ϕ. We show:

(a) *Any orbit of $\phi : \underline{G}^{sp} \to \underline{G}^{sp}$ is finite. In particular, $\phi : \underline{G}^{sp} \to \underline{G}^{sp}$ is bijective.*

Let $g \in G^{sp}$. We choose $k \geq 1$ such that $j^k = 1$ and $g_s^{q^k} = g_s$. Then we have $\phi^k(g) = j^k(g^{q^k}) = g_s g_u^{q^k}$ and this is conjugate to g. This proves (a). From the definitions we have

(b) *$\underline{G}^{sp}_*$ is exactly the fixed point set of $\phi : \underline{G}^{sp} \to \underline{G}^{sp}$.*

We show:

(c) *$\phi : G^{sp} \to G^{sp}$ is a bijection.*

Let $g \in G^{sp}$, $g' \in G^{sp}$ be such that $j(g^q) = j(g'^q)$. Then $g^q = g'^q$, hence $g_s^q = g_s'^q$ and $g_u^q = g_u'^q$. The last equality implies $g_u = g'_u$ since $p = 0$. From $g_s^q = g_s'^q$ we deduce using $g_s^{q^r} = g_s$, $g_s'^{q^{r'}} = g'_s$ (where we can assume $r = r' \geq 1$) that $g_s = g'_s$ hence $g = g'$. Thus the map in (c) is injective. Since $\phi : \underline{G}^{sp} \to \underline{G}^{sp}$ is surjective it follows that the map in (c) is surjective. This proves (c).

2.10

Let $C \in \underline{G}^{sp}$. Let $\tilde{C} = \sqcup_{g \in C} \bar{Z}_G(g)$. We can view $\tilde{C}$ as an algebraic variety which is a finite covering of C under the obvious map $\iota\colon \tilde{C} \to C$. Note that G acts on $\tilde{C}$ by $h\colon (g, zZ'_G(g)) \mapsto (hgh^{-1}, hzh^{-1}Z'_G(hgh^{-1}))$; here $g \in G$, $z \in Z_G(g)$. This is well defined since $Z'_G(hgh^{-1}) = hZ'_G h^{-1}$. Now the orbits of G on $\tilde{C}$ are exactly the connected components of $\tilde{C}$. Let $M(C)$ be the set of all pairs $(\mathcal{O}, \mathcal{E})$ where $\mathcal{O}$ is a G-orbit in $\tilde{C}$ and $\mathcal{E}$ is an irreducible G-equivariant local system on $\mathcal{O}$. For any $g \in C$ there is a canonical bijection $M(C) \xrightarrow{\sim} M(\bar{Z}_G(g))$ (notation of 1.1) which associates to $(\mathcal{O}, \mathcal{E})$ the equivalence class of (x, σ) where x belongs to $\mathcal{O} \cap \iota^{-1}(g)$ (a $\bar{Z}_G(g)$-conjugacy class in $\bar{Z}_G(g)$) and σ is the stalk of $\mathcal{E}$ at x viewed as an irreducible representation of $Z_{\bar{Z}_G(g)}(x)$. Thus the sets $M(\bar{Z}_G(g))$ with $g \in C$ are all in canonical bijection with $M(C)$. Let

$$CS(G) = \sqcup_{C \in \underline{G}^{sp}} M(C).$$

2.11

For any $g \in G^{sp}$, we define a homomorphism $\bar{Z}_G(g) \to \bar{Z}_G(\phi(g))$ by $zZ'_G(g)) \mapsto j(z)Z'_G(j(g^q))$ with $z \in Z_G(g)$. We show:

(a) *this map is a bijection.*

Since $\bar{Z}_G(g), \bar{Z}_G(\phi(g))$ are finite, it is enough to show that our map is injective. Let z, z' in $Z_G(g)$ be such that $j(z) = j(z') \bmod Z'_G(j(g^q))$. Then $z = z' \bmod Z'_G(g^q) = Z'_G(g)$, see 2.3(a). This proves injectivity, hence (a).

We define $\tilde{\phi}\colon \sqcup_{C \in \underline{G}^{sp}} \tilde{C} \to \sqcup_{C \in \underline{G}^{sp}} \tilde{C}$ by $(g, zZ'_G(g)) \mapsto (j(g^q), j(z)Z'_G(j(g^q)))$; here $g \in C$, $z \in Z_G(g)$. Using (a) and 2.9(a) we see that $\tilde{\phi}$ is a bijection.

For any $C \in \underline{G}^{sp}$, $\tilde{\phi}$ restricts to a (bijective) morphism $\tilde{C} \to \tilde{C}'$ where $C' = \phi(C)$. Note that the G-action on $\tilde{C}$ is compatible under $\tilde{\phi}$ with the G-action on $\tilde{C}'$ composed with the automorphism j of G; in particular, $\tilde{\phi}$ carries any G-orbit on $\tilde{C}$ to a G-orbit on $\tilde{C}'$. We see that $(\mathcal{O}, \mathcal{E}) \mapsto (\tilde{\phi}(\mathcal{O}), \tilde{\phi}_!\mathcal{E})$ defines a bijection $M(C) \xrightarrow{\sim} M(C')$. Taking disjoint unions we see that $\tilde{\phi}$ induces a bijection

(b) $$CS(G) \xrightarrow{\sim} CS(G).$$

The fixed points of this bijection are the pairs $(\mathcal{O}, \mathcal{E}) \in M(C)$ with $\phi(C) = C$ such that $\mathcal{O} = \tilde{\phi}(\mathcal{O})$, $\mathcal{E} \cong \tilde{\phi}_!\mathcal{E}$. The condition that $\phi(C) = C$ is the same as $C \in G_{j,q}$, see 2.9(b). Assuming that this is the case, let $g \in C$. Then $\Delta(g), \eta$ and $M^1_\eta(\Delta(g))$ are well defined and $\bar{Z}_G(g)$, $M(\bar{Z}_G(g))$ are well defined. Note that $\bar{Z}_G(g)$ is the subgroup $\eta^{-1}(0)$ of $\Delta(g)$. Hence $M^1_\eta(\Delta(g))$ can be viewed as the subset of $M(\bar{Z}_G(g)) = M(C)$ defined by the two conditions in 1.3(c). In terms of $M(C)$ those two conditions are exactly the conditions that $\mathcal{O} = \tilde{\phi}(\mathcal{O})$, $\mathcal{E} \cong \tilde{\phi}_!\mathcal{E}$ (as above). In this way we see that

(c) *$AC(G)$ (see 2.8) can be naturally identified with the fixed point set of the map* (b) *of $CS(G)$ into itself.*

2.12

In the reminder of this section we assume that q is a power of a prime number and that $\mathcal{G}$ is a connected reductive group over $\bar{\mathbf{F}}_q$ of a type dual to that of G with a fixed $\mathbf{F}_q$-rational structure which is twisted according to j. Let $\mathcal{F}$ be the vector space of class functions $\mathcal{G}(\mathbf{F}_q) \to \mathbf{C}$. Now $\mathcal{F}$ has a basis β given by the characters of the irreducible representations of $\mathcal{G}(\mathbf{F}_q)$. We define a function $\delta\colon \beta \to \{1,-1\}$ as follows: if χ is the character of an irreducible representation ε of $\mathcal{G}(\mathbf{F}_q)$ and $D(\varepsilon)$ is its dual (as in [DL]), then $\delta(\chi)$ is characterized by the property that $\delta(\chi)D(\varepsilon)$ is an irreducible representation.

Another basis β' of $\mathcal{F}$ (whose elements are defined only up to multiplication by roots of 1) is provided by the characteristic functions of the character sheaves on $\mathcal{G}$ which are isomorphic to their inverse image under the Frobenius map $\mathcal{G} \to \mathcal{G}$. It is known that β is indexed by $\mathcal{C}(G)$, the character sheaves on $\mathcal{G}$ are indexed by $CS(G)$ and the Frobenius-invariant character sheaves on $\mathcal{G}$ are indexed by the fixed point set of ϕ on $CS(G)$. Thus, β' is indexed by $AC(G)$.

The decompositions 2.8(a),(b) give rise to decompositions $\beta = \sqcup_C \beta_C$, $\beta' = \sqcup_C \beta'_C$. We now fix $C \in \underline{G}^{sp}_*$, $g \in C$ and we list the elements of β_C as $\{\mathbf{b}_{x,\sigma}; (x,\sigma) \in M^1_\eta(\Delta)\}$ and the elements of β'_C as $\{\mathbf{b}'_{x,\sigma}; (x,\sigma) \in M^0_\eta(\Delta)\}$ (where $\Delta = \Delta(g)$).

For any $(x,\sigma) \in M^0_\eta(\Delta)$ we set

(a)
$$\mathbf{b}''_{x,\sigma} = \sum_{(y,\tau)\in M^1_\eta(\Delta)} \sum_{w\in\Delta/\langle\xi\rangle; xwyw)^{-1}=wyw^{-1}x} \eta(\xi)|Z_\Delta(x)/\langle\xi\rangle|^{-1}|Z_\Delta(y)/\langle\xi\rangle|^{-1} \times \mathrm{tr}(w^{-1}x^{-1}w,\tilde{\tau})\mathrm{tr}(wyw^{-1},\tilde{\sigma})\delta(\mathbf{b}_{y,\tau})\mathbf{b}_{y,\tau}$$

where $\xi \in \Delta^*$ (notation of 1.3), $\tilde{\sigma}$ is a representation of $Z_\Delta(x)$ extending σ on which ξ acts as 1, $\tilde{\tau}$ is the representation of $Z_\Delta(y)$ extending τ on which y acts as 1. Note that $\mathbf{b}''_{x,\sigma}$ is independent (up to multiplication by a root of 1) of the choice of $\xi, \tilde{\sigma}$, see 1.3.

We expect that for any $(x,\sigma) \in M^0_\eta(\Delta)$ we have

(b)
$$\mathbf{b}'_{x,\sigma} = \mathbf{b}''_{x,\sigma}$$

up to multiplication by a root of 1.

Note that (b) is known in many cases but not in complete generality. It implies that:

(c) *for any $C \in \underline{G}^{sp}_*$, the subspace of $\mathcal{F}$ spanned by β_C coincides with the subspace of $\mathcal{F}$ spanned by β'_C.*

The functions $\mathbf{b}''_{x,\sigma}$ are called *almost characters*. Note that the almost characters (for various C) form a basis of $\mathcal{F}$ indexed by $AC(G)$.

References

[DL] P. Deligne and G. Lusztig, *Duality for representations of reductive groups over a finite field*, J. Alg **74** (1982), 284–291.

[Ko] B. Kostant, *The principal three dimensional subgroup and the Betti numbers of a complex simple Lie group*, Amer. J. Math. **81** (1959), 973–1032.

[L1] G. Lusztig, *Unipotent representations of a finite Chevalley group of type E_8*, Quart. J. Math. **30** (1979), 315–338.

[L3] G. Lusztig, *Characters of reductive groups over finite fields*, Proc. Int. Congr. Math. Warsaw 1983, North Holland, 1984, pp. 877–880.

[L2] G. Lusztig, *Characters of reductive groups over a finite field*, Ann. Math. Studies 107, Princeton U. Press, 1984.

[L4] G. Lusztig, *Families and Springer's correspondence*, Pacific J. Math. **267** (2014), 431–450.

Verlinde Formulas for Nonsimply Connected Groups

Eckhard Meinrenken

To the memory of Bert Kostant

Abstract In 1999, Fuchs and Schweigert proposed formulas of Verlinde type for moduli spaces of surface group representations in compact nonsimply connected Lie groups. In this paper, we will prove a symplectic version of their conjecture for surfaces with at most one boundary component. A key tool in our computations is Kostant's notion of a *maximal torus in apposition*.

Keywords: Verlinde formulas • group-valued momentum maps • maximal torus in apposition

Mathematics Subject Classification (2010): 53D30, 53D20, 57S15

1 Introduction

Let G be a compact, simple, simply connected Lie group, Σ_{g} a compact oriented surface of genus g without boundary, and

$$M_G(\Sigma_{\mathrm{g}}) = \mathrm{Hom}(\pi_1(\Sigma_{\mathrm{g}}), G)/G \tag{1}$$

the representation variety. Choosing standard generators of the fundamental group, we have $M_G(\Sigma_{\mathrm{g}}) = \Phi^{-1}(e)/G$, where

$$\Phi\colon G^{2\mathrm{g}} \to G, \;\; (a_1, b_1, \ldots, a_{\mathrm{g}}, b_{\mathrm{g}}) \mapsto \prod_{i=1}^{\mathrm{g}} a_i b_i a_i^{-1} b_i^{-1}\,. \tag{2}$$

By interpreting (1) as a moduli space of flat G-bundles over Σ_{g} modulo gauge transformations, Atiyah-Bott [5] defined a symplectic structure on the smooth part of $M_G(\Sigma_{\mathrm{g}})$, with 2-form depending on the choice of an invariant inner product B on

Mathematics Department, University of Toronto, 40 St George Street, Toronto Ontario M6S3W6

V. G. Kac, V. L. Popov (eds.), *Lie Groups, Geometry, and Representation Theory*, Progress in Mathematics 326, https://doi.org/10.1007/978-3-030-02191-7_14

the Lie algebra. (Taking into account the singularities, $M_G(\Sigma_g)$ is a stratified symplectic space in the sense of [28]; for the purpose of this introduction we will ignore the singularities.) If B is a k-th multiple of the *basic inner product* on $\mathfrak{g}$, for some integral level $k \in \mathbb{N}$, then $M_G(\Sigma_g)$ acquires a prequantum line bundle L. After suitable desingularization, one can define a quantization $\mathcal{Q}(M_G(\Sigma)) \in \mathbb{Z}$ as the index of a spin-c Dirac operator with coefficients in L. This index is given by the (symplectic) *Verlinde formula*

$$\mathcal{Q}(M_G(\Sigma_g)) = \sum_{\lambda \in P_k} S_{0,\lambda}^{2-2g}, \tag{3}$$

where P_k is the finite set of level k weights of G and $S_{0,\lambda}$ are components of the S-matrix. (The relevant notation is explained in Section 2 below.) The numbers (3) are known to compute the dimension of 'space of conformal blocks' [32] or 'generalized theta functions' [14], see [6] for background and references; this dimension formula is equivalent to the index formula in the presence of suitable vanishing theorems [30].

Let $Z \subseteq Z(G)$ be a finite subgroup of the center, and $G' = G/Z$ the resulting nonsimply connected simple Lie group. Fuchs-Schweigert [16] proposed a generalization of (3) to the space

$$M_{G'}(\Sigma_g) = \mathrm{Hom}(\pi_1(\Sigma_g), G')/G'.$$

In general, this space has several connected components $M_{G'}(\Sigma_g)_{(c)}$ indexed by the elements $c \in Z$. Fuchs-Schweigert suggested that the resulting formula would involve the natural action $\bullet_k$ of the center $Z(G)$ on P_k, and e.g. for $c = e$ is of the form

$$\mathcal{Q}\big(M_{G'}(\Sigma_g)_{(e)}\big) = \frac{1}{\# Z^{2g}} \sum_{c_1,\cdots c_{2g} \in Z} \varepsilon(c_1,\ldots,c_{2g}) \sum_{\substack{\lambda \in P_k \\ c_i \bullet_k \lambda = \lambda}} S_{0,\lambda}^{2-2g} \tag{4}$$

for suitable phase factors $\varepsilon(c_1,\ldots,c_{2g}) \in \mathrm{U}(1)$. Their paper is not specific on the precise conditions for prequantizability, or how to compute these phase factors. More general formulas of a similar kind are conjectured for surfaces with boundary, with prescribed holonomies around the boundary components. Some special cases of (4) were already known in the algebro-geometric context: Pantev [26] obtained the Verlinde numbers for $G' = \mathrm{SO}(3)$, and Beauville [7] for $G' = \mathrm{PU}(n)$ for n prime. In the symplectic framework, the case of $G' = \mathrm{SO}(3)$ had been worked out for an arbitrary number of boundary components in [23, 24].

In this article, we will prove the symplectic version of the Fuchs-Schweigert conjectures for arbitrary compact, simple G', for surfaces with at most one boundary component. The case with several boundary components is more difficult, and we hope to return to it elsewhere. Similar to the argument in [4] towards a proof of (3) as a fixed point formula, we will obtain these results by reduction from a suitable quasi-Hamiltonian G-space [1] (M, ω, Φ) with G-valued moment map $\Phi\colon M \to G$.

Given a prequantization of such a space, at integral level k, its quantization is an element

$$\mathcal{Q}(M) \in R_k(G)$$

of the level k fusion ring (Verlinde algebra). By the 'quantization commutes with reduction principle', the multiplicity of the zero weight computes the quantization of the symplectic quotient $\mathcal{Q}(M/\!/G) \in \mathbb{Z}$. On the other hand, the quantization of M is determined by its values $\mathcal{Q}(M)(t) \in \mathbb{C}$ at a certain finite collection of regular elements of the maximal torus T, which in turn are computed by a localization formula as a sum of contributions from the fixed point manifolds $F \subseteq M$ of t.

For the case at hand, $M_{G'}(\Sigma_\mathsf{g})_{(e)}$ will be obtained by reduction from $M \cong (G')^{2\mathsf{g}}$, with G-action by conjugation on each factor, and with G-valued moment map a lift of the product of Lie group commutators. By a result of Krepski [22] (which we will re-prove in this paper), M is prequantizable if and only if the bilinear form B is integer-valued on the lattice $\Lambda_Z := \exp_T^{-1}(Z) \subseteq \mathfrak{t}$. If $Z = \{1\}$, the fixed point manifolds are simply $T^{2\mathsf{g}} \subseteq G^{2\mathsf{g}}$, and the evaluation of the fixed point contributions is straightforward [4]. For general Z, the fixed point manifolds of t may be disconnected, and the evaluation of the fixed point data becomes much more involved. The main difficulty is the computation of the phase factors: since the stabilizer groups of the new fixed point components are disconnected, the phase factors are not directly determined by the values of the moment map.

Our main tool for resolving these difficulties is Kostant's notion of a *maximal torus in apposition*, introduced in his famous 1959 paper, *The Principal Three-Dimensional Subgroup and the Betti Numbers of a Complex Simple Lie Group* [19]. In this work, Kostant shows that among regular elements of the maximal torus T, there is a unique W-conjugacy class of elements t_* such that the adjoint action of t_* has minimal order, and this order coincides with the order of a Coxeter element of the Weyl group. To explain this coincidence, he constructs a second maximal torus U that is normalized by the conjugation action of t_*, and on which Ad_{t_*} acts as a Coxeter transformation. For our application to moduli spaces, we will use this second torus U to relate the various fixed point manifolds of the regular element t, using the observation that $U' = U/Z$ meets every component of the fixed point set of Ad_t on G'. In this way, we are able to study the prequantizations of $(G')^{2\mathsf{g}}$, and evaluate the fixed point data, in terms of their restrictions to $(U')^{2\mathsf{g}} \subseteq (G')^{2\mathsf{g}}$. In particular, we obtain an explicit formula for the phase factors $\varepsilon(c_1,\ldots,c_{2\mathsf{g}})$ in terms of the Coxeter transformation.

Acknowledgements. This project started out many years ago as a collaboration with Anton Alekseev and Chris Woodward. Following our approach [4] to Verlinde formulas as fixed point formulas, we had computed the quantization of moduli spaces for nonsimply connected groups, but only for specific levels k that were not expected to be optimal. Section 4.5 of this paper will use some material from our unfinished manuscript [3], and I thank my coauthors for letting me reproduce it here. I also thank the referee for helpful suggestions.

2 Lie-theoretic preliminaries

2.1 Notation

Throughout this section, G will denote a compact, simple, simply connected Lie group of rank l, with center $Z(G)$ and maximal torus T. Let $\mathfrak{g},\mathfrak{t}$ be their Lie algebras. We denote by $Q^\vee \subseteq P^\vee \subseteq \mathfrak{t}$ the coroot and coweight lattices, and by $Q \subseteq P \subseteq \mathfrak{t}^*$ the root and weight lattices so that $Q = \mathrm{Hom}(P^\vee,\mathbb{Z})$, $P = \mathrm{Hom}(Q^\vee,\mathbb{Z})$. (Note that we are working with 'real' weights as in [11, p. 185].) Then $Q^\vee$ is the integral lattice of G, while $P^\vee$ is the integral lattice of the adjoint group $G/Z(G)$:

$$Q^\vee = \exp_T^{-1}(\{e\}),\quad P^\vee = \exp_T^{-1}(Z(G)). \tag{5}$$

The coroot of any root α is denoted $\alpha^\vee$. Let

$$\{\alpha_1,\dots,\alpha_l\} \subseteq Q,\quad \{\alpha_1^\vee,\dots,\alpha_l^\vee\} \subseteq Q^\vee. \tag{6}$$

be the $\mathbb{Z}$-bases of the two lattices given by a choice of simple roots and coroots, with dual bases the fundamental coweights and weights,

$$\{\varpi_1^\vee,\dots,\varpi_l^\vee\} \subseteq P^\vee,\quad \{\varpi_1,\dots,\varpi_l\} \subseteq P. \tag{7}$$

We put $\rho = \sum_{i=1}^l \varpi_i \in P$, $\rho^\vee = \sum_{i=1}^l \varpi_i^\vee \in P^\vee$. The highest root is denoted θ; the Coxeter number and the dual Coxeter number are given by

$$h = 1 + \langle \theta, \rho^\vee \rangle,\quad h^\vee = 1 + \langle \rho, \theta^\vee \rangle. \tag{8}$$

The *basic inner product* B_{basic} on $\mathfrak{g}$ is the unique invariant inner product such that $B_{\mathrm{basic}}(\theta^\vee,\theta^\vee) = 2$. It has the property that

$$B_{\mathrm{basic}}(Q^\vee, P^\vee) \subseteq \mathbb{Z}. \tag{9}$$

Put differently, the isomorphism $\mathfrak{t} \to \mathfrak{t}^*$ given by the basic inner product takes $Q^\vee$ into Q and $P^\vee$ into P.

We denote by $\mathfrak{t}_+ \subseteq \mathfrak{t}$ the fundamental Weyl chamber, defined by the inequalities $\langle \alpha_i, \xi \rangle \geq 0$, $i = 1,\dots,l$, and by

$$\mathfrak{A} = \{\xi \in \mathfrak{t}_+ \mid \langle \theta, \xi \rangle \leq 1\} \tag{10}$$

the fundamental Weyl alcove. The maps

$$\mathfrak{t}_+ \to \mathfrak{g}/\mathrm{Ad}(G),\ \xi \mapsto G.\xi, \qquad \mathfrak{A} \to G/\mathrm{Ad}(G),\ \xi \mapsto G.\exp(\xi)$$

are bijections, identifying $\mathfrak{t}_+$ and $\mathfrak{A}$ with the set of adjoint orbits in $\mathfrak{g}$ and in G, respectively.

2.2 Coxeter elements

Let $W = N(T)/T$ be the Weyl group. It acts as a group of reflections on $\mathfrak{t}$, with $\mathfrak{t}_+$ as a fundamental domain. An element $w_* \in W$ is called a *Coxeter element* if it is W-conjugate to the product $s_l \cdots s_1$ of simple reflections. The Coxeter number (8) is the order of such a Coxeter element:

$$\mathrm{ord}(w_*) = h.$$

One of the key properties of the Coxeter element is that its action on $\mathfrak{t}$ has trivial fixed point set. More precisely, there is the following statement.

Proposition 2.1. [12, Theorem 1.6] *The transformation $1 - w_*$ of $\mathfrak{t}$ restricts to an isomorphism from the coweight lattice to the coroot lattice:*

$$(1 - w_*)P^\vee = Q^\vee. \tag{11}$$

Equivalently, the fixed point set of the action of a Coxeter element $w_* \in W$ on the maximal torus T is exactly the center:

$$T^{w_*} = Z(G). \tag{12}$$

More generally, we are interested in the fixed point set of w_* on the maximal torus $T' = T/Z$ of $G' = G/Z$, for any subgroup $Z \subseteq Z(G)$. Let $\Lambda_Z = \exp_T^{-1}(Z)$, so that Λ_Z is the integral lattice of G'.

Proposition 2.2. *The fixed point set of the action of a Coxeter element $w_* \in W$ on $T' = T/Z$ is the finite subgroup*

$$(T')^{w_*} = ((1 - w_*)^{-1}\Lambda_Z)/\Lambda_Z. \tag{13}$$

It fits into an exact sequence of groups

$$1 \to Z(G)/Z \to (T')^{w_*} \to Z \to 1; \tag{14}$$

in particular, $\#(T')^{w_*} = \#Z(G)$.

Proof. An element $\xi \in \mathfrak{t}$ satisfies $w_* \exp_{T'} \xi = \exp_{T'} \xi$ if and only if $(1 - w_*)\xi \in \Lambda_Z$, i.e., $\xi \in (1 - w_*)^{-1}\Lambda_Z$. This proves (13). The inclusion $P^\vee = (1 - w_*)^{-1}Q^\vee \hookrightarrow (1 - w_*)^{-1}\Lambda_Z$ defines an injective map $Z(G)/Z \hookrightarrow (T')^{w_*}$. On the other hand, we have a surjective map $(T')^{w_*} \to Z$ induced by

$$\begin{aligned}(1 - w_*)^{-1}\Lambda_Z/\Lambda_Z &\cong \Lambda_Z/(1 - w_*)\Lambda_Z \\ &\to \Lambda_Z/(1 - w_*)P^\vee = \Lambda_Z/Q^\vee \cong Z\end{aligned}$$

with kernel $(1 - w_*)^{-1}Q^\vee/\Lambda_Z = P^\vee/Z = Z(G)/Z$. This establishes exactness of (14). □

2.3 Kostant's maximal torus 'in apposition'

In [19], Kostant observed that the Coxeter number h also arises in the following context. For $g \in G$, let $\mathrm{ord}(\mathrm{Ad}_g)$ be the order of its adjoint action on $\mathfrak{g}$.

Theorem 2.3 (Kostant [19]). *Suppose $g \in G$ is regular. Then*

$$\mathrm{ord}(\mathrm{Ad}_g) \geq h. \tag{15}$$

Equality holds if and only if g is conjugate to $t_* = \exp(\zeta_*)$, where

$$\zeta_* = \frac{1}{h}\rho^\vee \in \mathrm{int}(\mathfrak{A}). \tag{16}$$

Proof. It is enough to prove (15) for elements of $\exp(\mathrm{int}(\mathfrak{A}))$. Let $t = \exp(\zeta)$ with $\zeta \in \mathrm{int}(\mathfrak{A})$. Write $\zeta = \sum_{i=1}^{l} r_i \varpi_i^\vee$ where $r_i > 0$. Then $\mathrm{ord}(\mathrm{Ad}_t) = m$ if and only if $m\zeta \in P^\vee$, if and only if all mr_i are integers. The condition $\langle \theta, \zeta \rangle < 1$ gives $\langle \theta, m\zeta \rangle \leq m - 1$, thus

$$m \geq 1 + \langle \theta, m\zeta \rangle = 1 + \sum_{i=1}^{l} mr_i \langle \theta, \varpi_i^\vee \rangle \geq 1 + \sum_{i=1}^{l} \langle \theta, \varpi_i^\vee \rangle = h.$$

Equality $m = h$ holds if and only if all mr_i are equal to 1, that is, $r_i = \frac{1}{h}$. □

The elements in the conjugacy class $\mathscr{C}_* = \mathrm{Ad}(G)t_*$ are called the *principal elements* of G. An explanation of why the order of the adjoint action by a principal element coincides with the order of a Coxeter element is provided by the following result.

Theorem 2.4 (Kostant [19]). *There exists a maximal torus $U \subseteq G$, invariant under the adjoint action of $t_* = \exp(\zeta_*)$, such that the action of Ad_{t_*} on U is given by the action of a Coxeter element of $N_G(U)/U$.*

The maximal torus U is called *in apposition to T* [19, Section 7]. Kostant gives the following construction: Let $e_1, \ldots, e_l \in \mathfrak{g}^\mathbb{C}$ be root vectors for the simple roots $\alpha_1, \ldots, \alpha_l$, with normalization $\langle e_i, e_i^* \rangle = 1$ (for some invariant metric on $\mathfrak{g}$), and let e_0 be a root vector for $\alpha_0 := -\theta$, with $\langle e_0, e_0^* \rangle = 1$. Write $\theta = \sum_{i=1}^{l} k_i \alpha_i$ with the Dynkin marks k_i. Then the *cyclic element*

$$z := e_0 + \sum_{i=1}^{l} \sqrt{k_i}\, e_i \in \mathfrak{g}^\mathbb{C}$$

is semi-simple [19, Lemma 6.3]. One checks that $[z, z^*] = 0$, which implies that the centralizer of z is invariant under complex conjugation. Let $\mathfrak{u} \subseteq \mathfrak{g}$ be the real part of the centralizer; then $U = \exp(\mathfrak{u})$ is the desired maximal torus.

Example 2.5. Let $G = \mathrm{SU}(l+1)$, with the standard maximal torus T consisting of diagonal matrices, and the standard choice of positive Weyl chamber. The Coxeter number is $h = l+1$. Letting q be the primitive $2l+2$-nd root of unity, the element $t_* = \exp(\zeta_*)$ is a diagonal matrix with entries

$$q^l,\, q^{l-2}, \ldots, q^{-l}$$

down the diagonal; note that t_*^{l+1} is $(-1)^l$ times the identity matrix. A maximal torus U in apposition to T is given by all those matrices $A \in \mathrm{SU}(l+1)$ that are constant along diagonals: That is, $A_{ij} = A_{i+1,j+1}$ for all i, j, where indices are taken modulo mod $l+1$. If $l = 1$, these are matrices of the form

$$A = \begin{pmatrix} \cos(\theta) & i\sin(\theta \\ i\sin(\theta) & \cos(\theta) \end{pmatrix}.$$

2.4 Action of the center on the alcove

The action of the center $Z(G)$ on G given by multiplication commutes with the G-action by conjugation. It thereby induces an action on the set of conjugacy classes, and hence on the alcove:

$$Z(G) \times \mathfrak{A} \to \mathfrak{A}. \tag{17}$$

As a consequence of Theorem 2.3, the distinguished conjugacy class $\mathscr{C}_* = \mathrm{Ad}_G\, t_*$ is invariant under left translation by $Z(G)$, thus ζ_* is a fixed point for the action (17). Put differently, for every $c \in Z(G)$ the element $c^{-1}t_*$ is W-conjugate to t_*. This defines an injective group homomorphism

$$Z(G) \to W, \quad c \mapsto w_c \tag{18}$$

such that $c^{-1}t_* = w_c(t_*)$. (See [31] for a characterization of its image.) Letting $\zeta_c \in P^\vee$ be the unique vertex of $\mathfrak{A}$ such that $\exp(\zeta_c) = c$, the action (17) is given by restriction of the affine action on $\mathfrak{t}$,

$$\zeta \mapsto w_c(\zeta - \zeta_{c^{-1}}). \tag{19}$$

Let $N(T)$ be the normalizer of T, and $N(T)^{(c)}$ its component mapping to w_c under the quotient map $N(T) \to W$. Equivalently,

$$N(T)^{(c)} = \{g \in G \mid \mathrm{Ad}_{t_*}(g) = cg\}. \tag{20}$$

Let $U \subseteq G$ be a maximal torus in apposition to T.

Proposition 2.6. *The maximal torus in apposition U intersects each of the components $N(T)^{(c)}$ for $c \in Z(G)$ in an orbit of the left action of $Z(G)$. It does not intersect any of the other components of $N(T)$.*

Proof. Consider the group homomorphism

(21) $$U \to U,\ u \mapsto u(\mathrm{Ad}_{t_*} u)^{-1}.$$

Its kernel is a finite subgroup $U^{t_*} = Z(G)$ (cf. Proposition 2.1); hence (21) is surjective. In particular, given $c \in Z(G)$, the equation $\mathrm{Ad}_{t_*} u = cu$ admits a solution $u \in U$, and this solution is unique up to elements of $Z(G)$. By (20), we have that $u \in N(T)^{(c)}$. Conversely, suppose that $u \in N(T) \cap U$ is given, representing a Weyl group element $w \in W$. Then $\mathrm{Ad}_u t_* = w(t_*)$ gives $u\,\mathrm{Ad}_{t_*}(u^{-1}) = w(t_*)t_*^{-1}$. The left-hand side is in U, the right-hand side is in T. Hence, both sides are in $T \cap U = Z(G)$. It follows that $w = w_c$ for some $c \in Z(G)$. □

Let $Z \subseteq Z(G)$ be a subgroup of the center. We are interested in the fixed point set of regular elements $t \in T$ on $G' = G/Z$. For $\zeta \in \mathfrak{t}$, let Z_ζ be the stabilizer under the affine action (19).

Proposition 2.7. *Let $\zeta \in \mathrm{int}(\mathfrak{A})$, and $t = \exp(\zeta)$. Then the fixed point set for Ad_t on $G' = G/Z$ is*
$$(G')^t = \bigcup_{c \in Z_\zeta} N(T)^{(c)}/Z.$$
In particular, every component of the fixed point set meets the maximal torus $U' = U/Z$, in an orbit of the translation action of $Z(G)/Z$.

Proof. Let $g \in G$ be a lift of $g' \in G'$. Then $g' \in (G')^t$ if and only if $\mathrm{Ad}_t g = cg$ for some $c \in Z$, that is, $g \in N(T)^{(c)}$. We obtain $w_c(t) = \mathrm{Ad}_g t = c^{-1}t$, thus $c \in Z_\zeta$. □

2.5 The level k fusion ring

For the material in this section, see e.g., [15]. We will use the basic inner product to identify $\mathfrak{g} \cong \mathfrak{g}^*$, hence also $\mathfrak{t} \cong \mathfrak{t}^*$. This identification takes $Q^\vee$ to the sublattice of Q spanned by the long roots. For $k \in \mathbb{N}$ we define a finite subgroup of the maximal torus,

(22) $$T_k = \left(\frac{1}{k}P\right) \Big/ Q^\vee \subseteq T.$$

Let $P_+ = P \cap \mathfrak{t}_+$ be the dominant weights, and $P_k = P \cap k\mathfrak{A}$ the *level k weights.* Denote by χ_μ the character of the irreducible representation of highest weight $\mu \in P_+$; the χ_μ are a $\mathbb{Z}$-basis of the representation ring $R(G)$ (viewed as a ring spanned by characters of finite-dimensional representations). It is well known that for all $k \in \mathbb{N} \cup \{0\}$, a weight of $P_{k+h^\vee}$ lies in $(k+h^\vee)\,\mathrm{int}(\mathfrak{A})$ if and only if it is of the form $\lambda + \rho$ with $\lambda \in P_k$. Put
$$\zeta_\lambda = \frac{\lambda+\rho}{k+h^\vee} \in \mathrm{int}(\mathfrak{A}),$$
and define the *special elements*

(23) $$t_\lambda = \exp \zeta_\lambda \in T_{k+h^\vee},\ \ \lambda \in P_k.$$

Then every regular element of $T_{k+h^\vee}$ is W-conjugate to a unique t_λ. The *level k fusion ideal* $\mathscr{I}_k(G) \subseteq R(G)$ is the ideal of characters vanishing at the special elements t_λ; the quotient ring

(24)
$$R_k(G) = R(G)/\mathscr{I}_k(G)$$

is the *level k fusion ring* (Verlinde algebra). It has an additive basis

(25)
$$\tau_\mu \in R_k(G),\ \mu \in P_k$$

given by the images of $\chi_\mu \in R(G),\ \mu \in P_k$ under the quotient map. For $\lambda \in P_k$, the evaluation map $\mathrm{ev}_{t_\lambda} : R(G) \to \mathbb{C}$ descends to $R_k(G)$, and any $\tau \in R_k(G)$ is uniquely determined by these values $\tau(t_\lambda)$. We hence obtain another additive basis (but *only over* $\mathbb{C}$), consisting of the elements $\widetilde{\tau}_\mu \in R_k(G) \otimes_{\mathbb{Z}} \mathbb{C}$ such that

(26)
$$\widetilde{\tau}_\mu(t_\lambda) = \delta_{\lambda,\mu},\quad \lambda,\mu \in P_k.$$

In this new basis, the product is diagonalized: $\widetilde{\tau}_\mu \widetilde{\tau}_{\mu'} = \delta_{\mu,\mu'}\widetilde{\tau}_\mu$. As a consequence of the Weyl character formula, one obtains the change-of-basis formulas

(27)
$$\widetilde{\tau}_\mu = \sum_{\lambda \in P_k} S_{\mu,0} S_{*\mu,\lambda} \tau_\lambda, \qquad \tau_\lambda = \sum_{\mu \in P_k} S_{\mu,0}^{-1} S_{\lambda,*\mu} \widetilde{\tau}_\mu.$$

Here the *S-matrix* is a symmetric, unitary $l \times l$-matrix given by (see e.g., [15])

(28)
$$S_{\mu,\lambda} = \frac{i^{\frac{1}{2}\dim(G/T)} J(t_\lambda)}{(\# T_{k+h^\vee})^{1/2}}\ \chi_\mu(t_\lambda)^*$$

where $J(t) = \sum_{w \in W} (-1)^{l(w)} t^{w\rho}$ is the W-skew symmetric Weyl denominator. For all $\mu, \lambda \in P_k$,

(29)
$$\tau_\lambda(t_\mu) = \frac{S_{\lambda,*\mu}}{S_{0,*\mu}}$$

by the Weyl character formula.

The action of $Z(G)$ on $\mathfrak{A}$ also induces an action on the set P_k of level k weights:

(30)
$$c \bullet_k \lambda = w_c(\lambda - k\zeta_{c^{-1}})$$

for $c \in Z(G),\ \lambda \in P_k$.

Lemma 2.8. *The map* $P_k \to \mathfrak{t},\ \lambda \mapsto \zeta_\lambda$ *is* $Z(G)$*-equivariant:*

$$c.\zeta_\lambda = \zeta_{c \bullet_k \lambda}.$$

Proof. The weight ρ is fixed under the level $h^\vee$ action: $c \bullet_{h^\vee} \rho = \rho$, since ρ is the unique weight in $h^\vee\,\mathrm{int}(\mathfrak{A})$. It follows that

$$c \bullet_{k+h^\vee} (\lambda + \rho) = w_c\big(\lambda + \rho - (k+h^\vee)\zeta_{c^{-1}}\big) = (c \bullet_k \lambda) + \rho,$$

and therefore

$$c.\zeta_\lambda = \frac{1}{k+h^\vee}\big(c \bullet_{k+h^\vee} (\lambda + \rho)\big) = \frac{1}{k+h^\vee}\big((c \bullet_k \lambda) + \rho\big) = \zeta_{c\bullet_k \lambda}. \qquad \square$$

3 Quasi-Hamiltonian spaces

We briefly recall the definition and basic properties of quasi-Hamiltonian G-spaces, and then discuss their prequantization and quantization.

3.1 Basic definitions

Let G be a compact Lie group, together with an invariant metric B on its Lie algebra $\mathfrak{g}$. Denote by $\theta^L, \theta^R \in \Omega^1(G,\mathfrak{g})$ the Maurer-Cartan-forms, and by

$$\eta = \frac{1}{12} B(\theta^L, [\theta^L, \theta^L]) \in \Omega^3(G) \tag{31}$$

the *Cartan* 3*-form*. A *quasi-Hamiltonian G-space* [1] is a G-manifold M, together with an invariant 2-form $\omega \in \Omega^2(M)$ and an equivariant *moment map* $\Phi\colon M \to G$ satisfying the following axioms:

(a) $\mathrm{d}\omega = -\Phi^*\eta$,
(b) $\iota(\xi_M)\omega = -\frac{1}{2}\Phi^* B(\theta^L + \theta^R, \xi)$,
(c) $\ker(\omega_m) \cap \ker(T_m\Phi) = 0$ for all $m \in M$.

Here ξ_M is the vector field on M defined by $\xi \in \mathfrak{g}$. There are many examples of such spaces; see e.g., [9, 13, 17, 18]. In this paper, we will need only two classes of examples. First, there are the G-conjugacy classes $\mathscr{C} \subseteq G$, with moment map the inclusion, and 2-form

$$\omega((\xi_1)_{\mathscr{C}}, (\xi_2)_{\mathscr{C}})|_g = \frac{1}{2} B\big((\mathrm{Ad}_g - \mathrm{Ad}_{g^{-1}})\xi_1, \xi_2\big).$$

Secondly, for a compact oriented surface Σ_{g}^1 with one boundary component, the space

$$M_G\big(\Sigma_{\mathrm{g}}^1\big) = \mathrm{Hom}(\pi_1(\Sigma), G)$$

is canonically a quasi-Hamiltonian G-space, with moment map given by restriction to the boundary circle $\partial\Sigma$ (and identifying $\mathrm{Hom}(\pi_1(\partial\Sigma), G) = \mathrm{Hom}(\mathbb{Z}, G) = G$.) If $G' = G/Z$ for some subgroup $Z \subseteq Z(G)$, then the quasi-Hamiltonian G'-space $M_{G'}\big(\Sigma_{\mathrm{g}}^1\big)$ can also be regarded as a quasi-Hamiltonian G-space, by viewing it as a quotient

$$M_{G'}(\Sigma_g^1) = M_G(\Sigma_g^1) \,/\, \mathrm{Hom}(\pi_1(\Sigma), Z)$$

(both ω and Φ is invariant under this action). For $g = 1$, one gets an explicit description of the space $M_G(\Sigma_g^1)$, by using standard generators of the fundamental group so that the boundary circle is their group commutator. One finds $M_G(\Sigma_g^1) \cong D(G)$ where $D(G) = G \times G$ is the *double*, with G acting diagonally by conjugation, with moment map the Lie group commutator $\Phi(a,b) = aba^{-1}b^{-1}$, and with the 2-form

$$\omega = \frac{1}{2}B(a^*\theta^L, b^*\theta^R) + \frac{1}{2}B(a^*\theta^R, b^*\theta^L) + \frac{1}{2}B((ab)^*\theta^L, (a^{-1}b^{-1})^*\theta^R). \tag{32}$$

(In this formula, the components $(a,b) \in D(G)$ are regarded as maps $a, b \colon D(G) \to G$.) Likewise $M_{G'}(\Sigma_g^1) = D(G')$, viewed as a quasi-Hamiltonian G-space

$$D(G') = D(G)/(Z \times Z). \tag{33}$$

Given two quasi-Hamiltonian G-spaces M_1, M_2, their direct product $M_1 \times M_2$ with the diagonal G-action, the 2-form $\omega_1 + \omega_2 + \frac{1}{2}B(\Phi_1^*\theta^L, \Phi_2^*\theta^R)$, and the moment map $\Phi_1\Phi_2$ is a quasi-Hamiltonian G-space. This so-called *fusion product* is associative. The moduli spaces for the higher genus surfaces with boundary are fusion products,

$$M_G(\Sigma_g^1) \cong D(G) \times \cdots \times D(G), \quad M_{G'}(\Sigma_g^1) \cong D(G') \times \cdots \times D(G'); \tag{34}$$

the second space is obtained from the first as a quotient by $\mathrm{Hom}(\pi_1(\Sigma), Z) \cong Z^{2g}$. Symplectic quotients (reductions) of quasi-Hamiltonian G-spaces are defined similar to ordinary moment maps by

$$M/\!/G = \Phi^{-1}(e)/G.$$

If e is a regular value of the moment map, then $M/\!/G$ is a symplectic orbifold, otherwise it is a stratified symplectic space as in [27]. Reductions with respect to conjugacy classes $\mathscr{C} \subseteq G$ are defined using the 'shifting trick', as $\Phi^{-1}(\mathscr{C})/G = (M \times \mathscr{C}^*)/\!/G$ where $\mathscr{C}^*$ denotes $\mathscr{C}$ with the opposite 2-form, and with the inverse of the inclusion map as a moment map. For the moduli space example, one has

$$\left(M_G(\Sigma_g^1) \times \mathscr{C}^*\right)/\!/G = M_G(\Sigma_g^1; \mathscr{C}),$$

the symplectic moduli space with boundary holonomy in the prescribed conjugacy class. As a special case $\mathscr{C} = \{e\}$, one finds $M_G(\Sigma_g^1)/\!/G = M_G(\Sigma_g^0)$.

Let us list the following basic properties of quasi-Hamiltonian G-spaces (M, ω, Φ). For proofs, see [1].

(a) If G is connected, then M is even-dimensional.
(b) If G is 1-connected, or more generally if it is connected and the map $\mathrm{Ad}\colon G \to \mathrm{SO}(\mathfrak{g})$ lifts to the spin group, then M has a canonical orientation.
(c) If $G = T$ is a torus, then ω is non-degenerate (i.e., symplectic).

(d) Let $\mathfrak{g}_0 \subseteq \mathfrak{g}$ be a convex neighborhood of 0 over which exp is a diffeomorphism, and $G_0 = \exp(\mathfrak{g}_0)$. Then $M_0 := \Phi^{-1}(G_0)$ has (canonically) the structure of a Hamiltonian G-space (M_0, ω_0, Φ_0), in such a way that $\Phi = \exp \circ \Phi_0$ and $\omega|_m = \omega_0|_m$ for all $m \in \Phi^{-1}(e)$. In fact, $\omega_0 = \omega|_{M_0} + \Phi_0^* \varpi$, where ϖ is the primitive of $\exp^* \eta|_{\mathfrak{g}_0}$ defined by the homotopy operator.
(e) For $a \in G$, with centralizer G^a, each component F of the fixed point set M^a is a quasi-Hamiltonian G^a-space. In particular, if G^a is abelian, then F is symplectic.
(f) If M is connected, and G is 1-connected, then the fibers of Φ are connected, and the image of $\Phi(M)$ under the quotient map $G \to G/\mathrm{Ad}_G \cong \mathfrak{A}$ is a convex polytope, referred to as the *moment polytope* of (M, ω, Φ).

3.2 Pre-quantization

Following Kostant [20] and Souriau [29], one defines a prequantization of a manifold M with closed 2-form ω to be a Hermitian line bundle $L \to M$ with connection, such that the Chern form of the connection equals ω. In particular, the Chern class $c_1(L) \in H^2(M, \mathbb{Z})$ is an integral lift of the cohomology class of the symplectic form. If a prequantization exists, then any two such differ by a flat Hermitian line bundle, or equivalently (if M is connected) by a homomorphism $\pi_1(M) \to \mathrm{U}(1)$. Given a Hamiltonian G-action on M, with equivariant moment map $\Phi\colon M \to \mathfrak{g}^*$, then the Kostant formula [20] gives a lift of the infinitesimal $\mathfrak{g}$-action, in such a way that the moment map measures the vertical part. If G is 1-connected, then this infinitesimal action integrates uniquely to a G-action.

A necessary and sufficient condition for prequantizability is the integrality of ω: for all closed oriented surfaces Σ and all smooth maps $f\colon \Sigma \to M$, the integral $\int_\Sigma f^* \omega$ must be an integer.

We will define prequantizations of quasi-Hamiltonian G-spaces only for the setting needed here, namely, G is simply connected and simple. (The generalization to the case of several simple factors is straightforward.) Then $H^j(G, \mathbb{Z})$ vanishes $j = 1, 2$ and is $\mathbb{Z}$ for $j = 3$, with generator represented in de Rham cohomology by the Cartan 3-form (31) for $B = B_{\mathrm{basic}}$. The conditions $\mathrm{d}\omega = -\Phi^* \eta$, $\mathrm{d}\eta = 0$ say that the pair (ω, η) is a closed form in the relative complex

$$\Omega^\bullet(\Phi) = \Omega^{\bullet - 1}(M) \oplus \Omega^\bullet(G).$$

We define:

Definition 3.1. [22] Suppose G is simply connected. Then a quasi-Hamiltonian G-space (M, ω, Φ) is *prequantizable* if the class $[(\omega, \eta)] \in H^3(\Phi, \mathbb{R})$ admits a lift $x \in H^3(\Phi, \mathbb{Z})$; the choice of such a lift is called a *prequantization*.

Here are some basic facts regarding prequantizations of connected quasi-Hamiltonian G-spaces (M, ω, Φ) for G simply connected. We will assume that G is simple; there are straightforward extensions to the case of several simple factors.

(a) The set of prequantizations is either empty, or is a torsor under the torsion subgroup of $H^2(M,\mathbb{Z})$, or equivalently the group $\mathrm{Hom}(\pi_1(M),\mathrm{U}(1))$.
(b) A necessary condition for the existence of a prequantization is that η is integral, hence B must be an integer multiple kB_{basic} of the basic inner product. We then say that M is *prequantized at level k*.
(c) For any quasi-Hamiltonian space (prequantized or not), there is a canonically defined class $y \in H^3(\Phi,\mathbb{Z})$ at level $2h^\vee$, which plays the role of the anti-canonical line bundle. Given a level k prequantization x, the class $2x+y \in H^3(\Phi,\mathbb{Z})$ at level $2(k+h^\vee)$ plays the role of a spin-c line bundle. See [2, 25]
(d) A conjugacy class $\mathscr{C} \subseteq G$ admits a level k prequantization, necessarily unique, if and only if $\mathscr{C} = G.\exp(\frac{1}{k}\mu)$ with $\mu \in P_k$.
(e) A fusion product of two quasi-Hamiltonian spaces is prequantizable if and only if both factors are prequantizable, and in this case a prequantization of the factors determines a prequantization of the product.
(f) Integrality of relative cocycles can be tested by pairing with relative cycles: The cocycle $(\omega,\eta) \in \Omega^3(\Phi)$ is integral if and only if for every smooth map $f\colon \Sigma \to M$ of a compact oriented surface Σ into M, and every smooth homotopy $h\colon [0,1]\times\Sigma \to G$ between $h_1 = \Phi \circ f$ and a constant map h_0, the sum

$$\int_\Sigma f^*\omega + \int_{[0,1]\times\Sigma} h^*\eta$$

is an integer. (This uses the fact that G is 2-connected.)
(g) If $H_2(M,\mathbb{Z}) = 0$, then integrality of η is sufficient for prequantizability. For example, $D(G)$ has a unique prequantization for any integer level.
(h) Let (M_0,ω_0,Φ_0) as in 3.1(d). Regarding Φ_0 as a map into the contractible subset $\mathfrak{g}_0 = \log(G_0) \subseteq \mathfrak{g}$, we have a pullback map

$$H^3(\Phi,\mathbb{Z}) \to H^3(\Phi_0,\mathbb{Z}) \cong H^2(M_0,\mathbb{Z}).$$

In particular, every prequantization $x \in H^3(\Phi,\mathbb{Z})$ determines a class $x_0 \in H^2(M_0,\mathbb{Z})$, which is an integral lift of the de Rham class of ω_0. Hence it corresponds to a prequantum line bundle $L_0 \to M_0$ for the Hamiltonian G-spaces (M_0,ω_0,Φ_0), with $x_0 = c_1(L_0)$. Similarly, the image of the class y (cf. (c)) under this map is the Chern class $y_0 = c_1(K^{-1})$ of the anti-canonical line bundle over M_0.
(i) Suppose $\iota_F : F \hookrightarrow M$ is a submanifold such that $\Phi(F) \subseteq T$. (For example, a component of the fixed point set of a regular element $a \in T$.) Denoting by η_r the Cartan 3-form (31) defined by rB_{basic}, there is a linear map in de Rham cohomology

$$H^3(\Phi) \to H^2(F), \quad [(\sigma,\eta_r)] \mapsto [\iota_F^*\sigma].$$

As explained in [25, Section 5.1], this map lifts to integral classes at *even* level: Letting $H^3(\Phi,\mathbb{Z})_r$ be elements at level r (i.e., the image in $H^3(G,\mathbb{Z}) = \mathbb{Z}$ is r), one obtains a canonical map $H^3(\Phi,Z)_r \to H^2(F,\mathbb{Z})$ for r *even*. (In the resulting line bundle over F is T_r-equivariant.) In particular, if M is prequantized at an even level, then F inherits a T_k-equivariant prequantization. If $F \subseteq M_0$, the restriction to even levels is not needed, since one can simply take the restriction of L_0.

In Appendix A, we will prove the following results about the descent of prequantizations to quotients. Suppose (M, ω, Φ) is a connected quasi-Hamiltonian G-space, such that the universal cover $\widetilde{M}$ (with $\widetilde{\omega}, \widetilde{\Phi}$ the pullback of 2-form and moment map) is prequantized. Similar to the case of manifolds with closed 2-forms, one then obtains a central extension of the fundamental group $\pi_1(M)$, and M itself admits a prequantization if and only if this central extension is trivial. If the fundamental group is *abelian*, then the central extension is described by its *commutator function* $q\colon \pi_1(M) \times \pi_1(M) \to \mathrm{U}(1)$ (see Proposition A.4). The latter has the following expression:

$$q(u,v) = \exp\Big(2\pi i\Big(\int_{S^1\times S^1} f^*\omega + \int_{S^1\times S^1\times[0,1]} h^*\eta\Big)\Big). \tag{35}$$

Here $f\colon S^1 \times S^1 \to M$ is any smooth map such that the induced map on fundamental groups takes the generators to u, v respectively, and h is a smooth homotopy between $h_1 = \Phi \circ f$ and a constant map h_0. That is, (ω, η) is integral if and only if $(\widetilde{\omega}, \eta)$ is integral and $q = 1$.

3.3 Quantization and the fixed point formula

Suppose G is simply connected and simple, and let M be a level k prequantized quasi-Hamiltonian G-space. As shown in [4] (using a fixed point formula as a definition) and [25] (more conceptually, as a K-homology push-forward and deriving the fixed point formula as a consequence), this then determines an element

$$\mathcal{Q}(M) \in R_k(G) \tag{36}$$

called its *quantization*. By [4, Section 2.3] (see also [25, Theorem 9.5]), the numbers $\mathcal{Q}(M)(t_\lambda)$ for $\lambda \in P_k$ are given as a sum of contributions from the fixed point manifolds of $t = t_\lambda$:

$$\mathcal{Q}(M)(t) = \sum_{F\subseteq M^t} \chi_F(t)^{1/2} \int_F \frac{\widehat{A}(F)\, \exp(\frac{1}{2}c_1(\mathscr{L}_F))}{D_{\mathbb{R}}(\nu_F, t)}. \tag{37}$$

Here each $F \subseteq M^t$ is a quasi-Hamiltonian $G^t = T$-space; in particular, the 2-form ω_F given as the pull-back of ω is symplectic, and endows F with an orientation. The class $\widehat{A}(F)$ is the usual $\widehat{A}$-class of F, while $D_{\mathbb{R}}(\nu_F, t)$ is an equivariant characteristic class for the normal bundle of F given by

$$D_{\mathbb{R}}(\nu_F, t) = i^{\mathrm{rank}(\nu_F)/2} \det{}^{1/2}(1 - A(t)^{-1} e^{R/2\pi}) \tag{38}$$

where $A(t)$ is the action of t on the normal bundle and R is the curvature 2-form of a t-invariant connection. (See [4]). The Chern class $c_1(\mathscr{L}_F)$ is the image of $2x+y$ under the map $H^3(\Phi,\mathbb{Z})_{2(k+h^\vee)} \to H^2(F,\mathbb{Z})$ described in 3.2(i).

The phase factor $\chi_F(t) \in \mathrm{U}(1)$ is given by a natural action of t on $\mathscr{L}_F$; it is constant on the fixed point components. The detailed definition of this phase factor, and the correct choice of square root is explained in [4]. If F contains a point m in $\Phi^{-1}(e)$ (or more generally a point in M_0), it may be computed as follows: Let $L_0 \to M_0$ be the prequantum line bundle as in 3.2(h), and denote by $\mu_F(t) \in \mathrm{U}(1)$ the weight the action of t on $L_0|_m$. Recall that the 2-form ω_m on T_mM is symplectic. Choose a compatible complex structure on the tangent space, invariant under the action of t, which hence becomes a unitary transformation $A_F(t)$. Then [4, Remark 4.7]

(39) $$\chi_F(t)^{1/2} = \mu_F(t)\,\det_{\mathbb{C}}(A_F(t)^{1/2}),$$

using the unique square root of $A_F(t)$ having all its eigenvalues in the set of $e^{i\theta}$ with $0 \le \theta < \pi$.

4 The space $M_{G'}(\Sigma_1^1)$

In this section, we will compute the prequantization and the quantization of the moduli space $M_{G'}(\Sigma_1^1) \cong D(G')$, where G is a compact, simple Lie group and $Z \subseteq G$ is a subgroup of the center $Z(G)$, and $G' = G/Z$. Throughout, we will denote the basic inner product by a dot, that is $B_{\mathrm{basic}}(\xi_1,\xi_2) = \xi_1 \cdot \xi_2$.

4.1 The basic level

As before, we denote $\Lambda_Z = \exp_T^{-1}(Z)$. Following Toledano-Laredo [31], we refer to the smallest $k_0 \in \mathbb{N}$ such that $k_0 B_{\mathrm{basic}}$ takes on integer values on Λ_Z, as the *basic level* (with respect to Z) . Equivalently, this is the smallest natural number such that

(40) $$k_0\,\Lambda_Z \cdot \Lambda_Z \subseteq \mathbb{Z}.$$

We will also be interested in another level k_1, defined to be the smallest natural number such that
$$k_1\,\Lambda_Z \cdot P^\vee \subseteq \mathbb{Z}.$$

Since $\Lambda_Z \subseteq P^\vee$, the level k_1 is a multiple of k_0. Some of the proofs below simplify whenever $k_1 = k_0$; in particular, this is the case when $Z = Z(G)$ so that $\Lambda_Z = P^\vee$.

The levels k_0, k_1 are easily worked out using the tables for compact Lie groups (e.g., Bourbaki [10]). For $Z = Z(G)$ one obtains [31]:

G	A_l	B_l	C_{2r}	C_{2r+1}	D_{2r}	D_{2r+1}	E_6	E_7
$Z(G)$	$\mathbb{Z}_{l+1}$	$\mathbb{Z}_2$	$\mathbb{Z}_2$	$\mathbb{Z}_2$	$\mathbb{Z}_2 \times \mathbb{Z}_2$	$\mathbb{Z}_4$	$\mathbb{Z}_3$	$\mathbb{Z}_2$
$k_0 = k_1$	$l+1$	1	1	2	2	4	3	2

If $Z = \{1\}$ one has $k_0 = k_1 = 1$, by the properties of the basic inner product. If Z is a non-trivial proper subgroup of $Z(G)$, we have:

- For A_l with $Z = \mathbb{Z}_m$, where m divides $l+1$, the basic level k_0 is the smallest natural number such that $k_0(l+1) \in m^2\mathbb{Z}$, while $k_1 = m$. (In particular, $k_1 = k_0$ if and only if m and $(l+1)/m$ are relatively prime.)
- For D_l, there are three different subgroups $Z \subseteq Z(G)$ isomorphic to $\mathbb{Z}_2$; with the labelings of fundamental coroots as in [10], they are generated by the exponentials of the coroots $\varpi_1^\vee, \varpi_{l-1}^\vee, \varpi_l^\vee$.

  (a) $Z = \{(e, \exp(\varpi_1^\vee)\}$. One checks that $\varpi_1^\vee \cdot \varpi_j^\vee$ equals 1 if $j \le l-2$ and $\frac{1}{2}$ for $j = l-1, l$. Hence $k_0 = 1$, $k_1 = 2$.
  (b) $Z = \{(e, \exp(\varpi_{l-1}^\vee)\}$. One finds that $\varpi_{l-1}^\vee \cdot \varpi_j^\vee$ equals $\frac{j}{2}$ for $j \le l-2$, $\frac{l}{4}$ for $j = l-1$, and $\frac{l-2}{4}$ for $j = l$. Hence $k_0 = k_1 = 4$ if l is odd, $k_0 = k_1 = 2$ if $l \in 4\mathbb{Z}+2$, and $k_0 = 1$, $k_1 = 2$ if $l \in 4\mathbb{Z}$.
  (c) $Z = \{(e, \exp(\varpi_l^\vee)\}$. The calculation is similar to that in (b), with the same results.

Notice that k_1 is different from k_0 only in a few exceptional cases.

4.2 *Prequantizability of the space* $D(G/Z)$

Let $D(G)$ be the double of G, with the quasi-Hamiltonian structure at level k (i.e., using the inner product $B = kB_{\mathrm{basic}}$). Since the moment map and 2-form of $D(G)$ are invariant under the $Z \times Z$-action, they descend to $D(G')$ regarded as a quasi-Hamiltonian G-space,

$$D(G') = D(G)/(Z \times Z).$$

The following result was proved in [22] using a case-by-case examination; we will give a new proof based on the criterion in Section 3.2.

Proposition 4.1 (Krepski [22]). *The quasi-Hamiltonian G-space $D(G/Z)$ is prequantizable at level k if and only if k is a multiple of the basic level k_0.*

Proof. We will compute the commutator map on $\pi_1(D(G')) = Z^2$ using the integral expression (35). Given $c, d \in Z^2$, write $c = (\exp_T(\xi_1), \exp_T(\xi_2))$ and $d = (\exp_T(\zeta_1), \exp_T(\zeta_2))$ with $\xi_i, \zeta_i \in \Lambda_Z$ for $i = 1, 2$. The map

$$\widetilde{f}\colon \mathbb{R} \times \mathbb{R} \to D(G'), \;\; (s,t) \mapsto \exp_{G'}(s\xi_1 + t\zeta_1,\; s\xi_2 + t\zeta_2)$$

descends to a map $f\colon S^1 \times S^1 \to D(G')$, and the induced map on fundamental groups $\mathbb{Z} \times \mathbb{Z} \to \pi_1(D(G'))$ takes $(1,0)$ and $(0,1)$ to $c, d \in \pi_1(D(G'))$, respectively. The map $h_1 = \Phi \circ f\colon D(G') \to G$ is constant; hence we may take the homotopy h to be a constant map, and (35) simplifies to

$$q(c,c') = \exp\Big(2\pi i \int_{S^1\times S^1} f^*\omega\Big) = \exp\Big(2\pi i \int_{[0,1]\times[0,1]} \widetilde{f}^*\omega\Big).$$

To compute the integral, note that $\widetilde{f}$ takes values in $D(T') = D(T)/Z \times Z$. But the pull-back of the 2-form on $D(G)$, given by (32), to $D(T)$ is simply

$$\omega_{D(T)} = k\, a^*\theta_T \cdot b^*\theta_T$$

where $\theta_T \in \Omega^1(T,\mathfrak{t})$ is the Maurer-Cartan form on T. Hence,

$$\widetilde{f}^*\omega = k(\xi_1\cdot\zeta_2 - \xi_2\cdot\zeta_1)\,\mathrm{d}s\wedge\mathrm{d}t.$$

Integrating over the square, we obtain

$$q(c,c') = \exp(2\pi i k(\xi_1\cdot\zeta_2 - \xi_2\cdot\zeta_1)).$$

In particular, $q = 1$ if and only if $k\xi\cdot\zeta \in \mathbb{Z}$ for all $\xi,\zeta \in \Lambda_Z$. □

By 3.2(e) and (34), Proposition 4.1 has the following immediate consequence.

Corollary 4.2. *For a compact oriented surface of genus g with one boundary component, the moduli space $M_{G'}(\Sigma_g^1)$ (regarded as a quasi-Hamiltonian G-space) is prequantizable at level k if and only if k is a multiple of the basic level k_0.*

4.3 Phase factors

In this section, we discuss the phase factors that we will encounter when evaluating the fixed point contributions later on. If k is a multiple of the basic level k_0, consider the map

(41) $$\Lambda_Z \times \Lambda_Z \to \mathrm{U}(1),\ \ (u,v) \mapsto e^{2\pi i k(1-w_*)^{-1}u\cdot v},$$

where w_* is the Coxeter transformation. Since $B = kB_{\mathrm{basic}}$ takes on integer values on Λ_Z, this expression takes on the value 1 whenever one of u or v lies in the sublattice $(1-w_*)\Lambda_Z$. That is, (41) descends to $\Lambda_Z/(1-w_*)\Lambda_Z$. Using the identification

$$(T')^{w_*} = (1-w_*)^{-1}\Lambda_Z/\Lambda_Z \cong \Lambda_Z/(1-w_*)\Lambda_Z,$$

this defines a bimultiplicative form

(42) $$\kappa\colon (T')^{w_*}\times(T')^{w_*} \to \mathrm{U}(1),\ (t_1',t_2') \mapsto \kappa(t_1',t_2')$$

given by (41) for $t_1' = \exp_{T'}((1-w_*)^{-1}u)$ and $t_2 = \exp_{T'}((1-w_*)^{-1}v)$.

If k is a multiple of the level k_1, this descends further to a bimultiplicative form on $Z = \Lambda_Z/Q^\vee$. Indeed, since $(1-w_*)^{-1}Q^\vee = P^\vee$ by Proposition 2.1, (41) takes on the value 1 whenever u or v are in $Q^\vee$. In particular, this is the case whenever $k_1 = k_0$.

If $k_1 > k_0$, the map (41) need not descend to Z in general. However, for our applications to the localization formula (37), the following more complicated statement suffices.

Lemma 4.3. *Suppose k is a multiple of k_0, and let $c_1, c_2 \in Z$ be given. If the action of c_1, c_2 on the set P_k of level k weights has at least one fixed point, then there is a well-defined phase factor*

$$\delta(c_1,c_2) = e^{2\pi i k\,(1-w_*)^{-1}u\cdot v}$$

where $u, v \in \Lambda_Z$ with $c_1 = \exp(u)$, $c_2 = \exp(v)$.

To prove this lemma, we must verify that the exponential does not depend on the choice of u, v. Equivalently, we must show that if the action of $c \in Z$ on P_k has a fixed point, and $c = \exp(u)$ for $u \in \Lambda_Z$, then $k_0\, u \cdot P^\vee \subseteq \mathbb{Z}$. Note that for $k_1 = k_0$, the existence of a fixed point is irrelevant, by definition of k_1. If $k_1 > k_0$, one can prove the lemma by a case-by-case calculation, which is done in the Appendix.

Remark 4.4. Note that the phase factor $\delta(c_1,c_2)$ does not depend on the choice of Coxeter element w_*. In fact, replacing w_* with its conjugate under a Weyl group element w amounts to replacing u, v with their images under w^{-1}. But $\exp(w^{-1}u) = \exp(u)$, since $\exp(u)$ is in the center, and similarly for v.

One may compute the phase factors δ case-by-case, using e.g., the tables in Bourbaki [10]. Here are some results of such computations (for details, see Appendix C):

Examples 4.5. (a) Let $G = A_l$, with $Z = \mathbb{Z}_m$ where m divides $l+1$. Let $c \in Z$ be a generator, and suppose k is a multiple of k_0, so that $k(l+1)$ is divisible by m^2. By explicit calculation, one obtains

$$\delta(c,c) = (-1)^{l\frac{k(l+1)}{m^2}},$$

while of course $\delta(c^r, c^s) = \delta(c,c)^{rs}$. In particular, if l is even, then $\delta = 1$ for all k.

(b) For $G = B_l$, thus $k_0 = 1$, and $c \in Z = \mathbb{Z}_2$ the non-trivial element, we have

$$\delta(c,c) = (-1)^k.$$

(c) For $G = C_l$, we have $\delta = 1$ for all k.

(d) For $G = D_l$ with l odd, $Z = Z(G) = \mathbb{Z}_4$, and k a multiple of $k_0 = 4$,

$$\delta(c,c) = (-1)^{\frac{k}{4}}$$

where c is the generator of $\mathbb{Z}_4$.

4.4 Classifying the prequantizations of $D(G/Z)$

Suppose that k is a multiple of k_0 so that $D(G')$ is prequantizable at level k. According to 3.2(c), the inequivalent prequantizations are a torsor under $\mathrm{Hom}(Z\times Z,\mathrm{U}(1))$. We would like to have a way of distinguishing these prequantizations. To this end, we consider the symplectic submanifold $D(T')\subseteq D(G')$. Since this is contained in $\Phi^{-1}(e)$, a prequantization of $D(G')$ gives rise to a W-equivariant prequantum line bundle over $D(T')$. We observe that the latter determines the former; in fact it suffices to look at the action of the Coxeter element w_*.

Proposition 4.6. *Any prequantization of $D(G')$ gives rise to a W-equivariant prequantum line bundle $L'\to D(T')$. Furthermore:*

(a) *If $m\in D(T')$ is a W-fixed point, then the W-action on the fiber $L'|_m$ is trivial.*
(b) *The prequantization of $D(G')$ is uniquely determined by the action of the Coxeter element w_* on the restriction of L' to the fixed point set $D(T')^{w_*}$.*

Proof. Let $D(G')_0$ be the Hamiltonian G-space as in 3.1(d). Given a prequantization of $D(G')$, we obtain by 3.2(h) a G-equivariant prequantum line bundle $L_0\to D(G')_0$. Elements of $D(Z(G)/Z)\subseteq D(G')_0$ are G-fixed points, and by Kostant's formula for prequantizations the G-action on the fiber at such points is trivial. Since $D(T')$ is contained in the identity level set of the moment map, it is contained in $D(G')_0$, and the pull-backs of ω_0 and of ω to $D(T')$ coincide. By restriction, the symplectic submanifold $D(T')$ inherits a W-equivariant prequantum line bundle $L'\to D(T')$, where the W-action at fibers over $D(Z(G)/Z))\subseteq D(T')$ is trivial. This proves (a).

As for (b), note that any two prequantizations of $D(G')$ differ by some $\phi\in \mathrm{Hom}\left(Z\times Z,\mathrm{U}(1)\right)$. The line bundle $L_0'\to D(G')_0$ changes by tensor product with the flat line bundle $D(G)_0\times_\phi\mathbb{C}\to D(G')_0$, and hence L' by tensor product with

$$D(T)\times_\phi\mathbb{C}\to D(T'). \tag{43}$$

Hence, it suffices to show that ϕ is uniquely determined by the action of w_* on the restriction of (43) to $D(T')^{w_*}$. Given $c_1,c_2\in Z$, choose $t_i'\in T'$ such that the lifts t_i satisfy $w_*(t_i)=t_ic_i^{-1}$. The action of w_* on an element $[(t_1,t_2,1)]$ in (43) is calculated as follows:

$$w_*[(t_1,t_2,1)]=[(w_*t_1,w_*t_2,1)]=[(t_1c_1^{-1},t_2c_2^{-1},1)]=[(t_1,t_2,\phi(c_1,c_2))].$$

This shows that we can recover ϕ from the action at fixed points. □

We are now in a position to classify the prequantizations of $D(G')$:

Proposition 4.7. *Suppose k is a multiple of the basic level k_0.*

(a) *There exists a unique level k prequantization of $D(G')$ with the property that the action of w_* on the fiber of $L'\to D(T')$ at $(t_1',t_2')\in D(T')^{w_*}$ is given by $\kappa(t_1',t_2')$.*

(b) *Any other prequantization of* $D(G')$ *is obtained from that in* (a) *by some* $\phi = (\phi_1, \phi_2) \in \mathrm{Hom}(Z \times Z, \mathrm{U}(1))$. *The phase factor for the action of* w_* *on the fiber at* (t_1', t_2') *changes to*

$$\phi_1(c_1)\ \phi_2(c_2)\ \kappa(t_1', t_2')$$

where $w_* t_i = t_i c_i^{-1}$.

Proof. The symplectic form on $D(T')$ is given by $k\,\mathrm{pr}_1^*\,\theta_{T'} \cdot \mathrm{pr}_2^*\,\theta_{T'}$; hence, its pull-back to $\mathfrak{t} \times \mathfrak{t}$ is the symplectic form $kdq \cdot dp$, where $\mathrm{d}q,\ \mathrm{d}p \in \Omega^1(\mathfrak{t}, \mathfrak{t})$ are the 'tautological' 1-forms for the two $\mathfrak{t}$-factors. It has a prequantum line bundle, unique up to isomorphism, given by the trivial line bundle $(\mathfrak{t} \times \mathfrak{t}) \times \mathbb{C}$ with the connection 1-form

$$\pi ik(q \cdot dp - p \cdot dq).$$

Given a prequantization of $D(G')$, resulting in a prequantum line bundle L' over $D(T') = T' \times T'$, there is a connection-preserving isomorphism between the pull-back of L' and this trivial line bundle. In particular, the latter inherits an action of the semi-direct product $(\Lambda_Z \times \Lambda_Z) \rtimes W$ by connection-preserving automorphisms. But the only connection-preserving lift of the W-action, in such a way that the action on the fiber at $(0,0)$ is trivial, is the obvious action

$$w.(q, p, z) = (wq, wp, z),$$

while the most general connection-preserving lift of the map $(q, p) \mapsto (q - u, p - v)$ for $u, v \in \Lambda_Z$ is of the form

$$(q, p, z) \mapsto \left(q - u, p - v, \sigma(u, v) e^{\pi ik\,(p \cdot u - q \cdot v)}\right)$$

for some scalars $\sigma(u, v) \in \mathrm{U}(1)$. The condition that this defines a group action of the semi-direct product $(\Lambda_Z \times \Lambda_Z) \rtimes W$ is that the map

$$\psi(u, v) = \sigma(u, v)\ e^{\pi ik\, u \cdot v}$$

defines a group homomorphism $\psi\colon \Lambda_Z \times \Lambda_Z \to \mathrm{U}(1)$, which furthermore is W-invariant. (In checking the homomorphism property, note that $e^{\pi iku \cdot v} = \pm 1$.) Write $\psi(u, v) = \psi_1(u)\psi_2(v)$ with W-invariant maps $\psi_i\colon \Lambda_Z \to \mathrm{U}(1)$. The W-invariance of the ψ_i means in particular that they descend to the quotient $\Lambda_Z/(1 - w_*)\Lambda_Z \cong (T')^{w_*}$.

Let $u, v \in \Lambda_Z$, and put

$$\xi = (1 - w_*)^{-1} u,\ (1 - w_*)^{-1} v$$

so that $(t_1', t_2') = (\exp_{T'} \xi, \exp_{T'} \zeta)$ is a w_*-fixed point. The point $(\xi, \zeta) \in \mathfrak{t} \times \mathfrak{t}$ is fixed under $((-u, -v), w_*) \in (\Lambda_Z \times \Lambda_Z) \rtimes W$, and its action on the fiber of $(\mathfrak{t} \times \mathfrak{t}) \times \mathbb{C}$ at that point coincides with the action of w_* on the fiber of its image in $D(T')$. We compute:

$$\begin{aligned}(-u,-v).w_*.(\xi,\zeta,1) &= (-u,-v).(\xi-u,\zeta-v,1)\\ &= \big(\xi,\zeta,\psi(-u,-v)\,e^{\pi ik((\zeta-v)\cdot(-u)-(\xi-u)\cdot(-v)-u\cdot v)}\big)\\ &= \big(\xi,\zeta,\psi_1(u)^{-1}\psi_2(v)^{-1}e^{2\pi ik\,\xi\cdot(1-w_*)\zeta}\big)\\ &= \big(\xi,\zeta,\psi_1(u)^{-1}\psi_2(v)^{-1}\kappa(t_1',t_2')\big).\end{aligned}$$

Hence, the action of w_* on the fiber at (t_1',t_2') is given by the scalar,

$$\psi_1(u)^{-1}\psi_2(v)^{-1}\kappa(t_1',t_2').$$

Writing $\phi_i=\psi_i^{-1}$, this proves the proposition. □

In the following section, we will consider a maximal torus U in apposition to T, with Ad_{t_*} acting as the Coxeter transformation. Proposition 4.7, with $U'=U/Z$ playing the role of T', shows that any prequantization of $D(G')$ defines a $N(U)/U$-equivariant prequantum line bundle $L'\to D(U')$, and that there is a unique prequantization such that for $(u_1',u_2')\in D(U')^{t_*}$, the weight for the action of t_* on L'_{u_1,u_2} is given by $\kappa(u_1',u_2')$.

4.5 Computation of the fixed point contributions

Suppose k is a multiple of the basic level k_0, and let $x\in H^3(\Phi,\mathbb{Z})$ be a given level k prequantization of the quasi-Hamiltonian G-space $D(G')$, with resulting quantization

$$\mathscr{Q}(D(G'))\in R_k(G').$$

Its values at $t=t_\lambda=\exp(\zeta_\lambda)$ with $\lambda\in P_k$ are given by the fixed point formula (37). In this section, we will work out the contributions

(44)
$$\chi_F(t)^{1/2}\int_F\frac{\widehat{A}(F)\,\exp(\frac{1}{2}c_1(\mathscr{L}_F))}{D_{\mathbb{R}}(\nu_F,t)}$$

from components $F\subseteq D(G')^t$. Proposition 2.7, together with Lemma 2.8 shows:

Proposition 4.8. *The fixed point set for the action of t on $D(G')$ is given by*

(45)
$$D(G')^t=\bigcup_{\substack{c_1,c_2\in Z\\ c_i\bullet_k\lambda=\lambda}}N(T')^{(c_1)}\times N(T')^{(c_2)},$$

where $N(T')^{(c)}=N(T)^{(c)}/Z$ for $c\in Z$.

Let $F\subseteq D(G')^t$ be the component indexed by c_1,c_2. Being a component of $D(N(T'))$, it is obtained from $D(T')\cong T'\times T'\subseteq D(G')$ by left translation. In particular,

$$\widehat{A}(F)=1.$$

Furthermore, the normal bundle is a trivial bundle with fiber $\mathfrak{g}/\mathfrak{t}$, with t acting diagonally by the adjoint action. By the definition (38) of $D_{\mathbb{R}}$, this gives

$$D_{\mathbb{R}}(\nu_F, t) = (-1)^{\frac{1}{2}\dim G/T} \det_{\mathfrak{g}/\mathfrak{t}}(1 - \mathrm{Ad}_t) = (-1)^{\frac{1}{2}\dim G/T} |J(t)|^2,$$

with the Weyl denominator $J(t) = \sum_{w \in W} (-1)^{l(w)} t^{w\rho}$. Hence, (44) simplifies to

$$(-1)^{\frac{1}{2}\dim G/T} \frac{\chi_F(t)^{1/2}}{|J(t)|^2} \int_F \exp\left(\frac{1}{2}c_1(\mathscr{L}_F)\right). \tag{46}$$

For the remaining integral we find:

Lemma 4.9. *We have*

$$\int_F \exp\left(\frac{1}{2}c_1(\mathscr{L}_F)\right) = \frac{\# T_{k+h^\vee}}{\# Z^2}.$$

Proof. (Cf. [3, Proposition 2.2].) The class $c_1(\mathscr{L}_F)$ is obtained from $2x + y$ by the map $H^3(\Phi, \mathbb{Z})_{\mathrm{ev}} \to H^2(F, \mathbb{Z})$ discussed in 3.2(h). Passing to real coefficients, the map $H^3(\Phi, \mathbb{R}) \to H^3(G, \mathbb{R})$ is an isomorphism since $H^2(D(G), \mathbb{R}) = 0$. Hence, there is a *unique* class in $H^3(\Phi, \mathbb{R})$ at any given level. It follows that the de Rham class of $c_1(\mathscr{L}_F)$ coincides with the class of the symplectic form on F, using the bilinear form at level $2(k + h^\vee)$. Letting $\omega_{F,1}$ be the symplectic structure for the 2-form defined by B_{basic}, it follows that

$$\int_F \exp\left(\frac{1}{2}c_1(\mathscr{L}_F)\right) = \int_F \exp((k + h^\vee)\omega_{F,1}).$$

Letting $\widetilde{F} = N(T)^{(c_1)} \times N(T)^{(c_2)} \subseteq D(G)$ be the pre-image of F, with $\omega_{\widetilde{F},1}$ the pull-back symplectic form, we obtain

$$\int_F \exp\left(\frac{1}{2}c_1(\mathscr{L}_F)\right) = \frac{1}{\# Z^2} \int_{\widetilde{F}} \exp((k + h^\vee)\omega_{\widetilde{F},1}).$$

We thus need to compute the symplectic volume of $\widetilde{F}$, for the 2-form defined by $(k + h^\vee) B_{\mathrm{basic}}$. We claim that this is the same as the symplectic volume of $D(T) \subseteq D(G)$. The latter equals $\# T_{k+h^\vee}$ (see e.g., [8, Proposition 1.2]), which hence completes the proof of the lemma.

To proceed, we need the explicit expression of the 2-form at points (a, b) of $\widetilde{F}$. Use left trivialization of the tangent bundles to identify $T_{(a,b)}(N(T)^{(c_1)} \times N(T)^{(c_2)}) \cong \mathfrak{t} \times \mathfrak{t}$. The transformations $\mathrm{Ad}_a, \mathrm{Ad}_b$ act on the two $\mathfrak{t}$-factors as *commuting* transformations $w_1 = w_{c_1}$, $w_2 = w_{c_2}$, respectively. For any given invariant bilinear form B, the 2-form (32) on $D(G)$ restricts to the skew-symmetric bilinear form on $\mathfrak{t} \times \mathfrak{t}$, given by

$$B\left(\begin{pmatrix} \xi_1 \\ \xi_2 \end{pmatrix}, \begin{pmatrix} a & b \\ c & d \end{pmatrix} \begin{pmatrix} \xi_1' \\ \xi_2' \end{pmatrix}\right),$$

for (ξ_1,ξ_2), $(\xi_1',\xi_2')\in\mathfrak{t}\times\mathfrak{t}$. with

$$a=-\frac{1}{2}(w_2-w_2^{-1}),\quad b=\frac{1}{2}(1+w_2+w_1^{-1}-w_2w_1^{-1}),$$

$$c=-\frac{1}{2}(1+w_1+w_2^{-1}-w_1w_2^{-1}),\quad d=\frac{1}{2}(w_1-w_1^{-1}).$$

Since the endomorphisms a,b,c,d commute, and since $ad-bc=I$ by elementary calculation, we have

$$\det\begin{pmatrix}a\ b\\ c\ d\end{pmatrix}=\det(ad-bc)=1.$$

It follows that the volume form defined by $\omega_{\widetilde{F}}$ coincides with the left translate of the volume form on $D(T)$. □

It remains to determine the scalar factor $\chi_F(t)^{1/2}$. This factor depends on the choice of prequantization. We have a distinguished prequantization, defined in terms of the maximal torus in apposition U as explained at the end of Section 4.4; the other prequantizations are related by a homomorphism $\phi=(\phi_1,\phi_2)\colon Z\times Z\to \mathrm{U}(1)$. Recall also the phase factor $\delta(c_1,c_2)\in\mathrm{U}(1)$ associated with the fixed point component F.

Lemma 4.10. *We have*

$$\chi_F(t)^{1/2}=(-1)^{\frac{1}{2}\dim(G/T)}\ \phi_1(c_2)\phi_2(c_2)\delta(c_1,c_2).$$

Proof. By Proposition 2.7, $F\cap D(U')$ is non-empty. Letting $m\in F\cap D(U')$, since $\Phi(m)=e$ we may compute the phase factor as a product

$$\chi_F(t)^{1/2}=\mu_F(t)\det{}_{\mathbb{C}}(A_F(t)^{1/2}),$$

see (39). Here $\mu_F(t)$ is the phase factor for the action of t on the prequantum line bundle at m. To compute it, let $T_F\subseteq F$ be the subgroup of elements acting trivially on F. We claim that the elements $t=t_\lambda$ and t_* are in the same component of T_F. Indeed, since the action of $Z(G)$ on $\mathfrak{t}$ is an affine action, the set of elements $\zeta\in\mathfrak{t}$ fixed by both c_1 and c_2 is an affine subspace, containing ζ_* and ζ_λ. By Proposition 2.7, for any ζ in this affine subspace, the element $\exp(\zeta)$ fixes F pointwise, proving the claim.

As a consequence, we obtain

$$\mu_F(t)=\mu_F(t_*)=\phi_1(c_2)\phi_2(c_2)\delta(c_1,c_2),$$

using Proposition 4.7 with U' playing the role of T'. To compute $\det_{\mathbb{C}}(A_F(t)^{1/2})$, we make some preliminary observations. Suppose V is a Euclidean vector space, and let $W=V\oplus V$ with the standard symplectic structure (and compatible complex structure) given by the matrix

$$\begin{pmatrix}0\ 1\\ -1\ 0\end{pmatrix}.$$

For $h \in \mathrm{O}(V)$, the transformation $g = h \oplus h$ of W is symplectic, and its eigenvalues other than 1 come in complex conjugate pairs $e^{i\theta}$, $e^{-i\theta}$, with $0 < \theta < \pi$. With our choice of square root, the eigenvalues of $g^{1/2}$ are then $e^{i\theta/2}$ and $e^{i(\pi/2-\theta/2)}$. Hence, $\det_{\mathbb{C}}(g^{1/2}) = (-1)^n$, where n is the codimension in V of the subspace where V acts trivially. The same result is obtained when we use *minus* the standard symplectic structure on $V \oplus V$.

By (32), the symplectic form on $T_m D(G') \cong \mathfrak{g} \oplus \mathfrak{g}$ (using left trivialization of the tangent bundle) is given in matrix form by

$$A = \begin{pmatrix} a & b \\ c & d \end{pmatrix},$$

with

$$a = -\frac{1}{2}(k_2 - k_2^{-1}), \quad b = \frac{1}{2}(1 + k_2 + k_1^{-1} - k_2 k_1^{-1}),$$
$$c = -\frac{1}{2}(1 + k_1 + k_2^{-1} - k_1 k_2^{-1}), \quad d = \frac{1}{2}(k_1 - k_1^{-1}).$$

Here $k_1 = \mathrm{Ad}_{u_1'}$, $k_2 = \mathrm{Ad}_{u_2'} \in \mathrm{O}(\mathfrak{g})$ for the given point $m = (u_1', u_2')$. Note that k_1, k_2 commute with each other (since the $u_i' \in U'$ commute), and they also commute with the transformation $h = \mathrm{Ad}_{t_*}$ (since the u_i' are fixed points of Ad_{t_*}). Let $V' \subseteq V = \mathfrak{g}$ be the subspace on which both k_1, k_2 act as minus the identity, and V'' its orthogonal complement. Then both $V' \oplus V'$ and $V'' \oplus V''$ are invariant under the diagonal action of Ad_{t_*}. On $V' \oplus V'$, the matrix A simplifies to minus the standard symplectic structure, hence by the discussion above, it contributes $(-1)^{\dim V'}$ to $\det_{\mathbb{C}}(A_F(t)^{1/2})$.

On $V'' \oplus V''$, we consider the homotopy

$$A_s = sA + (1-s)\begin{pmatrix} 0 & 1 \\ -1 & 0 \end{pmatrix}, \quad 0 \le s \le 1,$$

with the standard symplectic structure. We claim that A_s is symplectic on $V'' \oplus V''$ for all $s \in [0,1]$. Indeed, using $ad - bc = 1$ one finds

$$\det_{V'' \oplus V''}(A_s) = \det_{V''}(1 - s(1-s)(2 - b + c)).$$

We want to show that the symmetric operator $s(1-s)(2-b+c)$ has no eigenvalues ≥ 1 on V''. Since $s(1-s) \in [0, \frac{1}{4}]$ for $s \in [0,1]$, this is equivalent to showing that $1 - b + c$ has no eigenvalues ≥ 3 on V''. But

$$1 - b + c = -\frac{w_1 + w_1^{-1}}{2} - \frac{w_2 + w_2^{-1}}{2} + \frac{w_2 w_1^{-1} + w_1 w_2^{-1}}{2}$$

has norm ≤ 3, since each summand has norm ≤ 1. It can take on the value 3 only on those generalized eigenspaces of w_1, w_2 where all three summands are equal to 1. But this does not happen on V'', by definition. We conclude that $V'' \oplus V''$ contributes $(-1)^{\dim V''}$ to $\det_{\mathbb{C}}(A_F(t)^{1/2})$. We see that $\det_{\mathbb{C}}(A_F(t_*)^{1/2})$ is $(-1)^n$, where n is the codimension of the Ad_{t_*}-fixed space in $\mathfrak{g}$. Since t_* is regular, this is just $\mathfrak{t}$, and we obtain $(-1)^{\dim G/T}$ as desired. □

Putting everything together, and summing over fixed point components, we have shown that the fixed point contribution (44) from F is given by

$$\frac{1}{\#Z^2}\frac{\#T_{k+h^\vee}}{|J(t_\lambda)|^2}\,\phi_1(c_1)\phi_2(c_2)\delta(c_1,c_2).$$

We can use the S-matrix to write $\frac{\#T_{k+h^\vee}}{|J(t_\lambda)|^2}=S_{\lambda,0}^{-2}$. Summing over all fixed point contributions, we then obtain

$$\mathscr{Q}(D(G'))(t_\lambda)=\frac{1}{\#Z^2}S_{\lambda,0}^{-2}\sum_{\substack{c_1,c_2\in Z\\ c_i\bullet_k\lambda=\lambda}}\phi_1(c_1)\phi_2(c_2)\delta(c_1,c_2).$$

Therefore,

$$\mathscr{Q}(D(G'))=\frac{1}{\#Z^2}\sum_{\lambda\in P_k}\Big(\sum_{\substack{c_1,c_2\in Z\\ c_i\bullet_k\lambda=\lambda}}\phi_1(c_1)\phi_2(c_2)\delta(c_1,c_2)\Big)S_{\lambda,0}^{-2}\,\widetilde{\tau}_\lambda$$

as an element of $R_k(G)$. Note that the condition $c_i\bullet_k\lambda=\lambda$ in this formula comes from the structure of the fixed point set, Proposition 4.8.

5 Moduli spaces for nonsimply connected groups

5.1 *Fuchs-Schweigert formulas for* $M_{G'}(\Sigma_{\mathrm{g}}^1,\mu)$

Let Σ_{g}^1 be the compact, oriented surface of genus g with one boundary component, and $M_{G'}(\Sigma_{\mathrm{g}}^1)=\mathrm{Hom}(\pi_1(\Sigma_{\mathrm{g}}^1),G')$ viewed as a quasi-Hamiltonian G-space. We have

$$M_{G'}(\Sigma_{\mathrm{g}}^1)=D(G')\times\cdots\times D(G')$$

(the g-fold fusion product). Since the quantization of a fusion product is just the product of the quantizations, we arrive at the following formula for a surface of genus g with one boundary component

$$\mathscr{Q}(M_{G'}(\Sigma_{\mathrm{g}}^1))=\frac{1}{\#Z^{2\mathrm{g}}}\sum_{\lambda\in P_k}\sum_{\substack{c_1,\ldots,c_{2\mathrm{g}}\in Z\\ c_i\bullet_k\lambda=\lambda}}\prod_{i=1}^{2\mathrm{g}}\phi_i(c_i)\prod_{j=1}^{g}\delta(c_{2j-1},c_{2j})S_{\lambda,0}^{-2\mathrm{g}}\,\widetilde{\tau}_\lambda\,.$$

Interchanging the two summations, we obtain

$$\mathcal{Q}(M_{G'}(\Sigma_{\mathrm{g}}^1)) = \frac{1}{\#Z^{2\mathrm{g}}} \sum_{c_1,\ldots,c_{2\mathrm{g}}} \prod_{i=1}^{2\mathrm{g}} \phi_i(c_i) \prod_{j=1}^{\mathrm{g}} \delta(c_{2j-1},c_{2j}) \sum_{\substack{\lambda\in P_k\\ c_i\bullet_k\lambda=\lambda}} S_{\lambda,0}^{-2\mathrm{g}}\, \widetilde{\tau}_\lambda\,.$$

(To be precise, the first sum is only over those c_i whose action on P_k has at least one common fixed point.) The reduced space at a value $\exp(\mu/k)$ is the moduli space $\mathcal{Q}(M_{G'}(\Sigma_{\mathrm{g}}^1,\mu))$ of flat G-bundles with boundary holonomy in the conjugacy class $\mathscr{C} = G.\exp(\mu/k)$. It inherits a prequantization whenever $\mu \in P_k$. According to the 'quantization commutes with reduction theorem' for group-valued moment maps (see [4] and [25]), the corresponding quantization is the multiplicity of τ_μ in $\mathcal{Q}(M_{G'}(\Sigma_{\mathrm{g}}^1))$. We hence arrive at the following version of the Fuchs-Schweigert formulas.

Theorem 5.1. *Let G be a compact, simple, simply connected Lie group, $Z \subseteq Z(G)$ a finite subgroup of the center, and $G' = G/Z$. Let $k \in \mathbb{N}$ be such that $B = kB_{\mathrm{basic}}$ takes on integer values on the lattice Λ_Z. Let Σ_{g}^1 be a compact oriented surface of genus* g *with one boundary component, and consider the prequantization indexed by $\phi \in \mathrm{Hom}(Z^{2\mathrm{g}},\mathrm{U}(1))$. Then*

$$\mathcal{Q}(M_{G'}(\Sigma_{\mathrm{g}}^1,\mu)) = \frac{1}{\#Z^{2\mathrm{g}}} \sum_{c_1,\ldots,c_{2\mathrm{g}}} \prod_{i=1}^{2\mathrm{g}} \phi_i(c_i) \prod_{j=1}^{\mathrm{g}} \delta(c_{2j-1},c_{2j}) \sum_{\substack{\lambda\in P_k\\ c_i\bullet_k\lambda=\lambda}} S_{\lambda,0}^{1-2\mathrm{g}}\, S_{\lambda,*\mu}$$

for all $\mu \in P_k$.

The term where all $c_i = e$ may be regarded as the 'leading term'. It is just the corresponding term for the group G, divided by $\#Z^{2\mathrm{g}}$:

$$\frac{1}{\#Z^{2\mathrm{g}}}\mathcal{Q}(M_G(\Sigma_{\mathrm{g}}^1,\mu)).$$

The terms where at least one $c_i \neq e$ involve a summation over a proper subset of P_k, and may be seen as 'correction terms'.

Remark 5.2. In some very special cases, the correction terms are zero. In particular, this then means that the integers $\mathcal{Q}(M_G(\Sigma_{\mathrm{g}}^1,\mu))$ must be divisible by $\#Z^{2\mathrm{g}}$. One such example, discussed in Appendix B, occurs for $G = A_l$ and $Z = \mathbb{Z}_m$ where m is a prime number dividing $l+1$. Then $k_0 = 1$, but unless k is a multiple of m the action of Z on P_k has trivial stabilizers. There are similar examples $G = D_l$, and suitable k, l and $Z \cong \mathbb{Z}_2 \subseteq \mathbb{Z}(G)$, as explained in Appendix B.

5.2 *The case* $G' = \mathrm{PU}(n)$*, with* n *prime*

An interesting case, considered by Beauville [7] in the algebro-geometric setting, is that of $G' = \mathrm{PU}(n)$ with n an odd prime number. Thus $G = \mathrm{SU}(n)$ with $Z = Z(G) = \mathbb{Z}_n$. A special feature of this situation is that the action of Z on the alcove has $\zeta_* = \frac{\rho}{h^\vee}$ as its unique fixed point, and is *free* away from this fixed point. We recall that $k_0 = n$, and that $\delta = 1$ (since $G = A_l$ where $l = n-1$ is even.) Suppose k is a multiple of n, so that $M_{G'}(\Sigma^1_{\mathrm{g}})$ is prequantizable at level k; the inequivalent prequantizations are indexed by $\phi = (\phi_1, \dots, \phi_{2\mathrm{g}}) \in \mathrm{Hom}(Z^{2\mathrm{g}}, \mathrm{U}(1))$. In the general formula, the 'leading term' coming from $c_1 = \ldots = c_{2\mathrm{g}} = e$ is $(\#Z)^{-2\mathrm{g}} M_G(\Sigma)$. If any $c_i \neq e$, then the unique level k weight fixed under $c_1, \dots, c_{2\mathrm{g}}$ is

$$\lambda_* = (k+n)\zeta_* = \frac{k}{n}\rho \in P_k.$$

We hence obtain

$$\mathscr{Q}(M_{G'}(\Sigma^1_{\mathrm{g}})) = \frac{1}{n^{2\mathrm{g}}}\left(\mathscr{Q}(M_G(\Sigma)) + (x-1) S_{\lambda_*,0}^{-2\mathrm{g}} \widetilde{\tau}_{\lambda_*}\right)$$

where

$$x = \sum_{c_1,\dots,c_{2\mathrm{g}}} \prod_{i=1}^{2\mathrm{g}} \phi_i(c_i) = \prod_{i=1}^{2\mathrm{g}} \left(\sum_{c \in \mathbb{Z}_n} \phi_i(c) \right) = \begin{cases} n^{2\mathrm{g}} & \text{if all } \phi_i = 1, \\ 0 & \text{if some } \phi_i \neq 1. \end{cases}$$

To read off multiplicities, we still need to express $\widetilde{\tau}_{\lambda_*}$ in terms of the standard basis of the fusion ring. For the following, see Theorem 2, Lemma 3.4.2 and Lemma 3.5.2 in [21].

Proposition 5.3 (Kostant [21]). *Let G be simple, simply connected and simply laced (i.e., $Q = Q^\vee$). Then*

(47) $$|J(t_*)|^2 = \#Z(G)\, h^l.$$

Furthermore, for $\mu \in P$ we have that

(48) $$\varepsilon(\mu) := \chi_\mu(t_*) \in \{-1, 0, 1\};$$

in fact, $\varepsilon(\mu) = (-1)^{l(w)}$ if there exists $w \in W$ with $w(\mu+\rho) - \rho \in hQ$, and 0 otherwise.

Equation (47) allows us to rewrite the square of the S-matrix element S_{0,λ_*} as

$$S^2_{0,\lambda_*} = \frac{|J(t_*)|^2}{\#T_{k+h}} = \frac{h^l}{(k+h)^l} = \left(\frac{k}{h}+1\right)^{-l}.$$

Define an element

(49) $$\tau_\natural = \sum_{\mu \in P_k} \varepsilon(\mu) \tau_\mu \in R_k(G)$$

with $\varepsilon(\mu)$ as in (48).

Lemma 5.4. *Suppose $k \in \mathbb{N}$ is such that $\lambda_* = k\zeta_* = \frac{k}{h^\vee}\rho \in P_k$. Then*

$$\widetilde{\tau}_{\lambda_*} = S^2_{0,\lambda_*}\, \tau_{\natural}.$$

Proof. By definition, $\varepsilon(\mu) = \tau_\mu(t_*) = \tau_\mu(t_{\lambda_*})$. In terms of the S-matrix, this is (see (29)) $S_{\mu,\lambda_*}/\,S_{0,\lambda_*}$. Together with the change-of-basis formulas (27), it follows that

$$\tau_{\natural} = \sum_{\mu,\nu} \frac{S_{\mu,\lambda_*}}{S_{0,\lambda_*}} \frac{S_{*\mu,\nu}}{S_{0,\nu}} \widetilde{\tau}_{*\nu} = \sum_{\nu} \frac{\delta_{\nu,\lambda_*}}{S_{0,\lambda_*} S_{0,\nu}} \widetilde{\tau}_{*\nu} = \frac{\widetilde{\tau}_{\lambda_*}}{S^2_{0,\lambda_*}}$$

where we used that the S-matrix is symmetric and unitary. □

Consequently, we have found that

$$S^{-2g}_{\lambda_*,0}\, \widetilde{\tau}_{\lambda_*} = \left(\frac{k}{h}+1\right)^{l(g-1)} \tau_{\natural}.$$

Returning to the case at hand, with $G = \mathrm{SU}(n)$, $h = n$ and $l = n-1$, we obtain:

Proposition 5.5. *Let $G = \mathrm{SU}(n)$, where n is prime, and $Z = Z(G)$ so that $G' = \mathrm{PU}(n)$, and let k be a multiple of n. For the prequantization given by*

$$\phi = (\phi_1, \dots, \phi_{2g}) \in \mathrm{Hom}(Z^{2g}, \mathrm{U}(1)),$$

we have that

$$\mathscr{Q}(M_{G'}(\Sigma^1_g)) = \frac{1}{n^{2g}} \left(\mathscr{Q}(M_G(\Sigma^1_g)) + (n^{2g}-1)\left(\frac{k}{n}+1\right)^{(n-1)(g-1)} \tau_{\natural} \right)$$

if all $\phi_i = 1$, while

$$\mathscr{Q}(M_{G'}(\Sigma^1_g)) = \frac{1}{n^{2g}} \left(\mathscr{Q}(M_G(\Sigma^1_g)) - \left(\frac{k}{n}+1\right)^{(n-1)(g-1)} \tau_{\natural} \right)$$

if any $\phi_i \neq 1$.

For $\mu \in P_k$, the integers $\mathscr{Q}(M_{G'}(\Sigma^1_g, \mu))$ are obtained from this formula as the coefficient of τ_μ, i.e., replacing $\mathscr{Q}(M_G(\Sigma^1_g))$ with $\mathscr{Q}(M_G(\Sigma^1_g, \mu))$ and $\tau_{\natural}$ with $\varepsilon(\mu)$. For $\mu = 0$ this is exactly the formulas given by Beauville [7, Proposition 3.3].

Remark 5.6. Note the following consequence for the Verlinde numbers $N(\mu) = \mathscr{Q}(M_G(\Sigma^1_g, \mu))$ with respect to $G = \mathrm{SU}(n)$: Whenever n is prime and the level k is a multiple of n, we have that

$$N(\mu) = \left(\frac{k}{n}+1\right)^{(n-1)(g-1)} \varepsilon(\mu) \mod n^{2g}$$

for all $\mu \in P_k$.

Appendix A Prequantization of nonsimply connected manifolds

In this section, we discuss the prequantizability of Hamiltonian and quasi-Hamiltonian G-spaces with an abelian fundamental group. We begin with some well known results about the prequantizability of closed 2-forms.

A.1 Prequantization of closed 2-forms

Let M be a connected manifold with a closed 2-form $\omega \in \Omega^2(M)$. Following Kostant [20] and Souriau [29], one defines a *prequantization* of (M,ω) to be a Hermitian line bundle L with Hermitian connection ∇ whose Chern form agrees with ω. The quotient of any two prequantum line bundles of (M,ω) is a Hermitian line bundle with a flat connection; the holonomy map for the connection gives a classification of flat bundles by homomorphisms $\pi_1(M) \to \mathrm{U}(1)$, or equivalently by the torsion subgroup of $H^2(M,\mathbb{Z})$.

Remark A.1. Given a Hamiltonian G-action on M, with an equivariant moment map $\Phi\colon M \to \mathfrak{g}^*$ satisfying $\iota(\xi_M)\omega = -\mathrm{d}\langle\Phi,\xi\rangle$, there is a corresponding G-equivariant version of this condition. The infinitesimal lift of the G-action is given by Kostant's formula [20]. If G is simply connected, this infinitesimal action always integrates to a G-action on L. Hence, to discuss prequantizability of Hamiltonian G-spaces with simply connected G, it suffices to consider the non-equivariant situation.

By Chern-Weil theory, a prequantization of (M,ω) exists if and only if the 2-form is integral, in the sense that its de Rham cohomology class $[\omega]$ lies in the image of the coefficient homomorphism $H^2(M,\mathbb{Z}) \to H^2(M,\mathbb{R})$. Equivalently, for every closed oriented surface Σ and every map $f\colon \Sigma \to M$,

$$\int_\Sigma f^*\omega \in \mathbb{Z}. \tag{50}$$

It suffices to check this condition on any collection of maps such that the classes $f_*[\Sigma]$ span the second homology $H_2(M,\mathbb{Z})$. If M is 1-connected, it follows from the Hurewicz theorem that $H_2(M,\mathbb{Z})$ is generated by *spherical* homology classes; hence one only needs to check for all maps $f\colon S^2 \to M$ from 2-spheres. Furthermore, in this case the prequantization is unique up to isomorphism.

In [20], Kostant also discusses the situation where M is connected but not simply connected. Clearly, a necessary condition for prequantizability is that the pull-back of ω to the universal cover, $\widetilde{\omega} \in \Omega^2(\widetilde{M})$, is prequantizable. Letting $\widetilde{L}$ be the prequantum line bundle, he shows that the group of connection preserving automorphisms of $\widetilde{L}$ define a $\mathrm{U}(1)$-central extension of the group $\mathrm{Diff}_\omega(\widetilde{M})$ of 2-form preserving diffeomorphism of $\widetilde{M}$. Pullback under the inclusion of $\pi_1(M)$ defines a central extension

$$1 \to \mathrm{U}(1) \to \widehat{\pi_1(M)} \to \pi_1(M) \to 1;$$

its triviality is necessary and sufficient for $\widetilde{L}$ to descend to a prequantum line bundle L of M.

Let us consider the special case that the fundamental group $\pi_1(M)$ is *abelian*. Recall that equivalence classes of U(1)-central extensions $\widehat{A}$ of a finitely generated abelian group A are in 1-1 correspondence with *commutator maps*

$$q\colon A\times A\to \mathrm{U}(1), \tag{51}$$

that is, maps with the properties $q(a,a)=1$ and

$$q(a_1+a_2,a)=q(a_1,a)\,q(a_2,a),\quad q(a,a_1+a_2)=q(a,a_1)\,q(a,a_2)$$

for all $a,a_1,a_2\in A$. Given $\widehat{A}$, the associated commutator map is given by

$$q(a_1,a_2)=\widehat{a_1}\widehat{a_2}\widehat{a_1}^{-1}\widehat{a_2}^{-1}$$

where $\widehat{a_i}\in\widehat{A}$ are lifts of $a_i\in A$. (Given a commutator map q, choose generators $a_1,\dots,a_r$ for A, define $c\colon A\times A\to B$ by $c(a_i,a_j)=q(a_i,a_j)$ for $i>j$ and $c(a_i,a_j)=e$ for $i\le j$, and extend bi-additively. Then c is a cocycle, and the resulting central extension has commutator map q.)

Thus, if $\omega\in\Omega^2(M)$ is a closed 2-form whose pull-back to $\widetilde{M}$ is prequantizable, then the central extension defined by Kostant defines a commutator map for $\pi_1(M)$. We will use the following alternative description of this commutator map, directly in terms of ω.

Proposition A.2. *Let M be a connected manifold with abelian fundamental group $\pi_1(M)$, and let $\omega\in\Omega^2(M)$ be a closed 2-form whose pull-back to the universal cover $\widetilde{M}$ is integral.*

(a) *There is a well-defined commutator map $q\colon \pi_1(M)\times\pi_1(M)\to \mathrm{U}(1)$ given by*

$$q(u,v)=\exp\Big(2\pi i\int_{S^1\times S^1} f^*\omega\Big).$$

Here $f\colon S^1\times S^1\to M$ for given $u,v\in\pi_1(M)$ is any smooth map such that the induced map on fundamental groups $f_*\colon \pi_1(S^1\times S^1)=\mathbb{Z}\times\mathbb{Z}\to\pi_1(M)$ satisfies $f_*(k,l)=ku+lv$.

(b) *The pair (M,ω) is prequantizable if and only if $q=1$.*

Proof. (a) To show that f with the desired properties exists, let us regard the 2-torus $S^1\times S^1$ as obtained from a square $[0,1]^2$, by gluing the sides according to the pattern $\alpha,\beta,\alpha^{-1},\beta^{-1}$. Given $u,v\in\pi_1(M)$, choose based loops $\gamma_u,\gamma_v\in C^\infty(S^1,M)$ taking the generator of $\pi_1(S^1)=\mathbb{Z}$ to u,v, respectively. This defines a map f_0 from the boundary of the square to M, given as the concatenation of loops $\gamma_u\gamma_v\gamma_u^{-1}\gamma_v^{-1}$. Since $\pi_1(M)$ is abelian, this loop is contractible, hence f_0 extends to a continuous map from the square, defining a continuous map $f\colon S^1\times S^1\to M$ with the desired property on $\pi_1(M)$. By standard techniques, it can be deformed into a smooth map.

Given two maps $f, f' \colon S^1 \times S^1 \to M$ inducing the same map on π_1, we can smoothly deform to arrange that $f_0 = f_0'$. The maps from f, f' correspond to two different extensions to the square $[0,1]^2$. Since ω takes on integer values on spherical homology classes, the resulting integrals $\int_{S^1\times S^1} f^*\omega$, $\int_{S^1\times S^1} (f')^*\omega$ differ by an integer. It is clear from the definition that the map q is a commutator map.

(b) Since $\pi_1(M)$ is abelian, the homology group $H_2(M,\mathbb{Z})$ is generated by images of 2-tori. (Regard a surface Σ of genus g as a connected sum of 2-tori, glued along circles $S_j \subseteq \Sigma$. Then S_j represents an element of $[\pi_1(\Sigma), \pi_1(\Sigma)]$, hence its image under a map $f \colon \Sigma \to M$ is contractible. Choosing such retractions, we have deformed f into a map sending each S_j to a point, thus effectively to a map from a wedge of 2-tori.) Hence, to check integrality of ω it suffices to verify (50) for 2-tori, which is precisely the condition that $q = 1$. □

A.2 Integrality of closed relative forms

To describe the prequantizability of quasi-Hamiltonian G-spaces, we need an extension of this discussion to *relative* forms. Suppose M, N are manifold, $\Phi \colon M \to N$ is a smooth map, and $(\omega, \eta) \in \Omega^3(\Phi) = \Omega^2(M) \oplus \Omega^3(N)$ relatively closed, that is,

$$\mathrm{d}\omega = -\Phi^*\eta, \ \ \mathrm{d}\eta = 0.$$

The class of (ω, η) in relative cohomology is integral if its natural pairing with all classes in $H_3(\Phi, \mathbb{Z})$ is integral. If N is 2-connected, this condition means that for every map $f \colon \Sigma \to M$ from a compact oriented surface, and every smooth homotopy $h \colon \Sigma \times [0,1] \to N$ between $h_1 = \Phi \circ f$ and a constant map h_0 (where we write $h_t = h(\cdot, t)$), we have that

$$\int_\Sigma f^*\omega + \int_{\Sigma\times[0,1]} h^*\eta \in \mathbb{Z}. \tag{52}$$

(Observe that by Stokes' theorem, the left-hand side is invariant under homotopies of the pair of maps f and h.) Applying this criterion to a constant map f, we see in particular that the closed 3-form η must be integral. Using again that $H_2(M,\mathbb{Z})$ is generated by images of 2-spheres (resp., of 2-tori) if M is 1-connected (resp., $\pi_1(M)$ is abelian), we have:

Lemma A.3. *Let $\Phi \colon M \to N$ be a smooth map, where N is 2-connected, and $(\omega, \eta) \in \Omega^3(\Phi)$ a closed relative 3-form such that η is integral.*

(a) *If M is 2-connected, then (ω, η) is integral.*
(b) *If M is 1-connected, then (ω, η) is integral if and only if* (52) *holds for all maps from a 2-sphere $\Sigma = S^2$.*
(c) *If M is connected and $\pi_1(M)$ is abelian, then (ω, η) is integral if and only if* (52) *holds for all maps from a 2-torus $\Sigma = S^1 \times S^1$.*

Repeating the argument from the proof of Proposition A.2, we obtain:

Proposition A.4. *Let M be a connected manifold with an abelian fundamental group $\pi_1(M)$, and $\Phi\colon M \to N$ a smooth map into a 2-connected manifold N. Suppose that $(\omega,\eta) \in \Omega^3(\Phi)$ is a relative cocycle whose pull-back $(\widetilde{\omega},\eta) \in \Omega^3(\widetilde{\Phi})$ to the universal cover $\widetilde{M}$ is integral.*

(a) *There is a well-defined commutator map $q\colon \pi_1(M)\times\pi_1(M) \to \mathrm{U}(1)$ given by*

$$q(u,v) = \exp\Big(2\pi i\big(\int_{S^1\times S^1} f^*\omega + \int_{S^1\times S^1\times[0,1]} h^*\eta\big)\Big).$$

Here $f\colon S^1\times S^1 \to M$ is any smooth map such that the induced map on fundamental groups takes the generators to u,v respectively, and h is a smooth homotopy between $h_1 = \Phi\circ f$ and a constant map h_0.

(b) *The relative cocycle (ω,η) is integral if and only if $q=1$.*

The following observations sometimes simplify the use of this criterion:

Remark A.5. (a) The condition that h_0 be a constant map can be weakened. For example, it is enough to require that h_0 takes values in a 1-dimensional submanifold.

(b) If f can be chosen in such a way that $\Phi\circ f$ is constant, then one can take h to be the trivial extension to $S^1\times S^1\times[0,1]$, and the second integral disappears.

Appendix B On the action of Z on P_k

Let G be compact, simple and simply connected, $Z\subseteq Z(G)$, and k a multiple of the basic level k_0. In this section we will prove the following fact:

Lemma B.1. *Suppose $c\in Z$ is an element whose action on P_k has a fixed point, and $\zeta\in\Lambda_Z$ with $\exp\zeta = c$. Then*

$$k\,\zeta\cdot P^\vee \subseteq \mathbb{Z}. \tag{53}$$

Note that since $Q^\vee\cdot P^\vee\subseteq\mathbb{Z}$, it suffices to check the condition for *any* ζ exponentiating to c.

Proof. If $k_1 = k_0$, there is nothing to prove (by definition of k_1). We examine the cases where $k_1 > k_0$, using the explicit description of the action of $Z(G)$ on level k weights, as described in [31].

(1) Let $G = D_l$, with the standard identification $\mathfrak{t}\cong\mathfrak{t}^*\cong\mathbb{R}^l$. The lattice P consists of all $\sum_i \mu_i e_i$ with $\mu_i\in\frac{1}{2}\mathbb{Z}$, such that $2\mu_1,\ldots,2\mu_l$ are all either even or all odd. We will show that if k is a multiple of k_0 but not of k_1, then the action of $c\in Z$ on P_k has no fixed points.

(1a) Let Z be generated by $c=\exp(\varpi_1^\vee)$, so that $k_0=1,\ k_1=2$. We claim that unless k is even (and hence a multiple of k_1), the fixed point set of c on P_k is empty. Indeed, this action is given by

$$\mu = (\mu_1,\ldots,\mu_l)\mapsto(k-\mu_1,\mu_2,\ldots,\mu_{l-1},-\mu_l),$$

hence μ is fixed if and only if $2\mu_1 = k$ and $\mu_l = 0$. In particular, $2\mu_l = 0$ is even, $2\mu_1$ must be even (by the description of P). Hence, there is no solution when k is odd.

(1b) Suppose l is divisible by 4, and Z is generated by $c = \exp(\varpi_{l-1}^\vee)$. Again, $k_0 = 1,\ k_1 = 2$. This time, the action on level k weights reads as

$$(\mu_1,\ldots,\mu_l) \mapsto \left(\frac{k}{2}+\mu_l, \frac{k}{2}-\mu_{l-1},\ldots,\frac{k}{2}-\mu_2, \frac{k}{2}-\mu_1\right).$$

The fixed point condition reads as $\mu_1 = \frac{k}{2}+\mu_l,\ldots$. Again, since the integers $2\mu_i$ must have the same parity, there is no solution when k is odd.

(1c) Suppose l is divisible by 4, and Z is generated by $c = \exp(\varpi_l^\vee)$, hence $k_0 = 1,\ k_1 = 2$. The action on level k weights reads as

$$(\mu_1,\ldots,\mu_l) \mapsto \left(\frac{k}{2}-\mu_l, \frac{k}{2}-\mu_{l-1},\ldots,\frac{k}{2}-\mu_2, \frac{k}{2}-\mu_1\right);$$

the same argument as in (b) shows that there are no fixed points when k is odd.

(2) Consider next $G = A_l$. The weight lattice P is identified with the quotient of $\mathbb{Z}^{l+1}$ by the rank one sublattice $\mathbb{Z}(e_1+\ldots+e_{l+1})$. The center $Z(G)$ is generated by $c_0 = \exp(\varpi_1^\vee)$, using the labeling of coweights as in Bourbaki [10]. The action of the generator c_0 on P_k is given by

$$(\mu_1,\ldots,\mu_{l+1}) \mapsto (k+\mu_{l+1},\mu_1,\ldots,\mu_l) \mod \mathbb{Z}(e_1+\ldots+e_{l+1}). \tag{54}$$

The action of some power c_0^N on P_k involves N such shifts. Note that the sum of coefficients $\mu_1+\ldots+\mu_{l+1}$ of the expression in parentheses changes by Nk, while the sum of coefficients of $e_1+\ldots+e_{l+1}$ is $l+1$. Hence, if the equivalence class of $(\mu_1,\ldots,\mu_{l+1})$ is a fixed point for the action of c_0^N, then Nk must be a multiple of $l+1$. With these initial observations, let $Z = \mathbb{Z}_m \subseteq Z(G) = \mathbb{Z}_{l+1}$ where m divides $l+1$. Recall that k_0 is the smallest natural number such that $k_0(l+1)$ is a multiple of m^2, while $k_1 = m$. Put $n = (l+1)/m$, so that Z is generated by $c_0^n = \exp(n\varpi_1^\vee)$. By the above, the fixed point set of $c := c_0^{nr} \in Z$ on P_k is empty unless nrk is a multiple of $l+1$, hence suppose that this is the case. The element $\zeta = nr\varpi_1 \in \Lambda_Z$ exponentiates to c^{nr}, and we obtain

$$k\zeta\cdot P^\vee = knr\varpi_1^\vee\cdot P^\vee \subseteq (l+1)\,\varpi_1^\vee\cdot P^\vee \subseteq \mathbb{Z},$$

as desired. □

Remark B.2. The discussion for $G = A_l = \mathrm{SU}(l+1)$ shows in particular the following: Suppose m is a prime number such that m^2 divides $l+1$, and let $Z = \mathbb{Z}_m$ so that $k_0 = 1$. Then, unless k is a multiple of m, the action of Z on P_k has trivial stabilizers.

Appendix C Coxeter transformations

In this section, we carry out several case-by-case calculations of the phase factors $\delta(c_1,c_2)$. Since δ is multiplicative in both entries, it suffices to compute it on generators. Our calculations will follow the tables in Bourbaki [10], and we will largely use the conventions given there, e.g., for enumeration of simple roots and fundamental weights. We will take the Coxeter element as the product of simple reflections in the order $w_* = s_l \cdots\cdots s_1$.

(a) Let $G = A_l$, with its standard realization of $\mathfrak{t}$ realized as the subspace of $\mathbb{R}^{l+1}$ where the coordinates add to zero, and $Q = Q^\vee = \mathfrak{t}\cap\mathbb{Z}^{l+1}$. The center is generated by $c_0 = \exp(\varpi_1^\vee)$, with the standard enumeration of the fundamental (co)weights as in [10]. The Coxeter element w_* acts by cyclic permutation, $e_i \mapsto e_{i+1}$ for $i < l+1$ and $e_{l+1} \mapsto e_1$. One finds that $\zeta_* = \frac{1}{h^\vee}\rho$ satisfies $w_*\zeta_* = \zeta_* - \varpi_1$; hence

$$(1-w_*)^{-1}\varpi_1 = \frac{1}{l+1}\rho = \frac{1}{l+1}\Big(\frac{l}{2}\alpha_1 + \ldots + \frac{i}{2}(l-i+1)\alpha_i + \ldots + \frac{l}{2}\alpha_l\Big).$$

Consequently,

$$(1-w_*)^{-1}\varpi_1^\vee \cdot \varpi_1^\vee = \frac{l}{2(l+1)}.$$

If $Z = Z(G)$, and if k is a multiple of $k_0 = l+1$, this gives

$$\delta(c_0,c_0) = (-1)^{l\frac{k}{l+1}}.$$

More generally, if m divides $l+1$, with quotient $(l+1)/m = n$, and $Z = \mathbb{Z}_m$ generated by $c_1 = c_0^n$, and if k is a multiple of k_0 so that $k(l+1)$ is a multiple of m^2, we obtain:

$$\delta(c_1,c_1) = \exp\Big(2\pi i\frac{kl(l+1)}{m^2}\Big) = (-1)^{l\frac{k(l+1)}{m^2}}.$$

(b) Let $G = B_l$, with the standard realization of the root system in $\mathfrak{t} = \mathbb{R}^l$ such that $Q = \mathbb{Z}^l$, and $Q^\vee \subseteq \mathbb{Z}^l$ are the integral points whose coefficient sum is even. The center $Z(G) = \mathbb{Z}_2$ is generated by the element $c_0 = \exp(\varpi_1^\vee)$ with $\varpi_1^\vee = e_1$, and the Coxeter transformation w_* is given by $e_i \mapsto e_{i+1}$ for $i < l$ and $e_l \mapsto -e_1$. One finds $(1-w_*)^{-1}\varpi_1^\vee = \frac{1}{2}(e_1 + \ldots + e_l)$, hence

$$(1-w_*)^{-1}\varpi_1^\vee \cdot \varpi_1^\vee = \frac{1}{2}.$$

It follows that for any level k,

$$\delta(c_0,c_0) = (-1)^k.$$

(c) Let $G=C_l$. The root system is obtained from that of B_l by interchanging the roles of Q and $Q^\vee$; with the same Weyl group and the same Coxeter element. Note however that the basic inner product for C_l is twice of that of B_l, and $\varpi_1^\vee = e_1$. We obtain,

$$(1-w_*)^{-1}\varpi_1^\vee\cdot\varpi_1^\vee = 1.$$

It follows that $\delta(c_0,c_0)=1$ for all k.

(d) Let $G=D_l$, with $\mathfrak{t}\cong\mathbb{R}^l$. The lattice $Q=Q^\vee$ consists of elements of $\mathbb{Z}^l$ with even coefficient sum, while $P=P^\vee$ is spanned by $\mathbb{Z}^l$ together with $\varpi_l^\vee = \frac{1}{2}(e_1+\ldots+e_l)$. The Coxeter element w_* acts by

$$e_1\mapsto e_2,\ldots,e_{l-2}\mapsto e_{l-1},\ e_{l-1}\mapsto -e_1,\ e_l\mapsto -e_l.$$

If l is odd, then $Z=\mathbb{Z}_4$ is generated by $c_0=\exp(\varpi_l)$. One computes

$$(1-w_*)^{-1}\varpi_l = \frac{1}{2}(e_1+2e_2+\ldots+(l-1)e_{l-1})+\frac{1}{4}e_l-\frac{l-1}{4}(e_1+\ldots+e_{l-1}),$$

and hence

$$(1-w_*)^{-1}\varpi_l^\vee\cdot\varpi_l^\vee = \frac{1}{8}(-l^2+3l+1).$$

Hence, for $Z=\mathbb{Z}_4$ and k a multiple of $k_0=4$,

$$\delta(c_0,c_0)=\exp\Big(2\pi i\frac{k}{8}(-l^2+3l+1)\Big)=(-1)^{\frac{k}{4}(-l^2+3l+1)}=(-1)^{\frac{k}{4}}$$

(since $-l^2+3l$ is even). If $Z=\mathbb{Z}_2$, generated by $c_1=c_0^2$, one has $k_0=1$, and $\delta(c_1,c_1)=(-1)^k$.

If l is even, then the center if generated by $c_0=\exp(\varpi_l^\vee)$ and $c_0'=\exp(\varpi_1^\vee)$ with $\varpi_1^\vee=e_1$. The expression for $(1-w_*)^{-1}\varpi_l^\vee$ is as before, while $(1-w_*)^{-1}\varpi_1^\vee=\frac{1}{2}(e_1+\ldots+e_{l-1})$. Hence

$$(1-w_*)^{-1}\varpi_l^\vee\cdot\varpi_1^\vee = \frac{3-l}{4},\ (1-w_*)^{-1}\varpi_1^\vee\cdot\varpi_1^\vee=\frac{1}{2}.$$

Hence, if $Z=Z(G)=\mathbb{Z}_2\times\mathbb{Z}_2$, and k a multiple of $k_0=2$,

$$\delta(c_0,c_0)=i^{\frac{k}{2}}(-1)^{\frac{kl}{4}},\ \delta(c_0,c_0')=(-i)^k,\ \delta(c_0',c_0')=1.$$

References

1. A. Alekseev, A. Malkin, and E. Meinrenken, *Lie group valued moment maps*, J. Differential Geom. **48** (1998), no. 3, 445–495.
2. A. Alekseev and E. Meinrenken, *Dirac structures and Dixmier-Douady bundles*, International Mathematics Research Notices **2012** (2012), no. 4, 904–956.

3. A. Alekseev, E. Meinrenken, and C. Woodward, *Formulas of Verlinde type for non simply-connected groups*, unfinished manuscript, available at arXiv:math/0005047.
4. A. Alekseev, E. Meinrenken, and C. Woodward, *The Verlinde formulas as fixed point formulas*, J. Symplectic Geom. **1** (2001), no. 1, 1–46.
5. M. F. Atiyah and R. Bott, *The Yang-Mills equations over Riemann surfaces*, Phil. Trans. Roy. Soc. London Ser. A **308** (1982), 523–615.
6. A. Beauville, *Conformal blocks, fusion rules and the Verlinde formula*, Proceedings of the Hirzebruch 65 Conference on Algebraic Geometry (Ramat Gan, 1993), Israel Math. Conf. Proc., no. 9, 1996, pp. 75–96.
7. ———, *The Verlinde formula for PGL(p)*, The mathematical beauty of physics (Saclay, 1996), Adv. Ser. Math. Phys., Vol. 24, World Sci. Publishing, 1997, pp. 141–151.
8. J.M. Bismut and F. Labourie, *Symplectic geometry and the Verlinde formulas*, Surveys in differential geometry: differential geometry inspired by string theory, Int. Press, Boston, MA, 1999, pp. 97–311.
9. P. Boalch, *Quasi-Hamiltonian Geometry of Meromorphic Connections*, Duke Math. J. **139** (2007), 369–405.
10. N. Bourbaki, *Lie groups and Lie algebras. Chapters 4–6*, Elements of Mathematics (Berlin), Springer-Verlag, Berlin, 2002.
11. T. Bröcker and T. tom Dieck, *Representations of Compact Lie Groups*, Graduate Texts in Mathematics, Vol. 98, Springer-Verlag, Berlin-Heidelberg-New York, 1985.
12. W. G. Dwyer and C. W. Wilkerson, *Centers and Coxeter elements*, Homotopy methods in algebraic topology (Boulder, CO, 1999), Contemp. Math., Vol. 271, Amer. Math. Soc., Providence, RI, 2001, pp. 53–75.
13. A. Eshmatov, *A new example of a group-valued moment map*, Journal of Lie Theory **19** (2009), 395–407.
14. G. Faltings, *A proof for the Verlinde formula*, J. Alg. Geom. **3** (1990), 347–374.
15. P. Di Francesco, P. Mathieu, and D. Sénéchal, *Conformal field theory*, Graduate Texts in Contemporary Physics, Springer-Verlag, New York, 1997.
16. J. Fuchs and C. Schweigert, *The action of outer automorphisms on bundles of chiral blocks*, Comm. Math. Phys. **206** (1999), 691–736.
17. J. Hurtubise, L. Jeffrey, and R. Sjamaar, *Group-valued implosion and parabolic structures*, Amer. J. Math. **128** (2006), no. 1, 167–214.
18. F. Knop, *Multiplicity free quasi-hamiltonian manifolds*, Preprint (2016), arXiv:1612.03843.
19. B. Kostant, *The principal three-dimensional subgroup and the Betti numbers of a complex simple Lie group*, Amer. J. Math. **81** (1959), 973–1032.
20. ———, *Quantization and unitary representations*, Lectures in Modern Analysis and Applications III (Washington, D.C.) (C. T. Taam, ed.), Lecture Notes in Mathematics, vol. 170, Springer-Verlag, Berlin-Heidelberg-New York, 1970, pp. 87–208.
21. ———, *On Macdonald's η-function formula, the Laplacian and generalized exponents*, Advances in Math. **20** (1976), no. 2, 179–212.
22. D. Krepski, *Pre-quantization of the moduli space of flat G-bundles over a surface*, J. Geom. Phys. **58** (2008), no. 11, 1624–1637.
23. D. Krepski and E. Meinrenken, *On the Verlinde formulas for SO(3)-bundles*, Quarterly Journal of Mathematics **64** (2013), no. 1, 235–252.
24. E. Meinrenken, *Quantization of q-Hamiltonian SU(2)-spaces*, Geometric Aspects of Geometry and Mechanics (in honor of the 65th birthday of Hans Duistermaat), Birkhäuser, 2011, pp. 257–293.
25. ———, *Twisted K-homology and group-valued moment maps*, International Mathematics Research Notices **2012 (20)** (2012), 4563–4618.
26. T. Pantev, *Comparison of generalized theta functions*, Duke Math. J. **76** (1994), no. 2, 509–539.
27. R. Sjamaar, *Singular orbit spaces in Riemannian and symplectic geometry*, Ph.D. thesis, Rijksuniversiteit te Utrecht, 1990.
28. R. Sjamaar and E. Lerman, *Stratified symplectic spaces and reduction*, Ann. of Math. (2) **134** (1991), 375–422.

29. J.-M. Souriau, *Structure des systèmes dynamiques*, Dunod, Paris, 1970.
30. C. Teleman, *Borel-Weil-Bott theory on the moduli stack of G-bundles over a curve*, Invent. Math. **134** (1998), no. 1, 1–57.
31. V. Toledano Laredo, *Positive energy representations of the loop groups of non-simply connected groups*, Comm. Math. Phys. **207** (1999), 307–339.
32. A. Tsuchiya, K. Ueno, and Y. Yamada, *Conformal field theory on universal family of stable curves with gauge symmetries*, Integrable systems in quantum field theory and statistical mechanics, Adv. Stud. Pure Math., vol. 19, Academic Press, 1989, pp. 459–566.

Equivariant Index of Twisted Dirac Operators and Semi-classical Limits

Paul-Émile Paradan and Michèle Vergne

To the memory of Bertram Kostant

Abstract Let G be a compact connected Lie group with Lie algebra $\mathfrak{g}$. Let M be a compact spin manifold with a G-action, and $\mathscr{L}$ be a G-equivariant line bundle on M. Consider an integer k, and let $\mathscr{Q}_G^{\text{spin}}(M, \mathscr{L}^k)$ be the equivariant index of the Dirac operator on M twisted by $\mathscr{L}^k$. Let $m_G(\lambda, k)$ be the multiplicity in $\mathscr{Q}_G^{\text{spin}}(M, \mathscr{L}^k)$ of the irreducible representation of G attached to the admissible coadjoint orbit $G\lambda$. We prove that the distribution $\langle \Theta_k, \varphi \rangle = k^{\dim(G/T)/2} \sum_\lambda m_G(\lambda, k) \langle \beta_{\lambda/k}, \varphi \rangle$ has an asymptotic expansion when k tends to infinity of the form $\langle \Theta_k, \varphi \rangle \equiv k^{\dim M/2} \sum_{n=0}^{\infty} k^{-n} \langle \theta_n, \varphi \rangle$. Here φ is a test function on $\mathfrak{g}^*$ and $\langle \beta_\xi, \varphi \rangle$ is the integral of φ on the coadjoint orbit $G\xi$ with respect to the canonical Liouville measure. We compute explicitly the distribution θ_n in terms of the graded $\hat{A}$ class of M and the equivariant curvature of $\mathscr{L}$.

If M is noncompact, we use these asymptotic techniques to give another proof of the fact that the formal geometric quantization of a manifold with a spinc structure is functorial with respect to restriction to subgroups.

Keywords: Dirac operator • equivariant index • semi-classical limits

Mathematics Subject Classification (2010): 57S15, 53C27, 58J20

This article pursues the fundamental idea of Kostant that a line bundle $\mathscr{L}$ on a G-manifold M, equipped with a G-invariant connection, gives rise to a moment map $\phi_G : M \to \mathfrak{g}^*$, and thus hopefully to a relation between the quantization of M and

Paul-Émile Paradan
IMAG, Université de Montpellier, CNRS, Montpellier, France,
e-mail: paul-emile.paradan@umontpellier.fr

Michèle Vergne
IMJ-PRG, Université Paris 7, CNRS, Paris, France,
e-mail: michele.vergne@imj-prg.fr

V. G. Kac, V. L. Popov (eds.), *Lie Groups, Geometry, and Representation Theory*, Progress in Mathematics 326, https://doi.org/10.1007/978-3-030-02191-7_15

the representation theory of G, since $\widehat{G}$ may be considered as a subset of $\mathfrak{g}^*$. We investigate further the corresponding quantization, via Dirac operators, in the case where the group G is compact, and the moment map proper. Our new insight is that the asymptotic behavior when $\mathscr{L}$ is replaced by $\mathscr{L}^k$ is easy to compute classically, and determines completely the quantization at $k = 1$. Certainly, Bertram Kostant would have found this idea obvious, but maybe also beautiful. Anyway, here it is.

1 Introduction

Let M be a compact oriented spin manifold of even dimension $2d$. Let $\mathscr{L}$ be a line bundle over M equipped with a connection of curvature $-i\Omega$ and let $\hat{A}(M)$ be the $\hat{A}$ class of M (normalized as in [2]). We do not assume the curvature $-i\Omega$ of $\mathscr{L}$ to be nondegenerate. Define the integral

$$\mathscr{Q}^{\mathrm{geo}}(M,\mathscr{L}) = \frac{1}{(-2i\pi)^d}\int_M e^{-i\Omega}\hat{A}(M).$$

The Dirac operator $D_{\mathscr{L}}$ twisted by $\mathscr{L}$ is an elliptic operator on M, and let

$$\mathscr{Q}^{\mathrm{spin}}(M,\mathscr{L}) = \dim\mathrm{Ker}(D_{\mathscr{L}}) - \dim\mathrm{Coker}(D_{\mathscr{L}})$$

be the index of $D_{\mathscr{L}}$. By the Atiyah-Singer index theorem,

$$\mathscr{Q}^{\mathrm{spin}}(M,\mathscr{L}) = \mathscr{Q}^{\mathrm{geo}}(M,\mathscr{L}).$$

Assume now that a compact connected Lie group G, with Lie algebra $\mathfrak{g}$, acts on $(M,\mathscr{L})$. The choice of an invariant connection on the line bundle $\mathscr{L}$ determines a moment map $\phi_G : M \to \mathfrak{g}^*$. This is the Kostant moment map [12]. Let T be a maximal torus of G, $\mathfrak{t}$ its Lie algebra. We identify the space $\widehat{G}$ of irreducible finite-dimensional representations of G with the discrete set of elements $\lambda \in \mathfrak{t}^*$ that are dominant, admissible, and regular, and we denote by $\chi_\lambda(g)$ the trace of the action of $g \in G$ on the irreducible representation of G parametrized by λ.

If k is an integer, we denote by $\mathscr{L}^k$ the k-th power of the line bundle $\mathscr{L}$. Define the function $\mathscr{Q}^{\mathrm{spin}}_G(M,\mathscr{L}^k)$ on G to be the trace of the action of $g \in G$ in the virtual space $\mathrm{Ker}(D_{\mathscr{L}^k}) - \mathrm{Coker}(D_{\mathscr{L}^k})$. Define $m_G(\lambda,k) \in \mathbb{Z}$ such that

$$\mathscr{Q}^{\mathrm{spin}}_G(M,\mathscr{L}^k) = \sum_{\lambda\in\hat{G}} m_G(\lambda,k)\chi_\lambda.$$

Consider the geometric (rescaled) analogue

$$\Theta_k^{(M,\mathscr{L})} = k^{\frac{\dim(G/T)}{2}} \sum_{\lambda\in\hat{G}} m_G(\lambda,k)\beta_{\lambda/k},$$

the weighted sum of the canonical Liouville measures $\beta_{\lambda/k}$ on the coadjoint orbits $G\lambda/k$.

The aim of this article is to study the asymptotic behavior of $\Theta_k^{(M,\mathscr{L})}$ when k is large, and M possibly noncompact, and to explore a functorial consequence of this formula for reduced spaces.

We will work in the spin context. The same argument would adapt to manifolds with spinc structures, provided that we work with odd powers of k. This context is more general than the Hamiltonian context (which includes the Kähler context), and it is more natural since the Weyl character formula for χ_λ is the fixed point formula for a twisted Dirac operator on the coadjoint orbit $G\lambda$. We will return to the comparison between these two contexts in a forthcoming article. The article [25] determines the asymptotic expansion of $\Theta_k^{(M,\mathscr{L})}$ in the Hamiltonian context, when G is a torus.

Let us return to the case of a spin manifold M with a G-invariant line bundle with connection. We no longer assume M to be compact, but we assume the moment map ϕ_G proper. In this introduction, to simplify, we assume that all stabilizers of the action of G on M are abelian and connected. We also assume that ρ is in the weight lattice of T. For every $\lambda \in \widehat{G} \subset \widehat{T} \subset \mathfrak{t}^*$, consider the reduced space

$$M_{\lambda,G} := \phi_G^{-1}(\lambda)/T,$$

which may be nonconnected. If λ is a regular value of ϕ_G, $M_{\lambda,G}$ is a smooth compact spin manifold. Denote by $\mathbb{C}_{[-\lambda]}$ the vector space $\mathbb{C}$ with the action of T given by the character $t^{-\lambda}$ of T. The restriction of the line bundle $\mathbb{C}_{[-\lambda]} \otimes \mathscr{L}^k$ to $\phi_G^{-1}(\lambda)$ is the pull-back of a line bundle $\mathscr{L}_{\lambda,k}$ on $M_{\lambda,G}$. So define

$$m_G^{\mathrm{geo}}(\lambda,k) = \mathscr{Q}^{\mathrm{geo}}(M_{\lambda,G}, \mathscr{L}_{\lambda,k}).$$

If λ is not a regular value of ϕ_G, this number can be defined by deformation. When M is compact, the $[Q,R]=0$ theorem in the spin context [21] asserts that $m_G(\lambda,k) = m_G^{\mathrm{geo}}(\lambda,k)$. This suggests extending the definition of $\mathscr{Q}_G^{\mathrm{spin}}(M,\mathscr{L})$, when $\mathscr{L}$ is equipped with a connection (that we leave implicit in the notation) such that the moment map ϕ_G is proper, in the following way. Define

$$\mathscr{Q}_G^{\mathrm{geo}}(M,\mathscr{L}^k) := \sum_\lambda m_G^{\mathrm{geo}}(\lambda,k) V_\lambda.$$

We call this object (in the spirit of [26]) the formal geometric quantization of $(M,\mathscr{L})$.

Here are two simple examples.

Example 1.1. • $M = T^*S^1$, with coordinates (t,θ), and the natural action of the circle group $G = S^1$. Consider the trivial line bundle $\mathscr{L}$ with connection $d - it d\theta$. The moment map is $\phi_G(t,\theta) = t$. Thus $\mathscr{Q}_G^{\mathrm{geo}}(M,\mathscr{L}^k)$ is independent of k:

$$\mathscr{Q}_G^{\mathrm{geo}}(M,\mathscr{L}^k) = \sum_{n\in\mathbb{Z}} e^{in\theta}.$$

• $M = \mathbb{C}$, with coordinate z. We consider an integer a and the line bundle $\mathscr{L} = M \times \mathbb{C}$ with action $e^{i\theta}(z,v) = (e^{2i\theta}z, e^{ia\theta}v)$. Take the connection $\nabla = d - \frac{i}{2} Im(zd\bar{z})$. The moment map is $\phi_G(z) = a + |z|^2$ and is proper. Then

$$\mathscr{Q}_G^{\mathrm{geo}}(M,\mathscr{L}^k) = e^{ika\theta} \sum_{j\geq 0} e^{i(2j+1)\theta}.$$

As M is noncompact, it is not possible to define the index of $D_{\mathscr{L}}$ without introducing additional data. We use transversally elliptic operators. The Kirwan vector field κ_G associated to ϕ_G allows us to deform the symbol of the Dirac operator, and to obtain a G-invariant transversally elliptic operator $D_{\mathscr{L},\phi_G}$ on M if the set of zeroes of κ_G is compact, which we assume in this introduction. The index $\mathscr{Q}_G^{-\infty}(M,\mathscr{L}^k)$ of this operator provides a well-defined generalized function on G. So define $m_G(\lambda,k) \in \mathbb{Z}$ such that

$$\mathscr{Q}_G^{-\infty}(M,\mathscr{L}^k)(g) = \sum_{\lambda \in \hat{G}} m_G(\lambda,k)\chi_\lambda(g).$$

Then, we have again (see [10])

$$m_G(\lambda,k) = m_G^{\mathrm{geo}}(\lambda,k).$$

An important consequence of this geometric relation is the fact that the function $m_G(\lambda,k)$ is a piecewise quasi-polynomial function. In particular, the map $k \mapsto m_G(k\lambda,k)$ is entirely determined by its behavior for large values of k.

We define, as in the compact case, the weighted sum of measures of the orbits $G\lambda/k$:

$$\Theta_k^{(M,\mathscr{L})} = k^{\dim(G/T)/2} \sum_\lambda m_G(\lambda,k)\beta_{\lambda/k}.$$

As is well known, at least in the Hamiltonian case, the measure $\frac{1}{k^{\dim M/2}}\Theta_k^{(M,\mathscr{L})}$ tends to the Duistermaat-Heckman measure when k tends to ∞. Here we will prove that $\Theta_k^{(M,\mathscr{L})}$ has an asymptotic expansion, when $k \to \infty$, as a Laurent series (in $1/k$) of distributions, and we will determine all coefficients as twisted Duistermaat-Heckman distributions related to the $\hat{A}$ class of M.

Recall in this spin context the definition of the Duistermaat-Heckman measure. Let $\Omega(X) = \Omega - \langle \phi_G, X\rangle$, $X \in \mathfrak{g}$. This is a closed equivariant form of degree 2 on M and $e^{-i\Omega(X)}$ is the equivariant Chern character of $\mathscr{L}$. The form $\Omega^{\dim M/2}$ is a density on M (which might be not positive). The Duistermaat-Heckman measure is the signed measure on $\mathfrak{g}^*$ obtained by push-forward of $\Omega^{\dim M/2}$ by the proper map ϕ_G. If M is compact, its Fourier transform is the function $X \mapsto \frac{1}{(-2i\pi)^d}\int_M e^{-i\Omega(X)}$ on $\mathfrak{g}$. Similarly, if $\nu(X)$ is a closed G-equivariant form on M with polynomial coefficients, we can define the distribution $DH^G(M,\Omega,\nu)$ on $\mathfrak{g}^*$ by the formula:

$$\langle DH^G(M,\Omega,\nu),\varphi\rangle = \frac{1}{(-2i\pi)^d}\int_M \int_{\mathfrak{g}} e^{-i\Omega(X)}\nu(X)\widehat{\varphi}(X)dX. \qquad (1.1)$$

Here φ is a test function on $\mathfrak{g}^*$, with Fourier transform $\widehat{\varphi}$. It is easy to see that this distribution is well defined if ϕ_G is proper.

Consider the equivariant $\hat{A}$ class of M which belongs to the completion of the equivariant cohomology ring of M, and its expansion $\hat{A}(M)(X) = \sum_{n=0}^{\infty} \hat{A}_n(M)(X)$ in equivariant classes homogeneous of degree $2n$. Finally, let $j_{\mathfrak{g}}^{1/2}(X) = \det_{\mathfrak{g}}^{1/2}\left(\frac{e^{X/2}-e^{-X/2}}{X}\right)$, a G-invariant function of $X \in \mathfrak{g}$. It determines a formal series $j_{\mathfrak{g}}^{1/2}(i\partial/k)$ of G-invariant constant-coefficient differential operators on $\mathfrak{g}^*$.

Our main contribution is the following result (see Theorem 4.6).

Theorem 1.2. *When k tends to ∞,*

$$\Theta_k^{(M,\mathscr{L})} \equiv j_{\mathfrak{g}}^{1/2}(i\partial/k)\Big(k^d \sum_{n=0}^{\infty} \frac{1}{k^n} DH^G(M,\Omega,\hat{A}_n(M))\Big). \tag{1.2}$$

The leading term in k^d is the Duistermaat-Heckman measure.

Let us check this formula in our two basic examples.

Example 1.3. • For $M = T^*S^1$, the $\hat{A}$ class is identically equal to 1, so there exists only one term in the expansion (1.2). Here Theorem 1.2 is the well-known fact that

$$\sum_{\lambda\in\mathbb{Z}} \varphi(\lambda/k) \equiv k \int_{\mathbb{R}} \varphi(\xi)d\xi.$$

• For $M = \mathbb{C}$, the equivariant $\hat{A}$ class for our action is the invariant function $\frac{\theta}{\sin(\theta)}$. Theorem 1.2 is the following variation of the Euler-Maclaurin formula (see [7]):

$$\sum_{j\geq 0} \varphi(a+(2j+1)/k) \equiv \frac{k}{2}\int_a^{\infty} \varphi(\xi)d\xi - \sum_{n=1}^{\infty} (\frac{2}{k})^{n-1} \frac{B_n(1/2)}{n!} \varphi^{(n-1)}(a).$$

Here $B_n(t)$ is the n-th Bernoulli polynomial and only even integers n occur, since $B_n(\frac{1}{2}) = 0$ for all odd $n \geq 0$.

Let us see why this formula is natural, when M is compact. By the Kirillov formula, the Fourier transform of $\Theta_k^{(M,\mathscr{L})}$ is the function

$$X \longmapsto \mathscr{Q}_G^{\mathrm{spin}}(M,\mathscr{L}^k)(\exp X/k)\, j_{\mathfrak{g}}^{1/2}(X/k)$$

on $\mathfrak{g}$. The delocalized formula [2] for the index says that for $X \in \mathfrak{g}$ small enough,

$$\mathscr{Q}_G^{\mathrm{spin}}(M,\mathscr{L}^k)(\exp X/k) = \frac{1}{(-2i\pi)^d}\int_M e^{-ik\Omega(X/k)}\hat{A}(M)(X/k).$$

But, by simple inspection of the relation between the equivariant degree and the polynomial degree, this is also equal, for X small, to

$$\frac{1}{(-2i\pi)^d} k^d \int_M e^{-i\Omega(X)} \Big(\sum_{n=0}^{\infty} \frac{1}{k^n} \hat{A}_n(M)(X) \Big).$$

The formula for the asymptotic expansion of $\Theta_k^{(M,\mathscr{L})}$ follows easily. Note the dichotomy between the equivariant Chern character $e^{-i\Omega(X)}$, which is unchanged in this asymptotic equivalence, and which involves the exponential function $e^{i\langle\phi_G,X\rangle}$, and the $\hat{A}$ class that we expand as a series of homogeneous equivariant classes with polynomial coefficients. A more delicate analysis, using the delocalized formula ([20]) for the generalized function $\mathscr{Q}_G^{-\infty}(M,\mathscr{L}^k)(\exp X)$, leads us also naturally to Theorem 1.2. The Chern character $e^{-i\Omega(X)}$ has to be replaced by the Chern character $e^{-i\Omega(X)}P(X)$ where $P(X)$ is an equivariant form with generalized coefficients, supported in a compact neighborhood of the zeroes of κ_G, equivalent to 1 in the equivariant cohomology without compact support conditions.

In conclusion, Theorem 1.2 says that the formula

$$\mathscr{Q}_G^{\mathrm{spin}}(M,\mathscr{L})(\exp X) = \frac{1}{(-2i\pi)^d} \int_M e^{i\Omega(X)} \hat{A}(M)(X)$$

has a meaning in the asymptotic sense for a noncompact manifold M when $\mathscr{L}$ is replaced by $\mathscr{L}^k$ and X by X/k, and is a good ansatz for the equivariant index formula of $D_{\mathscr{L}}$.

Let us now explain a consequence of this asymptotic formula for reduced spaces.

Return for a moment to the case where M is compact. Plugging $g=1$ in the identity $\mathscr{Q}_G^{\mathrm{spin}}(M,\mathscr{L}^k)(g) = \sum_{\lambda\in\hat{G}} m_G(\lambda,k)\chi_\lambda(g)$ leads to the remarkable identity

$$\mathscr{Q}^{\mathrm{geo}}(M,\mathscr{L}^k) = \sum_{\lambda\in\widehat{G}} \mathrm{vol}(G\lambda)\, \mathscr{Q}^{\mathrm{geo}}(M_{\lambda,G},\mathscr{L}_{\lambda,k}) \tag{1.3}$$

relating an integral on M to a sum of integrals on the finite number of reduced spaces $M_{\lambda,G}$. We will see that this equality generalizes to reduction in stages, even when M is noncompact.

Let H be a connected compact subgroup of G with torus T_H. The H action on $(M,\mathscr{L})$ leads to the moment map $\phi_H : M \to \mathfrak{h}^*$. Assume ϕ_H is still proper. In this case we can define $\mathscr{Q}_G^{-\infty}(M,\mathscr{L})$, and $\mathscr{Q}_H^{-\infty}(M,\mathscr{L})$. When M is compact, from the description of $\mathscr{Q}_G^{\mathrm{spin}}(M,\mathscr{L})$ as the index of the elliptic operator $D_{\mathscr{L}}$ on M, we see that $\mathscr{Q}_H^{\mathrm{spin}}(M,\mathscr{L})$ is the restriction of the representation $\mathscr{Q}_G^{\mathrm{spin}}(M,\mathscr{L})$ to H. When M is noncompact, this relation is not obvious since our G-transversally elliptic operator $D_{\mathscr{L},\phi_G}$ is not (usually) H-transversally elliptic. The following theorem

$$\mathscr{Q}_G^{-\infty}(M,\mathscr{L})|_H = \mathscr{Q}_H^{-\infty}(M,\mathscr{L}) \tag{1.4}$$

was proved by the first author using cutting [19]. Here we will show that this also follows from comparing the large-k behavior of $\mathscr{Q}_G^{-\infty}(M,\mathscr{L}^k)$ and $\mathscr{Q}_H^{-\infty}(M,\mathscr{L}^k)$ and the fact that the multiplicity maps $k \to m_G(k\lambda,k)$ and $k \to m_H(k\lambda,k)$ are entirely determined by their behaviors for large values of k.

Let $\lambda \in \widehat{G}$, and $\mu \in \widehat{H}$. Denote by $c(\lambda,\mu)$ the multiplicity of V_μ in $V_\lambda|_H$. Theorem (1.4) is equivalent to the following equality of indices of Dirac operators on reduced spaces for H and G.

For any $\mu \in \widehat{H}$, we have

$$\mathscr{Q}^{\rm geo}(M_{\mu,H},\mathscr{L}_{\mu,k}) = \sum_{\lambda\in\widehat{G}} c(\lambda,\mu)\,\mathscr{Q}^{\rm geo}(M_{\lambda,G},\mathscr{L}_{\lambda,k}).$$

When M is compact, we can take $H=\{1\}$, and this is the mysterious equality (1.3). In general, this equality of indices of Dirac operators on reduced spaces for different groups is not clear, since $M_{\mu,H} = \phi_H^{-1}(\mu)/T_H$ do not carry any visible group action.

Let us briefly sketch a proof of this equality for the case where G,H are tori. This simple case could be treated by considering the action of G/H on the compact spin manifold $M_{\mu,H}$, but we treat it by a different method which will generalize to any pair of groups $H\subset G$.

In this case, $\widehat{G}$, $\widehat{H}$ are lattices in $\mathfrak{g}^*,\mathfrak{h}^*$, and we have to prove

$$m_H(\mu,k) = \sum_{\lambda\in\widehat{G},\lambda|_{\mathfrak{h}}=\mu} m_G(\lambda,k).$$

In a companion article [23], we proved the following easy result. Consider a distribution $D_k = \sum_{\mu\in\hat{H}} q(\mu,k)\delta_{\mu/k}$ associated to a piecewise quasi-polynomial function $q(\mu,k)$. Assume that for any element $\zeta \in H$ of finite order, the distribution $\sum_{\mu\in\hat{H}} q(\mu,k)\zeta^\mu\delta_{\mu/k} = O(k^{-\infty})$ when k tends to ∞. Then $q(\mu,k)=0$.

We have computed the asymptotic expansion of both distributions $\Theta_k^G = \sum_{\lambda\in\hat{G}} m_G(\lambda,k)\,\delta_{\lambda/k}$ and $\Theta_k^H = \sum_{\mu\in\hat{H}} m_H(\mu,k)\,\delta_{\mu/k}$. One has

$$\Theta_k^G \equiv k^d \sum_{n=0}^{\infty} \frac{1}{k^n} DH^G(M,\Omega,\hat{A}_n(M)),$$

$$\Theta_k^H \equiv k^d \sum_{n=0}^{\infty} \frac{1}{k^n} DH^H(M,\Omega,\hat{A}_n(M)).$$

Define $m'_H(\mu,k) = \sum_{\lambda\in\widehat{G},\lambda|_{\mathfrak{h}}=\mu} m_G(\lambda,k)$. The distribution

$$S_k^H = \sum_{\mu\in\hat{H}} m'_H(\mu,k)\delta_{\mu/k}$$

is the push-forward $r_*\Theta_k^G$ of the distribution Θ_k^G under the map $r:\mathfrak{g}^*\to\mathfrak{h}^*$.

The Duistermaat-Heckman distributions $DH^G(M,\Omega,\nu)$ behave very well under the push-forward map:

$$r_*DH^G(M,\Omega,\nu) = DH^H(M,\Omega,\nu).$$

Indeed, at least in the compact case, the Fourier transform of $DH^H(M,\Omega,\nu)$ is the restriction to $\mathfrak{h}$ of the function $\frac{-1}{(2\pi)^d}\int_M e^{-i\Omega(X)}\nu(X)$ on $\mathfrak{g}$. So we see that S_k^H and Θ_k^H have the same asymptotic expansion. This is not entirely sufficient to prove that $\Theta_k^H = S_k^H$. But we use the fact that both functions $m'_H(\mu,k)$ and $m_H(\mu,k)$ are piecewise quasi-polynomials and that a similar asymptotic descent formula holds for the distribution

$$\sum_{\mu\in\widehat{H}} m_H(\mu,k)\zeta^{\mu}\delta_{\mu/k}$$

when $\zeta \in T_H$ is of finite order.

The proof, sketched here for H and G abelian, works equally well for two subgroups $H \subset G$. We just have to use the formula for the push-forward of the Liouville measure $r_*(\beta_\lambda)$ of admissible coadjoint orbits. As expected, the full series for the $\hat{A}(M)$ class, as well as the Duflo operator $j_{\mathfrak{g}}^{1/2}(\partial)$, plays a role in the functoriality.

Asymptotic behavior of quantization when k tends to ∞ has been considered by many authors. Let us give a very limited and personal selection of influential works.

Asymptotic behavior of kernels of Laplacians twisted by $\mathscr{L}^k$ was used by Boutet de Monvel-Guillemin [6] to produce a formal star-product of functions on symplectic manifolds. In general, only the few first terms of the formal deformation of the star-product are computable (see for example [5], [13]).

Our asymptotic trace formula for the transversally elliptic operator $D_{\mathscr{L},\phi_G}$ has the same flavor as Fedosov's trace formula for deformation quantization [8].

The article of E. Meinrenken [15] where multiplicities $m_G(\lambda,k)$ are identified at the large-k limit to the index of reduced spaces is closely related to our setting. Asymptotic Riemann sums of values of smooth functions at equally spaced sample points of a Delzant polytope $\Delta \subset \mathfrak{g}^*$ were studied by Guillemin-Sternberg [9], where the full asymptotic formula is given in terms of the Todd class of the corresponding toric manifold. The natural geometric rescaling $\mu \to \mu/k$ in computing $\frac{1}{k^d}\sum_{\mu\in\mathbb{Z}^d\cap k\Delta}\varphi(\mu/k)$ leads to the consideration of the rescaling $X \to X/k$ in the variable $X \in \mathfrak{g}$. This was one of the inspiring examples.

Our application to the determination of restrictions to subgroups is a striking example of the phenomenon that an obvious property at the semi-classical level (functoriality of push-forward of distributions) can lead to a proof of a subtle relation at the quantum level.

2 Asymptotics and representations

2.1 Fourier transforms

When V is a finite-dimensional real vector space, we denote by $\mathscr{D}'(V^*)$ the vector space of complex-valued distributions on the dual vector space V^*. If $A \in \mathscr{D}'(V^*)$ and $\varphi \in \mathscr{C}^\infty_{cpt}(V^*)$, we denote by $\langle A,\varphi\rangle \in \mathbb{C}$ (or $\langle A(\xi),\varphi(\xi)\rangle$) the value of A on φ.

When $A \in \mathscr{D}'(V^*)$ has compact support, its Fourier transform $\mathscr{F}_{V^*}(A) \in \mathscr{C}^\infty(V)$ is defined by the relation

$$\mathscr{F}_{V^*}(A)(v) := \langle A(\xi), e^{i\langle \xi, v\rangle}\rangle, \quad v \in V.$$

When f is a smooth function on $\mathfrak{g}$ and has at most polynomial growth, its Fourier transform $\mathscr{F}_V(f) \in \mathscr{D}'(V^*)$ is defined by the relation

$$\langle \mathscr{F}_V(f), \varphi \rangle = \int_V f(v)\widehat{\varphi}(v)dv,$$

where $\widehat{\varphi}(v) = \frac{1}{(2\pi)^{\dim V}} \int_{V^*} \varphi(\xi) e^{-i\langle \xi, v\rangle} d\xi$. Here dv and $d\xi$ are dual volume forms on V and V^*.

For any $k \geq 1$, we define the operation

$$\text{rescaling}_{V^*}(k) : \mathscr{D}'(V^*) \to \mathscr{D}'(V^*) \tag{2.1}$$

by $\langle \text{rescaling}_{V^*}(k)(A), \varphi\rangle = \langle A(\xi), \varphi(\xi/k)\rangle$. We have also the operation $\text{rescaling}_V(k) : \mathscr{C}^\infty(V) \to \mathscr{C}^\infty(V)$ defined by $\text{rescaling}_V(k)(f)(v) = f(v/k)$.

2.2 The isomorphism $\mathscr{R}_\mathfrak{g}$

Let G be a connected compact Lie group with Lie algebra $\mathfrak{g}$. Let $T \subset G$ be a maximal torus with Lie algebra $\mathfrak{t}$. Let W_G be the Weyl group. We consider $\widehat{T}$ as a lattice Λ in $\mathfrak{t}^*$. If $\lambda \in \Lambda$, we denote by t^λ the corresponding character of T. If $t = \exp(X)$ with $X \in \mathfrak{t}$, then $t^\lambda = e^{i\langle \lambda, X\rangle}$. We choose a system Δ^+ of positive roots. In our convention, Δ^+ is contained in $\mathfrak{t}^*$. We choose an invariant scalar product on $\mathfrak{g}$.

Let $\mathscr{D}'(\mathfrak{g}^*)^G$ be the space of G-invariant distributions on $\mathfrak{g}^*$ and let $\mathscr{D}'(\mathfrak{t}^*)^{W_G-alt}$ be the subspace of W_G-anti-invariant distributions on $\mathfrak{t}^*$. Let $\Pi_{\mathfrak{g}/\mathfrak{t}}(X) = \prod_{\alpha \in \Delta^+} \langle \alpha, X\rangle$ (a W_G-anti-invariant function on $\mathfrak{t}$).

We will use the following classical fact.

Lemma 2.1. *There exists a unique linear isomorphism*

$$\mathscr{R}_\mathfrak{g} : \mathscr{D}'(\mathfrak{g}^*)^G \longrightarrow \mathscr{D}'(\mathfrak{t}^*)^{W_G-alt}$$

with the following two properties.

- *$\mathscr{R}_\mathfrak{g}(fA) = f|_{\mathfrak{t}^*}\mathscr{R}_\mathfrak{g}(A)$ for all $A \in \mathscr{D}'(\mathfrak{g}^*)^G$ and $f \in \mathscr{C}^\infty(\mathfrak{g}^*)^G$.*
- *For compactly supported distributions A, the Fourier transform of $\mathscr{R}_\mathfrak{g}(A)$ is given by*

$$\langle \mathscr{R}_\mathfrak{g}(A), e^{i\langle -, X\rangle}\rangle = (i)^r \Pi_{\mathfrak{g}/\mathfrak{t}}(X) \langle A, e^{i\langle -, X\rangle}\rangle, \qquad X \in \mathfrak{t},$$

with $r = \dim(G/T)/2$.

2.3 Asymptotics

Let $\mathscr{D}'(\mathfrak{g}^*)^G$ be the space of G-invariant distributions on $\mathfrak{g}^*$. We will study asymptotic expansions of sequences of distributions.

Definition 2.2. Let $(\Theta_k)_{k\geq 1}$ and $(\theta_n)_{n\in\mathbb{N}}$ be two sequences in $\mathscr{D}'(\mathfrak{g}^*)^G$. We write

$$\Theta_k \equiv k^{n_o} \sum_{n=0}^{\infty} \frac{1}{k^n} \theta_n \tag{2.2}$$

for some $n_o \in \mathbb{Z}$ if for any test function φ and any $N \in \mathbb{N}$, we have

$$\langle \Theta_k, \varphi\rangle = k^{n_o} \sum_{n=0}^{N} \frac{1}{k^n} \langle \theta_n, \varphi\rangle + o(k^{n_o - N}).$$

In this text, we will allow each distribution θ_n to depend periodically on the integers k. This means that for every n, there exist $D \geq 1$ and distributions (θ_ζ) parametrized by the roots of unity $\{\zeta^D = 1\}$ such that $\theta_n(k) = \sum_{\zeta^D=1} \zeta^k \theta_\zeta$ for all $k \geq 1$.

Example 2.3. Consider the distributions $\mathrm{T}_k = \sum_{\lambda\in\mathbb{N}} \delta_{\lambda/k}$ $(k \geq 1)$ on $\mathbb{R}$. The Euler-Maclaurin formula gives the expansion

$$\mathrm{T}_k \equiv k\, 1_{[0,\infty[} + \frac{1}{2}\delta_0 - \sum_{n=1}^{\infty} \frac{1}{k^{2n-1}} \frac{b_{2n}}{(2n)!} \delta_0^{(2n-1)},$$

where b_{2n} are the Bernoulli numbers and $\delta_0^{(2n-1)}$ is the $(2n-1)$-th derivation of the Dirac distribution δ_0.

Let d be a C^∞ function defined near the origin $0 \in \mathfrak{g}$. Consider the Taylor series $\sum_{n=0}^{\infty} d_n(X)$ of d at the origin. Thus, d_n is a homogeneous polynomial function of degree n on $\mathfrak{g}$. We associate to the function d the formal series

$$d(i\partial/k) = \sum_{n=0}^{\infty} k^{-n} d_n(i\partial)$$

of constant-coefficient differential operators $d_n(i\partial)$ on $\mathfrak{g}^*$. Thus, if $Q(k) = k^{n_o} \sum_{n=0}^{\infty} k^{-n}\theta_n$ is a formal series of distributions on $\mathfrak{g}^*$, the series $d(i\partial/k)Q(k)$ is the formal series of distributions $k^{n_o} \sum_{n=0}^{\infty} k^{-n} s_n$ on $\mathfrak{g}^*$, where $s_n = \sum_{l+m=n} d_l(i\partial)\theta_m$.

We have the following basic fact.

Lemma 2.4. *The expansion* $\Theta_k \equiv k^{n_o} \sum_{n=0}^{\infty} \frac{1}{k^n} \theta_n$ *holds in* $\mathscr{D}'(\mathfrak{g}^*)^G$ *if and only if we have the expansion* $\mathscr{R}_{\mathfrak{g}}(\Theta_k) \equiv k^{n_o} \sum_{n=0}^{\infty} \frac{1}{k^n} \mathscr{R}_{\mathfrak{g}}(\theta_n)$ *in* $\mathscr{D}'(\mathfrak{t}^*)^{W_G-alt}$.

Our asymptotic expansions will arise from Taylor series.

Let τ be a smooth function on $\mathfrak{g}$. Assume that τ and all its derivatives have at most polynomial growth. Consider the Taylor series $\sum_{n=0}^{\infty} \tau_n(X)$ of τ at the origin.

If $b(X)$ is a smooth function on $\mathfrak{g}$ with at most polynomial growth, then $f_k(X) = b(X)\tau(X/k)$ $(k \geq 1)$ defines a tempered distribution on $\mathfrak{g}$. Similarly, as $\tau_n(X)$ is a polynomial function, $g_n(X) = b(X)\tau_n(X)$ defines a tempered distribution on $\mathfrak{g}$. We note the following result.

Proposition 2.5. *When k tends to* ∞*, we have the asymptotic expansion*

$$\mathscr{F}_{\mathfrak{g}}(f_k) \equiv \sum_{n=0}^{\infty} \frac{1}{k^n} \mathscr{F}_{\mathfrak{g}}(g_n) = \tau(i\partial/k)\mathscr{F}_{\mathfrak{g}}(b).$$

Later, we will need a uniform version with parameters of this proposition. So we give the proof.

Proof. Let φ be a smooth function on $\mathfrak{g}^*$ with compact support. Thus $\langle \mathscr{F}_{\mathfrak{g}}(f_k), \varphi \rangle = \int_{\mathfrak{g}} b(X)\tau(X/k)\widehat{\varphi}(X)dX$.

Let N be a positive integer, and let $\tau_{\leq N}$ be the sum of the Taylor coefficients τ_n up to degree N, so $\tau = \tau_{\leq N} + \tau_{>N}$. We write $I_k = \int_{\mathfrak{g}} b(X)\tau(X/k)\widehat{\varphi}(X)dX$ as $I_k^0 + R_k$ with

$$I_k^0 = \int_{\mathfrak{g}} b(X)\tau_{\leq N}(X/k)\widehat{\varphi}(X)dX = \sum_{k=0}^{N} \frac{1}{k^n} \int_{\mathfrak{g}} b(X)\tau_n(X)\widehat{\varphi}(X)dX$$

and $R_k = \int_{\mathfrak{g}} b(X)\tau_{>N}(X/k)\widehat{\varphi}(X)dX$.

In multi-index notation, we may write (via an integral formula depending on τ and its derivatives) $\tau_{>N}(X) = \sum_{\alpha,|\alpha|=N+1} X^{\alpha} D_{\alpha}(X)$ with $D_{\alpha}(X)$ bounded by a polynomial function of X. If $\tau(X)$ depends smoothly on a parameter x, we can construct $D_{\alpha}(X)$ depending smoothly on x. Thus

$$R_k = \frac{1}{k^{N+1}} \sum_{\alpha,|\alpha|=N+1} \int_{\mathfrak{g}} b(X)X^{\alpha}D_{\alpha}(X/k)\widehat{\varphi}(X)dX.$$

Since $\|X/k\| \leq \|X\|$, we can bound $b(X)X^{\alpha}D_{\alpha}(X/k)$ by a polynomial function of X. Since $\widehat{\varphi}(X)$ is rapidly decreasing, we see that $|R_k| \leq \frac{c_N}{k^{N+1}}$. □

Let $b(X,k) = \sum_{m \in F} k^m b_m(X)$ be a sum over a finite set $F \subset \mathbb{Z}$ of smooth functions $b_m(X)$ with at most polynomial growth, and let $\tau(X)$ be a function with polynomial growth as well as all its derivatives. Then it is clear that we can obtain the asymptotic expansion of $\int_{\mathfrak{g}} b(X,k)\tau(X/k)\widehat{\varphi}(X)dX$ by summing over $m \in F$ the asymptotic expansions of $\int_{\mathfrak{g}} b_m(X)\tau(X/k)\widehat{\varphi}(X)dX$ multiplied by k^m. So this asymptotic expansion is given by the Laurent series (in $1/k$)

$$\sum_{m \in F} \sum_{n=0}^{\infty} k^m \frac{1}{k^n} \int_{\mathfrak{g}} b_m(X)\tau_n(X)\widehat{\varphi}(X)dX.$$

We write this somewhat informally as

$$\int_{\mathfrak{g}} b(X,k)\tau(X/k)\widehat{\varphi}(X)dX \equiv \int_{\mathfrak{g}} b(X,k) \sum_{n=0}^{\infty} \tau_n(X/k)\widehat{\varphi}(X)dX. \tag{2.3}$$

In short, we replace τ by its formal Taylor series $\sum_{n=0}^{\infty} \tau_n(X)$ and keep $b(X,k)$ as it is.

2.4 Kirillov formula

Recall that any coadjoint orbit $\mathscr{O} \subset \mathfrak{g}^*$ is provided with the Kirillov-Kostant-Souriau symplectic form $\Omega_{\mathscr{O}}$ that is normalized as follows. We have $\Omega_{\mathscr{O}}|_\xi(X\cdot\xi, Y\cdot\xi) = \langle \xi, [X,Y]\rangle$, for $\xi \in \mathscr{O}$ and $X,Y \in \mathfrak{g}$. Let $n_{\mathscr{O}} = \frac{1}{2}\dim \mathscr{O}$.

Definition 2.6. We denote by $\beta_{\mathscr{O}}$ the distribution on $\mathfrak{g}^*$ defined by the relation

$$\langle \beta_{\mathscr{O}}, \varphi \rangle = \int_{\xi \in \mathscr{O}} \varphi(\xi) \frac{(\Omega_{\mathscr{O}})^{n_{\mathscr{O}}}}{(2\pi)^{n_{\mathscr{O}}} n_{\mathscr{O}}!}$$

for any smooth function φ on $\mathfrak{g}^*$.

Recall that we have chosen a system $\Delta^+ \subset \mathfrak{t}^*$ of positive roots, and let $\rho \in \mathfrak{t}^*$ be the corresponding element. We consider the positive Weyl chamber $\mathfrak{t}^*_{\geq 0}$ with interior $\mathfrak{t}^*_{>0}$. We identify $\widehat{G}$ with the discrete set $A_G = (\rho + \Lambda) \cap \mathfrak{t}^*_{>0}$ of positive admissible regular elements of $\mathfrak{t}^*$.

If $\lambda \in A_G$, we denote by $\chi_\lambda(g)$ the trace of the action of $g \in G$ on the irreducible representation V_λ parametrized by λ. We have the fundamental identity

$$\chi_\lambda(e^X) j_{\mathfrak{g}}^{1/2}(X) = \langle \beta_\lambda, e^{i\langle -, X\rangle} \rangle \tag{2.4}$$

where $j_{\mathfrak{g}}(X) = \det_{\mathfrak{g}}\left(\frac{e^{X/2} - e^{-X/2}}{X}\right)$, and $\beta_\lambda := \beta_{G\lambda}$.

Let $\mathscr{R}_{\mathfrak{g}} : \mathscr{D}'(\mathfrak{g}^*)^G \longrightarrow \mathscr{D}'(\mathfrak{t}^*)^{W_G-alt}$ be the isomorphism introduced in Section 2.2. If $\lambda \in A_G$, we have

$$\mathscr{R}_{\mathfrak{g}}(\beta_\lambda) = \sum_{w \in W_G} \varepsilon(w) \delta_{w\lambda} \tag{2.5}$$

where $\delta_{w\lambda}$ is the δ function at $w\lambda$.

2.5 Quasi-polynomials and asymptotics

We recall the notion of piecewise quasi-polynomial functions (for more details see [23]).

Let $\tilde{\Lambda}$ be the lattice generated by Λ and ρ. We consider the vector space $E = \mathfrak{t}^* \times \mathbb{R}$, which is equipped with the lattice $\tilde{\Lambda} \times \mathbb{Z}$.

A function $m : \tilde{\Lambda} \times \mathbb{Z} \to \mathbb{C}$ is periodic if there exists a positive integer D such that $m(x_0 + Dx) = m(x_0)$ for $x, x_0 \in \tilde{\Lambda} \times \mathbb{Z}$. By definition, the algebra of quasi-polynomial functions on $\tilde{\Lambda} \times \mathbb{Z}$ is generated by polynomials and periodic functions on $\tilde{\Lambda} \times \mathbb{Z}$.

To each closed rational polyhedron $P \subset \mathfrak{t}^*$, we associate

- the subspace $E_P \subset E$ generated by $(\xi, 1), \xi \in P$,
- the lattice $\tilde{\Lambda}_P := (\tilde{\Lambda} \times \mathbb{Z}) \cap E_P$ in E_P,
- the cone $C_P := \{(t\xi, t), t \geq 0, \xi \in P\} \subset E_P$,
- $[C_P]$, the characteristic function of C_P.

If m_P is a quasi-polynomial function on the lattice $\tilde{\Lambda}_P$, we can form the product $m_P[C_P]$ that defines a function on $\tilde{\Lambda} \times \mathbb{Z}_{>0}$ as follows:

$$m_P[C_P](\lambda, k) = \begin{cases} 0 & \text{if } \lambda/k \notin P, \\ m_P(\lambda, k) & \text{if } \lambda/k \in P. \end{cases}$$

Definition 2.7. A function $m : \tilde{\Lambda} \times \mathbb{Z}_{>0} \to \mathbb{C}$ is called a piecewise quasi-polynomial function if there is a collection $\mathscr{A}$ of closed rational polyhedrons in $\mathfrak{t}^*$ and a collection of quasi-polynomial functions[1] $(m_P)_{P \in \mathscr{A}}$ such that

$$m = \sum_{P \in \mathscr{A}} m_P[C_P]. \tag{2.6}$$

Here $\mathscr{A}$ is not necessarily finite but, in order that (2.6) makes sense, we assume that for any compact $\mathscr{K} \subset \mathfrak{t}^*$, the set $\{P \in \mathscr{A}, P \cap \mathscr{K} \neq \emptyset\}$ is finite.

We denote by $\mathscr{S}(\tilde{\Lambda})$ the group of piecewise quasi-polynomial functions on $\tilde{\Lambda} \times \mathbb{Z}_{>0}$.

To a piecewise quasi-polynomial function $m \in \mathscr{S}(\tilde{\Lambda})$, we associate the family of invariant distributions on $\mathfrak{g}^*$,

$$\Theta_k(m) = k^r \sum_{\lambda \in A_G} m(\lambda, k) \beta_{\lambda/k}, \quad k \geq 1.$$

In a companion article [23], we proved the following result.

Proposition 2.8. *The family $\Theta_k(m)$ admits an asymptotic expansion*

$$\Theta_k(m) \equiv \mathrm{AS}_k(m) := \sum_{n \in \mathbb{Z}} \frac{1}{k^n} \theta_n(k).$$

Here $(\theta_n(k))_{n \in \mathbb{Z}}$ is a collection of invariant distributions on $\mathfrak{g}^*$, depending periodically on k, and such that $\theta_n(k) = 0$ if $n < n_0^m$.

For our computation, we need also a variation of Proposition 2.8. Let $\mathrm{j} : \mathfrak{g} \to \mathbb{C}$ be an invariant analytic function such that its Fourier transform $B_{\mathrm{j}} := \mathscr{F}_{\mathfrak{g}}(\mathrm{j})$ is a compactly supported measure. We define $B_{\mathrm{j}}^k = \mathrm{rescaling}_{\mathfrak{g}^*}(k) B_{\mathrm{j}}$.

We consider now the family of distributions $B_{\mathrm{j}}^k \star \Theta_k(m)$ where $\star$ denotes the convolution.

[1] Each quasi-polynomial m_P is defined on the sublattice $\tilde{\Lambda}_P \subset \tilde{\Lambda} \times \mathbb{Z}$.

Proposition 2.9. *The family $B_{\mathrm{j}}^k \star \Theta_k(m)$ admits the asymptotic expansion*

$$B_{\mathrm{j}}^k \star \Theta_k(m) \equiv \mathrm{j}(i\partial/k)\mathrm{AS}_k(m).$$

Proof. Let φ be a test function on $\mathfrak{g}^*$. Let $R > 0$ be such that the ball $\{\xi \in \mathfrak{g}^*, \|\xi\| \leq R/2\}$ contains the compact supports of φ and B_{j}.

We start with the relation

$$\langle B_{\mathrm{j}}^k \star \Theta_k(m), \varphi\rangle = \langle \Theta_k(m), \varphi_k\rangle$$

where

$$\varphi_k(\xi) = \int_{\mathfrak{g}^*} B_{\mathrm{j}}(\xi')\varphi(\xi + \xi'/k), \quad k \geq 1$$

are smooth functions supported on the ball $\{\xi \in \mathfrak{g}^*, \|\xi\| \leq R\}$. Let us write the Taylor series of φ at ξ: $\varphi(\xi + \eta) = \varphi(\xi) + \sum_{n=1}^N d_n\varphi(\xi)(\eta) + r_N(\xi, \eta)$. Here $\eta \mapsto d_n\varphi(\xi)(\eta)$ is a homogeneous polynomial of degree n depending smoothly on the variable ξ, and r_N is a smooth function of (ξ, η) such that

$$|r_N(\xi, \eta)| \leq c_N \|\eta\|^{N+1}, \quad \forall (\xi, \eta) \in \mathfrak{g}^* \times \mathfrak{g}^*, \tag{2.7}$$

for some $c_N > 0$.

If we write the Taylor series of j at 0, $\mathrm{j}(X) = \sum_{n=0}^\infty \mathrm{j}_n(X)$, we see that

$$\int_{\mathfrak{g}^*} B_{\mathrm{j}}(\eta) d_n\varphi(\xi)(\eta) = \mathrm{j}_n(-i\partial)\varphi(\xi).$$

So the functions φ_k admit the following description:

$$\varphi_k = \sum_{n=0}^N \frac{1}{k^n} \mathrm{j}_n(-i\partial)\varphi + R_k^N,$$

where $R_k^N(\xi) = \int_{\mathfrak{g}^*} B_{\mathrm{j}}(\xi') r_N(\xi, \xi'/k)$. Thanks to (2.7), we see that there exists C_N such that $|R_k^N(\xi)| \leq C_N/k^{N+1}, \forall \xi \in \mathfrak{g}^*$. We check also that for any $k \geq 1$, the support of R_N^k is contained in the ball $\{\xi \in \mathfrak{g}^*, \|\xi\| \leq R\}$.

Finally we obtain

$$\langle B_{\mathrm{j}}^k \star \Theta_k(m), \varphi\rangle = \sum_{n=0}^N \frac{1}{k^n} \langle \Theta_k(m), \mathrm{j}_n(-i\partial)\varphi\rangle + \mathrm{Rest}_k^N \tag{2.8}$$

where

$$\mathrm{Rest}_k^N = \langle \Theta_k(m), R_k^N\rangle = k^r \sum_{\|\lambda\|/k \leq R} m(\lambda, k) \langle \beta_{\lambda/k}, R_k^N\rangle.$$

The term $|\mathrm{Rest}_k^N|$ is bounded by $\frac{C_N}{k^{N+1}} \sum_{\|\lambda\|/k \leq R} |m(\lambda, k)| \mathrm{vol}(G\lambda)$. Since $m(\lambda, k)$ is a piecewise quasi-polynomial function, we see that for any $p \geq 1$, there exists $N \geq 1$ such that

$$\mathrm{Rest}_k^N = O(k^{-p}). \tag{2.9}$$

Identities (2.8) and (2.9) show that the family $B_{\mathrm{j}}^k \star \Theta_k(m)$ admits the asymptotic expansion $\mathrm{j}(i\partial/k)\mathrm{AS}_k(m)$. □

3 Spin quantization

Let M be an even-dimensional oriented spin manifold of dimension $2d$ and provided with an action of G. If $(\mathscr{L},\nabla)$ is a G-equivariant line bundle equipped with a G-invariant Hermitian connection ∇, we obtain a moment map $\phi_G : M \to \mathfrak{g}^*$ and a closed two-form Ω on M using Kostant's formula:

$$\nabla^2 = -i\Omega \quad \text{and} \quad L(X) - \nabla_X = i\langle \phi_G, X\rangle. \tag{3.1}$$

Here $X \in \mathfrak{g}$ and $L(X)$ is the infinitesimal action of $X \in \mathfrak{g}$ on smooth sections of $\mathscr{L}$.

Assume M compact. The spin quantization $\mathscr{Q}_G^{\mathrm{spin}}(M,\mathscr{L})$ is a virtual finite-dimensional representation of G, constructed as the index of the Dirac operator on M twisted by $\mathscr{L}$. If k is an integer, we denote by $\mathscr{L}^k$ the k-th power of the line bundle $\mathscr{L}$; we can consider the equivariant index

$$\mathscr{Q}_G^{\mathrm{spin}}(M,\mathscr{L}^k) = \sum_{\lambda\in A_G} m_G(\lambda,k)V_\lambda.$$

Taking traces, we also write, for $g \in G$,

$$\mathscr{Q}_G^{\mathrm{spin}}(M,\mathscr{L}^k)(g) = \sum_{\lambda\in A_G} m_G(\lambda,k)\chi_\lambda(g).$$

We will need the following basic fact.

Lemma 3.1. *There exists a polynomial $P(k)$ such that $\left|\mathscr{Q}_G^{\mathrm{spin}}(M,\mathscr{L}^k)(g)\right| \le P(k)$ for all $g \in G$ and $k \ge 1$.*

We associate to $(M,\mathscr{L})$ the distributions on $\mathfrak{g}^*$ given by

$$\Psi_k^{(M,\mathscr{L})} := \mathscr{F}_{\mathfrak{g}} \circ \mathrm{rescaling}_{\mathfrak{g}}(k)\Big(\mathscr{Q}_G^{\mathrm{spin}}(M,\mathscr{L}^k)(e^X)\Big) \tag{3.2}$$

and

$$\begin{aligned}\Theta_k^{(M,\mathscr{L})} &:= \mathscr{F}_{\mathfrak{g}} \circ \mathrm{rescaling}_{\mathfrak{g}}(k)\Big(\mathscr{Q}_G^{\mathrm{spin}}(M,\mathscr{L}^k)(e^X)\,j_{\mathfrak{g}}^{1/2}(X)\Big)\\ &= \mathrm{rescaling}_{\mathfrak{g}^*}(k)\Big(\sum_{\lambda\in A_G} m_G(\lambda,k)\beta_\lambda\Big)\\ &= k^r \sum_{\lambda\in A_G} m_G(\lambda,k)\beta_{\lambda/k},\end{aligned} \tag{3.3}$$

where $r = \dim(G/T)/2$. Thus $\Theta_k^{(M,\mathscr{L})}$ is the rescaled geometric analogue of $\mathscr{Q}_G^{\mathrm{spin}}(M,\mathscr{L}^k) = \sum_{\lambda \in A_G} m_G(\lambda,k)\chi_\lambda$.

3.1 Quasi-polynomial behavior of the multiplicities

We consider the multiplicity function $m_G : A_G \times \mathbb{Z}_{>0} \to \mathbb{Z}$ defined by the relation $\mathscr{Q}_G^{\mathrm{spin}}(M,\mathscr{L}^k) = \sum_{\lambda \in A_G} m_G(\lambda,k)\chi_\lambda$. We extend m_G to $\tilde{\Lambda} \times \mathbb{Z}_{>0}$ by defining $m_G(\lambda,k) = 0$ if $\lambda \notin A_G$.

The main objective of this section is the following result.

Theorem 3.2. *Suppose that the G-action on M admits* abelian infinitesimal stabilizers. *Then* $m_G \in \mathscr{S}(\tilde{\Lambda})$.

In particular, Theorem 3.2 implies that the function $k \mapsto m_G(k\lambda,k)$ is a quasi-polynomial function of $k \geq 1$.

Theorem 3.2 is a consequence of the geometric formulas for the multiplicities $m_G(\lambda,k)$ obtained in [21]. Let us first recall these geometric formulas.

Let $\tilde{T} \to T$ be the covering such that $\tilde{\Lambda}$ is the weight lattice of the torus $\tilde{T}$. Every element $\lambda \in \tilde{\Lambda}$ determines a character of $\tilde{T}$, and we denote by $\mathbb{C}_\lambda$ the corresponding 1-dimensional representation.

Recall that we have chosen a G-invariant Hermitian connection on $\mathscr{L}$, and $\phi_G : M \to \mathfrak{g}^*$ is the associated moment map. We consider the positive Weyl chamber $\mathfrak{t}^*_{\geq 0}$ with interior $\mathfrak{t}^*_{>0}$ and the following subset

$$Y = \phi_G^{-1}(\mathfrak{t}^*_{>0}).$$

Then Y is a T-invariant submanifold of M, not necessarily connected, but every connected component of Y is even-dimensional. We denote by $\phi_T : Y \to \mathfrak{t}^*$ the restriction of the map ϕ_G to Y. Thus $\phi_T(Y)$ is contained in $\mathfrak{t}^*_{>0}$.

Lemma 3.3. *The K-equivariant spin structure on M induces a $\tilde{T}$-equivariant spin structure on Y. We denote by $\mathscr{S}_Y$ the corresponding spinor bundle.*

Now we explain how we can attach a multiplicity function

$$m_j : \tilde{\Lambda} \times \mathbb{Z}_{>0} \to \mathbb{Z}$$

to a connected component Y_j of Y. Let $\mathfrak{t}_j \subset \mathfrak{t}$ be the generic infinitesimal stabilizer of the T-action on Y_j. We denote by $I_j \subset \mathfrak{t}^*$ the affine rational subspace with direction $\mathfrak{t}_j^\perp$ that contains $\phi_T(Y_j)$.

We consider a finite collection $\mathscr{B}_j$ of affine co-dimension-1 subspaces of I_j defined as follows: $A \in \mathscr{B}_j$ if $A = \phi_T(y) + \mathfrak{t}_y^\perp$ for some $y \in Y_j$ such that $\dim(\mathfrak{t}_y/\mathfrak{t}_j) = 1$.

Definition 3.4. A chamber of I_j is a connected component of the open subset $(I_j)_{\mathrm{reg}} := \left(I_j \setminus \bigcup_{E \in \mathscr{B}_j} E\right) \cap \mathfrak{t}^*_{>0}$.

Thus, the closure of a chamber is a rational polyhedron contained in $\mathfrak{t}^*_{\geq 0}$.

Let $\mathfrak{c}$ be a chamber of I_j and let $\xi \in \mathfrak{c}$ be a regular value of the map $\phi_T : Y_j \to I_j$. We consider the orbifold

$$\mathscr{Y}_{j,\xi} := (\phi_T^{-1}(\xi) \cap Y_j)/T.$$

Lemma 3.5. *Let $(\lambda, k) \in \tilde{\Lambda} \times \mathbb{Z}_{>0}$ such that $\frac{\lambda}{k} \in I_j$. The $\tilde{T}$-equivariant spinc bundle $\mathscr{S}_{Y_j} \otimes \mathscr{L}^k|_{Y_j} \otimes \mathbb{C}_{-\lambda}$ on Y_j induces a spinc bundle $\mathscr{S}^{\lambda,k}_{j,\xi}$ on $\mathscr{Y}_{j,\xi}$. We denote by $\mathscr{Q}(\mathscr{Y}_{j,\xi}, \mathscr{S}^{\lambda,k}_{j,\xi})$ the index of the corresponding Dirac operator.*

Let $E_j \subset \mathfrak{t}^* \times \mathbb{R}$ be the subspace generated by $(v, 1), v \in I_j$. We denote by $\tilde{\Lambda}_j := (\tilde{\Lambda} \times \mathbb{Z}) \cap E_j$ the corresponding lattice of E_j. The Kawasaki index theorem [11] tells us that the map

$$(\lambda, k) \longmapsto \mathscr{Q}(\mathscr{Y}_{j,\xi}, \mathscr{S}^{\lambda,k}_{j,\xi})$$

is a quasi-polynomial function $\mathrm{m}_{j,\xi}$ on $\tilde{\Lambda}_j$.

The following result is proved in [21].

Theorem 3.6. • *The quantity $\mathscr{Q}(\mathscr{Y}_{j,\xi}, \mathscr{S}^{\lambda,k}_{j,\xi})$ does not depend on the choice of $\xi \in \mathfrak{c}$. We denote by $\mathrm{m}_{j,\mathfrak{c}} : \tilde{\Lambda}_j \to \mathbb{C}$ the corresponding quasi-polynomial.*

• *Let $(\lambda, k) \in \tilde{\Lambda} \times \mathbb{Z}_{>0}$ and let $\mathfrak{c}_1, \mathfrak{c}_2$ be two chambers such that $\frac{\lambda}{k} \in \overline{\mathfrak{c}_1} \cap \overline{\mathfrak{c}_2}$. Then*

$$\mathrm{m}_{j,\mathfrak{c}_1}(\lambda, k) = \mathrm{m}_{j,\mathfrak{c}_2}(\lambda, k).$$

The first point of the previous theorem tells us that $\mathrm{m}_{j,\mathfrak{c}}$ is the zero map if the chamber $\mathfrak{c}$ is not contained in the image of the map $\Phi_T : Y_j \to I_j$. It is due to the fact that we can choose a regular value $\xi \in \mathfrak{c}$ with empty reduced space $\mathscr{Y}_{j,\xi}$.

Thus we can define multiplicity functions $(m_j)_{j \in J}$ parametrized by the set J of connected components of the manifold Y as follows.

Definition 3.7. The map $m_j : \tilde{\Lambda} \times \mathbb{Z}_{>0} \to \mathbb{Z}$ is defined as follows:

$$m_j(\lambda, k) = \begin{cases} 0 & \text{if } \frac{\lambda}{k} \notin I_j, \\ \mathrm{m}_{j,\mathfrak{c}}(\lambda, k) & \text{if } \frac{\lambda}{k} \in \overline{\mathfrak{c}} \subset I_j. \end{cases}$$

We can now state the main result of [21].

Theorem 3.8. *The following relation*

$$m_G(\lambda, k) = \sum_{j \in J} m_j(\lambda, k)$$

holds for any $(\lambda, k) \in \tilde{\Lambda} \times \mathbb{Z}_{>0}$.

Let us fix j. Let $\mathscr{A}_j$ be the collection of closed rational polyhedrons formed by the faces of the closures $\overline{\mathfrak{c}}$ of the chambers in I_j (they are contained in $\mathfrak{t}^*_{\geq 0}$). To each $P \in \mathscr{A}_j$, we associate

- the subspace $E_P \subset \mathfrak{t}^* \times \mathbb{R}$ generated by $(v,1), v \in P$ and the lattice $\tilde{\Lambda}_P := (\tilde{\Lambda} \times \mathbb{Z}) \cap E_P$,
- the cone $C_P := \{(t\xi, t), t \geq 0, \xi \in P\} \subset E_P$,
- the quasi-polynomial $m_P : \tilde{\Lambda}_P \to \mathbb{C}$ such that $m_P(\lambda,k) = \mathrm{m}_{j,\mathfrak{c}}(\lambda,k)$, for $(\lambda,k) \in \tilde{\Lambda}_P$, if P is a face of $\bar{\mathfrak{c}}$.

By inclusion-exclusion, we see that the multiplicity function m_j admits the decomposition

$$m_j = \sum_{P \in \mathscr{A}_j} \alpha_P m_P[C_P], \tag{3.4}$$

for some appropriate choice of constants $\alpha_P \in \mathbb{Z}$. In other words, the multiplicity function m_j is in the space $\mathscr{S}(\tilde{\Lambda})$. The relation $m_G = \sum_j m_j$ given by Theorem 3.8 shows that $m_G \in \mathscr{S}(\tilde{\Lambda})$. The proof of Theorem 3.2 is completed. □

Example 3.9. Let us give a simple example for the group $G = SU(2)$. Consider $\mathfrak{t}^*$ with basis ρ. In this basis $A_G = \{\lambda; \lambda \in \mathbb{Z}, \lambda > 0\}$. We consider the line bundle $\mathscr{L}$ associated to (ρ,ρ) on the spin manifold $M = G/T \times G/T$. We see that $\mathscr{Q}_G^{\mathrm{spin}}(M,\mathscr{L}^k)$ is the tensor product $V(k\rho) \otimes V(k\rho)$, and $V(k\rho)$ is the irreducible representation of dimension k of G. So $\mathscr{Q}_G^{\mathrm{spin}}(M,\mathscr{L}^k)$ is the sum of the representations $V(j\rho)$ with j odd and less than $2k$. If we write $\mathscr{Q}_G^{\mathrm{spin}}(M,\mathscr{L}^k) = \sum_{\lambda>0} m_G(\lambda,k) V(\lambda\rho)$,

$$m_G(\lambda,k) = \frac{1}{2}(1-(-1)^{\lambda}), \quad \text{for } 1 \leq \lambda \leq 2k.$$

3.2 Equivariant cohomology

Let N be a G-manifold and let $\mathscr{A}(N)$ be the space of differential forms on N, graded by its exterior degree. Following [3] and [27], an equivariant form is a G-invariant smooth function $\alpha : \mathfrak{g} \to \mathscr{A}(N)$, thus $\alpha(X)$ is a differential form on N depending smoothly on $X \in \mathfrak{g}$. Consider the operator

$$d_{\mathfrak{g}}\alpha(X) = d\alpha(X) - \iota(v_X)\alpha(X) \tag{3.5}$$

where $\iota(v_X)$ is the contraction by the vector field v_X generated by the action of $-X$ on N. Then $d_{\mathfrak{g}}$ is an odd operator with square 0, and the equivariant cohomology is defined to be the cohomology space of $d_{\mathfrak{g}}$. It is important to note that the dependence of α on X may be C^{∞}. If the dependence of α on X is polynomial, we denote by $H_G^*(N)$ the corresponding $\mathbb{Z}$-graded algebra. By definition, the grading of $P(X) \otimes \mu$, where P is a homogeneous polynomial and μ is a differential form on N, is the exterior degree of μ plus twice the polynomial degree in X.

The line bundle $\mathscr{L} \to M$ determines the closed equivariant form $\Omega(X) = \Omega - \langle \phi_G, X \rangle$ on M. Here we will not assume any nondegeneracy condition on Ω.

Choose a G-invariant Riemannian metric on M. Let

$$J(A) = \det_{\mathbb{R}^{2d}} \left(\frac{e^{A/2} - e^{-A/2}}{A} \right),$$

an invariant function of A. Then $J(0) = 1$. Consider $\frac{1}{J^{1/2}(A)}$ and its Taylor expansion at 0:

$$\frac{1}{J^{1/2}(A)} = \det{}^{1/2}_{\mathbb{R}^{2d}} \left(\frac{A}{e^{A/2} - e^{-A/2}} \right) = \sum_{n=0}^{\infty} B_n(A).$$

Each function $B_n(A)$ is an invariant polynomial of degree n (in fact B_n is nonzero only for n even) and by the Chern-Weil construction, and choice of connections on TM, the function B_n determines a closed equivariant form $\hat{A}_n(M)(X)$ on M of homogeneous equivariant degree $2n$. Note that $\hat{A}_0(M) = 1$. We define the formal series of equivariant forms:

$$\hat{A}(M)(X) = \sum_{n=0}^{\infty} \hat{A}_n(M)(X).$$

As M is compact, we can find a positive constant r_M such that for $\|X\| < r_M$, this series of equivariant forms is convergent. In particular, $\hat{A}(M)(0)$ is a closed differential form on M which represents the usual $\hat{A}$ class of M.

3.3 The equivariant index

Recall the "delocalized index formula" (see [2]). For $X \in \mathfrak{g}$ such that $\|X\| < r_M$, so that $\hat{A}(M)(X)$ is well defined, we have

$$\mathscr{Q}_G^{\mathrm{spin}}(M, \mathscr{L})(\exp X) = \frac{1}{(-2i\pi)^d} \int_M e^{-i\Omega(X)} \hat{A}(M)(X). \tag{3.6}$$

Here $d = \dim M/2$. Note that Kostant relations (3.1) imply that the equivariant form $\Omega(X) = \Omega - \langle \phi_G, X \rangle$ is closed. In other words, $\iota(X)\Omega + d\langle \phi_G, X \rangle = 0$ for any $X \in \mathfrak{g}$.

For every integer n, consider the analytic function on $\mathfrak{g}$ given by

$$\mathscr{I}_n(X) = \frac{1}{(-2i\pi)^d} \int_M e^{-i\Omega(X)} \hat{A}_n(M)(X).$$

There is a remarkable relation between the character associated to $\mathscr{L}^k$ and the dilation $X \to X/k$ on $\mathfrak{g}$.

Lemma 3.10. *When $X \in \mathfrak{g}$ is such that $\|X\| < r_M$, then, for any $k \geq 1$, one has*

$$\mathcal{Q}_G^{\text{spin}}(M,\mathcal{L}^k)(\exp(X/k)) = k^d \sum_{n=0}^{\infty} \frac{1}{k^n}\mathcal{I}_n(X).$$

Proof. Write $\hat{A}(M)(X) = \sum_{n=0}^{\infty} \tau_n(X)$ as a sum of forms with homogeneous polynomials in X of degree n as coefficients. Thus

$$\hat{A}_n(M)(X) = \sum_{q\leq n} (\tau_q(X))_{[n-2q]}$$

where $\omega_{[s]}$ is the component of exterior degree s of a differential form ω.

For $\|X\| < r_M$, $\sum_{n=0}^{\infty} \tau_n(X)$ is a convergent series whose sum is the equivariant $\hat{A}$ class. We obtain

$$\mathcal{Q}_G^{\text{spin}}(M,\mathcal{L}^k)(\exp(X/k)) = \frac{1}{(-2i\pi)^d}\int_M e^{-ik\Omega(X/k)}\Big(\sum_{n=0}^{\infty} \tau_n(X/k)\Big).$$

Let $b(X,k) = e^{-ik\Omega(X/k)} = e^{-ik\Omega}e^{i\langle\phi_G,X\rangle}$. Note that $b(X,k)$ depends polynomially on k. We rewrite

$$\mathcal{Q}_G^{\text{spin}}(M,\mathcal{L}^k)(\exp(X/k)) = \frac{1}{(-2i\pi)^d}\int_M b(X,k)\Big(\sum_{n=0}^{\infty} \frac{1}{k^n}\tau_n(X)\Big). \tag{3.7}$$

Only the top exterior degree term contributes to the integral. We compute it in terms of the equivariant forms $\hat{A}_n(M)$.

Lemma 3.11.

$$\Big[b(X,k)\Big(\sum_{n=0}^{\infty}\frac{1}{k^n}\tau_n(X)\Big)\Big]_{[2d]} = k^d\Big[e^{-i\Omega(X)}\Big(\sum_{n=0}^{\infty}\frac{1}{k^n}\hat{A}_n(M)(X)\Big)\Big]_{[2d]}.$$

Proof. The coefficient of $k^d\frac{1}{k^n}$ on the left-hand side is

$$\sum_{a=0}^{d}\frac{(-i\Omega)^a}{a!}(\tau_{n+a-d}(X))_{[2d-2a]}e^{i\langle\phi_G,X\rangle}.$$

This is the term of exterior degree $2d$ in the differential form $e^{-i\Omega(X)}\hat{A}_n(M)(X)$. □

This identity implies the lemma. □

3.4 *Twisted Duistermaat-Heckman distributions*

Let ν be a closed equivariant form on M with polynomial coefficients. Let $-i\Omega(X) = -i\Omega + i\langle\phi_G,X\rangle$ be our equivariant curvature. As $\nu(X)$ depends polynomially on X, $\int_M e^{-i\Omega(X)}\nu(X)$ is a function on $\mathfrak{g}$ of at most polynomial growth.

Definition 3.12. We denote by $DH^G(M,\Omega,\nu) \in \mathscr{D}'(\mathfrak{g}^*)^G$ the Fourier transform of $\frac{1}{(-2i\pi)^d}\int_M e^{-i\Omega(X)}\nu(X)$.

We thus have, for φ smooth with compact support,

$$\begin{aligned}\langle DH^G(M,\Omega,\nu),\varphi\rangle &= \frac{1}{(-2i\pi)^d}\int_M\int_{\mathfrak{g}} e^{-i\Omega(X)}\nu(X)\widehat{\varphi}(X)dX\\ &= \frac{1}{(-2i\pi)^d}\int_M e^{-i\Omega}[\nu(-i\partial)\varphi](\phi_G(m)).\end{aligned}$$

If $\nu(X)=\sum_a p_a(X)\nu_a$ with polynomial functions p_a on $\mathfrak{g}$ and differential forms ν_a on M, the expression

$$[\nu(-i\partial)\varphi](\phi_G(m)) = \sum_a [p_a(-i\partial)(\varphi)](\phi_G(m))\nu_a$$

is a differential form on M supported on the pull-back by ϕ_G of the support of φ. This shows, and this will be used in our application to a line bundle with a proper moment map, that $DH^G(M,\Omega,\nu)$ is a well-defined distribution supported on $\phi_G(M)$, provided ϕ_G is proper. Notice that the distribution $DH^G(M,\Omega,\nu)$ depends only on the class $[\nu]\in H^*_G(M)$.

Recall that we have associated to $(M,\mathscr{L})$ the distributions on $\mathfrak{g}^*$ given by

$$\begin{aligned}\Psi_k^{(M,\mathscr{L})} &:= \mathscr{F}_{\mathfrak{g}}\circ \mathrm{rescaling}_{\mathfrak{g}}(k)\Big(\mathscr{Q}_G^{\mathrm{spin}}(M,\mathscr{L}^k)(e^X)\Big),\\ \Theta_k^{(M,\mathscr{L})} &:= \mathscr{F}_{\mathfrak{g}}\circ \mathrm{rescaling}_{\mathfrak{g}}(k)\Big(\mathscr{Q}_G^{\mathrm{spin}}(M,\mathscr{L}^k)(e^X)j_{\mathfrak{g}}^{1/2}(X)\Big).\end{aligned}$$

The distribution $\Theta_k^{(M,\mathscr{L})}$ is the weighted sum $k^r\sum_{\lambda\in A_G} m_G(\lambda,k)\beta_{\lambda/k}$ of the canonical measures on the orbits $G\lambda/k$.

Theorem 3.13. *Let M be a compact even-dimensional oriented spin manifold with a G-equivariant line bundle $\mathscr{L}$. When k tends to ∞, we have the asymptotic expansions*

$$\Psi_k^{(M,\mathscr{L})} \equiv k^d\sum_{n=0}^{\infty}\frac{1}{k^n}DH^G(M,\Omega,\hat{A}_n(M)) \tag{3.8}$$

and

$$\Theta_k^{(M,\mathscr{L})} \equiv j_{\mathfrak{g}}^{1/2}(i\partial/k)\Big(k^d\sum_{n=0}^{\infty}\frac{1}{k^n}DH^G(M,\Omega,\hat{A}_n(M))\Big). \tag{3.9}$$

Proof. We present a detailed proof since we will have to adapt this proof to the case of a noncompact manifold M.

We fix a G-invariant function $h:\mathfrak{g}\to\mathbb{R}$ equal to 1 on a small ball $\|X\|\le r$ and with compact support contained in $\|X\|<r_M$, so that $X\mapsto h(X)\hat{A}(M)(X)$ is a smooth compactly supported function on $\mathfrak{g}$ whose values are differential forms on M.

Let φ be a smooth compactly supported function on $\mathfrak{g}^*$. We have $\langle\Psi_k^{(M,\mathscr{L})},\varphi\rangle = \mathrm{I}_k+\mathrm{J}_k$ where

$$\mathrm{I}_k = \int_{\mathfrak{g}} \mathscr{Q}_G^{\mathrm{spin}}(M,\mathscr{L}^k)(e^{X/k})h(X/k)\widehat{\varphi}(X)dX,$$

and

$$\mathrm{J}_k = \int_{\mathfrak{g}} \mathscr{Q}_G^{\mathrm{spin}}(M,\mathscr{L}^k)(e^{X/k})(1-h(X/k))\widehat{\varphi}(X)dX.$$

We will see that $\mathrm{J}_k = O(k^{-\infty})$. For estimating I_k, we will use the delocalized index formula (3.6).

Using the facts that $\left|\mathscr{Q}_G^{\mathrm{spin}}(M,\mathscr{L}^k)(g)\right|$ is uniformly bounded by a polynomial in k, that the support of $(1-h(X/k))$ is contained in the set $\|X\| \geq rk$, $r>0$, and that the function $\widehat{\varphi}$ is rapidly decreasing, we see that $\mathrm{J}_k = O(k^{-\infty})$.

We now analyze I_k. We can use Formula (3.6) since $\|X/k\| \leq r_M$. Thus

$$\mathrm{I}_k = \frac{1}{(-2i\pi)^d}\int_M \int_{\mathfrak{g}} e^{-ik\Omega(X/k)}\hat{A}(M)(X/k)h(X/k)\widehat{\varphi}(X)dX.$$

Proposition 2.5 extends to the case of vector-valued functions on $\mathfrak{g}$. Let E be a finite-dimensional space, and let $\tau : \mathfrak{g} \to \Lambda E$ be a smooth function with at most polynomial growth, as well as all of its derivatives. The coefficients $\tau_n(X)$ of its Taylor series $\sum_{n=0}^{\infty}\tau_n(X)$ are ΛE-valued polynomial functions homogeneous of degree n. on $\mathfrak{g}$. Let $b(X,k) = \sum_{m=0}^{d} k^m b_m(X)$ where $b_m(X)$ are smooth functions of X with values in ΛE with at most polynomial growth. Thus

$$b(X,k)\sum_{n=0}^{\infty}\tau_n(X/k) = \Big(\sum_{m=0}^{d} k^m b_m(X)\Big)\Big(\sum_{n=0}^{\infty}\frac{1}{k^n}\tau_n(X)\Big)$$

is a Laurent series in $\frac{1}{k}$ of functions on $\mathfrak{g}$ with values in ΛE.

We have

$$\int_{\mathfrak{g}} b(X,k)\tau(X/k)\widehat{\varphi}(X)dX \equiv \int_{\mathfrak{g}} b(X,k)\Big(\sum_{n=0}^{\infty}\tau_n(X/k)\Big)\widehat{\varphi}(X)dX. \tag{3.10}$$

In short, we replace τ by its Taylor series.

Consider the differential form $h(X)\hat{A}(M)(X)$. For any given $x \in M$, $\tau_x^h(X) = h(X)\hat{A}(M)|_x(X)$ is a smooth compactly supported function on $\mathfrak{g}$ with values in ΛT_x^*M. Let

$$b_x(X,k) = e^{i\langle \phi_G(x),X\rangle}e^{-ik\Omega_x}.$$

It depends polynomially on k and each coefficient of this polynomial is a bounded function of X (as it is proportional to $e^{i\langle \phi_G(x),X\rangle}$). So we can certainly apply the asymptotic formula (3.10).

Let $\mathrm{I}_k^x = \frac{1}{(-2i\pi)^d}\int_{\mathfrak{g}} b_x(X,k)\tau_x^h(X/k)\widehat{\varphi}(X)dX$, with values in ΛT_x^*M. So $x \mapsto \mathrm{I}_k^x$ is a differential form and $\mathrm{I}_k = \int_{x\in M}\mathrm{I}_k^x$.

The manifold M is compact. Proposition 2.5 gives us an asymptotic expansion for I_k^x, and the proof shows that the rest is bounded uniformly. So the asymptotic expansion of I_k is obtained by replacing (at each $x \in M$) $\tau_x^h(X/k)$ by its Taylor series. As h is identically 1 near zero, we obtain, in the notation of the proof of Lemma 3.10, that

$$\mathrm{I}_k \equiv \frac{1}{(-2i\pi)^d} \int_M \int_{\mathfrak{g}} b(X,k) \Big(\sum_{n=0}^{\infty} \frac{1}{k^n} \tau_n(X) \Big) \widehat{\varphi}(X) dX.$$

By Lemma 3.11, this is

$$k^d \frac{1}{(-2i\pi)^d} \int_M \int_{\mathfrak{g}} e^{-i\Omega(X)} \Big(\sum_{n=0}^{\infty} \frac{1}{k^n} \hat{A}_n(M)(X) \Big) \widehat{\varphi}(X) dX.$$

By definition of the twisted Duistermaat-Heckman measure, we obtain the expansion of formula (3.8).

The proof of the expansion formula (3.9) is identical. □

3.5 Asymptotics localized at $g \in G$

For applications to formal geometric quantization, we state an analogous asymptotic descent formula. Now let $g \in G$. Let $G(g)$ be the centralizer of g in G, and $\mathfrak{g}(g)$ its Lie algebra.

Consider the manifold M^g of fixed points of the action of g on M: it may have several connected components M^g_a, all of even dimension. Since M is spin, M^g admits an orientation (that is all connected components are oriented). Let $\mathcal{N}^g$ be the normal bundle of M^g in M and let

$$R_{\mathcal{N}^g}(X) = R_{\mathcal{N}^g} + \mu_{\mathcal{N}^g}(X)$$

be its equivariant curvature [2].

We denote by $\Omega_g(X)$ the restriction of $\Omega(X)$ to M^g. We consider the following equivariant form:

$$D_g(M)(X) = \det{}^{1/2}_{\mathcal{N}^g} \Big(I - (g^{-1})^{\mathcal{N}^g} e^{-R_{\mathcal{N}^g}(X)} \Big).$$

Here the square root is chosen such that $\det^{1/2}_{\mathcal{N}^g}(I - (g^{-1})^{\mathcal{N}^g}) > 0$.

For $X \in \mathfrak{g}(g)$ small enough, we have

$$\mathcal{Q}_G^{\mathrm{spin}}(M, \mathcal{L})(g \exp(X)) = \int_{M^g} c_g \frac{\hat{A}(M^g)(X)}{D_g(M)(X)} g^{\mathcal{L}} e^{-i\Omega_g(X)} \tag{3.11}$$

where

• $g^{\mathcal{L}}$ is the locally constant function on M^g with value a complex number of modulus 1 given by the action of g on the fiber of $\mathcal{L}|_{M^g}$. We write u_a for the restriction of $g^{\mathcal{L}}$ to a connected component M^g_a. This is a complex number of modulus 1. If g is of finite order, u_a is a root of unity.

• $c_g = \varepsilon_g(-2i\pi)^{-\dim M/2}(2\pi)^{\mathrm{rk}\mathcal{N}^g/2}$ is a locally constant function independent of $\mathcal{L}$. Here the sign $\varepsilon_g = \pm$ depends only on the action of g on the spinor bundle $\mathcal{S}|_{M^g}$.

Consider the equivariant form

$$V_g(\mathcal{L},k)(X) = c_g(g^{\mathcal{L}})^k \frac{\hat{A}(M^g)(X)}{D_g(M)(X)}$$

on M^g. Then we have $V_g(\mathcal{L},k)(X) = \sum_{n=0}^{\infty} V_g(n,k)(X)$ where $V_g(n,k)(X)$ are $G(g)$-equivariant forms on M^g, homogeneous of even degree $2n$.

Thus, for $X \in \mathfrak{g}(g)$ small, we have

$$\mathcal{Q}_G^{\mathrm{spin}}(M,\mathcal{L}^k)(g\exp(X)) = \sum_{n=0}^{\infty} \int_{M^g} e^{-ik\Omega_g(X)} V_g(n,k)(X).$$

Here Lemma 3.10 becomes

Lemma 3.14. *When $X \in \mathfrak{g}(g)$ is sufficiently small, then, for any $k \geq 1$, one has*[2]

$$\mathcal{Q}_G^{\mathrm{spin}}(M,\mathcal{L}^k)(g\exp(X/k)) = \sum_{n=0}^{\infty} \frac{1}{k^n} k^{\frac{\dim M^g}{2}} \mathcal{I}_g(n,k)(X)$$

where $\mathcal{I}_g(n,k)(X) := \int_{M^g} e^{-i\Omega_g(X)} V_g(n,k)(X)$.

Now consider the character χ_λ of the irreducible representation attached to the admissible orbit $\mathcal{O}_\lambda$. If $g \in G$, the fixed-point set $\mathcal{O}_\lambda^g$ is a union of coadjoint orbits of the connected component $G(g)^o$ of $G(g)$. They are all of the same even dimension $r_g = \dim G(g) - \dim T$.

Then we have the formula, for $X \in \mathfrak{g}(g)$,

$$\chi_\lambda(ge^X) j_{\mathfrak{g}(g)}^{1/2}(X) \det{}_{\mathfrak{g}/\mathfrak{g}(g)}^{1/2}(1-g^{-1}e^{-X}) = \langle \beta(g,\lambda)(\xi), e^{i\langle \xi, X\rangle} \rangle. \tag{3.12}$$

Here the square root is determined such that $\det_{\mathfrak{g}/\mathfrak{g}(g)}^{1/2}(1-g^{-1}) > 0$ and $\beta(g,\lambda)$ is a measure on $\mathcal{O}_\lambda^g$ entirely determined by this equation.

Let $\tilde{T}$ be the torus with weight lattice $\tilde{\Lambda}$ (the lattice generated by Λ and ρ). Then any $\lambda \in A_G$ defines a character $\tilde{t} \in \tilde{T} \to \tilde{t}^\lambda$. Suppose that $g \in T$ and denote by $W_{G(g)} \subset W_G$ the Weyl group of $G(g)^o$. Then

$$\mathcal{O}_\lambda^g = \bigcup_{\overline{w} \in W_{G(g)} \backslash W_G} G(g)^o w\lambda.$$

Let $\mathcal{R}_{\mathfrak{g}(g)} : \mathcal{D}'(\mathfrak{g}(g))^{G(g)^o} \to \mathcal{D}'(\mathfrak{t})^{W_{G(g)}-alt}$ be the canonical isomorphism.

Lemma 3.15. *Choose $\tilde{g} \in \tilde{T}$ above $g \in T$.*

---

[2] The term $k^{\frac{\dim M^g}{2}} \mathcal{I}_g(n,k)(X)$ must be understood as the sum $\sum_a k^{\frac{\dim M_a^g}{2}} \int_{M_a^g} e^{-i\Omega_g(X)} V_g(n,k)(X)$.

• *The measure* $\beta(g,\lambda)$ *satisfies the relation*

$$\beta(g,\lambda)=\gamma_{\tilde{g}}\sum_{\overline{w}\in W_{G(g)}\backslash W_G}\tilde{g}^{w\lambda}\beta_{G(g)^o w\lambda}$$

where $\gamma_{\tilde{g}}$ *is a complex number of modulus* 1.

• *We have* $\mathscr{R}_{\mathfrak{g}(g)}(\beta(g,\lambda))=\gamma_{\tilde{g}}\sum_{w\in W_G}\varepsilon(w)\tilde{g}^{w\lambda}\delta_{w\lambda}$.

Proof. The first point follows from the character formula and the second is a direct consequence of the first. □

Definition 3.16. We denote by $DH^{G(g)}(M^g,\Omega_g,V_g(n,k))\in\mathscr{D}'(\mathfrak{g}(g)^*)^{G(g)}$ the Fourier transform of the function $\mathscr{I}_g(n,k)$ (see Lemma 3.14).

Then for a test function φ, $\langle DH^{G(g)}(M^g,\Omega_g,V_g(n,k)),\varphi\rangle$ is of the form $\sum_a u_a^k\langle s_{a,n},\varphi\rangle$ with $\langle s_{a,n},\varphi\rangle=\int_{M_a^g}\int_{\mathfrak{g}(g)}e^{-i\Omega_g(Y)}\alpha_{a,n}(Y)\widehat{\varphi}(Y)dY$; here $\alpha_{a,n}(Y)$ is an equivariant form on M_a^g of degree $2n$, and u_a is the restriction of $g^{\mathscr{L}}$ to M_a^g.

The asymptotic expansions of distributions of the form $\sum_a\sum_{n=0}^{\infty}u_a^k k^{-n}h_{n,a}$ are thus well defined and the coefficients $H(n,k)=\sum_a u_a^k h_{n,a}$ are uniquely determined.

With the same proof as for Theorem 3.13, we obtain the following theorem.

Theorem 3.17. *Let M be a compact even-dimensional oriented spin manifold with a G-equivariant line bundle* $\mathscr{L}$. *Let* $g\in G$ *be of finite order. When k tends to* ∞, *we have the asymptotic equivalence*[3]

$$\text{rescaling}_{\mathfrak{g}(g)^*}(k)\Big(\sum_{\lambda\in\Lambda}m(\lambda,k)\beta(g,\lambda)\Big)\equiv$$

$$j_{\mathfrak{g}(g)}^{\frac{1}{2}}(i\partial/k)\det{}^{\frac{1}{2}}_{\mathfrak{g}/\mathfrak{g}(g)}\big(1-g^{-1}e^{-i\partial/k}\big)\sum_{n=0}^{\infty}\frac{1}{k^n}k^{\frac{\dim M^g}{2}}DH^{G(g)}(M^g,\Omega_g,V_g(n,k)).$$

4 Spin quantization in the noncompact setting

In this section we work with an even-dimensional oriented spin manifold M of dimension $2d$ provided with an action of G. We do not assume that M is compact. Let $(\mathscr{L},\nabla)$ be a G-equivariant line bundle equipped with a G-invariant Hermitian connection ∇.

We assume that the moment map $\phi_G: M\to\mathfrak{g}^*$ defined by the Kostant relation (3.1) is a proper map. In the next section we explain how the formal geometric quantization of the data $(M,\phi_G,\mathscr{L})$ is defined.

Notice that the twisted Duistermaat-Heckman distributions $DH^G(M,\Omega,\nu)\in\mathscr{D}'(\mathfrak{g}^*)^G$ are still defined by the relation

---

[3] The term $k^{\frac{\dim M^g}{2}}DH^{G(g)}(M^g,\Omega_g,V_g(n,k))$ must be understood as the sum $\sum_a k^{\frac{\dim M_a^g}{2}}DH^{G(g)}(M_a^g,\Omega_g,V_g(n,k))$.

$$\langle DH^G(M,\Omega,\nu),\varphi\rangle = \frac{1}{(-2i\pi)^d}\int_M e^{-i\Omega}[\nu(-i\partial)\varphi](\phi_G(m)). \tag{4.1}$$

Here ν is a closed equivariant form on M with polynomial coefficients and φ is a test function on $\mathfrak{g}^*$. The integral in (4.1) is well defined since the differential form $e^{-i\Omega}[\nu(-i\partial)\varphi](\phi_G(m))$ has compact support. The distribution $DH^G(M,\Omega,\nu)$ depends only on the class defined by ν in $H^*_G(M)$.

4.1 Formal geometric quantization: definition

The invariant scalar product on $\mathfrak{k}^*$ provides an identification $\mathfrak{k} \simeq \mathfrak{k}^*$.

Definition 4.1. • The *Kirwan vector field* associated to ϕ_G is defined by

$$\kappa_G(m) = -\phi_G(m)\cdot m, \quad m \in M. \tag{4.2}$$

• We denote by Z_G the set of zeroes of κ_G. Thus Z_G is a G-invariant closed subset of M.

The set Z_G, which is not necessarily smooth, admits the following description. Consider the closed Weyl chamber $\mathfrak{t}^*_{\geq 0}$. We see that

$$Z_G = \coprod_{\gamma\in\mathscr{B}_G} Z_\gamma, \tag{4.3}$$

where Z_γ corresponds to the compact set $G(M^\gamma \cap \phi_G^{-1}(\gamma))$, and $\mathscr{B}_G = \phi_G(Z_G)\cap \mathfrak{t}^*_{\geq 0}$. The properness of ϕ_G ensures that, for any compact subset $C \subset \mathfrak{t}^*$, the intersection $\mathscr{B}_G \cap C$ is finite. Here M^γ is the set of zeroes of the vector field on M defined by the infinitesimal action of γ.

Let $\mathscr{S} \to M$ be the spinor bundle on M. The principal symbol of the Dirac operator $D_{\mathscr{S}}$ is the bundle map $\sigma(M) \in \Gamma(\mathrm{T}^*M, \hom(\mathscr{S}^+,\mathscr{S}^-))$ defined by the Clifford action

$$\sigma(M)(m,\nu) = \mathbf{c}_m(\tilde{\nu}) : \mathscr{S}|_m^+ \to \mathscr{S}|_m^-,$$

where $\nu \in \mathrm{T}^*M \simeq \tilde{\nu} \in \mathrm{T}M$ is an identification associated to an invariant Riemannian metric on M.

Definition 4.2. The symbol $\sigma(M,\phi_G)$ shifted by the vector field κ_G is the symbol on M defined by

$$\sigma(M,\phi_G)(m,\nu) = \sigma(M)(m,\tilde{\nu} - \kappa_G(m))$$

for any $(m,\nu) \in \mathrm{T}^*M$.

For any G-invariant open subset $\mathscr{U} \subset M$ such that $\mathscr{U} \cap Z_G$ is compact in M, we see that the restriction $\sigma(M,\phi_G)|_{\mathscr{U}}$ is a transversally elliptic symbol on $\mathscr{U}$. If $E \to M$ is a complex vector bundle, we denote by $D^{\mathscr{U}}_{E,\phi_G}$ a pseudo-differential operator whose principal symbol defines the same class as $\sigma(M,\phi_G)\otimes E|_{\mathscr{U}}$ in the group of equivariant K-theory of $\mathrm{T}^*_G\mathscr{U}$ (see [1]).

Thus we can define the following localized equivariant indices (see [1,22]).

Definition 4.3. • A closed invariant subset $Z \subset Z_G$ is called a *component* of Z_G if it is a union of connected components of Z_G.

• If Z is a *compact component* of Z_G, and $E \to M$ is an equivariant vector bundle, we denote by

$$\mathcal{Q}_G^{\text{spin}}(M,E,Z) \in \hat{R}(K)$$

the equivariant index of the transversally elliptic operator $D^{\mathcal{U}}_{E,\phi_G}$. Here $\mathcal{U}$ is an invariant neighborhood of Z such that $\mathcal{U} \cap Z_G = Z$.

By definition, $Z = \emptyset$ is a component of $Z_{\mathcal{S}}$ and $\mathcal{Q}_G^{\text{spin}}(M,\mathcal{S},\emptyset) = 0$. For any $\gamma \in \mathcal{B}_{\mathcal{S}}$, Z_γ is a compact component of $Z_{\mathcal{S}}$.

When the manifold M is compact, the set $\mathcal{B}_G$ is finite and we have the decomposition $\mathcal{Q}_G^{\text{spin}}(M,\mathcal{L}^k) = \sum_{\gamma\in\mathcal{B}_G} \mathcal{Q}_G^{\text{spin}}(M,\mathcal{L}^k,Z_\gamma)$.

Definition 4.4. When the moment map ϕ_G is proper, we define the formal geometric quantization of the data $(M,\phi_G,\mathcal{L}^k)$ as

$$\mathcal{Q}_G^{-\infty}(M,\mathcal{L}^k) := \sum_{\gamma\in\mathcal{B}_G} \mathcal{Q}_G^{\text{spin}}(M,\mathcal{L}^k,Z_\gamma). \tag{4.4}$$

The sum of the right-hand side is not necessarily finite but it converges in $\hat{R}(G)$ (see [10,14,18,19]).

In the following example, for any $\lambda \in \mathbb{Z}$, we denote by $\mathbb{C}_{[\lambda]}$ the vector space $\mathbb{C}$ with the action of $S^1 : t\cdot z = t^\lambda z$, for $(t,z) \in S^1 \times \mathbb{C}_{[\lambda]}$.

Example 4.5. Consider the S^1-manifold $M = \mathbb{C}_{[2]}$. The S^1-equivariant spinor bundle on M is $\mathcal{S} = M \times \left(\mathbb{C}_{[-1]} \oplus \mathbb{C}_{[1]}\right)$.

Fix $a \in \mathbb{N}$. Consider the equivariant line bundle $\mathcal{L}(a) = M \times \mathbb{C}_{[a]}$ with connection $\nabla = d - \frac{i}{2} Im(zd\bar{z})$. The two-form is $\Omega^a = \frac{1}{2} Im(dzd\bar{z})$, the moment map $\phi^a_{S^1}(z) = a + |z|^2$ is proper, and the corresponding critical set $Z^a_{S^1}$ is reduced to $\{0\} \subset M$.

A small computation shows that $\mathcal{Q}_{S^1}^{-\infty}(M,\mathcal{L}(a)^k) = \mathcal{Q}_{S^1}^{\text{spin}}(M,\mathcal{L}(a)^k,\{0\})$ is equal to $\mathbb{C}_{[ka]} \otimes \sum_{j\geq 0} \mathbb{C}_{[2j+1]}$ (see [22]).

Let $m_G(\lambda,k)$ be the multiplicity of V_λ in $\mathcal{Q}_G^{-\infty}(M,\mathcal{L}^k)$. In other words,

$$\mathcal{Q}_G^{-\infty}(M,\mathcal{L}^k) = \sum_{\lambda\in A_G} m_G(\lambda,k) V_\lambda.$$

In this context, the multiplicities $m_G(\lambda,k)$ still have an interpretation in terms of reduced spaces (see [10]). Hence, when the generic infinitesimal stabilizer is abelian, Theorem 3.8 still holds, so $m_G(\lambda,k) \neq 0$ only if $\lambda/k \in \phi_G(M)$.

As in the previous section, we are interested in the asymptotic behavior of the following family of distributions:

$$\Theta_k^{(M,\mathcal{L})} := k^r \sum_{\lambda\in A_G} m_G(\lambda,k)\beta_{\lambda/k},$$

where $r = \dim(G/T)/2$. For any test function φ with compact support, the identity

$$\langle \Theta_k^{(M,\mathscr{L})}, \varphi \rangle := k^r \sum_{\lambda \in A_G} m_G(\lambda,k) \langle \beta_{\lambda/k}, \varphi \rangle$$

is well defined since there exists only a finite number of terms such that $\langle \beta_{\lambda/k}, \varphi \rangle \neq 0$.

Let us recall that we can associate the twisted Duistermaat-Heckman distribution $DH^G(M,\Omega,\hat{A}_n(M)) \in \mathscr{D}'(\mathfrak{g}^*)^G$ to the equivariant form $\hat{A}_n(M)$ for any $n \geq 0$ (see (4.1)).

The aim of this section is to prove the following extension of Theorem 3.13.

Theorem 4.6. *Let M be an even-dimensional oriented spin manifold with a G-equivariant line bundle $(\mathscr{L},\nabla)$. Suppose that the moment map ϕ_G is proper. When k tends to ∞, we have the asymptotic expansion*

$$\Theta_k^{(M,\mathscr{L})} \equiv j_{\mathfrak{g}}^{1/2}(i\partial/k)\Big(k^d \sum_{n=0}^{\infty} \frac{1}{k^n} DH^G(M,\Omega,\hat{A}_n(M))\Big). \tag{4.5}$$

Example 4.7. We continue Example 4.5 with $M = \mathbb{C}_{[2]}$ and $\mathscr{L}(a) = M \times \mathbb{C}_{[a]}$. Then $\Theta_k^{(M,\mathscr{L}(a))} = \sum_{j \geq 0} \delta_{a+(2j+1)/k}$. The equivariant $\hat{A}$-class is $\hat{A}(M)(X) = \frac{X}{\sin(X)}$ for $X \in \mathrm{Lie}(S^1) \simeq \mathbb{R}$. Identity (4.5) says that

$$\Theta_k^{(M,\mathscr{L}(a))} \equiv \frac{k}{2} \frac{i\partial/k}{\sin(i\partial/k)} 1_{[a,\infty[}.$$

This is the formula given in Example 1.3.

With the same proof as in Theorem 4.6, we obtain the following theorem.

Theorem 4.8. *Let M be an even-dimensional oriented spin manifold with a G-equivariant line bundle $\mathscr{L}$. Suppose that the moment map ϕ_G is proper. Let $g \in G$ of finite order. When k tends to ∞, we have the asymptotic equivalence*

$$\mathrm{rescaling}_{\mathfrak{g}(g)^*}(k)\Big(\sum_{\lambda \in \Lambda} m(\lambda,k)\beta(g,\lambda)\Big) \equiv$$

$$j_{\mathfrak{g}(g)}^{\frac{1}{2}}(i\partial/k) \det{}^{\frac{1}{2}}_{\mathfrak{g}/\mathfrak{g}(g)}\big(1 - g^{-1}e^{-i\partial/k}\big) \sum_{n=0}^{\infty} \frac{1}{k^n} k^{\frac{\dim M^g}{2}} DH^{G(g)}(M^g,\Omega_g,V_g(n,k)).$$

We prove Theorem 4.6 in the next two subsections.

4.2 Formal geometric quantization: delocalized formulas

We consider the generalized character

$$\mathscr{Q}_G^{\rm spin}(M,\mathscr{L}^k,Z_\gamma) = \sum_{\lambda\in A_G} m_\gamma(\lambda,k)\,V_\lambda.$$

Here $\mathscr{Q}_G^{\rm spin}(M,\mathscr{L}^k,Z_\gamma)(g) = \sum_{\lambda\in A_G} m_\gamma(\lambda,k)\,\chi_\lambda(g)$ defines a distribution on G because the multiplicity function $\lambda\mapsto m_\gamma(\lambda,k)$ has at most polynomial growth.

The function $m_\gamma(\lambda,k)$ has in fact at most polynomial growth in both variables (λ,k). This follows, for example, from the general multiplicity formula for transversally elliptic operators given in [24]. In this article, the multiplicity formula is given for a single general transversally elliptic symbol σ and multiplicities are obtained as values on A_G of a certain piecewise quasi-polynomial function on A_G. If we consider a family $\sigma_k = \sigma\otimes\mathscr{L}^k$, the formula is locally piecewise polynomial on a certain finite number of affine cones in $\mathfrak{t}^*_{\geq 0}\oplus\mathbb{R}$ intersected with $A_G\oplus\mathbb{Z}$.

We consider the corresponding distribution on $\mathfrak{g}$:

$$\mathscr{Q}_G^{\rm spin}(M,\mathscr{L}^k,Z_\gamma)(e^X) = \sum_{\lambda\in A_G} m_\gamma(\lambda,k)\,\chi_\lambda(e^X).$$

The distribution on $\mathfrak{g}^*$ defined by

$$\Theta_k^\gamma := k^r \sum_{\lambda\in A_G} m_\gamma(\lambda,k)\,\beta_{\lambda/k}$$

is the Fourier transform of $\mathscr{Q}_G^{\rm spin}(M,\mathscr{L}^k,Z_\gamma)(e^{X/k})j_{\mathfrak{g}}^{1/2}(X/k)$. It is tempered also because the multiplicity function $\lambda\mapsto m_\gamma(\lambda,k)$ has at most polynomial growth.

In order to give a formula for the asymptotic behavior of Θ_k^γ, we introduce an equivariant form $P_\gamma(X)$ with generalized coefficients on M that is supported in a small neighborhood of Z_γ.

Take $\chi: M\to\mathbb{R}$, a G-invariant function equal to 1 in a neighborhood of Z_γ, compactly supported, and such that $\mathrm{Support}(\chi)\cap Z_G = Z_\gamma$. Consider the invariant 1-form $\theta := (\kappa_G,-)$ where κ_G is the Kirwan vector field associated to ϕ_G, and $(-,-)$ is an invariant Riemannian metric.

We consider the equivariant form $D\theta(X) = d\theta - \langle\Phi_\theta,X\rangle$, and the equivariant form with generalized coefficients

$$P_\gamma(X) := \chi + i(d\chi)\theta\int_0^\infty e^{-itD\theta(X)}dt.$$

Thus, if $f(X)$ is in the Schwartz space of $\mathfrak{g}$, the integral

$$\int_{\mathfrak{g}} P_\gamma(X)f(X)dX := \chi\hat{f}(0) + i(d\chi)\theta\int_0^\infty e^{-itd\theta}\hat{f}(t\Phi_\theta)dt$$

is convergent. Indeed, $\langle\Phi_\theta,\phi_G\rangle(m) = \|\kappa_G(m)\|^2$, and so $\Phi_\theta\neq 0$ on the support of $d\chi$. It defines a differential form with compact support since it is equal to 0 outside of the support of χ. So if $X\mapsto f(X)$ is a function from $\mathfrak{g}$ to the space of differential forms on M such that $X\to f(X)$ is rapidly decreasing, then we can define $\int_M\int_{\mathfrak{g}} P_\gamma(X)f(X)dX$. Let φ be a test function on $\mathfrak{g}^*$, and $\nu(X)$ an equivariant form. Then $e^{-i\Omega(X)}\nu(X)\widehat{\varphi}(X)$ is rapidly decreasing if $\nu(X)$ is an equivariant form with polynomial coefficients. So we define $DH^G(M,\Omega,\nu P_\gamma) \in \mathscr{D}'(\mathfrak{g}^*)^G$ by

$$\langle DH^G(M,\Omega,\nu P_\gamma),\varphi\rangle = \frac{1}{(-2i\pi)^d}\int_M\int_{\mathfrak{g}} e^{-i\Omega(X)}\nu(X)P_\gamma(X)\widehat{\varphi}(X)dX.$$

An improved version of Witten's nonabelian localization theorem is the following formula:

$$DH^G(M,\Omega,\nu) = \sum_{\gamma\in\mathscr{B}_G} DH^G(M,\Omega,\nu P_\gamma), \tag{4.6}$$

which is somehow a consequence of the fact that $\sum_{\gamma\in\mathscr{B}_G} P_\gamma$ is equal to 1 in cohomology (see [16]).

We will prove the following theorem in the next section.

Theorem 4.9. *We have the following relation:*

$$\Theta_k^\gamma \equiv k^d\, j_{\mathfrak{g}}^{1/2}(i\partial/k)\sum_{n=0}^{\infty}\frac{1}{k^n}DH^G(M,\Omega,\hat{A}_n(M)P_\gamma).$$

We end this section with some observations.

We see that $P_\gamma(X) = \chi + \sum_{j=1}^d \omega_j \int_0^\infty t^{j-1}e^{it\langle\Phi_\theta,X\rangle}dt$, where ω_j is a differential form, compactly supported, of degree $2j$. Then

$$E(X,k) := P_\gamma(X/k) = \chi + \sum_{j=1}^d k^j\,\omega_j\int_0^\infty t^{j-1}e^{it\langle\Phi_\theta,X\rangle}dt, \quad k\geq 1,$$

is a polynomial in k whose values are equivariant forms with generalized coefficients.

Let $U_\gamma \subset M$ be a relatively compact invariant subset containing the support of χ. So we can choose $r_\gamma > 0$ so that $\hat{A}(M)(X)$ is well defined on U_γ when $\|X\| < r_\gamma$.

We will deduce Theorem 4.9 from the following result proved in [20].

Proposition 4.10. *If $\|X\| < r_\gamma$, we have the relation*

$$\mathscr{Q}_G^{\mathrm{spin}}(M,\mathscr{L}^k,Z_\gamma)(e^X) = \frac{1}{(-2i\pi)^d}\int_M \hat{A}(M)(X)P_\gamma(X)e^{-ik\Omega(X)}.$$

So, for $k\geq 1$, the generalized function $\mathscr{Q}_G^{\mathrm{spin}}(M,\mathscr{L}^k,Z_\gamma)(e^{X/k})j_{\mathfrak{g}}^{1/2}(X/k)$ coincides with

$$\frac{1}{(-2i\pi)^d}\int_M j_{\mathfrak{g}}^{1/2}(X/k)\hat{A}(M)(X/k)P_\gamma(X/k)e^{-ik\Omega(X/k)}$$

on the ball $\{\|X\|/k < r_\gamma\}$.

Let us compute the Laurent series of

$$\left[j_{\mathfrak{g}}^{1/2}(X/k)\hat{A}(M)(X/k)P_\gamma(X/k)e^{-ik\Omega(X/k)}\right]_{[2d]}$$

where $[-]_{[2d]}$ means the component of maximal degree $2d$ in $\mathscr{A}^\bullet(M)$. We write $j_\mathfrak{g}^{1/2}(X)\hat{A}(M)(X) = \sum_{n=0}^\infty d_n(X)$ as a sum of closed equivariant forms of equivariant degree n. Recall that $E(X,k) = P_\gamma(X/k)$ and $b(X,k) = e^{-ik\Omega(X/k)}$ depend polynomially on k.

We also write $j_\mathfrak{g}^{1/2}(X/k)\hat{A}(M)(X) = \sum_{n=0}^\infty \tau_n(X)$ where $\tau_n(X)$ is an equivariant form whose coefficients are homogeneous polynomial functions of X of degree n. We proceed as in the proof of Lemma 3.10 comparing the terms of top exterior degree $2d$. Arguing as in the proof of Lemma 3.11, we have the following formula.

Lemma 4.11.

$$\left[b(X,k)\Big(\sum_{n=0}^\infty \frac{1}{k^n}\tau_n(X)\Big)E(X,k)\right]_{[2d]} = \left[e^{-i\Omega(X)}\Big(\sum_{n=0}^\infty \frac{1}{k^n}d_n(X)\Big)P_\gamma(X)\right]_{[2d]}.$$

Thus we can write formally

$$\mathscr{Q}_G^{\mathrm{spin}}(M,\mathscr{L}^k,Z_\gamma)(e^{X/k})j_\mathfrak{g}^{1/2}(X) = k^d\sum_{n=0}^\infty \frac{1}{k^n}g_n(X)$$

where $g_n(X)$ is the distribution

$$g_n(X) := \frac{1}{(-2i\pi)^d}\int_M d_n(X)P_\gamma(X)e^{-i\Omega(X)}.$$

4.3 *Proof of Theorem 4.9*

We fix a G-invariant function $h : \mathfrak{g} \to \mathbb{R}$ equal to 1 in a neighborhood of 0 and with support contained in $\|X\| < r_\gamma$. If φ is a smooth function with compact support on $\mathfrak{g}^*$, we have $\langle \Theta_k^\gamma, \varphi\rangle = \mathrm{I}_k + \mathrm{J}_k$ where

$$\mathrm{I}_k = \int_\mathfrak{g} \mathscr{Q}_G^{\mathrm{spin}}(M,\mathscr{L}^k,Z_\gamma)(e^{X/k})j_\mathfrak{g}^{1/2}(X/k)h(X/k)\widehat{\varphi}(X)dX$$

and

$$\mathrm{J}_k = \int_\mathfrak{g} \mathscr{Q}_G^{\mathrm{spin}}(M,\mathscr{L}^k,Z_\gamma)(e^{X/k})j_\mathfrak{g}^{1/2}(X/k)(1-h(X/k))\widehat{\varphi}(X)dX.$$

For estimating J_k, we will use the estimation of the Fourier coefficients $m_\gamma(\lambda,k)$ and we will prove that $\mathrm{J}_k = O(k^{-\infty})$. For estimating I_k, we will use Proposition 4.10.

To analyze these expressions, we need the following technical lemma.

Lemma 4.12. *Let $u(X)$ be a C^∞ function of X. Assume that $u(X)$ and all its derivatives are of at most polynomial growth and that $u(X)$ vanishes at order N at $X = 0$.*

Let R be an integer. Then there exists a constant c_R such that

$$\left|\int_{\mathfrak{g}} u(X/k)e^{i\langle\xi,X\rangle}\widehat{\varphi}(X)dX\right| \leq \frac{1}{k^{N+1}}\frac{c_R}{(1+\|\xi\|^2)^R}$$

for all $(\xi,k) \in \mathfrak{g}^* \times \mathbb{N}\setminus\{0\}$.

Proof. We may write in multi-index notation

$$u(X) = \sum_{\alpha,|\alpha|=N+1} X^\alpha v_\alpha(X)$$

with $v_\alpha(X)$ (and derivatives) bounded by polynomial functions of X. We compute $I_R = k^{N+1}(1+\|\xi\|^2)^R \int_{\mathfrak{g}} u(X/k)e^{i\langle\xi,X\rangle}\widehat{\varphi}(X)dX$. Thus

$$I_R = (1+\|\xi\|^2)^R \int_{\mathfrak{g}} \sum_{|\alpha|=N+1} X^\alpha v_\alpha(X/k)e^{i\langle\xi,X\rangle}\widehat{\varphi}(X)dX.$$

Let $\Delta = -\sum_j \partial^2_{X_j}$ be the Laplacian on $\mathfrak{g}$. Then I_R is equal to

$$\sum_{|\alpha|=N+1} \int_{\mathfrak{g}} X^\alpha v_\alpha(X/k)\widehat{\varphi}(X)\left((1+\Delta^2)^R \cdot (e^{i\langle\xi,X\rangle})\right)dX = \int_{\mathfrak{g}} L_R(X,k)e^{i\langle\xi,X\rangle}dX$$

where $L_R(X,k) = (1+\Delta^2)^R \cdot \left(\sum_{|\alpha|=N+1} X^\alpha v_\alpha(X/k)\widehat{\varphi}(X)\right)$.

Using the facts that $\widehat{\varphi}(X)$ is rapidly decreasing (as well as are all its derivatives), that the derivatives of the functions v_α are bounded by polynomials, and that $1/k \leq 1$ for $k \geq 1$, we see that $L_R(X,k)$ can be bounded by a rapidly decreasing function of X independent of k. So $\left|\int_{\mathfrak{g}} L_R(X,k)e^{i\langle\xi,X\rangle}dX\right| \leq c_R$. □

We return to our proof. We start by checking that $\mathrm{J}_k = O(k^{-\infty})$. For this computation, we can assume that φ is G-invariant. Let

$$c(\lambda,k) = \int_{\mathfrak{g}} \chi_\lambda(e^{X/k}) j_{\mathfrak{g}}^{1/2}(X/k)(1-h(X/k))\widehat{\varphi}(X)dX.$$

This is equal to $\mathrm{vol}(G\lambda)\int_{\mathfrak{g}} e^{i\langle\lambda/k,X\rangle}(1-h(X/k))\widehat{\varphi}(X)dX$.

By definition, $\mathscr{Q}_G^{\mathrm{spin}}(M,\mathscr{L}^k,Z_\gamma)(e^X) = \sum_{\lambda\in A_G} m_\gamma(\lambda,k)\chi_\lambda(e^X)$, so we get

$$\mathrm{J}_k = \sum_{\lambda\in A_G} m_\gamma(\lambda,k)\mathrm{vol}(G\lambda)\int_{\mathfrak{g}} e^{i\langle\lambda/k,X\rangle}(1-h(X/k))\widehat{\varphi}(X)dX.$$

Now consider $u(X) = 1-h(X)$. It vanishes identically in a neighborhood of 0. As $h(X)$ is compactly supported, $u(X)$ and all derivatives are bounded. So we can apply Lemma 4.12 and obtain $|c(\lambda,k)| \leq \mathrm{vol}(G\lambda)\frac{1}{k^{N+1}}\frac{c_{R,N}}{(1+\|\lambda/k\|^2)^R}$ for any integers (R,N). Since the multiplicities $m_\gamma(\lambda,k)$ have at most a polynomial growth in the variable (λ,k), we can conclude that $\mathrm{J}_k = O(k^{-\infty})$.

We now estimate

$$\mathrm{I}_k = \int_{\mathfrak{g}} \mathscr{Q}_G^{\mathrm{spin}}(M, \mathscr{L}^k, Z_\gamma)(e^{X/k}) j_{\mathfrak{g}}^{1/2}(X/k) h(X/k) \widehat{\varphi}(X) dX.$$

As $h(X/k) = 0$ when $\|X\|/k \geq r_\gamma$, we may use Proposition 4.10. Let $\tau^h(X) = h(X)\hat{A}(M)(X) j_{\mathfrak{g}}^{1/2}(X)$, a smooth compactly supported function from $\mathfrak{g}$ to differential forms on U_γ. We fix $x \in U_\gamma$ and consider $\tau_x^h(X)$, a function from $\mathfrak{g}$ to $\Lambda T_x^* M$. Define

$$\mathrm{I}_k^x = \int_{\mathfrak{g}} \tau_x^h(X/k) E_x(X,k) b_x(X,k) \widehat{\varphi}(X) dX$$

with

$$E_x(X,k) = \chi(x) + \sum_{j=1}^{d} k^j (\omega_j)|_x \int_0^\infty t^{j-1} e^{it\langle \zeta, X\rangle} dt$$

if $\zeta = \Phi_\theta(x)$ and $b_x(X,k) = e^{-ik\Omega_x} e^{i\langle \xi, X\rangle}$ if $\xi = \phi_G(x)$. So

$$\mathrm{I}_k^x = \int_M \int_{\mathfrak{g}} \tau_x^h(X/k) E_x(X,k) b_x(X,k) \widehat{\varphi}(X) dX.$$

In view of Lemma 4.11, we only need to prove that the asymptotic expansion of I_k^x is obtained by replacing $\tau_x^h(X)$ by its Taylor series (we need to care about uniform estimates in x in the compact support of χ).

Thus we fix x and write $E_x(X,k) = \chi(x) + R_x(X,k)$. So $\mathrm{I}_k^x = \mathrm{S}_k^x + \mathrm{T}_k^x$ with

$$\mathrm{S}_k^x = \chi(x) \int_{\mathfrak{g}} \tau_x^h(X/k) b_x(X,k) \widehat{\varphi}(X) dX$$

and

$$\mathrm{T}_k^x = \int_{\mathfrak{g}} R_x(X,k) \tau_x^h(X/k) b_x(X,k) \widehat{\varphi}(X) dX.$$

The expression for S_k^x has been already analyzed when proving Proposition 3.13, and indeed it admits an asymptotic expansion obtained by replacing $\tau_x^h(X)$ by its Taylor series.

Consider $a(X,t) = e^{i\langle \xi, X\rangle} e^{it\langle \zeta, X\rangle} P(t)$ where $P(t)$ is a polynomial function on t with values in $\Lambda T_x^* M$. We only need to prove that

$$\mathrm{W}_k^x = \int_{t=0}^\infty \int_{\mathfrak{g}} a(X,t) \tau_x^h(X/k) \widehat{\varphi}(X) dX$$

admits an asymptotic expansion, also obtained by replacing $\tau_x^h(X)$ by its Taylor series. We fix $x \in M$, write $\tau_x^h(X) = \tau_{\leq N}(X) + \tau_{>N}(X)$. Then $\tau_{>N}(X)$ vanishes with order N at $X = 0$. As $\tau^h(X)$ was compactly supported, $\tau_{\leq N}(X)$ and $\tau_{>N}(X)$ are of at most polynomial growth, as well as derivatives. So we use Lemma 4.12 and obtain, for any positive integer R,

$$\left| \int_{\mathfrak{g}} a(X,t) \tau_{>N}(X/k) \widehat{\varphi}(X) dX \right| \leq P(t) \frac{1}{k^{N+1}} \frac{c_{N,R}}{(1 + |\xi + t\zeta|^2)^R}.$$

If R is sufficiently large, $\int_{t=0}^{\infty} P(t)\frac{c_{N,R}}{(1+|\xi+t\zeta|^2)^R} < \infty$. So we obtain our estimate for the rest. All our estimates can be done uniformly in x when x runs in the compact support of χ. This ends the proof of Theorem 4.9.

4.4 Proof of Theorem 4.6

The decomposition $\Theta_k^{(M,\mathscr{L})} = \sum_\gamma \Theta_k^\gamma$ is well defined since the distribution Θ_k^γ is supported in $\{\|\xi\| \geq \|\gamma\|\}$ (see [18]). If φ is a test function, then

$$\langle \Theta_k^{(M,\mathscr{L})}, \varphi \rangle = \sum_\gamma \langle \Theta_k^\gamma, \varphi \rangle$$

where the sum on the right-hand side has only a finite number of nonzero terms.

Thanks to Theorem 4.9, we have the asymptotic expansion

$$\Theta_k^\gamma \equiv k^d\, j_{\mathfrak{g}}^{1/2}(i\partial/k) \sum_{n=0}^{\infty} \frac{1}{k^n} DH^G(M,\Omega,\hat{A}_n(M)P_\gamma)$$

for any γ. Hence

$$\Theta_k^{(M,\mathscr{L})} \equiv k^d\, j_{\mathfrak{g}}^{1/2}(i\partial/k) \sum_{n=0}^{\infty} k^{-n}\theta_n$$

with θ_n equal to $\sum_\gamma DH^G(M,\Omega,\hat{A}_n(M)P_\gamma)$. The proof of Theorem 4.6 is complete since $\sum_\gamma DH^G(M,\Omega,\hat{A}_n(M)P_\gamma) = DH^G(M,\Omega,\hat{A}_n(M))$ by (4.6).

5 Functoriality

5.1 H and G

Let H be a connected compact subgroup of G. Let $r : \mathfrak{g}^* \to \mathfrak{h}^*$ be the projection. For θ a distribution with compact support on $\mathfrak{g}^*$, the push-forward $\langle r_*\theta, \varphi \rangle = \langle \theta, \varphi \circ r \rangle$ is well defined. The Fourier transform $\mathscr{F}_{\mathfrak{h}^*}(r_*\theta)$ is the restriction to $\mathfrak{h}$ of the Fourier transform $\mathscr{F}_{\mathfrak{g}^*}(\theta)$. We can define $r_*\theta$ more generally whenever θ is compactly supported along the fibers of r.

The twisted Duistermaat-Heckman distributions behave very well under the push-forward map r_*.

From (4.1) we immediately get the following:

Proposition 5.1. *Suppose that the moment map $\phi_H := r \circ \phi_G$ is proper. Then, for any $\nu \in H_G^*(M)$, we have*

$$r_* DH^G(M,\Omega,\nu) = DH^H(M,\Omega,\nu).$$

Rescaling also behaves very well under the push-forward map r_*:

$$\text{rescaling}_{\mathfrak{h}^*}(k) \circ r_* = r_* \circ \text{rescaling}_{\mathfrak{g}^*}(k).$$

In the rest of the article, objects associated to λ are associated to G, while objects associated to μ are associated to H.

For $\mu \in A_H$ and $\lambda \in A_G$, let $c(\mu,\lambda)$ be the multiplicity of the representation V_μ of H in the restriction of V_λ to H. In other words, for $h \in H$,

$$\chi_\lambda(h) = \sum_{\mu \in A_H} c(\mu,\lambda)\chi_\mu(h).$$

Consider the H-invariant function $j^{1/2}_{\mathfrak{g}/\mathfrak{h}}(Y) = \det^{1/2}_{\mathfrak{g}/\mathfrak{h}}\left(\frac{e^{Y/2}-e^{-Y/2}}{Y}\right)$ on $\mathfrak{h}$. Its Fourier transform is a compactly supported measure $B_{\mathfrak{g}/\mathfrak{h}}$ on $\mathfrak{h}^*$. We use the notation $B^k_{\mathfrak{g}/\mathfrak{h}}$ for its rescaling: $B^k_{\mathfrak{g}/\mathfrak{h}} = \text{rescaling}_{\mathfrak{h}^*}(k)B_{\mathfrak{g}/\mathfrak{h}}$. We have an exact relation between the push-forward of the measure β_λ on $\mathfrak{h}^*$ and measures of H-admissible coadjoint orbits. We denote by $B_{\mathfrak{g}(g)/\mathfrak{h}(g)}$ the compactly supported measure on $\mathfrak{h}(g)^*$ which is the Fourier transform of the $H(g)$-invariant function $j^{1/2}_{\mathfrak{g}(g)/\mathfrak{h}(g)}$. We denote by $C_{\mathfrak{q}/\mathfrak{q}(g)}$ the Fourier transform of the $H(g)$-invariant function $Y \mapsto \det^{1/2}_{\mathfrak{q}/\mathfrak{q}(g)}(1-g^{-1}e^{-Y})$. We use the notations $B^k_{\mathfrak{g}(g)/\mathfrak{h}(g)}$ and $C^k_{\mathfrak{q}/\mathfrak{q}(g)}$ for their rescaling.

$$B^k_{\mathfrak{g}(g)/\mathfrak{h}(g)} := \text{rescaling}_{\mathfrak{h}(g)^*}(k)B_{\mathfrak{g}(g)/\mathfrak{h}(g)},$$

$$C^k_{\mathfrak{q}/\mathfrak{q}(g)} := \text{rescaling}_{\mathfrak{h}(g)^*}(k)C_{\mathfrak{q}/\mathfrak{q}(g)}.$$

Lemma 5.2. *For $\lambda \in A_G$, we have the relations*

$$r_*(\beta_\lambda) = B_{\mathfrak{g}/\mathfrak{h}} \star \Big(\sum_{\mu \in A_H} c(\mu,\lambda)\,\beta_\mu\Big),$$

$$\text{rescaling}_{\mathfrak{h}^*}(k)\,(r_*(\beta_\lambda)) = B^k_{\mathfrak{g}/\mathfrak{h}} \star \Big(\text{rescaling}_{\mathfrak{h}^*}(k)\big(\sum_{\mu \in A_H} c(\mu,\lambda)\,\beta_\mu\big)\Big).$$

Here the $\star$ sign denotes the convolution.

Proof. The first identity follows immediately by Fourier transform of the formula $\chi_\lambda(e^Y)j^{1/2}_{\mathfrak{g}}(Y) = j^{1/2}_{\mathfrak{g}/\mathfrak{h}}(Y)\sum_{\mu \in A_H} c(\mu,\lambda)\,\chi_\mu(e^Y)j^{1/2}_{\mathfrak{h}}(Y)$, for $Y \in \mathfrak{h}$.

We get the second identity by applying the operator $\text{rescaling}_{\mathfrak{h}^*}(k)$ to the first one. □

We now prove a similar descent formula. We write $\mathfrak{g} = \mathfrak{h} \oplus \mathfrak{q}$ where $\mathfrak{q}$ is H-invariant. Let $g \in H$ and consider the function

$$d(Y) := \chi_\lambda(ge^Y)\,j^{1/2}_{\mathfrak{g}(g)}(Y)\det{}^{1/2}_{\mathfrak{g}/\mathfrak{g}(g)}(1-g^{-1}e^{-Y}), \quad Y \in \mathfrak{g}(g).$$

On the one hand $d(Y) = \langle \beta(g,\lambda), e^{i\langle -,Y\rangle}\rangle$ when $Y \in \mathfrak{g}(g)$, and on the other hand, we have

$$d(Y) = j^{1/2}_{\mathfrak{g}(g)/\mathfrak{h}(g)}(Y) \det{}^{1/2}_{\mathfrak{q}/\mathfrak{q}(g)}(1 - g^{-1}e^{-Y}) \sum_{\mu} c(\lambda,\mu)\langle \beta(g,\mu), e^{i\langle -,Y\rangle}\rangle$$

when $Y \in \mathfrak{h}(g)$.

If A is a compactly supported distribution on $\mathfrak{g}(g)^*$, we still denote by $r_*(A)$ the push-forward distribution on $\mathfrak{h}(g)^*$. We denote by $B_{\mathfrak{g}(g)/\mathfrak{h}(g)}$ the compactly supported measure on $\mathfrak{h}(g)^*$ which is the Fourier transform of the $H(g)$-invariant function $j^{1/2}_{\mathfrak{g}(g)/\mathfrak{h}(g)}$. We define $B^k_{\mathfrak{g}(g)/\mathfrak{h}(g)} := \text{rescaling}_{\mathfrak{h}(g)^*}(k) B_{\mathfrak{g}(g)/\mathfrak{h}(g)}$. Let $C_{\mathfrak{q}/\mathfrak{q}(g)}$ be the Fourier transform of the $H(g)$-invariant function $Y \mapsto \det^{1/2}_{\mathfrak{q}/\mathfrak{q}(g)}(1 - g^{-1}e^{-Y})$. We define $C^k_{\mathfrak{q}/\mathfrak{q}(g)} := \text{rescaling}_{\mathfrak{h}(g)^*}(k) C_{\mathfrak{q}/\mathfrak{q}(g)}$.

The previous identities give the following result.

Lemma 5.3. *Let $g \in H$. Then* $\text{rescaling}_{\mathfrak{h}(g)^*}(k)\Big(r_*\beta(g,\lambda)\Big)$ *is equal to*

$$B^k_{\mathfrak{g}(g)/\mathfrak{h}(g)} \star C^k_{\mathfrak{q}/\mathfrak{q}(g)} \star \left(\text{rescaling}_{\mathfrak{h}(g)^*}(k)\Big(\sum_{\mu \in A_H} c(\mu,\lambda)\beta(g,\mu)\Big)\right).$$

5.2 Functoriality relative to restrictions

Here we consider an oriented spin manifold M of even dimension with an action of a compact connected Lie group G. Let $\mathscr{L} \to M$ be a G-equivariant line bundle equipped with an invariant Hermitian connection ∇. We assume here that ϕ_G is proper. In this case we can define $\mathscr{Q}^{-\infty}_G(M, \mathscr{L}^k) \in \hat{R}(G)$ for any $k \geq 1$.

The main result of this section is the following theorem.

Theorem 5.4. *Suppose that $H \subset G$ is a closed connected subgroup such that ϕ_H is proper. Then*

(1) *$\mathscr{Q}^{-\infty}_G(M, \mathscr{L}^k)$ is H-admissible,*
(2) *$\mathscr{Q}^{-\infty}_G(M, \mathscr{L}^k)|_H = \mathscr{Q}^{-\infty}_H(M, \mathscr{L}^k)$.*

We will prove Theorem 5.4 under the assumption that the generic infinitesimal stabilizer of the G-action on M is abelian. It can be shown easily that this implies the general case (see Lemma 4.2 in [19]).

Let us denote by $c(\lambda,\mu)$ the multiplicity of V_μ in $V_\lambda|_H$. If we consider the generalized characters

$$\mathscr{Q}^{-\infty}_G(M, \mathscr{L}^k) = \sum_{\lambda \in \hat{G}} m_G(\lambda, k) V_\lambda$$

and

$$\mathcal{Q}_H^{-\infty}(M,\mathcal{L}^k) = \sum_{\mu\in\hat{H}} m_H(\mu,k)V_\mu,$$

we see that Theorem 5.4 is equivalent to the following theorem.

Theorem 5.5. *For any $\mu \in A_H$, we have*

$$m_H(\mu,k) = \sum_{\lambda\in A_G} m_G(\lambda,k)\,c(\lambda,\mu),$$

where the right-hand side is a finite sum.

We consider the sequences of distributions $\Theta_k^G = k^r\sum_\lambda m_G(\lambda,k)\,\beta_{\lambda/k}$ and $\Theta_k^H = k^{r'}\sum_\mu m_H(\mu,k)\,\beta_{\mu/k}$ associated to the formal geometric quantizations $\mathcal{Q}_G^{-\infty}(M,\mathcal{L}^k)$ and $\mathcal{Q}_H^{-\infty}(M,\mathcal{L}^k)$.

We start with the following proposition.

Proposition 5.6. • *We have $B^k_{\mathfrak{g}/\mathfrak{h}} \star \Theta_k^H \equiv j^{1/2}_{\mathfrak{g}/\mathfrak{h}}(i\partial/k)\Theta_k^H$.*

• *The push-forward $r_*(\Theta_k^G)$ is a well-defined sequence of distributions on $\mathfrak{h}^*$. We have*

$$r_*(\Theta_k^G) \equiv j^{1/2}_{\mathfrak{g}/\mathfrak{h}}(i\partial/k)\Theta_k^H.$$

Proof. Since $m_H(\mu,k)$ is a piecewise quasi-polynomial function (see Section 3.1), the first point is a consequence of Proposition 2.9.

We now prove the second point. Let $\varphi \in \mathcal{C}^\infty(\mathfrak{h}^*)$ be a function with compact support. Let $R_\varphi > 0$ be such that the compact set $\mathcal{K}_\varphi := \phi_G\big(\phi_H^{-1}(\text{Support}(\varphi))\big)$ is contained in $\{\xi \in \mathfrak{g}^*,\ \|\xi\| \le R_\varphi\}$. The expression

$$\langle r_*(\Theta_k^G),\varphi\rangle := k^r\sum_\lambda m_G(\lambda,k)\,\varphi\,(r(\lambda/k))$$

is well defined since the term $m_G(\lambda,k)\,\varphi\,(r(\lambda/k))$ is nonzero only for a finite number of λ. More precisely, since $m_G(\lambda,k)\neq 0$ only if $\lambda/k \in \phi_G(M)$, we see that $m_G(\lambda,k)\,\varphi\,(r(\lambda/k)) \neq 0$ only if $\lambda/k \in \mathcal{K}_\varphi$.

If $\varepsilon \in \mathcal{C}^\infty(\mathfrak{g}^*)$ is a function with compact support such that $\varepsilon(\xi) = 1$ if $\|\xi\| \le R_\varphi$, we define $\tilde{\varphi} := \varphi\circ r\times\varepsilon$. We see that

$$\big\langle r_*\big(\Theta_k^G\big),\varphi\big\rangle = \big\langle \Theta_k^G,\tilde{\varphi}\big\rangle = k^d\sum_{n=0}^N \frac{1}{k^n}\big\langle \theta_n^G,\tilde{\varphi}\big\rangle + o\big(k^{d-N}\big)$$

where (θ_n^G) is the family of distributions on $\mathfrak{g}^*$ such that

$$j_{\mathfrak{g}}^{1/2}(i\partial/k)\sum_{n=0}^\infty \frac{1}{k^n}DH^G\big(M,\Omega,\hat{A}_n(M)\big) = \sum_{n=0}^\infty \frac{1}{k^n}\theta_n^G.$$

We check easily that, for any polynomial P on $\mathfrak{g}$, we have

$$\big\langle DH^G\big(M,\Omega,\hat{A}_n(M)\big),P(\partial)\tilde{\varphi}\big\rangle = \big\langle DH^H\big(M,\Omega,\hat{A}_n(M)\big),P|_{\mathfrak{h}}(\partial)\varphi\big\rangle.$$

Hence the formal series $\langle j_{\mathfrak{g}}^{1/2}(i\partial/k)\sum_{n=0}^{\infty}\frac{k^d}{k^n}DH^G(M,\Omega,\hat{A}_n(M)),\tilde{\varphi}\rangle$ is equal to

$$\langle j_{\mathfrak{g}}^{1/2}(i\partial/k)\sum_{n=0}^{\infty}\frac{k^d}{k^n}DH^H(M,\Omega,\hat{A}_n(M)),\varphi\rangle \equiv \langle j_{\mathfrak{g}/\mathfrak{h}}^{1/2}(i\partial/k)\Theta_k^H,\varphi\rangle. \qquad \square$$

Consider the function $m'_H(\mu,k) := \sum_{\lambda\in A_G} m_G(\lambda,k)\,c(\lambda,\mu)$ and the difference $d(\mu,k) := m_H(\mu,k) - m'_H(\mu,k)$. Our aim is to show that $d = 0$.

Proposition 5.7. *We have* rescaling$_{\mathfrak{h}^*}(k)\left(\sum_{\mu\in A_H} d(\mu,k)\beta_\mu\right) \equiv 0$.

Proof. The previous proposition tells us that

$$r_*(\Theta_k^G) \equiv B^k_{\mathfrak{g}/\mathfrak{h}} \star \text{rescaling}_{\mathfrak{h}^*}(k)\Big(\sum_{\mu\in A_H} m_H(\mu,k)\,\beta_\mu\Big).$$

If we use Lemma 5.2, we can compute $r_*(\Theta_k^G)$ in another manner:

$$\begin{aligned} r_*(\Theta_k^G) &= \text{rescaling}_{\mathfrak{h}^*}(k)\Big(\sum_{\lambda\in A_G} m_G(\lambda,k) r_*(\beta_\lambda)\Big) \\ &= B^k_{\mathfrak{g}/\mathfrak{h}} \star \text{rescaling}_{\mathfrak{h}^*}(k)\Big(\sum_{\mu\in A_H} m'_H(\mu,k)\,\beta_\mu\Big). \end{aligned}$$

At this stage we obtain that $B^k_{\mathfrak{g}/\mathfrak{h}} \star \text{rescaling}_{\mathfrak{h}^*}(k)\big(\sum_{\mu\in A_H} d(\mu,k)\beta_\mu\big) \equiv 0$, or equivalently, $B^k_{\mathfrak{g}/\mathfrak{h}} \star \big(\sum_{\mu\in A_H} d(\mu,k)\beta_{\mu/k}\big) \equiv 0$. Since the function $d(\mu,k)$ is a piecewise quasi-polynomial function, we know that $\Theta_k := \sum_{\mu\in A_H} d(\mu,k)\beta_{\mu/k}$ admits an asymptotic expansion

$$\Theta_k \equiv k^{n_o}\sum_{n=0}^{\infty} k^{-n}\theta_n(k)$$

where the distributions $\theta_n(k)$ depend periodically on k (see Proposition 2.8). Thanks to Proposition 2.9, we know that $B^k_{\mathfrak{g}/\mathfrak{h}} \star \Theta_k$ admits the asymptotic expansion

$$0 \equiv B^k_{\mathfrak{g}/\mathfrak{h}} \star \Theta_k \equiv j^{1/2}_{\mathfrak{g}/\mathfrak{h}}(i\partial/k)\Big(k^{n_o}\sum_{n=0}^{\infty} k^{-n}\theta_n(k)\Big).$$

If we write the Taylor series $j^{1/2}_{\mathfrak{g}/\mathfrak{h}}(X) = \sum_{n=0}^{\infty} d_n(X)$, we obtain for any $n \geq 0$ the relation $\sum_{m+l=n} d_m(i\partial)\theta_l(k) = 0$. As $d_0 = 1$, we see that $\theta_n(k) = 0$ for all $n \geq 0$. $\square$

If we use Lemma 5.3 and Theorem 4.8, we can prove similarly the following extension of Proposition 5.7.

Proposition 5.8. *For any $g \in H$ of finite order, we have*

$$\text{rescaling}_{\mathfrak{h}(g)^*}(k)\Big(\sum_{\mu\in A_H} d(\mu,k)\beta(g,\mu)\Big) \equiv 0.$$

The function $d(\mu,k)$ is defined for $\mu \in A_H = (\rho_H + \Lambda_H) \cap \mathfrak{t}^*_{H,>0}$. Let $\tilde{\Lambda}_H$ be the lattice generated by Λ_H and ρ_H. Let W_H be the Weyl group of H. We can then extend the function $d(\mu,k)$ to a W_H-anti-invariant function $\tilde{d}$ on $\tilde{\Lambda}_H$: $\tilde{d}(w\mu,k) = \varepsilon(w)d(\mu,k)$ for $w \in W_H$, $\mu \in A_H$ and $\tilde{d}(\mu,k) = 0$ if $\mu \notin \rho_H + \Lambda_H$.

Let T_H be a Cartan subgroup of H. We have a covering $\tilde{T}_H \to T_H$ such that $\tilde{\Lambda}_H$ is the weight lattice of the torus $\tilde{T}_H$. Take $\tilde{g} \in \tilde{T}_H$ and its image $g \in T_H$.

We now identify $H(g)$-invariant distributions on $\mathfrak{h}(g)^*$ to $W_{H(g)}$-anti-invariant distributions on $\mathfrak{t}^*_H$ through the isomorphism $\mathscr{R}_{\mathfrak{h}(g)}$. Using Lemma 3.15, we have $\mathscr{R}_{\mathfrak{h}(g)}(\beta(g,\mu)) = \gamma_{\tilde{g}} \sum_{w\in W_G} \varepsilon(w)\tilde{g}^{w\mu}\delta_{w\mu}$.

We thus obtain the following proposition.

Proposition 5.9. *For any $\tilde{g} \in \tilde{T}_H$ of finite order, we have*

$$\sum_{\nu \in \tilde{\Lambda}_H} \tilde{d}(\nu,k)\,\tilde{g}^{\nu}\,\delta_{\nu/k} \equiv 0.$$

Since $\tilde{d}(\nu,k)$ is a piecewise quasi-polynomial function on $\tilde{\Lambda}_H \times \mathbb{Z}_{>0}$, Proposition 5.9 implies that $\tilde{d} = 0$ (see Proposition 3.1 in [23]). Hence $d = 0$; the proof of Theorem 5.4 is complete.

References

[1] M.F. Atiyah, *Elliptic operators and compact groups*, Lecture Notes in Mathematics, Vol. 401, Springer-Verlag, 1974.

[2] Nicole Berline, Ezra Getzler, and Michèle Vergne, *Heat Kernels and Dirac Operators*, Grundlehren Text Editions, Springer-Verlag, 2004.

[3] Nicole Berline and Michèle Vergne, *Classes caractéristiques équivariantes. Formules de localisation en cohomologie équivariante*, C.R.A.S. **295** (1982), 539–541.

[4] ______, *L'indice équivariant des opérateurs transversalement elliptiques*, Invent. Math. **124** (1996), 51–101.

[5] Martin Bordemann, Eckhard Meinrenken, and Martin Schlichenmaier, *Toeplitz quantization of Kähler manifolds and* gl(N), $N \to \infty$ *limits*, Comm. Math. Phys. **165** (1994), 281–296.

[6] L. Boutet de Monvel and V. Guillemin, *The spectral theory of Toeplitz operators*, Annals of Mathematics Studies, Vol. 99, Princeton University Press, Princeton, NJ; University of Tokyo Press, Tokyo, 1981.

[7] Henri Cohen, *Number theory. Vol. II. Analytic and modern tools*, Graduate Texts in Mathematics, Vol. 240, Springer, New York, 2007.

[8] Boris Fedosov, *On G-trace and G-index in deformation quantization*, Lett. Math. Phys. **52** (2000), 29–49. Conference Moshé Flato 1999 (Dijon).

[9] Victor Guillemin and Shlomo Sternberg, *Riemann sums over polytopes*, Ann. Inst. Fourier (Grenoble) **57** (2007), no. 7, 2183–2195 (English, with English and French summaries). Festival Yves Colin de Verdière.

[10] Peter Hochs and Yanli Song, *Equivariant indices of Spinc-Dirac operators for proper moment maps*, Duke Math J. **166** (2017), 1125–1178.

[11] Tetsuro Kawasaki, *The index of elliptic operators over V-manifolds*, Nagoya Math. Journal **84** (1981), 135–157.

[12] Bertram Kostant, *Quantization and unitary representations. I. Prequantization*, Lectures in modern analysis and applications, III, Springer, Berlin, 1970, pp. 87–208. Lecture Notes in Math., Vol. 170.

[13] Xiaonan Ma and George Marinescu, *Berezin-Toeplitz quantization and its kernel expansion*, Geometry and quantization, Trav. Math., Vol. 19, Univ. Luxemb., Luxembourg, 2011, pp. 125–166.
[14] Xiaonan Ma and Weiping Zhang, *Geometric quantization for proper moment maps: the Vergne conjecture*, Acta Mathematica **212** (2014), 11–57.
[15] Eckhard Meinrenken, *On Riemann-Roch formulas for multiplicities*, J. Amer. Math. Soc. **9** (1996), 373–389.
[16] Paul-Émile Paradan, *The moment map and equivariant cohomology with generalized coefficients*, Topology **39** (2000), 401–444.
[17] ———, *Formal geometric quantization*, Ann. Inst. Fourier (Grenoble) **59** (2009), no. 1, 199–238.
[18] ———, *Formal geometric quantization II*, Pacific J. Math. **253** (2011), 169–211.
[19] ———, *Formal Geometric Quantization III, Functoriality in the spin-c setting*, arXiv **1704.06034** (2017).
[20] Paul-Émile Paradan and Michèle Vergne, *Index of transversally elliptic operators*, Astéristique **328** (2009), 297–338.
[21] ———, *Equivariant Dirac operators and differentiable geometric invariant theory*, Acta Math. **218** (2017), 137–199.
[22] ———, *Witten non abelian localization for equivariant K-theory, and the $[Q,R]=0$ problem*, to appear in Memoirs of the A.M.S.
[23] ———, *Asymptotic distributions associated to piecewise quasi-polynomials*, arXiv **1708.08283** (2017).
[24] Michèle Vergne, *Formal equivariant $\hat{A}$ class, splines and multiplicities of the index of transversally elliptic operators*, Izvestiya: Mathematics **80** (2016), no. 5.
[25] ———, *The equivariant Riemann-Roch theorem and the graded Todd class*, Comptes Rendus Mathématique **355** (2017), no. 5, 563–570.
[26] Jonathan Weitsman, *Nonabelian symplectic cuts and the geometric quantization of noncompact manifolds*, Lett. Math. Phys. **56** (2001), 31–40.
[27] Edward Witten, *Supersymmetry and Morse theory*, J. Differential Geom. **17** (1982), 661–692.

Modality of Representations, and Packets for θ-Groups

Vladimir L. Popov

To the memory of Bertram Kostant

Abstract We first establish several general properties of modality of algebraic group actions. In particular, we introduce the notion of a modality-regular action and prove that every visible action is modality-regular. Then, using these results, we classify irreducible linear representations of connected simple algebraic groups of every fixed modality $\leqslant 2$. Next, exploring a finer geometric structure of linear actions, we generalize to the case of any cyclically graded semisimple Lie algebra the notion of a packet (or a Jordan/decomposition class) and establish the properties of packets.

Keywords: Algebraic group action • Modality • θ-group • Packet • Sheet

Mathematics Subject Classification (2010): 14L30 (Primary), 20G05

1 Introduction

The modality of a group action is the maximal number of parameters on which a family of orbits may depend. This notion, as a natural measure of complexity of a group action, goes back to V. I. Arnold's works on the theory of singularities in which the actions of diffeomorphism groups on the spaces of functions have been explored. V. I. Arnold and his collaborators succeeded in classifying the cases of a small modality (0 and 1); this led to the famous lists of singularities that enjoy remarkable properties; see [A75].

This work is supported by grant RFBR 15-01-02158 and the Program of the Presidium of the Russian Academy of Sciences No. 01 "Fundamental Mathematics and its Applications" under grant PRAS-18-01.

Steklov Mathematical Institute, Russian Academy of Sciences, Gubkina 8, Moscow 119991, Russia and National Research University, Higher School of Economics, Myasnitskaya 20, Moscow 101000, Russia, e-mail: popovvl@mi.ras.ru

V. G. Kac, V. L. Popov (eds.), *Lie Groups, Geometry, and Representation Theory*, Progress in Mathematics 326, https://doi.org/10.1007/978-3-030-02191-7_16

The concept of modality naturally adapts to the setting of algebraic group actions on algebraic varieties [V86], [PV94, Sect. 5.2]. As in the theory of singularities, this concept allows one to systematically approach the classification problem of algebraic group actions by their degree of complexity. Here we explore the concept of modality in this setting.

In Section 2, we first discuss general properties of modality of algebraic group actions; in particular, we introduce the notion of a modality-regular action and prove that every visible action is modality-regular. Then we consider the basic class of algebraic group actions, namely, that of the linear actions on (finite-dimensional) vector spaces. Guided by the analogy with V. I. Arnold's standpoint in the theory of singularities and with a view to obtaining the distinguished classes of linear actions, we consider the problem of classifying linear actions (representations) of a small modality and classify all irreducible representations of simple algebraic groups of every fixed modality $\leqslant 2$. They do, indeed, turn out to be remarkable because of allowing for some nice equivalent characterizations. Actually, for modality 0, the classification is not new: the definition of modality implies that the class of representations of modality 0 coincides with that of representations with finitely many orbits; much attention has been given to the latter in the literature, in particular, all irreducible representations of reductive groups from this class have been listed [SK77], [K80$_1$], [KKY86]. For simple groups, however, nice characterizations are possible, which for all reductive groups are no longer satisfied (see below Remark 2.17).

In Section 3, we explore the finer properties of the geometry of linear actions. Namely, a finer study of actions presupposes finding not only the maximal number of parameters on which a family of orbits may depend (i.e., the modality), but also on describing all maximal families of orbits (i.e., the sheets), and, where possible, presenting varieties as the disjoint unions of the finer families which have a standard structure and the better controlled geometric properties. For the adjoint representations of semisimple algebraic groups, and the isotropy representations of symmetric spaces, the solutions to these finer study problems are known, see [TY05, Chap. 39] and the references therein. In these cases, the latter finer families are the Jordan classes (also known as the decomposition classes and the packets). We generalize this notion (using the term "packet") to the case of any cyclically graded semisimple Lie algebra (or θ-group, in the terminology of [V76], [K80$_1$]). We describe all packets and explore their properties; in particular, we find their dimensions and modality.

As the base field, we fix an algebraically closed field k of characteristic zero. Below we freely use the standard notation and terminology of algebraic group theory and invariant theory from [B91] and [PV94], where also the proofs of unreferenced claims and/or the relevant references can be found. All considered actions of algebraic groups on algebraic varieties and all homomorphisms of algebraic groups (in particular, representations) are assumed to be algebraic (i.e., regular/morphic). Topological terms are related to the Zariski topology.

The results of Section 2 are partly announced in [P17$_1$].

2 Modality

2.1 Let G be a connected algebraic group. We call any irreducible algebraic variety F endowed with an action of G such that all G-orbits in F have the same dimension d a *family* of G-orbits depending on

$$\operatorname{mod}(G:F) := \dim F - d \tag{1}$$

parameters; the integer $\operatorname{mod}(G:F)$ is called the *modality* of F. If $F \dashrightarrow F/\!\!/G$ is a rational quotient of this action (which exists by the Rosenlicht theorem), then

$$\operatorname{mod}(G:F) = \dim F/\!\!/G = \operatorname{tr}\deg_k k(F)^G \tag{2}$$

and $F/\!\!/G$ may be informally viewed as the variety parametrizing typical G-orbits in F.

Given an algebraic variety X endowed with an action of G, we denote by $\mathscr{F}(X)$ the set of all locally closed G-stable subsets of X which are families. The integer

$$\operatorname{mod}(G:X) := \max_{F\in\mathscr{F}(X)} \operatorname{mod}(G:F), \tag{3}$$

is then called the *modality* of X. If X is a vector space and the action is linear determined by a representation $\rho\colon G\to \mathrm{GL}(X)$, then we call $\operatorname{mod}(G:X)$ the *modality of representation* ρ and denote it by $\operatorname{mod}\rho$.

If Y is an algebraic variety endowed with an action of a (not necessarily connected) algebraic group H and if H^0 is the identity component of H, then by definition[1],

$$\operatorname{mod}(H:Y) := \operatorname{mod}(H^0:Y).$$

Similarly, the modality of a representation of H is defined as the modality of its restriction to H^0.

Recall that for every integer d, the set $\{y\in Y \mid \dim H\cdot y \leqslant d\}$ is closed in Y. Whence, for every locally closed irreducible (not necessarily H-stable) subset Z in Y, the subset

$$Z^{\mathrm{reg}} := \{z\in Z \mid \dim H\cdot z \geqslant \dim H\cdot y \text{ for every } y\in Z\} \tag{4}$$

is dense and open in Z.

The definition of modality implies that equality (3) still holds if $\mathscr{F}(X)$ is replaced by the set of all maximal (with respect to inclusion) families in X, i.e., by the *sheets* of X [PV94, Sect. 6.10]. Recall that there are only finitely many sheets of X. If X is irreducible, then X^{reg} is a sheet, called *regular*, which is open and dense in X. It follows from (2) that

$$\operatorname{mod}(G:X^{\mathrm{reg}}) = \operatorname{tr}\deg_k k(X)^G. \tag{5}$$

[1] This definition fixes the inaccuracy in [V86], [PV94, Sect. 5.2], where $\operatorname{mod}(G:X)$ is defined by (3) for any G, not necessarily connected: as is easily seen, for a disconnected G, the set $\mathscr{F}(X)$ may be empty, so this definition should be corrected.

This implies that equality (3) still holds if $\mathscr{F}(X)$ is replaced by the set of all G-stable locally closed (or closed) subsets of X, and $\operatorname{mod}(G:F)$ by $\operatorname{tr}\deg_k k(F)^G$.

The aforesaid shows that $\operatorname{mod}(G:X)=0$ if and only if the set of all G-orbits in X is finite.

If G is reductive and X is affine, then

$$\begin{aligned}\operatorname{mod}(G:X)\geqslant \operatorname{mod}(G:X^{\mathrm{reg}}) &\overset{(5)}{=} \operatorname{tr}\deg_k k(X)^G\\ &\geqslant \operatorname{tr}\deg_k k[X]^G=\dim X/\!/G.\end{aligned} \tag{6}$$

2.2 The existence of regular sheets leads to defining the following naturally distinguished class of actions:

Definition 2.1. An action of a connected algebraic group G on an irreducible algebraic variety X is called *modality-regular* if $\operatorname{mod}(G:X)=\operatorname{mod}(G:X^{\mathrm{reg}})$. A linear representation $G\to \mathrm{GL}(V)$ is called *modality-regular* if it determines a modality-regular action of G on V.

There are affine algebraic groups G such that every action of G is modality-regular. Here is their complete classification:

Theorem 2.2 (groups all actions of which are modality-regular)**.** *The following properties of a connected affine algebraic group G are equivalent:*

(i) *all actions of G on irreducible algebraic varieties are modality-regular;*
(ii) *for every irreducible algebraic variety X and every action of G on X with a dense open G-orbit, there are only finitely many G-orbits in X;*
(iii) *G is one of the following groups:*

— *a torus,*
— *a product of a torus and a group isomorphic to* $\mathbf{G}_a$.

Proof. See [P17$_2$]. □

2.3 However, a restriction of the class of actions under consideration may lead to an extension of the class of those groups for which all actions of this class are modality-regular. Apparently for the first time, such a phenomenon was discovered in the following theorem:

Theorem 2.3. *Let G, B, and U be respectively a connected reductive algebraic group, a Borel subgroup of G, and a maximal unipotent subgroup of G. Then the restrictions to B and U of any action of G on an irreducible algebraic variety are modality-regular.*

Proof. See [V86, Thms. 2 and 3]. □

Remark 2.4. By Theorem 2.2, if G is not abelian, then there are actions of B which are not modality-regular. By Theorem 2.3, these B-actions cannot be extended up to G-actions. If $\operatorname{rk} G\geqslant 2$, the same holds for U.

2.4 The next example of this phenomenon, in particular, shows that apart from tori, there are other connected reductive algebraic groups for which every representation is modality-regular:

Theorem 2.5 (actions of SL_2). *Every action of $G = \mathrm{SL}_2$ on an irreducible quasi-affine algebraic variety X is modality-regular.*

Proof. Since X is quasiaffine, there is an equivariant open embedding of X in an affine algebraic variety endowed with an action of G; see [PV94, Thm. 1.6]. Therefore, we may (and shall) assume that X is affine. Given a sheet $S \neq X^{\mathrm{reg}}$, we need to show that

$$\operatorname{mod}(G:S) \leqslant \operatorname{mod}(G:X^{\mathrm{reg}}). \tag{7}$$

First, we note that

$$\dim X^G \leqslant \operatorname{mod}(G:X^{\mathrm{reg}}). \tag{8}$$

Indeed, since $k[X]^G$ separates closed orbits, the restriction of the quotient morphism $X \to X/\!/G$ to X^G is injective. Hence $\dim X^G \leqslant \dim X/\!/G$. This and (5) imply (8) because $\dim X/\!/G = \operatorname{tr deg}_k k[X] \leqslant \operatorname{tr deg}_k k(X)^G$.

In view of (8) we need to consider only the case where $S \cap X^G = \varnothing$. Assume that this equality holds. As is well-known, every one-dimensional homogeneous space of G is projective (actually, isomorphic to $\mathbf{P}^1$). Hence, since every G-orbit in X is quasiaffine, its dimension may be only $\dim G = 3$, 2, or 0. This and $S \neq X^{\mathrm{reg}}$ imply that

$$\dim G\cdot x = \begin{cases} 2 & \text{if } x \in S, \\ 3 & \text{if } x \in X^{\mathrm{reg}}, \end{cases} \quad \text{and} \quad \dim S \leqslant \dim X - 1. \tag{9}$$

From (1), (9) we get $\operatorname{mod}(G:S) \leqslant \dim X - 3 = \operatorname{mod}(G:X^{\mathrm{reg}})$, whence (7). □

Remark 2.6. By Theorem 2.2, the quasiaffinity condition in Theorem 2.5 cannot be dropped.

2.5 Using Theorem 2.5, one computes the modality of every representation of SL_2. Namely, let ρ_n be the $(n+1)$-dimensional linear representation of SL_2 determining the natural SL_2-module structure on the space of binary forms of degree n over k. It is irreducible and every linear representation of SL_2 is equivalent to a direct sum of such representations. Given a representation ρ and an integer $s > 0$, we denote $s\rho := \rho \oplus \cdots \oplus \rho$ (s summands).

Corollary 2.7 (modality of SL_2-representations). *Let $\rho\colon G \to \mathrm{GL}(V)$ be a linear representation of $G = \mathrm{SL}_2$. Then*

$$\operatorname{mod}\rho = \begin{cases} \dim\rho - 3 & \text{if } \rho \neq s\rho_0 \oplus \rho_1,\, s\rho_0 \oplus \rho_2,\, s\rho_0, \\ \dim\rho - 2 & \text{if } \rho = s\rho_0 \oplus \rho_1, s\rho_0 \oplus \rho_2, \\ \dim\rho & \text{if } \rho = s\rho_0. \end{cases}$$

Proof. By Definition 2.1 and Theorem 2.5, $\operatorname{mod}\rho = \dim\rho - \max_{v\in V} \dim G\cdot v$. The integer $\max_{v\in V} \dim G\cdot v$ is computed in [P74]. Whence the claim. □

2.6 The following example shows that for every integer $n \geqslant 3$, there are linear representations of $G = \mathrm{SL}_n$ which are not modality-regular. In particular, SL_2 in Theorem 2.5 cannot be replaced by SL_n for $n \geqslant 3$.

Example 2.8. Consider the natural action of G on k^n and the diagonal action of G on $V := k^n \oplus \cdots \oplus k^n$ (d summands). If $d \leqslant n-1$, there is an open G-orbit in V, so we have $\operatorname{mod}(G : V^{\mathrm{reg}}) = 0$. On the other hand, for every nonzero $v, u \in k^n$ and $\lambda_1, \ldots, \lambda_{d-1}, \mu_1, \ldots, \mu_{d-1} \in k^\times$, the elements $(v, \lambda_1 v, \ldots, \lambda_{d-1} v) \in V$ and $(u, \mu_1 u, \ldots, \mu_{d-1} u) \in V$ lie in the same G-orbit if and only if $\lambda_i = \mu_i$ for all i. This imples that $\operatorname{mod}(G : V) \geqslant d-1$.

2.7 We shall now prove that all the representations from a certain important class are modality-regular; this will be then used in the proof of the classification results in Subsection 2.10.

Recall from [K75], [K80$_1$] (see also [PV94, §8]) that a linear action of a reductive algebraic group G on a vector space V (and the corresponding representation $G \to \mathrm{GL}(V)$) is called *visible* if there are only finitely many G-orbits in the level variety of $k[V]^G$ in V containing 0. As a matter of fact, then automatically every level variety of $k[V]^G$ in V contains only finitely many G-orbits [PV94, Cor. 3 of Prop. 5.1]. Extending this terminology to a more general setting, we introduce the following:

Definition 2.9. An action of G on an affine algebraic variety X is called *visible* if every fiber of the categorical quotient

$$\pi_{G,X} : X \to X /\!/ G \tag{10}$$

contains only finitely many G-orbits.

Theorem 2.10 (modality of visible actions)**.** *Every visible action of a reductive algebraic group G on an irreducible affine algebraic variety X enjoys the following properties:*

(i) *it is modality-regular;*
(ii) $\operatorname{mod}(G : X) = \dim X /\!/ G$*;*
(iii) *the induced action of G on every closed G-stable subset of X is visible.*

Proof. First, we prove (iii). Let Y be a G-stable closed subset of X. Then the set $Z := \pi_{G,X}(Y)$ is closed in $X/\!/G$ and $\pi_{G,X}|_Y : Y \to Z$ is the categorical quotient for the action of G on Y; see [PV94, Sect. 4.4]. Every fiber of $\pi_{G,X}|_Y$ is the intersection of Y with a fiber of $\pi_{G,X}$; since the latter contains only finitely-many G-orbits, this intersection shares this property.

Now we prove (i) and (ii). Let $m_{G,X} := \max_{x\in X} \dim G\cdot x$. For every $x \in X^{\mathrm{reg}}$, we then have $\dim G{\cdot}x = m_{G,X}$. Since the fiber $\pi_{G,X}^{-1}(\pi_{G,X}(x))$ contains only finitely many orbits, the latter equality entails that the dimension of this fiber be equal to $m_{G,X}$. From this we infer that

$$\operatorname{mod}(G : X^{\mathrm{reg}}) = \dim X^{\mathrm{reg}} - m_{G,X} = \dim X - m_{G,X} = \dim X/\!/G; \tag{11}$$

(the second equality in (11) holds as X^{reg} is open in X, and the third by the fiber dimension theorem).

Let S be a sheet of X, let $\overline{S}$ be its closure in X, and let $Z := \pi_{G,X}(\overline{S}) = \overline{S}/\!/G$. As $\overline{S}$ is a G-stable closed subset in X, the action of G on $\overline{S}$ is visible by (iii). Therefore, replacing X in (11) by $\overline{S}$ and taking into account that $S = \overline{S}^{\mathrm{reg}}$, we obtain

$$\operatorname{mod}(G : S) = \dim Z. \tag{12}$$

Now the inclusion $Z \subseteq X/\!/G$ combined with (11) and (12), yields the inequality $\operatorname{mod}(G : X^{\mathrm{reg}}) \geqslant \operatorname{mod}(G : S)$. This completes the proof. □

2.8 Theorem 2.10 is applicable to the class of so-called θ-groups studied in [V76] (see also [K80$_1$]).

Namely, let m be either a positive integer or ∞. Denote by $\mathbf{Z}_m$ the following additively written cyclic group of order m: for $m < \infty$, the elements of $\mathbf{Z}_m$ are the integers between 0 and $m-1$, and the sum of i and j in $\mathbf{Z}_m$ is the remainder of dividing $i+j$ by m; by definition, $\mathbf{Z}_\infty = \mathbf{Z}$.

Consider a simply connected semisimple algebraic group G and assume that its Lie algebra $\mathfrak{g} = \operatorname{Lie} G$ is $\mathbf{Z}_m$-graded:

$$\mathfrak{g} = \bigoplus_{i\in\mathbf{Z}_m} \mathfrak{g}_i. \tag{13}$$

Then $\operatorname{Aut}\mathfrak{g}$ contains a subgroup θ, which, for $m < \infty$, is cyclic of order m, and for $m = \infty$, is a one-dimensional torus, such that (13) is the weight decomposition of $\mathfrak{g}$ with respect to the natural action of θ.

The component $\mathfrak{g}_0$ in (13) is a reductive subalgebra of $\mathfrak{g}$. Let G_0 be a closed connected subgroup of G with

$$\operatorname{Lie} G_0 = \mathfrak{g}_0. \tag{14}$$

Every $\mathfrak{g}_i$ in (13) is G_0-stable with respect to the adjoint action, so one can consider $\mathfrak{g}_i$ as the G_0-module with respect to this action. As any G_0-module $\mathfrak{g}_i$ coincides with the G_0-module $\mathfrak{g}_1$ for another appropriate cyclic grading of $\mathfrak{g}$, when studying the orbital decompositions it suffices to explore only the G_0-module $\mathfrak{g}_1$. By [V76, Thm. 1], any two maximal linear subspaces of $\mathfrak{g}_1$ consisting of semisimple pairwise commuting elements are transformed into each other by G_0. These subspaces are called *Cartan subspaces* and their mutual dimension is called the *rank of the graded*

Lie algebra (13). The image of the adjoint representation $G_0 \to \mathrm{GL}(\mathfrak{g}_1)$ is called the *θ-group* associated with (13).

Theorem 2.11 (modality of θ-group actions). *In the above notation, the adjoint representation $\rho\colon G_0 \to \mathrm{GL}(\mathfrak{g}_1)$ is modally-regular and its modality is equal to the rank r of the graded Lie algebra* (13).

Proof. By [V76, Thm. 4], ρ is visible, and by [V76, Thm. 5], $\dim \mathfrak{g}_1/\!/G_0 = r$. In view of this, the claim follows from Theorem 2.10. □

Corollary 2.12 (case $m = 1$). *The adjoint representation of every connected semisimple algebraic group G is modality-regular and its modality is equal to the rank of G.*

Corollary 2.13 (case $m = 2$). *The isotropy representation of every symmetric space X of a connected semisimple algebraic group is modality-regular and its modlity is equal to the rank of X.*

2.9 We now turn to classifying representations of a small modality. First, note that the following finiteness theorem can be considered as an argument in favor of considering the problem of classifying representations in terms of the magnitude of the modality.

Theorem 2.14 (finiteness for modality). *For every connected semisimple algebraic group G and every integer $m \geqslant 0$, there are only finitely many (up to equivalence) linear representations of G of modality m.*

Proof. First, note that G has only finitely many (up to equivalence) linear representations of any fixed dimension. Indeed, given the complete reducibility of representations, it suffices to prove this for irreducible representations. Denote by $\rho(\lambda)$ the irreducible representation of G with the highest weight λ regarding a fixed Borel subgroup and its torus T. We may (and shall) assume that G is simply connected; then let $\varpi_1, \ldots, \varpi_r$ be the fundamental weights of T. It follows from the Weyl formula for $\dim \rho(\lambda)$, see [J62, Chap. VIII, §4, (41)], that $\dim \rho(\sum_{i=1}^{r} n_i \varpi_i) < \dim \rho(\sum_{i=1}^{r} m_i \varpi_i)$ if $0 \leqslant n_i \leqslant m_i$ for all i and $n_{i_0} < m_{i_0}$ for some i_0. This implies the finiteness statement.

The claim of the theorem now follows from this statement combined with the inequality $\operatorname{mod} \rho \leqslant \dim \rho$ that stems from the definition of $\operatorname{mod} \rho$. □

2.10 In Theorem 2.15 below, we classify irreducible representations of connected simple algebraic groups of modalities $0, 1$, and 2. We use in this theorem the following agreements and notation, cf. [P15].

Let G be a a connected semisimple algebraic group and let $\pi\colon \widetilde{G} \to G$ be its universal covering. The map $\rho \mapsto \widetilde{\rho} := \rho \circ \pi$ is a bijection between the set of all representations of G and the set of all representations of $\widetilde{G}$ factoring through G. This allows one to specify ρ by specifying $\widetilde{\rho}$. If $\widetilde{\rho}$ is irreducible, it is uniquely up to equivalence determined by its highest weight λ (with respect to a fixed Borel subgroup B of $\widetilde{G}$ and its maximal torus T). Given this, we denote $\underset{\sim}{\rho}$ (considered up to equivalence) by (R,λ), where R is the type of the root system of $\widetilde{G}$. The fundamental weights of $\widetilde{G}$ with respect to the pair (B,T) are denoted by $\varpi_1,\dots,\varpi_r$; we use their Bourbaki numbering [B68]. For $\mathsf{R}=\mathsf{A}_r,\mathsf{B}_r,\mathsf{C}_r,\mathsf{D}_r$, we assume that, respectively, $r\geqslant 1,3,2,4$. The group of characters of T is considered in additive notation. The representation contragredient to ρ is denoted by ρ^*.

Theorem 2.15 (irreducible representations of simple algebraic groups of modalities $0,1,2$). *Let G be a connected simple algebraic group and let V be a G-module determined by a nontrivial irreducible representation $\rho\colon G\to \mathrm{GL}(V)$.*

(M_1) *The condition* $\operatorname{mod}\rho=0$ *is equivalent to either of the following*:

(i) $k[V]^G=k$;
(ii) *the action of G on V is nonstable;*
(iii) *ρ or ρ^* is contained in the following list:*

$$(\mathsf{A}_r,\varpi_1);(\mathsf{A}_r,\varpi_2),r\geqslant 4 \textit{ is even};(\mathsf{C}_r,\varpi_1);(\mathsf{D}_5,\varpi_5).$$

(M_2) *The condition* $\operatorname{mod}\rho=1$ *is equivalent to either of the following*:

(i) $\operatorname{tr}\deg_k k[V]^G=1$;
(ii) $V/\!/G=\mathbf{A}^1$;
(iii) *ρ or ρ^* is contained in the following list:*

$$(\mathsf{A}_r,2\varpi_1);(\mathsf{A}_r,\varpi_2),r\geqslant 3 \textit{ is odd};(\mathsf{B}_r,\varpi_1);(\mathsf{D}_r,\varpi_1);$$
$$(\mathsf{A}_1,3\varpi_1);(\mathsf{A}_5,\varpi_3);(\mathsf{A}_6,\varpi_3);(\mathsf{A}_7,\varpi_3);(\mathsf{B}_3,\varpi_3);(\mathsf{B}_4,\varpi_4);(\mathsf{B}_5,\varpi_5);$$
$$(\mathsf{C}_2,\varpi_2);(\mathsf{C}_3,\varpi_3);(\mathsf{D}_6,\varpi_6);(\mathsf{D}_7,\varpi_7);(\mathsf{G}_2,\varpi_1);(\mathsf{E}_6,\varpi_1);(\mathsf{E}_7,\varpi_7).$$

(M_3) *The condition* $\operatorname{mod}\rho=2$ *is equivalent to either of the following* :

(i) $\operatorname{tr}\deg_k k[V]^G=2$;
(ii) $V/\!/G=\mathbf{A}^2$;
(iii) *ρ or ρ^* is contained in the following list:*

$$(\mathsf{A}_1,4\varpi_1);(\mathsf{A}_2,\varpi_1+\varpi_2);(\mathsf{A}_2,3\varpi_1);(\mathsf{B}_6,\varpi_6);$$
$$(\mathsf{C}_2,2\varpi_1);(\mathsf{C}_3,\varpi_2);(\mathsf{F}_4,\varpi_4);(\mathsf{G}_2,\varpi_2).$$

If $\operatorname{mod}\rho\leqslant 2$, *then ρ is modality-regular.*

First, we prove the following:

Lemma 2.16. *Let G be a connected semisimple algebraic group and let V be a G-module such that* $\operatorname{mod}(G:V)\leqslant 2$. *Then $k[V]^G$ is a free k-algebra.*

Proof. By (6), we have $\dim V/\!/G\leqslant 2$. If $\dim V/\!/G=0$, then $k[V]^G=k$. If $\dim V/\!/G=1$, then it follows from the Lüroth theorem that $V/\!/G=\mathbf{A}^1$; see [P80, Prop. 12] (this is true for any reductive G). If $\dim V/\!/G=2$, then, as is proved in

[$K80_2$] (and first conjectured in [P77, Sect. 7, Rem. 2°]), the assumption that G is connected semisimple entails $V/\!/G = \mathbf{A}^2$ (cf. also [PV94, Sect. 8.4]). This completes the proof. □

Proof of Theorem 2.15. By Lemma 2.16, if $\operatorname{mod}\rho \leqslant 2$, then ρ is cofree (i.e., $k[V]^G$ is a free k-algebra). The list of all cofree irreducible representations of connected simple algebraic groups is obtained in [KPV76] (see also Summary Table in [PV94, pp. 259–262]). This leads to determining which representations from this list have modality $\leqslant 2$. Comparing this list with the list of all irreducible visible representations obtained in [$K80_1$, Thm. 1] shows that every representation ρ from the former list is visible. Hence, by Theorem 2.10, it is modality-regular and $\operatorname{mod}\rho = \dim V/\!/G$. Since the integers $\dim V/\!/G$ are known (they are specified in the fifth column of the Summary Table in [PV94, pp. 259–262]), this yields the lists in (M_1)(iii), (M_2)(iii), and (M_3)(iii), thereby proving all the claims except (M_1)(ii). By [P71, Thm. 1], ρ is nonstable if and only if the G-stabilizer of a point in general position in V is nonreductive. Since the stabilizers of points in general position for the representations from this list are known as well (they are specified in the fourth column of the Summary Table in [PV94, pp. 259–262]), applying this criterion yields that (M_1)(ii) and (M_1)(iii) are equivalent. This completes the proof. □

Remarks 2.17.

1. In the following statements, the assumptions of the irreducibility of ρ and the simplicity of G are essential:

(a) In (M_1), in the claims that $k[V] = k$ or the nonstability of the action of G on V implies $\operatorname{mod}\rho = 0$. Examples:

— If $\rho = 2(\mathsf{A}_r, \varpi_1)$ and $r \geqslant 2$, then $k[V]^G = k$, the action of G on V is nonstable, and $\operatorname{mod}\rho > 0$; see Example 2.8.

— If $\rho = (\mathsf{A}_r, \varpi_2) \otimes (\mathsf{A}_1, \varpi_1)$, where $r \geqslant 8$ is even, then $k[V]^G = k$, see [L89], and the action of G on V in nonstable, see [PV94, Thm. 3.3, Cor. of Thm. 2.3], but the number of G-orbits in V is infinite (equivalently, $\operatorname{mod}\rho > 0$); see [SK77], [$K80_2$].

(b) In (M_2), in the claim that $\operatorname{tr deg}_k k[V]^G = 1$ is equivalent to $\operatorname{mod}\rho = 1$. For example, if $\rho = (r+1)(\mathsf{A}_r, \varpi_1)$ and $r \geqslant 2$, then $\operatorname{tr deg}_k k[V]^G = 1$, and $\operatorname{mod}\rho \geqslant r$; see Example 2.8.

(c) In (M_3), in the claim that $\operatorname{mod}\rho = 2$ implies $\operatorname{tr deg}_k k[V]^G = 2$. For example, if $\rho = 3(\mathsf{A}_r, \varpi_1)$ and $r \geqslant 3$, then it is not difficult to see that $k[V]^G = k$ and $\operatorname{mod}\rho = 2$.

The representations ρ specified in (a), (b), (c) are not modality-regular.

2. Arguing along the same lines, one can extend the classifications obtained in Theorem 2.15 up to the classifications of all irreducible castling reduced representations of connected semisimple algebraic groups of every fixed modality $\leqslant 2$. The reason being that the complete list of cofree irreducible castling reduced representations of connected semisimple algebraic groups is known [L89]. Comparing it with that of visible representations obtained in [SK77], [$K80_1$], [KKY86], one ascertains that the majority of representations in this list (all but two) are visible, and therefore, for them, the modality is given by Theorem 2.10 (the integers $\dim V/\!/G$ are specified in the fifth column of [L89, Tabelle]). To the remaining two representations one applies the ad hoc considerations.

3. The arguments from the proof of Theorem 2.15 yield the modalities of all cofree irreducible representations of connected simple algebraic groups.

2.11 We conclude this section with a statement, which in some cases helps to practically determine the modality.

Lemma 2.18. *Let X be an algebraic variety endowed with an action of an algebraic group G. Let $\{C_i\}_{i\in I}$ be a collection of subsets of X such that*

(i) *I is finite;*
(ii) *$\bigcup_{i\in I} C_i = X$;*
(iii) *the closure $\overline{C_i}$ of C_i in X is irreducible for every $i \in I$;*
(iv) *every C_i is G-stable;*
(v) *all G-orbits in C_i have the same dimension d_i for every $i \in I$.*

Then the following hold:

(a) $\operatorname{mod}(G:X) = \max_{i\in I}(\dim\overline{C_i} - d_i)$*;*
(b) *if X is irreducible, then $X = \overline{C_{i_0}}$ for some i_0, and* $\operatorname{mod}(G:X^{\mathrm{reg}}) = \dim X - d_{i_0}$.

Proof. By (iii), we have a family $\overline{C_i}^{\mathrm{reg}}$, and (v) implies $C_i \subseteq \overline{C_i}^{\mathrm{reg}}$. Hence

$$\operatorname{mod}(G:\overline{C_i}^{\mathrm{reg}}) = \dim\overline{C_i} - d_i. \tag{15}$$

From (3) and (15), we infer that $\operatorname{mod}(G:X) \geqslant \max\limits_{i\in I}(\dim\overline{C}_i - d_i)$. To prove the opposite inequality, let $Z \in \mathscr{F}(X)$ be a family of s-dimensional G-orbits such that $\operatorname{mod}(G:X) = \dim Z - s$ and let $J := \{i \in I \mid Z \cap C_i \neq \varnothing\}$. By (ii), we have $Z = \bigcup_{j\in J}(Z \cap \overline{C_j})$. Since Z is irreducible and, by (i), J is finite, there is $j_0 \in J$ such that $Z \subseteq \overline{C_{j_0}}$. As $Z \cap C_{j_0} \neq \varnothing$, we have $s = d_{j_0}$. Therefore, $\operatorname{mod}(G:X) = \dim Z - s \leqslant \dim\overline{C_{j_0}} - d_{j_0}$. This proves (a).

By (ii), $\bigcup_{i\in I}\overline{C_i} = X$. If X is irreducible, then, in view of (i), this equality implies the existence of i_0 such that $X = \overline{C_{i_0}}$. This and (15) prove (b). □

Example 2.19. Consider the graded semisimple Lie algebra (13) and the adjoint representation $\rho\colon G_0 \to \mathrm{GL}(\mathfrak{g}_1)$. In Section 3 we define a collection $\{C_i\}_{i\in I}$ of subsets of $\mathfrak{g}_1$ called packets (see below Definition 3.6). In Propositions 3.8, 3.9, 3.13 below, we prove that this collection satisfies all conditions (i)–(v) from Lemma 2.18, and in Corollary 3.14 below we compute $\operatorname{mod}(G_0 : \overline{C_i})$ for every i. The integer $\max_{i\in I}\operatorname{mod}(G_0 : \overline{C_i})$ turns out then to be equal to the rank of the graded Lie algebra (13). By Lemma 2.18, this agrees with Theorem 2.11.

Corollary 2.20. *Let Y and F be the algebraic varieties endowed with the actions of an algebraic group G. Consider the diagonal action of G on $X := Y \times F$. If G acts on F trivially, then*

(i) $\operatorname{mod}(G:X) = \operatorname{mod}(G:Y) + \dim F$*;*
(ii) *the action of G on Y is modality-regular if and only if the action of G on X is modality-regular.*

Proof. Let $\{S_i\}_{i\in I}$ be the collection of all sheets of Y. The claim then follows by applying Lemma 2.18 to $\{C_i\}_{i\in I}$ where $C_i := S_i \times F$. □

2.12 It seems natural to use the concepts of modality and sheets for expanding the exploration undertaken in [P16].

3 Packets for θ-groups

In this section we explore the finer geometric properties of θ-group actions.

3.1 Besides the notation of Subsection 2.8 above, we also use the following:

- If $\mathfrak{a}$ and $\mathfrak{b}$ are nonempty subsets of $\mathfrak{g}$, then $\mathfrak{a}^{\mathfrak{b}}$ is the centralizer of $\mathfrak{b}$ in $\mathfrak{a}$,

$$\mathfrak{a}^{\mathfrak{b}} := \{x \in \mathfrak{a} \mid [x,y] = 0 \text{ for all } y \in \mathfrak{b}\},$$

  and $[\mathfrak{a},\mathfrak{b}]$ is the k-linear span of all $[x,y]$, where $x \in \mathfrak{a}$, $y \in \mathfrak{b}$.
- The center of $\mathfrak{a}$, i.e., $\mathfrak{a}^{\mathfrak{a}}$, is denoted by $\mathfrak{z}(\mathfrak{a})$.
- For any subset $\mathfrak{s}$ in $\mathfrak{g}$, we put $\mathfrak{s}_i := \mathfrak{s} \cap \mathfrak{g}_i$ (the case $\mathfrak{s} \cap \mathfrak{g}_i = \varnothing$ is not excluded).

The following facts, used below, are proved in [V76]:

There exists a nondegenerate G-invariant and θ-invariant scalar multiplication $\mathfrak{g} \times \mathfrak{g} \to k$, $(x,y) \mapsto \langle x,y \rangle$.

If $x = x_{\rm s} + x_{\rm n}$ is the Jordan decomposition of an element $x \in \mathfrak{g}$ with $x_{\rm s}$ semisimple and $x_{\rm n}$ nilpotent, then $x \in \mathfrak{g}_i$ entails that $x_{\rm s}, x_{\rm n} \in \mathfrak{g}_i$.

There are only finitely many nilpotent G_0-orbits in every $\mathfrak{g}_i$.

3.2 We first consider the following general construction and introduce the related terminology and notation.

Definition 3.1. Let M be a nonempty set and let F be a nonempty set of functions $M \to k$. Define the equivalence relation $\sim_F$ on M by

$$x \sim_F y \iff \text{for every } \alpha \in F, \text{ either } \alpha(x) = \alpha(y) = 0 \text{ or } \alpha(x)\alpha(y) \neq 0.$$

The equivalence classes of $\sim_F$ are called the *cells* of $\sim_F$. The set of all cells of $\sim_F$ is denoted by $\mathscr{C}_F(M)$.

Proposition 3.2. *Let V be a finite-dimensional vector space over k and let F be a finite subset of the dual space V^*.*

(i) *For every linear subspace L in V^*, the set*

$$C_L := \left\{ v \in V \;\middle|\; \begin{array}{l} \alpha(v) = 0 \textit{ for all } \alpha \in F \cap L, \\ \beta(v) \neq 0 \textit{ for all } \beta \in F \setminus L. \end{array} \right\} \tag{16}$$

is nonempty.

(ii) *A subset of V is a cell of $\sim_F$ if and only if it is C_L for some L.*

(iii) *The closure $\overline{C_L}$ of C_L in V is the linear subspace*

$$\{v \in V \mid \alpha(v) = 0 \textit{ for all } \alpha \in F \cap L\}, \tag{17}$$

and the complement of C_L in $\overline{C_L}$ is the union of hyperplanes

$$\bigcup_{\beta \in F \setminus L} \{v \in \overline{C_L} \mid \beta(v) = 0\}.$$

(iv) *The set $\mathscr{C}_F(V)$ is finite.*

Proof. (i) In view of (16), without changing C_L, we may (and shall) assume that L is the linear span of $L \cap F$. If $L = V^*$, then $C_L = 0$. Let $L \neq V^*$; then the dimension of the linear subspace (17) is positive. The restriction of every $\beta \in F \setminus L$ to it is nonzero. For, otherwise, $\operatorname{rk}\big((F \cap L) \cup \beta\big) = \operatorname{rk}(F \cap L)$, hence β lies in the linear span of $F \cap L$, i.e., in L,—a contradiction. Thus the locus of zeros of the restriction of β to the linear subspace (17) is its proper linear subspace. Hence C_L is nonempty.

(ii) It follows directly from the definitions of $\sim_F$ and C_L that C_L is a cell of $\sim_F$. Conversely, let C be a cell of $\sim_F$ in V. If L is the linear span of $\{\alpha \in F \mid \alpha|_C = 0\}$, then Definition 3.1 implies that every $\beta \in F \setminus L$ vanishes nowhere on C. Hence $C \subseteq C_L$. Since, by (i), C_L is a cell, this yields $C = C_L$.

(iii) This follows from (i).

(iv) In view of (i), this follows from the finiteness of F. □

Corollary 3.3. *Every cell of $\sim_F$ in V is an irreducible smooth rational affine algebraic variety locally closed in V.*

3.3 We now fix a maximal torus $\mathfrak{t}$ of $\mathfrak{g}$ such that $\mathfrak{t}_1$ is a Cartan subspace of $\mathfrak{g}_1$ (since the minimal algebraic subalgebra of a Cartan subspace is a torus, such a $\mathfrak{t}$ exists). Let $R \subset \mathfrak{t}^*$ be the root system of $\mathfrak{g}$ with respect to $\mathfrak{t}$. As usual, if $\alpha \in R$, then $\mathfrak{g}_\alpha := \{x \in \mathfrak{g} \mid [t, x] = \alpha(t)x \text{ for every } t \in \mathfrak{t}\}$.

We consider the cells of $\sim_R$ in $\mathfrak{t}$.

Proposition 3.4. *Let $\mathfrak{c}$ be a cell of $\sim_R$ in $\mathfrak{t}$ and let*

$$R_{\mathfrak{c}} := \{\alpha \in R \mid \alpha|_{\mathfrak{c}} = 0\}.$$

(i) *For every nonempty subset $\mathfrak{s} \subseteq \mathfrak{c}$, the following hold:*

(i_1) *$\mathfrak{g}^{\mathfrak{s}}$ is a reductive subalgebra of $\mathfrak{g}$ with the maximal torus $\mathfrak{t}$ and the $\mathfrak{t}$-root decomposition*

$$\mathfrak{g}^{\mathfrak{s}} = \mathfrak{t} \oplus \big(\textstyle\bigoplus_{\alpha \in R_{\mathfrak{c}}} \mathfrak{g}_\alpha\big). \tag{18}$$

(i_2) *the center $\mathfrak{z}(\mathfrak{g}^{\mathfrak{s}})$ of $\mathfrak{g}^{\mathfrak{s}}$ is $\overline{\mathfrak{c}}$.*

(i$_3$) *the commutator ideal* $[\mathfrak{g}^{\mathfrak{s}}, \mathfrak{g}^{\mathfrak{s}}]$ *of* $\mathfrak{g}^{\mathfrak{s}}$ *is* $\overline{\mathfrak{c}}^{\perp} \oplus \left(\bigoplus_{\alpha \in R_{\mathfrak{c}}} \mathfrak{g}_{\alpha}\right)$, *where* $\overline{\mathfrak{c}}^{\perp}$ *is the orthogonal complement of* $\overline{\mathfrak{c}}$ *in* $\mathfrak{t}$ *with respect to* $\langle \cdot, \cdot \rangle|_{\mathfrak{t}}$.

(ii) *If* $t \in \mathfrak{g}$ *is a semisimple element such that* $\mathfrak{g}^t = \mathfrak{g}^{\mathfrak{c}}$, *then* $t \in \mathfrak{c}$.

Proof. (i$_1$) Since the elements of $\mathfrak{s}$ are semisimple and pairwise commute, the Lie algebra $\mathfrak{g}^{\mathfrak{s}}$ is reductive. As $\mathfrak{g}^{\mathfrak{s}} = \bigcap_{c \in \mathfrak{s}} \mathfrak{g}^c$, to prove (18) it suffices to show that (18) holds for $\mathfrak{s}$ being a single element $c \in \mathfrak{c}$. Let $x \in \mathfrak{g}$. The root decomposition of $\mathfrak{g}$ with respect to $\mathfrak{t}$,

$$\mathfrak{g} = \mathfrak{t} \oplus \left(\bigoplus_{\alpha \in R} \mathfrak{g}_{\alpha}\right), \tag{19}$$

yields $x = t + \sum_{\alpha \in R} x_{\alpha}$ for some $t \in \mathfrak{t}$, $x_{\alpha} \in \mathfrak{g}_{\alpha}$. Hence $[c, x] = \sum_{\alpha \in R} \alpha(c) x_{\alpha}$. Therefore, $x \in \mathfrak{g}^c$ if and only if $\alpha(c) = 0$ for every α such that $x_{\alpha} \neq 0$. The condition $c \in \mathfrak{c}$ implies that $\alpha(c) = 0$ if and only if $\alpha \in R_{\mathfrak{c}}$. Hence $x \in \mathfrak{g}^c$ if and only if $x_{\alpha} = 0$ for all $\alpha \notin R_{\mathfrak{c}}$. This and (19) prove (i).

(i$_2$) Since the center of any reductive Lie algebra lies in every maximal torus of the latter, (i$_1$) and Proposition 3.2 entail

$$\begin{aligned} \mathfrak{z}(\mathfrak{g}^{\mathfrak{s}}) &= \{t \in \mathfrak{t} \mid [t, \mathfrak{g}_{\alpha}] = \alpha(t)\mathfrak{g}_{\alpha} = 0 \text{ for each } \alpha \in R_{\mathfrak{c}}\} \\ &= \{t \in \mathfrak{t} \mid \alpha(t) = 0 \text{ for each } \alpha \in R_{\mathfrak{c}}\} = \overline{\mathfrak{c}}. \end{aligned}$$

(i$_3$) This follows from (18) and (i$_2$).

(ii) As $\mathfrak{t}$ lies in $\mathfrak{g}^{\mathfrak{c}}$ by (i$_1$), the equality $\mathfrak{g}^t = \mathfrak{g}^{\mathfrak{c}}$ entails that t commutes with $\mathfrak{t}$. As t is semisimple and $\mathfrak{t}$ is a maximal torus, this shows that $t \in \mathfrak{t}$. Let $\mathfrak{c}'$ be the unique cell in $\mathfrak{t}$ containing t. From (i$_2$) we obtain $\overline{\mathfrak{c}'} = \mathfrak{z}(\mathfrak{g}^t) = \mathfrak{z}(\mathfrak{g}^{\mathfrak{c}}) = \overline{\mathfrak{c}}$. By Proposition 3.2(ii),(iii), this yields $\mathfrak{c}' = \mathfrak{c}$. □

Corollary 3.5. *Every two nonempty subsets of any cell of* $\sim_R$ *in* $\mathfrak{t}$ *have the same centralizers in* $\mathfrak{g}$.

3.4 The decomposition $\mathfrak{t} = \bigsqcup_{\mathfrak{c} \in \mathscr{C}_R(\mathfrak{t})} \mathfrak{c}$ yields the decomposition

$$\mathfrak{t}_1 = \bigsqcup_{\mathfrak{c} \in \mathscr{C}_R(\mathfrak{t})} \mathfrak{c}_1. \tag{20}$$

Definition 3.6. For every cell $\mathfrak{c} \in \mathscr{C}_R(\mathfrak{t})$ such that $\mathfrak{c}_1 \neq \varnothing$ and every nilpotent element $n \in \mathfrak{g}_1^{\mathfrak{c}}$, the set $G_0 \cdot (\mathfrak{c}_1 + n)$ is called a *packet* in $\mathfrak{g}_1$.

Remark 3.7. By [TY05, Prop. 39.1.5] (cf. also Proposition 3.10 below), for $m = 1$ in (13), Definition 3.6 is equivalent to the usual definition of a packet (also known as Jordan/decomposition class) in $\mathfrak{g}$; see [TY05, 39.1.3] and the references in [P08].

3.5 In this subsection we prove some basic properties of packets in $\mathfrak{g}$. Below, for every locally closed subset X in $\mathfrak{g}_1$, the notation X^{reg} refers to the action of G_0 on $\mathfrak{g}_1$ unless stated otherwise.

Proposition 3.8. *The union of all packets in* $\mathfrak{g}_1$ *coincides with* $\mathfrak{g}_1$.

Proof. Let $x \in \mathfrak{g}_1$; we should show that x lies in a packet. We have $x_{\rm s}, x_{\rm n} \in \mathfrak{g}_1$. Definition 3.6 shows that packets in $\mathfrak{g}_1$ are G_0-stable. On the other hand, since $\mathfrak{t}_1$ is a Cartan subspace in $\mathfrak{g}_1$ (by the choice of $\mathfrak{t}$), the G_0-orbit of $x_{\rm s}$ intersects $\mathfrak{t}_1$, see [V76, Cor. of Thm. 1]. Hence when proving that x lies in a packet, we can (and shall) assume that $x_{\rm s} \in \mathfrak{t}_1$. In view of (20), there exists a cell $\mathfrak{c} \in \mathscr{C}_R(\mathfrak{t})$ such that $x_{\rm s} \in \mathfrak{c}_1$. By the definition of the Jordan decomposition of x, we have $x_{\rm n} \in \mathfrak{g}_1^{x_{\rm s}}$. In view of Corollary 3.5, we have $\mathfrak{g}_1^{x_{\rm s}} = \mathfrak{g}_1^{\mathfrak{c}}$. Hence, by Definition 3.6, $G_0\cdot(\mathfrak{c}_1 + x_{\rm n})$ is a packet in $\mathfrak{g}_1$; clearly x lies in it. □

Proposition 3.9. *There are only finitely many packets in* $\mathfrak{g}_1$.

Proof. Since the set $\mathscr{C}_R(\mathfrak{t})$ is a finite by Proposition 3.2(iv), it suffices to show that, for any given cell $\mathfrak{c} \in \mathscr{C}_R(\mathfrak{t})$ with $\mathfrak{c}_1 \neq \varnothing$, there are only finitely many packets of the form $G_0\cdot(\mathfrak{c}_1 + n)$ where n is a nilpotent element of $\mathfrak{g}_1^{\mathfrak{c}}$.

To this end, consider the Lie subalgebra $\mathfrak{g}^{\mathfrak{c}_1}$ of $\mathfrak{g}$. By Proposition 3.4(i), it is reductive. As $\mathfrak{g}_1$ is a weight space of θ, the inclusion $\mathfrak{c}_1 \subseteq \mathfrak{g}_1$ entails that $\mathfrak{g}^{\mathfrak{c}_1}$ be θ-stable, i.e., $\mathfrak{g}^{\mathfrak{c}_1}$ is a graded subalgebra of the graded Lie algebra $\mathfrak{g}$. In view of Corollary 3.5, we have $\mathfrak{g}^{\mathfrak{c}_1} = \mathfrak{g}^{\mathfrak{c}}$.

Let S be a closed connected subgroup of G_0 such that

$$\operatorname{Lie} S = \mathfrak{g}_0^{\mathfrak{c}}. \tag{21}$$

By [V76, Prop. 2], there are only finitely many orbits of the adjoint action of S on the variety of nilpotent elements of $\mathfrak{g}_1^{\mathfrak{c}}$. Let $n_1, \ldots, n_d$ be the representatives of these orbits.

Now consider a packet $G_0\cdot(\mathfrak{c}_1 + n)$, where n is a nilpotent element of $\mathfrak{g}_1^{\mathfrak{c}}$. By the aforesaid, $S\cdot n = S\cdot n_i$ for some i. Since, in view of (21), every element of $\mathfrak{c}_1$ is a fixed point of S, this implies that

$$S\cdot(\mathfrak{c}_1 + n) = S\cdot(\mathfrak{c}_1 + n_i). \tag{22}$$

In turn, since S is a subgroup of G_0, from Definition 3.6 and (22) we obtain

$$G_0\cdot(\mathfrak{c}_1 + n) = G_0 S\cdot(\mathfrak{c}_1 + n) \overset{(22)}{=} G_0 S\cdot(\mathfrak{c}_1 + n_i) = G_0\cdot(\mathfrak{c}_1 + n_i).$$

This proves the required finiteness statement. □

Proposition 3.10. *For any $x, y \in \mathfrak{g}_1$ the following properties are equivalent:*

(P) *There exists a packet $G_0\cdot(\mathfrak{c}_1 + n)$ in $\mathfrak{g}_1$ containing both x and y.*
(J) *There exists an element $g \in G_0$ such that*

$$g\cdot\mathfrak{g}^{x_{\rm s}} = \mathfrak{g}^{g\cdot x_{\rm s}} = \mathfrak{g}^{y_{\rm s}} \quad \textit{and} \quad g\cdot x_{\rm n} = y_{\rm n}. \tag{23}$$

Proof. (P)⇒(J) Let $x, y \in G_0\cdot(\mathfrak{c}_1 + n)$. By Definition 3.6, there are $c_1, c_2 \in \mathfrak{c}_1$ and $g_1, g_2 \in G_0$ such that $g_1\cdot x = c_1 + n$, $g_2\cdot y = c_2 + n$. Hence

$$g_1{\cdot}x_{\rm s} = (g_1{\cdot}x)_{\rm s} = c_1, \quad g_1{\cdot}x_{\rm n} = (g_1{\cdot}x)_{\rm n} = n,$$
$$g_2{\cdot}y_{\rm s} = (g_2{\cdot}y)_{\rm s} = c_2, \quad g_2{\cdot}y_{\rm n} = (g_2{\cdot}y)_{\rm n} = n, \tag{24}$$

From (24) and Corollary 3.5 we obtain $\mathfrak{g}^{g_1\cdot x_{\rm s}} = \mathfrak{g}^{g_2\cdot y_{\rm s}}$ and $g_1{\cdot}x_{\rm n} = g_2{\cdot}y_{\rm n}$; hence (23) with $g = g_2^{-1}g_1$.

(J)⇒(P) Assume that (23) holds for $x, y \in \mathfrak{g}_1$. By Proposition 3.8, there is a packet $G_0{\cdot}(\mathfrak{c}_1 + n)$ in $\mathfrak{g}_1$ containing y; we have to show that $x \in G_0{\cdot}(\mathfrak{c}_1 + n)$. By Definition 3.6, there are $c \in \mathfrak{c}_1$ and $h \in G_0$ such that $y = h{\cdot}(c+n)$; hence $y_{\rm s} = h{\cdot}c$, $y_{\rm n} = h{\cdot}n$. Plugging this into (23), we obtain

$$\mathfrak{g}^{w\cdot x_{\rm s}} = \mathfrak{g}^{c}, \quad w{\cdot}x_{\rm n} = n, \qquad \text{where } w = h^{-1}g. \tag{25}$$

By Proposition 3.4(i$_1$), we have $\mathfrak{t} \subset \mathfrak{g}^c$. Hence, by the first equality in (25), the semisimple element $w(x_{\rm s})$ centralizes the maximal torus $\mathfrak{t}$. Hence $w{\cdot}x_{\rm s} \in \mathfrak{t}$ and, therefore, there is a unique cell $\mathfrak{c}'$ in $\mathfrak{t}$ such that

$$w{\cdot}x_{\rm s} \in \mathfrak{c}'. \tag{26}$$

From Proposition 3.4(i$_2$) we then infer that $\mathfrak{z}(\mathfrak{g}^{w\cdot x_{\rm s}}) = \overline{\mathfrak{c}'}$. On the other hand, by the same reason, $\mathfrak{z}(\mathfrak{g}^c) = \overline{\mathfrak{c}}$. Combining this with the first equality in (25), we conclude that $\overline{\mathfrak{c}'} = \overline{\mathfrak{c}}$. By Proposition 3.2(iii), this yields

$$\mathfrak{c}' = \mathfrak{c}. \tag{27}$$

From (26), (27), the second equality in (25), and Definition 3.6 we now conclude that $x \in G_0{\cdot}(\mathfrak{c}_1 + n)$. □

Corollary 3.11. *The packets in $\mathfrak{g}_1$ do not depend on the choice of a torus $\mathfrak{t}$.*

Corollary 3.12. *Every two packets in $\mathfrak{g}_1$ are either equal or disjoint.*

Proof. This is because property (J) is clearly an equivalence relation on $\mathfrak{g}_1$ and, by Proposition 3.10, packets in $\mathfrak{g}_1$ are precisely its equivalence classes. □

Proposition 3.13. *For every packet $G_0{\cdot}(\mathfrak{c}_1 + n)$ and its closure $\overline{G_0{\cdot}(\mathfrak{c}_1 + n)}$ in $\mathfrak{g}_1$, the following hold:*

(i) $\mathfrak{g}_i^x = \mathfrak{g}_i^{\mathfrak{c}} \cap \mathfrak{g}_i^n$ *for any $x \in \mathfrak{c}_1 + n$ and i;*
(ii) $[\mathfrak{g}_0, x] = [\mathfrak{g}_0, y]$ *for any $x, y \in \mathfrak{c}_1 + n$.*
(iii) $G_0{\cdot}(\mathfrak{c}_1 + n)$ *is irreducible and contains a dense open subset of* $\overline{G_0{\cdot}(\mathfrak{c}_1 + n)}$.
(iv) $\overline{G_0{\cdot}(\mathfrak{c}_1 + n)}$ *is contained in a sheet of* $\mathfrak{g}_1$.
(v) $\overline{G_0{\cdot}(\mathfrak{c}_1 + n)}$ *is an irreducible algebraic variety of dimension $d + \dim \mathfrak{c}_1$, where d is the dimension of G_0-orbits in a sheet containing $G_0{\cdot}(\mathfrak{c}_1 + n)$.*

Proof. If $c \in \mathfrak{c}_1$ and $x = c + n$, then $c = x_{\rm s}$ and $n = x_{\rm n}$. This, the uniqueness of the Jordan decomposition and Corollary 3.5 imply that

$$\mathfrak{g}^x = \mathfrak{g}^c \cap \mathfrak{g}^n = \mathfrak{g}^{\mathfrak{c}} \cap \mathfrak{g}^n. \tag{28}$$

Since c, n, x are the homogeneous elements, $\mathfrak{g}^c$, $\mathfrak{g}^n$, $\mathfrak{g}^x$ are the graded subalgebras of $\mathfrak{g}$. Taking the ith components, from (28) we obtain (i).

The subspace $\mathfrak{g}_{-1}$ is dual to $\mathfrak{g}_1$ with respect to $\langle\cdot,\cdot\rangle$, and the orthogonal complement in $\mathfrak{g}_1$ to $\mathfrak{g}^x_{-1}$ with respect to this duality is $[\mathfrak{g}_0, x]$; see [V76, Prop. 5]. Since $\mathfrak{g}^x_{-1}$, by (i), is the same for all $x \in \mathfrak{c}_1 + n$, this proves (ii).

Corollary 3.3 and the connectedness of G_0 yield that $G_0 \times \mathfrak{c}_1$ is an irreducible smooth affine algebraic variety. In view of Definition 3.6, the packet $G_0 \cdot (\mathfrak{c}_1 + n)$ is the image of the morphism:

$$\varphi\colon G_0 \times \mathfrak{c}_1 \to \mathfrak{g}_1, \quad (g, c) \mapsto g \cdot (c + n). \tag{29}$$

By the general properties of morphisms, this implies (iii).

In view of (14), we have $\dim G_0 \cdot x = \dim \mathfrak{g}_0 - \dim \mathfrak{g}_0^x$. This and (i) show that $\dim G_0 \cdot x$ is the same for all $x \in G_0 \cdot (\mathfrak{c}_1 + n)$, which, by (iii), means that $G_0 \cdot (\mathfrak{c}_1 + n) \subseteq \overline{G_0 \cdot (\mathfrak{c}_1 + n)}^{\mathrm{reg}}$. This implies (iv).

We may (and shall) consider φ as a dominant morphism $G_0 \times \mathfrak{c}_1 \to \overline{G_0 \cdot (\mathfrak{c}_1 + n)}$. By [H77, Lem. 10.5], there is a point $z = (g, c) \in G_0 \times \mathfrak{c}_1$ such that $\varphi(z)$ is a smooth point of $\overline{G_0 \cdot (\mathfrak{c}_1 + n)}$ and $d\varphi_z$ is a surjective map of the tangent spaces; this implies that $\dim \overline{G_0 \cdot (\mathfrak{c}_1 + n)}$ is equal to the dimension of the image of $d\varphi_z$. We shall now compute this latter dimension.

First, note that since φ is G_0-equivariant, we may (and shall) assume that g is the identity element. From Proposition 3.2(ii), we infer that the closure $\overline{\mathfrak{c}_1}$ of $\mathfrak{c}_1$ in $\mathfrak{t}_1$ is a linear subspace, and $\overline{\mathfrak{c}_1} \setminus \mathfrak{c}_1$ is a union of finitely many hyperplanes. This shows that $\mathfrak{g}_0 \oplus \overline{\mathfrak{c}_1}$ is the tangent space of $G_0 \times \mathfrak{c}_1$ at z. Given this, we deduce from (14) and (29) that

$$\text{the image of } d\varphi_z \text{ is } [\mathfrak{g}_0, c + n] + \overline{\mathfrak{c}_1}. \tag{30}$$

For the point $x := c + n \in \mathfrak{c}_1 + n$, we have $x_{\mathrm{s}} = c$, $x_{\mathrm{n}} = n$. Therefore,

$$[\mathfrak{g}, x] \cap \mathfrak{z}(\mathfrak{g}^c) = 0; \tag{31}$$

see [TY05, Lem. 39.2.8]. But $\mathfrak{z}(\mathfrak{g}^c) = \overline{\mathfrak{c}}$ by Proposition 3.4(i_2). This and (31) yield $[\mathfrak{g}, x] \cap \overline{\mathfrak{c}} = 0$; hence

$$[\mathfrak{g}_0, x] \cap \overline{\mathfrak{c}_1} = 0. \tag{32}$$

Combining (30), (32), (ii), and taking into account that $[\mathfrak{g}_0, x]$ is the tangent space to $G_0 \cdot x$ at x, we now obtain that the dimension of the image of $d\varphi_z$ is equal to $\dim[\mathfrak{g}_0, x] + \dim \overline{\mathfrak{c}_1} = d + \dim \mathfrak{c}_1$. This completes the proof of (v). □

Corollary 3.14. $\mathrm{mod}\,(G_0 : \overline{G_0 \cdot (\mathfrak{c}_1 + n)}) = \dim \mathfrak{c}_1$.

Lemma 3.15. $x \in (\mathfrak{z}(\mathfrak{g}^x)_1)^{\mathrm{reg}}$ *for every* $x \in \mathfrak{g}_1$.

Proof. If $y \in \mathfrak{z}(\mathfrak{g}^x)$, then $\mathfrak{g}^x \subseteq \mathfrak{g}^y$; see [TY05, 35.3.2]. As x is homogeneous, $\mathfrak{g}^x$ is graded. If y is homogeneous, $\mathfrak{g}^y$ is graded as well, hence the specified inclusion yields $\mathfrak{g}_i^x \subseteq \mathfrak{g}_i^y$ for every i. In particular, $\mathfrak{g}_0^x \subseteq \mathfrak{g}_0^y$ for every $y \in \mathfrak{z}(\mathfrak{g}^x)_1$, hence $\dim G_0 \cdot y \leqslant \dim G_0 \cdot x$ and, by (4), the claim. □

Lemma 3.16. *Let $x \in \mathfrak{g}_1$ be a nilpotent element. Then*

$$G_0{\cdot}x \cap \mathfrak{z}(\mathfrak{g}^x)_1 = (\mathfrak{z}(\mathfrak{g}^x)_1)^{\mathrm{reg}}.$$

Proof. By by (4) and Lemma 3.15, for every $z \in \mathfrak{z}(\mathfrak{g}^x)_1$, we have

$$z \in (\mathfrak{z}(\mathfrak{g}^x)_1)^{\mathrm{reg}} \iff \dim G_0{\cdot}z = \dim G_0{\cdot}x. \tag{33}$$

As x is nilpotent and $\mathfrak{g}$ is semisimple, all elements of $\mathfrak{z}(\mathfrak{g}^x)$ are nilpotent (see [TY05, 35.1.2]). Therefore $\mathfrak{z}(\mathfrak{g}^x)_1$ lies in the variety of nilpotent elements of $\mathfrak{g}_1$. By [V76, Prop. 2], this variety is the union of finitely many G_0-orbits. From this we infer that among them there is an orbit $\mathscr{O}$ such that $\mathscr{O} \cap \mathfrak{z}(\mathfrak{g}^x)_1$ is a dense open subset of $\mathfrak{z}(\mathfrak{g}^x)_1$. This and (33) entail that

$$\overline{\mathscr{O} \cap \mathfrak{z}(\mathfrak{g}^x)_1} = \mathfrak{z}(\mathfrak{g}^x)_1, \tag{34}$$

$$\mathscr{O} \cap \mathfrak{z}(\mathfrak{g}^x)_1 \subseteq (\mathfrak{z}(\mathfrak{g}^x)_1)^{\mathrm{reg}}, \tag{35}$$

$$\dim \mathscr{O} = \dim G_0{\cdot}x. \tag{36}$$

If $z \in (\mathfrak{z}(\mathfrak{g}^x)_1)^{\mathrm{reg}}$, then (33), (36) yield $\dim G_0{\cdot}z = \dim \mathscr{O}$. Since, by (34), $z \in \overline{\mathscr{O}}$, the latter equality implies that $G_0{\cdot}z = \mathscr{O}$. This equality and (35) yield

$$\mathscr{O} \cap \mathfrak{z}(\mathfrak{g}^x)_1 = (\mathfrak{z}(\mathfrak{g}^x)_1)^{\mathrm{reg}}. \tag{37}$$

From (37) and Lemma 3.15 we now infer that $\mathscr{O} = G_0{\cdot}x$. This equality and (37) complete the proof. □

Proposition 3.17. *For all $x, y \in \mathfrak{g}_1$, the following conditions are equivalent:*

(i) *x and y lie in the same packet in $\mathfrak{g}_1$.*
(ii) *There exists $g \in G_0$ such that $\mathfrak{g}^y = g{\cdot}\mathfrak{g}^x$.*
(iii) *There exists $g \in G_0$ such that $\mathfrak{z}(\mathfrak{g}^y) = g{\cdot}\mathfrak{z}(\mathfrak{g}^x)$.*

Proof. The implication (ii)⇒(iii) is clear.

Assume that (iii) holds. Then $\mathfrak{z}(\mathfrak{g}^y) = \mathfrak{z}(\mathfrak{g}^{g\cdot x})$, which implies that $\mathfrak{g}^y = \mathfrak{g}^{g\cdot x} = g{\cdot}\mathfrak{g}^x$; see [TY05, 35.3.2]. This proves the implication (iii)⇒(ii).

Assume that (i) holds. Then by Proposition 3.10 there exists $g \in G_0$ such that (23) holds. Therefore

$$\mathfrak{g}^y = \mathfrak{g}^{y_{\mathrm{s}}} \cap \mathfrak{g}^{y_{\mathrm{n}}} = g{\cdot}\mathfrak{g}^{x_{\mathrm{s}}} \cap g{\cdot}\mathfrak{g}^{x_{\mathrm{n}}} = g{\cdot}(\mathfrak{g}^{x_{\mathrm{s}}} \cap \mathfrak{g}^{x_{\mathrm{n}}}) = g{\cdot}\mathfrak{g}^x.$$

This proves the implication (i)⇒(ii).

Finally, let us prove that x and y lie same packet in $\mathfrak{g}_1$ assuming that (iii) holds. Since packets are G_0-stable, we may (and shall) assume that g is the identity element, i.e.,

$$\mathfrak{z}(\mathfrak{g}^x) = \mathfrak{z}(\mathfrak{g}^y). \tag{38}$$

According to [TY05, 39.1.1],

$$
\begin{aligned}
\mathfrak{z}(\mathfrak{g}^x) &= \mathfrak{z}(\mathfrak{g}^{x_s}) \oplus \mathfrak{z}\big([\mathfrak{g}^{x_s},\mathfrak{g}^{x_s}]^{x_n}\big),\\
\mathfrak{z}(\mathfrak{g}^y) &= \mathfrak{z}(\mathfrak{g}^{y_s}) \oplus \mathfrak{z}\big([\mathfrak{g}^{y_s},\mathfrak{g}^{y_s}]^{y_n}\big),
\end{aligned}
\tag{39}
$$

and the first (respectively, second) direct summand in each of the equalities (39) is the set of all semisimple (respectively, nilpotent) elements of the left-hand side of this equality. This and (38) then entail that

$$
\begin{aligned}
\mathfrak{z}(\mathfrak{g}^{x_s}) &= \mathfrak{z}(\mathfrak{g}^{y_s}),\\
\mathfrak{z}\big([\mathfrak{g}^{x_s},\mathfrak{g}^{x_s}]^{x_n}\big) &= \mathfrak{z}\big([\mathfrak{g}^{y_s},\mathfrak{g}^{y_s}]^{y_n}\big).
\end{aligned}
\tag{40}
$$

Since x and y are homogeneous elements, the left- and right-hand sides of (40) are the graded subalgebras of $\mathfrak{g}$. The subalgebras $[\mathfrak{g}^{x_s},\mathfrak{g}^{x_s}]$, $[\mathfrak{g}^{y_s},\mathfrak{g}^{y_s}]$ are semisimple and $x_n \in [\mathfrak{g}^{x_s},\mathfrak{g}^{x_s}]_1$, $y_n \in [\mathfrak{g}^{y_s},\mathfrak{g}^{y_s}]_1$.

Let H be the connected closed subgroup of G_0 such that $\operatorname{Lie} H = [\mathfrak{g}^{x_s},\mathfrak{g}^{x_s}]_0$. Applying Lemma 3.16, in which x and $\mathfrak{g}$ are replaced respectively by x_n and $[\mathfrak{g}^{x_s},\mathfrak{g}^{x_s}]$, and taking (40) into account, we obtain

$$
H\cdot x_n \cap \mathfrak{z}\big([\mathfrak{g}^{x_s},\mathfrak{g}^{x_s}]^{x_n}\big)_1 = \Big(\mathfrak{z}\big([\mathfrak{g}^{x_s},\mathfrak{g}^{x_s}]^{x_n}\big)_1\Big)^{\mathrm{reg}} \overset{(40)}{=\!=} \Big(\mathfrak{z}\big([\mathfrak{g}^{y_s},\mathfrak{g}^{y_s}]^{y_n}\big)_1\Big)^{\mathrm{reg}},
\tag{41}
$$

where the notation X^{reg} refers to the action of H on $[\mathfrak{g}^{x_s},\mathfrak{g}^{x_s}]_1$.

By Lemma 3.15, the element y_n lies on the right-hand side of (41). Therefore (41) implies that there exists $g \in H$ such that

$$
g\cdot x_n = y_n.
\tag{42}
$$

Since $[\mathfrak{g}^{x_s},\mathfrak{g}^{x_s}]$ and $\mathfrak{z}(\mathfrak{g}^{x_s})$ commute, H acts trivially on $\mathfrak{z}(\mathfrak{g}^{x_s})$. This and the first equality in (40) yield

$$
\mathfrak{z}(\mathfrak{g}^{g\cdot x_s}) = g\cdot\mathfrak{z}(\mathfrak{g}^{x_s}) = \mathfrak{z}(\mathfrak{g}^{x_s}) \overset{(40)}{=\!=} \mathfrak{z}(\mathfrak{g}^{y_s}).
\tag{43}
$$

From (43), using [TY05, 35.3.2], we infer that

$$
\mathfrak{g}^{g\cdot x_s} = \mathfrak{g}^{y_s}.
\tag{44}
$$

By Proposition 3.10, the equalities (42) and (44) imply (i), thereby completing the proof of the implication (iii)⇒(i). □

Using Proposition 3.17, we now can show that in fact a property stronger than Proposition 3.13(iii) holds:

Proposition 3.18. *Every packet P in $\mathfrak{g}_1$ is open in its closure in $\mathfrak{g}_1$.*

Proof. By Proposition 3.13, there exists an integer d such that P lies in the locally closed subset $X = \{x \in \mathfrak{g} \mid \dim \mathfrak{g}^x = d\}$ of $\mathfrak{g}$. Let $\operatorname{Grass}(\mathfrak{g},d)$ be the Grassmannian of d-dimensional linear subspaces in $\mathfrak{g}$ endowed with the natural action of G. The map

$$
\varphi\colon X \to \operatorname{Grass}(\mathfrak{g},d), \quad x \mapsto \mathfrak{g}^x.
$$

is G-equivariant. Let $x \in P$. Then Proposition 3.17 implies that

$$P = \mathfrak{g}_1 \cap \varphi^{-1}(G_0 \cdot \varphi(x)). \tag{45}$$

Since φ is a morphism (see [TY05, 19.7.6 and 29.3.1]) and orbits of algebraic transformation groups are open in their closures, (45) shows that P enjoys the latter property as well. □

3.6 We conclude with describing the relationship between the sheets and the packets in $\mathfrak{g}_1$. Below bar stands for the closure in $\mathfrak{g}_1$.

Proposition 3.19. *For every sheet S of $\mathfrak{g}_1$, there is a unique packet P in $\mathfrak{g}_1$ such that $P \subseteq S$ and $\overline{P} = \overline{S}$. Moreover, $S = \overline{P}^{\mathrm{reg}}$.*

Proof. If d is the dimension of G_0-orbits in S, then by Propositions 3.8, 3.13 and in view of the definition of a sheet, we have

$$\{x \in \mathfrak{g}_1 \mid \dim G_0(x) = d\} = \textstyle\bigcup_{i=1}^{p} S_i = \bigcup_{j=i}^{q} P_j,$$

where $S_1, \ldots, S_p$ and $P_1, \ldots, P_q$ are, respectively, some pairwise different sheets and packets in $\mathfrak{g}_1$; the sheet S is one of $S_1, \ldots, S_m$. Hence

$$\textstyle\bigcup_{i=1}^{p} \overline{S_i} = \bigcup_{j=i}^{q} \overline{P_j}. \tag{46}$$

As sheets and packets contain open subsets of their closures, we have $\overline{S_i} \not\subseteq \overline{S_j}$ and $\overline{P_i} \not\subseteq \overline{P_j}$ for all $i \neq j$. Since all $\overline{S_i}$ and $\overline{P_j}$ are irreducible, (46) then implies that $p = q$ and there is a permutation σ of $\{1, \ldots, p\}$ such that $\overline{S_i} = \overline{P_{\sigma(i)}}$.

The last claim of this proposition follows from the definition of a regular sheet. □

3.7 It is expected that the topic of this section may be linked with that of [P15], [O15], [O16$_1$], [O16$_2$].

References

[A75] V. I. Arnold, *Critical points of smooth functions*, in: *Proceedings of the International Congress of Mathematicians*, Vol. 1 (Vancouver, BC, 1974), Canad. Math. Congress, Montreal, Que., 1975, pp. 19–39.

[B68] N. Bourbaki, *Groupes et Algèbres de Lie*, Chap. IV, V, VI, Hermann, Paris, 1968.

[B91] A. Borel, *Linear Algebraic Groups*, 2nd enlarged ed., Graduate Texts in Mathematics, Vol. 126, Springer-Verlag, 1991.

[H77] R. Hartshorne, *Algebraic Geometry*, Springer-Verlag, New York, 1977.

[J62] N. Jacobson, *Lie Algebras*, Interscience Tracts in Pure and Applied Mathematics, No. 10, Interscience Publishers, New York, 1962.

[K75] V. G. Kac, *On the question of describing the orbit space of linear algebraic groups*, Uspekhi Mat. Nauk **30** (1975), no. 6(186), 173–174 (in Russian).

[K80$_1$] V. G. Kac, *Some remarks on nilpotent orbits*, J. Algebra **64** (1980), 190–213.

[K80$_2$] G. Kempf, *Some quotient surfaces are smooth*, Michigan Math. J. **27** (1980), 295–299.

[L89] P. Littelmann, *Koreguläre und äquidimensionale Darstellungen*, J. Algebra **123** (1989), 193–222.

[KPV76] V. G. Kac, V. L. Popov, É. B. Vinberg, *Sur les groupes linéaires algébriques dont l'algèbre des invariants est libre*, C. R. Acad. Sci. Paris Sér. A **283** (1976), A875–A878.

[KKY86] T. Kimura, S.-I. Kasai, O. Yasukura, *A classification of the representations of reductive algebraic groups which admit only a finite number of orbits*, Amer. J. Math. **108** (1986), no. 3, 643–691.

[O15] D. O. Orlov, *Geometric realizations of quiver algebras*, Proc. Steklov Inst. Math. **290** (2015), 70–83.

[O16$_1$] D. Orlov, *Smooth and proper noncommutative schemes and gluing of DG categories*, Adv. Math. **302** (2016), 59–105.

[O16$_2$] D. O. Orlov, *Gluing of categories and Krull–Schmidt partners*, Russian Math. Surveys **71** (2016), no. 3, 594–596.

[P71] V. L. Popov, *A stability criterion for an action of a semisimple group of a factorial variety*, Math USSR, Izv. **4** (1971), 527–535.

[P74] V. L. Popov, *Structure of the closure of orbits in spaces of finite-dimensional linear* SL(2) *representations*, Math. Notes of the Academy of Sciences of the USSR **16** (1974), no. 6, 1159–1162.

[P77] V. L. Popov, *Representations with a free module of covariants*, Funct. Anal. Appl. **10** (1977), 242–244.

[P80] V. L. Popov, *Classification of spinirs of dimension fourteen*, Trans. Moscow Math. Soc. (1980), no. 1, 181–232.

[P08] V. L. Popov, *Irregular and singular loci of commuting varieties*, Transformation Groups **13** (2008), no. 3–4, special issue dedicated to Bertram Kostant on the occasion of his 80th birthday (2008), 819–837.

[P15] V. L. Popov, *Number of components of the nullcone*, Proc. Steklov Inst. of Math. **290** (2015), 84–90.

[P16] V. L. Popov, *Algebras of general type: Rational parametrization and normal forms*, Proc. Steklov Inst. Math. **292** (2016), no. 1, 202–215.

[P17$_1$] V. L. Popov, *On modality of representations*, Doklady Mathematics **96** (2017), no. 1, 312–314.

[P17$_2$] V. L. Popov, *Algebraic groups whose orbit closures contain only finitely many orbits*, `arXiv:1707.06914` (2017).

[PV94] V. L. Popov, E. B. Vinberg, *Invariant Theory*, in: *Algebraic Geometry* IV, Enc. Math. Sci., Vol. 55, Springer, Berlin, 1994, pp. 123–284.

[SK77] M. Sato, T. Kimura, *A classification of irreducible prehomogeneous vector spaces and their relative invariants*, Nagoya Math. J. **65** (1977), 1–155.

[TY05] P. Tauvel, R. W. T. Yu, *Lie Algebras and Algebraic Groups*, Springer Manographs in Mathematics, Springer, Berlin, 2005.

[V76] É. B. Vinberg, *The Weyl group of a graded Lie algebra*, Math. USSR-Izv. **10** (1976), no. 3, 463–495.

[V86] É. B. Vinberg, *Complexity of action of reductive groups*, Funct. Analysis Appl. **20** 1986, no. 1, 1–11.

Distributions on Homogeneous Spaces and Applications

N. Ressayre

Dedicated to the memory of Professor Bertram Kostant

Abstract Let G be a complex semisimple algebraic group. In 2006, Belkale-Kumar defined a new product $\odot_0$ on the cohomology group $\mathrm{H}^*(G/P,\mathbb{C})$ of any projective G-homogeneous space G/P. Their definition uses the notion of Levi-movability for triples of Schubert varieties in G/P.

In this article, we introduce a family of G-equivariant subbundles of the tangent bundle of G/P and the associated filtration of the de Rham complex of G/P viewed as a manifold. As a consequence one gets a filtration of the ring $\mathrm{H}^*(G/P,\mathbb{C})$ and proves that $\odot_0$ is the associated graded product. One of the aims of this more intrinsic construction of $\odot_0$ is that there is a natural notion of a fundamental class $[Y]_{\odot_0} \in (\mathrm{H}^*(G/P,\mathbb{C}),\odot_0)$ for any irreducible subvariety Y of G/P.

Given two Schubert classes σ_u and σ_v in $\mathrm{H}^*(G/P,\mathbb{C})$, we define a subvariety Σ_u^v of G/P. This variety should play the role of the Richardson variety; more precisely, we conjecture that $[\Sigma_u^v]_{\odot_0} = \sigma_u \odot_0 \sigma_v$. We give some evidence for this conjecture, and prove special cases.

Finally, we use the subbundles of TG/P to give a geometric characterization of the G-homogeneous locus of any Schubert subvariety of G/P.

Keywords: Belkale-Kumar Schubert calculus • Kostant's harmonic forms

Mathematics Subject Classification (2010): 14N15, 14M15

The author is partially supported by the French National Agency (Project GeoLie ANR-15-CE40-0012) and the Institut Universitaire de France (IUF).

N. Ressayre
Université Lyon I - 43 boulevard du 11 novembre 1918 - 69622 Villeurbanne cedex - France - e-mail: ressayre@math.univ-lyon1.fr

V. G. Kac, V. L. Popov (eds.), *Lie Groups, Geometry, and Representation Theory*, Progress in Mathematics 326, https://doi.org/10.1007/978-3-030-02191-7_17

1 Introduction

Let G be a complex semisimple group and let P be a parabolic subgroup of G. In this paper, we are interested in the Belkale-Kumar product $\odot_0$ on the cohomology group of the flag variety G/P.

The Belkale-Kumar product. Fix a maximal torus T and a Borel subgroup B such that $T \subset B \subset P$. Let W and W_P denote respectively the Weyl groups of G and P. Let W^P be the set of minimal length representative in the cosets of W/W_P. For any $w \in W^P$, let X_w be the corresponding Schubert variety (that is, the closure of BwP/P) and let $[X_w] \in \mathrm{H}^*(G/P, \mathbb{C})$ be its cohomology class. The structure coefficients c_{uv}^w of the cup product are written as

$$[X_u].[X_v] = \sum_{w \in W^P} c_{uv}^w [X_w]. \tag{1}$$

Let L be the Levi subgroup of P containing T. This group acts on the tangent space $T_{P/P}G/P$ of G/P at the base point P/P. Moreover, this action is multiplicity free, and we have a unique decomposition

$$T_{P/P}G/P = V_1 \oplus \cdots \oplus V_s, \tag{2}$$

as the sum of irreducible L-modules. It turns out that for any $w \in W^P$, the tangent space $T_w := T_{P/P}w^{-1}X_w$ of the variety $w^{-1}X_w$ at the smooth point P/P decomposes as

$$T_w = (V_1 \cap T_w) \oplus \cdots \oplus (V_s \cap T_w). \tag{3}$$

Set $T_w^i := T_w \cap V_i$. Since $[X_w]$ has degree $2(\dim(G/P) - \dim(T_w))$ in the graded algebra $\mathrm{H}^*(G/P)$, if $c_{uv}^w \neq 0$, then

$$\dim(T_u) + \dim(T_v) = \dim(G/P) + \dim(T_w), \tag{4}$$

or equivalently

$$\sum_{i=1}^{s} \left(\dim(T_u^i) + \dim(T_v^i) \right) = \sum_{i=1}^{s} \left(\dim(V_i) + \dim(T_w^i) \right). \tag{5}$$

The Belkale-Kumar product requires the equality (5) to hold term by term. More precisely, the structure constants $\tilde{c}_{uv}^w$ of the Belkale-Kumar product [BK06],

$$[X_u] \odot_0 [X_v] = \sum_{w \in W^P} \tilde{c}_{uv}^w [X_w] \tag{6}$$

can be defined as follows (see [RR11, Proposition 2.4]):

$$\tilde{c}_{uv}^w = \begin{cases} c_{uv}^w & \text{if } \forall 1 \le i \le s \ \ \dim(T_u^i) + \dim(T_v^i) = \dim(V_i) + \dim(T_w^i), \\ 0 & \text{otherwise.} \end{cases} \tag{7}$$

The product $\odot_0$ defined in this way is associative and satisfies Poincaré duality. The Belkale-Kumar product was proved to be the more relevant product for describing the Littlewood-Richardson cone (see [BK06, Res10, Res11a]).

Motivations. If G/P is cominuscule, then $T_{P/P}G/P$ is an irreducible L-module (that is, $s = 1$). In this case, the Belkale-Kumar product is simply the cup product. This paper is motivated by the conjecture that several known results for cominuscule G/P could be generalized to any G/P using the Belkale-Kumar product. In particular, it might be a first step toward a positive geometric uniform combinatorial rule for computing the coefficients $\tilde{c}_{uv}^w$. Indeed, we define a subvariety Σ_u^v which is encoded by combinatorial datum (precisely a subset of roots of G). We also define a Belkale-Kumar fundamental class $[\Sigma_u^v]_{\odot_0}$ and conjecture that $[\Sigma_u^v]_{\odot_0} = [X_u] \odot_0 [X_v]$.

A geometric construction of the Belkale-Kumar ring. The first aim of this paper is to give a geometric construction of the Belkale-Kumar ring which does not deal with the Schubert basis. Consider the connected center Z of L and its character group $X(Z)$. The Azad-Barry-Seitz theorem (see [ABS90]) asserts that V_i in the decomposition (2) is an isotypical component for the action of Z associated to some weight denoted by $\alpha_i \in X(Z)$. The group P acts on $T_{P/P}G/P$ but does not stabilize the decomposition (2). But the group $X(Z)$ is endowed with a partial order $\succcurlyeq$ (see Section 3.1 for details), such that for any $\alpha \in X(Z)$, the sum

$$V^{\succcurlyeq\alpha} := \bigoplus_{\alpha_i \succcurlyeq \alpha} V_i \tag{8}$$

is P-stable. Since $V^{\succcurlyeq\alpha}$ is P-stable, it induces a G-homogeneous subbundle $T^{\succcurlyeq\alpha}G/P$ of the tangent bundle TG/P. We obtain a family of distributions indexed by $X(Z)$. This family forms a decreasing multi-filtration: if $\alpha \succcurlyeq \beta$, then $T^{\succcurlyeq\alpha}G/P$ is a subbundle of $T^{\succcurlyeq\beta}G/P$. Moreover, these distributions are globally integrable in the sense that

$$[T^{\succcurlyeq\alpha}G/P, T^{\succcurlyeq\beta}G/P] \subset T^{\succcurlyeq\alpha+\beta}G/P. \tag{9}$$

This allows us to define a filtration ("à la Hodge") of the de Rham complex, and therefore of the algebra $\mathrm{H}^*(G/P,\mathbb{C})$ indexed by the group $X(Z) \times \mathbb{Z}$. We consider the associated graded algebra.

Theorem 1 *The* $(X(Z) \times \mathbb{Z})$*-graded algebra* $\mathrm{Gr}\,\mathrm{H}^*(G/P,\mathbb{C})$ *associated to the* $(X(Z) \times \mathbb{Z})$*-filtration is isomorphic to the Belkale-Kumar algebra* $(\mathrm{H}^*(G/P,\mathbb{C}), \odot_0)$.

The first step to prove Theorem 1 is to give a precise sense defining the orders on $X(Z)$, $X(Z) \times \mathbb{Z}$, and the filtrations. The key point to prove the isomorphism is that the Schubert basis $([X_w])_{w \in W^P}$ of $\mathrm{H}^*(G/P,\mathbb{C})$ is adapted to the filtration. Indeed each subspace of the filtration is spanned by the Schubert classes it contains. To obtain this result, we use Kostant's harmonic forms [Kos61].

Theorem 1 is closed to [BK06, Theorem 43] obtained by Belkale-Kumar. In [BK06], the filtration on $\mathrm{H}^*(G/P,\mathbb{C})$ is defined using the Schubert basis. On the other hand, the filtration on $\mathrm{H}^*_{DR}(G/P,\mathbb{C})$ is defined using Kostant's K-invariant forms (where K is a compact form of G). Here, the filtration is defined independently of any basis or any choice of a compact form of G.

This "intrinsic" definition of the Belkale-Kumar also gives a pleasant interpretation of the functoriality result of [RR11, Theorem 1.1]. Indeed, let τ be a one-parameter subgroup of Z, such that

$$\forall \alpha \in X(Z) \qquad \alpha \succcurlyeq 0 \Rightarrow \langle \tau, \alpha \rangle \geq 0,$$

and

$$\forall 1 \leq i \neq j \leq s \qquad \langle \tau, \alpha_i \rangle \neq \langle \tau, \alpha_j \rangle. \tag{10}$$

Setting for any $n \in \mathbb{Z}$

$$V^{\geq n} := \bigoplus_{\langle \tau, \alpha \rangle \geq n} V_i,$$

one gets a globally integrable family of distributions on G/P indexed by $\mathbb{Z}$. Then one obtains a $\mathbb{Z}$-filtration of the ring $\mathrm{H}^*(G/P, \mathbb{C})$. By (6) and (10), the associated $\mathbb{Z}$-graded ring is isomorphic to $\mathrm{Gr}\,\mathrm{H}^*(G/P, \mathbb{C})$. Then, [RR11, Theorem 1.1] is a direct consequence of the immediate lemma 12 below.

A conjecture. The main motivation to show Theorem 1 is to define the fundamental class for the Belkale-Kumar product of any irreducible subvariety Y of G/P. This class $[Y]_{\odot_0}$ which belongs to $\mathrm{Gr}\,\mathrm{H}^*(G/P, \mathbb{C})$ is defined in Section 4.4.

Let w_0 and w_0^P be the longest elements of W and W_P respectively. If $v \in W^P$, then $v^\vee := w_0 v w_0^P$ belongs to W^P and $[X_{v^\vee}]$ is the Poincaré dual class of $[X_v]$. Consider the weak Bruhat order $\lessdot$ on W^P. We are interested in the product $[X_u] \odot_0 [X_v] \in \mathrm{H}^*(G/P, \mathbb{C})$, for u and v in W^P. Lemma 24 below shows that if $[X_u] \odot_0 [X_v] \neq 0$, then $v^\vee \lessdot u$. Assume that $v^\vee \lessdot u$ and consider the group

$$H_u^v := u^{-1} B u \cap w_0^P v^{-1} B v w_0^P. \tag{11}$$

It is a closed connected subgroup of G containing T; in particular, it can be encoded by its set Φ_u^v of roots. Let Σ_u^v denote the closure of the H_u^v-orbit of P/P:

$$\Sigma_u^v = \overline{H_u^v . P/P}. \tag{12}$$

Another characterization of this subvariety is given by the following statement.

Proposition 1 *The variety Σ_u^v is the unique irreducible component of the intersection $u^{-1} X_u \cap w_0^P v^{-1} X_v$ containing P/P. Moreover, this intersection is transverse along Σ_u^v.*

Our main conjecture can be stated as follows.

Conjecture 1 *If $v^\vee \lessdot u$, then*

$$[\Sigma_u^v]_{\odot_0} = [X_u] \odot_0 [X_v] \in \mathrm{Gr}\,\mathrm{H}^*(G/P, \mathbb{C}).$$

Write

$$[\Sigma_u^v]\odot_0 = \sum_{w\in W^P} d_{uv}^w [X_w].$$

By Proposition 11 below, d_{uv}^w are integers. Moreover, Conjecture 1 is equivalent to $\tilde{c}_{uv}^w = d_{uv}^w$ for any $w \in W^P$. The first evidence is the following weaker result.

Proposition 2 *Then*

(i) $d_{uv}^w \neq 0 \iff \tilde{c}_{uv}^w \neq 0$;
(ii) $d_{uv}^w \leq \tilde{c}_{uv}^w$.

Known cases. Conjecture 1 generalizes another case for G/B. Indeed, if $G/P = G/B$ is a complete flag variety, then Conjecture 1 is equivalent to the following:

Conjecture 2 *For G/B and any u, v, and w in W, the structure constant $\tilde{c}_{uv}^w$ is equal to 1 if for any $1 \leq i \leq s$, $\dim(T_u^i) + \dim(T_v^i) = \dim(V_i) + \dim(T_w^i)$ and 0 otherwise.*

In particular, Conjecture 2 implies that we have a uniform combinatorial and geometric model for the Belkale-Kumar product. Conjecture 2 was explicitly stated in [DR09]. E. Richmond proved in [Ric09] and [Ric12] this conjecture for $G = \mathrm{SL}_n(\mathbb{C})$ or $G = \mathrm{Sp}_{2n}(\mathbb{C})$. In Section 7, we prove it for $G = \mathrm{SO}_{2n+1}(\mathbb{C})$ (this proof is certainly known by some specialists, but I have shortly included it for convenience). Very recently, Dimitrov-Roth obtained also a proof for the classical groups and G2 [DR17]. Using [BK06, Corollary 44], we wrote a program [Res13] to check this conjecture: it is checked in type F_4 and E_6.

Combinatorial evidences. Consider the following degenerate version of Conjecture 1.

Conjecture 3 *The product $[X_u]\odot_0[X_v]$ only depends on the set Φ_u^v.*

The expression of the Belkale-Kumar structure coefficients as products, given in [Ric09], shows that Conjecture 3 holds in type A. Consider now the case $G = \mathrm{SO}_{2n+1}(\mathbb{C})$ or $\mathrm{Sp}_{2n}(\mathbb{C})$ and P maximal. In this case, in [Res12], it is proved that the set of triples $(u, v, w) \in W^P$ such that $\tilde{c}_{uv}^w = 1$ only depends on Φ_u^v, according to Conjecture 3. If G/P is cominuscule, $\tilde{c}_{uv}^w = c_{uv}^w$ for any $(u, v, w) \in W^P$. Then the Thomas-Young combinatorial rule [TY09] for c_{uv}^w implies that Conjecture 3 holds.

Distributions and Schubert varieties. In Section 3, we study the restriction of the distributions to the Schubert varieties X_u. More precisely, for any x in X_u and $\alpha \in X(Z)$ we are interested in $T_xX_u \cap T_x^{\succcurlyeq\alpha}G/P$. For α fixed, the dimension of $T_xX_u \cap T_x^{\succcurlyeq\alpha}G/P$ has a fixed value for $x \in X_u$ general and can increase for x in a strict subvariety of X_u. Consider the maximal open subset X_u^0 of X_u, such that for any $\alpha \in X(Z)$ the dimension of $T_xX_u \cap T_x^{\succcurlyeq\alpha}G/P$ does not depend on $x \in X_u^0$. Consider the global stabilizer Q_u of X_u, that is, the set of $g \in G$ such that $g.X_u = X_u$. Since X_u is B-stable, Q_u is a standard parabolic subgroup of G.

Proposition 3 *With the above notation, we have*

$$X_u^0 = Q_u.uP/P.$$

If G/P is cominuscule, the filtration is trivial and Proposition 3 asserts that Q_u acts transitively on the smooth locus of X_u. This was previously proved by Brion and Polo in [BP00]. Proposition 3 is in the philosophy of generalizing known results from cominuscule homogeneous spaces to any homogeneous space G/P using the Belkale-Kumar product.

Note that Proposition 3 is equivalent to [BKR12, Theorem 7.4]. Nevertheless, we think that the distributions give a pleasant interpretation of this result. In Section 3 we present a proof using the properties of the Peterson map.

Returning to the setting of Conjecture 1, we assume moreover that the intersection $u^{-1}X_u \cap w_0^P v^{-1}X_v$ is proper. Then Conjecture 1 is implied by the fact that Σ_u^v is the only irreducible component of this intersection that has the same $X(Z)$-dimension (see Section 2.3). Proposition 3 is clearly related to this version of Conjecture 1.

Acknowledgment. I thank M. Herzlich, P.E. Paradan, N. Perrin, C. Vernicos for useful discussions.

Contents

2 Infinitesimal filtrations

2.1 The case of a vector space

Ordered group. Let Γ be a finitely generated free abelian group denoted by $+$. Consider the vector space $\Gamma \otimes_{\mathbb{Z}} \mathbb{Q}$. Assume that a closed strictly convex cone $\mathscr{C}$ in $\Gamma \otimes_{\mathbb{Z}} \mathbb{Q}$ of a nonempty interior is given. Moreover, we assume that $\mathscr{C}$ is a rational polyhedral that is defined by finitely many linear rational inequalities, or equivalently generated, as a cone, by finitely many vectors in $\Gamma \otimes_{\mathbb{Z}} \mathbb{Q}$. We consider the partial order $\preccurlyeq$ on Γ defined by $\alpha \preccurlyeq \beta$ if and only if $\beta - \alpha$ belongs to $\mathscr{C}$. The group Γ endowed with the order $\preccurlyeq$ is an ordered group:

$$\forall \alpha, \beta, \gamma \in \Gamma \quad \alpha \preccurlyeq \beta \Rightarrow (\alpha + \gamma) \preccurlyeq (\beta + \gamma). \tag{13}$$

The order $\preccurlyeq$ satisfies the following version of the Ramsey theorem (see also the Bolzano-Weierstrass theorem).

Lemma 1 *Let* $(\alpha_n)_{n\in\mathbb{N}}$ *be a sequence of pairwise distinct elements of* Γ *such that* $\alpha_n \preccurlyeq 0$ *for any* n*. Then there exists a subsequence* $(\alpha_{\phi(n)})_{n\in\mathbb{N}}$*, such that for any* n

$$\alpha_{\phi(n+1)} \prec \alpha_{\phi(n)}.$$

Proof. Let $\varphi_1, \dots, \varphi_s$ be elements of $\mathrm{Hom}(\Gamma, \mathbb{Z})$, such that $x \in \mathscr{C}$ if and only if $\varphi_i(x) \geq 0$ for any $i = 1, \dots, s$.

Consider first the sequence $\varphi_1(\alpha_n)$ and set

$$I_1 = \{n \mid \forall m \geq n \;\; \varphi_1(\alpha_m) > \varphi_1(\alpha_n)\}.$$

Assume for a contradiction that I_1 is infinite. Denoting by $\phi(k)$ the kth element of I_1, we get an increasing subsequence of $(\varphi_1(\alpha_n))_{n\in\mathbb{N}}$. But $\varphi_1(\alpha_n) \in \mathbb{Z}$ and $\varphi_1(\alpha_n) \leq 0$: a contradiction. Hence I_1 is finite.

Up to taking a subsequence, we may assume that I_1 is empty; that is

$$\forall n \geq 0 \qquad \exists m > n \qquad \varphi_1(\alpha_m) \leq \varphi_1(\alpha_n).$$

This property allows us to construct a nonincreasing subsequence of $\varphi_1(\alpha_n)$. Hence, by considering such a subsequence, we may assume that

$$\forall n \geq 0 \qquad \varphi_1(\alpha_{n+1}) \leq \varphi_1(\alpha_n).$$

By successively proceeding similarly, for $i = 2, \dots, s$, one gets a subsequence $\alpha_{\psi(n)}$ such that

$$\forall i = 1, \dots, n \qquad \forall n \qquad \varphi_i(\alpha_{\psi(n+1)}) \leq \varphi_1(\alpha_{\psi(n)}).$$

Since the α_n are pairwise distinct, we deduce that $\alpha_{\psi(n+1)} \preccurlyeq \alpha_{\psi(n)}$. □

Remark. Consider the cone $\mathscr{C} = \{(x,y) \in \mathbb{R}^2 : y \geq 0 \text{ and } \sqrt{2}x - y \geq 0\}$ and the group $\Gamma = \mathbb{Z}^2$. Lemma 1 does not hold for the induced order $\succcurlyeq$ showing that the rationality assumption on $\mathscr{C}$ is necessary. Indeed, denote by $\pi : \mathbb{R}^2 \longrightarrow \mathbb{R}$ the linear projection on the line $y = 0$ with kernel $y = \sqrt{2}x$. Then $\pi(\mathbb{Z}^2)$ is dense as the group generated by 1 and $\frac{\sqrt{2}}{2}$. In particular, one can construct a sequence $(x_n, y_n)_{n \in \mathbb{N}}$ such that $y_{n+1} < y_n < 0$ and $0 > \sqrt{2}x_{n+1} - y_{n+1} > \sqrt{2}x_n - y_n$. Then the elements of the sequence are pairwise incomparable for the partial order $\succcurlyeq$.

Γ-filtration. The group Γ is used here to index filtrations.

Definition. Let V be a finite-dimensional real or complex vector space. A *Γ-filtration* of V is a collection $F^{\succcurlyeq \beta}V$ of linear subspaces of V indexed by $\beta \in \Gamma$ satisfying

(i) $\alpha \preccurlyeq \beta \Rightarrow F^{\succcurlyeq \beta}V \subset F^{\succcurlyeq \alpha}V$,
(ii) $\exists \beta_0 \in \Gamma$ s.t. $V = F^{\succcurlyeq \beta_0}V$,
(iii) if $F^{\succcurlyeq \alpha}V \neq \{0\}$ then $\alpha \preccurlyeq 0$.

Lemma 2 *Let $(F^{\succcurlyeq \beta}V)_{\beta \in \Gamma}$ be a Γ-filtration. Then the set $\{F^{\succcurlyeq \beta}V \mid \beta \in \Gamma\}$ of linear subspaces of V is finite.*

Proof. By contradiction, assume that there exists a sequence $F^{\succcurlyeq \alpha_n}V$ of pairwise distinct linear subspaces of V. By axiom (iii), $\alpha_n \preccurlyeq 0$ for any but eventually one n. Now Lemma 1 implies that there exists a decreasing subsequence $\alpha_{\phi(k)}$. Since the linear subspaces $F^{\succcurlyeq \alpha_n}V$ are pairwise distinct, the subsequence $F^{\succcurlyeq \alpha_{\phi(k)}}V$ is increasing. This contradicts the assumption that V is finite-dimensional. □

Γ-filtrations coming from decompositions. For $\beta \in \Gamma$, $\sum_{\alpha \succ \beta} F^{\succcurlyeq \alpha}V$ is a linear subspace of $F^{\succcurlyeq \beta}V$. Let us choose a supplementary subspace S^β:

$$F^{\succcurlyeq \beta}V = S^\beta \oplus \sum_{\alpha \succ \beta} F^{\succcurlyeq \alpha}V. \tag{14}$$

One motivation for axiom (iii) in the definition of Γ-filtration is the following lemma.

Lemma 3 *With the above notation,*

$$F^{\succcurlyeq \beta}V = \sum_{\alpha \succcurlyeq \beta} S^\alpha. \tag{15}$$

Proof. It is clear that the sum is contained in $F^{\succcurlyeq\beta}V$. Conversely, since V is finite-dimensional, we have

$$F^{\succcurlyeq\beta}V = S^{\beta} \oplus (F^{\succcurlyeq\alpha_1}V + \cdots + F^{\succcurlyeq\alpha_s}V),$$

for some $\alpha_i \in \Gamma$, such that $\alpha_i \succ \beta$. By axiom (iii), we may assume that for any $i = 1,\ldots,s$, we have $\alpha_i \preccurlyeq 0$. If $F^{\succcurlyeq\alpha_i}V$ satisfies the lemma, the lemma is proved for $F^{\succcurlyeq\beta}V$. Otherwise, we restart the proof with α_i in place of β. Since Γ is discrete, the set of $\alpha \in \Gamma$, such that $0 \succcurlyeq \alpha \succcurlyeq \beta$ is finite. In particular, the procedure ends by axiom (iii) of the definition of a Γ-filtration. □

Conversely, assume that a linear subspace S^{α} of V is given for any $\alpha \in \Gamma$. If these linear subspaces satisfy ($S^{\alpha} \neq \{0\} \Rightarrow \alpha \preccurlyeq 0$), and there exist $\alpha_1,\ldots,\alpha_s$ such that $V = S^{\alpha_1} + \cdots + S^{\alpha_s}$, then the formula (15) defines a Γ-filtration of V. The Γ-filtration of V is said to *come from a decomposition* if there exists a decomposition

$$V = \bigoplus_{\alpha\in\Gamma} S^{\alpha}, \text{ with } S^{\alpha} \neq \{0\} \Rightarrow \alpha \preccurlyeq 0, \tag{16}$$

such that (15) holds.

The *f-dimension vector* (f stands for filtered) of the Γ-filtration is the vector $(fd^{\beta}(V))_{\beta\in\Gamma}$ of $\mathbb{N}^{\Gamma}$ defined by

$$\Gamma \longrightarrow \mathbb{N},\ \beta \longmapsto fd^{\beta}(V) = \dim(F^{\succcurlyeq\beta}V),$$

for any $\beta \in \Gamma$. Define the *grading associated to the Γ-filtration* by setting

$$\mathrm{Gr}^{\beta} V = \frac{F^{\succcurlyeq\beta}V}{\sum_{\alpha\succ\beta} F^{\succcurlyeq\alpha}V}, \text{ and } \mathrm{Gr}\,V = \bigoplus_{\beta\in\Gamma} \mathrm{Gr}^{\beta} V. \tag{17}$$

The *g-dimension vector* (g stands for graded) $(gd^{\beta}(V))_{\beta\in\Gamma}$ of the Γ-filtration is defined by

$$\Gamma \longrightarrow \mathbb{N},\ \beta \longmapsto gd^{\beta}(V) := \dim(\mathrm{Gr}^{\beta} V).$$

Lemma 4 *The Γ-filtration comes from a decomposition if and only if*

$$\dim(\mathrm{Gr}\,V) = \dim(V). \tag{18}$$

In this case, the g-dimension vector of V only depends on the f-dimension vector of V.

Proof. Assume first that the Γ-filtration comes from a decomposition. Fix linear subspaces S^{α} satisfying the conditions (16) and (15). For any $\beta \in \Gamma$, the identity (14) holds and $\dim(\mathrm{Gr}^{\beta} V) = \dim(S^{\beta})$. Hence the lemma follows from the condition (16).

Conversely, assume that the condition (18) is satisfied and choose the linear subspaces S^β satisfying (14). Let $\beta_0 \in \Gamma$ such that $F^{\succcurlyeq\beta_0}V = V$. Lemma 3 implies that $V = \sum_{\beta\succcurlyeq\beta_0} S^\beta$. The condition (18) implies that the sum is direct. Moreover, it implies that $S^\gamma = \{0\}$ if $\gamma \not\succcurlyeq \beta_0$. Using Lemma 3 again, we deduce that the filtration comes from the decomposition $V = \bigoplus_\beta S^\beta$.

If $fd^\beta = 0$, then $F^{\succcurlyeq\beta}V = \{0\}$, $\mathrm{Gr}^\beta = \{0\}$ and $gd^\beta = 0$. Let $\Gamma_{\max}$ be the set of maximal elements among the elements β in Γ satisfying $F^{\succcurlyeq\beta}V \neq \{0\}$. For $\beta \in \Gamma_{\max}$, we have $gd^\beta(V) = fd^\beta(V)$, etc. □

Example. Consider the group $\mathbb{Z}^2$ endowed with the order $(a,b)\preccurlyeq(a',b')$ if and only if $a \leq a'$ and $b \leq b'$. Fix a two-dimensional vector space V and three pairwise distinct lines l_1, l_2, and l_3 in V. Consider the following family $(S^\beta)_{\beta\in\mathbb{Z}^2}$ of linear subspaces of V: $S^{(-2,0)} = l_1$, $S^{(0,-2)} = l_2$, $S^{(-1,-1)} = l_3$, and $S^\beta = \{0\}$ if $\beta \notin \{(-2,0),(0,-2),(-1,-1)\}$. The filtration defined by the formula (15) does not come from a decomposition. More precisely, $\mathrm{Gr}V \simeq l_1 \oplus l_2 \oplus l_3$ has dimension three, whereas V has dimension two.

Another useful notion is the *weight* $\rho(V)$ of the Γ-filtration of V defined by

$$\rho(V) = \sum_{\beta\in\Gamma} gd^\beta(V)\beta. \tag{19}$$

Filtrations induced on a linear subspace. Let W be a linear subspace of V. The Γ-filtration on V induces a Γ-filtration on W by setting

$$\forall\beta \in \Gamma \quad F^{\succcurlyeq\beta}W := W \cap F^{\succcurlyeq\beta}V. \tag{20}$$

Lemma 5 *If the Γ-filtration on V comes from a decomposition, then the induced Γ-filtration on W comes from a decomposition.*

Proof. Fix linear subspaces S_V^β and S_W^β of V such that

$$V = S_V^\beta \oplus F^{\succ\beta}V, \quad W = S_W^\beta \oplus F^{\succ\beta}W, \quad S_W^\beta \subset S_V^\beta.$$

Lemma 3 implies that

$$W = \sum_{\beta\in\Gamma} S_W^\beta. \tag{21}$$

Lemma 4 shows that

$$V = \bigoplus_{\beta\in\Gamma} S_V^\beta.$$

Since $S_W^\beta \subset S_V^\beta$, it follows that the sum (21) is direct. □

Filtrations induced on p-forms. Let p a nonnegative integer. A Γ-filtration of V induces a filtration on the space $\wedge^p V^*$ of skew-symmetric p-forms on V as follows.

Definition. Let $\beta \in \Gamma$. Denote by $F^{\preccurlyeq\beta} \wedge^p V^*$ the linear subspace of forms $\omega \in \wedge^p V^*$, such that for any $\alpha_1, \dots, \alpha_p \in \Gamma$, and for any $v_i \in F^{\succcurlyeq\alpha_i} V$, we have

$$\alpha_1 + \cdots + \alpha_p \not\preccurlyeq \beta \Rightarrow \omega(v_1, \dots, v_p) = 0. \tag{22}$$

The first properties of these linear subspaces are the following.

Proposition 4 (i) *If $\beta \preccurlyeq \gamma$, then $F^{\preccurlyeq\beta} \wedge^p V^* \subset F^{\preccurlyeq\gamma} \wedge^p V^*$.*
(ii) *Let $\beta_0 \in \Gamma$ be such that $F^{\succcurlyeq\beta_0} V = V$. If $F^{\preccurlyeq\gamma} \wedge^p V^* \neq \{0\}$, then $\gamma \succcurlyeq p\beta_0$.*
(iii) *We have $F^{\preccurlyeq 0} \wedge^p V^* = \wedge^p V^*$.*
(iv) *For β and γ in Γ, we have $F^{\preccurlyeq\beta} \wedge^p V^* \wedge F^{\preccurlyeq\gamma} \wedge^q V^* \subset F^{\preccurlyeq\beta+\gamma} \wedge^{p+q} V^*$.*

Proof. If $\beta \preccurlyeq \gamma$, then $\alpha_1 + \cdots + \alpha_p \not\preccurlyeq \gamma$ implies $\alpha_1 + \cdots + \alpha_p \not\preccurlyeq \beta$. Hence the conditions defining $F^{\preccurlyeq\gamma} \wedge^p V^*$ are conditions defining $F^{\preccurlyeq\beta} \wedge^p V^*$. The first assertion follows.

Let $\gamma \not\succcurlyeq p\beta_0$. The definition of $F^{\preccurlyeq\gamma} \wedge^p V^*$ with $\alpha_1 = \cdots = \alpha_p = \beta_0$ implies that $F^{\preccurlyeq\gamma} \wedge^p V^*$ is reduced to zero.

Let ω be any p-form. We want to prove that $\omega \in F^{\preccurlyeq 0} \wedge^p V^*$. Let $\alpha_1, \dots, \alpha_p$ be such that $\alpha_1 + \cdots + \alpha_p \not\preccurlyeq 0$. Then i_0 satisfies $\alpha_{i_0} \not\preccurlyeq 0$. In particular, $F^{\succcurlyeq\alpha_{i_0}} V = \{0\}$. This implies that ω is zero on $F^{\succcurlyeq\alpha_1} V \times \cdots \times F^{\succcurlyeq\alpha_s} V$.

Let ω_1 and ω_2 belong to $F^{\preccurlyeq\beta} \wedge^p V^*$ and $F^{\preccurlyeq\alpha} \wedge^q V^*$ respectively. Let $\alpha_1, \dots, \alpha_{p+q}$ be such that $\alpha_1 + \cdots + \alpha_{p+q} \not\preccurlyeq \beta + \gamma$. Let $v_i \in F^{\succcurlyeq\alpha_i} V$. Then

$$\begin{aligned}&(\omega_1 \wedge \omega_2)(v_1, \dots, v_{p+q}) \\ &= \frac{1}{(p+q)!} \sum_{\sigma \in \mathscr{S}_{p+q}} \varepsilon(\sigma) \omega_1(v_{\sigma(1)}, \dots, v_{\sigma(p)}) . \omega_2(v_{\sigma(p+1)}, \dots, v_{\sigma(p+q)}).\end{aligned} \tag{23}$$

It is sufficient to prove that any term in the sum (23) is zero. Since

$$(\alpha_{\sigma(1)} + \cdots + \alpha_{\sigma(p)}) + (\alpha_{\sigma(p+1)} + \cdots + \alpha_{\sigma(p+q)}) \not\preccurlyeq \beta + \gamma,$$

either $(\alpha_{\sigma(1)} + \cdots + \alpha_{\sigma(p)}) \not\preccurlyeq \beta$ or $(\alpha_{\sigma(p+1)} + \cdots + \alpha_{\sigma(p+q)}) \not\preccurlyeq \gamma$. In both cases, the product

$$\omega_1(v_{\sigma(1)}, \dots, v_{\sigma(p)}) . \omega_2(v_{\sigma(p+1)}, \dots, v_{\sigma(p+q)})$$

is equal to zero. □

Remark. The three first assertions of Proposition 4 mean that $(F^{\preccurlyeq\beta} \wedge^p V^*)_{\beta \in \Gamma}$ is a Γ-filtration of $\wedge^p V^*$ up to the changing of index $\beta \mapsto p\beta_0 - \beta$. Indeed, even for $p = 1$, taking the orthogonal reverses inclusions and exchanges $\{0\}$ with the whole space.

Filtrations coming from a decomposition.

Lemma 6 *Let p be a positive integer. If the Γ-filtration on V comes from a decomposition, then the induced Γ-filtration $F^{\succcurlyeq p\beta_0 - \beta} \wedge^p V^*$ on $\wedge^p V^*$ comes from a decomposition.*

Proof. Write

$$V = \bigoplus_{\alpha\in\Gamma} S^\alpha \quad \text{and} \quad F^{\succcurlyeq\beta}V = \bigoplus_{\alpha\succcurlyeq\beta} S^\alpha,$$

with $(S^\alpha V \neq \{0\} \Rightarrow \alpha\preccurlyeq 0)$. For any $\beta\in\Gamma$, denote by T^β the orthogonal of $\bigoplus_{\alpha\neq\beta} S^\alpha$ in V^*. It can be identified with the dual of S^β and

$$V^* = \bigoplus_{\beta\in\Gamma} T^\beta. \tag{24}$$

For a collection of subspaces $F_1,\dots,F_p$ of V^*, $\pi(F_1\otimes\cdots\otimes F_p)$ denotes the subspace of $\wedge^p V^*$ obtained by adding wedge products of the elements of subspaces F_i. For any $\theta\in\Gamma$, set

$$(\wedge^p V^*)^\theta := \sum_{\beta_1+\cdots+\beta_p=\theta} \pi(T^{\beta_1}\otimes\cdots\otimes T^{\beta_p}).$$

It is clear that (24) implies that

$$\wedge^p V^* = \bigoplus_{\theta\in\Gamma} (\wedge^p V^*)^\theta.$$

Moreover, for any $\theta\in\Gamma$, $(\wedge^p V^*)^\theta$ is the set of p-forms ω, such that for any $\alpha_i\in\Gamma$ and $v_i\in S^{\alpha_i}$ such that $\alpha_1+\cdots+\alpha_p\neq\theta$, we have $\omega(v_1,\dots,v_p)=0$.

We claim that

$$F^{\preccurlyeq\beta}\wedge^p V^* = \bigoplus_{\theta\preccurlyeq\beta} (\wedge^p V^*)^\theta. \tag{25}$$

Indeed $F^{\preccurlyeq\beta}\wedge^p V^*$ is the subspace of forms $\omega\in\wedge^p V^*$, such that for any $\alpha_1,\dots,\alpha_p\in\Gamma$, and $v_i\in S^{\alpha_i}$, we have

$$\alpha_1+\cdots+\alpha_p \not\preccurlyeq \beta \Rightarrow \omega(v_1,\dots,v_p)=0.$$

Let θ be such that $(\wedge^p V^*)^\theta\neq\{0\}$. Then there exist $\beta_1,\dots,\beta_p$ in Γ, such that $\beta_1+\cdots+\beta_p=\theta$ and $T^{\beta_i}\neq\{0\}$ for any i. Hence $S^{\beta_i}\neq\{0\}$ for any i and $\beta_i\preccurlyeq 0$. We deduce that $\theta\preccurlyeq 0$. □

2.2 *The case of manifolds*

Let M be a smooth connected manifold and let TM denote its tangent bundle. Here comes the central definition of this work.

Definition. An *infinitesimal Γ-filtration* of M is a collection $F^{\succcurlyeq\beta}TM$ of vector subbundles of TM indexed by $\beta\in\Gamma$ satisfying

(i) $\alpha\preccurlyeq\beta \Rightarrow F^{\succcurlyeq\beta}TM \subset F^{\succcurlyeq\alpha}TM$,

(ii) $\exists \beta_0 \in \Gamma$ s.t. $TM = F^{\succcurlyeq \beta_0} TM$,
(iii) if $F^{\succcurlyeq \alpha} TM \neq \{\underline{0}\}$, then $\alpha \preccurlyeq 0$.

The *f-rank vector* of the infinitesimal filtration is the map

$$\beta \longmapsto \mathrm{rk}(F^{\succcurlyeq \beta} TM), \tag{26}$$

and belongs to $\mathbb{N}^{\Gamma}$.

Definition. An infinitesimal Γ-filtration is said to *come from a decomposition* if for any $x \in M$, the Γ-filtration of T_xM comes from a decomposition.

Remark. We do not require a Γ-decomposition of the tangent bundle TM, but only for a punctual decomposition.

Lemma 7 *Consider an infinitesimal Γ-filtration on M coming from a decomposition. Then for any β, the sum $\sum_{\alpha \succ \beta} F^{\succcurlyeq \alpha} TM$ is a subbundle of TM.*

Proof. Fix x in M and a Γ-decomposition of $T_xM = \oplus_\alpha S^\alpha$ such that the identities (16) and (15) hold. Then $\sum_{\alpha \succ \beta} F^{\succcurlyeq \alpha} T_xM = \sum_{\alpha \succ \beta} S^\alpha$. In particular, its dimension only depends on the g-dimension vector of the filtration of T_yM. This g-dimension vector only depends on the f-dimension by Lemma 4. It follows that the dimension of $\sum_{\alpha \succ \beta} F^{\succcurlyeq \alpha} T_xM$ does not depend on x. Now, the lemma follows from classical properties of vector subbundles. □

Define the *grading associated to the infinitesimal Γ-filtration coming from a decomposition* by setting

$$\mathrm{Gr}^\beta TM = \frac{F^{\succcurlyeq \beta} TM}{\sum_{\alpha \succ \beta} F^{\succcurlyeq \alpha} TM} \quad \text{and} \quad \mathrm{Gr}\, TM = \bigoplus_{\beta \in \Gamma} \mathrm{Gr}^\beta TM. \tag{27}$$

Then, $\mathrm{Gr}^\beta TM$ and $\mathrm{Gr}\, TM$ are vector bundles on M. The *g-rank vector* $(gd^\beta(M))_{\beta \in \Gamma}$ of the Γ-filtration is defined by

$$\Gamma \longrightarrow \mathbb{N}, \; \beta \longmapsto gd^\beta(M) := \mathrm{rk}(\mathrm{Gr}^\beta TM).$$

2.3 The case of varieties

Let X be a smooth complex irreducible variety. Consider the complex tangent bundle TX.

Definition. An infinitesimal Γ-filtration of X is said to be *algebraic* if each $F^{\succcurlyeq \beta} TX$ is a complex algebraic vector subbundle of TX.

Let Y be an irreducible subvariety of X. For $y \in Y$, the Zariski-tangent space T_yY of Y at the point y is a complex subspace of T_yX. Set

$$F^{\succcurlyeq \beta} T_y Y = F^{\succcurlyeq \beta} T_y X \cap T_y Y. \tag{28}$$

Even if Y is smooth, $F^{\succcurlyeq \beta} T_y Y$ does not define a subbundle of TY since its dimension depends on y.

Lemma 8 *For any $\beta \in \Gamma$ and $y \in Y$, there exists an open neighborhood U of y' in Y such that for any $y' \in U$, we have*

$$\dim(F^{\succcurlyeq \beta} T_y Y) \geq \dim(F^{\succcurlyeq \beta} T_{y'} Y). \tag{29}$$

Proof. Locally in $y \in Y$, the subspace $F^{\succcurlyeq \beta} T_y Y$ of T_yX can be expressed as the kernel of a matrix whose coefficients depend algebraically on y. The lemma follows. □

The point $y \in Y$ is said to be *Γ-regular* if

$$\forall \beta \in \Gamma \quad \dim(F^{\succcurlyeq \beta} T_y Y) = \min_{y' \in Y} \dim(F^{\succcurlyeq \beta} T_{y'} Y). \tag{30}$$

Since Γ is countable, Lemma 8 shows that a very general point in Y is Γ-regular. More precisely, Lemma 2 implies that the set of Γ-regular points in Y is open. The open set of Γ-regular points of Y is denoted by $Y^{\Gamma\text{-reg}}$. If $x \in Y^{\Gamma\text{-reg}}$, the g-dimension of T_xY is called the Γ-dimension of Y.

3 Infinitesimal filtration of G/P and Schubert varieties

3.1 Infinitesimal filtration of G/P

As in the introduction, G is a complex semisimple group, P is a parabolic subgroup of G, $T \subset B \subset P$ are a fixed maximal torus and a Borel subgroup. Moreover, L denotes the Levi subgroup of P containing T and Z denotes the neutral component of its center. The group of multiplicative characters of Z is denoted by $X(Z)$. Set $\Gamma = X(Z)$. Our main example is an infinitesimal $X(Z)$-filtration of G/P.

Let S be any torus. If V is any S-module, then $\Phi(V,S)$ denotes the set of nonzero weights of S on V. For $\beta \in X(S)$, V_β denotes the eigenspace of weight β.

Denote by $\mathfrak{p}$ and $\mathfrak{g}$ the Lie algebras of P and G and consider the convex cone $\mathscr{C}$ generated by $\Phi(\mathfrak{p}, Z)$ in $X(Z) \otimes \mathbb{Q}$. Note that $\mathscr{C}$ is a closed strictly convex polyhedral cone of nonempty interior in $X(Z) \otimes \mathbb{Q}$. The associated order on $X(Z)$ is denoted by $\succcurlyeq$. The decomposition of $\mathfrak{g}/\mathfrak{p}$ under the action of Z:

$$\mathfrak{g}/\mathfrak{p} = \bigoplus_{\alpha \in X(Z)} (\mathfrak{g}/\mathfrak{p})_\alpha \tag{31}$$

is supported on $-\mathscr{C}\cap X(Z)$. The group P acts on $\mathfrak{g}/\mathfrak{p}$ by the adjoint action but does not stabilize the decomposition (31). For any $\beta\in X(Z)$, the linear subspace

$$F^{\succcurlyeq\beta}\mathfrak{g}/\mathfrak{p}=\bigoplus_{\substack{\alpha\in X(Z)\\ \alpha\succcurlyeq\beta}}(\mathfrak{g}/\mathfrak{p})_\alpha \tag{32}$$

is P-stable. More precisely, the set of $F^{\succcurlyeq\beta}\mathfrak{g}/\mathfrak{p}$ forms a P-stable $X(Z)$-filtration of $\mathfrak{g}/\mathfrak{p}$ coming from the decomposition (31). The tangent bundle $T(G/P)$ identifies with the fiber bundle $G\times_P\mathfrak{g}/\mathfrak{p}$ over G/P. These remarks allow us to define a G-equivariant infinitesimal $X(Z)$-filtration on G/P by setting for any $\beta\in X(Z)$

$$F^{\succcurlyeq\beta}T(G/P)=G\times_P F^{\succcurlyeq\beta}\mathfrak{g}/\mathfrak{p}. \tag{33}$$

Consider the set $\Phi(\mathfrak{g}/\mathfrak{p},T)$ of weights of T acting on $\mathfrak{g}/\mathfrak{p}$. Then $\Phi(\mathfrak{g}/\mathfrak{p},T)$ is a subset of Φ. Let w belong to W^P and consider the centered Schubert variety $w^{-1}X_w$. Then P/P belongs to the open $w^{-1}Bw$-orbit in $w^{-1}X_w$. In particular, it is $X(Z)$-regular. Denote by $\Phi(w)$ the set of weights of T acting on $T_{P/P}w^{-1}X_w$. Then $\Phi(w)=\Phi^-\cap w^{-1}\Phi^+$ is the inversion set of w. Moreover, $\Phi(w)$ is contained in $\Phi(\mathfrak{g}/\mathfrak{p},T)$. Since P/P is $X(Z)$-regular in $w^{-1}X_w$, the g-dimension of X_w is equal to the g-dimension of $T_{P/P}w^{-1}X_w$. The following result follows directly:

Lemma 9 *The g-dimension of $gd(X_w)$ of X_w is equal to*

$$\begin{aligned} X(Z) &\longrightarrow \mathbb{Z}_{\geq 0}\\ \alpha &\longmapsto \#\{\theta\in\Phi(w)\,:\,\theta_{|Z}=\alpha\},\end{aligned}$$

where θ belongs to $X(T)$ and $\theta_{|Z}$ denotes its restriction to Z.

3.2 Peterson's application

Let V' be any T-module without multiplicity and let $\beta\in X(T)$. Under the action of $\operatorname{Ker}\beta\subset T$, V' decomposes

$$V'=\bigoplus_{\alpha\in X(T)/\mathbb{Z}\beta}\left(\oplus_{k\in\mathbb{Z}}V'_{\alpha+k\beta}\right). \tag{34}$$

A subset Λ of $\Phi(V',T)$ is said to be β-*convex* if

$$\alpha\in\Lambda,\ \alpha+\beta\in\Phi(V',T)\Rightarrow\alpha+\beta\in\Lambda. \tag{35}$$

For any submodule V of V', V^β denotes the unique sub-T-module of V' isomorphic to V as a $\operatorname{Ker}(\beta)$-module, and such that $\Phi(V^\beta,T)$ is β-convex. In other words, on each line $\alpha+\mathbb{Z}\beta\cap\Phi(V',T)$, one pushes the elements of $\Phi(V,T)$ in the direction β to get $\Phi(V^\beta,T)$.

Let $w \in W$. The point wP/P is denoted by $\dot{w}$. Let V be a T-submodule of $T_{\dot{w}}G/P$. Let β be a root of (G,T). We are interested in the action of the unipotent one-parameter subgroup U_β associated to β on $\dot{w}$ and V. Consider the point $\dot{v} = \lim_{\tau\to\infty} U_\beta(\tau)\dot{w}$. For any $\tau \in \mathbb{C}$, $U_\beta(\tau)V$ is a linear subspace of $T_{U_\beta(\tau)\dot{w}}G/P$ of the same dimension as V. Hence it is a point of a bundle in Grassmannians over G/P. Consider the limit in this bundle

$$\tau(V,\beta) := \lim_{\tau\to\infty} U_\beta(\tau)V. \tag{36}$$

This limit $\tau(V,\beta)$ is a T-stable submodule of the T-module without multiplicity $T_{\dot{v}}G/P$.

We can now state Peterson's result (see [CK03, Section 8]).

Lemma 10 *The T-submodule $s_\beta\tau(V,\beta)$ of $T_{\dot{w}}G/P$ is equal to $V^{-\beta}$.*

Proof. The set $\{U_\beta(\tau)\dot{w} : \tau \in \mathbb{C}\} \cup \dot{v}$ is a T-stable curve isomorphic to $\mathbb{P}^1$. The computation of $\tau(V,\beta)$ lies in a bundle in Grassmannians over this line. This computation can be made quite explicitly by trivializing this bundle on the two T-stable open affine subsets of $\mathbb{P}^1$. □

3.3 A lemma on T-varieties

The following result is used in this paper to characterize Schubert varieties in terms of their tangent spaces among the irreducible T-stable subvarieties of G/P.

Lemma 11 *Let V be a T-module. Let $\mathscr{C}$ be a strictly convex cone in $X(T)\otimes\mathbb{Q}$. Let Σ be a closed T-stable subvariety of V such that*

(i) *Σ is smooth at* 0;
(ii) $T_0\Sigma = \bigoplus_{\chi\in\mathscr{C}} V_\chi$.

Then $\Sigma = \bigoplus_{\chi\in\mathscr{C}} V_\chi$.

Proof. Since $\mathscr{C}$ is strictly convex and $\Phi(V,T)$ is finite, there exist finitely many one-parameter subgroups $\lambda_1,\dots,\lambda_k$ of T, such that

$$\forall\chi \in X(T) \quad \chi \in \mathscr{C} \iff \forall i \ \langle\lambda_i,\chi\rangle > 0.$$

For any i, there exists a T-stable neighborhood of 0 in Σ, such that any point x in this neighborhood satisfies $\lim_{t\to 0}\lambda_i(t)x = 0$. Consider the set W of $v \in V$ such that $\lim_{t\to 0}\lambda_i(t)v = 0$, for any i. By the second condition, W is precisely $T_0\Sigma$. We just proved that $T_0\Sigma$ contains an open subset of Σ. But these two varieties are irreducible and of the same dimension (since Σ is assumed to be smooth at 0). Hence $\Sigma = T_0\Sigma$. □

3.4 Schubert varieties

Let Y be a subvariety of G/P. Let $G(X)$ denote the stabilizer of Y in G; this is the set of g in G such that $gY = Y$. If $G(Y)$ has an open orbit in Y, then this orbit is called the *homogeneous locus* of Y; otherwise, the homogeneous locus of Y is defined to be empty. In other words, the homogeneous locus of Y is the biggest open subset of Y that is homogeneous under a subgroup of G; it is denoted by Y^{hom}.

Recall that $X_w = \overline{BwP/P}$. If $Y = X_w$ for some $w \in W^P$, then the group $G(X_w)$ contains B and is a standard parabolic subgroup of G. In particular, this stabilizer is characterized by a subset Δ_w of simple roots. More precisely, we set

$$\Delta_w = \{\alpha \in \Delta \ :\ P_\alpha X_w = X_w\}.$$

Proposition 5 *We have*

$$X_w^{X(Z)\text{-reg}} = X_w^{\text{hom}}.$$

Proof. Since the infinitesimal filtration is G-invariant, it is clear that $X_w^{X(Z)\text{-reg}}$ is $G(X_w)$-stable and contains X_w^{hom}. Moreover Lemma 8 implies that $X_w^{X(Z)\text{-reg}}$ is open in X_w.

Assume that $X_w^{X(Z)\text{-reg}} - X_w^{\text{hom}}$ is nonempty. Choose an open B-orbit in $X_w^{X(Z)\text{-reg}} - X_w^{\text{hom}}$ and a T-fixed point $\dot{v}$ on it.

Obviously v is smaller than w for the Bruhat order. Since the Bruhat order is generated by T-stable curves, there exists a positive root β, such that $s_\beta v \in W^P$ and $v < s_\beta v < w$. Since $B.\dot{v}$ is dense in an irreducible component of $X_w^{X(Z)\text{-reg}} - X_w^{\text{hom}}$, $s_\beta \dot{v}$ belongs to X_w^{hom}.

Since $s_\beta \dot{v}$ is a T-fixed point in $G(X_w).\dot{w}$, it is equal to $u\dot{w}$ for some $u \in W(G(X_w))$.
We claim that

$$s_\beta \in G(X_w)/T. \tag{37}$$

Let us first explain how the claim leads to a contradiction. Since u belongs to $G(X_w)/T$, the claim implies that $s_\beta u^{-1} X_w^{\text{hom}} = X_w^{\text{hom}}$. But $\dot{v} = s_\beta u^{-1}\dot{w}$ and $\dot{w}$ belongs to X_w^{hom}. Hence $\dot{v} \in X_w^{\text{hom}}$, which is a contradiction.

Consider $\gamma = \pm w^{-1}u\beta$ where the sign is chosen to make γ negative. Since $u \in G(X_w)/T$, Claim (37) is equivalent to $s_\beta u^{-1} X_w = u^{-1} X_w$ or to $s_{u\beta} X_w = X_w$ or to

$$s_\gamma.(w^{-1}X_w) = w^{-1}X_w. \tag{38}$$

Look at these two varieties in a neighborhood of P/P. More precisely, consider the unique affine open T-stable neighborhood Ω of P/P in G/P. Then Ω is isomorphic as a T-variety to a T-module without multiplicity. Since the two varieties of (38) are irreducible, it is sufficient to prove that

$$\Omega \cap s_\gamma.(w^{-1}X_w) = \Omega \cap w^{-1}X_w. \tag{39}$$

Since $s_\gamma P/P \in w^{-1}X_w$, $\gamma \in \Phi(w)$. In particular, $w^{-1}X_w$ is U_γ-stable. But, $s_\gamma P/P$ and P/P are smooth points in $w^{-1}X_w$. Hence

$$\lim_{\tau\to\infty} U_\gamma(\tau) T_{P/P} w^{-1}X_w = T_{s_\gamma P/P} w^{-1}X_w.$$

Then Lemma 10 shows that

$$\begin{aligned}\Phi(T_{P/P} s_\gamma w^{-1}X_w, T) &= s_\gamma \Phi(T_{s_\gamma P/P} w^{-1}X_w, T)\\ &= s_\gamma \Phi\left(\lim_{\tau\to\infty} U_\gamma(\tau) T_{P/P} w^{-1}X_w, T\right)\\ &= \Phi\left((T_{P/P} w^{-1}X_w)^{-\gamma}, T\right).\end{aligned}$$

Since P/P is Γ-regular in $s_\gamma w^{-1}X_w$,

$$\forall \alpha \in X(Z) \quad \dim(F^{\succcurlyeq\alpha}(T_{P/P} w^{-1}X_w)^{-\gamma}) = \dim(F^{\succcurlyeq\alpha}(T_{P/P} w^{-1}X_w)). \qquad (40)$$

But $\gamma \notin \Phi(P)$, hence $\gamma_{|Z}$ is nontrivial. Then, equality (40) implies that

$$\Phi\left((T_{P/P} w^{-1}X_w)^{-\beta}, T\right) = \Phi\left((T_{P/P} w^{-1}X_w), T\right).$$

Equality (39) follows and the theorem is proved. □

4 Infinitesimal filtration and cohomology

4.1 Filtration of differential forms on a manifold

In this subsection, M is a smooth connected manifold of dimension d endowed with an infinitesimal Γ-filtration. The idea that allows controlling the differential, relative to the filtration, is the following.

Definition. An infinitesimal Γ-filtration of M is said to be *integrable*, if for any α and β in Γ we have

$$[F^{\succcurlyeq\alpha}TM, F^{\succcurlyeq\beta}TM] \subset F^{\succcurlyeq\alpha+\beta}TM. \qquad (41)$$

Example. Let L be an integrable distribution on M. We get an integrable infinitesimal $\mathbb{Z}$-filtration by setting

$$\begin{aligned} F^{\succcurlyeq a}TM &= TM \quad \forall a \in \mathbb{Z}_{<0},\\ F^{\succcurlyeq 0}TM &= L,\\ F^{\succcurlyeq a}TM &= \underline{0} \quad \forall a \in \mathbb{Z}_{>0}.\end{aligned}$$

Example. Let L be any distribution on M. We get an integrable infinitesimal $\mathbb{Z}$-filtration by setting

$$F^{\succcurlyeq a}TM = TM \quad \forall a \leq -2,$$
$$F^{\succcurlyeq -1}TM = L,$$
$$F^{\succcurlyeq a}TM = \underline{0} \quad \forall a \in \mathbb{Z}_{\geq 0}.$$

Consider the sheaf Ω^p of differential p-forms on M and the de Rham differential $d_p : \Omega^p \longrightarrow \Omega^{p+1}$. The de Rham cohomology group is

$$\mathrm{H}^p_{DR}(M,\mathbb{R}) := \frac{\mathrm{Ker}\ d_p(M)}{\mathrm{Im}\ d_{p-1}(M)}.$$

The exterior product

$$\wedge : \Omega^p \times \Omega^{p'} \longrightarrow \Omega^{p+p'}$$
$$(\omega, \omega') \longmapsto \omega \wedge \omega'$$

induces the product $\wedge$ on the cohomology since

$$d(\omega \wedge \omega') = (d\omega) \wedge \omega' + (-1)^p \omega \wedge d\omega'.$$

In particular, $\mathrm{H}^*_{DR}(M,\mathbb{R}) := \bigoplus_{k=0}^d \mathrm{H}^k_{DR}(M,\mathbb{R})$ is a graded algebra.

We now consider the Γ-filtration on the sheaf Ω^p induced by the infinitesimal Γ-filtration.

Definition. Let $\beta \in \Gamma$ and let U be an open subset of M. The subspace $F^{\preccurlyeq\beta}\Omega^p(U)$ of $\Omega^p(U)$ is defined to be the set of forms $\omega \in \Omega^p(U)$, such that for any $\alpha_1, \dots, \alpha_p \in \Gamma$, for any $x \in U$, and for any $\xi_i \in F^{\succcurlyeq\alpha_i}T_xM$, we have

$$\alpha_1 + \cdots + \alpha_p \not\preccurlyeq \beta \Rightarrow \omega_x(\xi_1, \dots, \xi_p) = 0. \tag{42}$$

A direct consequence of Proposition 4 is

Proposition 6 (i) *If $\beta \preccurlyeq \gamma$, then $F^{\preccurlyeq\beta}\Omega^p \subset F^{\preccurlyeq\gamma}\Omega^p$.*
(ii) *Let $\beta_0 \in \Gamma$ be such that $F^{\succcurlyeq\beta_0}TM = TM$. If $F^{\preccurlyeq\gamma}\Omega^p \neq \{0\}$, then $\gamma \succcurlyeq p\beta_0$.*
(iii) *We have $F^{\preccurlyeq 0}\Omega^p = \Omega^p$.*
(iv) *For β and γ in Γ, we have $F^{\preccurlyeq\beta}\Omega^p \wedge F^{\preccurlyeq\gamma}\Omega^q \subset F^{\preccurlyeq\beta+\gamma}\Omega^{p+q}$.*

Integrability is essential in the following result.

Proposition 7 *Assume that the infinitesimal filtration is Γ-integrable. Then for any $\beta \in \Gamma$*

$$d_p(F^{\preccurlyeq\beta}\Omega^p) \subset F^{\preccurlyeq\beta}\Omega^{p+1}.$$

Proof. Let U be an open subset of M and let $\omega \in F^{\preccurlyeq\beta}\Omega^p(U)$. Let $x \in U$ and let $\xi_i \in F^{\succcurlyeq\alpha_i}TM$ be defined in a neighborhood of x, such that $\alpha_1 + \cdots + \alpha_{p+1} \not\preccurlyeq \beta$. It remains to prove that

$$d_p(\omega)_x(\xi_1,\dots,\xi_{p+1}) = 0.$$

Cartan's formula implies that

$$d_p(\omega)_x(\xi_1,\dots,\xi_{p+1}) = \sum_i \pm\xi_i \cdot \omega(\xi_1,\dots,\hat{\xi_i},\dots,\xi_{p+1}) \\ + \sum_{i<j} \pm\omega_x([\xi_i,\xi_j],\xi_1,\dots,\hat{\xi_i},\dots,\hat{\xi_j},\dots,\xi_{p+1}).$$

Since $[\xi_i,\xi_j] \in F^{\succcurlyeq\alpha_i+\alpha_j}M$ and

$$(\alpha_i+\alpha_j)+\alpha_1+\cdots+\hat{\alpha_i}+\cdots+\hat{\alpha_j}+\cdots+\alpha_{p+1} \not\preccurlyeq \beta,$$

the term $\omega_x([\xi_i,\xi_j],\xi_1,\dots,\hat{\xi_i},\dots,\hat{\xi_j},\dots,\xi_{p+1})$ is zero.

Consider now the term

$$\xi_i \cdot \omega(\xi_1,\dots,\hat{\xi_i},\dots,\xi_{p+1}). \tag{43}$$

If $\alpha_i \not\preccurlyeq 0$, then $\xi_i = 0$ and the term (43) is zero. Assume now that $\alpha_i \preccurlyeq 0$. The weight of $\xi_1,\dots,\hat{\xi_i},\dots,\xi_{p+1}$ is $\theta := \sum_{j=1}^{p+1}\alpha_j - \alpha_i$. Since $\theta + \alpha_i \not\preccurlyeq \beta$ and $\alpha_i \preccurlyeq 0$, we have $\theta \not\preccurlyeq \beta$. Since ω belongs to $F^{\preccurlyeq\beta}\Omega^p(U)$, it follows that $\omega(\xi_1,\dots,\hat{\xi_i},\dots,\xi_{p+1}) = 0$. □

4.2 Filtration on the cohomology

The Γ-filtration on M induces an increasing Γ-filtration on the cohomology. Indeed, Propositions 6 and 7 show that the de Rham complex is Γ-filtered. Namely, we set

$$F^{\preccurlyeq\beta}\,\mathrm{H}^p(M,\mathbb{R}) := \frac{\mathrm{Ker}(d_p) \cap F^{\preccurlyeq\beta}\Omega^p(M,\mathbb{R})}{d_{p-1}(\Omega^{p-1}(M,\mathbb{R})) \cap F^{\preccurlyeq\beta}\Omega^p(M,\mathbb{R})}. \tag{44}$$

Propositions 6 and 7 show the following:

Proposition 8 *The sets $F^{\preccurlyeq\beta}\,\mathrm{H}^p(M,\mathbb{R})$ are canonically identified with subspaces of* $\mathrm{H}^p(M,\mathbb{R})$.

(i) *If $F^{\succcurlyeq\beta_0}M = TM$, then $F^{\succcurlyeq p\beta_0-\beta}\,\mathrm{H}^p(M,\mathbb{R})$ is a Γ-filtration of* $\mathrm{H}^p(M,\mathbb{R})$.

(ii) *The filtration respects the structure of algebra. Namely, for β and γ in Γ, we have*

$$F^{\preccurlyeq\beta}\,\mathrm{H}^p(M,\mathbb{R}) \wedge F^{\preccurlyeq\gamma}\mathrm{H}^q(M,\mathbb{R}) \subset F^{\preccurlyeq\beta+\gamma}\mathrm{H}^{p+q}(M,\mathbb{R}).$$

Remark. The Γ-filtration is defined at the level of the de Rham complex, not only at the level of the cohomology. In particular, it induces a spectral sequence which should be studied to understand the relations between the ordinary and the Belkale-Kumar cohomologies. Here we only study the Belkale-Kumar product.

Consider now the $(\Gamma \times \mathbb{Z})$-graded algebra associated to this Γ-filtration of the $\mathbb{Z}$-graded (by degree) algebra $\mathrm{H}^*(M,\mathbb{R})$ by setting

$$Gr^{\beta}\,\mathrm{H}^p(M,\mathbb{R}) := \frac{F^{\preccurlyeq\beta}\,\mathrm{H}^p(M,\mathbb{R})}{\sum_{\gamma\prec\beta} F^{\preccurlyeq\gamma}\mathrm{H}^p(M,\mathbb{R})} \tag{45}$$

and

$$Gr^{\bullet}\,\mathrm{H}^*(M,\mathbb{R}) := \bigoplus_{\beta\in\Gamma,\, p\in\mathbb{N}} Gr^{\beta}\,\mathrm{H}^p(M,\mathbb{R}). \tag{46}$$

Then $Gr^{\bullet}\,\mathrm{H}^*(M,\mathbb{R})$ is a ring graded by $\Gamma \times \mathbb{Z}$.

Now, we observe the following easy functoriality result.

Lemma 12 *Let M and N be two smooth manifolds endowed with integrable infinitesimal Γ-filtrations. Let ϕ $M \longrightarrow N$ be a smooth map such that*

$$\forall \alpha \in \Gamma \qquad T\phi(F^{\geq\alpha}TM) \subset F^{\geq\alpha}TN.$$

Then the pullback $\phi^* : \mathrm{H}^*(N,\mathbb{R}) \longrightarrow \mathrm{H}^*(M,\mathbb{R})$ respects the Γ-filtration. In particular, it induces a Γ-graded morphism $Gr\phi^* : Gr\mathrm{H}^*(N,\mathbb{R}) \longrightarrow Gr\mathrm{H}^*(M,\mathbb{R})$.

4.3 Cohomology with complex coefficients

Recall that M is a connected manifold. Consider the cohomology group $\mathrm{H}^*(M,\mathbb{C})$ with complex coefficients and consider the following complex vector bundle on M:

$$T^{\mathbb{C}}M := TM \otimes_{\mathbb{R}} \mathbb{C}.$$

A *complex infinitesimal Γ-filtration of M* is a family of complex subbundles

$$F^{\preccurlyeq\beta}T^{\mathbb{C}}M \subset T^{\mathbb{C}}M,$$

indexed by $\beta \in \Gamma$ and satisfying the three assertions of Definition 2.2. A complex infinitesimal Γ-filtration is said to be Γ-integrable if for any β and γ in Γ, we have

$$[F^{\preccurlyeq\beta}T^{\mathbb{C}}M, F^{\preccurlyeq\gamma}T^{\mathbb{C}}M] \subset F^{\preccurlyeq\beta+\gamma}T^{\mathbb{C}}M. \tag{47}$$

A complex infinitesimal integrable Γ-filtration induces a filtration on the de Rham complex and on the groups $\mathrm{H}^p(M,\mathbb{C})$.

Example. Let M be an holomorphic manifold. Let J denote the complex structure on the tangent bundle TM. Since $J^2 = -\mathrm{Id}$, its eigenvalues acting on $TM \otimes \mathbb{C}$ are $\pm\sqrt{-1}$. Let $T^{1,0}M$ (resp. $T^{0,1}M$) denote the complex subbundle of $TM \otimes \mathbb{C}$ associated to the eigenvalue $\sqrt{-1}$ (resp. $-\sqrt{-1}$). There is a natural $\mathbb{C}$-linear isomorphism $\iota^{1,0} : TM \longrightarrow T^{1,0}M$. It is well known that $T^{1,0}M$ is an integrable distribution in $T^{\mathbb{C}}M$. Then we get a complex infinitesimal integrable $\mathbb{Z}$-filtration by setting

$$\begin{aligned} F^{\succcurlyeq a}T^{\mathbb{C}}M &= T^{\mathbb{C}}M \qquad && \forall a \in \mathbb{Z}_{<0},\\ F^{\succcurlyeq 0}T^{\mathbb{C}}M &= T^{1,0}M, &&\\ F^{\succcurlyeq a}T^{\mathbb{C}}M &= \underline{0}, && \forall a \in \mathbb{Z}_{>0}. \end{aligned}$$

The $\mathbb{Z}$-filtration of $\mathrm{H}^p(M,\mathbb{C})$ is called the Hodge filtration of M (see for example [Voi07]).

4.4 The case of a smooth complex variety

Let M be a smooth complex irreducible variety endowed with an algebraic Γ-filtration. Assume that this filtration is integrable and comes from a decomposition (recall the definition in Section 2.2). Set $\tilde{\Gamma} := \Gamma \times \mathbb{Z}$ endowed with the order $(\beta,n)\succcurlyeq(\gamma,m)$ if and only if $\beta\succcurlyeq\gamma$ and $n \geq m$.

Define a complex $\tilde{\Gamma}$-filtration on $T^{\mathbb{C}}M$ by setting for any $\beta \in \Gamma$,

$$\begin{aligned} F^{\succcurlyeq(\beta,a)}T^{\mathbb{C}}M &= T^{\mathbb{C}}M \qquad && \forall a \in \mathbb{Z}_{<0},\\ F^{\succcurlyeq(\beta,0)}T^{\mathbb{C}}M &= \iota^{1,0}(F^{\succcurlyeq\beta}TM), &&\\ F^{\succcurlyeq(\beta,a)}T^{\mathbb{C}}M &= \underline{0}, && \forall a \in \mathbb{Z}_{>0}. \end{aligned}$$

Integration along subvarieties. Let N be an irreducible subvariety of M. Denote by n the dimension of M, and by d that of N. By Lemma 4, the dimension vector $(fd^\beta(T_xN))_{\beta\in\Gamma}$ does not depend on $x \in N$ general. This general value of the dimension vector is by definition the *f-dimension vector of N* and is denoted by $fd^\beta(N)$. For any x in N, the Γ-filtration of T_xN comes from a decomposition by Lemma 5. In particular, Lemma 4 shows that the g-dimensional vector of T_xN does not depend on x in N general. This remark allows us to define the g-dimension vector of N. Then the weight $\rho(N) \in \Gamma$ of N is defined by the formula

$$\rho(N) = \sum_{\beta\in\Gamma} gd^\beta(N)\beta. \tag{48}$$

Consider the extended notions to $\tilde{\Gamma}$: $\widetilde{gd}^{(\beta,0)}(N) = gd^{(\beta,0)}(T_xN \otimes \mathbb{C}) = gd^\beta(N)$, $\widetilde{gd}^{(0,-1)}(N) = d$ and $\widetilde{gd}^{(\beta,a)}(N) = 0$ otherwise. Note that $\tilde{\rho}(N) = (\rho(N), -d)$.

Consider now the linear map

$$\begin{aligned}\Omega^{2d}(M,\mathbb{C}) &\longrightarrow \mathbb{C}\\ \omega &\longmapsto \int_N \omega_{|N}.\end{aligned}$$

The following lemma relies on the filtration and the integration.

Lemma 13 *Let $\beta\in\Gamma$ and let $e\in\mathbb{Z}$, such that $(\beta,e)\not\succcurlyeq\tilde{\rho}(N)$. If $\omega\in F^{\preccurlyeq(\beta,e)}\Omega^{2d}(M,\mathbb{C})$ then*

$$\int_N \omega_{|N} = 0.$$

Proof. Let $x\in N$ be a general point. By Lemma 5, the Γ-filtration on T_xN comes from a decomposition. Then there exists a basis $(\xi_1,\dots,\xi_d)$ of T_xN such that for any $\beta\in\Gamma$, the set of ξ_i which belongs to $F^{\succcurlyeq\beta}T_xN$ spans $F^{\succcurlyeq\beta}T_xN$. Such a basis exists since by Lemma 5, the Γ-filtration on T_xN comes from a decomposition. Let α_i be the maximal element of Γ such that ξ_i belongs to $F^{\succcurlyeq\alpha_i}T_xN$.

Consider the basis $(\iota^{(1,0)}(\xi_1),\dots,\iota^{(1,0)}(\xi_d),\iota^{(0,1)}(\xi_1),\dots,\iota^{(0,1)}(\xi_d))$ of $T_xN\otimes\mathbb{C}$. Since x is any general point on N, it is sufficient to prove that

$$\omega(\iota^{(1,0)}(\xi_1),\dots,\iota^{(1,0)}(\xi_d),\iota^{(0,1)}(\xi_1),\dots,\iota^{(0,1)}(\xi_d)) = 0.$$

But $\iota^{(1,0)}(\xi_i)\in F^{\succcurlyeq(\alpha_i,0)}T^{\mathbb{C}}N$ and $\iota^{(0,1)}(\xi_i)\in F^{\succcurlyeq(0,-1)}T^{\mathbb{C}}N$. Hence the weight of $(\iota^{(1,0)}(\xi_1),\dots,\iota^{(1,0)}(\xi_d),\iota^{(0,1)}(\xi_1),\dots,\iota^{(0,1)}(\xi_d))$ is $\sum_{i=1}^d(\alpha_i,0)+d(0,-1)=\tilde{\rho}(N)$. The lemma follows. □

The restriction of the map $\omega\longmapsto\int_N\omega_{|N}$ to the closed $2d$-forms is zero on the exact forms and induces a linear map

$$\int_N : \mathrm{H}^{2d}(M,\mathbb{C}) \longrightarrow \mathbb{C}.$$

Consider now the restriction of this map to $F^{\preccurlyeq\tilde{\rho}(N)}\mathrm{H}^{2d}(M,\mathbb{C})$. By Lemma 13, this restriction induces a linear map

$$\int_N : \mathrm{Gr}^{\tilde{\rho}(N)}\mathrm{H}^{2d}(M,\mathbb{C}) \longrightarrow \mathbb{C}.$$

Poincaré pairing. Assume that M is compact and recall that it is orientable since it is holomorphic. Let p be an integer such that $0\le p\le 2d$. The integration allows us to define a pairing

$$\begin{aligned}\mathrm{H}^p(M,\mathbb{C})\times\mathrm{H}^{2d-p}(M,\mathbb{C}) &\longrightarrow \mathbb{C}\\ ([\omega_1],[\omega_2]) &\longmapsto \int_M \omega_1\wedge\omega_2.\end{aligned} \tag{49}$$

By Poincaré duality, this bilinear form is nondegenerate. In particular, $\mathrm{H}^p(M,\mathbb{C})$ and $\mathrm{H}^{2d-p}(M,\mathbb{C})$ have the same dimension.

Let $\tilde{\alpha}\in\tilde{\Gamma}$. Consider the following restriction of the bilinear form (49):

$$\begin{aligned} F^{\preccurlyeq\tilde{\alpha}}\,\mathrm{H}^p(M,\mathbb{C}) \times F^{\preccurlyeq\tilde{\rho}(M)-\tilde{\alpha}}\,\mathrm{H}^{2d-p}(M,\mathbb{C}) &\longrightarrow \mathbb{C} \\ ([\omega_1],[\omega_2]) &\longmapsto \int_M \omega_1\wedge\omega_2. \end{aligned} \tag{50}$$

Since

$$\begin{aligned} &\tilde{\alpha}\succcurlyeq\tilde{\rho}(M) \Rightarrow F^{\preccurlyeq\tilde{\alpha}}\Omega^{2d}(M) = \Omega^{2d}(M), \text{ and} \\ &\tilde{\alpha}\not\succcurlyeq\tilde{\rho}(M) \Rightarrow F^{\preccurlyeq\tilde{\alpha}}\Omega^{2d}(M) = 0, \end{aligned}$$

Lemma 13 shows that

$$\tilde{\alpha}+\tilde{\beta}\not\succcurlyeq\tilde{\rho}(M) \Rightarrow F^{\preccurlyeq\tilde{\alpha}}\,\mathrm{H}^p(M,\mathbb{C})\wedge F^{\preccurlyeq\tilde{\beta}}\,\mathrm{H}^{2d-p}(M,\mathbb{C}) = \{0\}. \tag{51}$$

In particular, the pairing (50) passes to the quotient and induces a pairing

$$\begin{aligned} \mathrm{Gr}^{\tilde{\alpha}}\,\mathrm{H}^p(M,\mathbb{C}) \times \mathrm{Gr}^{\tilde{\rho}(M)-\tilde{\alpha}}\,\mathrm{H}^{2d-p}(M,\mathbb{C}) &\longrightarrow \mathbb{C} \\ ([\omega_1],[\omega_2]) &\longmapsto \int_M \omega_1\wedge\omega_2. \end{aligned} \tag{52}$$

Definition. The $\tilde{\Gamma}$-filtration of $\mathrm{H}^*(M,\mathbb{C})$ is said to be *compatible with Poincaré duality* if for any integer $0\leq p\leq 2d$ and for any $\tilde{\alpha}\in\tilde{\Gamma}$, the pairing (52) is nondegenerate.

Lemma 14 *The $\tilde{\Gamma}$-filtration of $\mathrm{H}^*(M,\mathbb{C})$ is compatible with Poincaré duality if and only if for any nonnegative integer p and any $\tilde{\alpha}\in\tilde{\Gamma}$, we have*

$$\dim(\mathrm{Gr}^{\tilde{\alpha}}\,\mathrm{H}^p(M,\mathbb{C})) = \dim(\mathrm{Gr}^{\tilde{\rho}(M)-\tilde{\alpha}}\,\mathrm{H}^{2d-p}(M,\mathbb{C})) \tag{53}$$

Proof. If the $\tilde{\Gamma}$-filtration of $\mathrm{H}^*(M,\mathbb{C})$ is compatible with Poincaré duality, we obviously have the equalities of dimensions.

Assume now that (53) hold. In a basis adapted to the filtration, implication (51) implies that the matrix A of the pairing (49) is upper triangular. Moreover, the matrices (in the induced basis) of the pairings (52) are the diagonal blocs of A. But equalities (53) imply that these blocs are square. Since A is invertible, it follows that any bloc is invertible. □

Definition. Let N be an irreducible subvariety of a compact smooth irreducible complex variety M, endowed with an integrable infinitesimal Γ-filtration coming from a decomposition. Assume that the $\tilde{\Gamma}$-filtration is compatible with Poincaré duality. Define $[N]_{\odot_0}\in\mathrm{Gr}^{\tilde{\rho}(M)-\tilde{\rho}(N)}\,\mathrm{H}^{2(n-d)}(M,\mathbb{C}))$ to satisfy the following formula:

$$\int_N[\omega] = \int_M [N]_{\odot_0}\wedge[\omega], \tag{54}$$

for any $[\omega]\in\mathrm{Gr}^{\tilde{\rho}(N)}\,\mathrm{H}^{2d}(M,\mathbb{C})$.

One can refer to Proposition 11 for a more concerte characterization of $[N]_{\odot_0}$ and in particular its relation with $[N]$, in the case when $M=G/P$.

5 Isomorphism with the Belkale-Kumar product

5.1 The Belkale-Kumar product

In this section, we recall the Belkale-Kumar notion of Levi-movability (see [BK06]).

The cycle class of the Schubert variety X_w in $\mathrm{H}^*(G/P,\mathbb{C})$ is denoted by σ_w and it is called a Schubert class. The degree of σ_w is $2(\dim G/P - l(w)$, where $l(w) = \sharp\Phi(w)$ is the length of w. The Schubert classes form a basis of the cohomology of G/P:

$$\mathrm{H}^*(G/P,\mathbb{C}) = \bigoplus_{w\in W^P} \mathbb{C}\sigma_w. \tag{55}$$

The Poincaré dual of σ_w is denoted by $\sigma_w^\vee$. Note that σ_e is the class of the point. Let σ_u, σ_v, σ_w be three Schubert classes (with $u, v, w \in W^P$). If there exists an integer d such that $\sigma_u.\sigma_v.\sigma_w = d\sigma_e$, then we set $c_{uvw} = d$; we set $c_{uvw} = 0$ otherwise. These coefficients are the (symmetrized) structure coefficients of the cup product on $\mathrm{H}^*(G/P,\mathbb{C})$ in the Schubert basis in the following sense:

$$\sigma_u.\sigma_v = \sum_{w\in W^P} c_{uvw}\sigma_w^\vee$$

and $c_{uvw} = c_{vuw} = c_{uwv}$.

Consider the tangent space T_u of the orbit $u^{-1}BuP/P$ at the point P/P; and similarly, consider T_v and T_w. Using the transversality theorem of Kleiman, Belkale and Kumar showed in [BK06, Proposition 2] the following important lemma.

Lemma 15 *The coefficient c_{uvw} is nonzero if and only if there exist $p_u, p_v, p_w \in P$ such that the natural map*

$$T_P(G/P) \longrightarrow \frac{T_P(G/P)}{p_uT_u} \oplus \frac{T_P(G/P)}{p_vT_v} \oplus \frac{T_P(G/P)}{p_wT_w}$$

is an isomorphism.

Then Belkale-Kumar defined Levi-movability.

Definition. The triple $(\sigma_u, \sigma_v, \sigma_w)$ is said to be *Levi-movable* if there exist $l_u, l_v, l_w \in L$, such that the natural map

$$T_P(G/P) \longrightarrow \frac{T_P(G/P)}{l_uT_u} \oplus \frac{T_P(G/P)}{l_vT_v} \oplus \frac{T_P(G/P)}{l_wT_w}$$

is an isomorphism.

Belkale-Kumar set

$$c^{\odot_0}_{uvw} = \begin{cases} c_{uvw} & \text{if } (\sigma_u, \sigma_v, \sigma_w) \text{ is Levi-movable;} \\ 0 & \text{otherwise.} \end{cases}$$

They defined on the group $\mathrm{H}^*(G/P,\mathbb{C})$ a bilinear product $\odot_0$ by the formula

$$\sigma_u \odot_0 \sigma_v = \sum_{w\in W^P} c^{\odot_0}_{uvw} \sigma^\vee_w.$$

Theorem 2 (Belkale-Kumar 2006). *The product $\odot_0$ is commutative, associative, and satisfies Poincaré duality.*

[RR11, Proposition 2.4] gives an equivalent characterization of Levi-movability. It can be formulated as follows.

Proposition 9 *Let $u, v, w \in W^P$ be such that $c_{uvw} \neq 0$. Then $(\sigma_u, \sigma_v, \sigma_w)$ is Levi-movable if and only if*

$$2gd(G/P) = gd(X_u) + gd(X_v) + gd(X_w).$$

5.2 *The statements*

The first aim of this section is to prove (see Section 5.4.10) the following result of compatibility between the basis of Schubert classes and the $\tilde{\Gamma}$-filtration on $\mathrm{H}^*(G/P,\mathbb{C})$.

Proposition 10

For any $\tilde{\beta} \in \tilde{\Gamma}$ and any integer p, the linear subspace $F^{\preccurlyeq\tilde{\beta}}\,\mathrm{H}^p(G/P,\mathbb{C})$ is spanned by the Schubert classes it contains.

More precisely, $F^{\preccurlyeq\tilde{\beta}}\,\mathrm{H}^p(G/P,\mathbb{C})$ is spanned by the Schubert classes $\sigma_{w^\vee}$, where $w \in W^P$ satisfies $(\rho(X_w), -l(w))\preccurlyeq\tilde{\beta}$.

For any $w \in W^P$, denote by $\overline{\sigma_{w^\vee}}$

the class of $\sigma_{w^\vee} \in F^{\preccurlyeq(\rho(X_w),-l(w))}\,\mathrm{H}^{l(w)}(G/P,\mathbb{C})$ in $\mathrm{Gr}^{(\rho(X_w),-l(w))}\,\mathrm{H}^{l(w)}(G/P,\mathbb{C})$.

Proposition 10 implies that $(\overline{\sigma_{w^\vee}})_{w\in W^P}$ is a basis of $\mathrm{Gr}\,\mathrm{H}^*(G/P,\mathbb{C})$. Consider now the obvious linear isomorphism

$$\begin{aligned} \Psi : \mathrm{H}^*(G/P,\mathbb{C}) &\longrightarrow \mathrm{Gr}\,\mathrm{H}^*(G/P,\mathbb{C}) \\ \sigma_{w^\vee} &\longmapsto \overline{\sigma_{w^\vee}} \qquad\qquad \text{for any } w \in W^P. \end{aligned}$$

Theorem 3 *The linear isomorphism Ψ from the algebra $(\mathrm{H}^*(G/P,\mathbb{C}),\odot_0)$ onto the algebra $\mathrm{Gr}\,\mathrm{H}^*(G/P,\mathbb{C})$ is an isomorphism of algebras.*

The theorem is proved in Section 5.5 after some preparation. The first consequence concerns Poincaré duality (see Section 5.4.10).

Corollary 1 *The* $(X(Z)\times\mathbb{Z})$*-filtration of* $\mathrm{H}^*(G/P,\mathbb{C})$ *is compatible with Poincaré duality.*

This corollary allows us to define the graded Schubert classes by setting, for any $w\in W^P$,

$$\sigma_w^{\odot_0} := [X_w]_{\odot_0}. \tag{56}$$

Finally, by applying Proposition 11 to $Y = X_w$, we get the following result of compatibility.

Lemma 16 *For any* $w\in W^P$*, we have*

$$\Psi(\sigma_w) = \sigma_w^{\odot_0}.$$

5.3 *An upper bound for* $\dim(F^{\preccurlyeq\tilde{\alpha}}\mathrm{H}^p(G/P,\mathbb{C}))$

For any $w\in W$, as a consequence of the relation $\Phi^- = (\Phi^-\cap w^{-1}\Phi^+)\cup(\Phi^-\cap w^{-1}\Phi^-)$, we have (see [Kum02, 1.3.22.3])

$$\sum_{\alpha\in\Phi^-\cap w^{-1}\Phi^+}\alpha = w^{-1}\rho-\rho. \tag{57}$$

Assume that $w\in W^P$. Since P/P is $X(T)$-regular and T acts on $T_{P/P}w^{-1}X_w$ without multiplicities and with weights $\Phi^-\cap w^{-1}\Phi^+$, we have

$$\rho(X_w) = \rho(w^{-1}X_w) = \left(\sum_{\alpha\in\Phi^-\cap w^{-1}\Phi^+}\alpha\right)_{|Z} = \left(w^{-1}\rho-\rho\right)_{|Z}. \tag{58}$$

In particular

$$\rho(G/P) = 2\,(\rho_L-\rho)_{|Z} = -2\rho_{|Z}, \tag{59}$$

since ρ_L is trivial on Z. Hence

$$\rho(G/P)-\rho(X_w) = \left(-\rho-w^{-1}\rho\right)_{|Z}. \tag{60}$$

Lemma 17 *For any* $w\in W^P$*, we have*

$$\rho(G/P)-\rho(X_w) = \rho(X_{w^\vee}).$$

Proof. Remark that

$$((w^\vee)^{-1}\rho)_{|Z} = ((w_0^P w^{-1}w_0\rho)_{|Z} = -w_0^P(w^{-1}\rho)_{|Z} = -(w^{-1}\rho)_{|Z},$$

since w_0^P belongs to L and acts trivially on Z. The lemma follows. □

Lemma 18
Let n denote the dimension of G/P. The dimension of $F^{\preccurlyeq\beta}\mathrm{H}^{2(n-d)}(G/P,\mathbb{C})$ is less or equal to the number of $w \in W^P$ such that $\rho(G/P)-\rho(X_w)\preccurlyeq\beta$ and $l(w)=d$.

Proof. For $w \in W^P$ such that $\rho(G/P)-\rho(X_w)\not\preccurlyeq\beta$ and $l(w)=d$, consider the linear form

$$\int_{X_{w^\vee}} : \mathrm{H}^{2(n-d)}(G/P,\mathbb{C}) \longrightarrow \mathbb{C}.$$

By Lemmas 17 and 13, this linear form is zero on $F^{\preccurlyeq\beta}\mathrm{H}^{2(n-d)}(G/P,\mathbb{C})$. But by Poincaré duality these linear forms are linearly independent. This implies that the codimension of $F^{\preccurlyeq\beta}\mathrm{H}^{2(n-d)}(G/P,\mathbb{C})$ in $\mathrm{H}^{2(n-d)}(G/P,\mathbb{C})$ is at least the number of $w \in W^P$ such that $\rho(G/P)-\rho(X_w)\not\preccurlyeq\beta$ and $l(w)=d$. The lemma follows. □

5.4 Kostant's harmonic forms

5.4.1 The role of Kostant's harmonic forms in this paper

Let w in W^P. In 1963, B. Kostant constructed an explicit $\mathbb{C}$-valued closed differential form ω_w on G/P, such that the associated cohomology class $[\omega_w]$ is equal to σ_w up to scalar multiplication. Kostant's form ω_w is used here to localize the Schubert class relative to the filtration.

Lemma 19 *The Schubert class $\sigma_{w^\vee}$ belongs to $F^{\preccurlyeq(\rho(X_w),-l(w))}\mathrm{H}^{l(w)}(G/P,\mathbb{C})$.*

Before proving Lemma 19 in Section 5.4.10, we recall Kostant's construction.

5.4.2 Restriction to K-invariant forms

Let K be a maximal compact subgroup of G, such that $T\cap K$ is a maximal torus of K. Then K is a connected compact Lie group.

Consider the subcomplex of K-invariant forms:

$$d_p : \Omega^p(G/P,\mathbb{C})^K \longrightarrow \Omega^{p+1}(G/P,\mathbb{C})^K,$$

and its cohomology $\mathrm{H}^*_{DR}(G/P,\mathbb{C})^K$. The identity

$$d_{p-1}(\Omega^{p-1}(G/P,\mathbb{C})^K) = d_{p-1}(\Omega^{p-1}(G/P,\mathbb{C}))\cap\Omega^p(G/P,\mathbb{C})^K$$

allows us to define a morphism

$$\mathrm{H}^*_{DR}(G/P,\mathbb{C})^K \longrightarrow \mathrm{H}^*_{DR}(G/P,\mathbb{C}),$$

which is an isomorphism.

Since K acts transitively on G/P, the restriction map to the tangent space at P/P provides a linear isomorphism

$$\Omega^p(G/P,\mathbb{C})^K \longrightarrow (\wedge^p \mathrm{Hom}_{\mathbb{R}}(\mathfrak{g}/\mathfrak{p},\mathbb{C}))^{K\cap L}. \tag{61}$$

Let $\mathfrak{k}$ denote the Lie algebra of K. This compact form $\mathfrak{k}$ determines a real structure $\Box^*$ on $\mathfrak{g}$. More precisely, $\Box^*$ is a $\mathbb{C}$-antilinear endomorphism of $\mathfrak{g}$ such that $\mathfrak{k}$ is the set of $\xi \in \mathfrak{g}$ satisfying $\xi^* = -\xi$.

Consider now the complex dual $(\mathfrak{g}/\mathfrak{l})^*$ of the complex vector space $\mathfrak{g}/\mathfrak{l}$. Since $\mathfrak{l}$ is stable by $\Box^*$, $\mathfrak{g}/\mathfrak{l}$ is endowed with a real structure still denoted by $\Box^*$. Then $(\mathfrak{g}/\mathfrak{l})^*$ is also endowed with a real structure by setting $\varphi^* = \overline{\varphi(\Box^*)}$, for any $\varphi \in (\mathfrak{g}/\mathfrak{l})^*$. Define a morphism

$$\begin{aligned} \theta : \mathrm{Hom}_{\mathbb{R}}(\mathfrak{g}/\mathfrak{p},\mathbb{C}) &\longrightarrow (\mathfrak{g}/\mathfrak{l})^* \\ \varphi+\psi &\longmapsto \varphi+\psi(\Box^*), \end{aligned}$$

where φ is $\mathbb{C}$-linear and ψ is $\mathbb{C}$-antilinear. One checks that θ is a $\mathbb{C}$-linear isomorphism and that it commutes with the real structure and actions of $K\cap L$. Note that L also acts on $(\mathfrak{g}/\mathfrak{l})$. Since $K\cap L$ is Zariski dense in L, we have

$$(\wedge^p(\mathfrak{g}/\mathfrak{l})^*)^{K\cap L} = (\wedge^p(\mathfrak{g}/\mathfrak{l})^*)^L. \tag{62}$$

Finally we get an isomorphism

$$\Omega^p(G/P,\mathbb{C})^K \longrightarrow (\wedge^p(\mathfrak{g}/\mathfrak{l})^*)^L. \tag{63}$$

5.4.3 The Lie algebra $\mathfrak{r}$

Let $\mathfrak{u}$ and $\mathfrak{u}^-$ be the algebras of the unipotent radicals of P and its opposite parabolic subgroup P^-. Consider the sum

$$\mathfrak{r} = \mathfrak{u}^- \oplus \mathfrak{u}, \tag{64}$$

endowed with a Lie algebra structure $[\cdot,\cdot]_{\mathfrak{r}}$ defined by keeping the brackets on $\mathfrak{u}^-$ and $\mathfrak{u}$ unchanged and by setting $[\mathfrak{u}^-,\mathfrak{u}]_{\mathfrak{r}} = 0$. The L-equivariant linear isomorphism $\mathfrak{r} \simeq \mathfrak{g}/\mathfrak{l}$ and its transpose $(\mathfrak{g}/\mathfrak{l})^* \simeq \mathfrak{r}^*$ induce isomorphisms

$$\Omega^\bullet(G/P,\mathbb{C})^K \simeq (\mathrm{Hom}_{\mathbb{R}}(\mathfrak{g}/\mathfrak{p},\mathbb{C}))^L \simeq (\wedge^\bullet\mathfrak{r}^*)^L \simeq \left(\wedge^\bullet(\mathfrak{u}^-)^* \otimes \wedge^\bullet\mathfrak{u}^*\right)^L. \tag{65}$$

The term $\wedge^\bullet(\mathfrak{u}^-)^*$ corresponds to holomorphic forms on G/P and the term $\wedge^\bullet\mathfrak{u}^*$ corresponds to antiholomorphic forms.

Combining $\Box^*$ and the Killing form $(\cdot,\cdot)$, one obtains an Hermitian form $\{\cdot,\cdot\}$ on $\mathfrak{g}$. Explicitly,

$$\{\xi,\eta\} = -(\xi,\eta^*),$$

for any $\xi, \eta \in \mathfrak{g}$. Denote by $\{\cdot,\cdot\}_{\mathfrak{r}}$ its restriction to $\mathfrak{r}$. The decomposition $\mathfrak{u}^- \oplus \mathfrak{u} = \mathfrak{r}$ is orthogonal for $\{\cdot,\cdot\}_{\mathfrak{r}}$. Consider now the graded exterior algebra $\wedge^\bullet \mathfrak{r}^* = \bigoplus_p \wedge^p \mathfrak{r}^*$ and extend the bilinear form $\{\cdot,\cdot\}_{\mathfrak{r}}$ on $\wedge^\bullet \mathfrak{r}^*$. The decomposition $\mathfrak{r} = \mathfrak{u}^- \oplus \mathfrak{u}$ induces an $\mathbb{N}^2$-grading $\wedge^\bullet \mathfrak{r}^* = \bigoplus_{(p,q)\in\mathbb{N}^2} \wedge^{p,q}\mathfrak{r}^*$ by setting

$$\wedge^{p,q}\mathfrak{r}^* = \wedge^p(\mathfrak{u}^-)^* \otimes \wedge^q(\mathfrak{u})^*.$$

Moreover, the sum $\bigoplus_{(p,q)\in\mathbb{N}^2} \wedge^{p,q}\mathfrak{r}^*$ is orthogonal for $\{\cdot,\cdot\}_{\mathfrak{r}}$.

Let $b \in \mathrm{End}(\wedge^\bullet \mathfrak{r}^*)$ be the Chevalley-Eilenberg coboundary operator of the Lie algebra $\mathfrak{r}$. It has degree $+1$; more precisely

$$b(\wedge^{p,q}\mathfrak{r}^*) \subset \wedge^{p+1,q}\mathfrak{r}^* \oplus \wedge^{p,q+1}\mathfrak{r}^*.$$

Set $b = b^{1,0} + b^{0,1}$ according to this decomposition. Let $\partial \in \mathrm{End}(\wedge^\bullet \mathfrak{r})$ denote the Chevalley-Eilenberg boundary operator. Using the Killing form, we identify $\mathfrak{r}$ and $\mathfrak{r}^*$ and transport ∂ to an operation $\partial^* \in \mathrm{End}(\wedge^\bullet \mathfrak{r}^*)$ of degree -1. Decompose $\partial^* = \partial^{-1,0} + \partial^{0,-1}$ according to the decomposition $\wedge^{p-1,q}\mathfrak{r}^* \oplus \wedge^{p,q-1}\mathfrak{r}^*$. Set

$$\mathscr{L} = \partial^* \circ b + b \circ \partial^*. \tag{66}$$

[Kos63, Proposition 4.2] gives an alternative expression of $\mathscr{L}$:

$$\mathscr{L} = \frac{1}{2}(\partial^{0,-1} \circ b^{0,1} + b^{0,1} \circ \partial^{0,-1}). \tag{67}$$

5.4.4 The $(X(Z)\times\mathbb{Z})$-filtration of $\wedge^\bullet \mathfrak{r}^*$

Consider the action of $Z \times \mathbb{C}^*$ on $\mathfrak{r}$ given by

$$(z,\tau).(\xi^- + \xi) = (\tau z \xi^-, \xi), \qquad \forall z \in Z, \tau \in \mathbb{C}^*, \xi^- \in \mathfrak{u}^-, \xi \in \mathfrak{u}. \tag{68}$$

Then the group $Z \times \mathbb{C}^*$ acts on $\wedge^\bullet \mathfrak{r}^*$ and induces a $\tilde{\Gamma}$-decomposition

$$\wedge^\bullet \mathfrak{r}^* = \bigoplus_{\tilde{\beta}\in X(Z)\times\mathbb{Z}} (\wedge^\bullet \mathfrak{r}^*)_{\tilde{\beta}}. \tag{69}$$

Note that the weights of Z acting on $(\mathfrak{u}^-)^*$ are the weights of Z acting on $\mathfrak{u}$; in particular, they are positive for the order $\succcurlyeq$. As a consequence, we have

$$(\wedge^\bullet \mathfrak{r}^*)_{\tilde{\beta}} \neq \{0\} \Rightarrow \tilde{\beta} \succcurlyeq 0. \tag{70}$$

Set

$$F^{\preccurlyeq\tilde{\beta}}(\wedge^\bullet \mathfrak{r}^*) = \bigoplus_{\tilde{\alpha}\preccurlyeq\tilde{\beta}} (\wedge^\bullet \mathfrak{r}^*)_{\tilde{\alpha}}. \tag{71}$$

Consider now, as in the formula (65), the diagonal action of L on $\mathfrak{r}$:

$$l.(\xi^- + \xi) = l\xi^- + l\xi, \qquad \forall l \in L,\ \xi^- \in \mathfrak{u}^-,\ \xi \in \mathfrak{u}.$$

Since Z is contained in the center of L, the action (68) of $Z \times \mathbb{C}^*$ and the above action of L commute. In particular the decomposition (69) is L-stable. Set $C = (\wedge^\bullet \mathfrak{r}^*)^L$ and $C_{\tilde{\beta}} = C \cap (\wedge^\bullet \mathfrak{r}^*)_{\tilde{\beta}}$. The $(Z \times \mathbb{C}^*)$-module C decomposes as follows:

$$C := \bigoplus_{\tilde{\beta} \in \tilde{\Gamma}} C_{\tilde{\beta}}. \tag{72}$$

The associated filtration of C is

$$F^{\preccurlyeq \tilde{\beta}} C = F^{\preccurlyeq \tilde{\beta}} (\wedge^\bullet \mathfrak{r}^*) \cap C.$$

5.4.5 Action of L on $\wedge^\bullet (\mathfrak{u}^-)^*$

We now recall results of Kostant in [Kos61] on the action of T on $\wedge^\bullet (\mathfrak{u}^-)^*$.

Theorem 4 (i) *The set of vertices of the convex hull of the weights of T acting on $\wedge^\bullet (\mathfrak{u}^-)^*$ is the set of $\rho - w^{-1}\rho$ where $w \in W^P$.*

These weights are multiplicity free and the eigenline corresponding to $\rho - w^{-1}\rho$ is generated by

$$\phi_w := \phi_{\alpha_1} \wedge \cdots \wedge \phi_{\alpha_p},$$

where $\{\alpha_1, \ldots, \alpha_p\} = \Phi^+ \cap w^{-1}\Phi^-$; and $\phi_{\alpha_i} \in (\mathfrak{u}^-)^*$ is a vector of weight α_i.

(ii) *For any $w \in W^P$, the vector ϕ_w is an highest weight vector for L. Denote by M_w the simple L-module generated by ϕ_w.*

5.4.6 A first differential form

We are now ready to define a first K-invariant differential form on G/P. Set

$$h_w = \mathrm{Id}_w \in M_w \otimes M_w^* \subset \left(\wedge^p (\mathfrak{u}^-)^* \otimes \wedge^p \mathfrak{u}^*\right)^L, \tag{73}$$

where p is the length of w, that is s the codimension of $X_{w^\vee}$. Since Z is central in L, Z acts with weight $(\rho - w^{-1}\rho)_{|Z}$ on M_w. In particular,

$$h_w \in C_{((\rho - w^{-1}\rho)_{|Z}, -p)}. \tag{74}$$

If G/P is cominuscule, then h_w corresponds by the isomorphism (65) to the desired closed differential form representing σ_w. In general, more work is useful.

5.4.7 An Hermitian product on $\mathfrak{r}$

Recall that the Hermitian product $\{\cdot,\cdot\}_{\mathfrak{r}}$ on $\mathfrak{r}$ induces Hermitian inner products on $\wedge^\bullet\mathfrak{r}$ and $\wedge^\bullet\mathfrak{r}^*$ still denoted by $\{\cdot,\cdot\}_{\mathfrak{r}}$.

Lemma 20 *For any nonnegative integer p, the* $(X(Z)\times\mathbb{Z})$*-decomposition* (72) *is* $\{\cdot,\cdot\}_{\mathfrak{r}}$*-orthogonal.*

Proof. It is sufficient to prove that the decomposition

$$\mathfrak{r}=\mathfrak{u}\oplus\bigoplus_{\alpha\in X(Z)}\mathfrak{u}_\alpha^- \tag{75}$$

is $\{\cdot,\cdot\}_{\mathfrak{r}}$-orthogonal. Since $\mathfrak{u}^*=\mathfrak{u}^-$ and the Killing form vanishes on $\mathfrak{u}^-$, $\mathfrak{u}$ and $\mathfrak{u}^-$ are $\{\cdot,\cdot\}_{\mathfrak{r}}$-orthogonal. Now let $\xi\in\mathfrak{u}_\beta^-$ and $\eta\in\mathfrak{u}_{\beta'}^-$ with $\beta\neq\beta'\in X(Z)$. Consider the adjoint action of Z on $\mathfrak{g}$, the induced one on $\mathrm{End}(\mathfrak{g})$ and the corresponding decomposition

$$\mathrm{End}(\mathfrak{g})=\bigoplus_{\alpha\in X(Z)}\mathrm{End}(\mathfrak{g})_\alpha.$$

Note that for any $A\in\mathrm{End}(\mathfrak{g})_\alpha$ with $\alpha\neq 0$, we have $\mathrm{tr}(A)=0$. The endomorphism $\mathrm{Ad}(\eta^*)$ belongs to $\mathrm{End}(\mathfrak{g})_{-\beta'}$. It follows that $\mathrm{Ad}(\eta^*)\circ\mathrm{Ad}(\xi)$ belongs to $\mathrm{End}(\mathfrak{g})_{\beta-\beta'}$ and that $\{\xi,\eta\}=-(\xi,\eta^*)=-\mathrm{tr}(\mathrm{Ad}(\eta^*)\circ\mathrm{Ad}(\xi))=0$. □

5.4.8 Operators on $\wedge^\bullet(\mathfrak{r}^*)$

Recall from the formula (66), the definition of the operator $\mathscr{L}\in\mathrm{End}(\wedge^\bullet\mathfrak{r}^*)$.

Lemma 21 *The operator* $\mathscr{L}$ *stabilizes* $C_{(\alpha,p)}$ *for any integer p and any* $\alpha\in X(Z)$.

Proof. By [Kos63, Proposition 3.4], $b^{0,1}(C_{(\alpha,p)})$ is contained in $C_{(\alpha,p+1)}$. By [Kos63, formula 3.5.3], $\partial^{0,-1}(C_{(\alpha,p+1)})$ is contained in $C_{(\alpha,p)}$. We deduce that $(\partial^{0,-1}\circ b^{0,1})(C_{(\alpha,p)})$ is contained in $C_{(\alpha,p)}$. Similarly, $(b^{0,1}\circ\partial^{0,-1})(C_{(\alpha,p)})$ is contained in $C_{(\alpha,p)}$. We conclude using the formula (67). □

Note that $\mathscr{L}$ is an Hermitian operator. In particular, we have a $\{\cdot,\cdot\}_{\mathfrak{r}}$-orthogonal decomposition $\mathrm{Ker}\mathscr{L}\oplus\mathrm{Im}\mathscr{L}=\wedge^\bullet\mathfrak{r}^*$. Consider the quasi-inverse $\mathscr{L}_0$ of $\mathscr{L}$: $\mathscr{L}_0$ is the Hermitian endomorphism of $\wedge^\bullet\mathfrak{r}^*$ such that $\mathrm{Ker}\mathscr{L}_0=\mathrm{Ker}\mathscr{L}$ and $\mathscr{L}_{0|\mathrm{Im}\mathscr{L}}=(\mathscr{L}_{|\mathrm{Im}\mathscr{L}})^{-1}$.

Let $\pi:\mathfrak{r}\longrightarrow\mathrm{End}(\wedge^\bullet\mathfrak{r}^*)$ be induced by the coadjoint action. Let f_i be eigenvectors in $\mathfrak{u}^-$ for the action of Z that form a basis of $\mathfrak{u}^-$. Let g_j be the basis of $\mathfrak{u}$ defined by the conditions $(f_i,g_j)=\delta_i^j$ (the Kronecker symbol). Set

$$\mathscr{E}:=2\sum_i\pi(g_i)\circ\pi(f_i)\in\mathrm{End}(\wedge^\bullet\mathfrak{r}^*). \tag{76}$$

Kostant defined a third operator

$$\mathscr{R} := -\mathscr{L}_0 \circ \mathscr{E} \in \mathrm{End}(\wedge^\bullet \mathfrak{r}^*); \tag{77}$$

he proved that $\mathscr{R}$ is nilpotent and he defined

$$s_w = (\mathrm{Id} - \mathscr{R})^{-1}(h_w) = h_w + \mathscr{R}(h_w) + \mathscr{R}^2(h_w) + \cdots. \tag{78}$$

Here, we need the following improvement by [Kos63, Lemma 4.6] that proves the nilpotency of $\mathscr{R}$.

Lemma 22 *For any integer p and $\alpha \in X(Z)$, we have*

$$\mathscr{R}\left(C_{(\alpha,p)}\right) \subset \bigoplus_{\beta \prec \alpha} C_{(\beta,p)}.$$

Proof. Lemma 21 asserts that $\mathscr{L}$ stabilizes the $(X(Z)\times\mathbb{Z})$-decomposition of C. Since this decomposition is $\{\cdot,\cdot\}_\mathfrak{r}$-orthogonal by Lemma 20, this implies that $\mathscr{L}_0$ also stabilizes the $\tilde{\Gamma}$-decomposition of C. By the formula (77), it remains to prove that $\mathscr{E}(C_{\alpha,p)}) \subset \bigoplus_{\beta\prec\alpha} C_{(\beta,p)}$.

But each $\pi(f_i)$ vanishes on $\wedge^\bullet \mathfrak{u}^*$ and each $\pi(g_i)$ respects the degree. It follows that $\mathscr{E}(C_{(\alpha,p)}) \subset \bigoplus_{\beta\in X(Z)} C_{(\beta,p)}$. But $\pi(g_i)$ vanishes on $\wedge^\bullet(\mathfrak{u}^-)^*$. Moreover, f_i belongs to $\mathfrak{u}^-$ and has weight $\gamma \preccurlyeq 0$. It follows that $\pi(f_i)(\wedge^\bullet(\mathfrak{u}^-)^*_\beta) \subset \wedge^\bullet(\mathfrak{u}^-)^*_{\beta-\gamma}$. The claim follows. □

5.4.9 Kostant's theorem

Theorem 5 ([Kos63]). *Let $w \in W^P$. The element $s_w \in \wedge^\bullet \mathfrak{r}^*$ defined by (78) is L-invariant. In particular, s_w corresponds by the isomorphism (65) to a K-invariant form ω_w on G/P.*

Then the form ω_w is closed and its class $[\omega_w]$ in $\mathrm{H}^*_{DR}(G/P,\mathbb{C})$ is equal to the Schubert class $\sigma_w^\vee$, up to a positive real scalar.

5.4.10 Application

We can now prove Lemma 19.

Proof (of Lemma 19). By Theorem 5, it is sufficient to prove that ω_w belongs to $F^{\preccurlyeq\tilde{\rho}(w)}\Omega^{l(w)}(G/P,\mathbb{C})$. But ω_w and the filtration are K-invariant on the K-homogeneous space G/P. Hence it is sufficient to prove that s_w belongs to $F^{\preccurlyeq\tilde{\rho}(w)}C$. This is a consequence of the property (74) and Lemma 22. □

Proof (of Proposition 10). Let $\tilde{\beta} \in \tilde{\Gamma}$ and let p be an integer such that $0 \le p \le \dim(G/P)$. Consider $F^{\preccurlyeq\tilde{\beta}}\,\mathrm{H}^{2p}(G/P,\mathbb{C})$. On one hand, Lemma 18 shows that the dimension of $F^{\preccurlyeq\tilde{\beta}}\,\mathrm{H}^p(G/P,\mathbb{C})$ is not more than the cardinality of the set

$$W(\tilde{\beta}, p) = \{w \in W^P : \tilde{\rho}(G/P) - \tilde{\rho}(X_w) \preccurlyeq \tilde{\beta} \text{ and } l(w) = n - p\}.$$

On the other hand, Lemma 19 shows that $F^{\preccurlyeq\tilde{\beta}}\,\mathrm{H}^p(G/P, \mathbb{C})$ contains the classes $\sigma_{w^\vee}$ for w in the set

$$W'(\tilde{\beta}, p) = \{w \in W^P : \tilde{\rho}(X_w) \preccurlyeq \tilde{\beta} \text{ and } l(w) = p\}.$$

But Lemma 17 implies that the Poincaré duality $w \mapsto w^\vee$ induces a bijection between $W(\tilde{\beta}, p)$ and $W'(\tilde{\beta}, p)$. Since the family $(\sigma_{w^\vee})_{w \in W'(\tilde{\beta}, p)}$ is linearly independent the proposition follows. □

Proof (of Corollary 1). The corollary is a direct consequence of Lemma 14 and the above proof of Proposition 10. □

5.5 Proof of Theorem 3

Let u and v be elements of W^P. Consider the following product in the ordinary cohomology ring $\mathrm{H}^*(G/P, \mathbb{C})$

$$\sigma_u \,.\, \sigma_v = \sum_{w \in W^P} c_{uv}^w \sigma_w.$$

By Lemma 19 and Lemma 17, σ_u belongs to $F^{\preccurlyeq\tilde{\rho}(G/P)-\tilde{\rho}(X_u)}\,\mathrm{H}^{l(w_0 w_o^P)-l(u)}(G/P, \mathbb{C})$. Similarly, σ_v belongs to $F^{\preccurlyeq\tilde{\rho}(G/P)-\tilde{\rho}(X_v)}\,\mathrm{H}^{l(w_0 w_o^P)-l(v)}(G/P, \mathbb{C})$. Now Proposition 8 shows that

$$\sigma_u \,.\, \sigma_v \in F^{\preccurlyeq 2\tilde{\rho}(G/P)-\tilde{\rho}(X_u)-\tilde{\rho}(X_v)}\,\mathrm{H}^{2l(w_0 w_o^P)-l(u)-l(v)}(G/P, \mathbb{C}).$$

By Proposition 10, this means that

$$c_{uv}^w \neq 0 \;\Rightarrow\; \tilde{\rho}(G/P) - \tilde{\rho}(X_w) \preccurlyeq 2\tilde{\rho}(G/P) - \tilde{\rho}(X_u) - \tilde{\rho}(X_v), \tag{79}$$

$$\Rightarrow\; \tilde{\rho}(X_u) + \tilde{\rho}(X_v) \preccurlyeq \tilde{\rho}(X_w) + \tilde{\rho}(G/P). \tag{80}$$

Proposition 10 implies also that

$$\overline{\sigma_u} \,.\, \overline{\sigma_v} = \sum_{\substack{w \in W^P \\ \tilde{\rho}(X_u)+\tilde{\rho}(X_v)=\tilde{\rho}(X_w)+\tilde{\rho}(G/P)}} c_{uv}^w \overline{\sigma_w}. \tag{81}$$

On the other hand, Proposition 9 shows that

$$\sigma_u \odot_0 \sigma_v = \sum_{\substack{w \in W^P \\ gd(X_u)+gd(X_v)=gd(X_w)+gd(G/P)}} c_{uv}^w \sigma_w. \tag{82}$$

Comparing the identities (81) and (82), it remains to prove under the assumption $c_{uv}^w \neq 0$ that the equivalence

$$\tilde{\rho}(X_u) + \tilde{\rho}(X_v) = \tilde{\rho}(X_w) + \tilde{\rho}(G/P) \iff gd(X_u) + gd(X_v) = gd(X_w) + gd(G/P)$$

holds.

The implication "$\Leftarrow$" is an immediate consequence of the definition (19) of $\rho(\cdot)$. Conversely, assume that $\tilde{\rho}(X_u) + \tilde{\rho}(X_v) = \tilde{\rho}(X_w) + \tilde{\rho}(G/P)$. Since $c_{uv}^w \neq 0$, the Belkale-Kumar numerical criterion of Levi-movability (see [BK06, Theorem 15]) implies that $\sigma_u \odot_0 \sigma_v \odot_0 \sigma_{w^\vee} = c_{uv}^w[pt]$. In particular, Proposition 9 implies that $gd(X_u) + gd(X_v) = gd(X_w) + gd(G/P)$. The theorem is proved.

5.6 The Belkale-Kumar fundamental class

Recall from Section 4.4 the definition of the Belkale-Kumar fundamental class of any subvariety of G/P. We can now give a simple characterization of this class using the notion of $X(Z)$-dimension.

Proposition 11 *Let Y be an irreducible subvariety of G/P of dimension d. Consider the expansion of its fundamental class in the Schubert basis*

$$[Y] = \sum_{w \in W^P} d_w \sigma_w.$$

Then the expansion of its $\odot_0$-fundamental class in the Schubert basis is

$$[Y]_{\odot_0} = \sum_{\substack{w \in W^P \\ \rho(X_w) = \rho(Y)}} d_w \sigma_w^{\odot_0}.$$

Proof. It remains to prove that for any $[\omega] \in \mathrm{Gr}^{\tilde{\rho}(Y)} \mathrm{H}^{2d}(G/P, \mathbb{C})$,

$$\int_Y \omega = [\omega]_{\odot_0} \odot_0 \Big(\sum_{\substack{w \in W^P \\ \rho(X_w) = \rho(Y)}} d_w \sigma_w \Big).$$

Since the two members of the equality depend linearly on $[\omega]$, it is sufficient to prove it for $[\omega] = \sigma_{u^\vee}$, for any $u \in W^P$ such that $\rho(X_u) = \rho(Y)$ and $l(u) = d$. By ordinary Poincaré duality, this case is equivalent to the following equality

$$\sigma_{u^\vee} . \Big(\sum_{\substack{w \in W^P \\ l(w) = n-d}} d_w \sigma_w \Big) = \sigma_{u^\vee} \odot_0 \Big(\sum_{\substack{w \in W^P \\ \rho(X_w) = \rho(Y)}} d_w \sigma_w \Big).$$

Since the only product $\sigma_{u^\vee} . \sigma_w$ that is nonzero in the above formula is $\sigma_{u^\vee} . \sigma_u$, the proposition follows. □

6 Intersecting Schubert varieties

Given $u, v \in W^P$ such that $v^\vee < u$, we construct in this section a family of varieties containing both the Richardson variety $X_u \cap w_0 X_v$ (up to translation) and the variety Σ_u^v. We prove (see Proposition 13) that Conjecture 4 holds for Σ_u^v if and only if it holds for all these varieties. To end this section, we show that Conjecture 4 is equivalent to a formula using the Kostant harmonic forms that looks like a Fubini formula.

6.1 *Products on* $\mathrm{H}^*(G/P, \mathbb{C})$ *and Bruhat orders*

The Bruhat order on W^P is defined by

$$u < v \iff X_u \subset X_v.$$

This order is generated by $u < v$ if $l(v) = l(u) + 1$ and $v = s_\alpha u$ for some positive root α. The weak Bruhat order on W^P is generated by the relation $u \lessdot v$ if $l(v) = l(u) + 1$ and $v = s_\alpha u$ for some simple root α. The relation between these two orders is

$$u \lessdot v \Rightarrow u < v. \tag{83}$$

A useful characterization of the weak Bruhat order is given by the following result (see [Bou68]).

Lemma 23 *Let u and v in W^P. Then $u \lessdot v$ if and only if $\Phi(u)$ is contained in $\Phi(v)$.*

The following relation between the cup product and the Bruhat order is well known:

$$\sigma_u . \sigma_v \neq 0 \iff v^\vee < u.$$

We have the following relation between the Belkale-Kumar product and the weak Bruhat order.

Lemma 24 *Let u and v in W^P. If $\sigma_u \odot_0 \sigma_v \neq 0$, then $v^\vee \lessdot u$.*

Proof. By assumption, there exists $w \in W^P$ such that (u, v, w) is Levi-movable and $l(u) + l(v) + l(w) = l(w_0 w_0^P)$. Hence, for (l_1, l_2, l_3) in a nonempty open subset of L^3,

$$l_1 T_u \cap l_2 T_v \cap l_3 T_w = \{0\}.$$

In particular, $l_1 T_u + l_2 T_v = T_{P/P} G/P$. Since $\Delta L.(B, w_0^P B_L)$ is open in L^2, there exist $l \in L$, $b_1, b_2 \in B_L$ such that $l b_1 T_u + l w_0^P b_2 T_v = T_{P/P} G/P$. But T_u and T_v are B_L-stable and $T_{P/P} G/P$ is L-stable, hence

$$T_u + w_0^P T_v = T_{P/P} G/P.$$

It follows that $\Phi(u) \cup w_0^P \Phi(v) = \Phi(G/P)$. But $\Phi(v^\vee) = \Phi(G/P) - w_0^P \Phi(v)$. Hence $\Phi(v^\vee) \subset \Phi(u)$ and $v^\vee \lessdot u$. □

Remark. The converse of the implication of Lemma 24 does not hold. Indeed consider $\mathrm{SL}_3(\mathbb{C})$ with its usual maximal torus and Borel subgroup B. Denote the two simple reflections of W by s_1 and s_2. Then $\sigma_{s_1 s_2} \odot_0 \sigma_{s_2 s_1} = 0$ while $(s_2 s_1)^\vee = s_2 \lessdot s_1 s_2$.

6.2 Similar to Richardson's varieties

Let $u, v \in W^P$. The Richardson variety X_u^v is defined by

$$X_u^v = X_u \cap w_0 X_v.$$

It is well known that X_u^v is irreducible, normal and satisfies $[X_u^v] = \sigma_u . \sigma_v$. In particular, X_u^v is empty if and only if $v^\vee < u$.

Assume now that $v^\vee \lessdot u$. Choose $y \in W^P$ such that $v^\vee \lessdot y \lessdot u$. Consider the intersection

$$I_u^v(y) := y^{-1} X_u \cap w_0^P v^{-1} . X_v. \tag{84}$$

The first example $I_u^v(v^\vee) = (v^\vee)^{-1} X_u^v$ is just a translated Richardson variety.

By the relation (83), the point yP/P belongs to X_u. Then $P/P \in y^{-1} X_u$. Since $vP/P \in X_v$, P/P is a point in $w_0^P v^{-1} . X_v$. It follows that

$$P/P \in I_u^v(y). \tag{85}$$

The following lemma shows that the variety $I_u^v(y)$ contains a translated Richardson variety.

Lemma 25 *Let u, v, and y in W^P be such that $v^\vee \lessdot y \lessdot u$. Then $I_u^{y^\vee}(y)$ is contained in $I_u^v(y)$.*

Proof. It remains to prove that $y^{-1} X_u \cap w_0^P (y^\vee)^{-1} . X_{y^\vee}$ is contained in $y^{-1} X_u \cap w_0^P v^{-1} . X_v$. It is sufficient to prove that $(y^\vee)^{-1} . X_{y^\vee}$ is contained in $v^{-1} . X_v$. But $(y^\vee)^{-1} . X_{y^\vee} = \overline{((y^\vee)^{-1} B y^\vee) . P/P}$ and $v^{-1} . X_v = \overline{(v^{-1} B v) . P/P}$. Hence it is sufficient to prove that $\Phi(\mathfrak{g}/\mathfrak{p}, T) \cap (y^\vee)^{-1} \Phi^+$ is contained in $\Phi(\mathfrak{g}/\mathfrak{p}, T) \cap v^{-1} \Phi^+$. But $v^\vee \lessdot y$ and hence $y^\vee \lessdot v$. With Lemma 23, the proof is concluded. □

The fact that X_u and X_v are B-stable implies that the group $H_u^v(y) := y^{-1} B y \cap w_0^P v^{-1} B v w_0^P$ acts on $I_u^v(y)$. Set $y' = y (v^\vee)^{-1}$ in such a way that $y = y' v^\vee$. Note that $y w_0^P v^{-1} = y' w_0$ and that

$$H_u^v(y) = (v^\vee)^{-1} (y'^{-1} B y' \cap B^-) v^\vee. \tag{86}$$

The group $H_u^v(y)$ is a connected subgroup of G, containing T and acting on $I_u^v(y)$. Consider now the group $U(y') = y'^{-1}Uy' \cap U^-$.

Let $\overset{\circ}{G/P} = B^-P/P$ denote the open T-stable affine cell containing P/P. Set $\overset{\circ}{I_u^v}(y) = \overset{\circ}{G/P} \cap I_u^v(y)$; it is an open T-stable affine neighborhood of P/P in $I_u^v(y)$. The following statement describes the geometry of this neighborhood.

Theorem 6 *Let u, v, and y in W^P be such that $v^\vee \lessdot y \lessdot u$. Then the following morphism*

$$\begin{aligned} \Psi:\ U(y') \times \overset{\circ}{I_u^{y^\vee}}(y) &\longrightarrow \overset{\circ}{I_u^v}(y) \\ (u,x) &\longmapsto (v^\vee)^{-1}uv^\vee.x \end{aligned}$$

is an isomorphism.

Proof. The weights of T acting on the Lie algebra of the group $U(y) = U^- \cap y^{-1}Uy$ are $\Phi(y) = \Phi^- \cap y^{-1}\Phi^+$. The weights of T acting on the tangent space at the point P/P of the variety $w_0^P(y^\vee)^{-1}X_{y^\vee}$ are $\Phi(\mathfrak{g}/\mathfrak{p}, T) \cap y^{-1}\Phi^-$. But $\overset{\circ}{G/P}$ is isomorphic as a T-variety to the affine space $\mathfrak{g}/\mathfrak{p}$. It follows that the map

$$\begin{aligned} U(y) \times [w_0^P(y^\vee)^{-1}X_{y^\vee} \cap \overset{\circ}{G/P}] &\longrightarrow \overset{\circ}{G/P} \\ (u,x) &\longmapsto ux \end{aligned} \tag{87}$$

is an isomorphism. The variety $y^{-1}X_u$ is stable by $y^{-1}By$, and so by $U(y)$. It follows that the map

$$\begin{aligned} U(y) \times [w_0^P(y^\vee)^{-1}X_{y^\vee} \cap \overset{\circ}{G/P} \cap y^{-1}X_u] &\longrightarrow \overset{\circ}{G/P} \cap y^{-1}X_u \\ (u,x) &\longmapsto ux \end{aligned}$$

is an isomorphism.

Since $v^\vee \lessdot y$ and $y = y'v^\vee$, the set $\Phi(y)$ is the disjoint union of $\Phi(v^\vee)$ and $(v^\vee).\Phi(y')$ (see for example [Bou02]). Then the map

$$\begin{aligned} U(y') \times U(v^\vee) &\longrightarrow U(y) \\ (u',u) &\longmapsto (v^\vee)^{-1}u'v^\vee u \end{aligned}$$

is an isomorphism. Note that in the above expression, we have fixed the representative (still denoted by $v^\vee$) of $v^\vee$ in the normalizer of the torus T. Composing these isomorphisms gives the following:

$$\begin{aligned} U(y') \times U(v^\vee) \times \overset{\circ}{I_u^{y^\vee}}(y) &\longrightarrow \overset{\circ}{G/P} \cap y^{-1}X_u \\ (u',u,x) &\longmapsto (v^\vee)^{-1}u'v^\vee ux. \end{aligned}$$

Since $\Phi(y')$ is contained in $(v^\vee)^{-1}\Phi^-$, and $w_0^P v^{-1}X_v = \overline{((v^\vee)^{-1}B^- v^\vee).P/P}$, the variety $w_0^P v^{-1}X_v$ is stable under the action of $U(y')$. Hence

$$U(y') \times \left[(U(v^\vee)\cdot \mathring{I}_u^{y^\vee}(y)) \cap w_0^P v^{-1}X_v\right] \longrightarrow \mathring{I}_u^v(y)$$
$$(u',x) \longmapsto (v^\vee)^{-1}u'v^\vee x$$

is an isomorphism. It remains to prove that

$$(U(v^\vee)\cdot \mathring{I}_u^{y^\vee}(y)) \cap w_0^P v^{-1}X_v = \mathring{I}_u^{y^\vee}(y).$$

Let $u \in U(v^\vee)$ and $x \in \mathring{I}_u^{y^\vee}(y)$ be such that ux belongs to $w_0^P v^{-1}X_v$. It is sufficient to prove that $u = e$. Replacing $y^\vee$ by v in the morphism (87), we get an isomorphism

$$\Theta:\ U(v^\vee) \times \left[w_0^P v^{-1}X_v \cap \mathring{G/P}\right] \longrightarrow \mathring{G/P}$$
$$(u',x') \longmapsto u'x'.$$

One can easily check that x belongs to $w_0^P v^{-1}X_v \cap \mathring{G/P}$ and that $\Theta(u,x) = \Theta(e,ux)$. Now, the injectivity of Θ implies that $u = e$. □

An important consequence of Theorem 6 for our purpose is the following statement.

Corollary 2 *The variety $I_u^v(y)$ is normal at the point P/P. In particular, there exists an unique irreducible component $\Sigma_u^v(y)$ of $I_u^v(y)$ which contains P/P.*

Proof. The corollary follows from the theorem and the fact that the Richardson varieties are irreducible and normal (see [KWY13] for a short proof). □

If $y = v^\vee$, then Theorem 6 is trivial. In the opposite situation when $y = u$, it implies the following result.

Corollary 3 *Let u and v in W^P be such that $v^\vee \lessdot u$. The orbit $H_u^v(u).P/P$ is open in $I_u^v(u)$. In other words, $\Sigma_u^v(u)$ is the closure of $H_u^v(u).P/P$.*

Proof. If $y = u$, then the translated Richardson variety $I_u^{y^\vee}(y) = I_u^{u^\vee}(u)$ is reduced to the point P/P. The corollary follows immediately. □

6.3 A conjecture

Here comes our main conjecture.

Conjecture 4 *Let $u, v \in W^P$ such that $v^\vee \preccurlyeq u$. Then*

$$[\Sigma_u^v(u)]_{\odot_0} = \sigma_u^{\odot_0} \odot_0 \sigma_v^{\odot_0}.$$

Some observations about this conjecture are noted in the following propositions.

Proposition 12 *Expand $[\Sigma_u^v(u)]_{\odot_0}$ and $\sigma_u \odot_0 \sigma_v$ in the Schubert basis:*

$$[\Sigma_u^v(u)]_{\odot_0} = \sum_{w\in W^P} d_{uv}^w \sigma_w^{\odot_0}, \text{ and}$$

$$\sigma_u^{\odot_0} \odot_0 \sigma_v^{\odot_0} = \sum_{w\in W^P} \tilde{c}_{uv}^w \sigma_w^{\odot_0}.$$

Then, for any $w \in W^P$,

(i) $\tilde{c}_{uv}^w \geq d_{uv}^w$;
(ii) $\tilde{c}_{uv}^w \neq 0 \iff d_{uv}^w \neq 0$.

Proof. Write $[\Sigma_u^v(u)] = \sum_{w\in W^P} e_{uv}^w \sigma_w$ and $\sigma_u . \sigma_v = \sum_{w\in W^P} c_{uv}^w \sigma_w$ in ordinary cohomology. Since $\Sigma_u^v(u)$ is an irreducible component of the intersection $I_u^v(u)$ and this intersection is proper along this component, the inequality

$$c_{uv}^w \geq e_{uv}^w \tag{88}$$

holds for any $w \in W^P$. Consider now a coefficient d_{uv}^w for some fixed $w \in W^P$. If $d_{uv}^w = 0$, then the first assertion of the proposition is obvious. Assume $d_{uv}^w \neq 0$. By Proposition 11, $d_{uv}^w = e_{uv}^w$. Comparing the inequality (88) and the first assertion, one observes that it is sufficient to prove that $\tilde{c}_{uv}^w = c_{uv}^w$, namely, that $\tilde{c}_{uv}^w \neq 0$.

Since $d_{uv}^w \neq 0$, Proposition 11 implies that $\rho(X_w) = \rho(\Sigma_u^v(u))$. Since P/P belongs to the open $H_u^v(u)$-orbit in $\Sigma_u^v(u)$, P/P is an $X(Z)$-regular point of Σ_u^v. In particular

$$\rho(\Sigma_u^v(u)) = \sum_{\gamma\in X(Z)} \dim[(T_{P/P}\Sigma_u^v(u))_\gamma]\gamma,$$

where $(T_{P/P}\Sigma_u^v(u))_\gamma$ is the weight space of weight γ of the Z-module $T_{P/P}\Sigma_u^v(u)$. But $T_{P/P}\Sigma_u^v(u)$ is the transverse intersection of $T_{P/P}u^{-1}X_u$ and $T_{P/P}w_0^P v^{-1}X_v$. It follows that $\rho(\Sigma_u^v(u)) = \rho(u^{-1}X_u) + \rho(w_0^P v^{-1}X_v) = \rho(X_u) + \rho(X_v)$. Finally $\rho(X_w) = \rho(X_u) + \rho(X_v)$ and Proposition 11 shows that $\tilde{c}_{uv}^w = c_{uv}^w$.

Assuming that $d_{uv}^w \neq 0$, the first assertion implies that $\tilde{c}_{uv}^w \neq 0$. Assume conversely that $\tilde{c}_{uv}^w \neq 0$ in other words that $(u, v, w^\vee)$ is Levi-movable. Arguing as in the proof of Lemma 24, one can check that there exists $l \in L$ such that $u^{-1}X_u$, $w_0^P v^{-1}X_v$ and $l(w^\vee)^{-1}X_{w^\vee}$ intersect transversally at P/P. It follows immediately that $\Sigma_u^v(u)$ and $l(w^\vee)^{-1}X_{w^\vee}$ intersect transversally at P/P. Hence $e_{uv}^w \neq 0$.

It remains to prove that $e_{uv}^w = d_{uv}^w$. The condition $\tilde{c}_{uv}^w \neq 0$ in the $X(Z)$-graded algebra $\operatorname{Gr} \mathrm{H}^*(G/P, \mathbb{C})$ implies that $\rho(X_w) = \rho(X_u) + \rho(X_v)$. But $\rho(X_u) + \rho(X_v) = \rho(\Sigma_u^v(u))$. Proposition 11 shows that $e_{uv}^w = d_{uv}^w$. □

Proposition 13 *Let $u, v \in W^P$ such that $v^\vee \preccurlyeq u$.*

(i) *Conjecture* 4 *holds if* $\Sigma_u^v(u)$ *has dimension* 0, 1 *or* 2.
(ii) *Conjecture* 4 *holds if and only if for any* $y \in W^P$, *such that* $v^\vee \lessdot y \lessdot u$, *we have* $[\Sigma_u^v(y)]_{\odot_0} = \sigma_u \odot_0 \sigma_v$.

Proof. If $\Sigma_u^v(u)$ has dimension 0, then $u = v^\vee$. In particular $[\Sigma_u^v(y)]_{\odot_0} = [pt] = \sigma_u \odot_0 \sigma_v$.

If $\Sigma_u^v(u)$ has dimension 1, then $u = s_\alpha v^\vee$, for some simple root α. Moreover, $l(u) = l(v^\vee) + 1$. This implies that X_u is stable by the action of P_α (the minimal parabolic subgroup associated to α). In particular $s_\alpha X_u = X_u$. It follows that $u^{-1}X_u = u^{-1}s_\alpha X_u = (v^\vee)^{-1}X_u$. In particular $I_u^v(u) = I_u^v(v^\vee)$ is a translated Richardson variety and is irreducible. Moreover, $\sigma_u . \sigma_v = [I_u^v(u)] = [\Sigma_u^v(u)]$. Proposition 11 implies that $\sigma_u \odot_0 \sigma_v = [\Sigma_u^v(u)]_{\odot_0}$.

Assume now that $u = s_\alpha s_\beta v^\vee$ for some simple roots α and β, such that $l(u) = l(v^\vee) + 2$. Then (note that $s_\alpha X_u = X_u$)

$$\begin{aligned} I_u^v(u) &= u^{-1}X_u \cap w_0^P v^{-1} X_v \\ &= (v^\vee)^{-1}(s_\beta s_\alpha X_u \cap w_0 X_v) \\ &= (v^\vee)^{-1} s_\beta (s_\alpha X_u \cap w_0 s_{\beta^*} X_v), \end{aligned}$$

where $\beta^* = -w_0\beta$. But the condition $v^\vee \lessdot s_\beta v^\vee$ implies that $s_{\beta^*} v \lessdot v$ (see for example Lemma 23). Then $s_{\beta^*} X_v = X_v$ and $I_u^v(u)$ is obtained by translation from the Richardson variety $s_\alpha X_u \cap w_0 s_{\beta^*} X_v$. The first assertion of the proposition follows.

Let α be a simple root such that $y \lessdot s_\alpha y \lessdot u$. Set $\beta = -y^{-1}\alpha$ and set $U_\beta : \mathbb{C} \longrightarrow G$, the associated additive one-parameter subgroup. Consider the flat limit $\lim_{t\to\infty} U_\beta(t)\Sigma_u^v(y)$. Since $U_\beta(t)y^{-1}B/B$ tends to $y^{-1}s_\alpha B/B$ when t goes to infinity, $\lim_{t\to\infty} U_\beta(t)y^{-1}X_u = y^{-1}s_\alpha X_u$. Since $v^\vee \lessdot y \lessdot s_\alpha y$, $\beta \in \Phi(s_\alpha y) - \Phi(v^\vee)$ and $w_0^P\beta \in \Phi(v)$. In particular, $w_0^P v^{-1} X_v$ is U_β-stable. But $\Sigma_u^v(s_\alpha y)$ is an irreducible component of the intersection $y^{-1}s_\alpha X_u \cap w_0^P v^{-1} X_v$, and this intersection is transverse along this component. It follows that $\Sigma_u^v(s_\alpha y)$ is an irreducible component of $\lim_{t\to\infty} U_\beta(t)\Sigma_u^v(y)$. Writing

$$[\Sigma_u^v(y)] = \sum_{w\in W^P} d'_w \sigma_w \quad \text{and} \quad [\Sigma_u^v(s_\alpha y)] = \sum_{w\in W^P} d''_w \sigma_w,$$

we deduce that

$$d''_w \leq d'_w \qquad \forall w \in W^P. \tag{89}$$

Write now

$$[\Sigma_u^v(v^\vee)] = \sum_{w\in W^P} d_w \sigma_w \quad \text{and} \quad [\Sigma_u^v(u)] = \sum_{w\in W^P} e_w \sigma_w.$$

Since $\Sigma_u^v(v^\vee)$ is a translation of the Richardson variety $X_u \cap w_0 X_v$,

$$\sigma_u . \sigma_v = \sum_{w\in W^P} d_w \sigma_w.$$

By an immediate induction, we deduce from (89) that

$$e_w \leq d'_w \leq d_w \qquad \forall w \in W^P.$$

Conjecture 4 holds for $y = u$ if and only if for any $w \in W^P$ such that (u, v, w) is Levi-movable $e_w = d_w$. Then, $d'_w = d_w$ for any such $w \in W^P$ and $[\Sigma_u^v(y)]_{\odot_0} = \sigma_u \odot_0 \sigma_v$. □

6.4 Interpretation in terms of harmonic forms

Kostant's harmonic forms allow us to formulate Conjecture 4 as an identity of integrals.

Proposition 14 *Let u and v in W^P be such that $v^\vee \lessdot u$. Then $\sigma_u \odot_0 \sigma_v = [\Sigma_u^v(u)]_{\odot_0}$ if and only if for any w in W^P, such that (u, v, w) is Levi-movable, we have*

$$\int_{(u^\vee)^{-1}X_{u^\vee}} \omega_{u^\vee} \, . \int_{(v^\vee)^{-1}X_{v^\vee}} \omega_{v^\vee} \, . \int_{\Sigma_u^v(u)} \omega_{w^\vee} = \int_{(P^u)^-} \omega_{u^\vee} \wedge \omega_{v^\vee} \wedge \omega_{w^\vee}.$$

Proof. For any $w \in W^P$, consider the Kostant's harmonic form ω_w and the nonzero complex number λ_w (see Theorem 5) such that

$$[\omega_w] = \lambda_w \sigma_w^\vee. \tag{90}$$

Then

$$\lambda_w = \int_{w^{-1}X_w} \omega_w. \tag{91}$$

By Propositions 11 and 12, Conjecture 4 is equivalent to the fact that for any $w \in W^P$ such that (u, v, w) is Levi-movable, we have

$$\sigma_u \, . \, \sigma_v \, . \, \sigma_w = [\Sigma_u^v(u)] \, . \, \sigma_w. \tag{92}$$

But on the one hand

$$\sigma_u \, . \, \sigma_v \, . \, \sigma_w = \frac{1}{\lambda_{u^\vee} \lambda_{v^\vee}} [\omega_{u^\vee} \wedge \omega_{v^\vee}] \, . \, \sigma_w = \frac{\int_{w^{-1}X(w)} \omega_{u^\vee} \wedge \omega_{v^\vee}}{\lambda_{u^\vee} \lambda_{v^\vee}}. \tag{93}$$

And on the other hand

$$[\Sigma_u^v(u)] \, . \, \sigma_w = \frac{\int_{\Sigma_u^v(u)} \omega_{w^\vee}}{\lambda_{w^\vee}}. \tag{94}$$

In particular the equality (92) is equivalent to

$$\lambda_{w^\vee}.\int_{w^{-1}X_w}\omega_{u^\vee}\wedge\omega_{v^\vee}=\lambda_{u^\vee}.\lambda_{v^\vee}.\int_{\Sigma_u^v(u)}\omega_{w^\vee}; \tag{95}$$

which is, by (91), equivalent to

$$\lambda_{w^\vee}.\int_{w^{-1}X_w}\omega_{u^\vee}\wedge\omega_{v^\vee}=\int_{(u^\vee)^{-1}X_{u^\vee}}\omega_{u^\vee}.\int_{(v^\vee)^{-1}X_{v^\vee}}\omega_{v^\vee}.\int_{\Sigma_u^v(u)}\omega_{w^\vee}. \tag{96}$$

We claim that

$$\lambda_{w^\vee}.\int_{w^{-1}X_w}\omega_{u^\vee}\wedge\omega_{v^\vee}=\int_{(P^u)^-}\omega_{u^\vee}\wedge\omega_{v^\vee}\wedge\omega_{w^\vee}. \tag{97}$$

Let d be the positive integer such that $\sigma_u.\sigma_v.\sigma_w=d[pt]$. We have

$$d=\int_{G/P}\frac{\omega_{u^\vee}\wedge\omega_{v^\vee}\wedge\omega_{w^\vee}}{\lambda_{u^\vee}\lambda_{v^\vee}\lambda_{w^\vee}}.$$

Since $\sigma_u.\sigma_v=d\sigma_{w^\vee}$, we also have

$$d=\int_{w^{-1}X(w)}\frac{\omega_{u^\vee}\wedge\omega_{v^\vee}}{\lambda_{u^\vee}\lambda_{v^\vee}}.$$

Claim (97) is obtained by identifying these two expressions of d.

The proposition follows now from the equations (97) and (96). □

Remark. Observe that $(P^u)^-$ is isomorphic to the product of the three T-stable affine neighborhoods of P/P in $(u^\vee)^{-1}X_{u^\vee}$, $(v^\vee)^{-1}X_{v^\vee}$ and $\Sigma_u^v(u)$. With this observation the equality of Proposition 14 looks like a Fubini formula.

7 The case of the complete flag varieties

Given u in W, set $\Phi(u)^c:=\Phi^- - \Phi(u)$. Let u,v, and w in W. For the complete flag variety G/B, the Levi-movability is easy to understand. Indeed T_u, T_v, and T_w are $L=T$-stable. In particular, $(\sigma_u,\sigma_v,\sigma_w)$ is Levi-movable if and only if the natural map $T_{B/B}(G/B)\longrightarrow\frac{T_{B/B}(G/B)}{T_u}\oplus\frac{T_{B/B}(G/B)}{T_v}\oplus\frac{T_{B/B}(G/B)}{T_w}$ is an isomorphism. This is equivalent to the fact that Φ^- is the disjoint union of $\Phi(u)^c$, $\Phi(v)^c$, and $\Phi(w)^c$. Since $\Phi(w)^c=\Phi(w^\vee)$, one gets the following equivalence:

$$\tilde{c}_{uv}^w\neq 0 \iff \Phi(w)^c=\Phi(u)^c\sqcup\Phi(v)^c.$$

Conjecture 4 generalizes a classical one on G/B.

Proposition 15 *Let G be a semisimple group and consider the Belkale-Kumar cohomology of G/B. Let u and v belong to W. Then $\sigma_u\odot_0\sigma_v=[\Sigma_u^v(u)]_{\odot_0}$ if and only if $\sigma_u\odot_0\sigma_v$ is either equal to zero or to σ_w for some $w\in W$.*

Proof. Assume that $\sigma_u \odot_0 \sigma_v = [\Sigma_u^v(u)]_{\odot_0}$.

Case 1. Suppose there exists $w \in W$ such that $\Phi(w) = \Phi(H_u^v(u))$. Then (see for example Lemma 11) $\Sigma_u^v(u) = w^{-1}X_w$; hence $[\Sigma_u^v(u)]_{\odot_0} = \sigma_w$. In particular $\sigma_u \odot_0 \sigma_v = \sigma_w$.

Case 2. Suppose there exists no $w \in W$ such that $\Phi(w) = \Phi(H_u^v(u))$. Since $\Phi(H_u^v(u)) = \Phi(u) \cap \Phi(v)$, there is no $w \in W$ such that $\Phi(w)^c = \Phi(u)^c \sqcup \Phi(v)^c$. Hence there is no $w \in W$, such that $(\sigma_u, \sigma_v, \sigma_{w^\vee})$ is Levi-movable. Then $\sigma_u \odot_0 \sigma_v = 0$. Moreover Proposition 10 implies that $\mathrm{Gr}^{\tilde{\rho}(G/P)-\tilde{\rho}(\Sigma_u^v(u))}\mathrm{H}^*(G/P,\mathbb{C}) = \{0\}$. In particular, $[\Sigma_u^v(u)]_{\odot_0} = 0$.

Assume now that $\sigma_u \odot_0 \sigma_v = \sigma_w$ for some $w \in W$. Since $\Phi(w)^c = \Phi(u)^c \sqcup \Phi(v)^c$, Lemma 11 shows that $w^{-1}X_w = \Sigma_u^v(u)$. Hence $\sigma_u \odot_0 \sigma_v = [\Sigma_u^v(u)]_{\odot_0}$.

Assume finally that $\sigma_u \odot_0 \sigma_v = 0$. It remains to prove that $[\Sigma_u^v(u)]_{\odot_0} = 0$. Since $\Phi(H_u^v(u)) = \Phi(u) \cap \Phi(v)$, $[\Sigma_u^v(u)]_{\odot_0}$ belongs to $\mathrm{Gr}^{\rho(X_u)+\rho(X_v)}\mathrm{H}^*(G/B,\mathbb{C})$. If there is no w in W such that $\rho(X_w) = \rho(X_u) + \rho(X_v)$, then Proposition 10 shows that $\mathrm{Gr}^{\rho(X_u)+\rho(X_v)}\mathrm{H}^*(G/B,\mathbb{C}) = \{0\}$. In particular $[\Sigma_u^v(u)]_{\odot_0} = 0$. Assume now that there exists w in W, such that $\rho(X_w) = \rho(X_u) + \rho(X_v)$. Then $[\Sigma_u^v] = d\sigma_w + \cdots$ for some integer d. If $d = 0$ there is nothing to prove. If $d \neq 0$, then $\sigma_u . \sigma_v = e\sigma_w + \cdots$ for some integer $e \geq d$. The numerical criterion [BK06, Theorem 15] shows that $\sigma_u \odot_0 \sigma_v = e\sigma_w$. This contradicts the assumption $\sigma_u \odot_0 \sigma_v = 0$. $\square$

Proposition 15 shows that for G/B, Conjecture 4 is equivalent to the following.

Conjecture 5 *Let u, v, and w in W such that $\Phi(w)^c = \Phi(u)^c \sqcup \Phi(v)^c$. Then*

$$\sigma_u \odot_0 \sigma_v = \sigma_w$$

in $\mathrm{H}^*(G/B,\mathbb{C})$.

Conjecture 5 was stated by Dimitrov and Roth in [DR09]. If $G = \mathrm{SL}_n(\mathbb{C})$ then Conjecture 5 was proved by Richmond in [Ric09]. If $G = \mathrm{Sp}_{2n}(\mathbb{C})$ then Conjecture 5 was proved independently in [Ric12] and [Res11b]. Dimitrov and Roth have a proof for each simple classical G, but it is not published. Here we include a proof for the group $\mathrm{SO}_{2n+1}(\mathbb{C})$.

Proposition 16 *Conjecture 5 holds for the group* $\mathrm{SO}_{2n+1}(\mathbb{C})$.

Proof. Let V be a $(2n+1)$-dimensional complex vector space and let $\mathscr{B} = (x_1, \ldots, x_{2n+1})$ be a basis of V^*. Let G be the special orthogonal group associated to the quadratic form $Q = x_{n+1}^2 + \sum_{i=1}^n x_i x_{2n+2-i}$. Consider the maximal torus $T = \{\mathrm{diag}(t_1, \ldots, t_n, 1, t_n^{-1}, \ldots, t_1^{-1}) : t_i \in \mathbb{C}^*\}$ of G. Let B be the Borel subgroup of G consisting of upper triangular matrices in the dual base of $\mathscr{B}$. Consider W, Φ, Φ^+ associated to $T \subset B \subset G$.

Let u, v, and w in W such that $\sigma_u \odot_0 \sigma_v \odot_0 \sigma_w = d[pt]$ for some positive integer d. It remains to prove that $d = 1$. The Levi-movability implies that $\Phi^- = \Phi(u)^c \sqcup \Phi(v)^c \sqcup \Phi(w)^c$.

Consider the linear group $\hat{G} = \mathrm{GL}(V)$. Let $\hat{T}$ denote the subgroup of $\hat{G}$ consisting of diagonal matrices and let $\hat{B}$ denote the subgroup of $\hat{G}$ consisting of upper triangular matrices in $\hat{G}$. Consider $\hat{W}$, $\hat{\Phi}$, $\hat{\Phi}^+$ associated to $\hat{T} \subset \hat{B} \subset \hat{G}$. Since T is a regular torus in $\hat{G}$, the group W identifies with a subgroup of $\hat{W}$. In particular, u, v, and w belong to $\hat{W}$. One can easily check that the similar property of Φ^- implies that $\hat{\Phi}^- = \hat{\Phi}(u)^c \sqcup \hat{\Phi}(v)^c \sqcup \hat{\Phi}(w)^c$. Consider now the three Schubert varieties $\hat{X}_u$, $\hat{X}_v$, and $\hat{X}_w$ in $\hat{G}/\hat{B}$. The fact that Conjecture 5 holds for $\hat{G}$ implies that

$$u^{-1}\hat{X}_u \cap v^{-1}\hat{X}_v \cap w^{-1}\hat{X}_w = \{\hat{B}/\hat{B}\}. \tag{98}$$

Consider now the inclusion $G/B \subset \hat{G}/\hat{B}$. Then X_u is contained in $\hat{X}_u$ (and similar inclusions hold for v and w). In particular, the condition (98) implies that

$$u^{-1}X_u \cap v^{-1}X_v \cap w^{-1}X_w = \{B/B\}. \tag{99}$$

Moreover, the condition on Φ^- implies that the intersection in (99) is transverse. It follows that $d = 1$. □

Proposition 17 *Conjecture 5 holds for the groups of type F_4 and E_6.*

Proof. For $w \in W$, set

$$p(w) = \prod_{\alpha \in \Phi^+ \cap w\Phi^+} (\rho, \alpha),$$

where $(\cdot, \cdot)$ is a W-invariant scalar product and ρ is the half sum of the positive roots. Let u, v, and w in W be such that $\Phi(w)^c = \Phi(u)^c \sqcup \Phi(v)^c$. By [BK06, Corollary 44],

$$\sigma_u \odot_0 \sigma_v = \frac{p(u).p(v)}{p(w)} \sigma_w$$

in $\mathrm{H}^*(G/B, \mathbb{C})$. To prove the proposition, it is sufficient to check that $p(w) = p(u).p(v)$. This is checked by a Sage program (see [Res13]). For example, in type F_4, if

$$u^\vee = s_3s_2s_3s_2, \quad v^\vee = s_1s_2s_3s_4s_2s_3s_1s_2s_3s_4 \quad \text{and}$$
$$w^\vee = s_1s_2s_3s_4s_2s_3s_1s_2s_3s_4s_3s_2s_3s_2,$$

then

$$p(u) = \frac{3}{2} \quad p(v) = 113400 \quad p(w) = 170100.$$

And, in type E_6, if

$$u = s_6s_5s_4s_3s_2s_4s_5s_6s_5s_3 \quad v = s_4s_3s_2s_4s_5s_4s_3s_2s_4s_2$$
$$w = s_6s_5s_4s_3s_2s_4s_5s_6s_5s_4s_3s_2s_4s_5s_4s_3s_2s_4s_3s_2,$$

then

$$p(u) = 20160 \quad p(v) = 4320 \quad p(w) = 87091200.$$

□

References

[ABS90] H. Azad, M. Barry, and G. Seitz, *On the structure of parabolic subgroups*, Com. in Algebra **18** (1990), no. 2, 551–562.

[BK06] Prakash Belkale and Shrawan Kumar, *Eigenvalue problem and a new product in cohomology of flag varieties*, Invent. Math. **166** (2006), no. 1, 185–228.

[BKR12] Prakash Belkale, Shrawan Kumar, and Nicolas Ressayre, *A generalization of Fulton's conjecture for arbitrary groups*, Math. Ann. **354** (2012), no. 2, 401–425.

[Bou68] Nicolas Bourbaki, *Éléments de mathématique. Fasc. XXXIV. Groupes et algèbres de Lie. Chapitre VI: systèmes de racines*, Hermann, Paris, 1968.

[Bou02] ———, *Lie groups and Lie algebras. Chapters 4–6*, Elements of Mathematics (Berlin), Springer-Verlag, Berlin, 2002, Translated from the 1968 French original by Andrew Pressley.

[BP00] M. Brion and P. Polo, *Large Schubert varieties*, Represent. Theory **4** (2000), 97–126.

[CK03] James B. Carrell and Jochen Kuttler, *Smooth points of T-stable varieties in G/B and the Peterson map.*, Invent. Math. **151** (2003), no. 2, 353–379.

[DR09] Ivan Dimitrov and Mike Roth, *Geometric realization of PRV components and the Littlewood-Richardson cone*, 2009, pp. 1–13.

[DR17] I. Dimitrov and M. Roth, *Intersection multiplicity one for classical groups*, ArXiv e-prints (2017), 1–13.

[Kos61] Bertram Kostant, *Lie algebra cohomology and the generalized Borel-Weil theorem*, Ann. of Math. (2) **74** (1961), 329–387.

[Kos63] ———, *Lie algebra cohomology and generalized Schubert cells*, Ann. of Math. (2) **77** (1963), 72–144.

[Kum02] Shrawan Kumar, *Kac-Moody groups, their flag varieties and representation theory*, Progress in Mathematics, Vol. 204, Birkhäuser Boston Inc., Boston, MA, 2002.

[KWY13] Allen Knutson, Alexander Woo, and Alexander Yong, *Singularities of Richardson varieties*, Math. Res. Lett. **20** (2013), no. 2, 391–400.

[Res10] Nicolas Ressayre, *Geometric invariant theory and generalized eigenvalue problem*, Invent. Math. **180** (2010), 389–441.

[Res11a] ———, *Geometric Invariant Theory and generalized eigenvalue problem II*, Ann. Inst. Fourier (Grenoble) **61** (2011), no. 4 (2012), 1467–1491.

[Res11b] ———, *Multiplicative formulas in Schubert calculus and quiver representation*, Indag. Math. (N.S.) **22** (2011), no. 1–2, 87–102.

[Res12] ———, *A cohomology-free description of eigencones in types A, B, and C*, Int. Math. Res. Not. IMRN (2012), no. 21, 4966–5005.

[Res13] ———, *Homepage*, October 2013.

[Ric09] Edward Richmond, *A partial Horn recursion in the cohomology of flag varieties*, J. Algebraic Combin. **30** (2009), no. 1, 1–17.

[Ric12] ———, *A multiplicative formula for structure constants in the cohomology of flag varieties*, Michigan Math. J. **61** (2012), no. 1, 3–17.

[RR11] Nicolas Ressayre and Edward Richmond, *Branching Schubert calculus and the Belkale-Kumar product on cohomology*, Proc. Amer. Math. Soc. **139** (2011), 835–848.

[TY09] Hugh Thomas and Alexander Yong, *A combinatorial rule for (co)minuscule Schubert calculus*, Adv. Math. **222** (2009), no. 2, 596–620.

[Voi07] Claire Voisin, *Hodge theory and complex algebraic geometry. I*, english ed., Cambridge Studies in Advanced Mathematics, vol. 76, Cambridge University Press, Cambridge, 2007, Translated from the French by Leila Schneps.

On the mod p reduction of orthogonal representations

Jean-Pierre Serre

to the memory of Bertram Kostant

Abstract We show that the reduction mod p of an orthogonal, or symplectic, linear representation of a group (or of an algebra with involution) is also orthogonal, or symplectic. The proof makes use of a (perhaps new) construction: the lower and upper "middles" of a pair of lattices.

Keywords: Brauer-Nesbitt • lattice • middle • orthogonal • semisimplification

Mathematics Subject Classification (2010): 20Cxx, 13F30

Introduction

The following fact is probably known :

(F) *Let G be a finite group and let ρ be a linear representation of G in characteristic* 0 *which is orthogonal* (resp. symplectic). *Let p be a prime number* $\neq 2$. *Then the reduction mod p of ρ is orthogonal* (resp. symplectic).

[Recall that a linear representation is said to be orthogonal (resp. symplectic) if it fixes a nondegenerate symmetric (resp. alternating) bilinear form. As for the precise meaning of "reduction mod p", see below.]

In what follows, we give a detailed proof of (F), and we extend it to linear representations of algebras with involution, including the group algebra of an infinite group.

Let us state the theorem explicitly, in the group algebra case:

Let R be a discrete valuation ring, with the ring of fractions K and residue field $k = R/\pi R$, where π is a uniformizer. Let V be a finite dimensional K-vector space.

Jean-Pierre Serre
Collège de France, 3 rue d'Ulm, 75005 Paris, France, e-mail: jpserre691@gmail.com

V. G. Kac, V. L. Popov (eds.), *Lie Groups, Geometry, and Representation Theory*,
Progress in Mathematics 326, https://doi.org/10.1007/978-3-030-02191-7_18

Let G be a group and let $\rho : G \to \mathrm{GL}(V)$ be a homomorphism. Assume that there exists a lattice L of V which is G-stable (such a lattice always exists when G is finite).

[Recall that a *lattice* is a free R-submodule L of V such that $K.L = V$, cf. §1.1.]

Let L be a G-stable lattice of V. The k-vector space $V_L = L/\pi L$ is a $k[G]$-module. The structure of this module may depend on the choice of L. However, by a theorem of Brauer-Nesbitt (see §2), its *semisimplification* V_L^{ss} is well-defined, up to isomorphism (cf. §2.2). It is V_L^{ss} that we call "the reduction mod π" of V, and we denote it by V_k. The precise form of (F) is :

Theorem A. *Assume that there exists a G-invariant symmetric* (resp. alternating) *nondegenerate K-bilinear form on V. Then there exists a k-bilinear form on V_k with the same properties.*

Assume now that 2 is invertible in R, i.e., $\mathrm{char}(k) \neq 2$, and hence $\mathrm{char}(K) \neq 2$. We may identify symmetric bilinear forms with quadratic forms. The symmetric part of th.A can then be restated, and made more precise, as follows:

Theorem B. *Let q be a nondegenerate G-invariant quadratic form on V. There exists a nondegenerate G-invariant quadratic form on V_k, whose class in the Witt ring $W(k)$ of k is the sum of the two Springer residues $\partial_1(q), \partial_2(q)$ of q.*

[For the definition of the Springer residues $\partial_1, \partial_2 : W(K) \to W(k)$, see §3.3.]

As mentioned above, we shall prove a generalized form of these two theorems, the generalization consisting in replacing the action of a group by the action of an R-algebra with involution (Th.5.1.4 and Th.5.1.7).

The first three sections are about lattices; we define and give the main properties of what we call the *lower middle* and the *upper middle* of two lattices. The last two sections give the proof of theorems A and B.

Acknowledgment. I want to thank R. Guralnick for his comments on a preliminary version of this paper, and for pointing out a paper of J. Thompson ([Th 84]), which contains all the necessary ingredients for proving Th.A. We shall use some of Thompson's proofs in §4 and §5.

§0. Notation

0.1. Lower middle and upper middle of a pair of integers

Let x, y be two integers.

The *lower middle* $m_-(x,y)$ of (x,y) is defined as the largest integer $\leqslant \frac{x+y}{2}$:

(0.1.1) $\quad m_-(x,y) = \lfloor \frac{x+y}{2} \rfloor.$

The *upper middle* $m_+(x,y)$ of (x,y) is the smallest integer $\geqslant \frac{x+y}{2}$:

(0.1.2) $\quad m_+(x,y) = \lceil \frac{x+y}{2} \rceil.$

[This is similar to calling Wednesday and Thursday the *middle days* of the week.]

We have:

(0.1.3) $m_-(x,y) = \frac{x+y}{2} = m_+(x,y)$ if $x \equiv y \pmod 2$

(0.1.4) $m_-(x,y) < \frac{x+y}{2} < m_+(x,y)$ and $m_+(x,y) = m_-(x,y)+1$ if $x \equiv y+1 \pmod 2$.

(0.1.5) $m_+(-x,-y) = -m_-(x,y)$.

(0.1.6) $m_-(x,y) = \sup_{n\in\mathbf{Z}} \inf(x-n, y+n)$.

(0.1.7) $m_+(x,y) = \inf_{n\in\mathbf{Z}} \sup(x-n, y+n)$.

(0.1.8) $m_-(x+1,y) = m_+(x,y)$ and $m_+(x+1,y) = m_-(x,y)+1$.

0.2. Discrete valuations

We keep the same notation as in the introduction: K is a field with a discrete valuation $v : K^\times \to \mathbf{Z}$, which is extended to K by putting $v(0) = +\infty$. The valuation ring R is the set of all $x \in K$ with $v(x) \geqslant 0$; the maximal ideal $\mathfrak{m}$ of R is the set of all $x \in K$ with $v(x) \geqslant 1$; we choose a generator π of $\mathfrak{m}$; we have $v(\pi) = 1$. The residue field is $k = R/\mathfrak{m} = R/\pi R$.

The letter V denotes a finite dimensional K-vector space.

§1. The lower and upper middles of a pair of lattices

1.1. Definitions

Recall that a *lattice* of V is a free R-submodule L of V such that the natural map $K \otimes_R L \to V$ is an isomorphism.

Let L and M be two lattices. Then $L \cap M$ is a lattice, and so is:

$$L + M = \text{set of all } x+y, \text{ with } x \in L, y \in M.$$

We now define two lattices $m_-(L,M)$ and $m_+(L,M)$, which are sandwiched between $L \cap M$ and $L+M$; we call them the *lower middle* and *upper middle* of L and M. (We shall see in Prop.1.1.5 below how they are related to the "middles" of §0.1.) They are defined as follows :

(1.1.1) $m_-(L,M) = +_{n\in\mathbf{Z}} (\pi^n L \cap \pi^{-n} M)$

(1.1.2) $m_+(L,M) = \bigcap_{n\in\mathbf{Z}} (\pi^n L + \pi^{-n} M)$.

[In these formulas, the $+$ symbol means "submodule generated by". For instance $m_-(L,M)$ is the R-submodule of V generated by the lattices $\pi^n L \cap \pi^{-n} M$.]

Note that L and M play a symmetric role: we have $m_\pm(L,M) = m_\pm(M,L)$.

Remark. In (1.1.1) and (1.1.2), it is not necessary to run n through the full set $\mathbf{Z}$; a finite subset suffices. For instance, in the case of (1.1.1), if n is large enough, then

$\pi^n L$ is contained in $L \cap M$, hence the nth term $\pi^n L \cap \pi^{-n} M$ is contained in the 0th-one; we may delete it from (1.1.1) without changing the sum. The same is true if $-n$ is large enough, by a similar argument. Same thing for (1.1.2). Hence the above infinite sums and intersections can be replaced by finite ones; this shows that $m_+(L,M)$ and $m_+(L,M)$ are lattices.

As an example, suppose that $\pi^2 L \subset M$ and $\pi^2 M \subset L$. Then the terms with $|n| > 1$ can be deleted, and the formulas reduce to:

$$m_-(L,M) = \pi L \cap \pi^{-1} M + L \cap M + \pi^{-1} L \cap \pi M,$$

$$m_+(L,M) = (\pi L + \pi^{-1} M) \cap (L+M) \cap (\pi^{-1} L + \pi M).$$

More generally, if $a \geqslant 0$ is such that $\pi^a L \subset M$ and $\pi^a M \subset L$, the terms with $|n| > a/2$ can be deleted.

Example : *middles of twisted lattices*

If L is a lattice, and if $a \in \mathbf{Z}$, the *a-twist* $L(a)$ of L is defined as :

(1.1.3) $\quad L(a) = \pi^{-a} L.$

We have $a \leqslant b \Rightarrow L(a) \subset L(b)$, and:

(1.1.4) $\quad L(a) + L(b) = L(\sup(a,b))$ and $L(a) \cap L(b) = L(\inf(a,b))$.

Proposition 1.1.5. *If* $x, y \in \mathbf{Z}$, *then* :

(1.1.6) $\quad m_-(L(x), L(y)) = L(m_-(x,y)),$

(1.1.7) $\quad m_+(L(x), L(y)) = L(m_+(x,y))$

[In other words: on the set of all twists of a given lattice, the "middle" operations coincide with those defined on $\mathbf{Z}$ in §0.1.]

Proof. This follows from (1.1.4), combined with (0.1.6) and (0.1.7).

1.2. Basic properties of the lower and upper middles of two lattices

As above, let L and M be two lattices of V.

Proposition 1.2.1. *We have* :

(1.2.2) $\quad L \cap M \subset m_-(L,M) \subset m_+(L,M) \subset L+M.$

Proof. The inclusions $L \cap M \subset m_-(L,M)$ and $m_+(L,M) \subset L+M$ are clear. Let us show that $m_-(L,M) \subset m_+(L,M)$. Note first that we have :

(1.2.3) $\quad \pi^a L \cap \pi^{-a} M \subset \pi^b L + \pi^{-b} M \quad$ for every $a, b \in \mathbf{Z}$.

Indeed, if $a \geqslant b$, this follows from :

$$\pi^a L \cap \pi^{-a} M \subset \pi^a L \subset \pi^b L \subset \pi^b L + \pi^{-b} M;$$

similarly, if $b \geqslant a$, (1.2.3) follows from :

$$\pi^a L \cap \pi^{-a} M \subset \pi^{-a} M \subset \pi^{-b} M \subset \pi^b L + \pi^{-b} M.$$

Since $m_-(L,M)$ is generated by the $\pi^a L \cap \pi^{-a} M$, formula (1.2.3) shows that it is contained in every $\pi^b L + \pi^{-b} M$, hence also in their intersection, which is $m_+(L,M)$.

Remark. The R-modules $(L+M)/m_+(L,M)$ and $m_-(L,M)/(L\cap M)$ are (non-canonically) isomorphic; this is proved by the method of reduction to dimension 1 used in the proof of Th. 1.2.4 below.
Similarly, we have $(L+M)/m_-(L,M) \simeq m_+(L,M)/(L\cap M)$.

Theorem 1.2.4. *We have* $\pi.m_+(L,M) \subset m_-(L,M)$.

Proof. Suppose first that $\dim V = 1$. In that case, all the lattices of V are twists of one of them. By Prop.1.1.5, the formula $\pi.m_+(L,M) \subset m_-(L,M)$ is equivalent to the obvious formula $-1+m_+(x,y) \leqslant m_-(x,y)$ for $x,y \in \mathbf{Z}$.

We now reduce the general case to that one. To do so, let $V = \oplus V_i$ be a splitting of V as a direct sum of subspaces V_i. We say that this splitting is *compatible* with a lattice P if $P = \oplus P_i$, where $P_i = P\cap V_i$. If $V = \oplus V_i$ is compatible with two lattices L and M, then the same is true for $L\cap M$, and we have $(L\cap M)_i = L_i \cap M_i$; same for $L+M$, $\pi^a L\cap \pi^b M$, $m_\pm(L,M)$. Hence, *if Th. 1.2.4 is true for the* V_i, *it is true for* V.

It only remains to prove :

Lemma 1.2.5. *There exists a splitting* $V = \oplus V_i$, *with* $\dim V_i = 1$ *for every i, which is compatible with both L and M.*

Proof. By replacing M by a suitable $\pi^n M$, we may assume that $M \subset L$. Since R is a principal ideal domain, there exists an R-basis (x_i) of L, and nonzero elements a_i of R such that the $(a_i x_i)$ make up a basis of M, cf. e.g., [A VII, §4, Th.1]. The splitting $V = \oplus Ke_i$ has the required properties: one has $L = \oplus Re_i$ and $M = \oplus Ra_ie_i$.

Remark. Let us mention three formulas which can also be proved by reduction to dimension 1:

(1.2.6) $\quad m_-(\pi L,M) = \pi.m_+(L,M)$,

(1.2.7) $\quad m_+(\pi L,M) = m_-(L,M)$,

(1.2.8) $\quad m_\pm(L,M) = m_\pm(L+M, L\cap M)$.

Building interpretation.

When R is complete for the π-adic topology, the lattices of V may be viewed as the vertices of the *affine building* X associated by Goldman-Iwahori to $\mathrm{GL}(V)$, cf. [GI 63] and [Ge 81]; the space X is isomorphic to the product of $\mathbf{R}$ by the Bruhat-Tits building of $\mathrm{SL}(V)$, cf. [BT 84]; the apartments of X correspond to the splittings of V as direct sums of 1-dimensional subspaces, and Lemma 1.2.5 is a special case of the fact that any two points of X are contained in an apartment, cf. [Ge, 2.3.4]. If L, M are two lattices, and $[L], [M]$ are the corresponding vertices of X, any barycenter $x[L]+y[M]$, with x,y real $\geqslant 0$, $x+y=1$, makes sense as a point of X (such barycenters make up the *geodesic segment* joining $[L]$ to $[M]$). In particular, $\frac{1}{2}[L]+\frac{1}{2}[M]$ makes sense; it is the middle (in the standard meaning of the word) of that geodesic segment. This middle point is not always a vertex, i.e., it does not always correspond to a lattice[1]. One can check that it is indeed a vertex if and only if $m_+(L,M)$ and $m_-(L,M)$ coincide, and in that case we have the good-looking formula:

$$\tfrac{1}{2}[L]+\tfrac{1}{2}[M] = [m_+(L,M)] = [m_-(L,M)].$$

---

[1] But it does after extension of scalars to $K(\sqrt{\pi})$.

1.3. The lower and upper middle submodules of a torsion R-module

[The content of this section will not be used in the rest of this paper.]

Let T be a torsion R-module of finite exponent, i.e., an R-module such that there exists $N \geqslant 1$ with $\pi^N T = 0$. Let us define its *lower middle* and *upper middle* submodules by the following formulas :

(1.3.1) $\quad m_-(T) = +_{n\geqslant 0}\,(\mathrm{Im}\ \pi_T^n \cap \mathrm{Ker}\ \pi_T^n),$

(1.3.2) $\quad m_+(T) = \bigcap_{n\geqslant 0}\,(\mathrm{Im}\ \pi_T^n + \mathrm{Ker}\ \pi_T^n),$

where π_T is the endomorphism of T defined by π. In these formulas, it is enough to let n run from 1 to $N-1$: larger n's contribute nothing, since then $\pi_T^n = 0$, and $n = 0$ does not give anything either.

The properties of $m_-(T)$ and $m_+(T)$ are analogous to the ones of §1.1 and §1.2. One may sum them up by :

Proposition 1.3.3. *We have* :

(1.3.4) $\quad \pi.m_+(T) \subset m_-(T) \subset m_+(T).$

(1.3.5) $\quad m_+(T) \simeq T/m_-(T).$

(1.3.6) $\quad m_-(T) \simeq T/m_+(T).$

The proof is analogous to that of Th.1.2.4: check first the case where T is cyclic, i.e., $T \simeq R/\pi^m R$ for some $m \leqslant N$, and then use the well known fact that T is a direct sum (finite or infinite) of cyclic submodules.

This construction is related to that of §1.1 and §1.2: if L and M are two lattices of V, we have $m_\pm(T) = m_\pm(L,M)/(L\cap M)$, where $T = (L+M)/(L\cap M)$.

Application to isogenies between abelian varieties

Let A and B be abelian varieties over a field F, and let $\varphi : A \to B$ be an isogeny. Assume that $\deg(\varphi)$ is a power of a prime number ℓ distinct from $\mathrm{char}(F)$. Let F^s be a separable closure of F, and let $T = \mathrm{Ker}\, A(F^s) \to B(F^s)$. By assumption, T is a finite abelian ℓ-group of order $\deg(\varphi)$, hence is a torsion module over $\mathbf{Z}_\ell$. Let $m_-(T), m_+(T)$ be the middles of T. The Galois group $\mathrm{Gal}(F^s/F)$ acts on T, and stabilizes $m_-(T)$ and $m_+(T)$. Thus, there is a factorization of φ as $A \to A_- \to A_+ \to B$, where A_- and A_+ are abelian varieties over F, and the kernel of $A(F^s) \to A_\pm(F^s)$ is $m_\pm(T)$; by (1.3.4), the kernel of $A_- \to A_+$ is killed by ℓ.

An interesting special case is when B is the dual variety of A, and φ is a polarization of degree a power of ℓ. In that case, one can show that A_- and A_+ are dual of each other. Hence, *if an abelian variety over F has a polarization of ℓ-power degree, there is an F-isogenous one which has the same property, and for which the polarization kernel is killed by ℓ.*

§2. The Brauer-Nesbitt theorem

2.1. Notation

Let A be an R-algebra and let $(a,x) \mapsto ax$ be an R-bilinear map $A \times V \to V$ which makes V into an A-module.

[If A is the group algebra $R[G]$ of a group G, this means that G acts linearly on V.]

Lemma 2.1.1. *The following properties are equivalent:*

(2.1.2) *There exists a lattice L of V with is stable under the action of A, i.e., $A.L = L$.*

(2.1.3) *The image of A in* $\mathrm{End}(V)$ *is a finitely generated R-module .*

Proof. Let A_V be the image of A in $\mathrm{End}(V)$. If L is as in (2.1.2), then we have $A_V \subset \mathrm{End}_R(L)$, and (2.1.3) follows. Conversely, if A_V is finitely generated, let M be a lattice, and let $L = A_V.M$; then L is a finitely generated R-module which contains M; hence it is a lattice such that $A.L = L$.

If (2.1.2) and (2.1.3) hold, we shall say that the action of A on V is *bounded*. A lattice L with $A.L = L$ will be called an *A-lattice*.

2.2. Semisimplification

Let L be an A-lattice; then $E_L = L/\pi L$ is a module over the k-algebra $A_k = A/\pi A$. Let $S_1, ..., S_m$ be the successive quotients of a Jordan-Hölder filtration of E_L; the direct sum $E_L^{\mathrm{ss}} = S_1 \oplus ... \oplus S_m$ is a semisimple A_k-module which is independent (up to isomorphism) of the chosen Jordan-Hölder filtration; that module is called the *semisimplification* of E_L.

Theorem 2.2.1 (Brauer-Nesbitt). *Let L and M be two A-lattices of V. Then the A_k-modules E_L^{ss} and E_M^{ss} are isomorphic.*

This was proved by Brauer and Nesbitt when A is the R-algebra of a finite group. The proof in the general case is similar. There are two steps:

(2.2.2) *The special case where $\pi L \subset M \subset L$.*

Let $T = L/M$. We have an exact sequence of A_k-modules :

(2.2.3) $$0 \to T \to E_M \to E_L \to T \to 0,$$

where the map $T \to E_M$ is induced by $x \mapsto \pi x$. This implies an isomorphism :

(2.2.4) $$T^{\mathrm{ss}} \oplus E_L^{\mathrm{ss}} \simeq E_M^{\mathrm{ss}} \oplus T^{\mathrm{ss}};$$

hence $E_L^{\mathrm{ss}} \simeq E_M^{\mathrm{ss}}$.

(2.2.5) *Reduction of the general case to the special case.*

Replacing M by a multiple $\pi^a M$ does not change E_M^{ss}; hence we may assume that $M \subset L$. There exists $n \geqslant 0$ such that $\pi^n L \subset M$; choose such an n and use induction on n. The case $n = 0$ is trivial since $M = L$; if $n > 0$, define $N = \pi^{n-1}L + M$. Since

$\pi^{n-1}L \subset N \subset L$, the induction hypothesis implies that $E_L^{ss} \simeq E_N^{ss}$. Since $\pi N \subset M \subset N$, we have $E_M^{ss} \simeq E_N^{ss}$ by (2.2.2). Hence $E_L^{ss} \simeq E_M^{ss}$.

Alternate proof. Since E_L^{ss} and E_M^{ss} are semisimple, to prove that they are isomorphic it is enough to show that, for every $a \in A$, the characteristic polynomials of a, acting on these two modules, are the same (this criterion, for group algebras of finite groups, is due to Brauer–for the general case, see Bourbaki [A VIII.377, §20, th.2]); but this is clear, since these polynomials are the reduction mod π of the characteristic polynomial of a acting on V.

Notation. The A_k-module E_L^{ss} will be called the *reduction* mod π of V; we shall denote it by V_k. It is defined up to a non-canonical isomorphism.

§3. Bilinear forms and lattices

Let $B(x,y)$ be a nondegenerate K-bilinear form on V, which is ε-symmetric, with $\varepsilon = \pm 1$, i.e., $B(x,y) = \varepsilon B(y,x)$ if $x, y \in V$.

3.1. Dual lattices

Let L be a lattice of V, and let L' be the *dual lattice* of L, namely the set of all $x \in V$ such that $B(x,y) \in R$ for every $y \in L$. If $x \in L'$, the map $y \mapsto B(x,y)$ is R-linear; this gives an isomorphism $L' \to \mathrm{Hom}_R(L,R)$; we may thus identify L' with the usual "R-dual" of L. We have $(L')' = L$.

A lattice L is *self-dual* (or *unimodular*) if $L' = L$; if $(e_1, ..., e_n)$ is an R-basis of L, this means that both the matrix $\mathsf{B} = (B(e_i, e_j))$ and its inverse have coefficients in R, i.e., we have $\mathsf{B} \in \mathrm{GL}_n(R)$.

We say that L is *almost self-dual* if $\pi L' \subset L \subset L'$, i.e., if the matrices B and $\pi \mathsf{B}^{-1}$ have coefficients in R.

Theorem 3.1.1 *Let L be a lattice and let L' be its dual. Then the lower middle $m_-(L,L')$ of L and L' is an almost self-dual lattice whose dual is $m_+(L,L')$.*

Proof. The functor "lattice $\mapsto$ dual lattice" transforms finite intersections into finite sums, and conversely. By (1.1.1) and (1.1.2), this shows that $m_-(L,L')$ and $m_+(L,L')$ are dual of each other. By Prop.1.2.1 and Th.1.2.4, we have

$$\pi . m_+(L,L') \subset m_-(L,L') \subset m_+(L,L'),$$

hence $m_-(L,L')$ is almost self-dual.

3.2. The bilinear forms b_1 and b_2 associated with an almost self-dual lattice

Let L be an almost self-dual lattice. The inclusions $\pi L' \subset L \subset L'$ give an exact sequence of k-vector spaces :

(3.2.1) $\quad 0 \to L'/L \to L/\pi L \to L/\pi L' \to 0,$

where the map $L'/L \to L/\pi L$ is given by $x \mapsto \pi x$.

By passage to quotients, the bilinear form B gives a k-bilinear form b_1 on $L/\pi L'$; similarly, πB defines a k-bilinear form b_2 on L'/L. These forms are ε-symmetric, and nondegenerate.

3.3. The quadratic case, and the Springer residues

Let us assume now that $\varepsilon = 1$, i.e., that B is symmetric. Assume also that $\mathrm{char}(k) \neq 2$, i.e., that 2 is invertible in R. We may thus identify symmetric bilinear forms and quadratic forms. Let $q(x) = B(x,x)$ be the quadratic form defined by B, and let $[q]$ be its image in the Witt ring $W(K)$ of the field K (for the definition and basic properties of the Witt ring, see [La 05, Chap.I-II]). Similarly, let q_1, q_2 be the quadratic forms defined by the k-bilinear forms b_1, b_2 of §3.2, and let $[q_1], [q_2]$ be their images in the Witt ring $W(k)$.

Recall (cf. [Sp 55], [La 05, chap.VI]) that Springer has defined two "residue" maps

$$\partial_1, \partial_2 : W(K) \; \to \; W(k).$$

The map ∂_1 is a ring homomorphism; the map ∂_2 is additive (and depends on the choice of the uniformizing element π). They are characterized by these properties, together with their values for 1-dimensional quadratic forms, which are as follows :

(3.3.1) If $u \in R^\times$ has image $\overline{u}$ in k, then the images of the 1-dimensional form $\langle u \rangle$ by ∂_1, ∂_2 are $\langle \overline{u} \rangle, 0$, and those of $\langle u\pi \rangle$ are $0, \langle \overline{u} \rangle$.

The map $(\partial_1, \partial_2) : W(K) \to W(k) \times W(k)$ is surjective; it is bijective if K is complete.

Theorem 3.3.2. *Let L be an almost self-dual lattice, and let q_1, q_2 be the corresponding quadratic forms over k. Then $\partial_1([q]) = [q_1]$ and $\partial_2([q]) = [q_2]$.*

Proof. We use the same method as for th.1.2.4, namely reduction to dimension 1 (in which case the formulas are obvious). What is needed is the following orthogonal analogue of lemma 1.2.5 :

Lemma 3.3.3. *If M is any lattice of V, there exists an orthogonal splitting $V = \oplus V_i$, with $\dim V_i = 1$ for every i, which is compatible with M.*

[The orthogonality assumption means that V_i and V_j are orthogonal for the bilinear form B if $i \neq j$. It implies that the splitting is compatible with the dual M' of M.]

Proof of the lemma. Use induction on $\dim V$. Let $m = \inf_{x\in M} v(q(x))$; we have $m > -\infty$. Choose $x \in M$ with $v(q(x)) = m$. For $y,z \in M$, we have $v(B(y,z)) \geqslant m$: this follows from the formula $2B(y,z) = q(y+z) - q(y) - q(z)$ since the valuations of $q(y+z), q(y), q(z)$ are $\geqslant m$. If $y \in V$, put $\ell(y) = B(x,y)/B(x,x)$. The linear form $\ell : V \to K$ is such that $\ell(x) = 1$, and it maps M onto R. It thus gives a splitting of V as $Kx \oplus \operatorname{Ker} \ell$, namely $y \mapsto (\ell(y)x, y - \ell(y)x$; this splitting is compatible with M. It is an orthogonal splitting, since $B(y - \ell(y)x, x) = 0$ for every y. The lemma follows by applying the induction assumption to the vector space $\operatorname{Ker} \ell$ and its lattice $M \cap \operatorname{Ker} \ell$.

Corollary 3.3.4. *The classes in $W(k)$ of the two quadratic forms associated with an almost self-dual lattice do not depend on the choice of that lattice.*

Remark. The fact that the class of q_1 in $W(k)$ is the same for any two almost self-dual lattices L and M does not imply that these two quadratic forms have the same dimension: they may differ by hyperbolic factors. But, if they do have the same dimension, a theorem of Bayer-Fluckiger and First shows that L and M are *isomorphic as quadratic R-modules* ([BF 17], th.4.1); hence they are conjugate of each other by an element of the orthogonal group of (V,q).

§4. Semisimplification of symplectic and quadratic modules over an algebra with involution

This section is essentially independent of §§1,2,3: we work over a field, and not over a discrete valuation ring.

4.1. Notation

Let k be a field, and let A_k be a k-algebra with a k-linear involution denoted by $a \mapsto a^*$; we have $a^{**} = a$ and $(ab)^* = b^*a^*$ for every $a, b \in A_k$.

Let E be an A_k-module which is finite dimensional over k (i.e., a "linear representation" of A_k), and let b be a nondegenerate bilinear form on E with the following property:

(4.1.1) $b(ax,y) = b(x,a^*y)$ for every $a \in A_k$, $x,y \in E$.

We then say that b is *compatible* with the A_k-module structure of E. When A_k is the group algebra $k[G]$ of a group G, with its canonical involution $g^* = g^{-1}$ for every $g \in G$, this means that b is invariant by G.

We assume one of the following :

(4.1.2) b is alternating.

(4.1.3) b is symmetric and $\operatorname{char}(k) \neq 2$.

In the first case, E is called a *symplectic A_k-module*, and in the second case, it is called an *orthogonal A_k-module*.

4.2. Semisimplification

Let A_k, E, b be as above, and let E^{ss} be the semisimplification of the A_k-module E.

Theorem 4.2.1. *In the case* (4.1.3) (*resp. in the case* (4.1.2)), *there exists a symmetric* (*resp. alternating*) *bilinear form on* E^{ss} *with the following two properties*:

(4.2.2) *It is compatible with the* A_k*-module structure of* E^{ss}.

(4.2.3) *It is isomorphic to* b.

One may sum up (4.2.2) by saying that the semisimplification of a quadratic (resp. alternating) module is quadratic (resp. alternating). As for (4.2.3), it is obvious in the symplectic case, since all nondegenerate alternating forms of a given rank are isomorphic; in the orthogonal case, it means that the corresponding quadratic forms have the same class in the Witt ring $W(k)$.

Proof. We use the method of [Th.84, §2]. Let S be an A_k-submodule of E, which is totally isotropic for b (i.e., $b(x,y) = 0$ for all $x, y \in E$), and is maximal for that property. Let $S_\perp$ be its orthogonal relative to b; because of (4.2.2), it is a submodule of E, and we have $0 \subset S \subset S_\perp \subset E$. The form b defines a nondegenerate form b_1 on $S_\perp/S$.

Lemma 4.2.4. (i) *The* A_k*-module* $X = S_\perp/S$ *is semisimple, and the only totally isotropic submodule of* X *is* 0.

(ii) *In the orthogonal case* (4.1.3), *the form* b *is isomorphic to the direct sum of* b_1 *and an hyperbolic form of rank* $2\dim S$.

Proof of (i). If Y is a totally isotropic submodule of X, its inverse image in $S_\perp$ is totally isotropic, hence equal to S, i.e., $Y = 0$. If Z is a submodule of X, then $Y = Z \cap Z_\perp$ is totally isotropic, hence 0; we have $X = Z \oplus Z_\perp$; this shows that every submodule of X is a direct summand, i.e., X is semisimple.

Proof of (ii). Let M be a subvector space of $S_\perp$ such that $S_\perp = S \oplus M$. The restriction of b to M is nondegenerate, and the projection $M \to S_\perp/S$ is an isomorphism of quadratic spaces; we have $E = M \oplus M_\perp$, and $\dim M_\perp = 2\dim S$. The form b splits as $b_1 \oplus h$, where h is the quadratic form of $M_\perp$; the form h is hyperbolic, since that space contains the totally isotropic subspace S, of dimension $\frac{1}{2}\dim M$. This proves (ii).

End of the proof of Th.4.2.1. The bilinear form b defines a duality between S and $E/S_\perp$; hence we may identify $E/S_\perp$ with the k-dual S' of S (with its natural A_k-structure). We have

$$E^{ss} = X \oplus (S^{ss} \oplus S'^{ss}),$$

since X is semisimple, cf. 4.2.4 (i). It is easy to see that S'^{ss} is isomorphic to the dual of S^{ss}. Hence, $E^{ss} \simeq X \oplus (Y \oplus Y')$, where $Y = S^{ss}$. We then put on E^{ss} the bilinear form which is the direct sum of the form b_1 on X and the natural bilinear form (symmetric or alternating, as needed) on $Y \oplus Y'$. By Lemma 4.2.4, that form has properties (4.2.2) and (4.2.3).

§5. Proof of theorem B

We now prove the theorems stated in the introduction; the proofs will merely consist in putting together the results of §§2,3,4.

5.1. The setting

We go back to the standard notation (K,R,V) and we assume that V is a module over an R-algebra A with involution. We also assume that this action is *bounded* (cf. §2.1), i.e., that there exists a lattice of V which is stable under A.

Let B be a nondegenerate K-bilinear form on V, which is compatible with the action of A; this means (as in (4.1.1)) :

(5.1.1) $B(ax,y) = B(x,a^*y)$ for every $a \in A,\ x,y \in V$.

As in §4, we assume one of the following:

(5.1.2) B is alternating.

(5.1.3) B is symmetric, and $\mathrm{char}(k) \neq 2$.

By the Brauer-Nesbitt theorem (see §2.2), the semisimplification V_k of V is well-defined; it is a module over the k-algebra $A_k = A/\pi A$.

Theorem 5.1.4. *There exists a nondegenerate k-bilinear form b on V_k with the following properties*:

(5.1.5) *It is compatible with the action of A_k, i.e., it satisfies condition* (4.1.1).

(5.1.6) *It is alternating* (*resp. symmetric*) *if B is.*

[More shortly: if V is a symplectic (resp. orthogonal) module, so is V_k.]

Note that condition (5.1.5) alone would be easy to satisfy: since V is isomorphic to its dual, the same is true for V_k, and that is equivalent to the existence of a bilinear form b compatible with the action of A_k.

In the orthogonal case (5.1.3), one may ask more of the form b. To state it, let us denote by q (instead of $[q]$) the Witt class in $W(K)$ of the quadratic form $q(x) = B(x,x)$, and let $q_1, q_2 \in W(k)$ be its images by the Springer residue maps, cf. §3.3. Then:

Theorem 5.1.7. *The form b of Th.5.1.4 can be chosen to have the following property*:

(5.1.8) *There exists an A_k-orthogonal splitting $V_k = E_1 \oplus E_2$ such that the Witt class of the restriction to E_i $(i = 1,2)$ of $x \mapsto b(x,x)$ is q_i.*

Note that (5.1.8) implies that the Witt class of the quadratic form $b(x,x)$ on V_k is $q_1 + q_2$. When A is a group algebra $R[G]$, we recover Th.B of the introduction. Similarly, Th.5.1.4 applied to $R[G]$ gives Th.A.

Theorems 5.1.4 and 5.1.7 will be proved in §5.3.

5.2. Existence of almost self-dual A-lattices

Theorem 5.2.1. *There exists an A-lattice of V which is almost self-dual* (cf. §3.1) *relative to the bilinear form B.*

Proof. Let L be an A-lattice of V. Condition (5.1.1) shows that the dual L' of L is also an A-lattice, and so are the lattices $\pi^n L, \pi^n L'$, and hence also the lower middle $m_-(L,L')$, which is almost self-dual by th.3.1.1.

Alternate proof. The proof above uses the "middle" notions of §§1,2. Here is a direct proof, taken from [Th.84]:

Choose an A-lattice L, with $L \subset L'$, and which is maximal for that property. Let m be the smallest integer such that $\pi^m L' \subset L$. If $m = 0$, or 1, then L is self-dual, or almost self-dual. Let us show that $m \geqslant 2$ is impossible. Define $M = \pi^{m-1}L' + L$; we have $M' = \pi^{1-m}L \cap L'$. The inequality $m \geqslant 2$ implies $\pi^{m-1}L' \subset \pi^{1-m}L$, hence $\pi^{m-1}L' \subset M'$; since $L \subset M'$, this shows that M is contained in M'. The maximality of L then implies $M = L$, hence $\pi^{m-1}L' \subset L$, which contradicts the minimality of m.

Remark. This short proof is not in fact very different from the first one. Indeed, it amounts to constructing an almost self-dual A-lattice, starting with any lattice L contained in its dual, by choosing m with $\pi^m L' \subset L$, then replacing L by $\pi^{m-1}L' + L$, and iterating until one gets $m = 0$ or $m = 1$. But, if one writes down the end result of that process, a simple computation shows that one finds the lower middle lattice $m_-(L,L')$, exactly as in the first proof.

There is however a definite advantage in using a maximal lattice: in that case, L'/L does not contain any nonzero totally isotropic submodule, hence it is a semi-simple A_k-module, cf. Lemma 4.2.4. This simplifies the proofs of the next section.

5.3. Proof of Th.5.1.4 and Th.5.1.7

Let L be an almost self-dual A-lattice, cf. th.5.2.1. Let $F_1 = L/\pi L'$, $F_2 = L'/L$. By (3.2.1), we have an exact sequence of A_k-modules :

(5.3.1) $0 \to F_2 \to L/\pi L \to F_1 \to 0.$

Let $E_1 = F_1^{\mathrm{ss}}$, $E_2 = F_2^{\mathrm{ss}}$, and let $V_k = (L/\pi L)^{\mathrm{ss}}$. The exact sequence above gives a splitting :

(5.3.2) $V_k = E_1 \oplus E_2.$

As explained in §3.2, the bilinear form B defines k-bilinear forms b_1 and b_2 on F_1 and F_2; these forms are compatible with the action of A_k and they are alternating (resp. symmetric) if B is alternating (resp. symmetric). By Th.4.2.1, applied to F_1 and F_2, there exist A_k-compatible forms b'_1, b'_2 on E_1, E_2 which are isomorphic (as bilinear forms) to b_1, b_2; in the orthogonal case, Th.3.3.1 shows that the Witt classes of these forms are the Springer residues of q. Using (5.3.2), we define a bilinear form on V_k as the orthogonal sum of b'_1 on E_1 and b'_2 on E_2. All the properties of Th.5.1.4 and Th.5.1.7. hold.

This concludes the proof.

References

[A VII] N. Bourbaki, *Algèbre, Chapitre 7, Modules sur les anneaux principaux*, Paris, Masson, 1981; English translation, *Algebra* II, Springer-Verlag, 1989.

[A VIII] ——, *Algèbre, Chapitre 8, Anneaux et modules semi-simples*, new revised edition, Springer-Verlag, 1998.

[BF 17] E. Bayer-Fluckiger & U.A. First, *Rationally isomorphic hermitian forms and torsors of some non-reductive groups*, Adv. Math. 312 (2017), 150–184.

[BT 84] F. Bruhat & J. Tits, *Schémas en groupes et immeubles des groupes classiques sur un corps local*, Bull. S.M.F. 112 (1984), 259–301 (= J. Tits, C.P., vol.IV, n°128, 1–43).

[Ge 81] P. Gérardin, *Immeubles des groupes linéaires généraux*, Lect. Notes Math. 880 (1981), 138–178.

[GI 63] O. Goldman & N. Iwahori, *The space of* $\mathfrak{p}$*-adic norms*, Acta Math. 109 (1963), 137–177.

[La 04] T-Y. Lam, *Introduction to Quadratic Forms over Fields*, A.M.S., GSM 67, 2004.

[Sp 55] T.A. Springer, *Quadratic forms over fields with a discrete valuation*, Indag. Math. 17 (1955), 352–362 and 18 (1956), 238–246.

[Th 86] J.G. Thompson, *Bilinear forms in characteristic p and the Frobenius-Schur indicator*, second part of *Some finite groups which appear as* $\mathrm{Gal}\, L/K$, *where* $K \subset \mathbf{Q}(\mu_n)$, in *Group Theory*, Lect. Notes Math. 1185 (1986), 221–230.

Printed by Printforce, the Netherlands